Atomic Weights (IUPAC, 1971; based on ^{12}C = 12.000)[a]

Aluminum	26.98154	Hydrogen	1.0079[b]	Rhenium	186.2
Antimony	121.7_5	Indium	114.82	Rhodium	102.9055
Argon	39.94_8[b]	Iodine	126.9045	Rubidium	85.467_8
Arsenic	74.9216	Iridium	192.2_2	Ruthenium	101.0_7
Barium	137.3_4	Iron	55.84_7	Samarium	150.4
Beryllium	9.01218	Krypton	83.80	Scandium	44.9559
Bismuth	208.9804	Lanthanum	138.905_5	Selenium	78.9_6
Boron	10.81[b]	Lead	207.2[b]	Silicon	28.08_6[b]
Bromine	79.904	Lithium	6.94_1[b,c]	Silver	107.868
Cadmium	112.40	Lutetium	174.97	Sodium	22.98977
Calcium	40.08	Magnesium	24.305	Strontium	87.62
Carbon	12.011[b]	Manganese	54.9380	Sulfur	32.06[b]
Cerium	140.12	Mercury	200.5_9	Tantalum	180.947_9
Cesium	132.9054	Molybdenum	95.9_4	Technetium	98.9062[d]
Chlorine	35.453	Neodymium	144.2_4	Tellurium	127.6_0
Chromium	51.996	Neon	20.17_9	Terbium	158.9254
Cobalt	58.9332	Neptunium	237.0482	Thallium	204.3_7
Copper	63.54_6[b]	Nickel	58.7_1	Thorium	232.0381
Dysprosium	162.5_0	Niobium	92.9064	Thulium	168.9342
Erbium	167.2_6	Nitrogen	14.0067	Tin	118.6_9
Europium	151.96	Osmium	190.2	Titanium	47.9_0
Fluorine	18.9984	Oxygen	15.999_4[b,c]	Tungsten	183.8_5
Gadolinium	157.2_5	Palladium	106.4	Uranium	238.029[c]
Gallium	69.72	Phosphorus	30.97376	Vanadium	50.941_4
Germanium	72.5_9	Platinum	195.0_9	Xenon	131.30
Gold	196.9665	Potassium	39.09_8	Ytterbium	173.0_4
Hafnium	178.4_9	Praseodymium	140.0977	Yttrium	88.9059
Helium	4.00260	Protactinium	231.0359	Zinc	65.38[e]
Holmium	164.9304	Radium	226.0254[d]	Zirconium	91.22

[a] Elements in materials of terrestrial origin. Reliable to ±1 in last digit or ±3 when subscript.
[b] Variation in natural isotopic abundance limits precision.
[c] Variations possible owing to artificial isotopic separation.
[d] Most commonly available long-lived isotope.
[e] A new coulometric determination gives 65.377 ± 0.003 (*Chem. Eng. News*, 1971, Sept. 27, p. 32).

basic inorganic chemistry

basic inorganic chemistry

F. ALBERT COTTON
Robert A. Welch Distinguished Professor of Chemistry
Texas A and M University
College Station, Texas, USA

GEOFFREY WILKINSON
Professor of Inorganic Chemistry
Imperial College of Science and Technology
University of London, England

JOHN WILEY & SONS
New York • Chichester • Brisbane • Toronto

Library of Congress Cataloging in Publication Data:

Cotton, Frank Albert, 1930-
Basic inorganic chemistry.
Includes bibliographies and index.
1. Chemistry, Inorganic. I. Wilkinson, Geoffrey, 1921– joint author. II. Title.
QD151.2.C69 546 75–26832
ISBN 0–471–17557–9

Printed in the United States of America

20 19 18 17 16 15 14 13 12 11 10

preface

Those who aspire not to guess and divine, but to discover and know, who propose not to devise mimic and fabulous worlds of their own, but to examine and dissect the nature of this very world itself, must go to facts themselves for everything."

F. Bacon, 1620

There are already several textbooks of inorganic chemistry that treat the subject in considerably less space than our comprehensive text, *Advanced Inorganic Chemistry*. Moreover, most of them include a great deal of introductory theory, which we omitted from our larger book because of space considerations. The net result is that these books contain very little of the real content of inorganic chemistry—namely, the actual facts about the properties and behavior of inorganic compounds.

Our purpose in *Basic Inorganic Chemistry*, is to meet the needs of teachers who present this subject to students who do not have the time or perhaps the inclination to pursue it in depth, but who may also require explicit coverage of basic topics such as the electronic structure of atoms and elementary valence theory. We therefore introduce material of this type, in an elementary fashion, and present only the main facts.

The point, however, is that this book does present the facts, in a systematic way. We have a decidedly Baconian philosophy about all chemistry, but particularly inorganic chemistry. We are convinced that inorganic chemistry *sans* facts (or nearly so), as presented in other books, is like a page of music with no instrument to play it on. One can appreciate the sound of music without knowing anything of musical theory, although of course one's appreciation is enhanced by knowing some theory. However, a book of musical theory, even if it is illustrated by audible snatches of themes and a few chord progressions, is quite unlike the hearing of a real composition in its entirety.

We believe that a student who has read a book on "inorganic chemistry" that consists almost entirely of theory and so-called principles, with but sporadic mention of the hard facts (only when they "nicely" illustrate the "principles") has not, in actual fact, had a course in inorganic chemistry. We deplore the current trend toward this way of teaching students who are not expected to specialize in the subject, and believe that even the nonspecialist ought to get a straight dose of the subject as it really is—"warts and all." This book was written to encourage the teaching of inorganic chemistry in a Baconian manner.

At the end of each chapter, there is a study guide. Occasionally this includes a few remarks on the scope and purpose of the chapter to help the student place it in the context of the entire book. A supplementary reading list is included in all chapters. This consists of relatively recent articles in the secondary (monograph and review) literature, which will be of interest to those who wish to pursue the subject matter in more detail. In some instances there is little literature of this kind available. However, the student—and the instructor—will find more detailed treatments of all the elements and classes of compounds, as well as further references, in our *Advanced Inorganic Chemistry*, third edition, Wiley, 1972, and in *Comprehensive Inorganic Chemistry*, J. C. Bailar, Jr., H. J. Emeléus, R. S. Nyholm, and A. F. Trotman-Dickinson, eds., Pergamon, 1973.

F. ALBERT COTTON
GEOFFREY WILKINSON

contents

basic inorganic chemistry

1

First Principles

1

some physical chemical preliminaries

1-1
Units

1-1a There is now an internationally accepted set of Units for the physical sciences. It is called the SI (for *Système International*) units. Based on the metric system, it is designed to achieve maximum internal consistency. However, since it requires the abandonment of many familiar units and numerical constants in favor of new ones, its adoption in practice will take time. In this book, we shall take a middle course, adopting some SI units (e.g., joules for calories) but retaining some non-SI units (e.g., angstroms).

1-1b The SI Units. The SI system is based on the following set of defined units:

Physical Quantity	Name of Unit	Symbol for Unit
Length	meter	m
Mass	kilogram	kg
Time	second	s
Electric current	ampere	A
Temperature	kelvin	K
Luminous intensity	candela	cd

Multiples and fractions of these are specified using the following prefixes:

Multiplier	Prefix	Symbol
10^{-1}	deci	d
10^{-2}	centi	c
10^{-3}	milli	m
10^{-6}	micro	μ
10^{-9}	nano	n
10^{-12}	pico	p
10	deka	da
10^{2}	hecto	h
10^{3}	kilo	k
10^{6}	mega	M
10^{9}	giga	G
10^{12}	tera	T

In addition to the defined units, the system includes a number of derived units, of which the following are the main ones.

Physical quantity	SI unit	Unit symbol
Force	newton	$N = kg\,m\,s^{-2}$
Work, energy, quantity of heat	joule	$J = Nm$
Power	watt	$W = J\,s^{-1}$
Electric charge	coulomb	$C = As$
Electric potential	volt	$V = WA^{-1}$
Electric capacitance	farad	$F = A\,s\,V^{-1}$
Electric resistance	ohm	$\Omega = V\,A^{-1}$
Frequency	hertz	$Hz = s^{-1}$
Magnetic flux	weber	$Wb = Vs$
Magnetic flux density	tesla	$T = Wb\,m^{-2}$
Inductance	henry	$H = V\,s\,A^{-1}$

1-1c Units To Be Used in This Book.

Energy. Joules and kilojoules will be used exclusively. Most of the chemical literature to date employs calories, kilocalories, electron volts and, to a limited extent, wave numbers (cm^{-1}). See Section 1-1d for the conversion of these units to joules.

Bond lengths. The angstrom, Å, will be used. This is defined as 10^{-8} centimeter. The nanometer (10 Å) and picometer (10^{-2} Å) are to be found in recent literature. The C—C bond length in diamond has the value:

1.54 angstrom
0.154 nanometers
154 picometers

Pressure. Atmospheres, atm, and torr (1/760 atm) will be used.

1-1d Some Useful Conversion Factors and Numerical Constants

Conversion Factors:

1 calorie = 4.184 joules(j)
1 electron volt per molecule = 96.5 kilojoules per mole (kJ mol^{-1})
1 kilojoule per mole = 83.54 wave numbers (cm^{-1})

Important Constants:

Avogadro's number (C^{12} = 12.0000...), $N_A = 6.02252 \times 10^{23}\ mol^{-1}$
Electron charge, $e = (4.8030 \pm 0.0001) \times 10^{-10}$ abs esu $= 1.602 \times 10^{-19}$ C
Electron mass, $m = 9.1091 \times 10^{-31}$ kg = 0.00054860 mu = 0.5110 Mev
Gas constant, R = 1.9872 defined cal $deg^{-1}\ mol^{-1}$ = 8.3143 $JK^{-1}\ mol^{-1}$ = 0.082057 liter atm $deg^{-1}\ mol^{-1}$
Ice point: 273.150 ± 0.01 K
Molar volume (ideal gas, 0°C, 1 atm) = $22.414 \times 10^3\ cm^3\ mol^{-1} = 2.241436 \times 10^{-2}\ m^3\ mol^{-1}$

Planck's constant, $h = 6.6256 \times 10^{-27}$ erg sec $= 6.6256 \times 10^{-34}$ Js
Boltzmann's constant, $\mathbf{k} = 1.3805 \times 10^{-23}$ JK^{-1}
Velocity of light in vacuum $= 2.99795 \times 10^8$ msec^{-1}
$\pi = 3.14159$
$e = 2.7183$
ln 10 = 2.3026

1-1e Coulombic Force and Energy Calculations in SI Units. Although SI units do, for the most part, lead to simplification, one computation that is important to inorganic chemistry becomes slightly more complex. We explain that point in detail here. It traces back to the concept of the dielectric constant, ε, which relates the intensity of an electric field induced within a substance, D, to the intensity of the field applied, E, by the equation

$$D = \varepsilon E$$

The same parameter appears in the Coulomb equation for the force, F, between two charges, q_1 and q_2, separated by a distance, d, and immersed in a medium with a dielectric constant ε:

$$F = \frac{q_1 \times q_2}{\varepsilon d^2}$$

In the old (cgs) system of units, which the SI system replaces, units and magnitudes were so defined that ε was a dimensionless quantity and for a vacuum we had $\varepsilon_0 = 1$.

For reasons that we shall not pursue here, coulomb's law of electrostatic force, in SI units, must be written

$$F = \frac{q_1 \times q_2}{4\pi\varepsilon d^2}$$

The charges are expressed in Coulombs, C, the distance in meters, m, and the force is obtained in newtons, N. It now develops that ε has units (i.e., is no longer a dimensionless quantity), namely, $C^2m^{-1}J^{-1}$. Moreover, the dielectric constant of a vacuum (the permittivity, as it should formally be called) is no longer unity. It is, instead,

$$\varepsilon_0 = 8.854 \times 10^{-12} C^2m^{-1}J^{-1}$$

Thus, to calculate a Coulomb energy, E, in joules, J, we must employ the expression

$$E = \frac{q_1 \times q_2}{4\pi\varepsilon d}$$

with all quantities being as defined above for the Coulomb force.

1-2 Thermochemistry

1-2a Standard States. To have universally recognized and understood values for energy changes in chemical processes, it is first necessary to define standard states for all substances.

The standard state for any substance is that phase in which it exists at 25°C (298.15 K) and one atmosphere (101.325 newtons per square meter) pressure. Substances in solution are at a concentration of 1 mole per liter.

1-2b Heat Content or Enthalpy. Virtually all physical and chemical changes either produce or consume energy, and generally this energy takes the form of heat. The gain or loss of heat may be attributed to a change in the "heat content," of the substances taking part in the process. "Heat content" is called *enthalpy*, symbolized H. The change in heat content is called the enthalpy change, ΔH.

$$\Delta H = (H \text{ of products}) - (H \text{ of reactants}) \qquad (1\text{-}2\text{-}1)$$

For the case where all products and reactants are in their standard states, the enthalpy change is represented by ΔH°, the *standard enthalpy* change of the process. For example, although the formation of water from H_2 and O_2 cannot actually be carried out at an appreciable rate under standard conditions, it is nevertheless useful to know, indirectly, that

$$H_2(\text{g, 1 atm, 25°C}) + \tfrac{1}{2}O_2(\text{g, 1 atm, 25°C}) = H_2O\,(\text{l, 1 atm, 25°C})$$
$$\Delta H^\circ = -285.7 \text{ kJ mol}^{-1} \qquad (1\text{-}2\text{-}2)$$

The heat contents of all elements in their standard states are arbitrarily set equal to zero for thermochemical purposes.

1-2c The Signs of ΔH's. In the above equation, ΔH° has a negative value. The heat content of the products is lower than that of the reactants, which means, heat is released. In general, we have

$$\text{Heat released: } \Delta H < 0$$
$$\text{Heat absorbed: } \Delta H > 0$$

The same convention will apply to changes in free energy, ΔG, to be discussed shortly.

Processes in which heat—or another form of energy—is released ($\Delta H < 0$) are called *exothermic* or *exoergic*. Those that consume energy ($\Delta H > 0$) are called *endothermic* or *endoergic*.

1-2d Standard Heats (Enthalpies) of Formation. The standard enthalpy change for any reaction can be calculated if the standard heat of formation, ΔH_f°, of each reactant and product is known. It is therefore useful to have tables of ΔH_f° values. The ΔH_f° value for a substance is the ΔH value for the process in which it is formed in its standard state from the elements, each in its standard state. Equation 1-2-2 describes such a process, and the ΔH° given is ΔH_f° of water.

The reason ΔH° for any reaction can be calculated from ΔH_f° values is because the set of equations for the ΔH_f°'s will always add up to the equation of the desired process with the elements, other than those which may necessarily appear in the final equation, themselves canceling out. This is illustrated by Eqs. 1-2-3 to 1-2-7.

$$LiAlH_4(s) = Li(s) + Al(s) + 2H_2(g) \qquad -\Delta H_f^\circ = 117.2 \qquad (1\text{-}2\text{-}3)$$

$$4H_2O(l) = 4H_2(g) + 2O_2(g) \qquad -4\,\Delta H_f^\circ = 1143.0 \qquad (1\text{-}2\text{-}4)$$

$$Li(s) + \tfrac{1}{2}O_2(g) + \tfrac{1}{2}H_2(g) = LiOH(s) \qquad \Delta H_f^\circ = -487.0 \qquad (1\text{-}2\text{-}5)$$

$$Al(s) + \tfrac{3}{2}O_2(g) + \tfrac{3}{2}H_2(g) = Al(OH)_3(s) \qquad \Delta H_f^\circ = -1272.8 \qquad (1\text{-}2\text{-}6)$$

$$LiAlH_4(s) + 4H_2O(l) = LiOH(s) + Al(OH)_3(s) + 4H_2(g) \qquad \Delta H^\circ = -734.0 \qquad (1\text{-}2\text{-}7)$$

Inspection shows that all elements cancel as the Eqs. 1-2-3 to 1-2-6 are added to give 1-2-7, and it is clear that the net ΔH° is simply the sum of the ΔH_f° values of the products minus the sum of the ΔH_f° values of the reactants, each ΔH_f° being multiplied by the coefficient required by the balanced chemical equation.

1-2e Other Special Enthalpy Changes. Aside from formation of a compound from the elements, there are several other physical and chemical processes of special importance for which the ΔH, or ΔH°, values are frequently required. Among these are the processes of melting (fusion), and vaporization (of either the solid or the liquid).

Ionization Enthalpies. Of particular interest is the process of ionization. For example,

$$Na(g) = Na^+(g) + e^-(g) \qquad \Delta H^\circ = 502 \text{ kJ mol}^{-1}$$

Thus, we speak of ionization enthalpies. Unfortunately, the more common tabulations of these quantities list them in units of the electron volt (see Section 1.1) and call them *ionization potentials*, but that terminology will not be followed here.

For many atoms, the enthalpies of removal of a second, third, etc., electron are also of chemical interest, and for most elements these enthalpies are known. For example, the first three ionization enthalpies of aluminum, and the overall energy to produce the $Al^{3+}(g)$ ion are

$$\begin{aligned} Al(g) &= Al^+(g) + e^- & \Delta H^\circ &= 577.5 \text{ kJ mol}^{-1} \\ Al^+(g) &= Al^{2+}(g) + e^- & \Delta H^\circ &= 1817 \text{ kJ mol}^{-1} \\ Al^{2+}(g) &= Al^{3+}(g) + e^- & \Delta H^\circ &= 2745 \text{ kJ mol}^{-1} \\ \hline Al(g) &= Al^{3+}(g) + 3e^- & \Delta H^\circ &= 5140 \text{ kJ mol}^{-1} \end{aligned}$$

Ionization enthalpies are also defined for molecules, for example

$$NO(g) = NO^+(g) + e^- \qquad \Delta H^\circ = 890.7 \text{ kJ mol}^{-1}$$

It is to be noted that for molecules and atoms the ionization enthalpies are always positive. Energy must be supplied to detach electrons. Also, the increasing magnitudes of successive ionization steps, as shown above for aluminum, is completely general. The more positive the atom or molecule becomes, the more difficult it is to ionize it further.

Electron Attachment Enthalpies. Consider the following processes:

$$Cl(g) + e^- = Cl^-(g) \qquad \Delta H^\circ = -348 \text{ kJ mol}^{-1}$$
$$O(g) + e^- = O^-(g) \qquad \Delta H^\circ = -142 \text{ kJ mol}^{-1}$$
$$O^-(g) + e^- = O^{2-}(g) \qquad \Delta H^\circ = 844 \text{ kJ mol}^{-1}$$

The $Cl^-(g)$ ion forms exothermically. The same is true of the other halide ions. Observe that the formation of the oxide ion, $O^{2-}(g)$ requires first an exothermic step and then an endothermic one. This is understandable because the O^- ion, which is already negative will tend to repel another electron.

In most of the chemical literature the negative of the enthalpy change, $-\Delta H^\circ$, for such a process is called the *electron affinity*, A, of the atom. In this book, however, we shall use only the systematic notation illustrated above.

Direct measurement of ΔH_{EA}'s is difficult, and indirect methods tend to be inaccurate. To give an idea of their magnitudes, some of those known, with those which are mere estimates in parentheses, are listed below:

H						
−73						
Li	Be	B	C	N	O	F
−58	(+60)	(−30)	−120	(+10)	−142	−333
Na			Si	P	S	Cl
(−50)			(−135)	(−75)	−200	−348
					Se	Br
					(−160)	−324
						I
						−295

1-2f Bond Energies. Consider the following processes and their ΔH°'s:

$$HF(g) = H(g) + F(g) \qquad \Delta H_{298} = 566 \text{ kJ mol}^{-1}$$
$$H_2O(g) = H(g) + OH(g) \qquad \Delta H_{298} = 497 \text{ kJ mol}^{-1}$$
$$OH(g) = H(g) + O(g) \qquad \Delta H_{298} = 421 \text{ kJ mol}^{-1}$$
$$H_2O(g) = 2H(g) + O(g) \qquad \Delta H_{298} = 918 \text{ kJ mol}^{-1}$$

The energy required in the first process has a simple, unambiguous significance. It is the energy required to break the H—F bond. It can be unambiguously called "the H—F bond energy." We can, if we prefer, think of 566 kJ mol^{-1} as the energy released when the H—F bond is formed; that is a perfectly equivalent, and equally unambiguous statement.

Consider the next three equations however. The actual processes of breaking the O—H bonds, *one after the other*, in H_2O have different energies. How then can we define "*the* O—H bond energy"? This requires us to refine the definition. If we take the mean of the two, which is one half of the sum, $\frac{918}{2} = 459$ kJ mol^{-1}, we can call this the *mean* O—H bond energy. This is also unambiguous, but we must remember that, knowing only this mean value, we cannot predict the actual enthalpy of either of the separate bond-breaking (or bond-forming) processes. Thus, there is a certain artificiality to the concept of the "mean" bond energy.

When we consider molecules containing more than one kind of bond the problem of defining bond energies becomes even more subtle. For example, we can think of the total enthalpy of the process

$$H_2N{-}NH_2(g) = 2\,N(g) + 4\,H(g) \qquad \Delta H_{298} = 1724 \text{ kJ mol}^{-1}$$

as consisting of the sum of the N—N bond energy, E_{N-N}, and four times the N—H bond energy, E_{N-H}. But is there any unique or rigorous way to divide the total energy into these component parts? The answer is no. The practical approach used is as follows.

We know from experiment that

$$NH_3(g) = N(g) + 3\,H(g) \qquad \Delta H_{298} = 1172 \text{ kJ mol}^{-1}$$

and thus we can say that

$$E_{N-H} = \frac{1172}{3} = 391 \text{ kJ mol}^{-1}$$

If we make the *assumption* that this value can be transferred to H_2NNH_2, then we can evaluate the N—N bond energy:

$$\begin{aligned} E_{N-N} + 4E_{N-H} &= 1724 \text{ kJ mol}^{-1} \\ E_{N-N} &= 1724 - 4E_{N-H} \\ &= 1724 - 4(391) \\ &= 160 \text{ kJ mol}^{-1} \end{aligned}$$

By proceeding in this way it is possible to build up a table of bond energies from which the enthalpies of forming molecules from their constituent gaseous atoms can be calculated fairly accurately in many cases. That this is possible shows the important fact that the energy of the bond between a given pair of atoms is fairly independent of the molecular environment in which that bond occurs. This is only approximately true, but it is a good enough approximation to be useful in understanding and interpreting many chemical processes.

Thus far only single bonds have been discussed. Certain pairs of atoms, for example, C with C, N with N, C with N, C with O, can form single bonds, double bonds, and even triple bonds. The bond energy increases as the bond order increases, in all cases. However, the increase is not, in general, linear. *Figure 1-1* gives some representative data.

A list of some useful bond energies is presented in Table 1-1.

Table 1-1 *Some Average Thermochemical Bond Energies at 25° in kJ mol^{-1}*

A. Single bond energies

	H	C	Si	Ge	N	P	As	O	S	Se	F	Cl	Br	I
H	436	416	323	289	391	322	247	467	347	276	566	431	366	299
C		356	301	255	285	264	201	336	272	243	485	327	285	213
Si			226	—	335	—	—	368	226	—	582	391	310	234
Ge				188	256	—	—	—	—	—	—	342	276	213
N					160	~200	—	201	—	—	272	193	—	—
P						209	—	~340	—	—	490	319	264	184
As							180	331	—	—	464	317	243	180
O								146	—	—	190	205	—	201
S									226	—	326	255	213	—
Se										172	285	243	—	—
F											158	255	238	—
Cl												242	217	209
Br													193	180
I														151

B. Multiple bond energies

C=C 598	C=N 616	C=O 695	N=N 418
C≡C 813	C≡N 866	C≡O 1073	N≡N 946

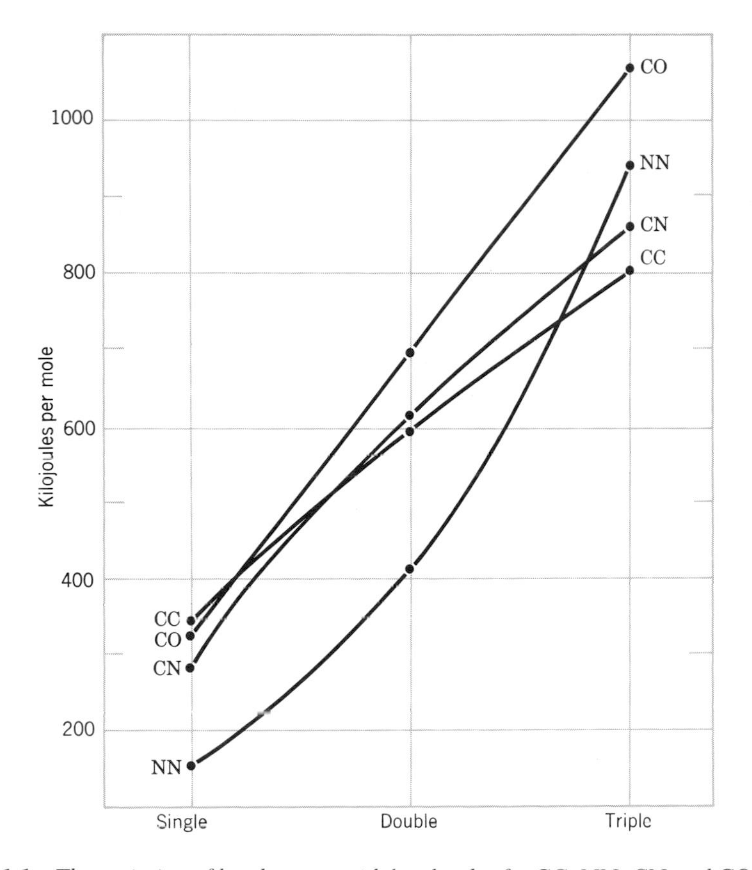

Figure 1-1 *The variation of bond energy with bond order for* CC, NN, CN *and* CO *bonds.*

1-3
Free Energy and Entropy

The direction in which a chemical reaction will go, or the point at which equilibrium will be reached depend on two factors: (1) The tendency to give off energy; exothermic processes are favored. (2) The tendency to attain a state that is statistically more probable, crudely describable as a "more disordered" state.

We already have a measure of the energy change of a system, namely, the magnitude and sign of ΔH.

The statistical probability of a given state of a system is measured by its *entropy*, denoted S. The greater the value of S, the more probable (and, generally, more disordered) is the state. Thus we can rephrase the two statements made in the first paragraph as follows. The likelihood of a process occurring increases as (1) ΔH becomes more negative, or (2) ΔS becomes more positive.

Only in rare cases, an example being racemization,

$$2d\text{-}[\mathrm{Coen}_3]^{3+} = d\text{-}[\mathrm{Coen}_3]^{3+} + l\text{-}[\mathrm{Coen}_3]^{3+}$$

does a reaction have $\Delta H = 0$. In such a case, the direction and extent of reaction depends solely on ΔS. In the case where $\Delta S = 0$, ΔH would alone determine the extent and direction of reaction. However, both cases are exceptional and it is, therefore, necessary to know how these two quantities combine to influence the direction and extent of a reaction. Thermodynamics provides the necessary relationship, which is

$$\Delta G = \Delta H - T\,\Delta S \tag{1-3-1}$$

in which T represents the absolute temperature in kelvins.

The letter G stands for the *free energy*, which is measured in kJ mol^{-1}. The units of entropy are joules per kelvin, JK^{-1}, but for use with ΔG and ΔH in kJ mol^{-1}, ΔS must be expressed as kJ K^{-1}.

1-4
Chemical Equilibrium

For any chemical reaction,

$$aA + bB + cC + \cdots = kK + lL + mM + \cdots$$

the position of equilibrium, for given temperature and pressure, is expressed by the equilibrium constant, K, This is defined as follows:

$$K = \frac{[K]^k[L]^l[M]^m\cdots}{[A]^a[B]^b[C]^c\cdots} \tag{1-4-1}$$

where $[A]$, $[B]$, etc. represent the thermodynamic *activities* of A, B, etc. For reactants in solution, the activities are approximated by the concentrations in moles per liter so long as the solutions are not too concentrated. For gases, the activities are approximated by the pressures. For a pure liquid or solid phase, X, the activity is defined as unity and $[X]^x$ can, therefore, be omitted from the expression for the equilibrium constant.

1-5
ΔG° as a Predictive Tool

For any reaction, the position of the equilibrium at 25° C is determined by the value of ΔG°. ΔG° is defined in a similar way to ΔH°, namely, at 298.15 K (25° C),

$$\Delta G^\circ = \sum \Delta G_f^\circ\,(\text{products}) - \sum \Delta G_f^\circ\,(\text{reactants})$$

In terms of the enthalpy and entropy components, at 25° C,

$$\Delta G^\circ = \Delta H^\circ - 298.15\,\Delta S^\circ$$

The standard entropy change, ΔS°, is defined similarly to ΔG° and ΔH°.

The following relationship exists between ΔG° and the equilibrium constant, K:

$$\Delta G^\circ = -RT \ln K \tag{1-4-2}$$

where R is the gas constant and has the value

$$R = 8.314 \text{ JK}^{-1} \text{ mol}^{-1}$$

in units appropriate to this equation. At 25°C we have

$$\Delta G^\circ = -5.69 \log K \tag{1-4-3}$$

For a reaction with $\Delta G^\circ = 0$, the equilibrium constant is unity. The more negative the value of ΔG° the more the reaction proceeds in the direction written, that is, to produce substances on the right and consume those on the left.

When ΔG° is considered as the net result of enthalpy (ΔH°) and entropy (ΔS°) contributions, a number of possibilities must be considered. Reactions which proceed as written, that is, from left to right, have $\Delta G^\circ < 0$. There are three main ways that this can happen.

1. Both ΔH° and ΔS° favor the reaction. That is, $\Delta H^\circ < 0$ and $\Delta S^\circ > 0$.

2. ΔH° favors the reaction while ΔS° does not, but ΔH° (<0) has a greater absolute value than $T\,\Delta S^\circ$ and thus gives a net negative ΔG°.

3. ΔH° disfavors the reaction (is > 0) but ΔS° is positive and sufficiently large that $T\,\Delta S^\circ$ has a larger absolute magnitude than ΔH°.

There are actual chemical reactions that belong to each of these categories.

Case 1 is fairly common. The formation of CO from the elements is an example:

$$\tfrac{1}{2}O_2(g) + C(s) = CO(g)$$
$$\Delta G^\circ = -137.2 \text{ kJ mol}^{-1}$$
$$\Delta H^\circ = -110.5 \text{ kJ mol}^{-1}$$
$$T\,\Delta S^\circ = 26.7 \text{ kJ mol}^{-1}$$

as are a host of combustion reactions, for example,

$$S(s) + O_2(g) = SO_2(g)$$
$$\Delta G^\circ = -300.4 \text{ kJ mol}^{-1}$$
$$\Delta H^\circ = -292.9 \text{ kJ mol}^{-1}$$
$$T\,\Delta S^\circ = 7.5 \text{ kJ mol}^{-1}$$

$$C_4H_{10}(g) + \tfrac{13}{2}O_2(g) = 4CO_2(g) + 5H_2O(g)$$
$$\Delta G^\circ = -2705 \text{ kJ mol}^{-1}$$
$$\Delta H^\circ = -2659 \text{ kJ mol}^{-1}$$
$$T\,\Delta S^\circ = 46 \text{ kJ mol}^{-1}$$

The reaction used in industrial synthesis of ammonia is an example of case 2:

$$N_2(g) + 3H_2(g) = 2NH_3(g)$$
$$\Delta G^\circ = -16.7 \text{ kJ mol}^{-1}$$
$$\Delta H^\circ = -46.2 \text{ kJ mol}^{-1}$$
$$T\,\Delta S^\circ = -29.5 \text{ kJ mol}^{-1}$$

The negative entropy term can be attributed to the greater "orderliness" of a product system that contains only two moles of independent particles compared with the reactant system in which there are four moles of independent molecules.

Case 3 is the rarest. Examples are provided by substances that dissolve endothermically to give a saturated solution greater than 1 *M* in concentration. This happens with sodium chloride:

$$\begin{aligned} NaCl(s) &= Na^+(aq) + Cl^-(aq) \\ \Delta G^\circ &= -2.7 \\ \Delta H^\circ &= +1.9 \\ T\,\Delta S^\circ &= +4.6 \end{aligned}$$

It must be stressed that the ΔG° value does not *necessarily* predict the *actual result* of a reaction, but only the result that corresponds to the attainment of equilibrium at 25°. It tells what is *possible*, but not what will actually *occur*. Thus, none of the first four reactions cited above which all have $\Delta G^\circ < 0$ actually occurs to a detectable extent at 25° simply on mixing the reactants. Activation energy and/or a catalyst (see page 18) must be supplied. Moreover, there are many compounds that are perfectly stable in a practical sense with positive values of ΔG_f°. They do not spontaneously decompose into the elements, although that would be the equilibrium situation. Common examples are benzene, CS_2, and hydrazine, H_2NNH_2.

The actual occurrence of a reaction requires not only that ΔG° be negative but that the *rate* of the reaction be appreciable. We shall discuss reaction rates and kinetics in Sec. 1-7.

1-6 Temperature Dependence of Chemical Equilibrium

The equilibrium constant for a reaction (or any other process) depends on temperature. That dependence is determined by, and can therefore be used as a way to measure, the magnitude of ΔH°. If the value of the equilibrium constant is K_1 at temperature T_1 and K_2 at T_2, we can write (from Eq. 1-4-2)

$$\ln K_1 = -\frac{\Delta H^\circ}{RT_1} + \frac{\Delta S^\circ}{R} \qquad (1\text{-}6\text{-}1a)$$

$$\ln K_2 = -\frac{\Delta H^\circ}{RT_2} + \frac{\Delta S^\circ}{R} \qquad (1\text{-}6\text{-}1b)$$

Subtracting, we obtain

$$\ln K_1 - \ln K_2 = \ln\frac{K_1}{K_2} = \frac{\Delta H^\circ}{R}\left(\frac{1}{T_1} - \frac{1}{T_2}\right) \qquad (1\text{-}6\text{-}2)$$

Thus we see that if the standard enthalpy change ΔH° is negative (an exothermic reaction) $K_1 < K_2$ if $T_1 > T_2$, and vice versa for ΔH° positive. This is, of course, in qualitative accord with Le Chatelier's principle, and states the result in quantitative form. The importance of Eq. 1-6-2 is that it allows us to calculate ΔH° if we know the equilibrium constant at two different temperatures.

In practice, we might secure greater accuracy by measuring the equilibrium constant at several temperatures and plotting $\ln K$ versus $1/T$. Such a plot should be a straight line, with a slope of $-(\Delta H^\circ/R)$ as can be seen from Eq. 1-6-1.

1-7 Reaction Kinetics

The experimental study of the rate of a reaction normally involves determining its dependence on two sets of factors: (1) The concentration or pressure of each reactant. (2) The reaction conditions, namely, temperature, solvent polarity, catalytic impurities, exposure to light, and the like. Of these, temperature is generally of greatest interest.

1-7a The Rate Law. This is an algebraic equation, deduced from experimental results, which tells how the rate depends on concentrations (or pressures) of reactants, other conditions being fixed. For example, the reaction

$$4\,HBr(g) + O_2(g) = 2\,H_2O(g) + 2\,Br_2(g)$$

has the rate law

$$\frac{d[O_2]}{dt} = -k[HBr][O_2]$$

which states that the concentration of oxygen decreases at a rate proportional to the first power of the HBr concentration and the first power of the oxygen concentration. Note that the rate law expression, $[HBr][O_2]$, does not involve exponents derived from the chemical equation; even though the latter contains the term $4\,HBr$, we have only $[HBr]$ and not $[HBr]^4$ in the rate law. Although a total of five molecules must react to complete the process, the rate law implies that this occurs in a series of steps, of which the slowest or rate-limiting step is one in which O_2 and a single HBr must react.

The above reaction is called a *second-order* reaction because the sum of the exponents in the rate law is 2. A total of two reactive molecules must come together in the rate-limiting step. There are a very few cases of *third-order* reactions, but they are rare because the probability of a 3-particle collision is very low.

The other common and important type of reaction, kinetically speaking, is *first order*. The decomposition of N_2O_5 is an example;

$$2\,N_2O_5(g) = 4\,NO_2(g) + O_2(g)$$

$$\frac{d[N_2O_5]}{dt} = -k[N_2O_5] \qquad (1\text{-}7\text{-}1)$$

The first-order rate law has additional importance because it describes all processes of radioactive decay. A few algebraic manipulations show that this rate law implies certain useful regularities. Equation 1-7-1 can be rearranged and then integrated as follows:

$$\frac{d[\mathrm{N_2O_5}]}{[\mathrm{N_2O_5}]} = -kdt$$

$$d\{\ln[\mathrm{N_2O_5}]\} = -kdt$$

$$\ln \frac{[\mathrm{N_2O_5}]_t}{[\mathrm{N_2O_5}]_0} = -kt \qquad (1\text{-}7\text{-}2)$$

$[N_2O_5]_0$ denotes the initial concentration and $[N_2O_5]_t$ denotes the concentration after t seconds.

An equivalent form of Eq. 1-7-2, written in general for any substance, X, whose decomposition follows first-order kinetics is

$$\frac{[\mathrm{X}]_t}{[\mathrm{X}]_0} = e^{-kt} \qquad (1\text{-}7\text{-}3)$$

From this it can be seen that the time required for *any* given quantity, $[X]_0$, to decrease to a certain fraction of that value is always the same for a particular substance. It depends only on k, the rate constant for the process

$$\mathrm{X} \longrightarrow \text{products}$$

For the particular case where half the original quantity decomposes we have, from Eq. 1-7-2

$$\ln \tfrac{1}{2} = -kt_{1/2}$$

or

$$t_{1/2} = \frac{1}{k} \ln 2 = 0.693/k \qquad (1\text{-}7\text{-}4)$$

Thus, the half-life, $t_{1/2}$, is a valid measure of the rate of the process. The faster the process, the shorter is $t_{1/2}$.

In dealing with radioactive decay, it is customary to use $t_{1/2}$ values and not rate constants as such.

1-7b The Effect of Temperature on Reaction Rates. The rates of chemical reactions increase with increasing temperature. Generally, the dependence of the rate constant, k on temperature T (in kelvins), follows the Arrhenius equation, at least over moderate temperature ranges (*ca.* 100 K):

$$k = Ae^{-E_a/RT} \qquad (1\text{-}7\text{-}5)$$

The coefficient A is called the frequency factor and E_a is called the activation energy. The higher the activation energy the slower the reaction at any given temperature.

By plotting log k against T the value of E_a (as well as A) can be determined. These E_a values are often useful in interpreting the reaction mechanism.

An alternative approach to interpreting the temperature dependence of reaction rates, especially for reactions in solution, is based on the so-called absolute reaction rate theory. In essence, this theory postulates that in the rate-limiting step, the reacting species, A and B, combine reversibly to form an "activated complex," $AB^{\ddagger}$, which can then decompose to products. Thus the following pseudo equilibrium constant is written

$$K^{\ddagger} = \frac{[AB^{\ddagger}]}{[A][B]}$$

The activated complex, $AB^{\ddagger}$, is treated as a normal molecule except that one of its vibrations is considered to have little or no restoring force and to allow dissociation into products. The frequency, ν, with which dissociation to products takes place is assumed to be given by equating the "vibrational" energy $h\nu$, to thermal energy, $\mathbf{k}T$. Thus we write

$$-\frac{d[A]}{dt} = \nu[AB^{\ddagger}] = \left(\frac{\mathbf{k}T}{h}\right)[AB^{\ddagger}]$$

The measureable rate constant is defined by

$$-\frac{d[A]}{dt} = k[A][B]$$

so that we have

$$k = \left(\frac{\mathbf{k}T}{h}\right)\frac{[AB^{\ddagger}]}{[A][B]} = \frac{\mathbf{k}T}{h}K^{\ddagger} \tag{1-7-6}$$

The formation of this activated complex is governed by the thermodynamic considerations similar to those of ordinary chemical equilibria. Thus we have

$$\Delta G^{\ddagger} = -RT \ln K^{\ddagger}$$

and therefore

$$k = \left(\frac{\mathbf{k}T}{h}\right)e^{-\Delta G^{\ddagger}/RT}$$

Furthermore, since

$$\Delta G^{\ddagger} = \Delta H^{\ddagger} - T\,\Delta S^{\ddagger}$$

we obtain

$$k = \left(\frac{\mathbf{k}T}{h}\right)e^{\Delta S^{\ddagger}/R}e^{-\Delta H^{\ddagger}/RT} \tag{1-7-7}$$

This equation can be related directly to the Arrhenius expression by noting that (from thermodynamic considerations not explained in this book)

$$E = \Delta H + RT$$

Making the appropriate substitution we get

$$k = (\mathbf{k}T/h)e^{\Delta S^{\ddagger}/R}e^{-(E-RT)/RT}$$
$$= \left(\frac{e\mathbf{k}T}{h}\right)e^{\Delta S^{\ddagger}/R}e^{-E_a/RT}$$

Thus we see that the Arrhenius frequency factor is a function of the entropy of activation.

1-7c Graphical Representation. The course of a chemical reaction as described in the absolute reaction rate theory can be conveniently depicted in a graph of free energy versus the "reaction coordinate." The latter is simply the pathway along which the changes in various interatomic distances progress as the system passes from reactants to activated complex to products. A representative graph is shown in *Fig. 1-2* for the unimolecular decomposition of formic acid.

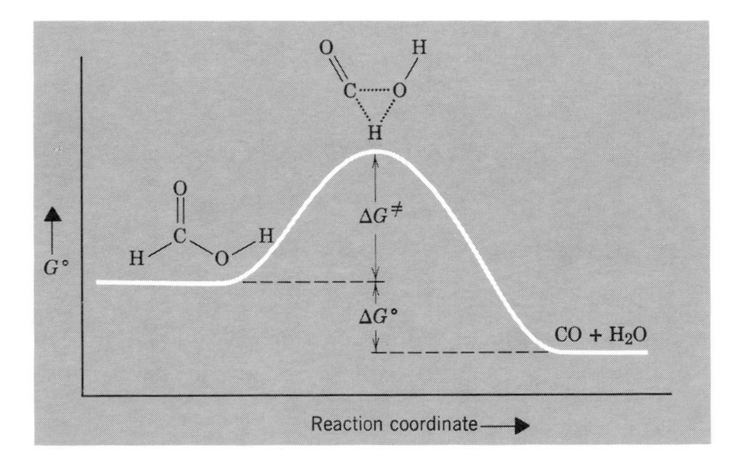

Figure 1-2 *The free energy versus reaction coordinate for the decomposition of formic acid. $\Delta G^{\ddagger}$ is the free energy of activation. ΔG° is the standard free energy change for the complete reaction.*

1-7d The Effect of Catalysts. A catalyst is a substance that causes a reaction to proceed more rapidly to equilibrium. It does not change the value of the equilibrium constant, and it does not itself undergo any net change. In terms of the absolute reaction rate theory, the role of a catalyst is to lower the free energy of activation, $\Delta G^{\ddagger}$. Some catalysts do this by simply assisting the reactants to attain basically the same activated complex as they do in the absence of a catalyst, but most of them appear to provide a different sort of pathway, in which they are temporarily bound, which has a lower free energy.

An example of acid catalysis, in which protonated intermediates play a role, is provided by the catalytic effect of protonic acids on the decomposition of formic acid. *Figure 1-3*, when compared with *Fig. 1-2* (the uncatalyzed reaction pathway) shows how the catalyst modifies the reaction pathway so that the highest value of the free energy that must be reached is diminished.

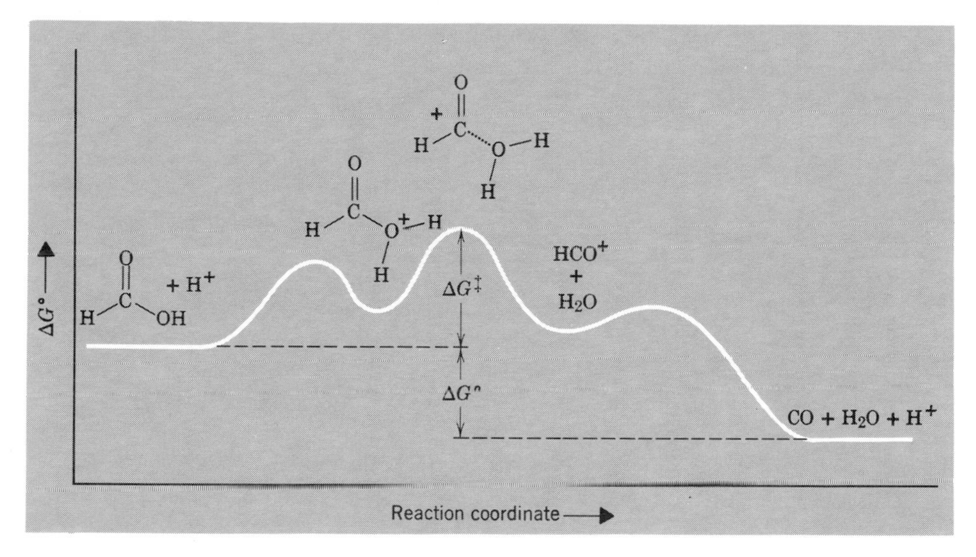

Figure 1-3 *Acid catalyzed decomposition of formic acid.* ΔG° *is the same as in Figure 1-2, but* $\Delta G^\ddagger$ *is now smaller.*

Catalysis may be either homogeneous or heterogeneous. In the example just cited, it is homogeneous. The strong acid is added to the solution of formic acid and the whole process proceeds in the one liquid phase. On the other hand, especially in the majority of industrially important reactions, the catalyst is a solid surface and the reactants, either as gases or in solution, flow over the surface. Many reactions can be catalyzed in more than one way, and in some cases both homogeneously and heterogeneously.

The hydrogenation of olefins affords an example where both heterogeneous and homogeneous catalyses are effective. The simple, uncatalyzed reaction,

$$RCH{=}CH_2 + H_2 \longrightarrow RCH_2CH_3$$

is impractically slow unless very high temperatures are used, and that gives rise to other difficulties, such as the expense and difficulty of maintaining the temperature and the occurrence of other, undesired reactions. If the gases are allowed to come in contact with certain forms of noble metals, for example, platinum, supported on high surface area materials such as silica or alumina, catalysis occurs. It is believed that both reactants are absorbed by the metal surface, possibly with dissociation of the hydrogen, as indicated in *Fig 1-4*. Homogeneous catalysis (one of many examples to be discussed in detail in Chapter 30) proceeds somewhat similarly but entirely on one metal ion that is present in solution as a complex. The

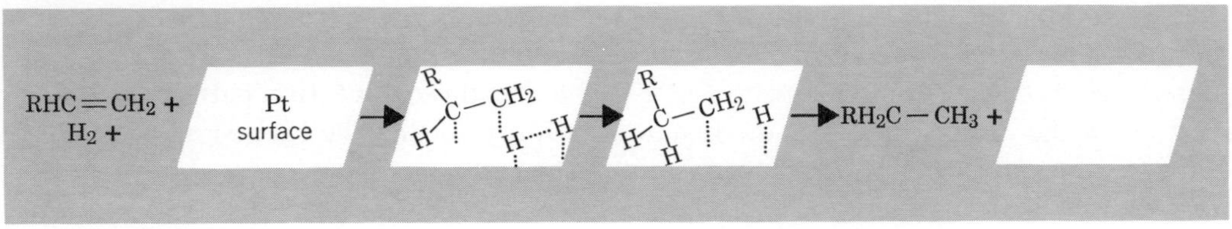

Figure 1-4 *Schematic indication of how a suitable platinum surface can catalyze olefin hydrogenation by binding and bringing together the reactants.*

important steps are believed to be the following, where M represents a metal atom such as Rh:

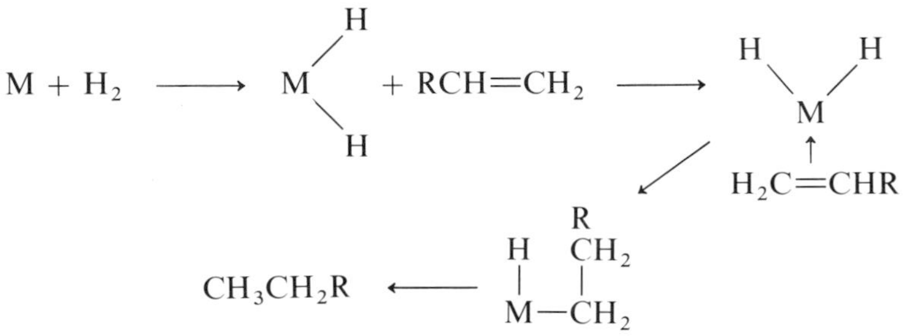

$$M + H_2 \longrightarrow M(H)_2 \xrightarrow{+\,RCH{=}CH_2} (H)_2M \leftarrow H_2C{=}CHR \longrightarrow H{-}M{-}CH_2CH_2R \longrightarrow CH_3CH_2R$$

1-8 Cell and Electrode Potentials

Although it is true that the direction and extent of a reaction are indicated by the sign and magnitude of ΔG°, this is not generally an easy quantity to measure. There is one class of reactions, redox reactions in solution, that frequently allows straightforward measurement of ΔG°. The quantity actually measured is the potential difference, ΔE, in volts, between two electrodes. Under the proper conditions, this can be related to ΔG° beginning with the following equation:

$$\Delta E = \Delta E^\circ - \frac{RT}{nF} \log Q \qquad (1\text{-}8\text{-}1)$$

ΔE° is the so-called standard potential, which will be discussed more fully below. n is the number of electrons in the redox reaction as written, and F is the faraday, 96,500 coulombs.

Q is an expression with the same algebraic form as the equilibrium constant for the reaction, into which the actual activities that exist when ΔE is measured are inserted. Clearly, when each concentration equals unity, $\log Q = \log 1 = 0$ and the measured ΔE equals ΔE°, the standard potential for the cell.

To illustrate, the reaction between zinc and hydrogen ions may be used

$$Zn(s) + 2H^+(aq) = Zn^{2+}(aq) + H_2(g)$$

For this, $n = 2$ and Q has the form:

$$\frac{A_{H_2} \cdot A_{Zn^{2+}}}{A_{H^+}^{\ 2}} \qquad (A_{Zn} = 1)$$

Now, suppose the reaction of interest is allowed to run until equilibrium is reached. The numerical value of Q is then equal to the equilibrium constant K. Moreover, at equilibrium there is no longer any tendency for electrons to flow from one electrode to the other: $\Delta E = 0$. Thus, we have

$$0 = \Delta E^\circ - \frac{RT}{nF} \ln K$$

or

$$\Delta E^\circ = \frac{RT}{nF} \ln K \qquad (1\text{-}8\text{-}2)$$

However, we already know that

$$\Delta G^\circ = -RT \ln K \qquad (1\text{-}4\text{-}2)$$

We, therefore, have a way of relating cell potentials to ΔG° values, that is

$$\frac{nF}{RT} \Delta E^\circ = -\frac{1}{RT} \Delta G^\circ$$

or

$$\Delta G^\circ = -nF\, \Delta E^\circ \qquad (1\text{-}8\text{-}3)$$

Just as ΔG° values for a series of reactions may be added algebraically to give ΔG° for a reaction which is the sum of those added so, too, may ΔE° values be combined. But, remember that it is $n\, \Delta E^\circ$, not simply ΔE°, which must be used for each reaction. The factor F will, of course, cancel out in such a computation. For instance,

$$\begin{array}{llr} (n = 2) & Zn(s) + 2H^+(aq) = Zn^{2+}(aq) + H_2(g) & \Delta E_1^\circ = +0.763 \\ (n = 2) & 2Cr(aq)^{3+} + H_2(g) = 2Cr^{2+}(aq) + 2H^+(aq) & \Delta E_2^\circ = -0.408 \\ \hline (n = 2) & Zn(s) + 2Cr^{3+}(aq) = Zn^{2+}(aq) + 2Cr^{2+}(aq) & \Delta E_3^\circ = +0.355 \end{array}$$

The correct relationship is

$$2\, \Delta E_3^\circ = 2\, \Delta E_1^\circ + 2\, \Delta E_2^\circ$$

In this example, we have added balanced equations to give a balanced equation. This automatically ensures that the coefficient n is the same for each ΔE°. However, in dealing with electrode potentials (see below) instead of potentials of balanced reactions the cancellation is not automatic, as we shall learn presently.

Signs of ΔE° Values. Physically, there is no absolute way to associate algebraic signs with measured ΔE° values. Yet a convention must be adhered to since, as illustrated above, the signs of some are opposite to those of others. For actual, balanced chemical reactions, the sign convention for ΔE° values is defined by Eq. 1-6-3. Negative values of ΔG° correspond to reactions for which the equilibrium state favors products, that is, reactions that proceed in the direction written. Therefore, reactions that "go" also have positive ΔE° values. The reduction of Cr^{3+} by elemental zinc ($\Delta E^\circ = +0.355$) therefore goes as written in the above example.

Half-cells and Half-cell (or Electrode) Potentials. Any complete, balanced chemical reaction can be artificially separated into two "half reactions." Correspondingly, any complete electrochemical cell can be separated into two hypothetical half-cells. The potential of the actual cell, ΔE°, can then be regarded as the algebraic sum of the two half-cell potentials. In the three reactions above, there are a total of three distinct half-cells. Let us consider first the reaction of zinc and $H^+(aq)$:

$$\begin{array}{rlr} Zn(s) = Zn^{2+}(aq) + 2e^- & & E_1^\circ = +0.763 \text{ V} \\ 2H^+(aq) + 2e^- = H_2(g) & & E_2^\circ = 0.000 \text{ V} \\ \hline Zn(s) + 2H^+(aq) = Zn^{2+}(aq) + H_2(g) & & \Delta E^\circ = +0.763 \text{ V} \end{array}$$

E_1° and E_2° must be chosen to give the sum $+0.763$ V. The only solution to this or any similar problem is to assign an arbitrary *conventional* value to one such half-cell potential. All others will then be determined. The conventional choice is to assign the hydrogen half-cell a standard potential of zero. The zinc half-cell reaction as written then has $E^\circ = +0.763$ V. In an exactly analogous way we get

$$Cr^{3+}(aq) + e^- = Cr^{2+}(aq) \qquad E^\circ = -0.408 \text{ V}$$

These two half-cell potentials may then be used directly to calculate the standard potential for reduction of Cr^{3+} by Zn(s):

$$\begin{array}{rlr} Zn(s) = Zn^{2+}(aq) + 2e^- & & E^\circ = +0.763 \text{ V} \\ 2e^- + 2Cr^{3+}(aq) = 2Cr^{2+}(aq) & & E^\circ = -0.408 \text{ V} \\ \hline Zn(s) + 2Cr^{3+}(aq) = Zn^{2+}(aq) + 2Cr^{2+}(aq) & & \Delta E^\circ = +0.355 \text{ V} \end{array}$$

Since each reaction involves the same number of electrons, the factor n in the expression $\Delta G^\circ = -nFE^\circ$ is the same in case and cancels out.

When two half-cell reactions are added to give a third half-cell reaction, then the n's cannot cancel out and must be explicitly employed in the computation.

For example:

$$\begin{array}{lll} Cl^- + 3H_2O = ClO_3^- + 6H^+ + 6e^- & E^\circ = -1.45 & 6E_1^\circ = -8.70 \text{ V} \\ e^- + \frac{1}{2}Cl_2 = Cl^- & E^\circ = +1.36 & 1E_2^\circ = +1.36 \text{ V} \\ \hline \frac{1}{2}Cl_2 + 3H_2O = ClO_3^- + 6H^+ + 5e^- & E^\circ = -1.47 & 5E_3^\circ = -7.34 \text{ V} \end{array}$$

$$5E_3^\circ = 6E_1^\circ + 1E_2^\circ$$
$$E_3^\circ \neq E_1^\circ + E_2^\circ$$

Table 1-2 *Some Half-cell Reduction Potentials*

Reaction Equation	E° (volts)
$Li^+ + e = Li$	−3.04
$Cs^+ + e = Cs$	−3.02
$Rb^+ + e = Rb$	−2.99
$K^+ + e = K$	−2.92
$Ba^{2+} + 2e = Ba$	−2.90
$Sr^{2+} + 2e = Sr$	−2.89
$Ca^{2+} + 2e = Ca$	−2.87
$Na^+ + e = Na$	−2.71
$Mg^{2+} + 2e = Mg$	−2.34
$\frac{1}{2}H_2 + e = H^-$	−2.23
$Al^{3+} + 3e = Al$	−1.67
$Zn^{2+} + 2e = Zn$	−0.76
$Fe^{2+} + 2e = Fe$	−0.44
$Cr^{3+} + e = Cr^{2+}$	−0.41
$H_3PO_4 + 2H^+ + 2e = H_3PO_3 + H_2O$	−0.20
$Sn^{2+} + 2e = Sn$	−0.14
$H^+ + e = \frac{1}{2}H_2$	0.00
$Sn^{4+} + 2e = Sn^{2+}$	0.15
$Cu^{2+} + e = Cu^+$	0.15
$S_4O_6^{2-} + 2e = 2S_2O_3^{2-}$	0.17
$Cu^{2+} + 2e = Cu$	0.34
$Cu^+ + e = Cu$	0.52
$\frac{1}{2}I_2 + e = I^-$	0.53
$H_3AsO_4 + 2H^+ + 2e = H_3AsO_3 + H_2O$	0.56
$O_2 + 2H^+ + 2e = H_2O_2$	0.68
$Fe^{3+} + e = Fe^{2+}$	0.76
$\frac{1}{2}Br_2 + e = Br^-$	1.09
$IO_3^- + 6H^+ + 6e = I^- + 3H_2O$	1.09
$IO_3^- + 6H^+ + 5e = \frac{1}{2}I_2 + 3H_2O$	1.20
$\frac{1}{2}Cl_2 + e = Cl^-$	1.36
$\frac{1}{2}Cr_2O_7^{2-} + 7H^+ + 3e = Cr^{3+} + \frac{7}{2}H_2O$	1.36
$MnO_4^- + 8H^+ + 5e = Mn^{2+} + 4H_2O$	1.52
$Ce^{4+} + e = Ce^{3+}$	1.61
$H_2O_2 + 2H^+ + 2e = 2H_2O$	1.77
$\frac{1}{2}S_2O_8^{2-} + e = SO_4^{2-}$	2.05
$O_3 + 2H^+ + 2e = O_2 + H_2O$	2.07
$\frac{1}{2}F_2 + e = F^-$	2.85
$\frac{1}{2}F_2 + H^+ + e = HF$	3.03

Tables of Half-cell or Electrode Potentials. The International Union of Pure and Applied Chemistry has agreed that half-cell and electrode potentials shall be written as reductions and the terms "half-cell potential" or "electrode potential" shall mean values carrying the sign appropriate to the reduction reaction. For example, the zinc electrode reaction is tabulated as

$$Zn^{2+}(aq) + 2e^- = Zn(s) \qquad E^\circ = -0.763 \text{ V}$$

Zinc is said to have an electrode potential of *minus* 0.763 V.

This convention is most easily remembered by noting that a half-cell reaction with a *negative* potential is *electron* rich. When two half-cells are combined to produce a complete electrolytic cell, the electrode having the more negative standard half-cell potential will be, physically, the negative electrode (electron source) if the cell is to be operated as a battery.

A list of some important standard half-cell or electrode potentials is given in Table 1-2.

1-9 Atomic Nuclei and Nuclear Reactions

Although chemical processes depend essentially on how the electrons in atoms and molecules interact with each other, the internal nature of nuclei, and changes in nuclear composition (nuclear reactions) play an important role in the study and understanding of chemical processes. Conversely, the study of nuclear processes constitutes an important area of applied chemistry, particularly inorganic chemistry.

Atomic nuclei consist of a certain number, N, of protons (p) called the *atomic number*, and a certain number of neutrons (n). The masses of these particles are each approximately equal to one mass unit and the total number of nucleons (protons and neutrons) is called the *mass number*, A. The two numbers, N and A, completely designate a given nuclear species (neglecting the excited states of nuclei). It is the number of protons, that is, the atomic number that tells us which *element* we are dealing with, and for a given N, the different values of A, resulting from different numbers of neutrons, are responsible for the existence of different *isotopes* of that element.

When it is necessary to specify a particular isotope of an element, the mass number is placed as a left superscript. Thus the isotopes of hydrogen are ^{1}H, ^{2}H, and ^{3}H. In this one case, there are generally used separate symbols and names for the less common isotopes: ^{2}H = D, deuterium, and ^{3}H = T, tritium.

All isotopes of an element have the same chemical properties except insofar as the mass differences may alter the exact magnitudes of reaction rates and thermodynamic properties.* These mass effects are virtually insignificant for elements other than hydrogen where the percentage variation in the masses of the isotopes is uniquely large.

Most elements are found in nature as a mixture of two or more isotopes. Tin occurs as a mixture of nine, from ^{112}Sn (0.96 %) through the most abundant

* A few other minor effects, for instance the existence of ortho and para forms of some molecules such as H_2, and the effect that radioactive decay may have on chemical properties, will not be considered here.

^{118}Sn (24.03%) to ^{124}Sn (5.94%). A few common elements that are terrestrially monoisotopic are ^{27}Al, ^{31}P and ^{55}Mn. Because the exact masses of protons and neutrons differ, and neither is precisely equal to one atomic mass unit, and for other reasons to be mentioned below, the masses of nuclei are not equal to their mass numbers. The actual atomic mass of ^{55}Mn, for example, is 54.9381 atomic mass units.

Usually, the isotopic composition of an element is constant all over the earth and thus its practical atomic weight, as found in the usual tables, is invariant. In a few instances, lead being most conspicuous, isotopic composition varies from place to place because of different parentage of the element in radioactive species of higher atomic number. Also, for elements that do not occur in nature, the "atomic weight" depends on which isotope, or isotopes, are made in nuclear reactions. In tables, it is customary to give for these the mass number of the longest-lived isotope known.

Spontaneous Decay of Nuclei. Only certain nuclear compositions are stable indefinitely. All others spontaneously decompose by emitting alpha particles (2p2n) or β particles (negative electrons) or, in rare cases in a few other ways. Emission of high energy photons (γ-rays) generally accompanies nuclear decay. Alpha emission reduces the atomic number by 2 and the mass number by 4. An example is

$$^{238}U \longrightarrow {}^{234}Th + \alpha$$

β-decay advances the atomic number by one unit without changing the mass number. Effectively, a neutron becomes a proton. An example is

$$^{60}Co \longrightarrow {}^{60}Ni + \beta^-$$

These two decay processes follow first-order kinetics (page 15) and are insensitive to the physical or chemical conditions surrounding the atom. The half-lives are unaffected by temperature, which is an important distinction from first-order chemical reactions. In short, the half-life of an unstable isotope is one of its fixed, characteristic properties.

All elements have some unstable (i.e., radioactive) isotopes. Of particular importance is the fact that some elements have no stable isotopes. No element with atomic number 84 (polonium) or higher has *any* stable isotope. Some, for instance, uranium and thorium, are found in substantial quantities in nature because they have, at least, one very long-lived isotope. Others, for instance, Ra, Rn, are found only in small quantities in a steady state as intermediates in radioactive decay chains. Others, for instance, At, Fr, have no single isotope stable enough to be present in macroscopic quantities. There are also two other elements, Tc and Pm, which not only have no stable isotope but none sufficiently long-lived that any detectable quantities of these elements occur in nature. Both are recovered from fission products.

Nuclear Fission. Many of the heaviest nuclei can be induced to break up into two fragments of intermediate size, a process called nuclear fission. The stimulus for this is the capture of a neutron by the heavy nucleus. This creates an excited state which splits. In the process, several neutrons and a great deal of energy are released. Because the process generates more neutrons than are required to stimulate it, a chain reaction is possible. Each individual fission can lead to an average of more than one subsequent fission. Thus, the process can become self-sustaining (nuclear reactor) or even explosive (atomic bomb). A representative example of a nuclear fission process (shown schematically in *Fig. 1-5*) is the following

	^{235}U +	n →	^{141}Ba +	^{92}Kr +	3n
Mass number	235	1	141	92	3
Atomic number	92	0	56	36	0
Neutrons	143	1	85	56	3

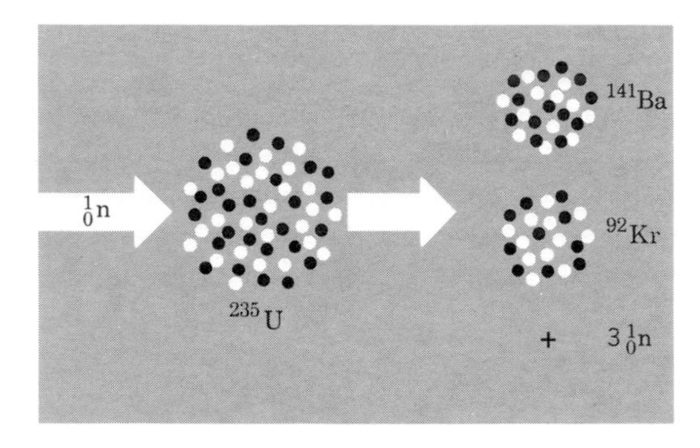

Figure 1-5 *A schematic equation for a typical nuclear fission process.*

Nuclear Fusion. In principle, very light nuclei can combine to form heavier ones and release energy as they do so. Such processes are the main source of the energy generated in the Sun and other stars. They also form the basis of the hydrogen bomb. At present, engineering research is underway to adapt nuclear fusion processes to the controlled, sustained generation of energy, but practical results cannot be expected in the near future.

Nuclear Binding Energies. The reasons why fission and fusion processes are sources of nuclear energy can be understood by reference to a plot of the binding energy per nucleon as a function of mass number (*Fig. 1-6*). Binding energy is

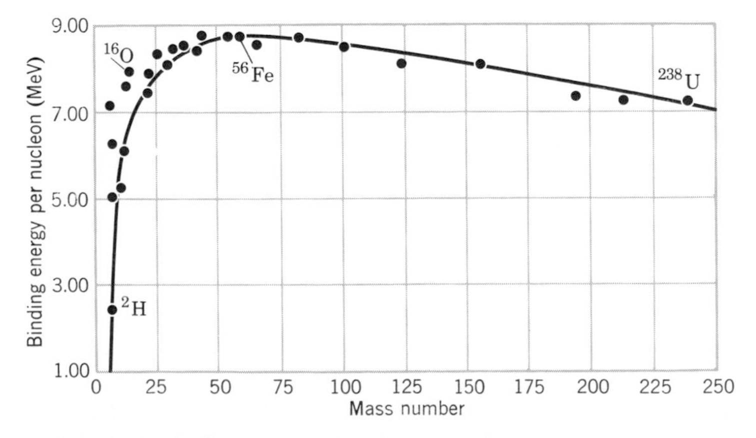

Figure 1-6 *Binding energy of nucleons as a function of mass number.*

figured by subtracting the actual nuclear mass from the sum of the individual masses of the constituent neutrons and protons and converting that mass difference into energy using Einstein's equation, $E = mc^2$. The usual unit for nuclear energies is one million electron volts (MeV), which is equal to 96.5×10^6 kJ mol^{-1}.

For example, for ^{12}C we have

1.	Actual mass:	12.000000 amu
2.	6x proton mass:	6.043662 amu
3.	6x neutron mass:	6.051990 amu
	(2) + (3) − (1) =	0.095652 amu

One amu = 931.4 MeV. Hence:
Total binding energy = (931.4)(0.095652) = 89.09 MeV.
Binding energy per nucleon = (89.09)/12 = 7.42 MeV.

Since the formation of nuclei of intermediate masses releases more energy per nucleon than the formation of very light or very heavy ones, energy will be released when very heavy nuclei split (fission) or when very light ones coalesce (fusion).

Nuclear Reactions. For many purposes the chemist often requires particular isotopes not available in nature, or even elements not found in nature. These can be made in nuclear reactors. In general, they are formed when the nucleus of a particular isotope of one element captures one or more particles (α-particles or neutrons) to form an unstable intermediate. This decays, ejecting one or more particles, to give the product. The more common changes are indicated in *Fig. 1-7*.

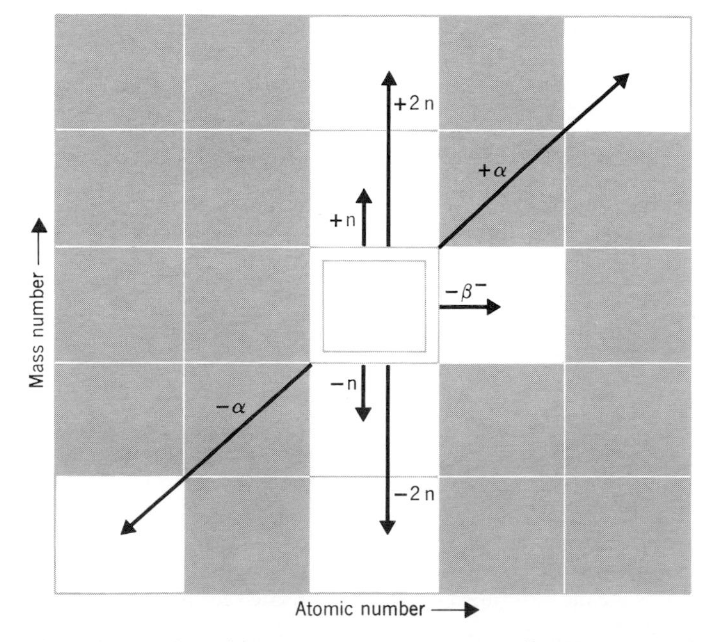

Figure 1-7 *Chart showing how the more important processes of capture or ejection of nuclear particles changes the nuclei.*

A convenient shorthand for writing nuclear reactions is illustrated by the following for the process used to prepare an isotope of astatine.

$$^{209}\text{Bi}(\alpha, 2\,\text{n})^{211}\text{At}$$

This equation says that ^{209}Bi captures an α-particle, and the resulting nuclear species, which is not isolable, promptly emits two neutrons to give the astatine isotope of mass number 211. The mass number increases by 4 (for α) minus 2 (for 2n) = 2 units and the atomic number increases by two units due to the two protons in the α-particle. Another representative nuclear reaction is

$$^{209}\text{Bi}(\text{n}, \gamma)^{210}\text{Bi} \longrightarrow {}^{210}\text{Po} + \beta$$

Study Questions

A

1. What are the prefixes for each of the multipliers 10^n for $n = -1, -2, -3, -6, -9, -12, 1, 2, 3, 6, 9$ and 12.
2. What are the SI units for length, mass, time, electric current, and temperature? What are their abbreviations?
3. What is an angstrom? Is it an SI unit? How is it related to some SI units?
4. What is the SI unit for energy? What are some commonly used non-SI energy units?
5. Define exothermic and endothermic. What are the signs of ΔH for each type of process?
6. How is the standard enthalpy of formation of a compound defined? Write the appropriate chemical equation for ΔH_f° of CF_3SO_3H.
7. What are the more common, trivial names for ionization enthalpy and electron attachment enthalpy?
8. Write an equation that can be used to define the mean S—F bond energy in SF_6. How is this value likely to be related to the energy of the process $SF_6(g) = SF_5(g) + F(g)$?
9. How do bond energies vary as a function of bond order?
10. Complete the following equations using the indicated quantities:

$$\Delta G = \Delta H, \Delta S, T$$
$$\Delta G^\circ = K, R, T$$
$$\Delta G^\circ = n, F, \Delta E$$

$$k = A, E_a, R, T$$

$$\Delta E = \Delta E^\circ, R, T, n, F, Q$$
$$\Delta E^\circ = R, T, n, F, K$$

11. Give a qualitative definition of entropy.
12. What significance do ΔG° values < 0, 0 and > 0 have?
13. What is meant by a rate law? How is it related to the balanced equation for the reaction?
14. How is the half-life ($t_{1/2}$) defined in terms of the rate constant? To what order of reaction only does the concept of a half-life apply?

15. What is the conventional, arbitrary assumption in defining the magnitudes of half-cell potentials? What is the convention for their signs?
16. Define mass number, atomic number, isotope, monoisotopic, α-decay, β-decay, nuclear fission, nuclear fusion, nuclear binding energy.
17. What element has different names for its isotopes? What are they?
18. Sketch the curve of nuclear binding energy per nucleon versus mass number.

B

1. How might one determine the absolute potential of a half-cell, rather than the arbitrary value based on E° for the hydrogen electrode being zero.
2. Name as many elements as you can whose electron attachment enthalpies might be negative.
3. The N—N bond energy in F_2NNF_2 is only about 80 kJ mol^{-1} compared to 160 kJ mol^{-1} in H_2NNH_2. Suggest a reason.
4. What would be the signs of the entropy changes for the following processes:

$$H_2O(l) = H_2O(g)$$
$$P_4(g) + 10F_2(g) = 4PF_5(g)$$
$$I_2(s) + Cl_2(g) = 2ICl(g)$$
$$BF_3(g) + NH_3(g) = H_3NBF_3(g)$$

5. Assuming bond energies are essentially independent of temperature, suggest a reason why there are many chemical reactions that are thermodynamically favored at high temperatures but not at low ones?
6. Using Table 1-1, estimate ΔH_f° values for the following molecules $HNCl_2$, CF_3SF_3, Cl_2NNH_2, $(CH_3)_6Ge_2$.
7. What do you suppose is the main thermodynamic reason that the following reaction has an equilibrium constant >1:

$$BCl_3(g) + BBr_3(g) = BCl_2Br(g) + BClBr_2(g)$$

8. Some chemical reactions have two-term rate laws; for example, in aqueous solution.

$$Pt(NH_3)_3Cl^+ + Br^- = Pt(NH_3)_3Br^+ + Cl^-$$
$$\text{Rate} = k[Pt(NH_3)_3Cl^+] + k'[Pt(NH_3)_3Cl^+][Br^-]$$

What, in general, do you suppose causes this; discuss specifically, then, the above reaction.
9. Sometimes rate laws for aqueous reactions are written with $[H^+]^{-1}$. What would be an equivalent way to write this to eliminate the negative exponent? How would the magnitude of the rate constant change when the rate law is thus rewritten?

Chapter 1
Study Guide

Scope and Purpose. This chapter summarizes physical chemistry that is helpful and sometimes essential to an understanding of the rest of the book. It is not intended that this chapter provide an *ab initio* introduction to this material but, instead, a summary or review. It is assumed that the student of this book will have already taken an introductory course in basic physical chemistry, or will be doing so as he studies this book.

It is not necessary that this chapter be covered in its entirety—or even in part—before proceeding to the rest of the book. The student should, however, be sure he is acquainted with Section 1-1 (units) and the sign conventions for energy changes, which are discussed in Section 1-2.

Supplementary Reading. Any good, modern textbook of physical chemistry will provide a more detailed and didactic presentation of the various topics covered in this chapter.

Thermochemistry is well covered by W. E. Dasent in *Inorganic Energetics*, Penguin Books, 1970.

Ionization enthalpies and electron attachment enthalpies have recently been reviewed by J. F. Liebman, *J. Chem. Educ.*, *50*, 831 (1973).

2

atomic structure

2-1
Introduction

The term atomic structure customarily is used to include not only the possible distributions of electrons about the nucleus but also the energies of the electrons and their magnetic properties, ionization enthalpies, and the like, which depend on the distribution.

Prior to 1913, a great deal of experimental work had been done to measure the frequencies of light that could be absorbed or emitted by atoms. It was found to be characteristic of atoms that they absorb and emit light only at certain sharply defined frequencies. The obvious questions raised were: Why is this true, and what determines the exact frequencies, which vary from one kind of atom to another? Efforts to answer these questions naturally tended to center on the hydrogen atom, which is the smallest and simplest one, and has the simplest spectra.

It had been found that the spectral lines for the hydrogen atom consisted of several converging series of lines (*Fig. 2-1*). The patterns conformed precisely to the following equations:

1. Ultraviolet (Lyman) series:

$$\bar{\nu} = R\left(\frac{1}{1^2} - \frac{1}{n^2}\right) \qquad n = 2, 3, 4, 5 \ldots$$

2. Visible (Balmer) series:

$$\bar{\nu} = R\left(\frac{1}{2^2} - \frac{1}{n^2}\right) \qquad n = 3, 4, 5, 6 \ldots$$

3. Infrared (Paschen) series:

$$\bar{\nu} = R\left(\frac{1}{3^2} - \frac{1}{n^2}\right) \qquad n = 4, 5, 6, 7 \ldots$$

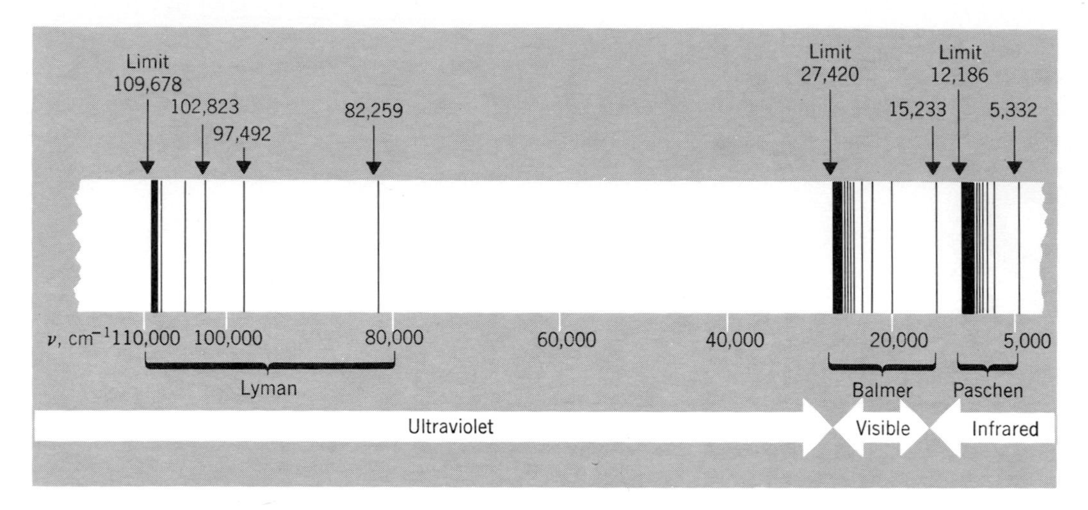

Figure 2-1 *The appearance of the emission spectrum of atomic hydrogen as recorded on a strip of film. In each series the lines become more and more closely spaced from right to left. As each limit is approached, they get too close together to be drawn separately. There are two more well-defined series, further into the infrared, which are not shown.*

These formulas, developed by Rydberg, contain an empirical constant, R, which equals 109, 678 cm^{-1}. Clearly they are all special cases of a general formula:

$$\bar{\nu} = R\left(\frac{1}{m^2} - \frac{1}{n^2}\right) \qquad \begin{array}{l} m = 1, 2, 3, 4 \ldots \\ n = (m+1), (m+2), (m+3), \ldots \end{array}$$

These observations seem neat and simple, just the sort of clear, orderly results scientists generally love to work with. However, in the first decade of this century scientists found these data extremely frustrating, since they could not devise any explanation for them.

In 1913, a young Danish physicist, Niels Bohr, made a bold proposal. He accepted the fact that the observations defied explanation in terms of the theories then in use; he then broke with tradition. Bohr proposed that the electron could revolve about the proton in orbits of certain radii indefinitely, whereas classical physics declared this to be impossible, asserting that the electron would spiral inward, radiating energy all the while and finally crash into the nucleus.

Bohr then found a way to select his orbits so that he could derive Rydberg's formulas. He accomplished this by imposing one condition. In each orbit, the angular momentum of the electron, mvr, must be quantized, that is, have a value given by the formula:

$$mvr = n\frac{h}{2\pi}, \qquad \text{where } n = 1, 2, 3, 4, \ldots$$

where m, v, r and h are the mass and velocity of the electron, the radius of the orbit, and Planck's constant. n, called the *quantum number* of the orbit, was required to be an integer. These two points—stable orbits and the quantization condition were, respectively, in conflict with and utterly outside of the accepted physical theory of the time. However, by using these assumptions, and treating all the rest of the physics in a perfectly traditional way, it was determined that for each orbit the energy and radius were given by

$$E = -\frac{2\pi^2 m Z^2 e^4}{n^2 h^2} \tag{2-1-1}$$

$$r = \frac{n^2 h^2}{4\pi^2 m e^2 Z} \tag{2-1-2}$$

Z is the nuclear charge. For the hydrogen atom itself it has the value 1 and could be omitted, but to make the equations general it is included here.

The collection of constants in the expression for E is equal to the value of Rydberg's constant. In short, Bohr had obtained

$$E = -\frac{R}{n^2}$$

The explanation for the spectral series was now at hand. The lowest energy (most negative) is $-R/1^2 = -R$. If the electron is excited to an orbit with higher

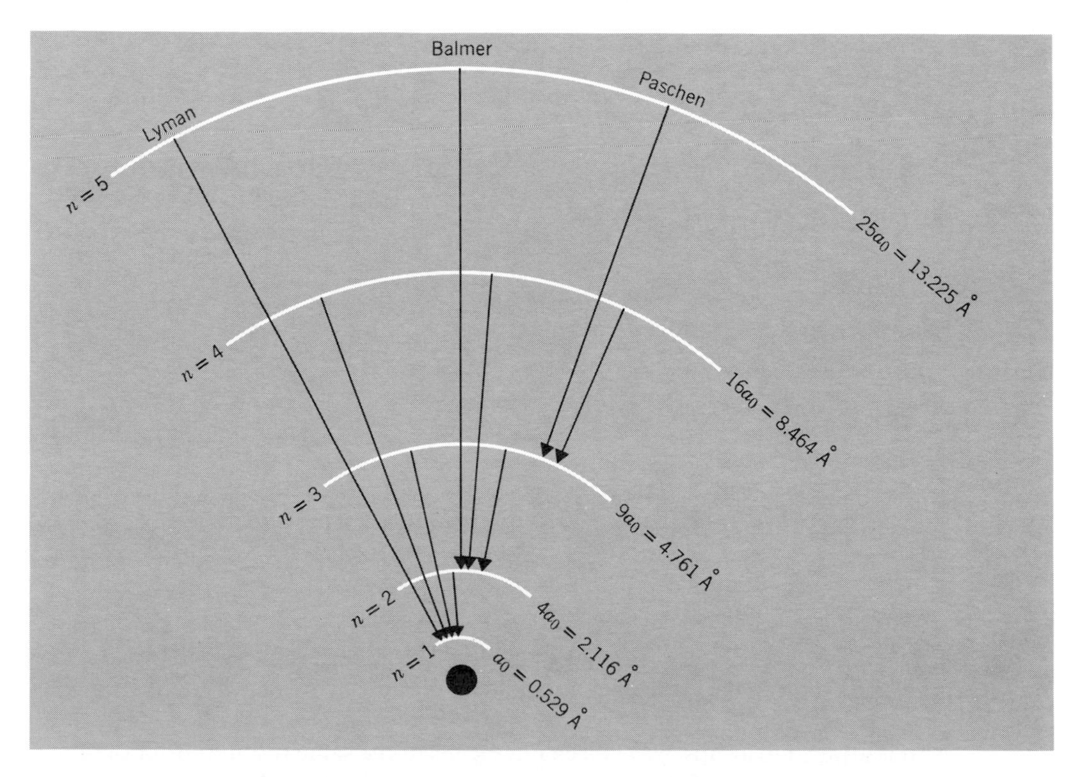

Figure 2-2 *A diagramatic indication of the Bohr orbits and energies of the hydrogen atom, and how they contribute to the spectral series discussed in the text. Each arc represents a portion of an orbit. The quantum number, n, and radius of each one is given.*

energy ($n > 1$) and then drops back to the orbit with $n = 1$, the ultraviolet series of lines will be obtained, since the electron will lose energy equal to $R(1/1^2 - 1/n^2)$. The other series arise when the electron drops from upper orbits to those with $n = 2$ (visible series), and $n = 3$ (infrared series). These relations are illustrated in *Fig. 2-2*.

In all of the above, Bohr was making use of Planck's earlier postulate, also in conflict with classical physics, that radiation is quantized, and each quantum of radiation of frequency $\bar{\nu}$ has an energy, E, given by

$$E = h\bar{\nu}$$

The radius calculated for the most stable orbit is 0.529 Å. This is called the Bohr radius.

The Bohr model was extended by Sommerfeld, who showed that finer features of the hydrogen atom spectrum, which were observed on application of a magnetic field, could also be accounted for if elliptical as well as circular orbits were used. This introduced another quantum number which gave the ellipticity of the orbit.

Unfortunately, the Bohr–Sommerfeld theory had to be abandoned for a number of reasons. For one thing, it could not deal with atoms more complex than hydrogen. Perhaps more important, other work showed that it is simply not correct to regard the electrons in atoms as discrete particles with precisely defined positions and velocities, as is done in Bohr's model.

2-2 Wave Mechanics

As we noted above, one reason for abandoning the Bohr model of the atom was the fact that other experimental and theoretical work showed that electrons have wavelike properties. In 1924 the French physicist de Broglie suggested that all matter could at times exhibit wavelike properties. In particular, small particles, such as electrons or nucleons, traveling with a velocity v, were postulated to have an associated wave character such that the wavelength, λ, is related to their mass, m, and velocity by Eq. 2-2-1.

$$\lambda = \frac{h}{mv} \tag{2-2-1}$$

A few years later, two Americans, Davisson and Germer, showed experimentally that a beam of electrons is diffracted by a crystal in much the same way as a beam of X rays, the effective wavelength of the electrons being just that required by de Broglie's formula.

Following very closely on these developments was the proposal of the German, Schrödinger that all such "moving particles" be described by an equation he had developed called the "wave equation." The complete theory of the behavior of subatomic particles according to the wave equation is called wave mechanics. As it is applied to atoms, wave mechanics leads to the same result as Bohr obtained for the energy levels of the electron in the hydrogen atom while it gives a correct account of the properties of other, more complex, atoms as well.

In wave mechanics one of the basic ideas is that the position of the electron, as a discrete particle, cannot be precisely specified. Instead, one has a *wave function*,

ψ, which has a precise value at each point in space. The wave function can be interpreted in one of two ways:

1. The electron is regarded as a discrete particle; the square of the wave function, ψ^2, tells the relative probability of its being at any given point.

2. The electron is regarded as a smeared out distribution of negative charge whose density varies from place to place according to the magnitude of ψ^2.

The second interpretation will be used here. To illustrate it, and compare it with the Bohr picture, *Fig. 2-3* shows the orbital of lowest energy for the hydrogen atom according to each theory. According to wave mechanics the electron density distribution has spherical symmetry about the nucleus. The electron density has its highest value at the nucleus and falls away exponentially with increasing radius. More interesting is a plot of the fraction of the electron density to be found in each successive thin spherical shell from the nucleus out. This is called the radial density distribution function. It is equal to $4\pi r^2\psi^2$ and is shown in *Fig. 2-4*. It is 0 at the nucleus where $r = 0$ and goes through a maximum. This maximum is at precisely the orbit radius in Bohr's theory, $a_0 = 0.529$ Å.

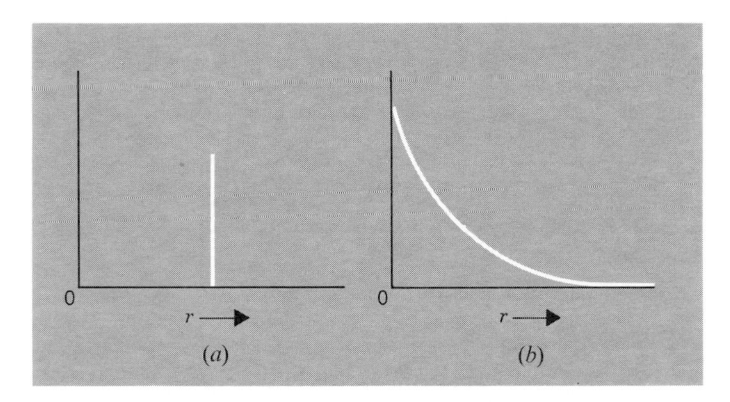

Figure 2-3 *Comparison of electron distribution as a function of distance, r, from the nucleus in the most stable orbital of the hydrogen atom for (a) the Bohr atom, and (b) the wave-mechanical atom. In (a) the electron is found entirely at one sharply defined distance (a_0) from the nucleus. In (b) the electron is spread over a range of distances.*

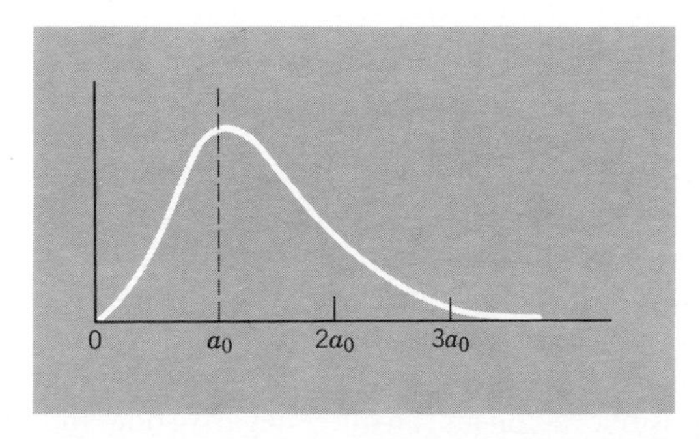

Figure 2-4 *The radial density distribution function for the most stable orbital of the hydrogen atom.*

2-3
Atomic Orbitals in Wave Mechanics

We now consider the entire set of wave functions for the electron in the hydrogen atom. These are called orbitals, since they are the wave-mechanical analogs of Bohr's orbits. In wave mechanics there is again a quantum number n, which takes integral values from 1 to ∞. For each value of n there is a second quantum number l, which takes values of $0, 1, 2 \ldots n - 1$. Finally, there is a quantum number m_l, which takes values from $-l$ to $+l$. For the more stable orbitals, the pattern is as shown in Table 2-1.

Table 2-1 Quantum Numbers and Symbols of Atomic Orbitals

n	l	m_l	
1	0	0	1*s*
2	0	0	2*s*
2	1	−1, 0, 1	2*p*
3	0	0	3*s*
3	1	−1, 0, 1	3*p*
3	2	−2, −1, 0, 1, 2	3*d*
4	0	0	4*s*
4	1	−1, 0, 1	4*p*
4	2	−2, −1, 0, 1, 2	4*d*
4	3	−3, −2, −1, 0, 1, 2, 3	4*f*

There are different types of orbitals, called *s*, *p*, *d*, *f* (followed alphabetically by *g*, *h*, etc.), which have l values of 0, 1, 2, 3, etc., respectively. For each value of n there is an *s* orbital. For each value of n beginning with 2 there is a set of three *p* orbitals, and so on.

From a chemist's viewpoint the shapes of these orbitals are very important. *Figure 2-5* shows in a diagrammatic way the shapes of orbitals. In each case, a contour is drawn that encloses an appreciable fraction of the total electron density. Each lobe carries an algebraic sign that is the sign of the wave function, ψ, in that region of space. It should be remembered that the electron density itself, which must always be positive, is given by ψ^2.

***s* Orbitals.** Every *s* orbital is spherically symmetrical. The 1*s* orbital is everywhere positive. Beginning with 2*s*, there are alternating positive and negative regions. This is best seen in *Fig. 2-6*, which shows how the amplitude varies with r for 1*s*, 2*s*, and 3*s* orbitals. Observe from the radial density distributions that the greatest concentration of electron density lies at progressively greater distances from the nucleus as n increases. Other types of orbitals (*p*, *d*, etc.) follow a similar pattern. The first such orbital (2*p*, 3*d*, etc.) has a simple exponential radial factor, the next has an inner positive region and an outer negative one, and so on. The spherical surfaces where a change of sign of ψ occurs (and $\psi^2 = 0$ as well) are called radial nodes.

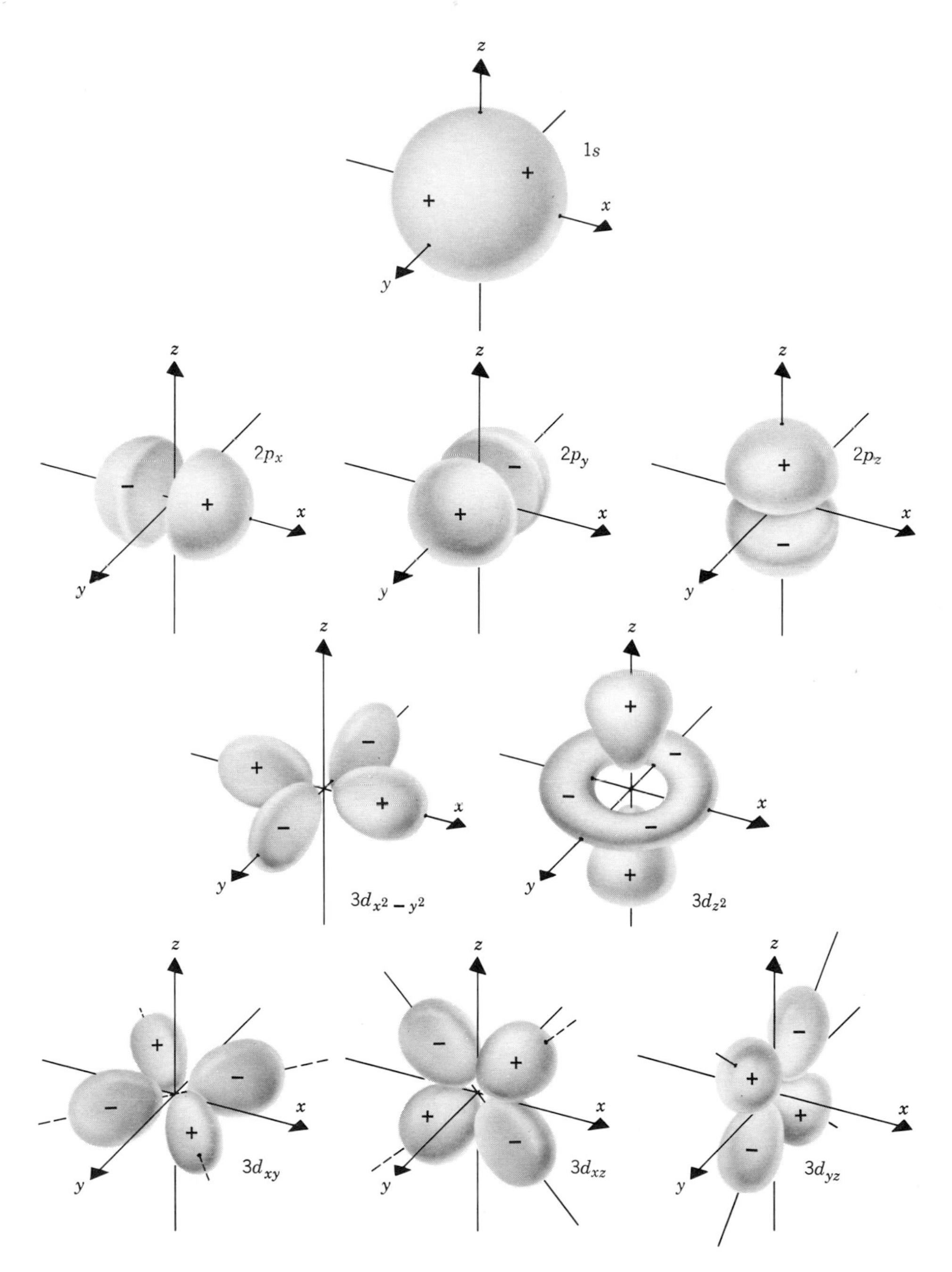

Figure 2-5 *Hydrogenic wave functions or orbitals. The d_{z^2} orbital and the p orbitals are figures of revolution about the axis of orientation, but the lobes of the remaining d orbitals are not circular in section. Note also that the two lobes of the p orbitals are of different sign, and the lobes of the d orbitals alternate in sign, that is, opposing pairs have the same sign. The orbitals are not drawn to the same scale.*

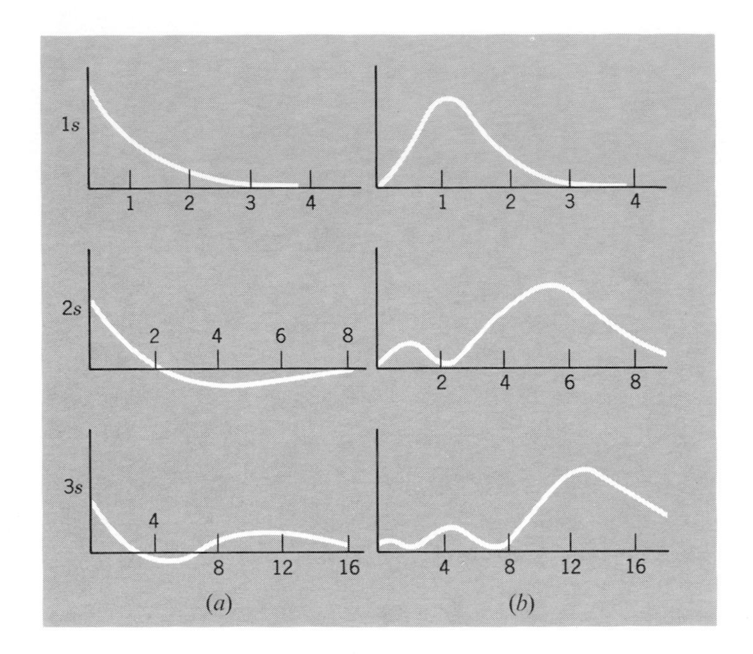

Figure 2-6 *(a) Plots of radial wave function, R, against r/a_0. (b) Plot of density distribution functions, $4\pi r^2\psi^2$, against r/a_0. Ordinates only relative; note different abscissa scales.*

***p* Orbitals.** Each *p* orbital consists of a positive lobe along the positive extension of a Cartesian axis and a negative lobe along the negative extension of the same axis. There are three orbitals in each set, one along the *x* axis, p_x, one along the *y* axis, p_y, and one along the *z* axis, p_z. The 2*p* orbitals have no radial nodes, but beginning with 3*p* there are radial nodes as well.

***d* Orbitals.** Each set of *d* orbitals consists of five members. There are many equally correct ways to represent them, but the particular set shown in *Fig. 2-5* is the conventional one. The following features are important.

1. The d_{z^2} orbital is symmetrical about the *z* axis.

2. The d_{xy}, d_{yz}, and d_{zx} orbitals are exactly alike except that they have their maximal amplitudes in the *xy*, *yz*, and *zx* planes, respectively.

3. The $d_{x^2-y^2}$ orbital has exactly the same shape as the d_{xy} orbital, but it is rotated by 45° about the *z* axis so that its lobes are directed along the *x* and *y* axis.

***f* Orbitals.** For every principal quantum number beginning with *4* there is a set of *f* orbitals. The 4*f* orbitals play only a slight role in chemical bonding, although the 5*f* orbitals are doubtless extensively employed in bonding in various compounds formed by the actinide elements. However, the need for detailed consideration of the shapes of *f* orbitals is insufficient to justify their inclusion here.

Finally, with respect to the hydrogen atom, the order of increasing energy of these orbitals is of importance. As we have already observed, the energy is given to a very good approximation by Eq. 2-2-1. Thus we have the pattern shown at

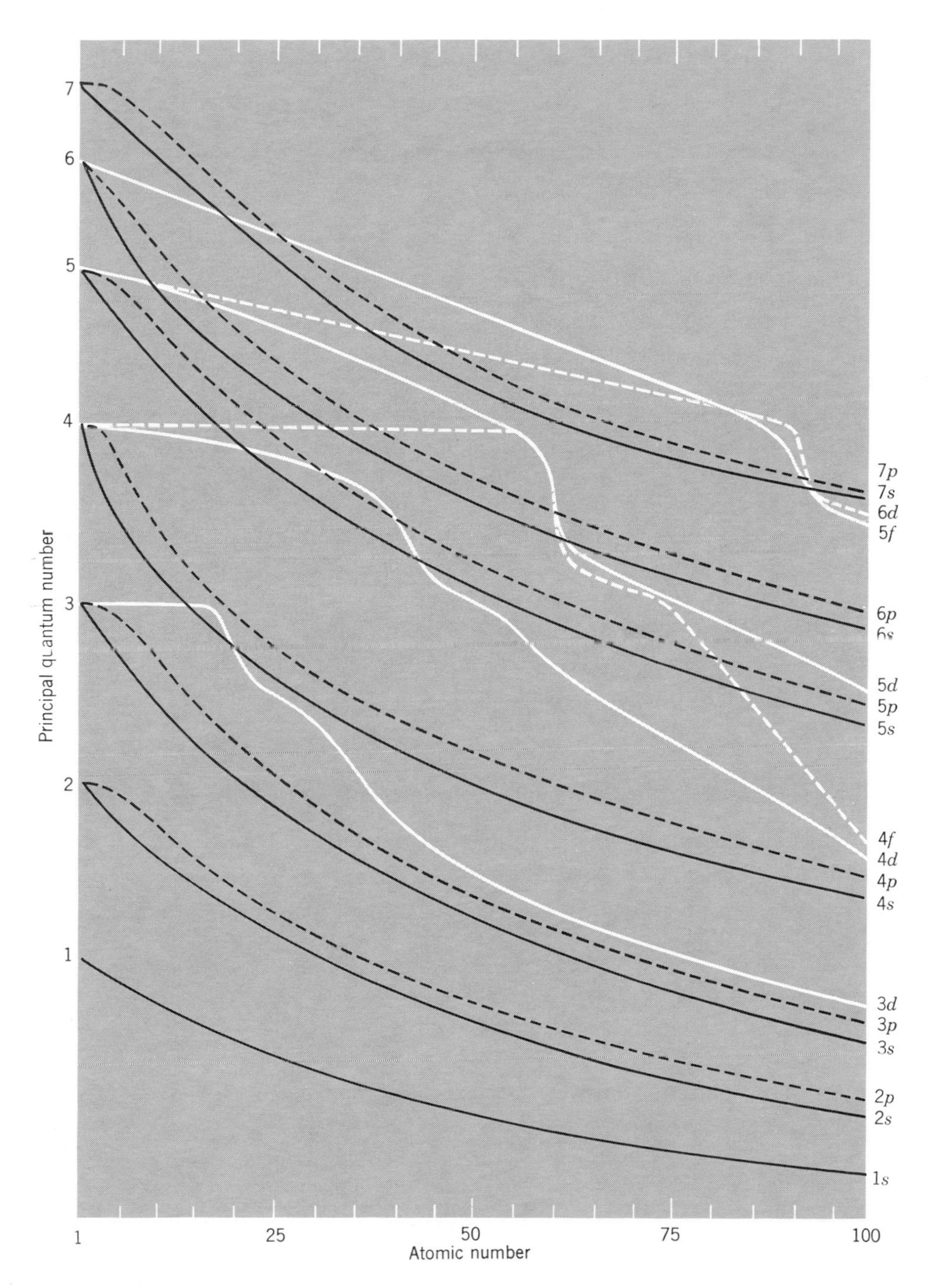

Figure 2-7 *The variation of the energies of atomic orbitals with increasing atomic number in neutral atoms (energies not strictly to scale).*

the left side of *Fig. 2-7*. There is no significant difference in the energies of the orbitals of given n, even though they differ in their l and m_l values, *in the hydrogen atom*. It will soon be seen, however, that this simple pattern changes in atoms with many electrons.

2-4
Electron Spin, The Exclusion Principle, Electron Configurations

Every electron has a property called *spin*, which can be thought of as though the electron, as a discrete particle, were spinning like a top about an axis passing through itself. There are two important consequences of electron spin. One is that it must be specified by an additional quantum number, m_s. This can take values of $+\frac{1}{2}$ or $-\frac{1}{2}$, depending on whether the direction of spin is clockwise or counterclockwise. Another is that the rotation of electric charge about an axis gives rise to a magnetic dipole moment that may point up or down depending on the direction of the spin. This leads to important magnetic properties of substances with more spins in one direction than in the other, as we explain later (page 48). The important point now is that the electron spin quantum number of $+\frac{1}{2}$ or $-\frac{1}{2}$ plays an important role in the way electrons can occupy orbitals when two or more electrons are present.

The Exclusion Principle. When an electron is assigned to an orbital, it can be fully described by listing four quantum numbers.

1. The principle quantum number of the orbital, n.

2. The l quantum number of the orbital, which tells whether it is an s, p, d, f, etc., orbital.

3. The m_l value of the orbital, which is equivalent to stating the subscript, such as x, y, or z for p orbitals, z^2, xy, yz, zx, $x^2 - y^2$ for a d orbital.

4. The spin quantum number $\pm\frac{1}{2}$ for the electron itself.

A fundamental rule which must be obeyed in placing two or more electrons in the orbitals of an atom is the exclusion principle:

No two electrons in one atom can have an identical set of quantum numbers.

The most important effect of this principle is that no orbital (as specified by n, l, and m_l) can ever contain more than two electrons, and these must have different spins. When two electrons occupy the same orbital with opposite spins, they also have oppositely directed magnetic moments and there is, therefore, no net magnetic moment from a filled orbital.

Electron Configurations. The way electrons occupy the orbitals in an atom is called the *electron configuration* of that atom. For the hydrogen atom, this is simply a statement of which orbital is occupied, for example, $1s$ for the ground state, $2s$, $3d$, $5f$, etc., for various excited states. For many-electron atoms, the same notation is extended. Double occupancy of an orbital is denoted with a superscript 2. Thus the ground state configuration of the lithium atom, with $Z = 3$, is $1s^2 2s$.

2-5
Structures of Atoms with Many Electrons

We can now consider the question of what electron configuration constitutes the ground state for the atom of each element. This is determined systematically by building up the configurations in order of increasing atomic number. In doing this, the exclusion principle must be observed (no more than two electrons per orbital), and each additional electron must be assigned to the lowest-energy orbital not yet filled.

The ground state for hydrogen is $1s$. For helium, another electron can be assigned to the same orbital giving $1s^2$. The first principal shell ($n = 1$) is now filled. The next element, lithium, has the third electron assigned to the next lowest orbital, $2s$, and the configuration is $1s^2 2s$.

Here we have encountered for the first of many times the important fact that orbitals of the same principal quantum number do *not* have the same energy in many-electron atoms, even though they do in the hydrogen atom. We shall examine the reason for this in some detail for the case of the $2s$ and $2p$ orbitals. The basic reason for splitting apart of the different types of orbitals with the same n (e.g., $3s$ from $3p$ and $3d$) is always the same, and once it is understood for any particular case, the others will, in principle, be understood also.

Figure 2-8 shows the radial distribution functions, $4\pi r^2 \psi^2$, for $1s$, $2s$, and $2p$ orbitals. Both the $2s$ and the $2p$ orbitals *penetrate* the $1s$ orbital. That is, a substantial fraction of the electron density of either a $2s$ or a $2p$ electron lies inside some of the electron density of the $1s$ electrons. It develops, when accurate calculations are made, that the $2s$ orbital penetrates somewhat more than $2p$. Thus, the electron density of a $2s$ electron is somewhat *less* screened from the nuclear charge than is the electron density of a $2p$ electron. The $2s$ orbital is, therefore, more stable

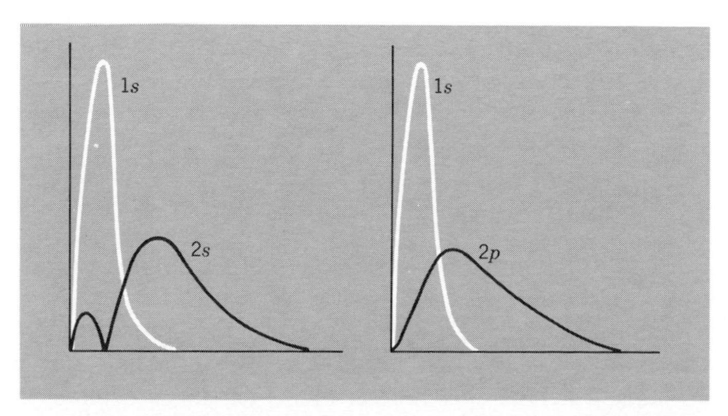

Figure 2-8 *Radial density distribution functions for hydrogen atom orbitals showing how both the 2s and the 2p orbitals penetrate the 1s orbital.*

than the $2p$ *when the* $1s$ *orbital is occupied*. For helium and lithium, the energy differences are:

	Energy Difference
He $1s2s$ $1s2p$	102 kJ mol^{-1}
Li $1s^22s$ $1s^22p$	178 kJ mol^{-1}

A difference of about 200 kJ mol^{-1} remains in effect from lithium through all succeeding elements. This is shown in *Fig. 2-7*.

After Li with $1s^22s$, we must add another electron for Be and this, too, enters the $2s$ orbital, giving the configuration $1s^22s^2$. The next set of stable orbitals are the $2p$ orbitals. There are three of them, each capable of holding two electrons, or six electrons in all. Thus, through the next six elements, the $2p$ levels are successively filled:

B	C	N
$1s^22s^22p$	$1s^22s^22p^2$	$1s^22s^22p^3$
O	F	Ne
$1s^22s^22p^4$	$1s^22s^22p^5$	$1s^22s^22p^6$

At neon the second principal shell ($n = 2$) is complete.

The next step must be to place electrons in the third principal shell, and through the next eight elements the $3s$ and then the $3p$ orbitals become filled, as we reach the next noble gas, argon. As with $2s$ and $2p$, once the inner shells have been filled, the $3s$ and $3p$ orbitals have different energies, with $3s$ the more stable.

After the $3s$ and $3p$ orbitals are filled, the next electrons enter the $4s$ orbital. The reason for this can be seen in *Fig. 2-7*. As the first 18 electrons have been added, the energies of the $4s$ and $4p$ orbitals have been dropping steeply, because they penetrate this core extensively. The $3d$ orbitals have remained fairly constant in energy, since they penetrate the core very little. The result is that when argon is reached, the $4s$ orbital has become more stable than the $3d$. Accordingly the next electron enters the $4s$ orbital. This situation continues to prevail, and the next electron enters the $4s$ orbital also. Thus for the configurations of K and Ca we have

$$\text{K: [Ar]}4s$$
$$\text{Ca: [Ar]}4s^2$$

where the symbol [Ar] stands for the complete argon configuration, $1s^22s^22p^6 3s^23p^6$.

The $3d$ orbitals do penetrate the $4s$ orbital and, therefore, through K and Ca drop sharply so as to become about as stable as the $4s$ orbitals and more stable than the $4p$ orbitals. Thus the next 10 electrons enter the $3d$ orbitals, in the series of elements Sc, Ti, V, Cr, Mn, Fe, Co, Ni, Cu and Zn. Thereafter, the $4p$ orbitals are filled, through Ga, Ge, As, Se, Br, to the next noble gas, Kr.

As *Fig.* 2-7 shows, an analogous pattern is now repeated, with electrons being added to the 5*s*, then 4*d*, then 5*p* orbitals until the noble gas Xe is reached.

At Xe the next available orbitals are the 6*s*. The 4*f* and 5*d* orbitals are only slightly penetrating and are so well shielded from the nuclear charge by the electrons of the Xe configuration that they remain high in energy. However, after the two 6*s* electrons have been added, both the 4*f* and 5*d* orbitals drop sharply in energy, with the 4*f* orbitals eventually becoming more stable than the 5*d* orbitals. As a result, for the next few elements, the configurations are $[Xe]4s^2 5d$, $[Xe]4s^2 4f^2$, $[Xe]4s^2 4f^3$, etc., until the 4*f* orbitals become filled. The 5*d* orbitals then get the next electrons until they are filled and, finally, the 6*p* electrons are added. This brings us to the heaviest noble gas atom, Rn.

The sequence of orbital filling following Rn becomes quite complex. As before, the next *s* orbital, 7*s* is filled, but then the 6*d* and 7*f* orbitals become very close in energy and the addition of electrons is controlled by interelectronic forces. It is not worthwhile to be concerned with the exact $7f^n 6d^m$ arrangement, since in each case two or more configurations differing in *n* and *m* differ so little in energy that the exact arrangement in the neutral atom has little to do with the chemical properties of the element.

2 6
The Periodic Table

More than a century ago chemists began to search for a tabular arrangement of the elements that would group together those with similar chemical properties and also arrange them in some logical sequence. The sequence was generally the order of increasing atomic weights. As is well known, these efforts culminated in the type of periodic table devised by Mendeleyev, in which the elements were arranged in horizontal rows with row lengths chosen so that like elements would form vertical columns.

It was Moseley who showed that the proper sequence criterion was not atomic weight but atomic number (although the two are only rarely out of register). It then followed that not only did the vertical columns contain chemically similar elements, but electronically similar atoms. We devote all of Chapter 8 to a discussion of the practical chemical aspects of the periodic table. Here, since we have just studied how the electron configurations of atoms are built up, it is appropriate to point out that these configurations lead logically to the same periodic arrangement as Mendeleyev deduced from strictly chemical observations.

To build up a periodic table based on similarities in electron configuration, a convenient point of departure is to require all atoms with outer ns^2np^6 configurations to fall in a column. It is convenient to place this column at the extreme right, and to include also He ($1s^2$). This column thus contains those elements called the *noble gases*, He, Ne, Ar, Kr, Xe, Rn.

If now the elements that each follow a noble gas and have a single electron in the *ns* orbital following the noble gas configuration, that is, [ ]*ns*, are placed in a column at the extreme left of the table, the form of the table is basically established. These elements, Li, Na, K, Rb, Cs, Fr are called the *alkali metals*. Each of them is immediately followed by an element with a [ ]ns^2 configuration and these naturally fall into a column. These elements, Be, Mg, Ca, Sr, Ba, and Ra, are called the *alkaline earth metals*.

If now we return to the noble gas column and begin to work back from right to left, it is clear that we shall get columns of elements with outer electron configurations $ns^2np^5, ns^2np^4, \ldots, ns^2np$. The ns^2np^5 elements F, Cl, Br, I, At, are called the *halogens* (meaning salt-formers). Those with the ns^2np^4 configurations are O, S, Se, and Te; they are given the family name *chalcogens*. The other three columns, that is the ns^2np (B, Al, Ga, In, Tl), ns^2np^2 (C, Si, Ge, Sn, Pb) and ns^2np^3 (N, P, As, Sb, Bi) elements have no trivial group names.

Thus far, we have developed a rational arrangement for nearly half of the elements. These elements, which involve outer shells consisting solely of s and p electrons are called the *main group elements*. Most of the remaining ones are called *transition elements*. They occupy the central region of the table in its usual form (*Fig. 2-9*). The two-dimensional array of the main group elements has a form that is uniquely dictated by their electronic structures. The way in which the remaining elements are fitted in and arranged has, at times, been done in different ways. However, the arrangement shown in *Fig. 2-9* is the most common one and is very logical.

Between Ca ($Z = 20$) and Ga ($Z = 31$) there are 10 elements, whose occurrence at this position is due to the filling of the $3d$ orbitals. With two minor deviations to be discussed later, they all have configurations $[Ar]4s^33d^x$, where x runs from 1 to 10. Similarly, between Sr and In there is another series whose electron configurations are $[Kr]5s^24d$ to $[Kr]5s^24d^{10}$. Finally there is a series beginning with La with the configuration $[Xe]6s^25d$ and ending with Hg, $[Xe]6s^25d^{10}$, where, for the present we omit 14 elements, from Ce to Lu through which the $4f$ shell is being filled. By making the latter omission, we are able to arrange 30 elements in a rectangular array, with columns such that in each column the configuration is of the form $[\quad]ns^2(n-1)d^x$.

With the exception of the group comprised of Zn, Cd, and Hg, these elements are called the transition elements or sometimes the *d-block elements*. Their common characteristic is that either the neutral atom or some important ion it forms, or both, have an incomplete set of d electrons. As we discuss in detail in Part III, the incomplete d shell leads to many characteristic physical and chemical properties. A few further points of notation are the following. The set of elements with incomplete $3d$ shells are called the *first transition* series and those with partial $4d$ and $5d$ shells are called the *second* and *third* transition series, respectively.

The elements Zn, Cd, and Hg have unique properties. While they resemble the alkaline earths in giving no oxidation state higher than $+2$, they differ because the configuration immediately underlying their valence orbitals is a rather polarizable nd^{10} shell instead of a more tightly bound noble gas shell. Their chemistry will be studied in a later chapter, following the other main group elements and preceding the transition elements.

Finally, the 14 elements between La and Hf, in which the $4f$ orbitals are being filled, are placed at the bottom of the table, to avoid making it excessively wide and unwieldy. These elements are called the *lanthanides* because of their chemical resemblence to lanthanum. A somewhat similar set of elements, called the *actinides*. have partially filled $5f$ orbitals. They, too, are placed at the bottom, under the corresponding lanthanides. These two series are sometimes called collectively, the *f-block elements*.

Period	Group Ia	Group IIa	Group IIIa	Group IVa	Group Va	Group VIa	Group VIIa	Group VIII			Group Ib	Group IIb	Group IIIb	Group IVb	Group Vb	Group VIb	Group VIIb	Group O
1 Is	1 H																1 H	2 He
2 2*s*2*p*	3 Li	4 Be											5 B	6 C	7 N	8 O	9 F	10 Ne
3 3*s*3*p*	11 Na	12 Mg											13 Al	14 Si	15 P	16 S	17 Cl	18 Ar
4 4*s*3*d* 4*p*	19 K	20 Ca	21 Sc	22 Ti	23 V	24 Cr	25 Mn	26 Fe	27 Co	28 Ni	29 Cu	30 Zn	31 Ga	32 Ge	33 As	34 Se	35 Br	36 Kr
5 5*s*4*d* 5*p*	37 Rb	38 Sr	39 Y	40 Zr	41 Nb	42 Mo	43 Tc	44 Ru	45 Rh	46 Pd	47 Ag	48 Cd	49 In	50 Sn	51 Sb	52 Te	53 I	54 Xe
6 6*s* (4*f*) 5*d* 6*p*	55 Cs	56 Ba	57* La	72 Hf	73 Ta	74 W	75 Re	76 Os	77 Ir	78 Pt	79 Au	80 Hg	81 Tl	82 Pb	83 Bi	84 Po	85 At	86 Rn
7 7*s* (5*f*) 6*d*	87 Fr	88 Ra	89** Ac															

*Lanthanide series 4*f*	58 Ce	59 Pr	60 Nd	61 Pm	62 Sm	63 Eu	64 Gd	65 Tb	66 Dy	67 Ho	68 Er	69 Tm	70 Yb	71 Lu
**Actinide series 5*f*	90 Th	91 Pa	92 U	93 Np	94 Pu	95 Am	96 Cm	97 Bk	98 Cf	99 Es	100 Fm	101 Md	102 No	103 Lr

Figure 2-9 *The conventional "long form" periodic table.*

The columns of chemically similar elements are formally called "groups." The groups are numbered to correspond with the number of valence shell electrons, which is often also the valence number—or one of the common valence numbers—of the group elements. It is necessary to use two sets of group numbers, Ia, IIa, etc., and Ib, IIb, etc., as is shown in *Fig. 2-9*, to handle the number of groups. The notation is somewhat arbitrary, but rather widely used.

2-7
Hund's First Rule; Variations of Ionization Enthalpies with Atomic Number

Thus far, for atoms in which there are partly filled p or d shells, the configurations have simply been written as p^n or d^n. However, it is possible, and often important, to specify them in greater detail. This is illustrated for the p^n configurations in *Fig. 2-10*. There are two important features of the pattern shown there:

1. Electrons continue to enter different orbitals as long as possible.

2. Two or more electrons each occupying a different orbital have *parallel spins*, that is, the same direction of spin.

The first of these features is a simple consequence of the charge of the electron. Since the three p orbitals are of equal energy but are concentrated in different regions of space, the electrons can minimize the repulsive forces between themselves by occupying different orbitals. The preference of single electrons in different orbitals of the set for parallel orientations of their spins is not explainable in any simple way, but it is always observed.

The two features just discussed can be summarized in a statement known as *Hund's first rule*: The most stable configuration, among several possible ones with

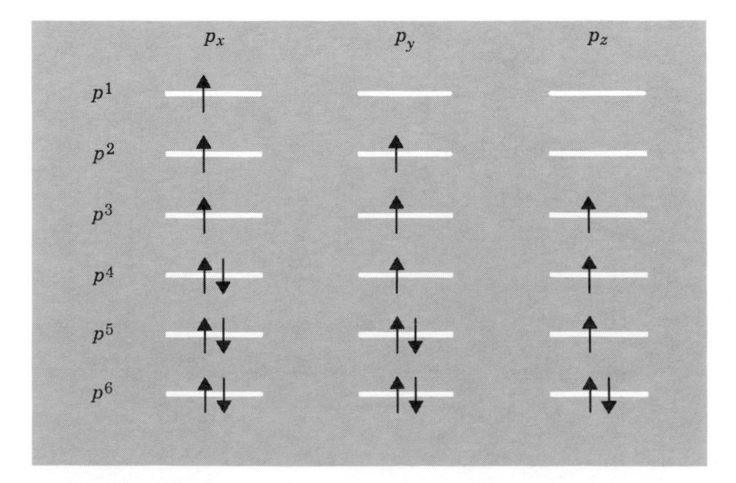

Figure 2-10 *The sequence of filling of a set of p orbitals in the ground states of atoms. The arrows represent electrons; the upward and downward directions of the arrows correspond to the two directions of spin.*

the same orbital energy, is that with the largest number of unpaired electron spins. This rule immediately implies the spreading out of the electrons, since any two in the same orbital are required by the exclusion principle to have their spins in opposite directions, a situation described as *pairing of spins*.

When Hund's rule is applied to the transition metal atoms, it leads to the predictions that atoms with $ns^2(n-1)d^x$ configurations will have x unpaired spins for $x = 1 - 5$ and then $10 - x$ unpaired spins for $x = 6 - 10$. These predictions are in exact agreement with observation.

The complete pattern of variation of first ionization enthalpies for the elements H to Rn is shown in *Fig. 2-11*. There are three major trends that merit comment.

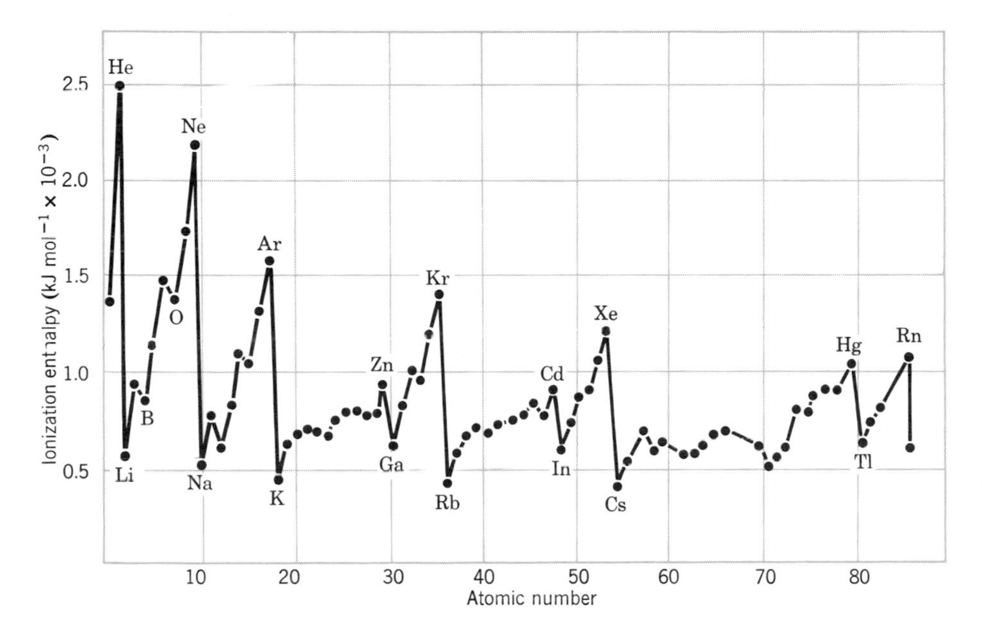

Figure 2-11 *Variation of first ionization enthalpies with atomic number.*

First, the maxima occur at the noble gases and the minima occur at the alkali metals. This is easily understandable, since the closed shell configurations of the noble gases are very stable and resist disruption, either to form chemical bonds or to become ionized. In the alkali metal atoms, there is an electron outside the preceding noble gas configuration that is well shielded from the attraction of the nucleus by all the inner shells; it is therefore relatively easy to remove.

Second, from each alkali metal to the next noble gas there is an overall increase in the ionization enthalpies. This is because electrons with the same principal quantum number have about the same average distance from the nucleus. One electron, therefore, only partly screens another from the attractive force of the nucleus. Thus, as we proceed through such a series of elements, the total nuclear charge increases by +1 at each step, but the added electron only partly screens the others from it. Hence, there is an overall increase in ionization enthalpies through a series s, s^2, s^2p, s^2p^2 . . . s^2p^6.

Third, the increase just discussed is not smooth. Instead, there are two well-defined jogs which occur at corresponding positions in each series, that is, from Li to Ne, from Na to Ar and, with some differences due to intervention of the transition elements, in subsequent periods of the table. In each case, the ionization enthalpy drops from the s^2 to the s^2p configuration and again from the s^2p^3 to the s^2p^4 configuration. The explanation is most readily apparent if the facts are stated in a slightly different way, where the elements of the Li–Ne period are used as an example. The ionization enthalpies of B, C, and N increase regularly but they are all lower than values that would be extrapolated from Li and Be. This is because *p* electrons are less penetrating than *s* electrons. They are, therefore, more shielded and more easily removed than extrapolation from the behavior of *s* electrons would predict. Again, the ionization enthalpies of O, F, and Ne increase regularly, but all are lower than would be expected by extrapolation from B, C, and N. This is because the 2*p* shell is half full at N and each of the additional 2*p* electrons enters an orbital already singly occupied. They are partly repelled by the electron already present in the same orbital and are thus less tightly bound.

One more observation which is pertinent here is that filled and half-filled shells give the appearance of having a specially high stability, and this leads to slight "anomalies" in the electron configurations of some of the transition elements. This is well illustrated by the ground state configurations of the atoms in the first transition series, which are, omitting the underlying [Ar] core:

Sc	Ti	V	Cr	Mn
$4s^23d$	$4s^23d^2$	$4s^23d^3$	$4s3d^5$	$4s^23d^5$
Fe	Co	Ni	Cu	Zn
$4s^23d^6$	$4s^23d^7$	$4s^23d^8$	$4s3d^{10}$	$4s^23d^{10}$

The anomalies occur at Cr and Cu, where a 4*s* electron seems to be "borrowed" to complete half-filled and filled *d* shells, d^5 and d^{10}.

2-8
Magnetic Properties of Atoms and Ions

Any atom, ion, or molecule that has one or more unpaired electrons is *paramagnetic*, meaning, that it, or any material in which it is found, will be attracted into a magnetic field. In cases where paramagnetic atoms or ions are very close together they interact cooperatively and other more intense and more complicated forms of magnetism, ferromagnetism and antiferromagnetism, in particular, are observed. We do not discuss these forms here as they have no direct chemical significance. Substances that contain no unpaired electrons (with certain exceptions that need not concern us here) are *diamagnetic*, meaning that they are repelled, weakly, by a magnetic field. Thus, the measurement of paramagnetism affords a powerful tool for detecting the presence and number of unpaired electrons in chemical elements and compounds.

The full power of magnetic measurements comes from the fact that the magnitude of the *magnetic susceptibility*, which is a measure of the force exerted by the field on a unit mass of the specimen, is related to the number of unpaired electrons present per unit weight—and, hence, per mole.

Actually, the paramagnetism of a substance containing unpaired electrons receives a contribution from the orbital motion of the unpaired electrons as well as from their spins. However, there are important cases where the spin contribution is so predominant that measured susceptibility values can be interpreted in terms of how many unpaired electrons are present. This correlation is best expressed by using a quantity called the *magnetic moment*, μ, which may be calculated from the measured susceptibility per mole, χ_M. It is best to use χ_M^{corr}, where a correction has been applied to the measured χ_M to allow for the diamagnetic effect that is always present, and that may be estimated from measurements on similar substances which lack the atom or ion which has the unpaired electrons.

Curie's Law. It was shown some 70 years ago by Pierrė Curie that for most paramagnetic substances, the magnetic susceptibility varies inversely with absolute temperature. In other words the product $\chi_M^{corr} \times T$ is a constant, called the Curie constant for the substance. From the theory of electric and magnetic polarization it can be shown that, if the paramagnetic susceptibility is due to the presence of individual, independent paramagnetic atoms or ions within the substance, each with a magnetic dipole moment, μ, the following equation holds true:

$$\mu = 2.84\sqrt{\chi_M^{corr} T} \tag{2-8-1}$$

It is seen that this expression incorporates Curie's law.

Now it can also be shown from the quantum theory for atoms (and ions) that the magnetic moment due entirely to n unpaired electrons on the atom or ion is given by

$$\mu = 2\sqrt{S(S + 1)} \tag{2-8-2}$$

where S equals the sum of the spins of all the unpaired electrons, that is, $n \times \frac{1}{2}$. From Eq. 2-8-2 it can easily be calculated that for 1 to 5 unpaired electrons the magnetic moments should be those shown in Table 2-2. The unit for atomic magnetic moments is the Bohr magneton, BM.

To illustrate the application of these ideas, consider copper(II) sulfate, $CuSO_4 \cdot 5H_2O$. From the magnetic susceptibility the magnetic moment is found to be 1.95 BM. This value is only a little higher than the calculated value for one unpaired electron, and the discrepancy can be attributed to the contribution made by orbital motion of the electron. Thus the magnetic properties of $CuSO_4 \cdot 5H_2O$

Table 2-2

Number of unpaired electrons	S	μ (BM)
1	$\frac{1}{2}$	1.73
2	1	2.83
3	$\frac{3}{2}$	3.87
4	2	4.90
5	$\frac{5}{2}$	5.92

are in accord with the presence of a Cu^{2+} ion which should have a [Ar]$3d^9$ configuration, with one unpaired electron. For comparison, $MnSO_4 \cdot 4H_2O$ has a magnetic moment of 5.86 BM, which is approximately the number expected for a Mn^{2+} ion with the electron configuration [Ar]$3d^5$.

2-9 Electronegativity

Electronegativity is an empirical measure of the tendency of an atom in a molecule to attract electrons. It will, naturally, vary with the oxidation state of the atom, and for a number of reasons the numerical values that have been assigned should not be taken too literally. As a semiquantitative notion, it is useful.

It should be stressed that electronegativity,* χ, is not the same thing as the enthalpy of electron attachment (Section 1-2e), ΔH_{EA}, although it is related to it. Mulliken has shown that the typical empirical values of χ are roughly proportional to the average of $-\Delta H_{EA} + \Delta H_{ION}$. That is, electronegativity is determined, in part, by the tendency of the atom to bind an additional electron and, in part, by its tendency to hold on to those it already has. A complete electronegativity scale cannot be set up using Mulliken's idea, however, if for no other reason than that ΔH_{EA}'s are known for only a few atoms.

Many ways of computing electronegativities have been suggested. The first general method was proposed by Pauling. He suggested that if two atoms, A and B, had the same electronegativity, the strength of the A—B bond would be equal to the geometric mean of the A—A and B—B bond energies, since the electrons in the bond would be equally shared in purely covalent bonds in all three cases. He observed, however, that for the majority of A—B bonds the energy exceeds that geometric average because, in general, two different atoms have different electronegativities, and there is an ionic contribution to the bond in addition to the covalent one. He proposed that the "excess" A—B bond energies could be used as an empirical basis to determine electronegativity differences. For instance, the H—F bond energy is 566 kJ mol^{-1}, whereas the H—H and F—F bond energies are 436 kJ mol^{-1} and 158 kJ mol^{-1}, respectively. Their geometric mean is $(158 \times 436)^{1/2} = 244$ kJ mol^{-1}. The difference, Δ, is 322 kJ mol^{-1}. He then found that to get a consistent set of electronegativities, so that $\chi_A - \chi_B = (\chi_C - \chi_B) - (\chi_C - \chi_A)$, etc., the electronegativity differences would have to obey the equation:

$$\chi_A - \chi_B = 0.102\ \Delta^{1/2} \qquad (2\text{-}9\text{-}1)$$

Pauling originally assigned the most electronegative of the elements, fluorine, $\chi = 4.00$. From the above data, we could calculate

$$\chi_H = 4.00 - 0.102(244)^{1/2} = 1.59\ldots$$

A more recently proposed method of calculating electronegativities, which produces a similar although not identical scale, is that of Allred and Rochow. It has the advantage of being more easily applied to more elements. Table 2-3

* Chi, χ, is conventionally used for electronegativity as well as for magnetic susceptibility.

Table 2-3 *Electronegativities of the Elements*

H 2.1																
Li 1.0	Be 1.5											B 2.0	C 2.5	N 3.1	O 3.5	F 4.1
Na 1.0	Mg 1.3											Al 1.5	Si 1.8	P 2.1	S 2.4	Cl 2.9
K 0.9	Ca 1.1	Sc 1.2	Ti 1.3	V 1.5	Cr 1.6	Mn 1.6	Fe 1.7	Co 1.7	Ni 1.8	Cu 1.8	Zn 1.7	Ga 1.8	Ge 2.0	As 2.2	Se 2.5	Br 2.8
Rb 0.9	Sr 1.0	Y 1.1	Zr 1.2	Nb 1.3	Mo 1.3	Tc 1.4	Ru 1.4	Rh 1.5	Pd 1.4	Ag 1.4	Cd 1.5	In 1.5	Sn 1.7	Sb 1.8	Te 2.0	I 2.2
Cs 0.9	Ba 0.9	La 1.1	Hf 1.2	Ta 1.4	W 1.4	Re 1.5	Os 1.5	Ir 1.6	Pt 1.5	Au 1.4	Hg 1.5	Tl 1.5	Pb 1.6	Bi 1.7	Po 1.8	At 2.0
Fr 0.9	Ra 0.9	Ac 1.0	Lanthanides: 1.0–1.2 Actinides: 1.0–1.2													

lists Allred–Rochow values. The rationale is that an atom will attract an electron in its valence shell according to Coulomb's law:

$$\text{Force} = \frac{(Z^*e)(e)}{r^2} \tag{2-9-2}$$

where (Z^*e) is the effective nuclear charge felt by the electron, e is the charge on the electron itself, and r is the mean radius of the orbital, which can be taken equal to the covalent radius (see Section 3-11) for the atom. The magnitude of Z^* can be calculated according to a set of rules developed long ago by Slater to take account of how much a given electron is screened from the nuclear charge by all the other electrons in the atom. On this basis equation (2-9-3) is obtained

$$\chi = 0.359\frac{Z^*}{r^2} + 0.744 \tag{2-9-3}$$

where the numerical constants are arbitrarily chosen to give the desired range. This equation gives the values in Table 2-3.

The variation of the χ values with position in the periodic table is qualitatively reasonable. The least electronegative atom is the largest alkali metal, Cs, and the most electronegative one is the smallest halogen, F.

One common application of electronegativities is to estimate the direction of polarity of chemical bonds. For example, the polarities of C—H and Si—H bonds should be opposite, and this can account for the distinct differences in their chemical properties:

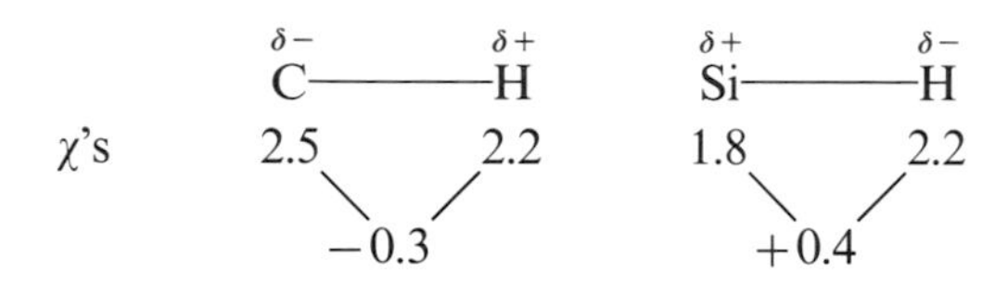

Study Questions

A

1. The emission lines of the hydrogen atom come in related sets. What is the form of the equations for these sets. An equation of this type is named for whom?
2. What were the two bold postulates made by Bohr which allowed him to derive an equation for the energies of the electron in a hydrogen atom?
3. Write and explain the meaning of the equation relating the energy and frequency of radiation. What is the constant in it called?
4. What does the term Bohr radius mean?
5. What is de Broglie's equation for the wavelength associated with a moving particle of mass m and velocity v? What physical effect first showed directly that the wave character of the electon really exists?
6. State the relationship between the Bohr orbit with $n = 1$ and the wave mechanical orbital with $n = 1$ for the hydrogen atom.
7. Specify the set of quantum numbers used to describe an orbital and state what values of each are possible.

8. State the quantum numbers for each of the following orbitals: 1*s*, 2*s*, 2*p*, 4*d*, 4*f*.
9. Draw diagrams of each of the following orbitals: 1*s*, $2p_x$, $2p_y$, $2p_z$, $3d_{z^2}$, $3d_{xy}$, $3d_{yz}$, $3d_{zx}$, $3d_{x^2-y^2}$.
10. State the exclusion principle in the form relevant to atomic structure. Show how it leads to the conclusion that in a given principal shell there can be only two *s*, six *p*, ten *d*, and fourteen *f* electrons.
11. What does the term penetration mean, and why is it important in understanding the relative energies of the *s*, *p*, *d*, and *f* electrons with the same principal quantum number?
12. Define each of the following: alkali metals; alkaline earth metals, halogens; noble gases; main group elements; *d*-block elements; *f*-block elements; lanthanides; transition elements.
13. What is Hund's first rule? Show how it is used to specify in detail the electron configurations of the elements from Li to Ne.
14. Why is the first ionization enthalpy of the oxygen atom lower than that of the nitrogen atom?
15. How is the magnetic moment of a substance containing an ion with unpaired electrons (e.g., $CuSO_4 \cdot 5H_2O$) related to its magnetic susceptibility at various temperatures if the substance follows Curie's law?
16. How is the magnetic moment, μ, related to the number of unpaired electrons if the magnetism is due solely to the electron spins? Calculate μ for an ion with three unpaired electrons.
17. R. S. Mulliken showed that electronegativity is related to both ΔH_{EA} and ΔH_{ION}. What is the relationship he gave?

B

1. For the He^+ ion the energy levels will be given by an equation similar to Eq. 2-1-1 with $Z = 2$ instead of 1. Calculate the frequencies, in cm^{-1}, for the first and last lines in each of the three series corresponding to those discussed for the H atom.
2. The first ionization enthalpy for Li is 520 kJ mol^{-1}. Calculate from this the effective charge felt by the 2*s* electron of Li. Why is this less than the actual nuclear charge of +3?
3. A consistent set of units which may be used in de Broglie's equation (2-2-1) is: λ in cm, m in g, v in cm sec^{-1}, and h in g cm sec^{-1} ($\equiv$erg sec^{-1}). What is the wavelength in centimeters and in angstroms of: (a) an electron traveling at 10^6 cm sec^{-1} (a velocity attainable in an electron microscope)? (b) A "thermal" neutron in an atomic reactor (you can calculate v by setting the kinetic energy, $\frac{1}{2}mv^2$, equal to **k**T at 300 K)? (c) A baseball or cricket ball thrown by a good pitcher or bowler, traveling at 10^3 cm sec^{-1}.
4. Write the most stable electron configurations of the atoms with the following atomic numbers: 7, 20, 26, 32, 37, 41, 85, 96. How many unpaired electrons would each one have? What would its magnetic moment, in Bohr magnetons, be? What would be the most highly charged stable ion each would be likely to form under normal chemical circumstances?
5. Test Mulliken's proposal that χ values are proportional to $-\Delta H_{EA} + \Delta H_{diss}$ using data for the halogen atoms. What is the mean deviation of χ values from the best straight line. ΔH_{EA} values are given on page 8. First ionization enthalpies in kJ mol^{-1} are 1681 (F), 1251 (Cl), 1140 (Br), 1008 (I).
6. As noted in the caption to *Fig. 2-1*, there are two more line series in the hydrogen emission spectrum that occur further into the infrared. Calculate the positions of the first line and the limit for each of them.

Chapter 2 **Study Guide**

Scope and Purpose. This chapter covers those fundamental facts and principles of atomic structure, as well as an introduction to the periodic table, that are indispensible in the subsequent discussions of bonding and general inorganic chemistry.

Supplementary Reading

Adamson, A. W., "Domain Representation of Orbitals," *J. Chem. Ed.*, *42*, 141 (1965).

Berry, R. S., Advisory Council on College Chemistry Resource Paper on "Atomic Orbitals," *J. Chem. Ed.*, *43*, 283 (1966).

Cohen, I. and Bustard, T., "Atomic Orbitals: Limitations and Variations," *J. Chem. Ed.*, *43*, 187 (1966).

Encyclopedia Britannica 1968, Vol. 17, page 616 has a good account of development of the period law and table.

Greenwood, N. N., *Principles of Atomic Orbitals*, Monograph for Teachers Series No. 8, The Royal Institute of Chemistry, London, 1968.

Guillemin, V., *The Story of Quantum Mechanics*, Charles Schribner's Sons, New York, 1968.

Hochstrasser, R. M., *Behavior of Electrons in Atoms*, Benjamin, Menlo Park, Calif., 1964.

Johnson, R. C. and Rettew, R. R., "Shapes of Atoms," *J. Chem. Ed.*, *42*, 145 (1965).

Karplus, M. and Porter, R. N., *Atoms and Molecules: An Introduction for Students of Physical Chemistry*, Benjamin, Menlo Park, Calif., 1970.

Lagowski, J. J., *The Structure of Atoms*, Houghton-Mifflin, Boston, 1964.

Ogryzlo, E. A. and Porter, G. B., "Contour Surfaces for Atomic and Molecular Orbitals," *J. Chem. Ed.*, *40*, 256 (1963).

Perlmutter-Hayman, B., "The Graphical Representation of Hydrogen-Like Functions," *J. Chem. Ed.*, *46*, 428 (1969).

Powell, R. E., "The Five Equivalent *d* Orbitals," *J. Chem. Ed.*, *45*, 1 (1968).

Price, W. C., Chissick, S. S., and Ravensdale, T., eds., *Wave Mechanics, The First 50 Years*, Butterworths, 1973.

Pritchard, H. O. and Skinner, H. A., "Electronegativity Scales," *Chem. Rev.*, *55*, 745 (1955).

3

chemical bonding

3-1
Introduction

To organize the subject, three main types of bonding are considered:

1. Covalent bonding between atom pairs (two-center bonds).
2. Delocalized (multicenter) covalent bonding.
3. Ionic Bonding.

The first two types are discussed in this chapter, while ionic bonding and related topics are considered in Chapter 4. In addition, a few special forms of bonding are discussed elsewhere, such as metallic bonding (Section 8-6), the hydrogen bond (Section 9-3), and ligand field theory (Chapter 23).

There is surely no bonding that is literally and completely *ionic* but, for practical purposes, a great many compounds can be treated to a reasonable approximation as if the attractive forces were just the electrostatic attractions between ions of opposite charge. The treatment of these substances, for example, NaCl, MgO, $NiBr_2$, and the like, takes a different form from that used for covalent bonding, where electron sharing between atoms is considered the dominant factor. It is, therefore, appropriate to discuss the two subjects separately.

3-2
Overlap of Orbitals

The detailed nature of chemical bonds is a complex matter. Chemists must, for practical purposes, employ simplified but useful descriptions of bonds. One of the simplest but essentially correct and broadly applicable ideas involved in such descriptions is that a chemical bond can exist when outer orbitals on different atoms overlap so as to concentrate electron density between the atomic cores. As a basic, qualitative guide to whether bonding will occur or not, *the criterion of net positive overlap of atomic orbitals* is of unparalleled usefulness. Consequently, the examination of these overlaps will be our first consideration.

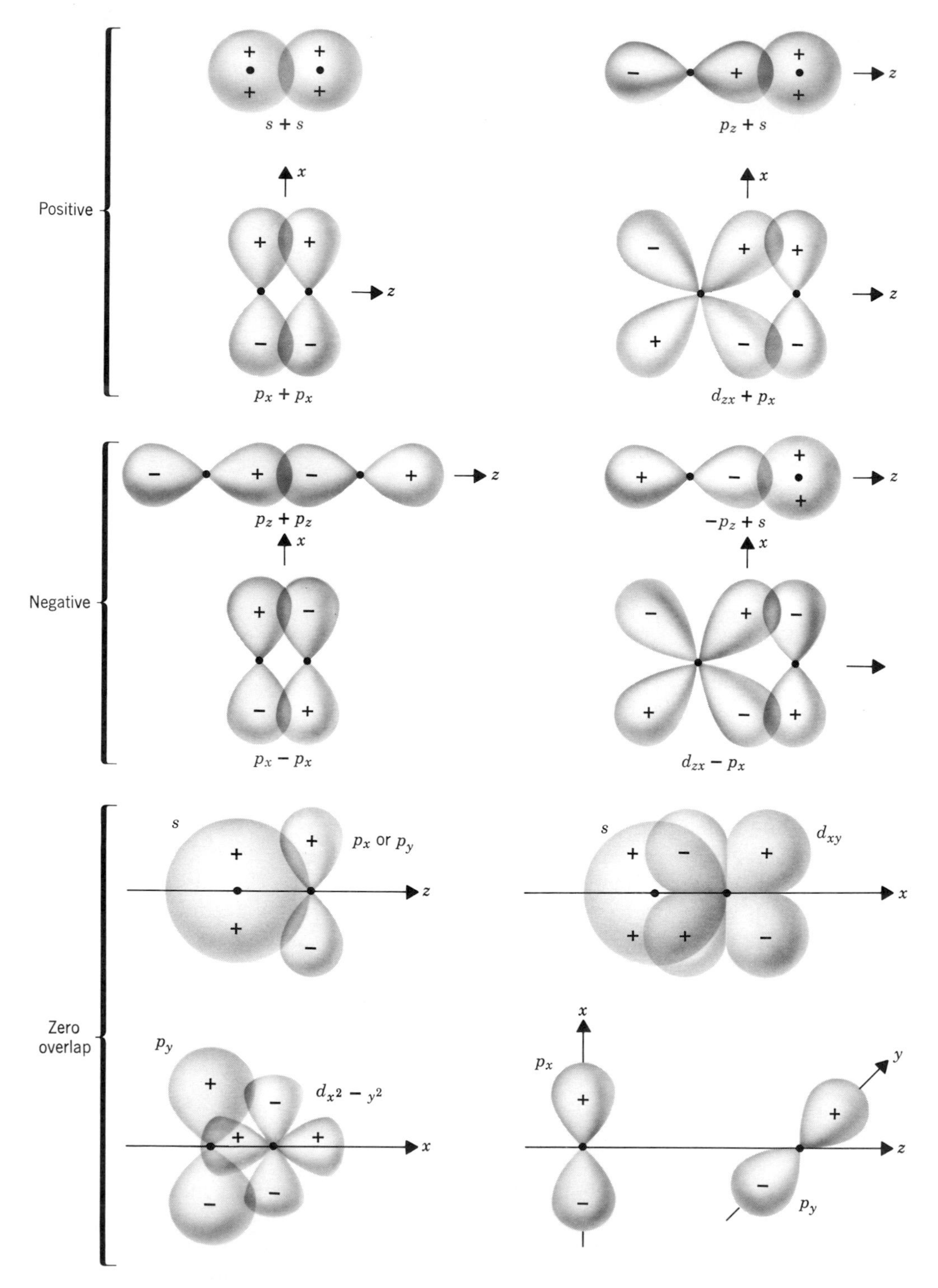

Figure 3-1 *Some common types of orbital overlap with positive, negative and precisely zero magnitudes.*

If two atoms approach each other closely enough for one orbital on each atom to have appreciable amplitude in a region of space common to both of them, the orbitals are said to overlap. The magnitude of the overlap may be positive, negative, or zero, according to the properties of the orbitals concerned. Examples of these three cases are illustrated in *Fig. 3-1.*

Overlap has a positive sign when the superimposed regions of the two orbitals have the same sign, both + or both −. Overlap has a negative sign when the superimposed regions of the two orbitals have opposite signs. Precisely zero overlap results when there are precisely equal regions of overlap with opposite signs.

The physical reason for the validity of the overlap criterion is straightforward. In a region where two orbitals, ϕ_1 and ϕ_2, have positive overlap, the electron density is higher than the mere sum of the electron densities of the two separate orbitals. That is, $(\phi_1 + \phi_2)^2$ is greater than $\phi_1{}^2 + \phi_2{}^2$, by $2\phi_1\phi_2$. More electron density is shared between the two atoms. The attraction of both nuclei for these electrons is greater than the mutual repulsion of the nuclei, and a net attractive force or bonding interaction therefore results.

This is shown in *Fig. 3-2* for the $H_2{}^+$ ion. The full light lines (1) show the electron distributions in the 1*s* orbitals for each atom, $\phi_A{}^2$ and $\phi_B{}^2$. The light dash line (2) shows the simple sum of these, $\phi_A{}^2 + \phi_B{}^2$. If these two orbitals are brought together with the same sign, they give a positive overlap and the electron density will be given by $(\phi_A + \phi_B)^2$. This is shown as line (3) which lies above (2) throughout the region between the nuclei. In other words, the electron becomes concentrated between the nuclei where it is simultaneously attracted to both of them and the $H_2{}^+$ ion is more stable than $H^+ + H$ or $H + H^+$.

Clearly, in the case of negative overlap, shared electron density is reduced by $2\phi_1\phi_2$ and internuclear repulsion increases. This causes a net repulsive or *antibonding* interaction between the atoms. This is also illustrated for $H_2{}^+$ in *Fig. 3-2.* The electron density distribution given by $(\phi_A - \phi_B)^2$ is given by the solid curve (4). The electron density is now much lower between the nuclei, actually reaching zero at the midpoint, and the nuclei repel each other strongly.

When the net overlap is zero there is neither an increase nor a decrease in shared electron density and, therefore, neither a repulsive nor an attractive interaction. This situation is described as a *nonbonding* interaction.

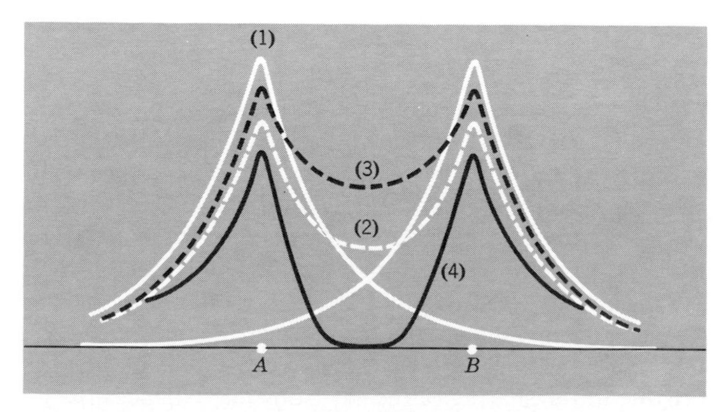

Figure 3-2 *Electron density distributions for the* $H_2{}^+$ *ion, with* H_A *at point* A *and* H_B *at point* B. *(1) For each orbital,* $\phi_A{}^2$ *and* $\phi_B{}^2$*, separately. (2) The sum,* $(\phi_A{}^2 + \phi_B{}^2)/2$. *(3) The bonding function* $(\phi_A + \phi_B)^2/\sqrt{2}$. *(4) The antibonding function* $(\phi_A - \phi_B)^2/\sqrt{2}$.

3-3
Why the H_2 Molecule Is Stable While He_2 Is Not

Once the sign and magnitude of the overlap between a particular pair of orbitals is known, the result, in terms of the energy of interaction, may be expressed in a diagram, called an energy level diagram. This is best explained by using an example, the hydrogen molecule, H_2. Each atom has only one orbital, namely its 1*s* orbital, which is stable enough to be used in bonding. Thus we examine the possible ways in which the two 1*s* orbitals, ϕ_1 and ϕ_2, may overlap as two H atoms approach each other.

There are only two possibilities, as is illustrated in *Fig. 3-3*. If the two 1*s* orbitals are combined with positive overlap, a bonding interaction results. The positively overlapping combination, $\phi_1 + \phi_2$, can be regarded as an orbital in itself, called a *molecular orbital* (MO), and denoted Ψ_b. The subscript *b* stands for *bonding*. Similarly, the negatively overlapping combination $\phi_1 - \phi_2$, also constitutes a molecular orbital, Ψ_a, where the subscript *a* stands for antibonding.

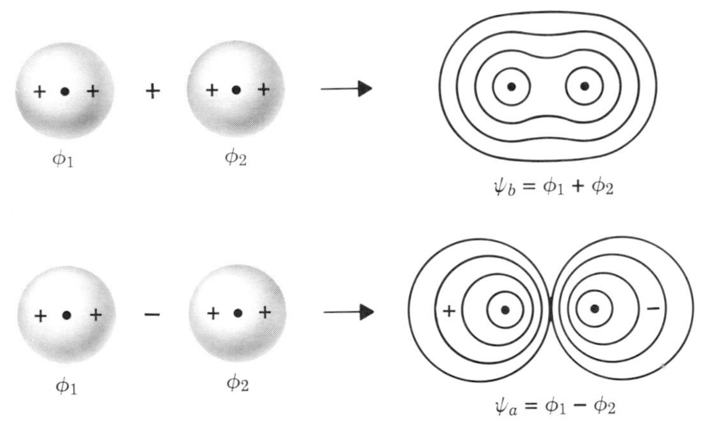

Figure 3-3 *The* 1*s orbitals on two hydrogen, or helium, atoms,* ϕ_1 *and* ϕ_2 *may combine to form either a bonding* MO, Ψ_b, *or an antibonding* MO, Ψ_a.

Let us now imagine that two hydrogen atoms approach each other so that the molecular orbital, MO, Ψ_b is formed. A molecular orbital, like an atomic orbital, is subject to the exclusion principle, which means that it may be occupied by no more than two electrons, and then only if these two electrons have opposite spins. Assuming that the two electrons present, one from each H atom, pair their spins and occupy Ψ_b a bond will be formed. The energy of the system will decrease as, *r*, the internuclear distance decreases following the curve labeled *b* in *Fig. 3-4*. At a certain value of the internuclear distance, r_e, the energy will reach a minimum and then begin to rise again, very steeply. At the minimum the attractive force due to the sharing of the electrons just balances the forces due to repulsions between particles of like charge. At shorter distances the repulsive forces increase very rapidly. It is this rapid increase in repulsive forces at short distances that causes the H_2 molecule (and all other molecules) to have a minimum

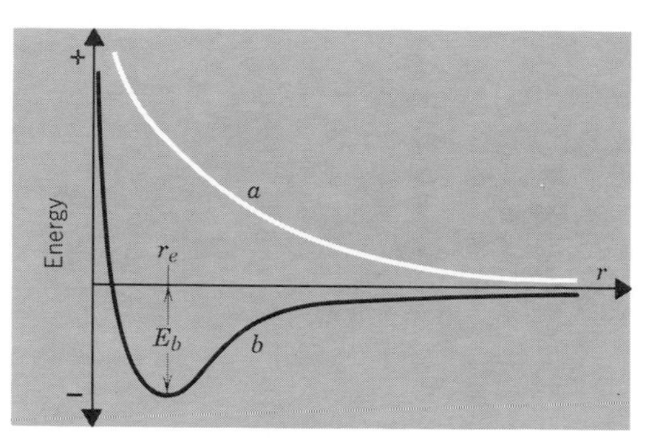

Figure 3-4 *A diagram showing how energy will vary with r as two* H *atoms approach and form a bonding* (*a*) *or an antibonding* (*b*) MO.

energy at a particular internuclear distance and prevents the atoms from coalescing. This minimum energy, relative to the energy of the completely separated ($r = \infty$) atoms is called the bond energy and is denoted E_b in *Fig. 3-4*.

Now if the two H atoms approach each other so as to form the antibonding orbital, Ψ_a, with both electrons occupying that orbital, the energy of the system would vary as is shown in curve *a*. The energy would continuously increase, because at all values of r the interaction is repulsive.

We may now consider the possible formation of an He_2 molecule by using the same basic considerations, represented in *Figs. 3-3* and *3-4*, as for the H_2 molecule. Again, only the 1*s* orbitals are stable enough to be potentially useful in bonding. The He atom differs from the H atom in having two electrons, and this is crucial because in the He_2 molecule there are then four electrons. This means that Ψ_b and Ψ_a must each be occupied by an electron pair. Therefore, whatever stabilization results from the occupation of Ψ_b, it is offset (actually outweighed), by the antibonding effect of the electrons in Ψ_a. The result is that no net, appreciable bonding occurs and the He atoms are more stable apart than together.

3-4
MO Theory of Homonuclear Diatomics in General

The foregoing explanation of why H_2 is a stable molecule and He_2 is not, when coupled with the previous results concerning orbital overlaps, provides all the essential features needed to discuss the bonding in all homonuclear diatomic molecules. We shall consider explicitly the molecules that might be formed by the elements of the first short period, that is, Li_2, Be_2, . . . , F_2, Ne_2.

Before we do so, however, we introduce a different type of energy diagram from that in *Fig. 3-4*—one more suitable to molecules with many molecular orbitals. Instead of trying to represent the energy as a function internuclear distance, we select one particular distance, namely r_e (or the estimated value thereof). The energies of the molecular orbitals at that distance are then shown in the center of the diagram, and the energies of the atomic orbitals are shown for the separate atoms on each side of the diagram. The presence of electrons in the orbitals can then be

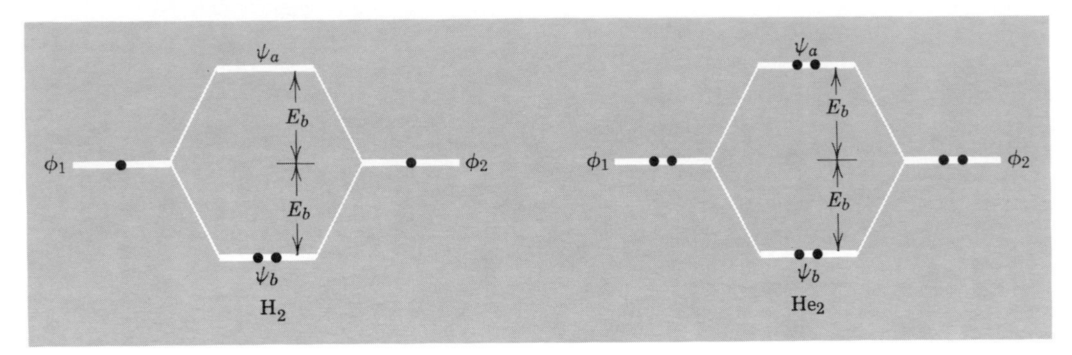

Figure 3-5 MO *energy diagrams for the* H_2 *and* He_2 *molecules. The orbitals marked* ϕ_1 *and* ϕ_2 *are the* 1*s atomic orbitals on the two* H *or two* He *atoms.*

represented by dots (or sometimes arrows). For H_2 and He_2 the appropriate diagrams are shown in *Fig. 3-5*.

Similar diagrams can be used when the two atomic orbitals are not of identical energy, in which case the appearance will be as is shown in *Fig. 3-6*. Two important features must be emphasized for this case. (1) The more the two atomic orbitals differ in energy to begin with, the less they interact and the smaller are the potential bonding energies. (2) While the MO's, Ψ_a and Ψ_b, in *Fig. 3-5* contain equal contributions from ϕ_1 and ϕ_2, this is not true when ϕ_1 and ϕ_2 differ in energy. In that case, Ψ_b has more ϕ_2 character than ϕ_1 character while, conversely, Ψ_a has a preponderance of ϕ_1 character. When ϕ_1 and ϕ_2 differ very greatly in energy, the interaction becomes so small that Ψ_a is virtually identical in form and energy with ϕ_1 and Ψ_b with ϕ_2, as is shown in *Fig. 3-6b*.

Diagrams of this type can be used to show the formation of bonding and antibonding molecular orbitals from any two atomic orbitals, or from two entire sets of atomic orbitals. We are interested here in the interactions of the entire set of $2s2p_x2p_y2p_z$ orbitals on one atom with the equivalent set on another.

If we define the internuclear axis as the z axis, we first note that, according

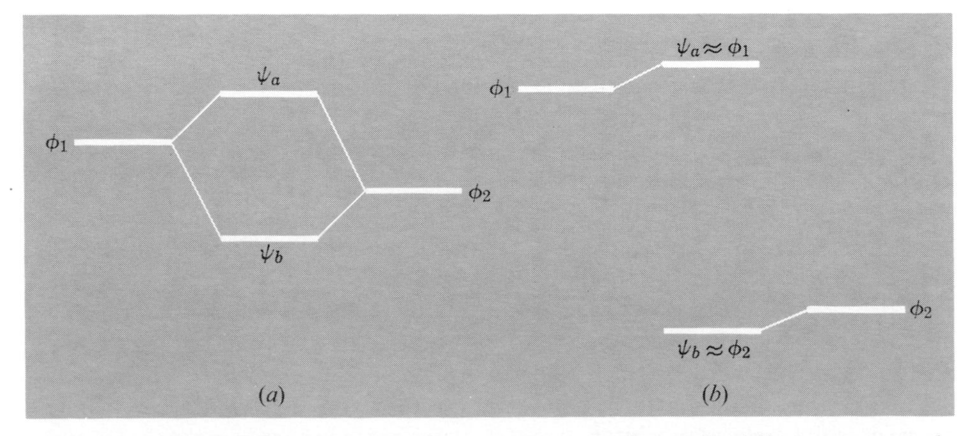

Figure 3-6 *Diagrams for cases where the two atomic orbitals initially differ in energy. In the case at the right the difference is so great that the* MO*'s are only slightly different in either energy or composition from the initial atomic orbitals.*

to the kind of considerations discussed in Section 3-2, only certain overlaps can be nonzero, namely,

$$\begin{array}{rcl} 2s & \text{with} & 2s' \\ 2s & \text{with} & 2p_z' \\ 2p_z & \text{with} & 2s' \\ 2p_z & \text{with} & 2p_z' \\ 2p_x & \text{with} & 2p_x' \\ 2p_y & \text{with} & 2p_y' \end{array}$$

All the remaining 10 (e.g., $2s$ with $2p_x'$, $2p_x$ with $2p_y'$, etc.) are rigorously zero and need not be further considered.

Figure 3-7 shows the overlaps just mentioned in more detail, and indicates how the resulting MO's are symbolized. The first four types of overlap, whether positive (to give a bonding MO) or negative (to give an antibonding MO) give rise to MO's that are designated σ. The $p_x - p_x$ and $p_y - p_y$ overlaps give rise to orbitals designated π. The last two, $s \pm p_z'$, also give σ MO's. The basis for this notation will now be explained.

σ, π, δ Notation. If we view a MO between two atoms along the direction of the bond, that is, we look at it end-on, the following possibilities must be considered, as shown in *Fig. 3-8*.

(*a*) We shall see a wave function that has the same sign, either + or −, all the way around. In other words, as we trace a circle about the bond axis, no change in sign occurs throughout the entire circle. An MO of this kind is called a σ (sigma) MO. Such an MO can only be formed by overlap (either + or −) of two atomic orbitals that also have the same property with respect to the axis in question. Thus these atomic orbitals can also be designated σ. Only the s and p_z orbitals in the sets we are using have this property. The symbol σ is used because σ is the letter s in the Greek alphabet, and a σ MO is analogous to an atomic s orbital, although it need not be formed from atomic s orbitals.

(*b*) We may see a wave function that is separated into two regions of opposite sign. With respect to the entire MO, there is a *nodal plane*. Precisely in this plane the wave function has an amplitude of zero, over the entire length of the bond. The symbol π, the Greek letter p, is used because this type of MO is analogous to an atomic p orbital. As is shown in *Fig. 3-7*, it can be formed by overlap of two suitably oriented p orbitals. In the simple case of a diatomic molecule, or any other linear molecule, π orbitals always come in pairs because there are always two similar p orbitals, p_x and p_y, on each atom. They are equivalent to each other and thus two equivalent π bonding MO's and two equivalent π antibonding MO's are formed.

(*c*) Although we shall not encounter this possibility until much later when we discuss certain transition metal compounds, there are MO's that have two nodal planes. These are called δ (Greek d) orbitals. The δ MO's cannot be formed with s and p atomic orbitals but the overlap of suitable atomic d orbitals, for example, two d_{xy} or two $d_{x^2-y^2}$ orbitals, will form a δ MO.

The F_2 Molecule. We can now consider energy level diagrams for the diatomic molecules. It is easiest to consider F_2 first, rather than Li_2, as will become apparent. Each fluorine atom has a configuration $1s^22s^22p^7$. The $1s$ electrons

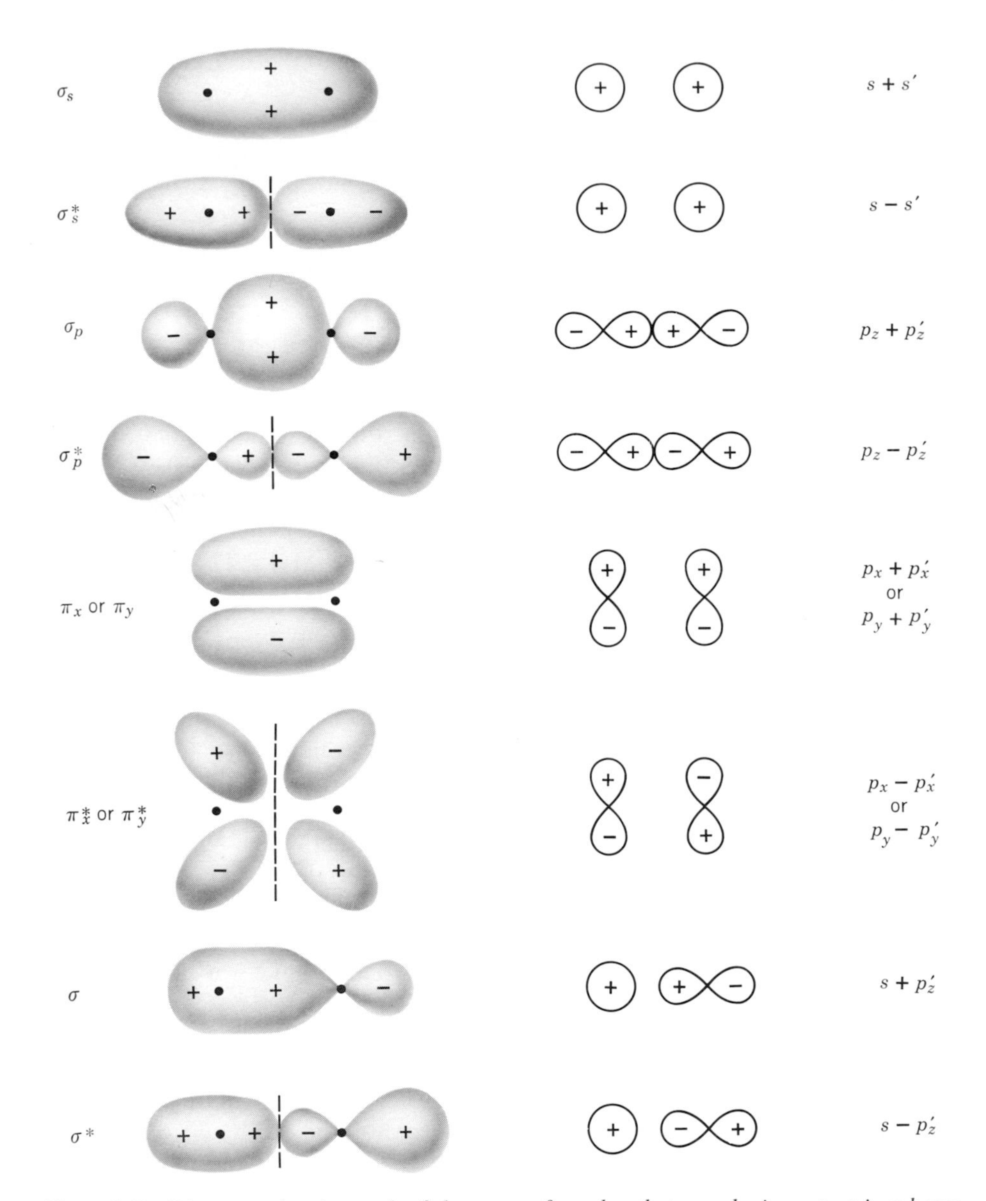

Figure 3-7 *Diagrams showing each of the types of overlap that may be important in a homonuclear diatomic molecule. Notice that the positive direction of the z axis for each atom is directed toward the other nucleus. Plus and minus signs are the signs of the wave function, and pictures are only a qualitative idea of the shape of the orbital. Dashed lines indicate nodal planes perpendicular to the internuclear axis. The σ and π notation is explained in the text. The * indicates an antibonding orbital.*

are so close to the nucleus and so low in energy that they play no significant role in bonding; this is almost always true of so-called inner shell electrons. Thus only the 2*s* and 2*p* orbitals and their electrons need be considered.

For the fluorine atom, the energy difference between the 2*s* and 2*p* orbitals is sufficiently great that the interaction of the 2*s* orbital on one atom with the $2p_z$ orbital on the other is very slight, as is indicated in *Fig. 3-6*, and can be ignored

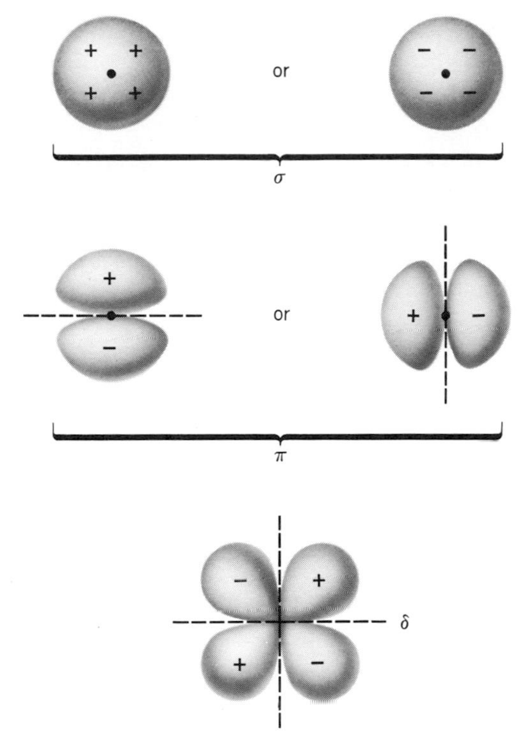

Figure 3-8 *Some orbitals seen along the internuclear axis, showing how σ, π, and δ are defined.*

in this approximate treatment. Thus only $2s - 2s$, $2p_x - 2p_x$, $2p_y - 2p_y$, and $2p_z - 2p_z$ interactions need be included, and the diagram in *Fig. 3-9* is obtained.

In *Fig. 3-9* the pairs of π and π^* orbitals, formed by overlap of the p_x and p_y orbitals, have identical energies, since they differ only in their orientation around the internuclear axis. The lowest σ orbital, σ_1, is simply a σ_s orbital, in the sense of *Fig. 3-7*. Similarly, σ_2 is σ_s^*, σ_3 is σ_p, etc. For F_2 there is now a total of $7 + 7 = 14$ electrons to occupy these MO's. By adding two electrons to each one, beginning at the lowest energy, we get the occupation shown in *Fig. 3-9*.

Bond Order. We can see in *Fig. 3-9* that while two electrons have dropped from the energy level of the 2*s* orbitals to that of σ_1, two others have been raised by roughly the same amount to the energy of σ_2. These two changes effectively cancel each other. Similarly, the electrons in the π_1 and π_2 orbitals have bonding and antibonding effects that approximately cancel each other. Hence, only the electron pair in σ_3 gives a net bonding effect, and we conclude that the F_2 molecule has a single bond.

This illustrates how bond order is defined in general in molecular orbital theory. If we take the number of electron pairs in bonding MO's (n_b) and subtract the number of pairs in antibonding MO's (n_a) we have the bond order, namely $n_b - n_a$.

The Li_2 Molecule. For the Li_2 molecule the diagram is somewhat different because the $2s - 2p$ separation in the Li atom is smaller and the $2s - 2p_z'$ and $2p_z - s'$ interactions are large enough that they cannot be ignored. The diagram

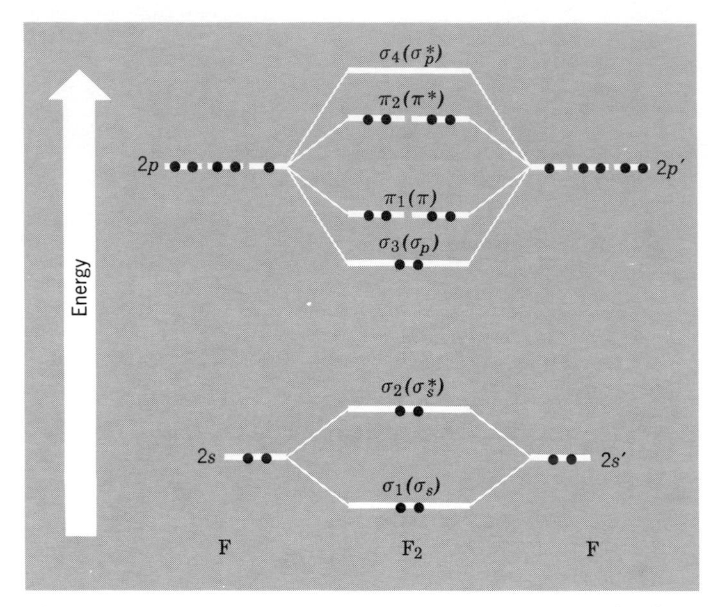

Figure 3-9 *A molecular orbital diagram for the fluorine molecule,* F_2.

is as shown in *Fig. 3-10.* As a result of the $s - p_z'$ and $p_z - s'$ interactions, σ_2 and σ_3 have both p_z and s character, and there is an upward displacement of σ_3 so that it lies above π_1. Although this has practically no importance for the stability of Li_2 itself, it will become very important as we proceed to molecules with more electrons. For Li_2 the 2s electrons occupy the σ_1 orbital and we have a σ bond. It is a weak bond because with Li atoms the 2s orbitals are diffuse and do not overlap very strongly. *Figure 3-11* shows a computer-drawn electron density map ($\sigma_1{}^2$) for this bonding electron pair. This quantitative representation of the $s + s'$ overlap for this case may be compared with the purely schematic depictions of such overlaps shown in *Figs. 3-3* and *3-7*.

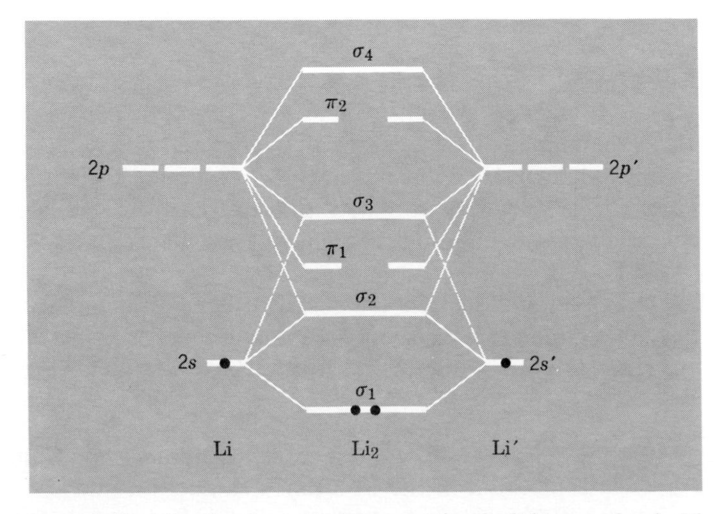

Figure 3-10 *A molecular orbital diagram for the lithium molecule,* Li_2.

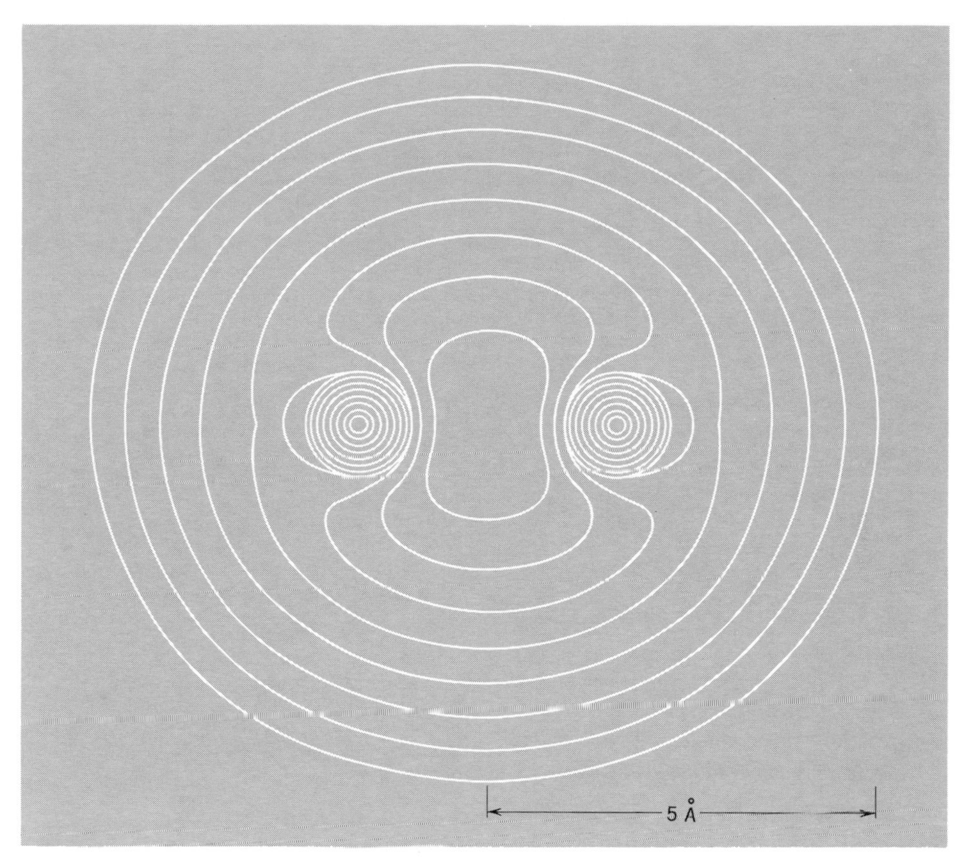

Figure 3-11 *Electron density contours for the filled bonding orbital, σ_1, in Li_2. Successive contours, from the outside in, represent increases by a factor of two.*

We can now consider the entire series of molecules from Li_2 to F_2. The progressive changes in orbital energies and electron populations from one extreme to the other are shown in *Fig. 3-12*, along with the bond distances and bond energies. Li_2 has the longest and weakest bond of all because it is only a single bond formed by overlap of two fairly diffuse orbitals. The fact that the $1s^2$ shells of each Li atom are still rather large because the nuclear charge is not high (+3), may also introduce a repulsive force that weakens the bond.

The beryllium atom has a $1s^2 2s^2$ configuration. Thus there would be four valence electrons in the molecule, and electron pairs occupy both σ_1 and σ_2. As with He_2, the bonding and antibonding effects cancel. In terms of bond order one has $n_b - n_a = 1 - 1 = 0$. Thus there is no stable Be_2 molecule.

For B_2 there are six electrons to occupy the MO's. The last two enter the π_1 orbital and behave just as would two p electrons in an atom according to Hund's first rule. They occupy different orbitals in the set and keep their spins in the same direction. The B_2 molecule is, therefore, paramagnetic with two unpaired electrons. The bond order is 1, since the σ_1 and σ_2 pairs cancel each other, but the π_1 electrons are occupying a bonding MO. Notice it is not necessary that they have their spins paired in order to serve as bonding electrons. The bond distance is shorter

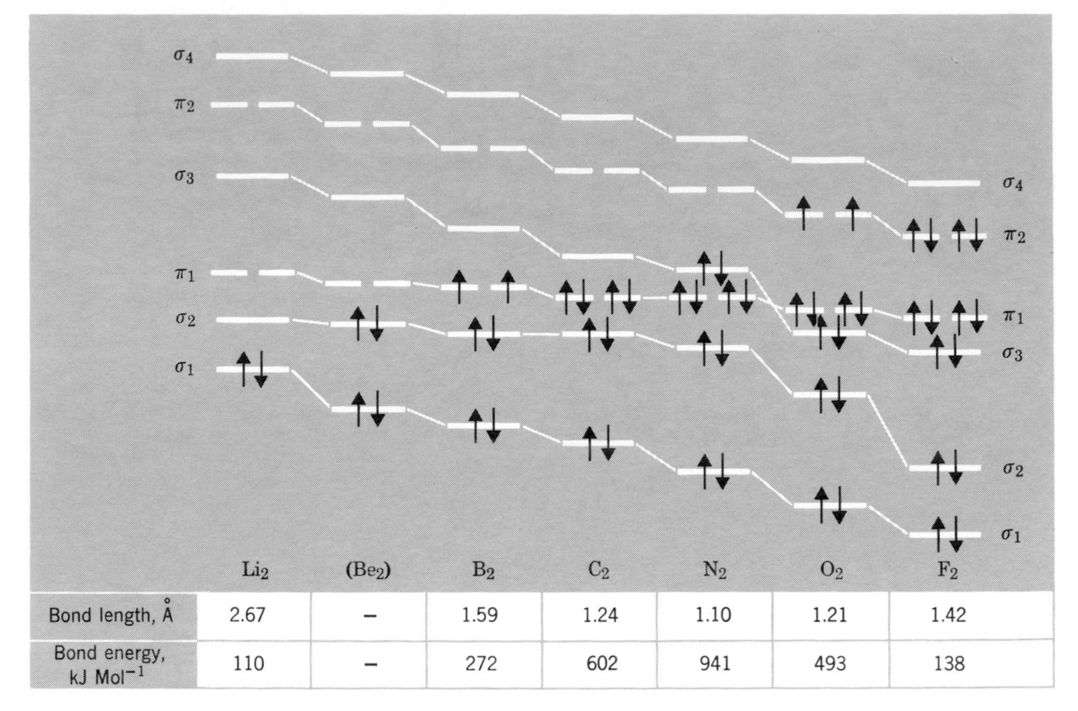

	Li_2	(Be_2)	B_2	C_2	N_2	O_2	F_2
Bond length, Å	2.67	–	1.59	1.24	1.10	1.21	1.42
Bond energy, kJ Mol^{-1}	110	–	272	602	941	493	138

Figure 3-12 *The diatomic molecules from* Li_2 *to* F_2 *showing the changes in energies, electron configurations, bond lengths, and bond energies.*

and the bond energy is higher than in Li_2 because of the smaller size of the boron atom.

At C_2 the π_1 orbitals are only slightly lower in energy than σ_3 but, nevertheless, by enough to give the configuration shown, with no unpaired electrons and a bond order of 2. In accord with this, the C_2 molecule in its ground state is diamagnetic and has a considerably shorter and stronger bond than does B_2. However, an excited state with the electron configuration $\sigma_1{}^2\sigma_2{}^2\pi_1{}^3\sigma_3$ lies only ~ 10 kJ mol^{-1} higher in energy.

The nitrogen molecule has the highest bond order, 3, the shortest bond, and the strongest bond of any molecule in the series. The net bond order of 3 is in agreement with the conventional representation of N_2 as a triply bonded molecule, :N≡N:.

With the oxygen molecule, bond order and bond strength begin to decrease since, following N_2, only antibonding orbitals remain to be occupied. The two additional electrons enter π_2, an antibonding orbital, thus reducing the bond order of 3 found in N_2 to 2 in O_2. Accordingly, the bond length increases and the bond energy decreases. A most important fact is that the electrons in π_2 (like those in π_1 in B_2) are unpaired, and this accords with the experimental fact that O_2 is paramagnetic with two unpaired electrons. The correct prediction of this by the simple MO theory is in contrast to the difficulty of explaining it in ordinary, simple terms where it is assumed that the atoms will share electrons so as to give a complete octet about each one. On this basis we would write

$$:\ddot{\underset{\cdot}{O}}\cdot + \cdot\ddot{\underset{\cdot}{O}}: \longrightarrow :\ddot{O}::\ddot{O}:$$

This predicts a double bond correctly, but not the presence of two unpaired electrons. This so-called singlet state is actually an excited state of the molecule where the two π_2 electrons have their spins paired, and lies about 92 kJ mol^{-1} above the ground state. Molecules in the singlet state have quite different reactivity from those in the paramagnetic ground state (see page 302).

The element following fluorine is neon. The Ne_2 molecule is not stable, and the reason for this is clear. Addition of two more electrons to the F_2 configuration would mean that all antibonding as well as all bonding orbitals would be occupied. Thus the bond order would be zero.

3-5 Heteronuclear Diatomic Molecules

The extension of the qualitative theory for homonuclear diatomics to heteronuclear ones, such as CO and NO, is not difficult. It depends on making allowance for the fact that the two sets of atomic orbitals that interact have different energies. This is shown in *Fig. 3-13* where the isoelectronic molecules N_2 and CO are contrasted. The most important features to be noted in this comparison are: (1) All orbitals of the oxygen atom lie at lower energies than the corresponding ones of the C atom, because oxygen has a nuclear charge which is two units higher. This is in keeping with *Fig. 2-11*, which indicates that the first ionization enthalpy of O is several hundred kJ mol^{-1} greater than that of C. (2) The $2s - 2p$ energy separation is greater for O than for C.

It can be shown that as a result of these initial atomic orbital energy differences, the MO's for CO are significantly different from those for N_2. For example, the highest filled orbital for N_2 is a σ orbital of fairly bonding character. Hence, loss of one electron from N_2 weakens the N—N bond. In CO, the highest filled orbital is a σ orbital that is slightly antibonding in character. Hence, the CO^+ ion has a slightly stronger bond than does CO.

Another important heteronuclear diatomic molecule is nitric oxide, NO. Since N and O differ by only one atomic number, the energy level diagram for this is rather similar to that for N_2. The additional electron occupies the antibonding π_2 orbital (*Fig. 3-13a*), from which it is relatively easily removed to give the NO^+ ion, which has a stronger bond than does neutral NO. The electronic structure of NO might, of course, equally well have been derived qualitatively by removing one electron from the configuration of the O_2 molecule.

3-6 Molecular Orbital Theory for Polyatomic Molecules

The molecular orbital method can be generalized to larger molecules. To illustrate, let us consider the simplest linear triatomic molecule, BeH_2. Let us choose the z axis as the molecular axis. We first note that only σ MO's can be formed because the hydrogen atoms have only their $1s$ orbitals to use in bonding. These orbitals are themselves of σ character with respect to any axis that passes through the nucleus, and therefore they can contribute only to σ MO's. On the Be atom then,

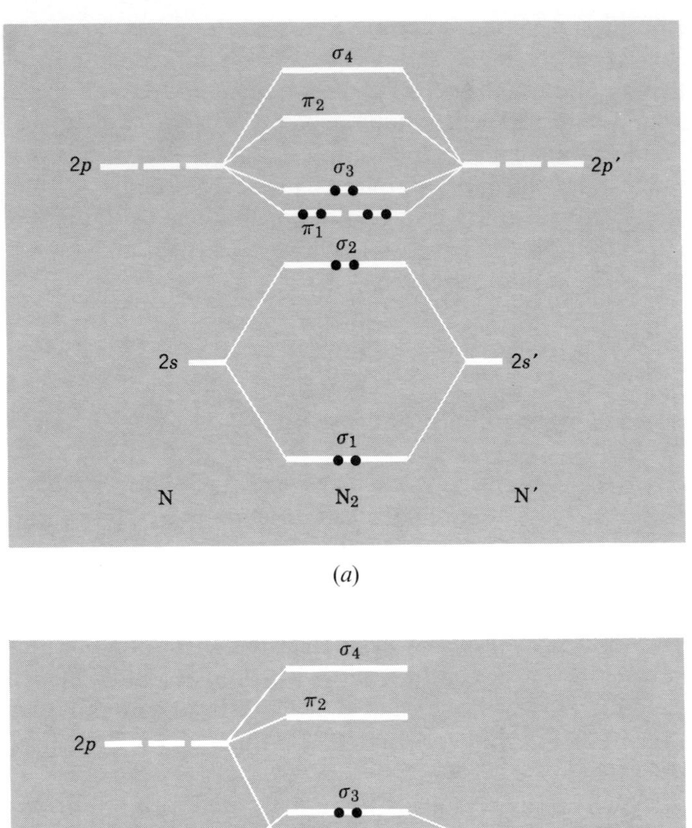

(a)

(b)

Figure 3-13 MO *diagram for* (*a*) *nitrogen molecule and* (*b*) *carbon monoxide molecule.*

only the 2*s* and $2p_z$ orbitals can participate in bonding. The p_x and p_y orbitals, which have π character and zero overlap with any σ orbital, will play no role in bonding in BeH_2.

The 2*s* orbital of beryllium can combine with the two 1*s* orbitals of the hydrogen atoms to form bonding and antibonding MO's as is shown in *Fig. 3-14*. In these, the signs of the two 1*s* orbitals are "in phase" with each other and either "in phase" or "out of phase" with the beryllium 2*s* orbital.

The $2p_z$ orbital of beryllium also combines with the hydrogen 1*s* orbitals, as is shown in *Fig. 3-14*, to form bonding and antibonding σ MO's. In these, the 1*s* orbitals are "out of phase" with each other.

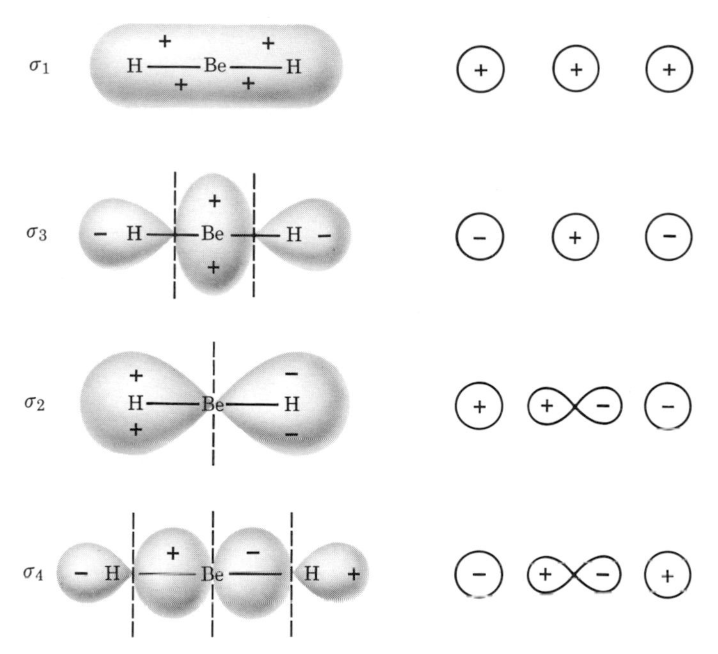

Figure 3-14 *The four σ-molecular orbitals for linear* BeH_2. *The dashed vertical lines are nodal planes perpendicular to the molecular axis.*

The important points to keep in mind about these four σ MO's are as follows.

1. In each bonding MO electron density is large and continuous between adjacent atoms, while in the antibonding MO's there is a node between each adjacent pair of nuclei.

2. In each one, the wave function indicates that an electron pair occupying it is "spread out" over the whole molecule, and is shared by all of the atoms, not just a particular adjacent pair. In other words, in MO's electrons are *delocalized* over the whole extent of the MO.

The MO treatment of the bonding in BeH_2 can be expressed in terms of an energy-level diagram, as shown in *Fig. 3-15*. The main features here are that the hydrogen 1*s* orbitals lie at much lower energy ($\sim$400 kJ mol^{-1}) than the beryllium 2*s* orbital and that the p_x and p_y orbitals of Be carry over completely unchanged into the center column, because they do not overlap with any other orbitals. The four valence electrons, $2s^2$ from Be and 1*s* from each H, occupy σ_1 and σ_2. The total bond order of the Be—H bonds is 2. Since each Be—H pair participates equally in the molecule, this is equivalent to saying that there are two equivalent B—H single bonds.

A particularly important and more general application of MO theory in polyatomic molecules deals with π bonding in planar species. One important group, which are qualitatively similar, though they differ in detail, are planar, symmetrical species of general formula AB_3. Important representatives are BF_3, CO_3^{2-}, NO_3^-. If such a molecule or ion is placed in the *xy* plane of a coordinate

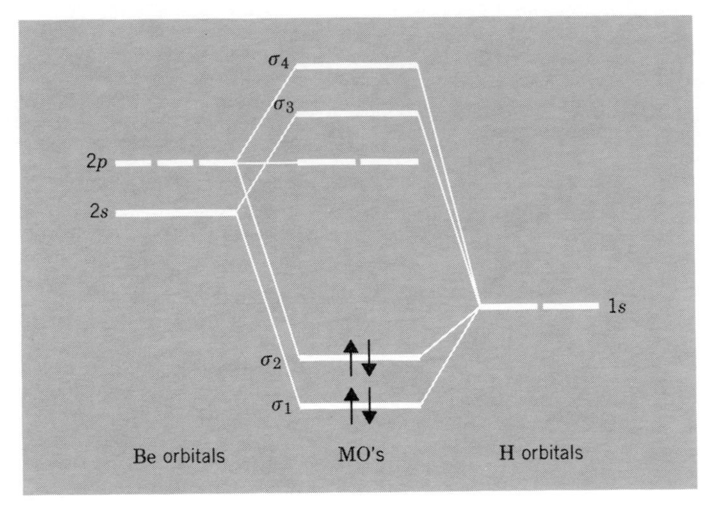

Figure 3-15 *A molecular orbital energy-level diagram for* BeH_2.

system, then the π bonding will be formed entirely by the p_z orbitals of the four atoms.

According to certain mathematical requirements which are an inherent part of quantum mechanics, but which cannot be explained here, the p_z orbitals on the outer atoms can be combined in the three ways shown in *Fig. 3-16*. Two of these, *a* and *b*, have a net overlap of zero with a p_z orbital on the central atom, $p_z^{(4)}$. For example, $p_z^{(1)}$ overlaps in a positive sense with $p_z^{(4)}$, but $p_z^{(2)}$ and $p_z^{(3)}$ each have a negative overlap and the sum of these exactly offsets the positive $p_z^{(1)}p_z^{(4)}$ overlap. Similarly, in *b*, the $p_z^{(2)}p_z^{(4)}$ and $p_z^{(3)}p_z^{(4)}$ overlaps are equal but opposite in sign and thus give a net overlap of zero. Combinations *a* and *b* are therefore nonbonding orbitals. The remaining one, *c*, in which the three p_z orbitals are "in phase" can have a finite overlap, either positive or negative, with the p_z orbital on the central atom. In this way, a bonding MO and an antibonding MO are formed (*Fig. 3-17*).

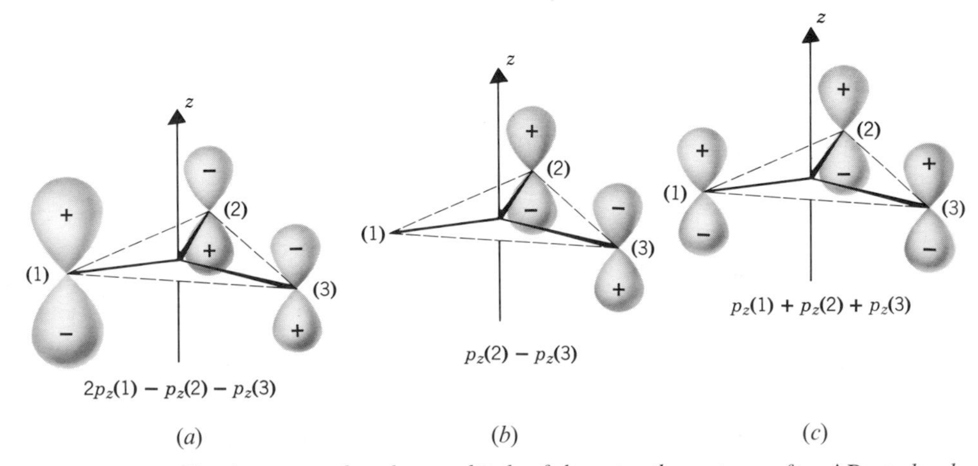

Figure 3-16 *The three ways that the* p_z *orbitals of the outer three atoms of an* AB_3 *molecule can be combined into sets for making molecular orbitals.*

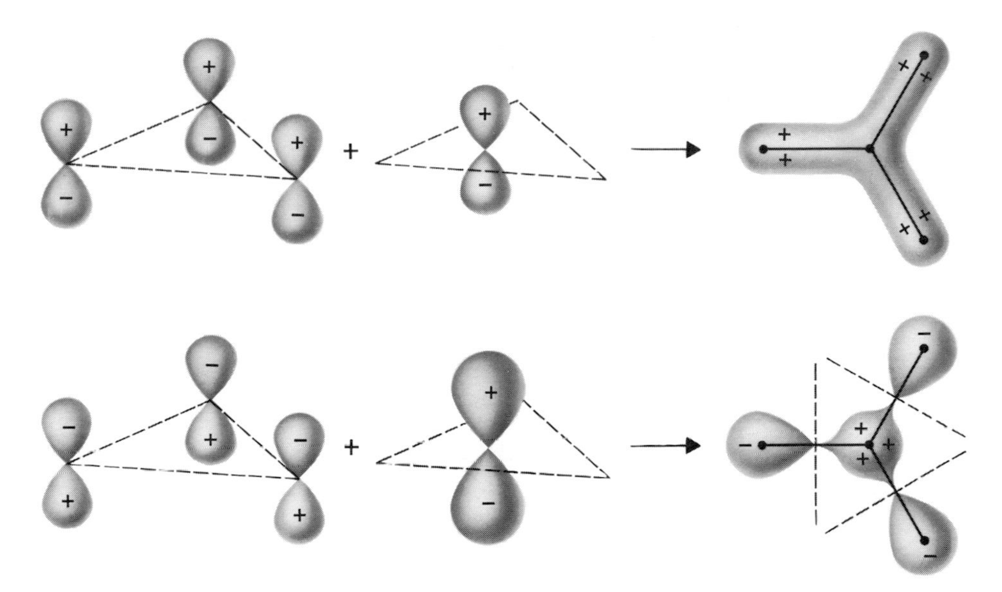

Figure 3-17 Diagrams showing how bonding (upper) and antibonding (lower) π molecular orbitals are formed in an AB_3 molecule. The MO's themselves, at the right, are viewed from above. The portions below the molecular plane have all signs reversed

The results of the foregoing analysis are depicted in an energy-level diagram in *Fig. 3-18.* For each of the species, BF_3, CO_3^{2-}, and NO_3^-, there are 6 electrons to occupy the π MO's; the other 18 electrons are in various σ orbitals which lie in the molecular plane. The 6 π electrons are distributed as is shown in *Fig. 3-18.*

The four electrons in the two, degenerate (meaning of equal energy) orbitals, π_{2a}, π_{2b}, neither contribute to nor detract from the bonding. Thus, the π bonding is provided entirely by the one electron pair in the π_1 orbital. The total π-bond order is therefore 1. However, the π-bond order per pair of atoms, B—F, C—O, or N—O, is $\frac{1}{3}$, since there are three such pairs, all sharing equally in the π bonding.

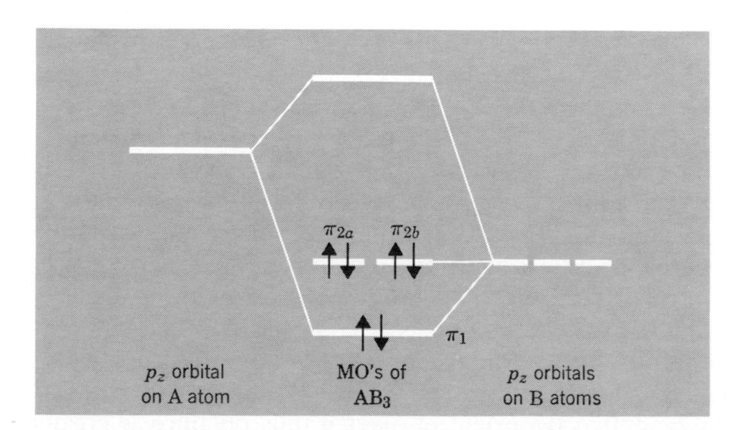

Figure 3-18 Molecular orbital energy-level diagram for the π bonding in a planar, symmetrical AB_3 molecule.

This is an example of fractional bond order and its formulation in MO theory. A different way of handling the same problem, involving the concept of resonance, will be discussed in Section 3-8.

3-7
The Localized Bond Approach; Valence States and Hybridization

The bonding in polyatomic molecules can also be treated as a collection of localized bonds between adjacent pairs of atoms. Although the concept of molecular orbitals is fundamentally correct, in many instances the actual extent of delocalization is small, and the idea of bonds localized between pairs of atoms is a useful approximation.

In BeH_2, for example, it is perfectly satisfactory, for nearly any purpose, to consider that there is one electron pair localized between each adjacent pair of atoms. Thus, we have the simple, familiar representation, H:Be:H. An electron pair bond of the type indicated can be thought of as arising from the overlap of two orbitals, one from each of the atoms bonded, with the electrons concentrated in the region of overlap between the atoms. In the case of BeH_2, which is linear, this raises the question of how to account for the linearity. In answering that question, two new concepts, the valence state and hybridization, are introduced.

The Valence State. The beryllium atom has the electron configuration $1s^2 2s^2$. Thus its valence shell has only one occupied orbital and the electrons are paired. On the other hand, if it is to form two bonds, by sharing one electron with each of two other atoms, it must first be put into a state where each electron is in a different orbital, and each spin is uncoupled from the other and, thus, ready to be paired with the spin of an electron on the atom to which the bond is to be formed. When it is in this condition, the atom is said to be in a valence state.

For the particular case of BeH_2, the valence state of lowest energy is obtained by promoting one of the electrons from the $2s$ orbital to one of the $2p$ orbitals, and decoupling their spins. This requires the expenditure of about 323 kJ mol^{-1}.

Hybridization. Although the promotion of the Be atom to the valence state prepares it to form the two bonds to the H atoms, it does not provide an explanation or a reason why the molecule should be linear, rather than bent. The $2s$ orbital of Be has the same amplitude in all directions. Therefore, whichever of the $2p$ orbitals is used to form one Be—H bond, the other bond in which the $2s$ orbital is used could make any angle with it, insofar as overlap of the H$1s$ and Be$2s$ orbitals is concerned. However, the preference for a linear structure can be attributed to the fact that if a $2s$ and $2p$ orbital are mixed so as to form two *hybrid* (i.e., mixed) orbitals, better total overlap with the H$1s$ orbitals can be obtained. The results of mixing the $2s$ and $2p_z$ orbitals are shown in *Fig. 3-19.*

Each of the hybrid orbitals has a large positive lobe concentrated in a particular direction and is, therefore, able to overlap very strongly with an orbital on another atom located at an appropriate distance in that direction. Actual calculations show that the extent of overlap thus obtained is greater than that obtainable by using either a pure $2s$ or pure $2p_z$ orbital. This is not difficult to see without calculation, if we note that only half of the p_z orbital is found in the $+z$

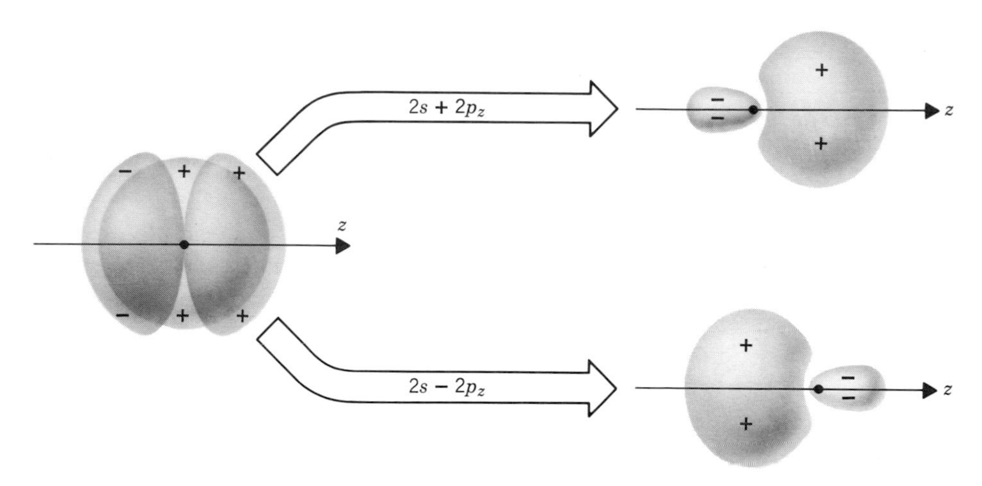

Figure 3-19 *The formation of hybrid orbitals by the mixing of a 2s and a $2p_z$ orbital.*

direction and only half in the $-z$ direction. The 2*s* orbital is uniformly distributed in all directions. The hybrid orbitals, however, are each strongly concentrated in just one direction.

The linearity of the BeH_2 molecule follows automatically from the use of the hybrid orbitals. It can be easily determined in *Fig. 3-19* that the best hybrids are obtained in the $+z$ and $-z$ directions, and in no others, because of the spatial properties of the *s* and *p* orbitals themselves. The best Be to H overlaps are then obtained by placing the H atoms along the $+z$ and $-z$ directions, as is shown in *Fig. 3-20*.

The hybrid orbitals just described are called *sp* hybrids, to indicate that they are formed from one *s* orbital and one *p* orbital. There are other ways of mixing *s* and *p* orbitals to form hybrids. The element boron forms many hydrogen compounds. The simplest, BH_3, which is not stable as such but dimerized (page 79) illustrates another important case of hybridization.

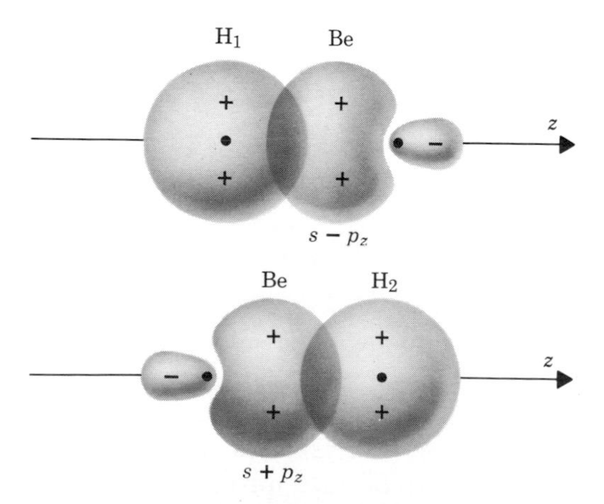

Figure 3-20 *The overlap of the sp hybrid orbitals on* Be *with the* 1*s orbitals of the* H *atoms in* BeH_2

The boron atom has a ground state electron configuration, $1s^2 2s^2 2p$. To form three bonds it must first be promoted to a valence state based on a configuration $2s2p_x2p_y$, in which the three valence electrons have decoupled their spins. The choice of $2p_x$ and $2p_y$ is arbitrary; any two $2p$ orbitals would be satisfactory. The ability of the central atom to form three bonds is now taken care of, but the question of securing maximum overlap must be dealt with. Again, it develops straightforwardly that by mixing the s and the two p orbitals equally, hybrid orbitals, called sp^2 hybrids can be formed, and they give superior overlap in certain definite directions, as is shown in *Fig. 3-21*. The three hybrid orbitals lie in the xy plane, and their maxima lie along lines that are 120° apart. Thus, the BH_3 molecule which would result has a planar, triangular structure.

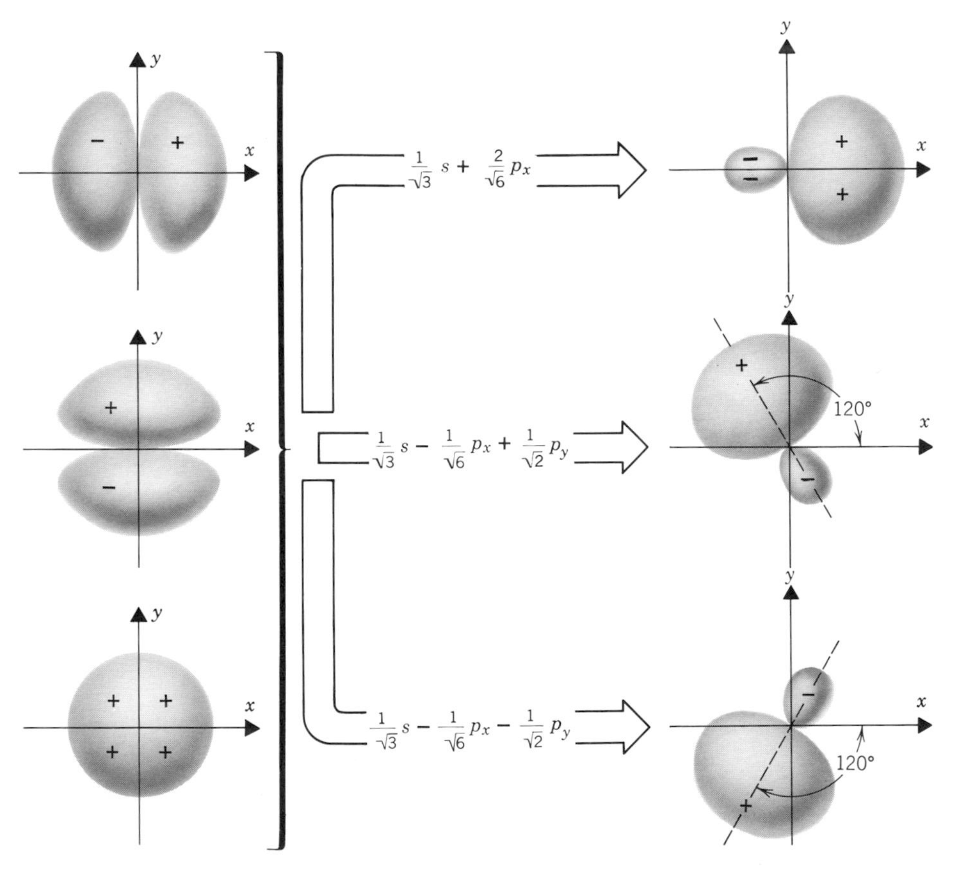

Figure 3-21 *The formation of the three equivalent sp^2 hybrid orbitals.*

The last type of hybridization involving only s and p orbitals can be discovered by considering how the carbon atom combines with four hydrogen atoms to form methane. Again, promotion from a ground state that does not have a sufficient number of unpaired electrons, $1s^2 2s^2 2p^2$, to a valence state, $2s2p_x2p_y2p_z$, is required first. Then, the four orbitals are mixed to give a set of four equivalent orbitals called sp^3 hybrids. This is shown in *Fig. 3-22*. These are directed to the vertices of a tetrahedron, and thus CH_4 has a tetrahedral configuration.

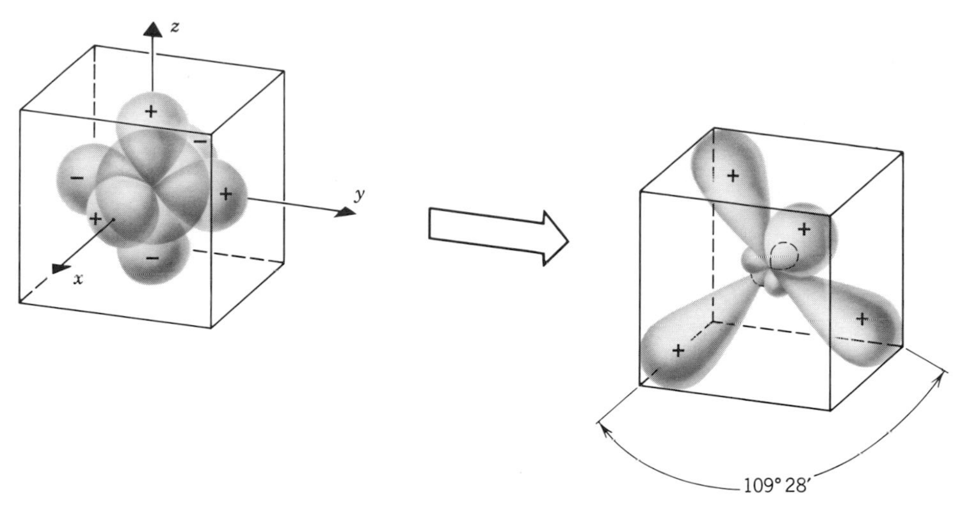

Figure 3-22 *The formation of four equivalent sp^3 hybrid orbitals. A tetrahedron is defined by the four alternating corners of the cube to which the hybrids are directed. The angle between the axes of any two hybrids is* 109°28′.

In summary, an atom that has only s and p orbitals in its valence shell can form three types of hybrid orbitals, depending on the number of electrons available to form bonds:

sp hybrids give a linear molecule
sp^2 hybrids give a plane triangular molecule
sp^3 hybrids give a tetrahedral molecule

When d orbitals as well as s and p orbitals are available, the following important sets of hybrids, each illustrated in *Fig. 3-23*, can arise.

1. d^2sp^3, *Octahedral hybridization.* When the $d_{x^2-y^2}$ and d_{z^2} orbitals are combined with an s orbital and a set of p_x, p_y, and p_z orbitals, a set of equivalent orbitals with lobes directed to the vertices of an octahedron can be formed.

2. dsp^2, *Square-planar hybridization.* A $d_{x^2-y^2}$ orbital, an s orbital and p_x and p_y orbitals can be combined to give a set of equivalent hybrid orbitals with lobes directed to the corners of a square in the xy plane.

3. sd^3, *Tetrahedral hybridization.* An s orbital and the set d_{xy}, d_{yz}, d_{zx} may be combined to give a tetrahedrally directed set of orbitals.

4. dsp^3, *Trigonal-bipyramidal hybridization.* The orbitals s, p_x, p_y, p_z, and d_{z^2} may be combined to give a nonequivalent set of five hybrid orbitals directed to the vertices of a trigonal bipyramid.

5. dsp^3, *Square-pyramidal hybridization.* The orbitals s, p_x, p_y, p_z, and $d_{x^2-y^2}$ may be combined to give a nonequivalent set of five hybrid orbitals directed to the vertices of a square pyramid.

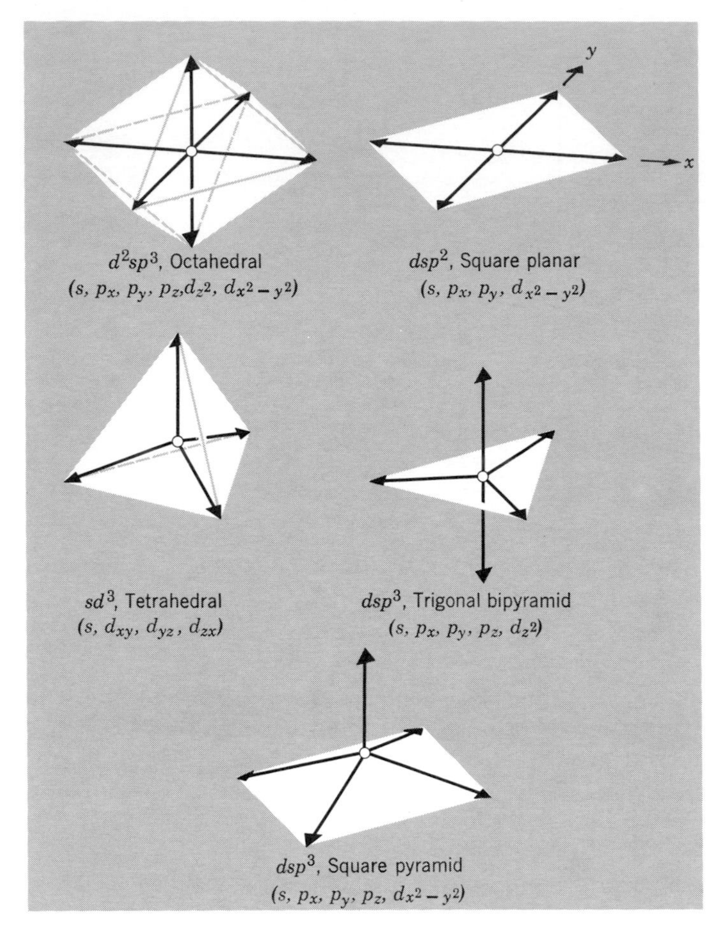

Figure 3-23 *Five important hybridization schemes involving d orbitals. Heavy arrows show the directions in which the lobes point.*

The use of hybridized orbitals to explain bonding and correlate structures has become less common in recent years, giving way to the more general use of MO theory. The main reasons for this are that the MO approach lends itself more readily to quantitative calculations employing digital computers and because, with such calculations, it is possible to account for molecular spectra more easily. Nevertheless the concept of hybrid orbitals retains certain advantages of simplicity and, in many instances, affords a very easy way to correlate and "explain" molecular structures. This latter point will be considered further in Section 3-10.

It is also worthwhile to point out, although we do not try to demonstrate it here, that the MO and hybrid orbital treatments give results that are, in most cases, not greatly different. In BeH_2, for example, it may be noted that the same atomic orbitals are employed in either treatment. The MO description suggests considerable delocalization of the bonding electrons, while the hybrid orbital description localizes them. However, the truth doubtless lies in between.

3-8
Resonance

This concept, like that of hybridization, is one whose popularity has declined in recent years, but still it retains utility for certain purposes. Again, it is a way of describing certain features of chemical bonding which is not really as different from the analogous molecular orbital description as superficially it might appear to be.

Let us consider again, for example, a molecule such as BF_3, CO_2^{2-}, or NO_3^-. If we try to write a formula for such a molecule, in which each atom has an octet of electrons, we obtain 3-I:

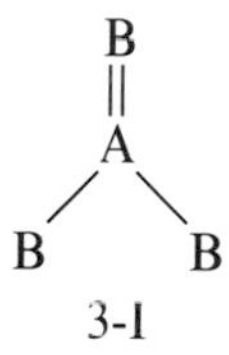

3-I

This representation implies that there are two A—B single bonds and one A=B double bond, whereas, experimental data show conclusively that all A—B bonds and all B—A—B angles are equal. To bring theory and experiment into accord, the former is modified by the postulate that structure 3-I alone does not describe the actual molecule but only one of three equivalent, hypothetical structures, I, II, and III. The real molecule has an electron distribution corresponding to the average of these three contributing structures, and is said to be a *resonance hybrid* of them. The double-headed arrow is used to indicate that the structures are mixing to give a resonance hybrid.

3-I ⟷ 3-II ⟷ 3-III

Care is required to avoid misinterpretation of the resonance concept. At no instant does the molecule actually have any one of the canonical structures. Each of these implies that one bond is stronger and thus presumably is shorter than the other two, whereas all three bonds are always entirely equivalent. The canonical structures have no real existence, in any way or sense, but their average corresponds to the actual structure. The analogy of a mule as a hybrid of a horse and a donkey is helpful. A mule does not alternate between being a horse and a donkey. It is always a mule but has characteristics that are intermediate between those of its parents.

We can see that the resonance description of the AB_3 molecule leads to the same conclusion as was previously obtained from MO theory regarding the A—B bond order. In MO theory it was shown that the π bond order per A—B pair is $\frac{1}{3}$. This, when added to a σ bond order of 1 for each pair, gives a total bond order for each A—B bond of $1\frac{1}{3}$. The resonance picture also gives a bond order of $(1 + 1 + 2)/3 = \frac{4}{3} = 1\frac{1}{3}$.

The concept of resonance can be justified from an energy point of view. It can be shown that a resonance hybrid must have a lower energy, that is, be more stable, than any single contributing structure. It is this which accounts for the fact that the molecule exists in the hybrid structure rather than any one of the contributing structures.

Other, familiar cases in which a resonance description can be used to account for fractional bond orders are the following:

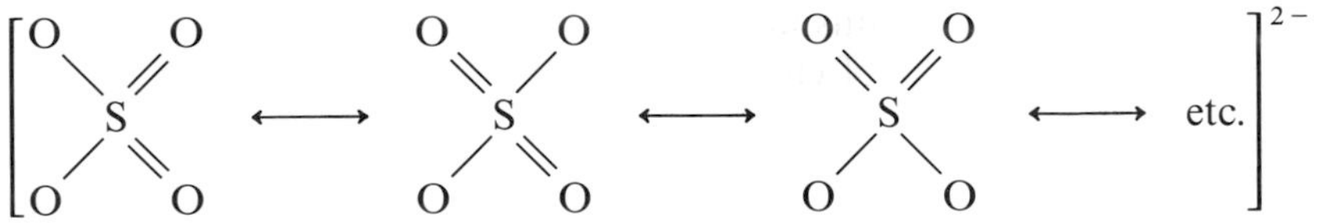

One particular type of resonance requires special mention, namely, *ionic-covalent resonance.* We pointed out in Section 2-9 that a bond between unlike atoms (A—B) is always more or less stronger than the average of the A—A and B—B bond strengths. This was used for calculating electronegativity differences, on the basis that an ionic or polar contribution to the bond made it stronger than the purely covalent bond alone. Actually, the situation is a little different, because it is resonance rather than simple additivity that Pauling proposed to account for the extra bond energy.

If A is more electronegative than B, the A—B bond can be represented by a resonance hybrid of 3-IVa and 3-IVb.

$$\underset{\text{3-IVa}}{\text{A—B}} \longleftrightarrow \underset{\text{3-IVb}}{\overset{-}{\text{A}}\overset{+}{\text{B}}}$$

As we explain above, the actual A—B bond will then (1) combine the properties of both contributing structures, and (2) be more stable than either one alone. Thus, the actual A—B bond will be polar to an extent depending on how much 3-IVb contributes to the average structure, and the increased strength of the bond, when it is compared with the strength expected for a purely covalent bond, will be proportional to the square of the electronegativity difference, since it is that difference which determines the importance of 3-IVb compared with 3-IVa.

3-9 Multicenter Bonding in Electron Deficient Molecules

We have thus far considered only molecules in which there are enough electrons to allow the use of, at least, one electron pair between each two adjacent atoms that are bonded to each other. In BeH_2, for example, the formation of two localized bonds, each formed by overlap of an *sp* hybrid orbital of Be with a 1*s* orbital of an H atom was described. There was one electron pair for each B—H bond. The same molecule was also discussed by using an MO treatment in which localization of the electron pairs is not required. However, here too, there was one bonding electron pair available per pair of adjacent atoms. In other cases, such as the diatomic molecules O_2 and N_2 we found two or three electron pairs between a given pair of atoms.

In some molecules, however, there are not enough electrons present to allow one (or more) electron pair between each adjacent pair of atoms. Such molecules are called *electron deficient*, and two examples are shown in 3-V and 3-VI. In 3-V and in the Al_2C_6 skeleton of 3-VI there are

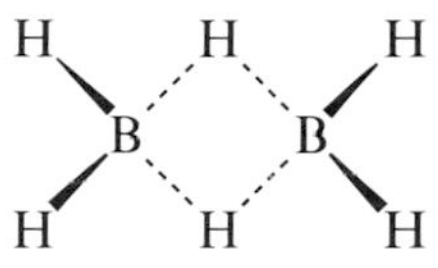

3-V 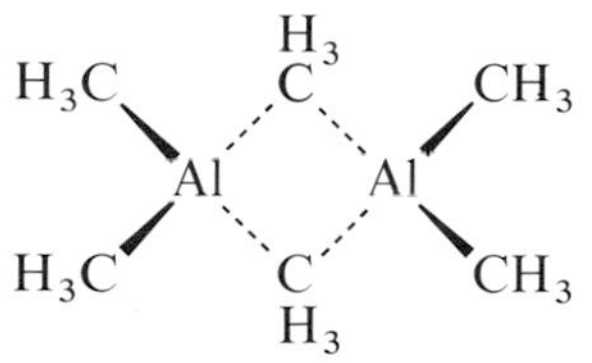 3-VI

eight adjacent pairs of atoms, but there are only six pairs of electrons available. How can there be eight bonds? We note, before proceeding, that structures 3-V and 3-VI are the actual ones for molecules with empirical formulas BH_3 and $Al(CH_3)_3$.

Both 3-V and 3-VI present the same basic problem. We concentrate first on 3-V, since it is less cumbersome. We could try to account for structure 3-V by invoking a resonance description, that is, by mixing of 3-Va and 3-Vb:

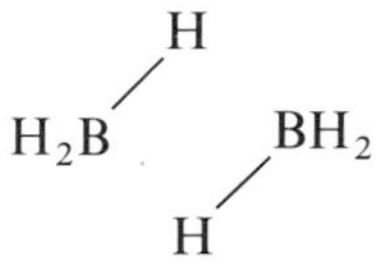

3-Va 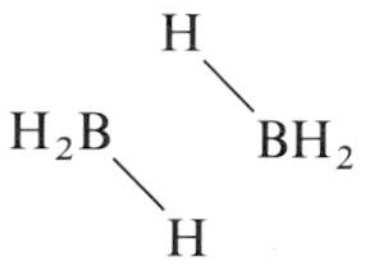 3-Vb

This would imply that in each B···H···B bridge one electron pair is shared or distributed over two B···H bonds, giving each one a bond order of $\frac{1}{2}$. An analogous description could be used for the central Al—C bonds in 3-VI.

The type of resonance shown in 3-Va and 3-Vb, called bond-no bond resonance, while formally acceptable, seems somewhat artificial. Another way to describe the bonding is to use an MO treatment that encompasses only the bridging system. The remaining B—H bonds are adequately described as ordinary electron-pair bonds.

In the case of 3-V, we begin with two BH_2 units, in which there are ordinary B:H bonds formed by sp^3 hybrid orbitals on the B atom. If these two BH_2 units are brought together as shown in *Fig. 3-24a*, so as to make the $H_2B\cdots BH_2$ set of atoms coplanar, the remaining two sp^3 hybrid orbitals on each B atom point toward each other. If now the remaining two H atoms are placed in their proper positions, then, as is shown in *Fig. 3-24b*, each of the 1*s* orbitals of these H atoms overlaps with two of the sp^3 orbitals from the B atoms. In this way an orbital extending over each B—H—B set is formed. This orbital has no nodes and is

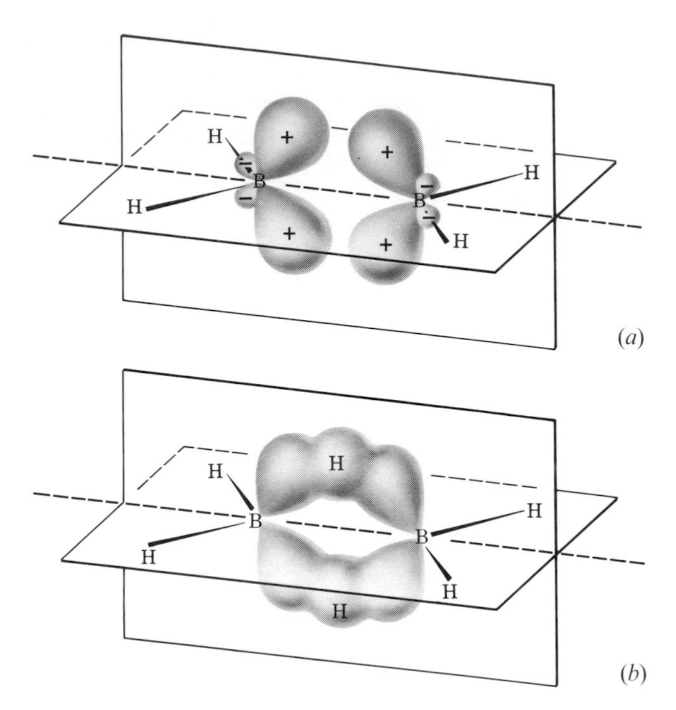

Figure 3-24 *The formation of 3c–2e bonds in* B_2H_6. *The orientation of two coplanar* H_2B *groups, with* sp^3 *hybrids on the* B *atoms is shown in* (*a*). *When the bridging* H *atoms are put in their places,* (*b*), *continuous overlap of the orbitals leading to the* 3*c*–2*e* *bonding results.*

therefore capable of bonding all three atoms together. Since each boron atom and each bridging H atom supplies one electron, there are a total of four electrons, or two for each B—H—B set. Thus, one electron pair can be used for each. In this way, we establish a type of bond called a 3-center, 2-electron bond, abbreviated 3*c*–2*e* bond. Since one electron pair is shared between three atoms instead of two, the bonds have only about one half the strength of a normal 2-center, 2-electron (2*c*–2*e*) bond. This is equivalent to the bond orders of $\frac{1}{2}$ obtained in the resonance treatment.

To appreciate and utilize more fully the concept of 3*c*–2*e* bonding, it is necessary to examine it in more detail. Suppose we consider only the sp^3 hybrid orbital on each B atom and the 1*s* orbital of the bridging H atom. These three atomic orbitals can be combined into three molecular orbitals as shown in *Fig. 3-25*. One of these, Ψ_b, is a bonding orbital; it is the same one already discussed. There is also an antibonding orbital, Ψ_a, which has a node between each adjacent pair of atoms. The third orbital, Ψ_n, has the signs of the two sp^3 orbitals out of phase and cannot have any net overlap with the H1*s* orbital. It is a *nonbonding* orbital.

We can now draw an energy level diagram that expresses these results, as in *Fig. 3-26*. By placing an electron pair in Ψ_b, the bonding MO, we have a complete picture of the 3*c*–2*e* bonding situation.

In the case of $Al_2(CH_3)_6$, 3-VI, the 3*c*–2*e* bridge bonding can be described in a very similar fashion. Each Al atom provides sp^3 hybrid orbitals, as do the boron

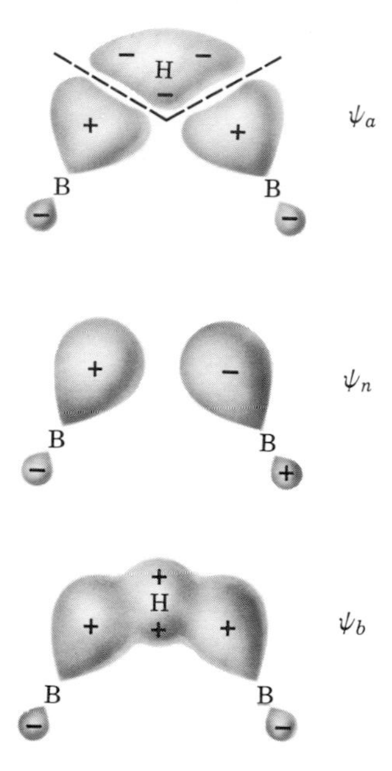

Figure 3-25 *The formation of three three-center* MO's *in a* B—H—B *bridge system.*

atoms in B_2H_6. Instead of the 1*s* orbital of the H atom, we now have the large positive lobe of a carbon sp^3 orbital at the center.

The energy-level diagram in *Fig. 3-26* can be applied in other cases as well. It can be extended to cover some cases of 3*c*–4*e* bonding. In the FHF^- ion, which is symmetrical (although most hydrogen bonds are weaker and unsymmetrical; see page 211), each F atom supplies a σ orbital and an electron pair. Thus a set of orbitals essentially similar to that in the BHB system is used, and an energy level diagram essentially like that in *Fig. 3-26* is applicable. However, there are now two electron pairs. One pair occupies Ψ_b and the other Ψ_n. The pair in Ψ_n has no significant effect on the bonding because Ψ_n is a nonbonding orbital. The net result is that here, too, the bond orders are $\frac{1}{2}$.

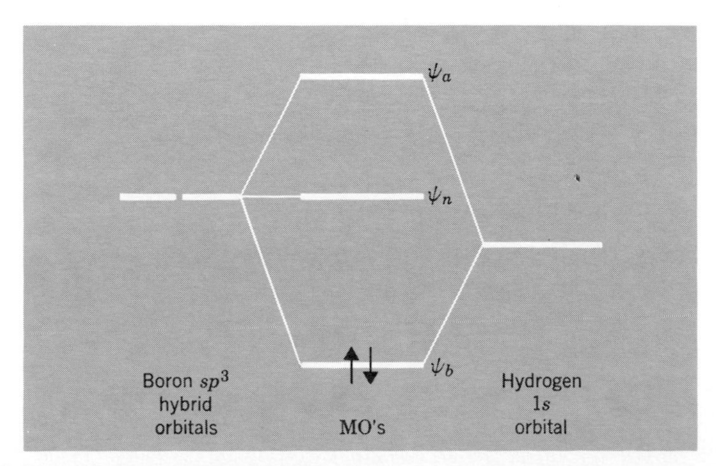

Figure 3-26 *An energy level diagram for the* 3*c*–2*e* B—H—B *bridge bonding in* B_2H_6.

One other type of $3c$–$4e$ bonding must also be discussed since it is essential to the discussion of molecular shapes in the next section. Suppose we have a set of three atoms, B—A—B, most probably linear but possibly bent to some extent, such that the central atom uses a *p* orbital rather than an *s* orbital. The situation is shown in *Fig. 3-27a*. Again, it is possible to form three multicenter orbitals as shown in *Fig. 3-27b*. The result turns out to be very similar to that already seen where the central atom uses an *s* orbital, in that bonding, Ψ_b, nonbonding, Ψ_n, and antibonding, Ψ_a, orbitals are formed and the energy level diagram is analogous, as is shown in *Fig. 3-27c*.

The interesting result, in either *Fig. 3-26* or *Fig. 3-27*, is that even if two electron pairs are available, the A—B bonds will have orders of only $\frac{1}{2}$, because one electron pair occupies the nonbonding orbital, Ψ_n. Here we are dealing with an orbitally deficient system rather than an electron deficient one. If the central atom in either case had an additional σ-type atomic orbital the system would be equivalent to that in BeH_2 and two bonds, each of order 1, could be formed.

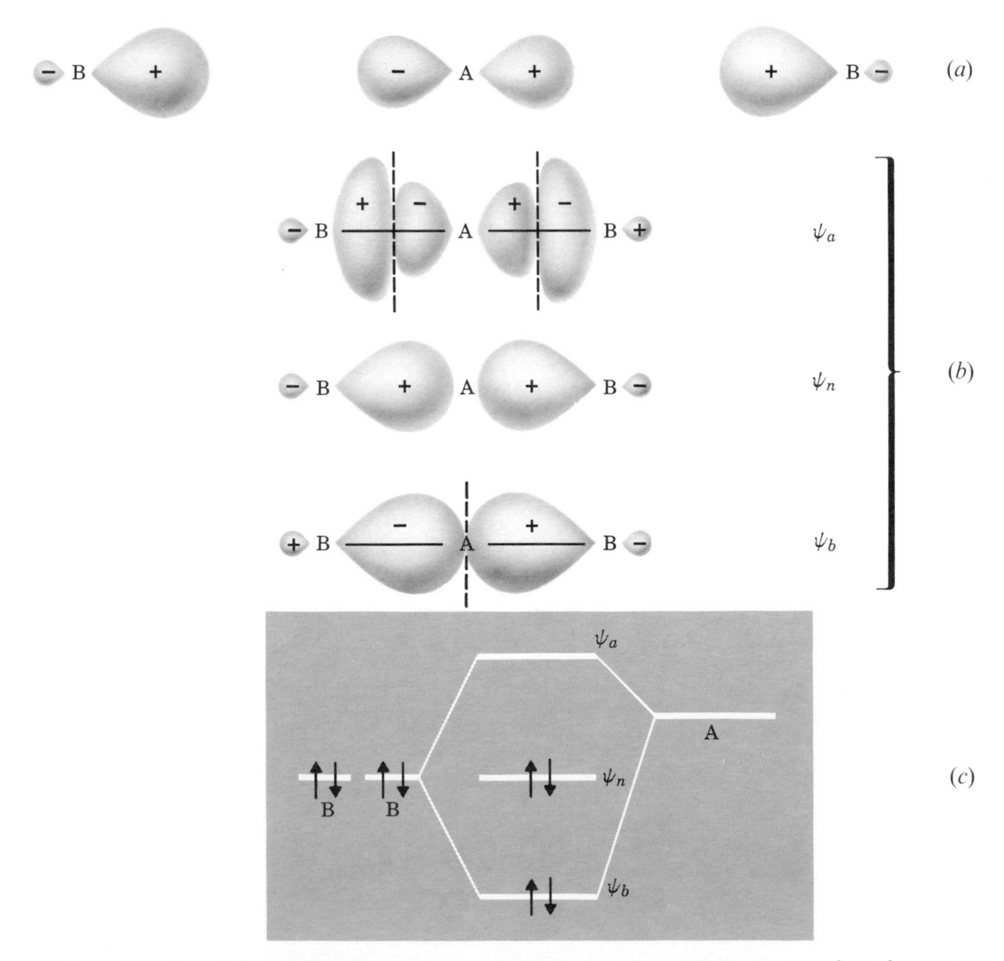

Figure 3-27 *The formation of three-center orbitals in a* B—A—B *system when the center atom uses a p orbital.* (*a*) *The atomic orbitals used.* (*b*) *The shapes of the* MO's *formed.* (*c*) *An energy-level diagram showing the occupation of the orbitals for a* $3c$–$4e$ *bond.*

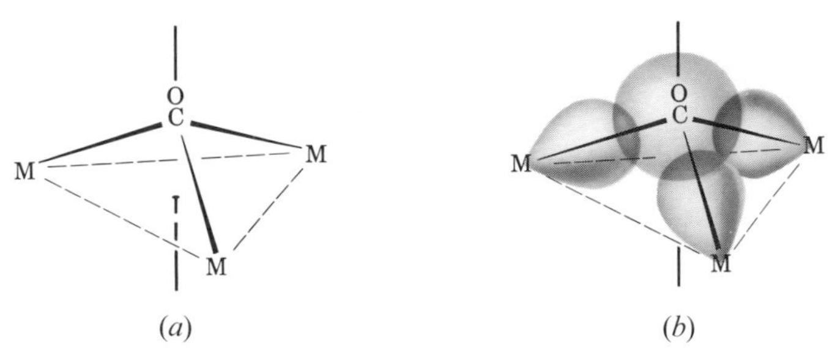

Figure 3-28 (*a*) *The spatial relationship of a* CO *to three metal atoms which is found in some metal carbonyl compounds.* (*b*) *The overlap of three metal atom orbitals with the* σ_3 *orbital of* CO (*and, to some extent with each other as well*) *to form a four-center orbital.*

Multicenter bonding can occur in larger groups of atoms. There are, for example, compounds in which a single carbon monoxide ligand lies perpendicularly over the center of a triangular set of metal atoms, as in *Fig. 3-28a*. In such cases, the best, simple way to describe the bonding is in terms of a $4c$–$2e$ bond. If one orbital on each metal atom is directed toward the carbon lone pair orbital of CO (σ_3 in *Fig. 3-13b*), there will be mutual overlap of all four orbitals, as is shown in *Fig. 3-28b*, and the resulting 4-center orbital will be occupied by the electron pair initially in σ_3 of the CO molecule.

3-10
Molecular Shapes

Although this subject includes molecules of all kinds, we restrict our discussion here to the specially important case of molecules in which a central atom A has various other atoms attached to it, but not to one another. Many common and important molecules as well as complex ions are of this type, and it is therefore useful to have simple ways of predicting and correlating their structures. Several superficially diverse ways of handling this problem have been proposed. None is rigorous, but each one has its virtues and, in many instances, any one of them can be used to obtain the same answer.

The Valence Shell Electron Pair Repulsion (VSEPR) Model. This model is based on the simple idea that the electrons about central atom A will form pairs (with opposed spins) and that the pairs will tend to stay as far from one another as possible to minimize their mutual electrostatic repulsions. Some or all of these pairs may be involved in bonds to the atoms B, while others may be so-called unshared or lone pairs. In the first approximation all pairs are treated alike.

For molecules with two to four electron pairs, the results are shown in *Fig. 3-29*. Two pairs adopt a linear arrangement, three a triangular one, and four occupy the vertices of a tetahedron. In the formulas used, E represents a lone pair and A and B represent the atoms. It is seen that the bent shapes of H_2O and SCl_2 as well as the pyramidal shapes of NH_3, H_3O^+, PF_3, etc., are unambiguously predicted.

The model can be refined by making the physically reasonable postulate that unshared pairs take up more space than shared pairs. This would be true because

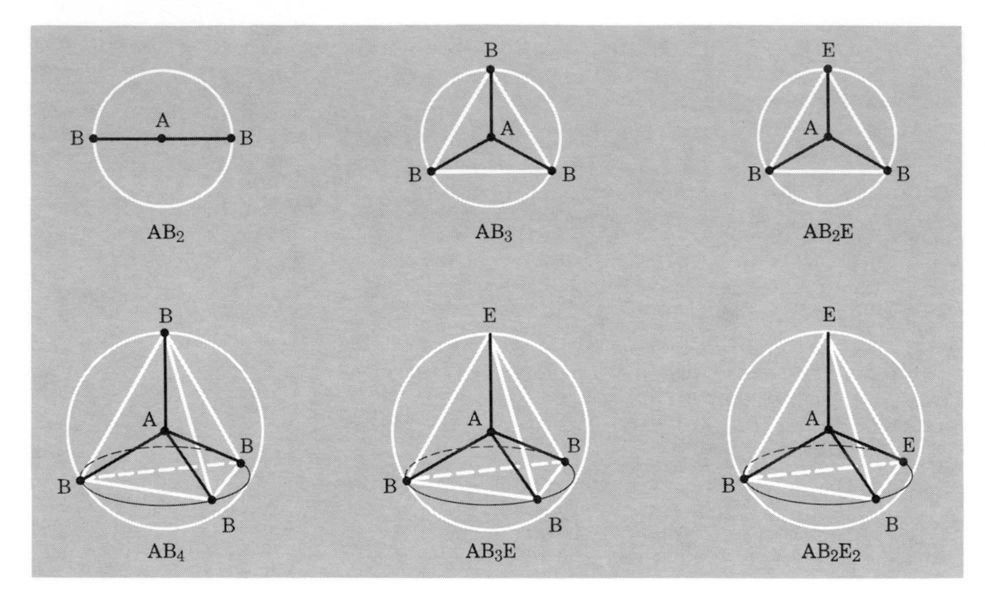

Figure 3-29 *Diagrams showing how the shapes of molecules are predicted when the valence shell of the central atom contains 2, 3, or 4 electron pairs.*

the shared pairs are attracted by two nuclei but the lone pairs are attracted by only one. This leads to the expectation that B—A—B angles will be smaller than the tetrahedral value because the E—A—E and E—A—B angles must be larger. This is, in fact, what is observed in almost every case, as the following interbond angles illustrate:

H_2O	105°	NH_3	107°
F_2O	102°	PF_3	98°
SCl_2	102°	AsI_3	100°

The model can readily be extended to cases where there are five, six, or more electron pairs surrounding atom A. The predicted stable geometries are listed in Table 3-1.

For five pairs the preferred arrangement is the trigonal bipyramid (**tbp**), although the square pyramid (**sp**) is only a little less stable (*Fig. 3-30*). Practically every AB_5 molecule does, in fact, have a **tbp** structure. For molecules of the types

Table 3-1 *Predicted Arrangements of Electron Pairs in One Valence Shell*

Number of Pairs	Polyhedron Defined
Two	Linear
Three	Equilateral triangle
Four	Tetrahedron
Five	Trigonal bipyramid, **tbp**
Six	Octahedron
Seven	Monocapped octahedron
Eight	Square antiprism
Nine	Tricapped trigonal prism

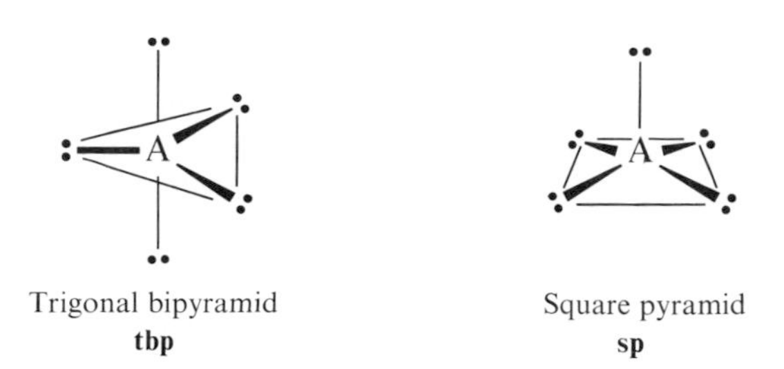

Figure 3-30 *The two most symmetrical arrangements for five electron pairs around a central atom. The* **tbp** *is somewhat the more stable.*

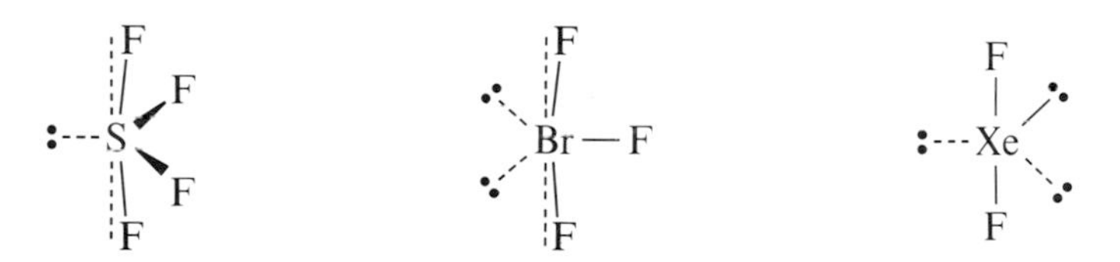

Figure 3-31 *The structures of some molecules of the* AB_4E, AB_3E_2 *and* AB_2E_3 *types, all based on a* **tbp** *with lone pairs in equatorial positions.*

AB_4E, AB_3E_2, and AB_2E_3, where some of the electron pairs are lone pairs, it is always found that the lone pairs are in equatorial positions. As examples we have SF_4, BrF_3, and XeF_2, *Fig. 3-31*.

Molecules with six electron pairs are mostly of the AB_6, AB_5E, and AB_4E_2 types. The first ones are, obviously, ordinary octahedral types. Those of the last two types have square pyramidal and square configurations, respectively, as are illustrated by BrF_5 and XeF_4, *Fig. 3-32*. In BrF_5 the F—Br—F angles are all less than 90° because, as noted above, the lone pair takes up more space than each of the shared pairs.

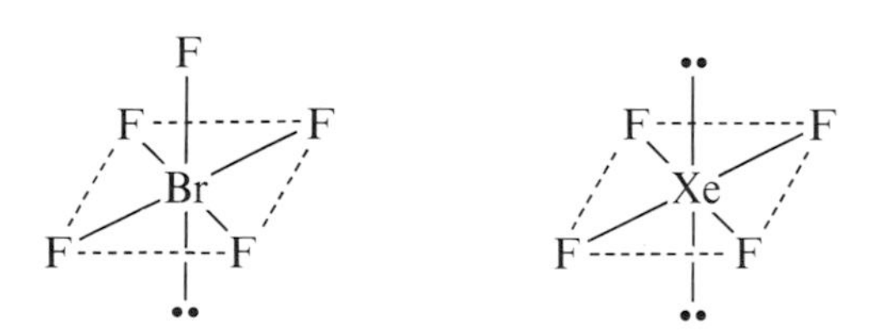

Figure 3-32 *The shapes of* AB_5E *and* AB_4E_2 *molecules according to the VSEPR model.*

The Hybridization or Directed Valence Theory. According to this treatment, bond directions are determined by a set of hybrid orbitals on the central atom which are used to form bonds to the ligand atoms and to hold unshared pairs. Thus AB_2 molecules are linear owing to the use of linear *sp* hybrid orbitals. AB_3 and AB_2E molecules should be equilateral triangular and angular, respectively, owing to use of trigonal sp^2 hybrids. AB_4, AB_3E, and AB_2E_2 molecules should be tetrahedral, pyramidal, and angular, respectively, because sp^3 hybrid orbitals are used. These cases are, of course, very familiar and involve no more than an octet of electrons.

For the AB_5, AB_4E, AB_3E_2, and AB_2E_3 molecules, the hybrids must now include *d* orbitals in their formation. The hybrid orbitals used must obviously be of the sp^3d type, but an ambiguity arises, since there are two such sets: $sp^3d_{z^2}$ leading to **tbp** geometry and $sp^3d_{x^2-y^2}$ leading to **sp** geometry. It is impossible to predict with certainty which set should give the more stable molecules and, hence, a decision must be made empirically. As we have already observed, all AB_5 molecules whose structures are known, with at most three exceptions, are **tbp**. It is therefore assumed that the $sp^3d_{z^2}$ hybrids and **tbp** geometry will generally be appropriate. Once this assumption is made, a consistent correlation of structures follows, exactly as in the VSEPR model where the basic **tbp** arrangement was adopted for a different reason. Again, however, the preferential allocation of lone pairs to equatorial orbitals is essentially arbitrary.

For AB_6 molecules octahedral sp^3d^2 hybrids are used. AB_5E molecules must, naturally, be **sp**. For AB_4E_2 molecules there is nothing in the directed valence theory itself to show whether the lone pairs should be *cis* or *trans*. The assumption that they must be *trans* leads to consistent results.

A Three-Center Bond Model. Interestingly, it is possible to use a model based on directed valence orbitals but that totally *omits* the *d* orbitals. For molecules involving only two, three or four electron pairs, *sp*, sp^2, and sp^3 hybrids are used just as previously, and the two models are identical. They diverge for all cases where the previous directed valence model employs one or more *d* orbitals. For an AB_4E, molecule such as SF_4, where five electron pairs are present around A, the reasoning is as follows: The S atom uses its p_x and p_y orbitals to form ordinary 2*c*–2*e* bonds to the two "equatorial" F atoms, but the "axial" F atoms are bonded only through the sulfur p_z orbital. The nearly linear F—S—F group then is described as a 3*c*–4*e* bond system (cf. Section 3-9). The sulfur 3*s* orbital is assumed to hold the remaining electron pair but to play no significant role in forming bonds. In an AB_5E type molecule (e.g., BrF_5) the axial A—B bond is supposed to be a 2*c*–2*e* bond that employs the p_z orbital of atom A. The remaining four A—B bonds are then regarded as two pairs of 3*c*–4*e* bonds, which are formed using the p_x and p_y orbitals on atom A. Again the unshared pair is placed in the *s* orbital of atom A.

This simple model is generally reliable in its predictions and is thus useful. It is probably an oversimplification to ignore *d* orbitals altogether, but so also is their full inclusion in a set of hybrids as in the previous model. The real situation is probably intermediate.

3-11
Bond Lengths and Covalent Radii

If we consider a single bond between like atoms, say Cl—Cl, we can define the single bond covalent radius of the atom as half of the bond length. Thus the Cl—Cl distance, 1.988 Å yields a covalent radius of 0.99 Å for the chlorine atom. In a similar way, radii for other atoms (e.g., 0.77 for carbon by taking $\frac{1}{2}$ the C—C bond length in diamond) are obtained. It is then gratifying to find that the lengths of heteronuclear bonds can often be predicted with useful accuracy. For example, from Table 3-2 we can predict the following bond lengths, which agree pretty well

Table 3-2 *Single Bond Covalent Radii, angstroms*

H	C	N	O	F
0.28	0.77	0.70	0.66	0.64
	Si	P	S	Cl
	1.17	1.10	1.04	0.99
	Ge	As	Se	Br
	1.22	1.21	1.17	1.14
	Sn	Sb	Te	I
	1.40	1.41	1.37	1.33

with the measured values given in parentheses:

C—Si	1.94 (1.87)	P—Cl	2.09 (2.04)
C—Cl	1.76 (1.77)	Cl—Br	2.13 (2.14)

The agreement is not perfect, and that cannot be expected, since bond properties (including length) vary somewhat with the environment.

Multiple bonds are always shorter than corresponding single bonds. This is illustrated by bonds between nitrogen atoms:

$$N{\equiv}N\ (1.10\ \text{Å}),\ N{=}N\ (1.25\ \text{Å}),\ N{-}N\ (1.45\ \text{Å})$$

Consequently, double- and triple-bond radii can also be defined. For the elements C, N, and O, which form most of the multiple bonds, the double- and triple-bond radii are approximately 0.87 and 0.78 times the single-bond radii, respectively.

The hybridization of an atom affects its covalent radius; since *s* orbitals are more contracted than *p* orbitals, the radius decreases with increasing *s* character. For carbon we have the following single bond radii: C(sp^3), 0.77; C(sp^2), 0.73; C(sp), 0.70.

When there is a great difference in the electronegativities (Section 2-9) of two atoms, the bond length is usually less than the sum of covalent radii, sometimes by a considerable amount. Thus, from Table 3-2, the C—F and Si—F distances are calculated to be 1.44 and 1.81 Å, whereas the actual distances in CF_4 and SiF_4 are 1.32 and 1.54 Å. In the case of the C—F bond it is believed that the shortening can be attributed to ionic-covalent resonance, which strengthens and, hence, shortens (by 0.12 Å) the bond. For SiF_4 only part of the very pronounced shortening can be thus explained. Much of it is thought to be due to π bonding using filled fluorine $p\pi$ and empty silicon $d\pi$ orbitals.

3-12 Molecular Packing; van der Waals Radii

When molecules pack together in the liquid and solid states, their approach to one another is limited by short-range repulsive forces, which result from overlapping of the diffuse outer regions of the electron clouds around the atoms.

The actual distance apart at which any two molecules would come to rest is determined by the equalization of attractive and repulsive forces. There are also weak, short-range attractive forces between molecules which result from permanent dipoles, dipole-induced dipole, and so-called London forces. The latter arise from interaction between fluctuating dipoles whose time-average value in any one molecule is zero.

Collectively, all these attractive and repulsive forces that are neither ionic nor covalent are called *van der Waals forces.*

It develops that both the attractive and repulsive forces are of roughly constant magnitude over the vast majority of molecules and thus the distances between molecules in condensed phases do not vary a great deal. As a result it is possible to compile a list of van der Waals radii, which give the typical internuclear distances between nearest neighbor atoms in different molecules in condensed phases. Van der Waals radii for some common atoms are listed in Table 3-3.

Table 3-3 *van der Waals Radii of Non-metallic Atoms (in Å)*

H	1.1–1.3					He	1.40
N	1.5	O	1.40	F	1.35	Ne	1.54
P	1.9	S	1.85	Cl	1.80	Ar	1.92
As	2.0	Se	2.00	Br	1.95	Kr	1.98
Sb	2.2	Te	2.20	I	2.15	Xe	2.18

Radius of a methyl group, 2.0 Å
Half-thickness of an aromatic ring, 1.85 Å

Van der Waals radii are far greater than covalent radii and are roughly constant for isoelectronic species. Thus, in crystalline Br_2, the covalent radius of Br is 1.15 Å, whereas the van der Waals radius (half the shortest intermolecular Br ··· Br distance) is 1.95 Å. The latter differs little from the Kr ··· Kr packing distance of 1.98 Å in solid krypton, since Br when bonded to another atom is isoelectronic with the Kr atom.

Study Questions

A

1. Why are the sign and magnitude of overlap between orbitals on adjacent atoms good indications of whether and how strongly the atoms are bonded?
2. Show with drawings how an *s* orbital, each of the three *p* orbitals, and each of the five *d* orbitals on one atom would overlap with the *s* orbital, one of the *p* orbitals, and any two of the *d* orbitals on another atom close to it. Characterize each overlap as positive, negative, or exactly zero.
3. Draw an energy level diagram for the interaction of two atoms each with an *s* orbital. Show how the MO's would be occupied if the two atoms in question were H atoms and if they were He atoms. What conclusions are to be drawn about the formation of a bond in each case?
4. When a bond is formed between two atoms, they are drawn together. What limits their internuclear distance so that they do not coalesce?
5. State the defining characteristics of σ, π, and δ MO's.

6. What is meant by a node? A nodal plane?
7. How is bond order defined for a diatomic molecule in MO theory?
8. Show with an energy level diagram why the C_2 molecule has a bond order of 2 and no unpaired electrons, but has a low-lying excited state in which there are two unpaired electrons.
9. Show how the electronic structure of the NO molecule can be inferred from that of O_2. Explain why NO^+ has a stronger bond than NO itself.
10. True or False: The set of valence shell orbitals (2*s*, 2*p*) for N are of higher energy than those for C. Explain the reason for your answer.
11. Write the electron configurations for the ground states and the valence states of Be, B, C, and N atoms so that each one can form the maximum number of 2*c*–2*e* (ordinary electron-pair) bonds.
12. What are the three important types of hybrid orbitals that can be formed by an atom with only *s* and *p* orbitals in its valence shell? Describe the molecular geometry which each of these produces.
13. State the geometric arrangement of bonds produced by each of the following sets of hybrid orbitals: dsp^2, d^2sp^3, dsp^3. For each one state explicitly which *d* and *p* orbitals are required for each geometric arrangement.
14. Explain in detail, using both the MO approach and the resonance theory why the NO bonds in NO_3^- have a bond order of $1\frac{1}{3}$.
15. Why is the use of hybrid orbitals preferable to the use of single atomic orbitals in forming bonds? Illustrate.
16. What does the term "electron deficient molecule" mean?
17. Why does not B_2H_6 have the same kind of structure as C_2H_6? Draw the structure that B_2H_6 does have and describe the nature of the two sorts of BH bonds therein.
18. Using the VSEPR model, predict the structures of the following ions and molecules: BeF_2, CH_2, OF_2, PCl_4^+, SO_2, ClF_2^+, BrF_3, BrF_5, SbF_5, ICl_4.
19. For the last five molecules and ions in question 18, predict their structures using the three-center bond model.
20. How and why do the lengths of the C—C single bonds differ in $HCCCH_3$, H_2CCHCH_3, and C_2H_6?
21. Why are the Kr · · · Kr and intermolecular Br · · · Br distances in the solid forms of the two elements practically identical? How would you expect the Br · · · Br distances in solid CBr_4 to be related to the above distances?

B

1. For the series of diatomics O_2^+, O_2, O_2^-, O_2^{2-}, determine from an MO energy level diagram how the bond strengths will vary and how many unpaired electrons each one should have.
2. The ionization enthalpies of H and F are 1299 and 1679 kJ mol^{-1}, respectively. Draw a molecular orbital energy level diagram for the HF molecule. What does it indicate about the polarity of the molecule?
3. Draw a qualitatively correct energy level diagram for the CO_2 molecule. Show that it accounts correctly for the presence of double bonds.
4. The ozone molecule is bent and symmetrical. Work out a MO energy level diagram for it and predict the O—O bond orders.
5. Consider the O_3 molecule again in the localized bond approach. How does the description obtained compare with that from the MO treatment?
6. Draw the principal contributing (Kekule-like) structures for the naphthalene molecule. There are four nonequivalent CC bonds in this molecule. On the basis of the resonance picture, what do you predict the bond order of each one to be? Assume that all contributing structures have equal weights in forming the resonance hybrid.

Chapter 3
Study Guide

Scope and Purpose. Only the briefest introduction to the subject can be given in one chapter and, in a book of this nature, the discussion must be qualitative rather than mathematical. The material in this chapter should be viewed as merely a nucleus to be enlarged as much as possible. At the same time, bonding theory and molecular structure should not be taken as ends in themselves, but only as important tools in understanding the actual properties and reactions of chemical compounds.

Supplementary Reading

Ballhausen, C. J. and Gray, H. B., *Molecular Orbital Theory*, Benjamin, Menlo Park, Calif., 1964.

Cartmell, E. and Fowles, G. W. A., *Valency and Molecular Structure*, 3rd ed., Butterworths, London, 1966.

Companion, A., *Chemical Bonding*, McGraw-Hill, New York, 1964.

Coulson, C. A., *Valence*, 2nd ed., Oxford University Press, New York, 1961.

Ferguson, J. E., *Stereochemistry and Bonding in Inorganic Chemistry*, Prentice-Hall, 1974.

Gillespie, R. J., *Molecular Geometry*, Van Nostrand Reinhold, London, 1972.

Karplus, M. and Porter, R. N., *Atoms and Molecules: An Introduction for Students of Physical Chemistry*, Benjamin, Menlo Park, Calif., 1970.

Murrell, J. N., Kettle, S. F. A., and Tedder, J. M., *Valence Theory*, 2nd Ed., John Wiley and Sons, Ltd., London, 1970.

Pimentel, G. C. and Spratley, R. D., *Chemical Bonding Clarified Through Quantum Mechanics*, Holden-Day, San Francisco, 1969.

Wade, K., *Electron Deficient Compounds*, Thomas Nelson and Sons, Ltd., London, 1971.

Wahl, A. C., "*Electron Density Maps*," *Science*, *151*, 961 (1966).

4

ionic solids

4-1 Introduction

A great many inorganic solids, and even a few organic ones, can usefully be thought of as consisting of a three-dimensional array of ions. This ionic model can be developed in further detail in two main ways.

First, it is assumed that the energy of this array of ions can be treated as the sum of the following contributions:

1. Coulombic (electrostatic) attractive and repulsive energies

2. Additional repulsive energy which results from repulsion between the overlapping outer electron density of adjacent ions.

3. A variety of minor energy terms, mainly van der Waals energy, and zero point vibrational energy.

The important point to note here is that no explicit account is taken of covalent bonding. This is doubtless an oversimplification in *every* case, but evidently in many substances the pure ionic description leads to fairly accurate estimates of the enthalpies of formation of the compounds. There is probably a certain approximate compensation so that covalent bond energy, which may actually be present, arises at the expense of a nearly equal amount of coulomb energy. Thus, so long as the covalence is small, the error involved in assuming that one form of energy exactly offsets the other is an acceptable approximation.

Second, by treating a substance as an efficiently packed array of ions, we can understand the main features of the structures they adopt. Aside from the overriding qualitative idea that the packing must maximize the number of contacts between oppositely charged ions while simultaneously keeping those with the same sign as far apart as possible, many more detailed insights are possible. Thus, by first determining a set of radii for the different ions, a combination of geometrical and electrostatic analysis can enable us to understand why, for example, CsCl, NaCl, and CuCl all have different structures.

4-2
The Lattice Energy of Sodium Chloride

We begin by considering how to calculate the enthalpy of forming a solid ionic compound from a dilute gaseous collection of the constituent ions. For definiteness, we shall first consider a specific example, NaCl. X-ray study shows that the atoms are arranged as in *Fig. 4-1.* If we assume that the atoms are in fact ions, Na^+ and Cl^-, the energy of the array can be calculated in the following way. The shortest Na^+—Cl^- distance is called r_0. The electrostatic energy between two neighboring

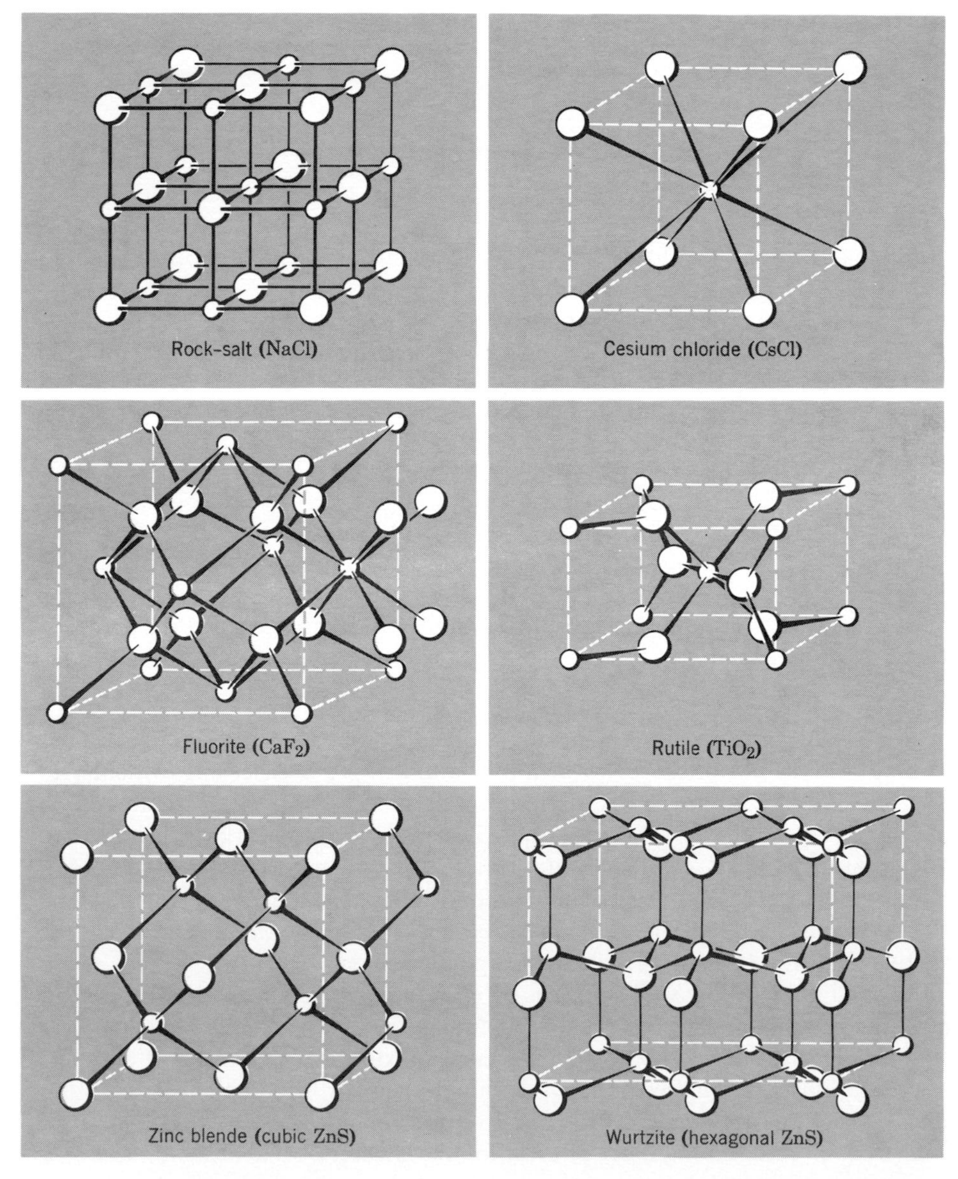

Figure 4-1 *Six important ionic structures. Small circles denote metal cations, large circles denote anions.*

ions is given (see Section 1-1e) by Eq. 4-1, where e is the electronic charge in coulombs.

$$E(\text{joules}) = \frac{e^2}{4\pi\varepsilon_0 r_0} \qquad (\varepsilon_0 = 8.854 \times 10^{-2}\ \text{C}^2\ \text{m}^{-1}\ \text{J}^{-1}) \tag{4-1}$$

Each Na^+ ion is surrounded by six Cl^- ions at the distance r_0 (in meters) giving an energy term $6e^2/4\pi\varepsilon_0 r_0$. The next closest neighbors to a given Na^+ ion are 12 Na^+ ions which, by simple trigonometry, lie $\sqrt{2}r_0$ away. Thus, another energy term, with a minus sign because it is repulsive, is $-12e^2/\sqrt{2}r_0 4\pi\varepsilon_0$. By repeating this sort of procedure, successive terms are found, leading to the expression:

$$\begin{aligned} E &= \frac{1}{4\pi\varepsilon_0}\left(\frac{6e^2}{r_0} - \frac{12e^2}{\sqrt{2}r_0} + \frac{8e^2}{\sqrt{3}r_0} - \frac{6e^2}{2r_0} + \cdots\right) \\ &= \frac{e^2}{4\pi\varepsilon_0 r_0}\left(6 - \frac{12}{\sqrt{2}} + \frac{8}{\sqrt{3}} - \frac{6}{2} + \frac{24}{\sqrt{5}} - \cdots\right) \end{aligned} \tag{4-2}$$

It is possible to derive a general formula for the infinite series and to find the numerical value to which it converges. That value is characteristic of the structure and independent of what particular ions are present. It is called the *Madelung constant*, M_{NaCl}, for the NaCl structure. It is actually an irrational number, whose value can be given to as high a degree of accuracy as needed, for example, 1.747 . . . , or 1.747558 . . . , or better. Madelung constants for many common ionic structures have been evaluated, and a few are given in Table 4-1 for illustrative purposes. The structures themselves, *Fig. 4-1*, will be discussed presently.

Table 4-1 *Madelung Constants for Several Structures*

Structure Type	M
NaCl	1.74756
CsCl	1.76267
CaF_2	5.03878
Zinc blende	1.63805
Wurtzite	1.64132

A unique Madelung constant is defined only for those structures in which all ratios of interatomic vectors are fixed by symmetry. In the case of the rutile structure there are two crystal dimensions that can vary independently. There is a different Madelung constant for each ratio of the two independent dimensions.

When a mole (N ions of each kind, where N is Avogadro's number) of sodium chloride is formed from the gaseous ions, the total electrostatic energy released is given by

$$E_e = NM_{NaCl}\left(\frac{e^2}{4\pi\varepsilon_0 r_0}\right) \tag{4-3}$$

This is true because the expression for the electrostatic energy of one Cl^- ion would be the same as that for an Na^+ ion. If we were to add the electrostatic energies for the two kinds of ions, the result would be twice the true electrostatic energy because each pairwise interaction would have been counted twice.

The electrostatic energy given by Eq. 4-3 is not the actual energy released in the process

$$Na^+(g) + Cl^-(g) = NaCl(s) \tag{4-4}$$

Real ions are not rigid spheres. The equilibrium separation of Na^+ and Cl^- in NaCl is fixed when the attractive forces are exactly balanced by repulsive forces. The attractive forces are coulombic and follow strictly a $1/r^2$ law. The repulsive forces are more subtle and follow an inverse r^n law where n is > 2 and varies with the nature of the particular ions. We can write, in a general way, that the total repulsive energy per mole at any value of r is

$$E_{\text{rep}} = \frac{NB}{r^n}$$

where B is a constant.

At the equilibrium distance, the net energy, U, for process (4-4) (where we now use algebraic signs in accord with convention; see Section 1-2c) is

$$U = -NM_{\text{NaCl}}\left(\frac{e^2}{4\pi\varepsilon_0 r_0}\right) + \frac{NB}{r_0{}^n} \tag{4-5}$$

Observe that the attractive forces produce an exothermic contribution and the repulsive ones an endothermic term.

The constant B can now be eliminated if we recognize that at equilibrium, when $r = r_0$ the energy, U, is a minimum by definition. The derivative of U with respect to r, evaluated at $r = r_0$ must equal zero. Differentiating Eq. 4-5 we get

$$\left(\frac{dU}{dr}\right)_{r=r_0} = \frac{NM_{\text{NaCl}}e^2}{4\pi\varepsilon_0 r_0{}^2} - \frac{nNB}{r_0{}^{n+1}} = 0$$

which can be rearranged and solved for B:

$$B = \frac{e^2 M_{\text{NaCl}}}{4\pi\varepsilon_0 n} r_0{}^{n-1}$$

When this expression for B is substituted into Eq. 4-5, we obtain

$$U = -\frac{NM_{\text{NaCl}}e^2}{4\pi\varepsilon_0 r_0}\left(1 - \frac{1}{n}\right) \tag{4-6}$$

The value of n can be estimated as 9.1 from the measured compressibility* of NaCl.

* Fractional change in volume per unit change in pressure, that is, $(\Delta V/V)/P$.

In a form suitable for calculating numerical results, in kJ mol^{-1}, using r_0 in angstroms Eq. 4-6 becomes

$$U = -1389 \frac{M_{\text{NaCl}}}{r_0}\left(1 - \frac{1}{n}\right)$$

and inserting appropriate values of the parameters we obtain

$$U = -1389 \frac{1.747}{2.82}\left(1 - \frac{1}{9.1}\right)$$

$$U = -860 + 95 = -765 \text{ kJ mol}^{-1}$$

Notice that the repulsive energy equals about 11% of the Coulombic energy. The net result is not very sensitive to the exact value of n. If a value of $n = 10$ had been used, an error of only 9 kJ mol^{-1} or 1.2% would have been made.

4-3 Generalization and Refinement of Lattice Energy Calculations

As noted, the Madelung constant is determined by the geometry of the structure only. Thus, for a case, such as MgO, where the structure is the same but each ion has a charge of ± 2, the only modification required is to replace $-e^2$ by $(2e)(-2e) = -4e^2$ in the Coulombic energy term. In general, Eq. 4-6 becomes

$$U = -\frac{NM_{\text{NaCl}}Z^2e^2}{4\pi\varepsilon_0 r_0}\left(1 - \frac{1}{n}\right)$$

for any structure whose Madelung constant is M with ions of charges Z^+ and Z^-.

The value of n can be estimated for alkali halides by using the average of the following numbers:

He	5	Kr	10
Ne	7	Xe	12
Ar	9		

where the noble gas symbol denotes the noble-gas-like electron configuration of the ion. Thus for LiF an average of the He and Ne values $(5 + 7)/2 = 6$ would be used.

To make very accurate calculations of crystal energies, sometimes called lattice energies, certain refinements must be introduced. The main ones are as follows.

1. A more accurate, quantum expression for the repulsion energy
2. A correction for van der Waals energy
3. A correction for the "zero-point energy," the vibrational energy present even at 0K.

The last two are opposite in sign and often of similar magnitude. Thus for NaCl a refined calculation gives:

Coulomb energy	−860
Repulsion energy	+99
van der Waals energy	−13
Zero-point energy	+8
	−766 kJ mol^{-1}

4-4 The Born–Haber Cycle

One test of whether an ionic model is a useful description of a substance such as sodium chloride is its ability to produce an accurate value of the enthalpy of formation. Note that we cannot make this test simply by measuring the enthalpy of reaction 4-4, or its reverse. The former is possible in principle but not experimentally feasible. The latter is not possible because sodium chloride does not vaporize cleanly to Na^+ and Cl^- but to NaCl, which then dissociates into atoms.

To handle the energy problem, a *thermodynamic cycle*, called the Born–Haber cycle is used. This is illustrated in *Fig. 4-2*. The basic idea is that the formation of NaCl(s) from the elements, Na(s) + $\frac{1}{2}Cl_2$(g), whose enthalpy is by definition the

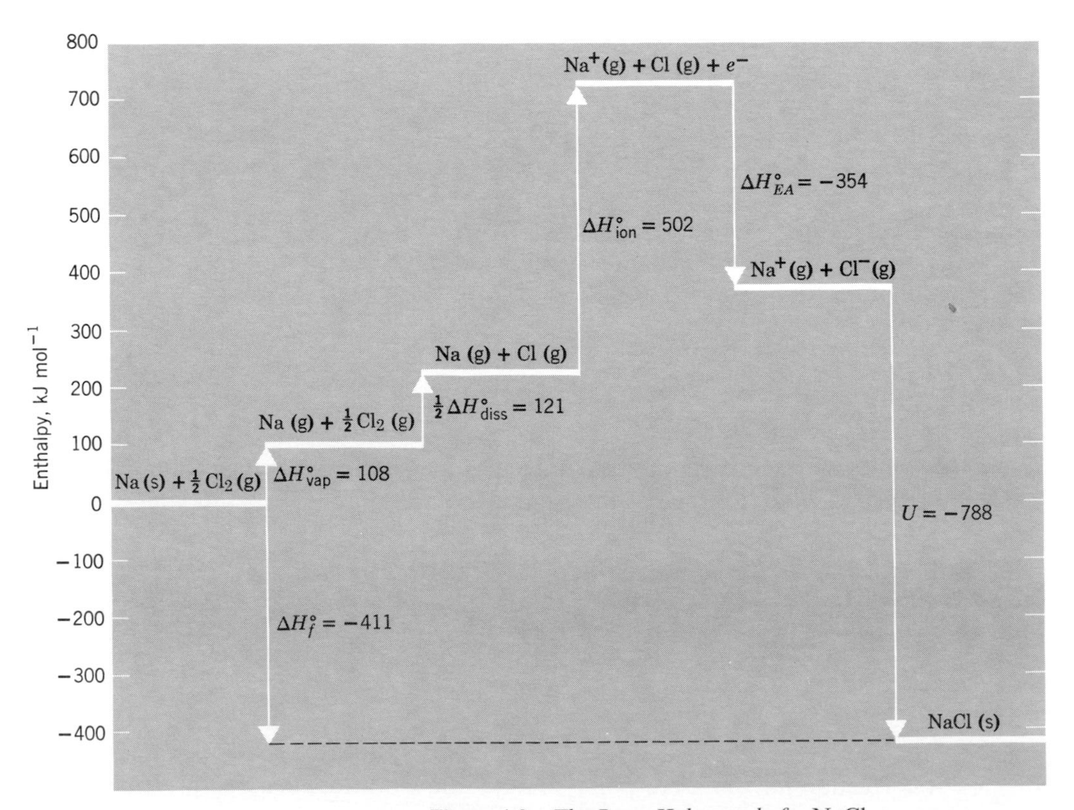

Figure 4-2 *The Born–Haber cycle for* NaCl.

enthalpy of formation of NaCl(s), can be broken down into a series of steps. If the enthalpies of these steps are added, algebraically, the result must be equal to ΔH_f° according to the Law of Conservation of Energy, the First Law of Thermodynamics. We thus have the equation

$$\Delta H_f^\circ = \Delta H_{\text{vap}}^\circ + \tfrac{1}{2}\,\Delta H_{\text{diss}}^\circ + \Delta H_{EA}^\circ + \Delta H_{\text{ion}}^\circ + U \qquad (4\text{-}8)$$

where the enthalpy terms are for vaporization of sodium ($\Delta H_{\text{vap}}^\circ$), dissociation of Cl_2(g) into gaseous atoms ($\Delta H_{\text{diss}}^\circ$), electron attachment to Cl(g) to give Cl^-(g)(ΔH_{EA}°), ionization of Na(g) to Na^+(g) + e ($\Delta H_{\text{ion}}^\circ$) and the formation of NaCl(s) from gaseous ions (U).

More generally, any one of these energies can be calculated if all the others are known. For NaCl all of the enthalpies except U in Eq. 4-7 have been measured independently. The following summation can thus be made:

$$\begin{array}{rcr} \Delta H_f^\circ & & -411 \\ -\Delta H_{\text{vap}}^\circ & & -108 \\ -\tfrac{1}{2}\,\Delta H_{\text{diss}}^\circ & = & -121 \\ -\Delta H_{\text{EA}}^\circ & & 354 \\ -\Delta H_{\text{ion}}^\circ & & -502 \\ \hline U & = & -788 \text{ kJ mol}^{-1} \end{array}$$

This result is very close (within 2.7%) of the value of U calculated by using the refined ionic model (-766 kJ mol^{-1}). Actually, even more precise calculations give a value that agrees to within less than 1%. This good agreement supports (but does not prove) the idea that the ionic model for NaCl is a useful one.

Often, the Born–Haber cycle, or a similar one, is used differently. If it is assumed that U calculated on the ionic model is correct, the cycle can be used to estimate some other energy term. For example, there is no convenient direct way to measure the enthalpy of formation of the gaseous CN^- ion. From a Born–Haber cycle for NaCN, where values of all the other enthalpies are available and U is calculated, it is found that ΔH_f for CN^-(g) is 29 kJ mol^{-1}.

4-5
Ionic Radii

In a manner similar in principle to that in which covalent radii were estimated, it is possible to assign radii to ions. The internuclear distance, d, between two ions in an ionic structure is assumed to be equal to the sum of the radii of the ions:

$$d = r^+ + r^-$$

By comparing distances in different compounds with an ion in common, it can be shown first that the radii of ions are substantially constant. For example, the difference in the radii of K^+ and Na^+ can be evaluated in four different halides:

$$\begin{aligned} r_{K^+} - r_{Na^+} &= d_{KF} - d_{NaF} = 0.35\text{ Å} \\ &= d_{KCl} - d_{NaCl} = 0.33\text{ Å} \\ &= d_{KBr} - d_{NaBr} = 0.32\text{ Å} \\ &= d_{KI} - d_{NaI} = 0.30\text{ Å} \end{aligned}$$

Table 4-2 *Some Ionic Radii, in Angstroms*[a]

Main group elements:													
												H^-	2.08
Li^+	0.60	Be^{2+}	0.31							O^{2-}	1.40	F^-	1.35
Na^+	0.96	Mg^{2+}	0.65	Al^{3+}	0.50					S^{2-}	1.84	Cl^-	1.81
K^+	1.33	Ca^{2+}	0.99	Ga^{3+}	0.62					Se^{2-}	1.98	Br^-	1.95
Rb^+	1.48	Sr^{2+}	1.13	In^{3+}	0.81	Sn^{4+}	0.71			Te^{2-}	2.21	I^-	2.16
Cs^+	1.69	Ba^{2+}	1.35	Tl^{3+}	0.95	Pb^{4+}	0.84	Pb^{2+}	1.21				
Transition metal ions:[b]										*Others:*			
Ti^{4+}	0.68	Fe^{3+}	0.53	Mn^{2+}	0.80					Zn^{2+}	0.74		
Zr^{4+}	0.80	Cr^{3+}	0.55	Fe^{2+}	0.75					Cd^{2+}	0.97		
Ce^{4+}	1.01			Co^{2+}	0.72					Hg^{2+}	1.10		
				Ni^{2+}	0.69								

[a] These are so-called Pauling radii, estimated using the essential assumption explained in the text but incorporating a number of additional small corrections.

[b] Lanthanide and actinide ion radii are discussed in Chapters 26 and 27.

Actually, the apparent trend as the halide ion size increases is a real effect that can be understood in terms of packing considerations, but we shall not discuss that topic further. Suffice it to say that if $r_{K^+} - r_{Na^+}$ is substantially constant, it is reasonable to assume that r_{K^+} and r_{Na^+} are themselves substantially constant.

It is easy to work out extensive sets of sums and differences of ionic radii. Then, provided that the actual radius of any one ion can be evaluated, the radii of all of the ions will be determined. This problem has no rigorous solution, but Pauling proposed a practical one, namely, that for two ions with the same noble gas configuration, say Na^+ and F^-, the ratio of the radii should be inversely proportional to the ratio of the nuclear charges felt by the outer electrons. These "effective" charges can be calculated by using empirical shielding constants which were developed by Slater. According to Slater's rules, an electron in the filled second principal shell is shielded by all the other electrons to the extent that it experiences a nuclear charge 4.15 units less than the actual one.

For Na^+, with an actual nuclear charge of 11, the effective charge is $11.0 - 4.15 = 6.85$. For F^- we have $9.00 - 4.15 = 4.85$. Hence, according to Pauling's proposal:

$$\frac{r_{Na^+}}{r_{F^-}} = \frac{4.85}{6.85} = 0.71$$

Since the interionic distance in NaF is 2.31 Å we have

$$r_{Na^+} + r_{F^-} = 2.31$$

which, together with the equation for the ratio, yields

$$r_{F^-} = 1.35$$
$$r_{Na^+} = 0.96$$

In this basic way, the ionic radii in Table 4-2 were derived.

4-6 Ionic Crystal Structures

Figure 4-1 shows six of the most important structures formed by essentially ionic substances. The common qualitative feature of all these structures is that the ions are packed to maximize the contacts between those of opposite charge and to minimize repulsions between those of the same charge. In a three-dimensional sense, ions of opposite charge alternate. The nearest neighbors of one ion are ions of opposite charge. However, this qualitative idea alone does not account for all of the features that can be seen in *Fig. 4-1*. For AB-type compounds we see four structure types. Consider first those of NaCl and CsCl. The difference is that in the NaCl structure each cation has six nearest neighbor anions, whereas in the CsCl structure each cation has eight such neighbors. We say that the *coordination numbers* of the cations are 6 and 8, respectively. In both the zinc blende and wurzite structures the cation has a coordination number of only 4. Again, for AB_2-type compounds there is a fluorite structure where the cation coordination number is 4 and a rutile structure where it is 6. Why does a particular AB or AB_2 compound adopt one and not another of these structures?

The answer lies partly in a consideration of the relative sizes of the ions. Anions are almost always larger than cations, since the net excess of nuclear charge on cations draws their electron clouds in, while the excess of negative charge on anions causes the electron clouds to expand. The optimum arrangement should allow the maximum number of oppositely charged ions to be neighbors without unduly squeezing together ions of the same charge. Thus the greater the ratio of cation to anion size, the higher the coordination number of the cation can—and should—be. That is why the relatively large Cs^+ surrounds itself with eight Cl^- ions, but for the smaller Na^+ there are only six.

It is possible to treat this idea in a semiquantitative way, by finding, for each structure, that ratio, r^-/r^+, for which the anions just touch one another while making contact with the cation, a situation we shall call perfect packing. For the CsCl structure, the relevant geometric relations (*Fig. 4-3*) are as follows.

1. The anions define a cube of edge length a.

2. The cation to anion distance equals half the length ($a\sqrt{3}$) of the body diagonal of the cube.

Therefore, for perfect packing

$$2r^- = a$$

$$r^+ + r^- = \frac{\sqrt{3}}{2}a$$

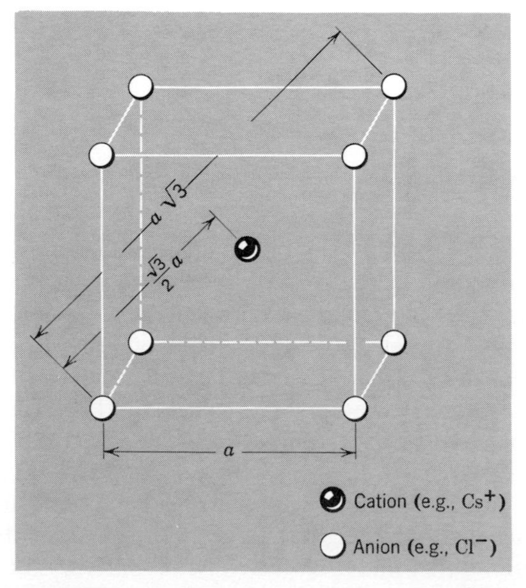

Figure 4-3 *Diagram showing the geometric relations in the* CsCl*-type structure.*

These equations are satisfied by any values of r^+ and r^- such that

$$r^-/r^+ = 1.37$$

Similarly, it is found that perfect packing requires $r^-/r^+ = 2.44$ for octahedral coordination number 6 and 4.44 for tetrahedral coordination number 4.

It should be stressed, however, that the foregoing analysis based on ion sizes is only a part of the picture. It works best for compounds that are most truly ionic, namely, alkali and alkaline earth halides, oxides, and sulfides; but even some of these do not obey predictions based solely on the radius ratio. With compounds in which the ions are highly polarizable [e.g., cuprous and zinc compounds] coordination numbers often are lower than expected.

In a case where the cation is very small relative to the anion ($r^-/r^+ \gg 4.44$), it will be impossible to achieve good cation-anion contact, even when anion-anion contacts are very close. Thus, ionic salts of this type are relatively unstable. Salts of the small cations Li^+, Be^{2+}, Al^{3+}, and Mg^{2+} with large polyatomic anions (e.g., ClO_4^-, CO_3^{2-}, NO_3^-, O_2^-) or even monatomic anions such as Cl^-, Br^-, and I^- are cases in point. The consequences of this are threefold.

1. In some cases, the anhydrous compounds are unstable relative to hydrates in which the cations surround themselves with water molecules. Thus $Mg(ClO_4)_2$ is a powerful absorbant for water, and lithium perchlorate forms a stable hydrate, $LiClO_4 \cdot 3H_2O$, whereas the other alkali metal perchlorates do not.

2. In other cases, the result of the bad packing is thermal instability such that the large polyatomic anion decomposes to leave behind a smaller one that can pack better with the small cation. Examples are

$$\begin{aligned} Li_2CO_3 &\longrightarrow Li_2O + CO_2 \\ 2NaO_2 &\longrightarrow Na_2O + \tfrac{3}{2}O_2 \\ 4Be(NO_3)_2 &\longrightarrow Be_4O(NO_3)_6 + N_2O_4 + \tfrac{1}{2}O_2 \end{aligned}$$

3. Related to point (1) are solubility relations. Thus $LiClO_4$ is about 10 times as soluble as $NaClO_4$ which, in turn, is about 10^3 times as soluble as $KClO_4$, $RbClO_4$, and $CsClO_4$. This trend is partly because the solvation enthalpies of the cations decrease as they increase in size, but it is enhanced by the fact that poor packing of the small Li^+ and Na^+ cations with the large ClO_4^- ions decreases the intrinsic stability of the crystals.

4-7 Structures with Close-packed Anions

Many ionic compounds, particularly those in which the cations are small relative to the anions, have structures based on *close-packing* of the spherical anions, with the cations occupying one or more sets of interstices. The NaCl structure is actually of this type, although it is not ordinarily useful to look at it that way.

Close packing of spheres in a single layer is shown in *Fig. 4-4*. The pattern produced is an array of contiguous equilateral triangles. Close packing in three

Figure 4-4 Close packing of spheres in a single layer.

dimensions is achieved by stacking such layers. *Figure 4-5a*, in which the pattern is represented only by the centers of the spheres and connecting lines, shows how this is done so that the spheres of each layer nestle into depressions between spheres in the other layer.

The most important thing to note in *Fig. 4-5a* is that the spheres of one layer (1) cover only *one half* the depressions in the other (2); thus, when a third layer (3) is added, there are two ways to do it. Spheres of layer 3 may be placed directly above those of 1, an arrangement that makes the first and third layers equivalent. The stacking pattern may therefore be called ABA. If this pattern is repeated, that is spheres in the fourth layer are placed over those in the second, and so on, the long-range pattern is ABABABA.... This is called *hexagonal closest packing* (*hcp*), because the hexagonal symmetry of each layer is retained by the entire stack.

When the third layer is added, it may be placed as is shown in *Fig. 4-5b*, that is, over those depressions in the first layer not covered by spheres in the second. This pattern may be described as ABC. If it is repeated indefinitely, ABCABCABCA..., we have *cubic closest packing* (*ccp*), so called not because the hexagonal pattern is destroyed but because a cubic arrangement is produced. This can be recognized by standing a cube on one vertex with a body diagonal vertical. If now one sphere is placed at each vertex and one at the center of each face, it can be seen that portions of closepacked layers lie in horizontal planes, as is indicated in *Fig. 4-6*.

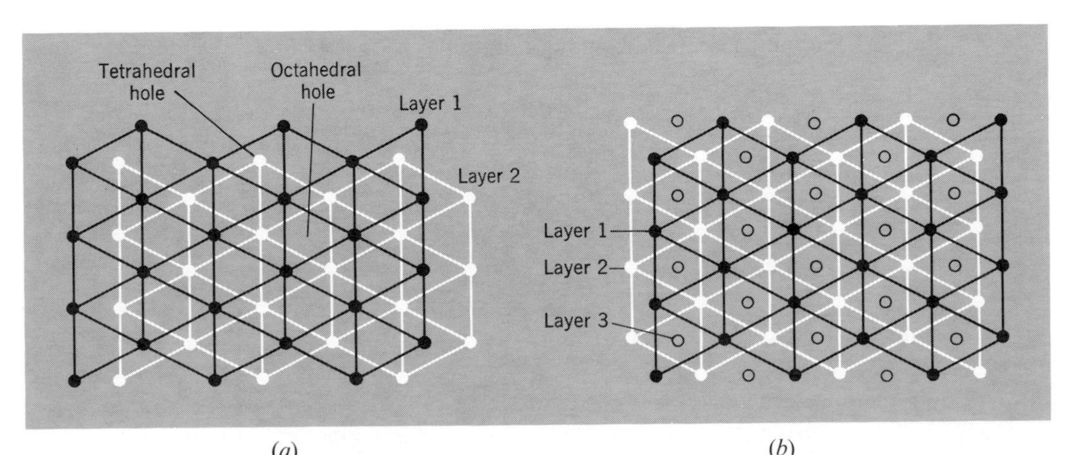

Figure 4-5 *(a) A diagram showing how two close-packed layers can be stacked to form both tetrahedral and octahedral holes between them. (b) A diagram showing three close-packed layers stacked so no two are superposed, that is the* ABC *or cubic close-packed (ccp) arrangement.*

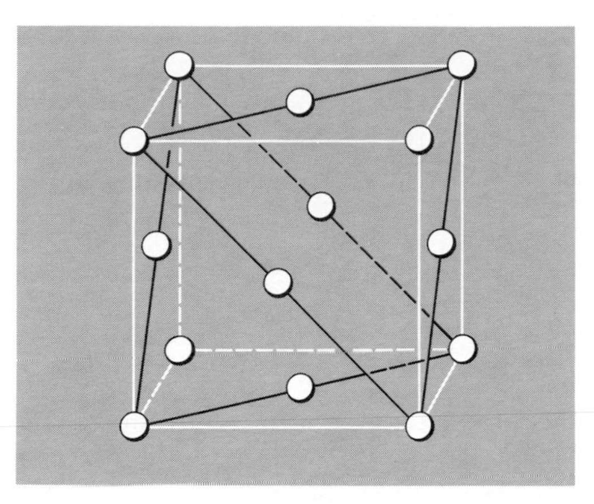

Figure 4-6 *A view of a ccp array which emphasizes its cubic symmetry.*

Of course, random forms of close packing, ABACBCACB..., etc., are possible, but occur only rarely.

It can be seen in *Fig. 4-5a* that between two layers of a close-packed structure there are two kinds of interstices: octahedral and tetrahedral. It is these that are occupied by cations in many close-packed halide and oxide structures. An example is the CdI_2 structure, *Fig. 4-7*, which is adopted by a number of MX_2 compounds. The anions are *hcp* and the metal ions occupy octahedral holes but only in every other layer. $CdCl_2$ has a *ccp* array of anions with, again, every other layer of octahedral holes occupied. The NaCl structure is related to the $CdCl_2$ structure by filling the octahedral holes in every layer. In the BI_3 structure, adopted by many

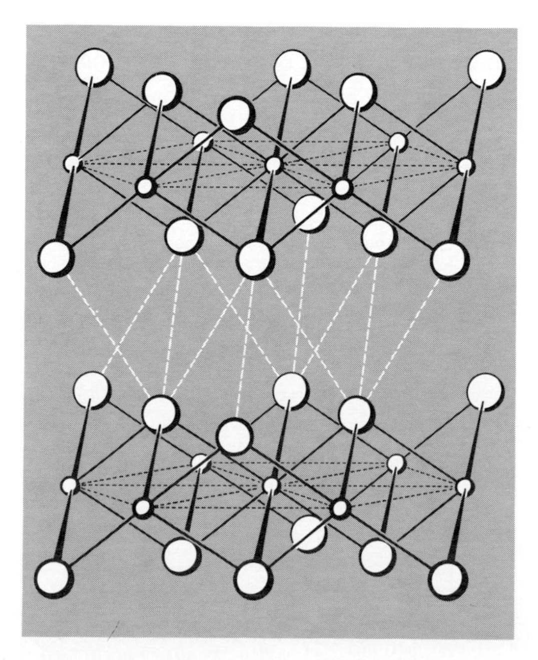

Figure 4-7 *A portion of the* CdI_2 *structure. Small spheres represent metal cations.*

MX_3 compounds, every other layer of octahedral holes in a *ccp* array is $\frac{2}{3}$'s occupied by cations.

Corundum, the α form of Al_2O_3, has an *hcp* array of oxide ions with $\frac{2}{3}$'s of the octahedral holes occupied, but not in a layered fashion. This important structure is adopted by many other M_2O_3 compounds such as Fe_2O_3, V_2O_3, and Rh_2O_3.

4-8 Mixed Metal Oxides

There is a large number of metal oxides, of great scientific and technical importance, which are essentially ionic substances. Many contain two or more different kinds of metal ions. They tend to adopt one of a few basic, general structures, the names of which are derived from the first compound—or an important one—found to have that structure.

The Spinel Structure. Spinel is a mineral, $MgAl_2O_4$. The structure is based on a *ccp* array of oxide ions, with Mg^{2+} ions in a set of tetrahedral holes and Al^{3+} ions in a set of octahedral holes. Many substances of the types $M^{2+}M^{3+}{}_2O_4$, $M^{4+}M^{2+}{}_2O_4$ and $M^{6+}M^{+}{}_2O_4$ have this structure. More highly charged cations tend to prefer the octahedral holes so that in $M^{4+}M^{2+}{}_2O_4$ compounds the octahedral holes are occupied by all the M^{4+} ions and one half of the M^{2+} ions.

The Ilmenite Structure. Ilmenite is the mineral $FeTiO_3$. Its structure is closely related to the corundum structure except that the cations are of two kinds. In ilmenite the cations are Fe^{2+} and Ti^{4+}, but many substances with the ilmenite structure have cations with charges of (+1, +5), or (+3, +3).

The Perovskite Structure. Perovskite is the mineral $CaTiO_3$. Its structure, shown in *Fig. 4-8*, is based on a *ccp* array of oxide ions together with large cations,

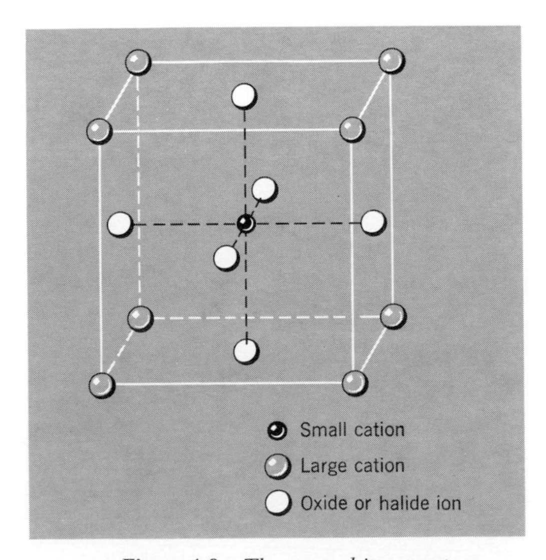

Figure 4-8 *The perovskite structure.*

similar in size to the oxide ion. The smaller cations lie in octahedral holes formed entirely by oxide ions. Again, the individual cation charges are not important so long as their sum is +6. The structure is adopted by many fluorides with cations of disparate sizes, such as $KZnF_3$.

Study Questions

A

1. What are the two main contributions to the cohesive energy of an ionic solid?
2. The pure ionic model neglects the possibility of any covalent bonding. Is this strictly correct or is it merely a useful approximation?
3. What is a Madelung constant?
4. Why do NaCl and MgO have the same Madelung constant?
5. In the Born expression for non-Coulomb repulsive energy, r^n occurs in the numerator. True or false?
6. In what range do values of n in the Born expression generally lie?
7. Using the same layout as in *Fig. 4-2*, show what the Born–Haber cycle would look like for CrN. Write a balanced equation for each step.
8. State in words the basic idea proposed by Pauling to estimate the ratio of the radii of a cation/anion pair.
9. What is a coordination number?
10. Why are we generally concerned with the coordination number of the cation rather than the anion?
11. Describe a close-packed layer of spherical atoms or ions.
12. Show with a drawing why there are two different ways to stack three such layers.
13. Explain the difference between cubic and hexagonal close packing.
14. What is corundum? Describe its structure.
15. What is the ilmenite structure, especially its relationship to the corundum structure?
16. What is the mineral of formula $MgAl_2O_4$ called? Describe its structure. What other cation charges besides 2+, 3+, and 3+ can occur in mixed oxides of this structure?
17. For the perovskite ($CaTiO_3$) structure to occur, what must be true of the sizes of the cations?

B

1. Krypton crystallizes in the cubic close-packed structure and has a density of 3.5 gcm^{-3}. What is the diameter of a krypton atom?
2. Consider a linear array of alternating cations and anions. Evaluate the Madelung constant to within 1%.
3. Show a Born–Haber type thermodynamic cycle from which the enthalpy of the reaction $NH_3(g) + H^+(g) = NH_4^+(g)$ could be calculated. *Hint*: consider the reaction $NH_3(g) + HCl(g) = NH_4Cl(g)$ as a starting point.
4. Using the same argument, due to Pauling, that is used to evaluate the radii of Na^+ and F^-, evaluate the radii of Mg^{2+} and O^{2-}, given that the unit cell edge of MgO (which has the NaCl structure) is 4.21 Å. Compare your results with those in Table 4-2.
5. What is the coordination number of each atom in a hexagonal close-packed structure?

6. As noted, MgO has the NaCl structure with a unit cell edge of 4.21 Å. Using a Born–Haber cycle calculate the enthalpy of electron attachment to form O^{2-}(g) from O(g). Additional data (in kJ mol^{-1}) required are: ΔH_f° of MgO(s) $= -801.7$, ΔH_{vap} of Mg $= 150.2$; ΔH_{diss} of $O_2 = 497.4$; $\Delta H_{ion}^{(1)} + \Delta H_{ion}^{(2)}$ for Mg $= 2188.1$.
7. By using radii for Mg^{2+} and F^-, decide whether it is likely to have a structure with 4-, 6-, or 8-coordinated Mg^{2+} ions.
8. In the vapor phase, lithium fluoride tends to dimerize to a planar structure. Assuming the Li-F distance to be the same for both monomer, LiF, and dimer $(LiF)_2$, namely 1.55 Å, and considering only electrostatic energies and Born repulsion energies, estimate the enthalpy of dimerization per mole of dimer. (Answer: 438.2 kJ mol^{-1}.)
9. Why do you think the value of n in the Born repulsion expression can be evaluated from compressibility data?
10. Demonstrate that r^-/r^+ for perfect packing in the wurzite structure is 4.44.

Chapter 4
Study Guide

Supplementary Reading

Adams, D. M., *Inorganic Solids*, John Wiley and Sons, 1974.
Dasent, W. E., *Inorganic Energetics*, Penguin Books, 1970.
Galasso, F. S., *Structure and Properties of Inorganic Solids*, Pergamon Press, 1970.
Greenwood, N. N., *Ionic Crystals, Lattice Defects and Non-Stoichiometry*, Butterworths, 1968.
Hannay, N. B., *Solid-State Chemistry*, Prentice-Hall, 1967.
Krebs, H., *Fundamentals of Inorganic Crystal Chemistry*, McGraw-Hill, 1968.
Wells, A. F., *Structural Inorganic Chemistry*, 3rd ed., Oxford University Press, 1962.

5

the chemistry of anions

5-1 Introduction

We have thus far discussed covalent bonding and some of the characteristics of simple ionic compounds, that is, compounds consisting mainly of monatomic cations, such as Na^+ or Ca^{2+}, and monatomic anions, such as F^- or O^{2-}. However, much of inorganic chemistry deals with ionic compounds of more elaborate types. In these, either the cation, or the anion, or both of them are polyatomic species, within which there are bonds and stereochemical relationships quite analogous to those within the uncharged polyatomic species that we call molecules.

In the next two chapters, the properties of anions and cations are considered in more detail, with particular though not exclusive reference to the more complex, polyatomic members of each group. The chemistry of cations is generally called *coordination chemistry* and is discussed in the next chapter. Here, the general properties of anions, as well as the specific chemistry of some of the more important ones, are outlined.

One term that must be defined here, in a preliminary way, although the subject will be covered in detail in Chapter 6, is *ligands*. When an anion (or other group) is bonded to a metal ion, it is called a ligand.

We may classify anions as follows.

1. Simple anions, such as O^{2-}, F^-, or CN^-
2. Discrete oxo anions, such as NO_3^- or SO_4^{2-}.
3. Polymeric oxo anions, such as silicates, borates, or condensed phosphates.
4. Complex halide anions, for example, TaF_6^- and anionic complexes containing multi-basic anions such as oxalate, for example, $[Co(C_2O_4)_3]^{3-}$.

Some of these, such as the oxide ion, O^{2-}, or most silicate anions, can exist only in the solid state. Others such as chloride ion, Cl^-, can exist also in aqueous solution. Further, some elements that form anions, notably the halogens, O and S, may be bound to other elements by covalent bonds as in PCl_3, CS_2, or NO_2.

More complex anions such as dithiocarbamate, $R_2NCS_2^-$, or acetylacetonate $CH_3COCHCOCH_3^-$, which occur mainly in coordination compounds are discussed in Chapter 6. Also described separately, since they constitute a quite different class of compounds, are those involving carbanions such as CH_3^-, $C_6H_5^-$, or $C_5H_5^-$ (Chapter 29). Hydride, H^-, and complex hydrido ions such as BH_4^- and AlH_4^- are also more conveniently treated separately (Chapters 9, 12, and 13). The most extensive, important, and varied classes of anions are those containing oxygen, and we discuss them first.

OXYGEN-CONTAINING ANIONS

5-2
The Oxide and Hydroxide Ions

The nature of several important oxide lattices has been discussed in Chapter 4. Discrete O^{2-} ions exist in many oxides but the ion cannot exist in aqueous solutions owing to the hydrolytic reaction

$$O^{2-}(s) + H_2O = 2OH^-(aq) \qquad K > 10^{22}$$

Thus only those ionic oxides that are insoluble in water are inert to it. Ionic oxides function as *basic anhydrides.* When insoluble in water, they usually dissolve in dilute acids, for example,

$$MgO(s) + 2H^+(aq) \longrightarrow Mg^{2+}(aq) + H_2O$$

In some cases, MgO being one, high-temperature ignition produces a very inert material, quite resistant to acid attack.

The covalent oxides of the nonmetals are usually acidic, dissolving in water to produce solutions of acids. Insoluble oxides of some less electropositive metals of this class generally dissolve in bases. Thus

$$N_2O_5 + H_2O \longrightarrow 2H^+(aq) + 2NO_3^-(aq)$$

$$Sb_2O_5(s) + 2OH^- + 5H_2O \longrightarrow 2Sb(OH)_6^-$$

Basic and acidic oxides will often combine directly to produce salts, such as

$$Na_2O + SiO_2 \xrightarrow{\text{Fusion}} Na_2SiO_3$$

Amphoteric Oxides. These behave as bases toward strong acids and acidically toward strong bases

$$ZnO + 2H^+(aq) \longrightarrow Zn^{2+} + H_2O$$

$$ZnO + 2OH^- + H_2O \longrightarrow Zn(OH)_4^{2-}$$

There are other oxides, some of which are relatively inert, dissolving in neither acids nor bases, for instance, N_2O, CO, and MnO_2; when MnO_2 (or PbO_2)

does react with acids, for example, concentrated HCl, it is a redox, not an acid-base, reaction giving Mn^{2+} and Cl_2.

There are also many ionic oxides that are nonstoichiometric, and there is an extensive chemistry of mixed metal oxides. When an element forms several oxides, the oxide with the element in the highest formal oxidation state is more acidic, for example, CrO, basic; Cr_2O_3 amphoteric; and CrO_3, fully acidic.

The Hydroxide Ion. Discrete OH^- ions exist only in the hydroxides of the more electropositive elements such as Na or Ba. For such an ionic material, dissolution in water results in formation of aquated metal and hydroxide ions:

$$M^+OH^-(s) + nH_2O \longrightarrow M^+(aq) + OH^-(aq)$$

and the substance is a strong base. In the limit of an extremely covalent M—O bond, dissociation will occur to varying degrees as follows:

$$MOH + nH_2O \rightleftharpoons MO^-(aq) + H_3O^+(aq)$$

and the substance must be considered an acid. Amphoteric hydroxides are those in which there is the possibility of either kind of dissociation, the one being favored by the presence of a strong acid:

$$M{-}O{-}H + H^+ = M^+ + H_2O$$

the other by a strong base:

$$M{-}O{-}H + OH^- = MO^- + H_2O$$

because the formation of water is so highly favored, that is

$$H^+ + OH^- = H_2O \qquad K_{25^\circ} = 10^{14}$$

Similarly, hydrolytic reactions of many metal ions can be written as

$$M^{n+} + H_2O = (MOH)^{(n-1)+} + H^+$$

However, because such ions are coordinated by H_2O molecules, a more realistic equation is

$$M(H_2O)_x^{n+} = [M(H_2O)_{x-1}(OH)]^{(n-1)+} + H^+$$

The higher the positive charge on the metal, the more acidic are the hydrogen atoms of the coordinated water molecules.

The OH^- ion has the ability to form bridges between metal ions. Thus there are various compounds of the transition and other metals containing OH bridges between pairs of metal atoms, as in (5-I). Although bridges of the type (5-I) are most common, there are also triply bridging hydroxo groups (5-II). The formation

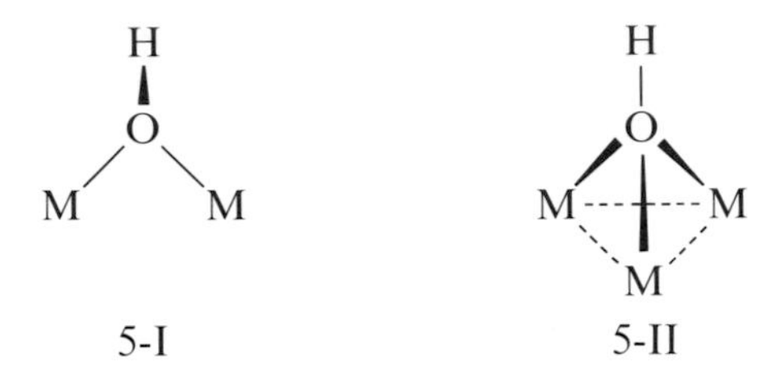

5-I 5-II

of hydroxo bridges occurs at an early stage in the precipitation of hydrous metal oxides. In the case of Fe^{3+}, precipitation of $Fe_2O_3 \cdot nH_2O$—commonly, but incorrectly, written $Fe(OH)_3$—proceeds through the following stages on adding OH^-:

$$
\begin{aligned}
&\underset{pH < 0}{[Fe(H_2O)_6]^{3+}} \longrightarrow \underset{0 < pH < 2}{[Fe(H_2O)_5OH]^{2+}} \\
&\longrightarrow \underset{\sim 2 < pH < \sim 3}{[(H_2O)_4Fe(OH)_2Fe(H_2O)_4]^{4+}} \\
&\longrightarrow \underset{\sim 3 < pH < \sim 5}{\text{colloidal } Fe_2O_3 \cdot xH_2O} \\
&\longrightarrow \underset{pH \sim 5}{Fe_2O_3 \cdot nH_2O \text{ ppt}}
\end{aligned}
$$

Note that many metal "hydroxides" are actually hydrous oxides; there is no hydroxide ion in the lattice.

Analogous to the OH^- ion are the *alkoxide ions*, OR^-. These are even stronger bases as a rule, being immediately hydrolyzed by water:

$$OR^- + H_2O = OH^- + ROH$$

Many *alkoxides* formally analogous to hydroxides are known, for example, $Ti(OH)_4$ and $Ti(OR)_4$. They are generally polymeric owing to the existence of OR bridge groups.

5-3
Mononuclear Oxo Anions

Discrete oxo anions are formed principally by the elements

B	C	N		
	Si	P	S	Cl
		As	Se	Br
			Te	I

and the transition metals

V	Cr	Mn	Fe
	Mo	Tc	Ru
	W	Re	Os

Some oxo ions can be bound covalently in compounds such as ethylene carbonate,

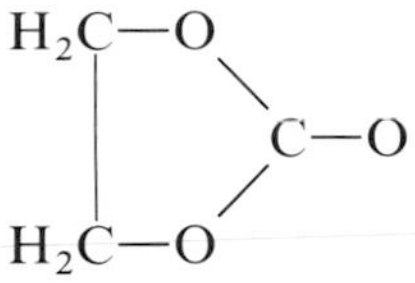

$$\begin{array}{l} H_2C{-}O \\ \quad\;\; | \qquad \diagdown \\ \quad\;\; | \qquad\;\; C{-}O \\ \quad\;\; | \qquad \diagup \\ H_2C{-}O \end{array}$$

and dimethyl sulfate $O_2S(OCH_3)_2$. They can also act as ligands in complex compounds.

Oxoacids of Carbon. The relation between CO_2 and "carbonic acid," H_2CO_3 is discussed on page 260. Both the carbonate, CO_3^{2-}, and bicarbonate, HCO_3^-, ions exist in crystalline ionic solids and in neutral or alkaline solutions. There are many naturally occurring carbonates, some of great importance, such as limestone, $CaCO_3$.

The ions are planar (5-III, 5-IV). In CO_3^{2-} because of delocalized π-bonding, (Section 3-6) the bond lengths are equal and the bond angles are 120°.

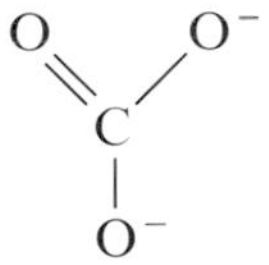

5-III 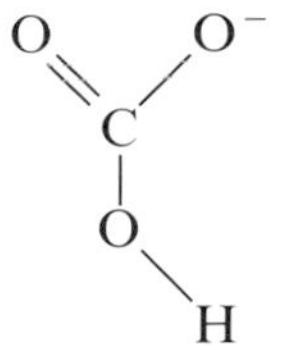 5-IV 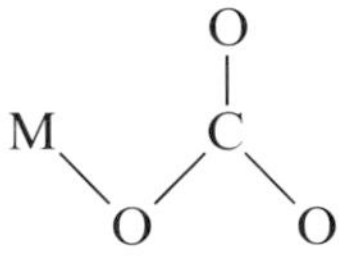 5-V 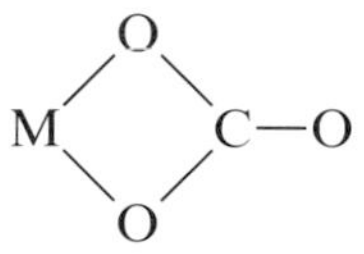 5-VI

Solutions of soluble carbonates such as those of the alkalis are alkaline as a result of hydrolysis

$$CO_3^{2-} + H_2O = HCO_3^- + OH^-$$

and, hence, the precipitation by sodium carbonate solutions of other insoluble carbonates, particularly those of transition metals, may lead to products contaminated with hydroxide.

With the exception of salts of the alkalis, Tl^+ and NH_4^+, carbonates are usually sparingly soluble in water. Like other oxo anions discussed below, carbonate can act as a ligand, for example, in $[Co(NH_3)_5CO_3]^{2+}$, forming either one bond (5-V) or two bonds (5-VI) to the cation.

Oxalate. $C_2O_4^{2-}$, gives insoluble salts with +2 ions such as Cu^{2+}. It is frequently found as a ligand, usually forming two bonds to the same cation, as in $[Cr(C_2O_4)_3]^{3-}$, but it can act as a bridge also.

Carboxylates. The carboxylate anions have several ways in which they can behave as ligands, as distinct from ionic behavior in say, sodium acetate. The main possibilities are 5-VII to 5-IX. The type of structure shown in 5-VIII is quite common and occurs in $Na[UO_2(RCO_2)_3]$. Symmetrical bridging (5-IX) occurs in the binuclear carboxylates $M_2(COOR)_4$ of Cu^{II}, Cr^{II}, Mo^{II}, and Rh^{II} where four carboxylato bridges are formed.

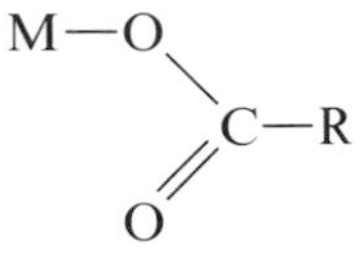

5-VII 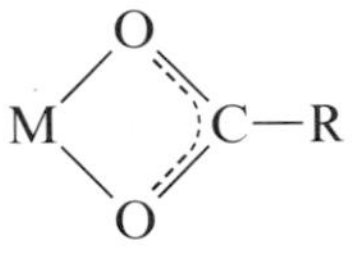 5-VIII 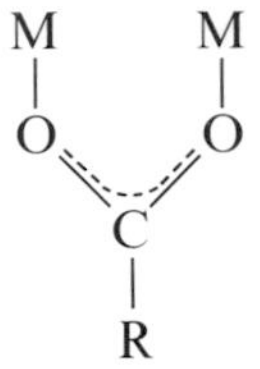 5-IX

Nitrogen Oxo Anions. Nitrite, NO_2^-, occurs normally as an anion only in $NaNO_2$ or KNO_2. It can act as a ligand in several ways (5-X, 5-XI, and 5-XII):

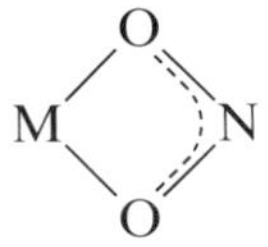

5-X 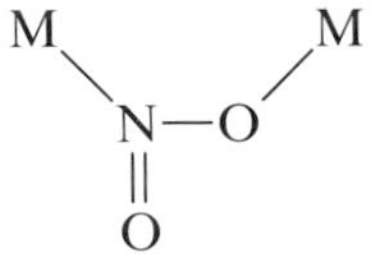 5-XI 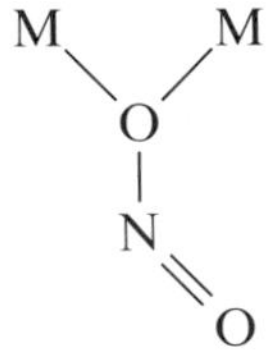 5-XII

the occurrence of a particular form can often be deduced from infrared spectra. Finally, there are tautomers, *nitrito*, M—ONO, and *nitro*, $M—NO_2$. Such tautomers occur for organic compounds, and the first inorganic example was discovered by S. M. Jørgensen in 1894 when he isolated $[Co(NH_3)_5ONO]Cl_2$ and $[Co(NH_3)_5NO_2]Cl_2$. The nitro isomer is always the more stable one.

Nitrates. Nitrates are made by dissolving the metals, oxides, or hydroxides in HNO_3. The crystalline salts are frequently hydrated and soluble in water. Alkali metal nitrates give nitrites on strong heating; others decompose to the metal oxides, water, and nitrogen oxides.

Like nitrite, nitrate may bond in several ways in complexes, 5-XIII to 5-XVI. The symmetrical structure (5-XVI) is quite common. Nitrate ion is a relatively weak ligand in aqueous solutions but cations of charge +3 or more are often complexed in solution as MNO_3^{2+}.

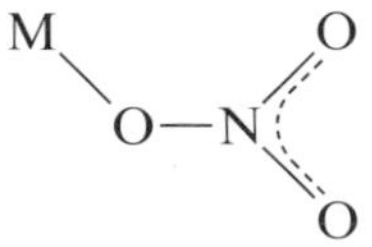

5-XIII 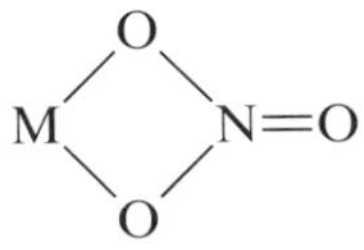 5-XIV 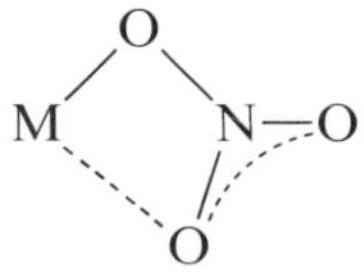 5-XV 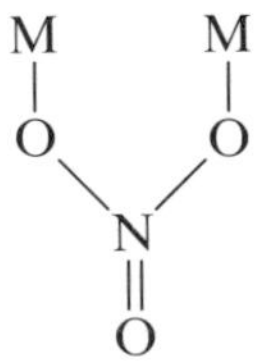 5-XVI

Phosphorus Oxoanions. There is a large number of these of which those of P^V, derived from the tribasic acid *orthophosphoric acid*, H_3PO_4 or $O{=}P(OH)_3$, are most important. Orthophosphates have tetrahedral PO_4 groups. Phosphates

PO_4^{3-}, HPO_4^{2-}, $H_2PO_4^-$ of most metal ions are known. Some are of practical importance, for example, ammonium phosphate fertilizers, alkali phosphate buffers in analysis, and the like. Natural phosphorus minerals are all orthophosphates, a major one being *fluoroapatite*. $Ca_9(PO_4)_6 \cdot CaF_2$. Hydroxy apatites, partly carbonated, make up the mineral part of teeth.

The precipitation of insoluble phosphates from 3–6 *M* nitric acid is a characteristic of the +4 ions of Ce, Th, Zr, and U.

Phosphates also form complexes in aqueous solution with many of the metal ions.

Arsenates. Arsenates generally resemble phosphates and the salts are often isomorphous. However, *antimony* differs in giving crystalline antimonates of the type $KSb(OH)_6$.

Sulfur Oxoanions. The common anions are pyramidal sulfite, SO_3^{2-}, or bisulfite, SO_3H^-, and tetrahedral sulfate, SO_4^{2-}, and bisulfate, SO_4H^- (XVII–XX). The sulfate ion forms many complexes in which it may coordinate to the metal ion through one oxygen atom (5-XXI), through two oxygen atoms (5-XXII), or it may serve as a bridge between two metal atoms (5-XXIII).

5-XVII  5-XVIII 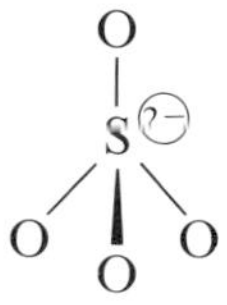 5-XIX 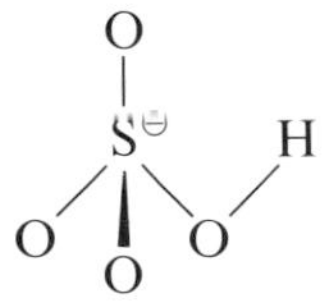 5-XX

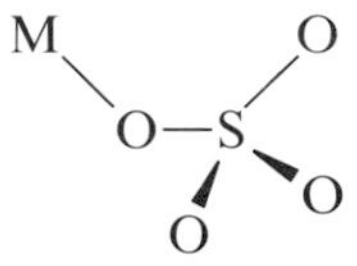

5-XXI 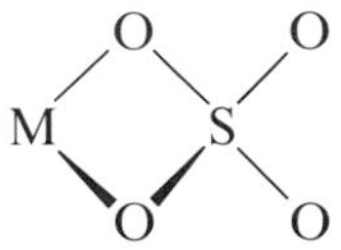 5-XXII 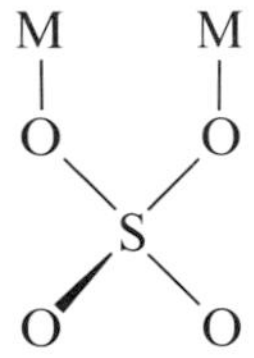 5-XXIII

Selenates. These are generally similar to the salts of SO_4^{2-} or SO_4H^- and are often isomorphous with them. *Tellurates* are invariably octahedral as in Hg_3TeO_6, $K[TeO(OH)_5] \cdot H_2O$ and the parent acid is best regarded as $Te(OH)_6$.

Halogen Oxoanions

Chlorates, Bromates, and Iodates. These pyramidal ions, XO_3^-, are known almost exclusively in alkali metal salts.

Iodates of +4 ions, Ce, Zr, Hf, Th, etc., can be precipitated from 6 *M* HNO_3 and provide a useful separation of these elements.

Perhalate Ions, XO_4^-. The most important is the perchlorate ion, ClO_4^-. It forms soluble salts with virtually all metal ions except the larger alkali ions, K^+, Rb^+, and Cs^+. It is often used to precipitate salts of other large +1 cations, for example $[Cren_2Cl_2]^+$, but this is highly inadvisable for organometallic ions such as $(\eta^5\text{-}C_5H_5)_2Fe^+$, as these compounds are often treacherously explosive. It is safer to employ $CF_3SO_3^-$, BF_4^-, or PF_6^- ions. The perchlorate ion has only a small tendency to serve as a ligand and is often used to minimize complex formation. It does, however, have some ability to coordinate, and a few perchlorate complexes are known.

Perbromate ion is a laboratory curiosity. Periodates are of the two types: tetrahedral IO_4^- and the octahedral species $IO_2(OH)_4^-$ and $IO_3(OH)_3^{2-}$. Both perbromates and periodates are chiefly important as oxidants.

Transition Metal Oxo Ions. Tetrahedral oxo anions, MO_4^{n-} are formed by V^{V}, Cr^{VI}, Mo^{VI}, W^{VI}, Mn^{VI}, Mn^{VII}, Tc^{VII}, Re^{VII}, Fe^{VIII}, Ru^{VIII}, and Os^{VIII} and can exist in solutions and in crystalline salts. They are not of general utility as anions. The best known are the permanganate, MnO_4^-, and chromate CrO_4^{2-}, ions that are widely used not as anions but as oxidants. We consider their chemistry elsewhere under the appropriate elements.

5-4
Polynuclear Oxo Anions

The oxo anions just discussed have 2, 3, or 4 oxygen atoms attached to a central atom to give a discrete anion. However, it is possible for one or more of these oxygen atoms to be shared between two atoms to give an ion with a bridge oxygen. One example of the simplest type is dichromate (5-XXIV) formed from CrO_4^- on acidification.

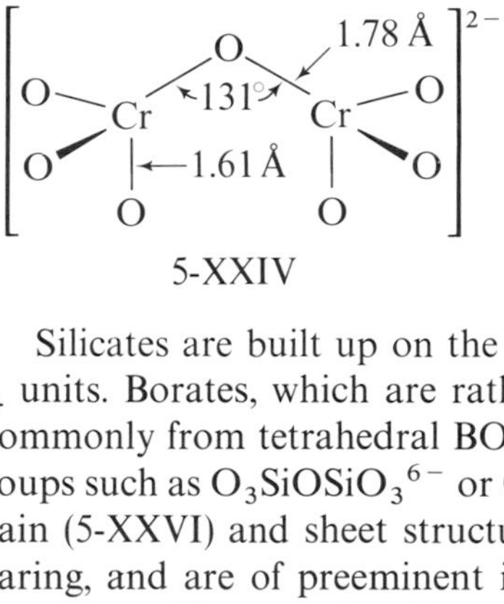

1.78 Å; 131°; 1.61 Å

5-XXIV

Silicates and Borates. Silicates are built up on the basis of sharing oxygen atoms of tetrahedral SiO_4 units. Borates, which are rather similar, are built up from planar BO_3 or less commonly from tetrahedral BO_4 units. Linking of such units can produce small groups such as $O_3SiOSiO_3^{6-}$ or $O_2BOBO_2^{4-}$. However, cyclic (5-XXV), infinite chain (5-XXVI) and sheet structures can be formed also by appropriate oxygen-sharing, and are of preeminent importance for silicates. The charges on the anions can be readily ascertained by regarding them as derived from acids where the oxygen atoms *not* involved in sharing can carry a negative charge as if derived from an —OH group by loss of H^+.

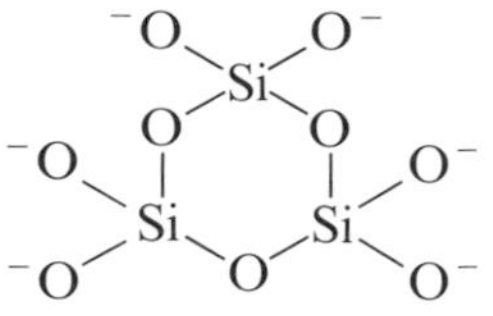

Ring anion
$Si_3O_9^{6-}$
5-XXV

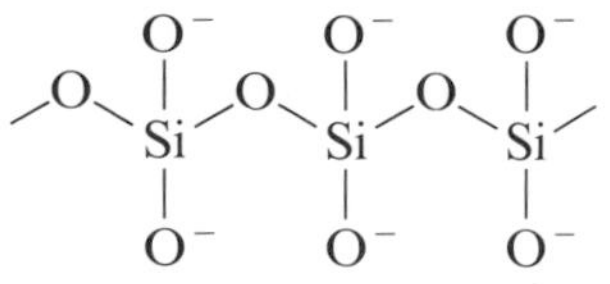

Infinite chain anion (pyroxene)
$(SiO_3^{2-})_n$
5-XXVI

An infinite sheet of SiO_4 units tetrahedrally linked in a 2-dimensional network is shown in *Fig. 5-1*; the stoichiometry is $(Si_2O_5^{2-})_n$.

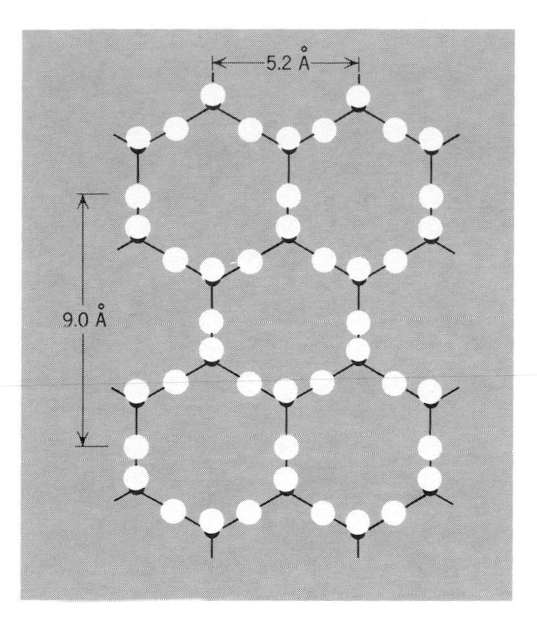

Figure 5-1 *The hexagonal arrangement of* SiO_4 *linked tetrahedra giving an infinite sheet of composition* $(Si_2O_5^{2-})_n$ ● = Si, ○ = O.

In silicate or borate structures, the specific nature of the cations or even their charges are relatively unimportant so long as the total positive charge is equivalent to the total negative charge. Thus for the pyroxene structure which occurs in many minerals, we can have $MgSiO_3$, $CaMg(SiO_3)_2$, $LiAl(SiO_3)_2$, and so forth. The cations lie between the chains so that their specific identity is of minor importance in the structure, so long as the required positive charge is supplied. Similarly, for sheet anions, the cations lie *between* sheets. Such substances could be expected to cleave readily and this is found to be so in *micas*, which are sheet silicates.

The final extension to complete sharing of oxygens of each tetrahedron leads, of course, to the structure of SiO_2—*silica* (page 270). However, if some of the formally Si^{4+} "ions" are replaced by Al^{3+}, then the framework must have a negative charge—and positive counter ions must be distributed through it. Such *framework minerals* are the *aluminosilicates*. They are among the most diverse, widespread, and useful natural silicate minerals. Many synthetic aluminosilicates can be made, and several are manufactured industrially for use as ion-exchangers (when wet) and "molecular sieves" (when dry).

Among the most important framework aluminosilicates are the *zeolites*. Their chief characteristic is the openness of the $[(Al, Si)O_2]_n$ framework (*Figs. 5-2* and *5-3*). The composition is always of the type $M_{x/n}[(AlO_2)_x(SiO_2)_y] \cdot zH_2O$ where n is the charge of the metal cation, M^{n+}, which is usually Na^+, K^+, or Ca^{2+}, and the z is the number of moles of water of hydration, which is highly variable. The openness of these structures results in the formation of channels and cavities of different sizes ranging from 2 to 11 Å in diameter. Molecules of appropriate sizes may thus be trapped in the holes, and it is this property that makes possible their use as selective adsorbents. Such zeolites are called "molecular sieves."

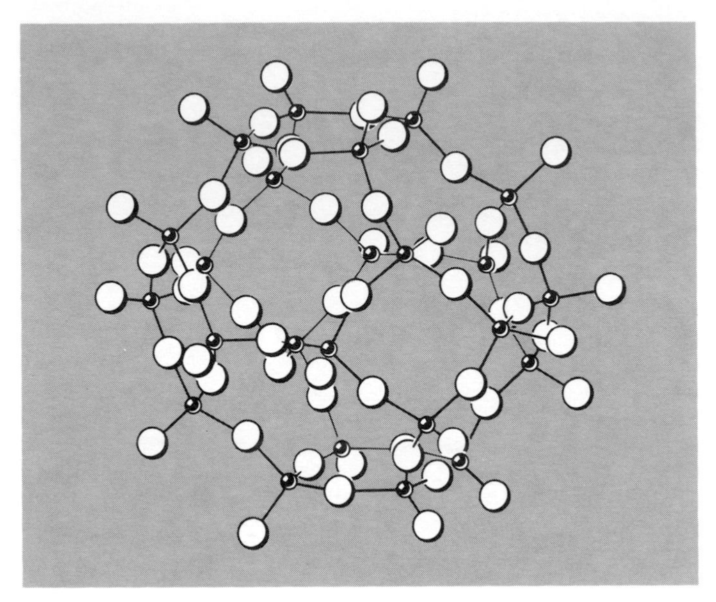

Figure 5-2 *The arrangement of* AlO_4 *and* SiO_4 *tetrahedra that gives the cubo-octahedral cavity in some zeolites and felspathoids. The* ● *represents* Si *or* Al.

They are used also as supports for metals or metal complexes used in heterogeneous catalytic reactions. The zeolites used are mainly synthetic. For example, slow crystallization under precisely controlled conditions of a sodium aluminosilicate gel of proper composition gives the crystalline compound $Na_{12}[(AlO_2)_{12}(SiO_2)_{12}]\cdot 27H_2O$. This hydrated form can be used as a cation-exchanger in basic solution.

In the hydrate all the cavities contain water molecules. In the anhydrous state obtained by heating in vacuum to *ca.* 350°, the same cavities may be occupied by other molecules brought into contact with the zeolite, providing such molecules are able to squeeze through the apertures connecting cavities. Molecules within the cavities then tend to be held there by attractive forces of electrostatic and van der Waals types. Thus the zeolite will be able to absorb and strongly retain molecules just small enough to enter the cavities. It will not absorb at all those too big to enter, and it will absorb weakly very small molecules or atoms that can enter but also leave easily. For example, straight-chain hydrocarbons but not branched-chain or aromatic ones may be absorbed.

Some *germanates* corresponding to silicates are known but Ge, Sn, and Pb more usually form octahedral anions $[M(OH)_6]^{2-}$. *Borates* do not form frameworks and are ring or chain polymeric anions. The most common boron mineral, *borax*, $Na_2B_4O_7\cdot 4H_2O$, contains an anion with the structure 5-XXVII.

```
[              OH              ]2-
               |
            O—B—O
           /    \   \
    HO—B        O   B—OH
           \    /   /
            O—B—O
               |
               OH              ]
```

5-XXVII

Figure 5-3 *Model of the zeolite edingtonite showing channels in the structure. The spheres represent oxygen atoms. The* Si *and* Al *atoms lie at the centers of* O_4 *tetrahedra and cannot be seen.*

Polymeric or Condensed Phosphates. Orthophosphate anions can also be linked by oxygen bridges. Three types of building blocks occur (5-XXVIII to 5-XXX). The resulting polymeric anions are called metaphosphates if they are cyclic (5-XXXI) or polyphosphates if they are linear (5-XXXII). Sodium salts of

$\diagdown O-P(=O)(O^-)-O^-$	$\diagup O-P(=O)(O^-)-O\diagup$	$\diagup O-P(=O)(O\diagup)-O\diagup$
$PO_{3.5}{}^{2-}$	$PO_3{}^-$	$PO_{2.5}$
End unit	Middle unit	Branching unit
5-XXVIII	5-XXIX	5-XXX

condensed phosphates are widely used as water softeners as they form soluble complexes with calcium and other metals. Their use has led to some ecological problems, since phosphates also act as fertilizers and in lakes can lead to abnormally high growths of algae.

$P_3O_9^{3-}$ 5-XXXI

$P_3O_{10}^{5-}$ 5-XXXII

Condensed phosphates are usually prepared by dehydration of orthophosphates under various conditions of temperature (300–1200°) and also by appropriate dehydration of hydrated species as, for example,

$$(n-2)NaH_2PO_4 + 2Na_2HPO_4 \xrightarrow{\text{Heat}} \underset{\text{Polyphosphate}}{Na_{n+2}P_nO_{3n+1}} + (n-1)H_2O$$

$$nNaH_2PO_4 \xrightarrow{\text{Heat}} \underset{\text{Metaphosphate}}{(NaPO_3)_n} + nH_2O$$

They can also be prepared by controlled addition of water to P_4O_{10}. The resulting complex mixtures of anions can be separated by ion-exchange or chromatography.

The most important *cyclic* phosphate is *tetrametaphosphate*, which can be prepared by heating copper nitrate with slightly more than an equimolar amount of H_3PO_4 (75%) slowly to 400°. The sodium salt can be obtained by treating a solution of the copper salt with Na_2S. Slow addition of P_4O_{10} to ice water gives ~75% of the P as tetrametaphosphate. Condensed arsenates exist only in the solid state, being rapidly hydrolyzed by water.

Transition Metal Polyanions. Although we cannot discuss them in detail, the oxo anions of V^{V}, Nb^{V}, Ta^{V}, Mo^{VI} and W^{VI} form extensive series of what

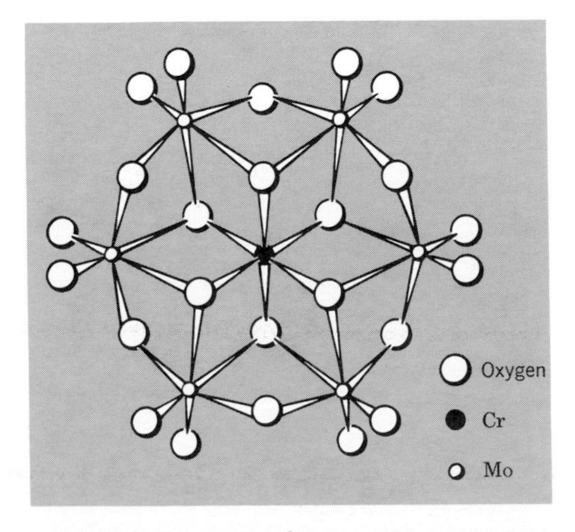

Figure 5-4 *The structure of* $[CrMo_6O_{24}H_6]^{3-}$; *the hydrogen atoms are probably bound to oxygen atoms of the central octahedron.*

are called *isopoly* and *heteropoly* anions. Both are built up by sharing oxygen atoms in MO_6 octahedra, where corners and edges, but not faces may be shared. An example is shown in *Figure 5-4*.

Isopoly anions, which contain only the element and oxygen, have stoichiometries such as $Nb_6O_{19}{}^{8-}$ and $Mo_7O_{24}{}^{6-}$. In heteropoly anions an additional metal or nonmetal atom is present. One example is $[Co_2{}^{II}W_{12}O_{42}]^{8-}$. The heteropoly salt ammonium phosphomolybdate $(NH_4)_3[P^{V}Mo_{12}O_{40}]$, is used in the determination of phosphorus while silico-tungstate, a large -1 anion, is sometimes used for precipitation of large $+1$ cations.

HALIDE, HALOGENO COMPLEX, PSEUDOHALIDE AND SULFIDE ANIONS

5-5 Ionic Halides

Most halides of metals in $+1$, $+2$, and $+3$ oxidation states are predominantly ionic in character. Of course, there is a uniform gradation from halides that are for all practical purposes purely ionic, through those of intermediate character, to those that are essentially covalent. Covalent halides and the preparation of halides are discussed in Chapter 20.

Many metals show their highest oxidation state in the fluorides. For very high oxidation states, which are formed notably with transition metals, for example, WF_6 or OsF_6, the compounds are generally gases, volatile liquids, or solids resembling closely the covalent fluorides of the nonmetals. The question as to whether a metal fluoride will be ionic or molecular cannot be reliably predicted, and the distinction between the types is not always sharp.

Fluorides in high oxidation states are often hydrolyzed by water, for example,

$$4RuF_5 + 10H_2O \longrightarrow 3RuO_2 + RuO_4 + 20\,HF$$

The driving force for such reactions results from the high stability of the oxides and the low dissociation of HF in aqueous solution.

The halides of the alkali and alkaline earth elements (with the exception of Be) and of most of the lanthanides and a few halides of the *d*-group metals and actinides can be considered as mainly ionic materials. As the charge/radius ratio of the metal ions increases, however, covalence increases. Consider, for instance, the sequence KCl, $CaCl_2$, $ScCl_3$, $TiCl_4$. KCl is completely ionic, but $TiCl_4$ is an essentially covalent molecular compound. Similarly, for a metal with variable oxidation state, the lower halides will tend to be ionic, whereas the higher ones will tend to be covalent. As examples we can cite $PbCl_2$ and $PbCl_4$, and UF_4, which is an ionic solid, while UF_6 is a gas.

Most ionic halides dissolve in water to give hydrated metal ions and halide ions. However, the lanthanide and actinide elements in the $+3$ and $+4$ oxidation states form fluorides insoluble in water. Fluorides of Li, Ca, Sr, and Ba also are sparingly soluble. Lead gives a sparingly soluble salt, PbClF, which can be used for gravimetric determination of F^-. The chlorides, bromides and iodides of

Ag^I, Cu^I, Hg^I, and Pb^{II} are also insoluble. The solubility through a series of mainly ionic halides of a given element, $MF_n \rightarrow MI_n$ may vary in either order. In cases where all four halides are essentially ionic, the solubility order will be iodide > bromide > chloride > fluoride, since the governing factor will be the lattice energies, which increase as the ionic radii decrease. This order is found among the alkali, alkaline-earth, and lanthanide halides. On the other hand, if covalence is fairly important, it can invert the trend, making the fluoride most and the iodide least soluble, as in the cases of Ag^+ and Hg_2^{2+} halides.

5-6 Halide Complex Anions

Complex halogeno anions, especially of fluoride and chloride, are of considerable importance. Halogeno anions may be formed by interaction of a metallic or non-metallic halide acting as a Lewis acid toward the halide acting as a base:

$$AlCl_3 + Cl^- = AlCl_4^-$$

$$FeCl_3 + Cl^- = FeCl_4^-$$

$$BF_3 + F^- = BF_4^-$$

$$PF_5 + F^- = PF_6^-$$

Many such halogeno anions can be formed in aqueous solution. The relative affinities of F^-, Cl^-, Br^-, and I^- for a given metal ion is not fully understood. For crystalline materials, lattice energies are important. For BF_4^-, BCl_4^-, BBr_4^-, the last two of which are known only in crystalline salts of large cations, lattice energies are governing. In considering the stability of the complex ions *in solution*, it is important to recognize that (a) the stability of the complex involves not only the bond strength of the M—X bond but also its stability relative to the stability of ion-solvent bonds, and (b) in general an entire series of complexes will exist, $M^{n+}(aq)$, $MX^{(n-1)+}(aq)$, $MX_2^{(n-2)+}(aq)$, . . . , $MX_x^{(n-x)+}(aq)$, where x is the maximum coordination number of the metal ion. These two points, of course, apply to all types of complexes in solution (page 138).

Generally the stability decreases in the series F > Cl > Br > I, but with some metal ions the order is the opposite, namely, F < Cl < Br < I. This is one of several problems involving acid-base interactions to be discussed in Chapter 7. It is to be emphasized that all complex fluoro "acids" such as HBF_4 and H_2SiF_6 are *necessarily strong*, since the proton can be bound *only* to a solvent molecule.

Halogeno anions are important in several ways. They are involved in many important reactions in which Lewis acids, particularly $AlCl_3$ and BF_3, take part; one example is the Friedel-Crafts reaction. For several elements, they are among the most accessible source materials; a good example is platinum as chloroplatinic acid, $(H_3O^+)_2PtCl_6$, or potassium chloroplatinite K_2PtCl_4. Large or undeformable anions like BF_4^- or PF_6^- can be used to obtain sparingly soluble salts of appropriate cations. Finally, halide complex formation can be used for separations with anion-exchange resins. To take an extreme example, Co^{2+} and Ni^{2+}, can be separated by passing a strong HCl solution through an anion-exchange column. Co^{2+} readily forms $CoCl_3^-$ and $CoCl_4^{2-}$ whereas nickel does not give chloro complexes in aqueous solutions. Effective separation usually depends on properly exploiting the *difference* in complex-formation between two cations *both* of which have some tendency to form anionic halide complexes.

5-7
Pseudohalides

Pseudohalides are substances containing two or more atoms that have halogenlike properties. Thus cyanogen, NC—CN, gives the *cyanide* ion, CN^-, and shows halogenlike behavior. Compare

$$Cu^{2+} + 2CN^- = CuCN + \tfrac{1}{2}(CN)_2$$

$$Cu^{2+} + 2I^- = CuI + \tfrac{1}{2}I_2$$

Other pseudohalide ions are *cyanate,* OCN^-, and *thiocyanate* SCN^-. These are formed, respectively, from CN^- by oxidation, for example, by PbO, and by fusing, say, KCN with S_8. Their Ag^+ salts like those of the halides are insoluble in water.

The pseudohalide ions are very good ligands. For cyanate and thiocyanate there are two binding possibilities—through N or through O or S. For OCN, most nonmetals seem to be N-bonded in covalent compounds such as $P(NCO)_3$ while the corresponding thiocyanates are S-bonded.

Cyanate and the more numerous thiocyanate complexes usually have stoichiometries similar to the analogous halide complexes.

Cyanide is somewhat different in that the formation of cyanide *complexes* is restricted to transition metal *d* block elements and Zn, Cd, and Hg. This suggests that π acceptor bonding is important in the binding of CN^- to the metal, which is almost invariably through carbon. The π acceptor character of CN^- is not nearly so high as for CO, RNC, or similar ligands (Chapter 28). This is clearly reasonable in view of its negative charge. Indeed, CN^- is a strong nucleophile, so that back-bonding need not be invoked to explain the stability of its complexes with metals in +2 and +3 oxidation states. However, CN^- does have the ability to stabilize metal ions in low oxidation states as in, for example, $[Ni(CN)_4]^{4-}$. Here, some acceptance of electron density into π^* orbitals of CN^- is likely.

The majority of cyanide complexes are anionic, typical being $[Fe^{II}(CN)_6]^{4-}$, $[Ni(CN)_4]^{2-}$, and $[Mo(CN)_8]^{3-}$. By contrast with the similar halide complexes, the free acids of many cyano anions are known, for example, $H_4[Fe(CN)_6]$ and $H_3[Rh(CN)_6]$. The reason for this is that the proton can be located in hydrogen bonds between the cyano anions, that is, $M—CN \cdots H \cdots NC—M$.

5-8
The Sulfide and Hydrosulfide Ions

Only the alkalis and alkaline earths form sulfides that contain the S^{2-} ion. Only these sulfides dissolve in water. Although S^{2-} is not as extensively hydrolyzed as O^{2-}, nevertheless essentially only SH^- ions are present in aqueous solutions owing to the low second dissociation constant of H_2S. The S^{2-} ion is present in strongly alkaline solution, but it cannot be detected in solution less alkaline than 8 M NaOH owing to the reaction

$$S^{2-} + H_2O = SH^- + OH^- \qquad K \sim 1$$

Polysulfide ions S_n^{2-} are formed when solutions of alkali sulfides are boiled with sulfur. Salts can be crystallized. The ions contain kinked chains of sulfur atoms as illustrated by the S_4^{2-} structure (5-XXXIII).

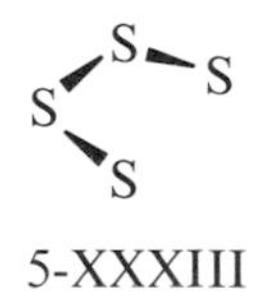

5-XXXIII

Study Questions

A

1. Why does the ion O^{2-} exist only in ionic lattices?
2. List the ways in which OH^- can act as a ligand.
3. List the elements that form oxoanions.
4. Many oxoanions can act as ligands in more than one way. Give the ways for (a) CO_3^{2-}, (b) SO_4^{2-}, (c) NO_3^-, (d) CH_3COO^-, (e) NO_2^-.
5. Draw the structures of $Cr_2O_7^{2-}$, $Si_2O_7^{6-}$, $B_2O_5^{4-}$.
6. How are 2-dimensional silicate networks built up?
7. What is the composition of zeolites? What are molecular sieves?
8. How do the oxoanions of Ge, Sn, and Pb differ from silicates?
9. Draw structures for cyclic and linear condensed phosphates.
10. What is meant by the terms iso- and hetero-poly anions?
11. What happens when 6 M HCl is added dropwise to a purple aqueous solution of $[Fe(H_2O)_6](ClO_4)_3$?
12. Why does the free acid HBF_4 not exist although the pure acid HNO_3 can be obtained?
13. What is a pseudohalogen? Give examples of the chemical similarity between a halogen and a pseudohalogen.
14. Why does the S^{2-} ion exist only in very strongly alkaline solution?
15. What anions can be used to precipitate +4 cations from nitric acid solutions?

B

1. Compare the properties of the oxides of Mg, B, Si, and Sb^V.
2. Why is the oxide of an element most acidic in the highest oxidation state?
3. Titanium ethoxide is a tetramer $[Ti(OEt)_4]_4$. Write a plausible structure for this molecule. What happens when water is added to it?
4. Compare the structures of the simpler oxoacids of S, Se, and Te.
5. What are the structures of the anions in $K_3B_3O_6$, CaB_2O_4 and $KB_5O_8 \cdot 4H_2O$?
6. Describe the pyroxene $(SiO_3^{2-})_n$ and amphibole $(Si_4O_{11}^{6-})_n$ structures.
7. Discuss the nature of multiple bonding in the oxo anions SO_4^{2-}, NO_3^-, and ClO_4^-.

Chapter 5
Study Guide

Supplementary Reading. Further details concerning individual anions and classes of anions will be found later in this book and in *Advanced Inorganic Chemistry*, 3rd ed., by F. A. Cotton and G. Wilkinson, John Wiley—Interscience, 1972.

6

coordination chemistry

6-1
Introduction

In all their compounds, cations are surrounded by anions or neutral molecules. The groups immediately surrounding a cation are called *ligands* and the branch of inorganic chemistry concerned with the combined behavior of cations and their ligands is called *coordination chemistry*. There is, of course, no sharp dividing line between coordination chemistry and the chemistry of covalent molecules on the one hand and that of ionic solids on the other. We traditionally consider methane and SF_6 as covalent molecules, while treating BH_4^- and AlF_6^{3-} as coordination compounds (i.e., $B^{3+} + 4H^-$ and $Al^{3+} + 6F^-$, respectively), but in terms of fundamental electronic properties, these distinctions would not be easy to defend. Similarly, the metal to ligand bonding in Na_3AlF_6 and in AlF_3(s) cannot be qualitatively very different even though we traditionally call the former a coordination compound (and the AlF_6^{-3} group a *complex* ion) and the latter a salt.

The main justification for classifying many substances as coordination compounds is that their chemistry can conveniently be described in terms of an essentially constant, cationic central species M^{n+} about which a great variety of ligands L, L′, L″, etc., may be placed in an essentially unlimited number of combinations. The overall charge on the resulting complex, $[ML_xL'_yL''_z\ldots]$ is determined by the charge on M, and the sum of the charges on the ligands. For example, the Pt^{2+} ion forms a great many complexes, studies of which have provided much of our basic knowledge of coordination chemistry. Examples of its complexes, all of which can be interconverted by varying the concentrations of the different ligands are

$$[Pt(NH_3)_4]^{2+},\quad [Pt(NH_3)_3Cl]^+,\quad [Pt(NH_3)_2Cl_2],\quad [Pt(NH_3)Cl_3]^-,\quad [PtCl_4]^{2-}$$

For complexes of Pt^{2+} the set of four ligands lie at the vertices of a square with the Pt^{2+} ion at the center. Thus, structurally, four of the five complexes in the above series are, unambiguously:

H_3N --- NH_3 / Pt / H_3N --- NH_3 H_3N --- NH_3 / Pt / H_3N --- Cl

H_3N --- Cl / Pt / Cl --- Cl Cl --- Cl / Pt / Cl --- Cl

Notice that the structure of the middle member of the series $[Pt(NH_3)_2Cl_2]$, is ambiguous from the formula. Two isomers, *cis* and *trans* are possible and both are well known:

Cl --- NH_3 / Pt / Cl --- NH_3 Cl --- NH_3 / Pt / H_3N --- Cl

cis *trans*

This is one of the simplest examples of the occurrence of isomers among coordination compounds. A number of other important cases will be discussed in Section 6-4.

The fundamental and classical investigations in coordination chemistry were carried out between about 1875 and 1915 by the Danish chemist S. M. Jørgensen (1837–1914) and the Swiss Alfred Werner (1866–1919). When they began their studies the nature of coordination compounds was a huge puzzle which the contemporary ideas of valence and structure could not accommodate. How, for example, could a stable metal salt, for example, MCl_n, combine with a group of stable, independently existing molecules, for example, xNH_3, to form a compound $M(NH_3)_xCl_n$ with wholly new properties? How were bonds formed? What was the structure? Jørgensen and Werner prepared thousands of new compounds, seeking to find regularities and relationships which would suggest answers to the above questions. Finally, Werner developed the concept of ligands surrounding a central metal ion—the concept of a coordination complex—and deduced the geometrical structures of many of them. His structure deductions were based on the study of isomers such as those just discussed. In this very instance, he reasoned that the arrangement had to be planar to give the two isomers; a tetrahedral structure could not account for their existence. Werner received the Nobel prize in Chemistry for his work in 1913.

6-2
Coordination Numbers and Geometries

The term coordination number has already been introduced (page 99) in discussing the packing of ions in ionic crystals. It is widely used in discussing the structures of the complexes formed by cations. In addition to the number of ligands surrounding a cation, it is important to know their spatial arrangement, the coordination geometry. We now discuss the coordination numbers, in increasing order, and under each the most common geometrical arrangements.

Coordination Number Two. This is relatively rare, occurring mainly with the +1 ions of Cu, Ag, and Au and with Hg^{2+}. The geometry is linear, and ions such as $[H_3N—Ag—NH_3]^+$, $[NC—Ag—CN]^-$, and $[Cl—Au—Cl]^-$ are examples.

Coordination Number Three. The most important geometries are planar and pyramidal. It is a relatively rare coordination number. Examples are the planar HgI_3^- ion and the pyramidal $SnCl_3^-$ ion. In most instances where the stoichiometry might suggest it, for example, $AlCl_3$, $FeCl_3$, $PtCl_2PR_3$, etc., no 3-coordinate mononuclear species occur. Instead, there are dinuclear species in which two ligands are shared so as to give each cation a coordination number of four, 6-I, 6-II.

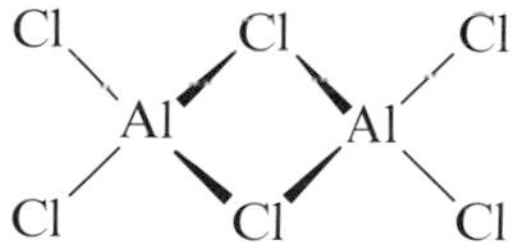

6-I

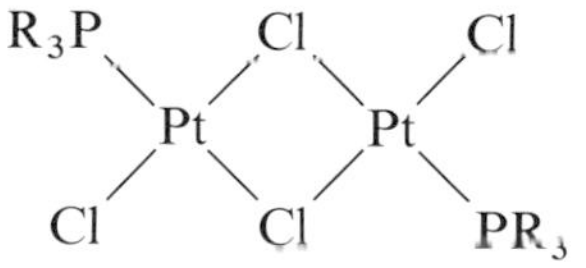

6-II

Coordination Number Four. This is one of the most important and gives two main geometries, tetrahedral and square. Tetrahedral complexes are most common, being formed almost exclusively by nontransition metal cations as well as by transition metal ones other than those to the right side of the *d* block. Examples of tetrahedral complexes include $Li(H_2O)_4^+$, BeF_4^{2-}, BF_4^-, $AlCl_4^-$, $FeCl_4^-$, $CoBr_4^{2-}$, ReO_4^-, and many others. Square complexes are particularly common for Cu^{2+}, Ni^{2+}, Pd^{2+}, Pt^{2+}, Au^{3+}, Rh^+, and Ir^+. The cations that characteristically form square complexes are those with eight *d* electrons. These eight electrons form four electron pairs that fill all of the *d* orbitals except $d_{x^2-y^2}$. This orbital together with the *s*, p_x, and p_y orbitals forms a set of hybrids, dsp^2, directed toward the vertices of a square.

Coordination Number Five. This is less common than 4 or 6, but still very important. There are two symmetrical geometric arrangements, the trigonal bipyramid (**tbp**) and the square pyramid (**sp**), 6-III and 6-IV. Although these two

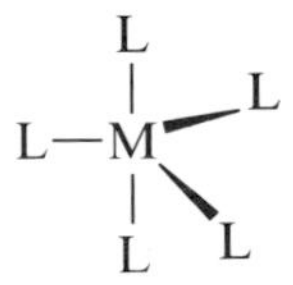

6-III

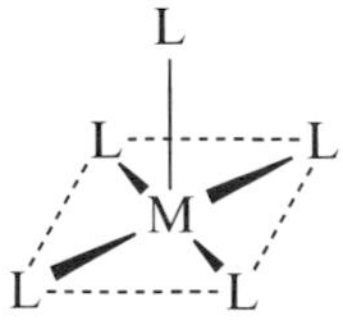

6-IV

may appear quite different, they usually do not differ greatly in energy, and one can be converted into the other by rather small changes in bond angles. For these reasons, many 5-coordinate complexes do not have either structure precisely, but a structure that is intermediate between the two. Moreover, even those that do have a structure approximating closely to one of these ideal forms are *stereochemically nonrigid*. This means that the ligands do not remain fixed in their places, but interchange places rapidly, as explained in Section 6-13.

An interesting illustration of the similar stabilities of the two geometries is afforded by the $Ni(CN)_5{}^{3-}$ ion, which forms one crystalline salt in which both the **tbp** and **sp** structures are present.

Coordination Number Six. This is enormously important, since nearly all cations form 6-coordinate complexes. Practically all of these have one geometrical form, the octahedron, 6-V. It is essential to recognize that the octahedron is an extremely symmetrical figure, even though some of the stylized ways of drawing it might not show this clearly. All six ligands, and all six M—L bonds are equivalent in a regular octahedral ML_6 complex.

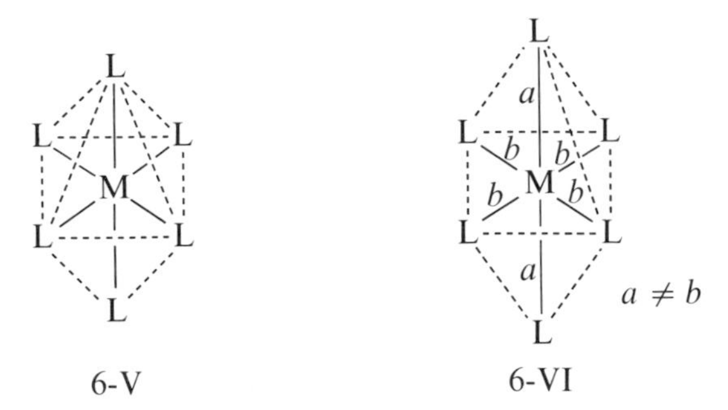

6-V 6-VI

As with other prototype geometries, we continue to describe complexes as "octahedral" even when different kinds of ligands are present and, hence, the full symmetry of the true octahedron cannot be retained. Even in cases where all ligands are chemically the same, octahedra are often distorted, either by electronic effects inherent in the metal ion (cf. Section 23-8) or by forces in the surroundings. A compression or elongation of one L—M—L axis relative to the other two is called a *tetragonal* distortion, 6-VI, whereas a complete breakdown of the equality of the axes gives a *rhombic* distortion, 6-VII. If the octahedron is compressed or elongated on an axis connecting the centers of two opposite triangular faces, the distortion is called *trigonal*, 6-VIII.

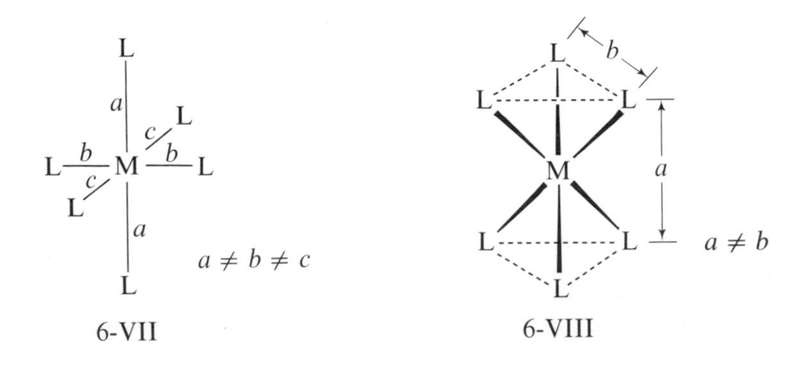

6-VII 6-VIII

There are a very few cases in which six ligands lie at the vertices of a trigonal prism, 6-IX. The prism is related to the octahedron in a simple way: if one triangular face of an octahedron is rotated 60° relative to the one opposite to it, a prism is formed. The superior stability of the octahedron compared with the prism has, at least, two causes. The most evident is steric: the octahedron allows the ligands to stay further away from each other, on the average, than does the prism for any given M—L distance. It is also likely that in most cases the metal ion can form stronger bonds to an octahedral set of ligands. The cases where a trigonal prism is found mostly involve either a set of six sulfur atoms, which may interact directly with each other to stabilize the prism, or some sort of rigid cage ligand which forces the prismatic arrangement.

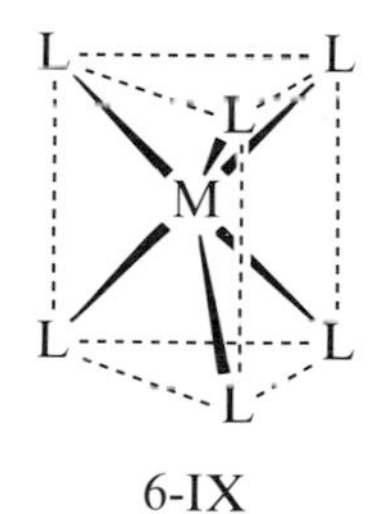

6-IX

Higher Coordination Numbers. Coordination numbers of 7, 8, and 9 are not infrequently found for some of the larger cations. In each of these cases there are several geometries which generally do not differ much in stability. Thus complexes with high coordination numbers are characteristically stereochemically nonrigid (Section 6-13).

For 7-coordination there are three fairly regular geometries, the pentagonal bipyramid (6-X), an arrangement derived from the octahedron by spreading one face to make room for the seventh ligand, and an arrangement similarly derived from a trigonal prism (6-XI).

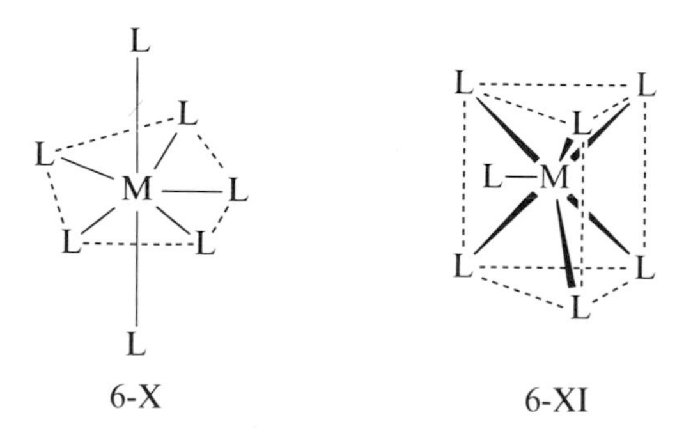

6-X 6-XI

Coordination number eight also has three important geometries, all of which are shown in *Fig. 6-1*. The cube itself is rare, since by distorting to either the antiprism or the triangular dodecahedron, inter-ligand repulsions can be diminished while still maintaining close M—L contacts.

For 9-coordination the only symmetrical arrangement is that shown in *Fig. 6-2*. This is observed in many lanthanide compounds in the solid state.

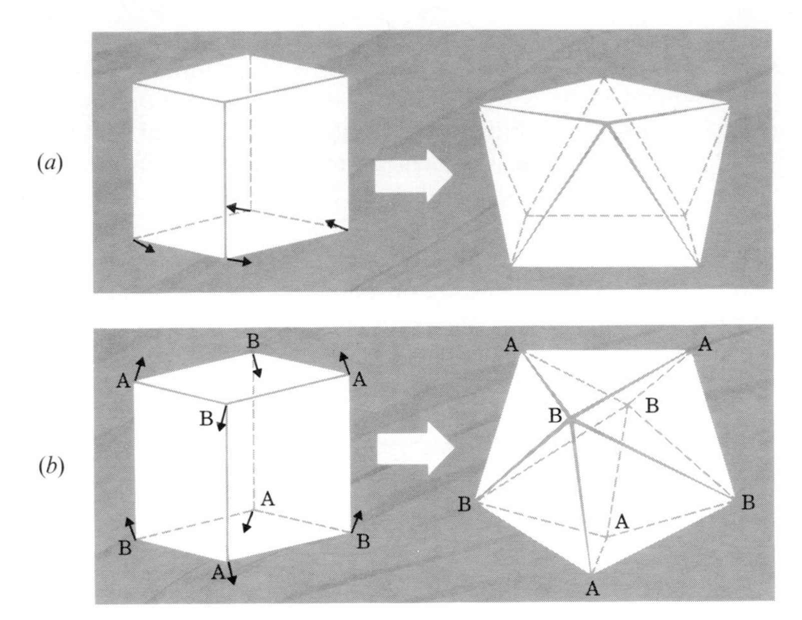

Figure 6-1 *The most important ways of distorting the cube: (a) to produce a square antiprism; (b) to produce a dodecahedron.*

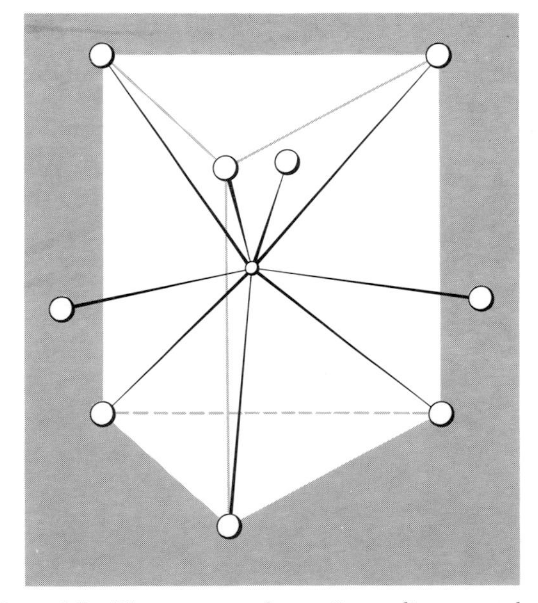

Figure 6-2 *The structure of many 9-coordinate complexes.*

6-3 Types of Ligands

The majority of ligands are anions or neutral molecules that can be thought of as electron pair donors. Common ones are F^-, Cl^-, Br^-, CN^-, NH_3, H_2O, CH_3OH, and OH^-. Ligands such as these, when they donate one electron pair to one metal atom, are called *monodentate* (literally, one-toothed) ligands. The five complexes of Pt^{2+} mentioned in the introduction contain only monodentate ligands, Cl^- and NH_3.

Ligands that contain two or more atoms, each of which can simultaneously form a two-electron donor bond to the same metal ion, are called *polydentate* ligands. They are also called *chelate* (from the Greek for claw) ligands since they appear to grasp the cation between the two, or more donor atoms.

Bidentate Ligands. These are the most common of the polydentate ligands. Neutral ones include diamines, diphosphines, diethers, and β-kctoenolate anions, of which the commonest are

$H_2\ddot{N}CH_2CH_2\ddot{N}H_2$ ethylenediamine, en,

$(C_6H_5)_2\ddot{P}CH_2CH_2\ddot{P}(C_6H_5)_2$ diphos

$CH_3\ddot{O}CH_2CH_2\ddot{O}CH_3$ glyme

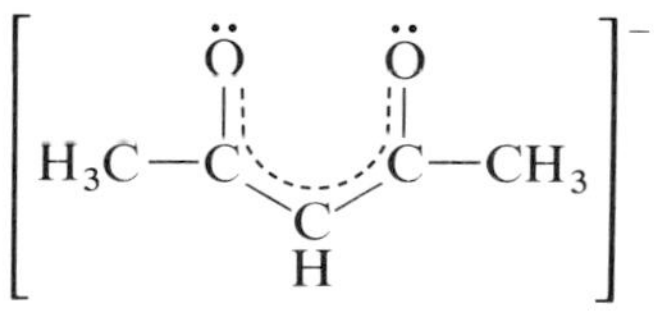

$[H_3C{-}C(\ddot{O}){\cdots}CH{\cdots}C(\ddot{O}){-}CH_3]^-$ acetylacetonate, acac

all of which form five-membered rings with the metal atom, as well as a number of anions that form four-membered rings, such as

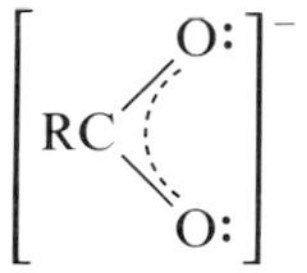

$[RC(O:)(O:)]^-$ carboxylates

$[O{=}N(O:)(O:)]^-$ nitrate

$[R_2NC(S:)(S:)]^-$ dithiocarbamates

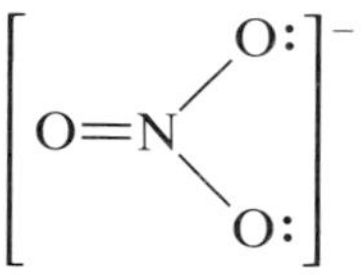

$[O_2S(O:)(O:)]^{2-}$ sulfate

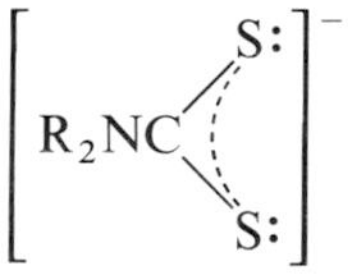

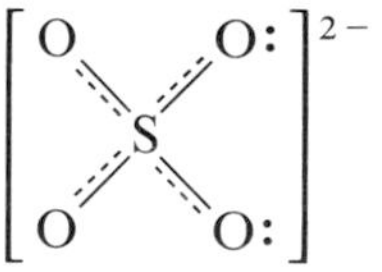

and a number of others.

Polydentate Ligands. These include tri-, quadri-, penta-, and hexadentate ligands. Examples of tridentate ligands are

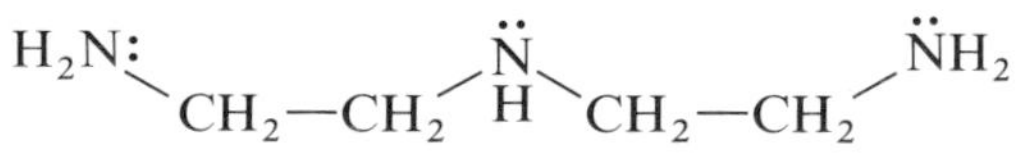

$H_2N{:}{-}CH_2{-}CH_2{-}\ddot{N}H{-}CH_2{-}CH_2{-}\ddot{N}H_2$ diethylene triamine, dien

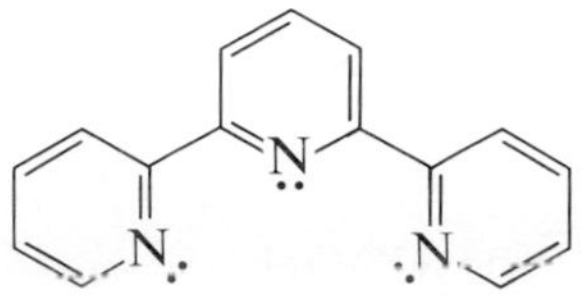

terpyridyl, terpy

Quadridentate ligands may be of the open chain type, such as the following Schiff base derived from acetylacetone, but more important are the many macro-

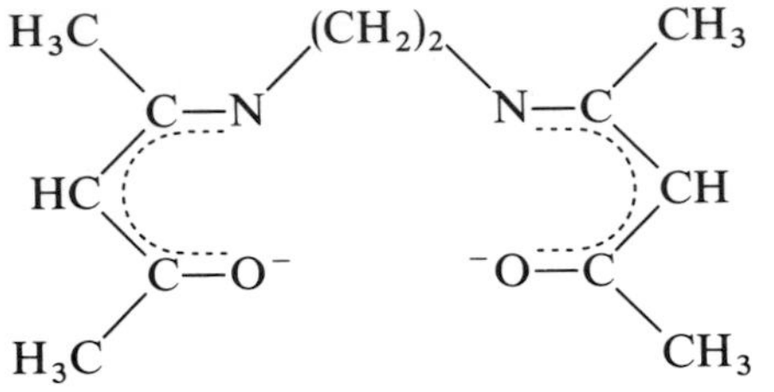

cyclic ligands, such as porphyrin (6-XII) and its derivatives, phthalocyanine (6-XIII), and a host of similar molecules that can be synthesized readily, for example, 6-XIV.

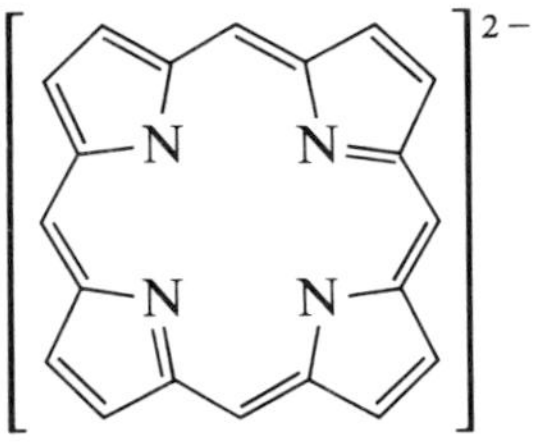

6-XII

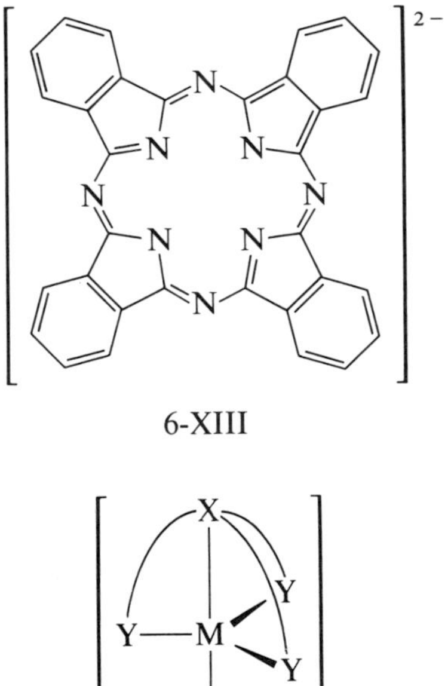

6-XIII

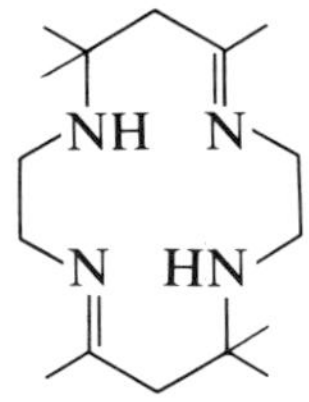

6-XIV

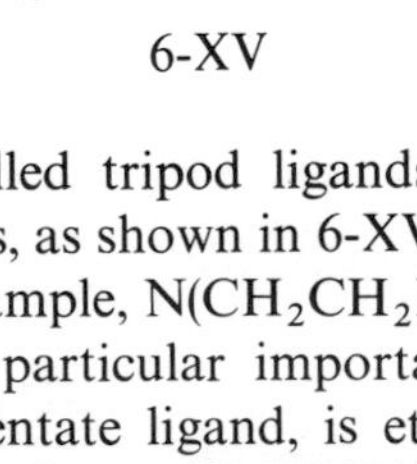

6-XV

There are also the so-called tripod ligands that favor the formation of trigonal bipyramidal complexes, as shown in 6-XV. The donor atoms in these are generally N, P, S, or As, for example, $N(CH_2CH_2PPh_2)_3$.

A hexadentate ligand of particular importance, which can also function as a pentadentate or quadridentate ligand, is ethylenediamine tetraacetic acid (H_4EDTA) in one of its anionic forms (H_2EDTA^{2-}, $HEDTA^{3-}$, $EDTA^{4-}$):

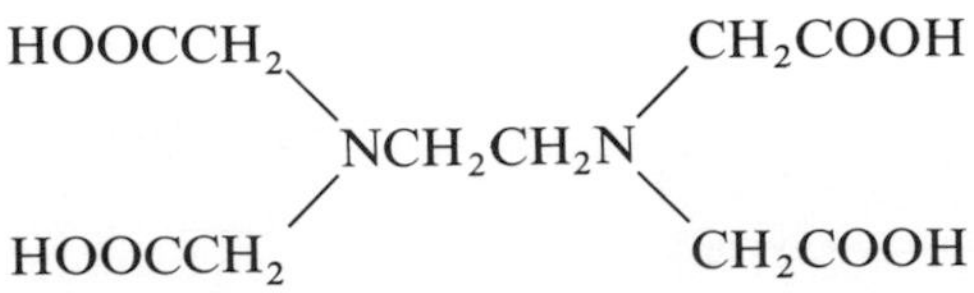

$(HOOCCH_2)_2NCH_2CH_2N(CH_2COOH)_2$ H_4EDTA

6-4
Isomerism in Coordination Compounds

One of the reasons coordination chemistry can become quite complicated is that there are many ways in which isomers can arise. We have already observed that square complexes of the type ML_2X_2 can exist as *cis* and *trans* isomers. Other important forms of geometrical isomerism are illustrated in 6-XVI to 6-XIX. Isomers of octahedral complexes that are of particular importance are the *trans*, 6-XVI, and *cis*, 6-XVII, isomers of the ML_4X_2 species and the facial (6-XVIII) and meridional (6-XIX) isomers of ML_3X_3 species.

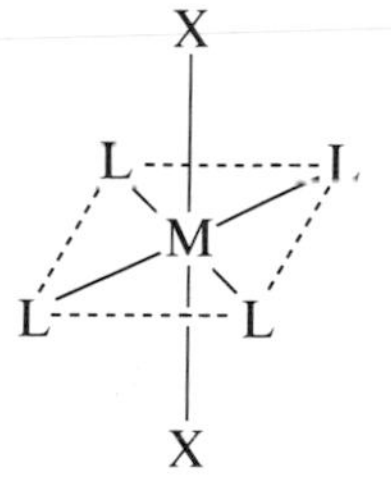

6-XVI

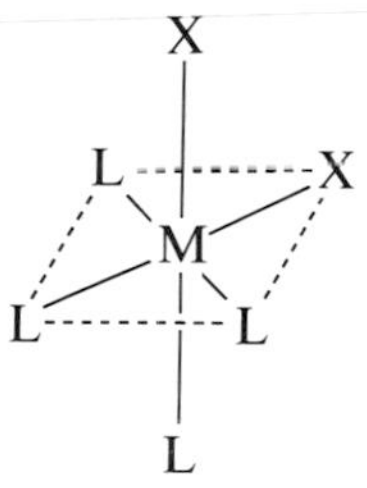

6-XVII

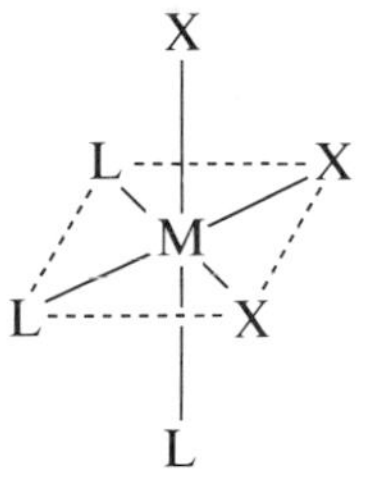

6-XVIII

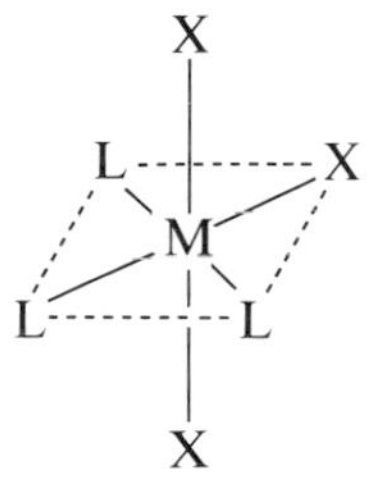

6-XIX

Optical isomers are molecules that are mirror images of each other that cannot be superimposed. Since they cannot be superimposed, they are not identical, even though all their internal distances and angles are identical. They also react identically unless the reactant is also one of a pair of optical isomers. Their most characteristic difference, which gives rise to the term *optical*, is that each one causes the plane of polarization of plane-polarized light to be rotated, but in opposite directions.

Two molecules that are optical isomers in the above sense are called enantiomorphs. Their existence was first recognized among organic compounds when a tetrahedral carbon atom was bonded to four different groups, as in lactic acid:

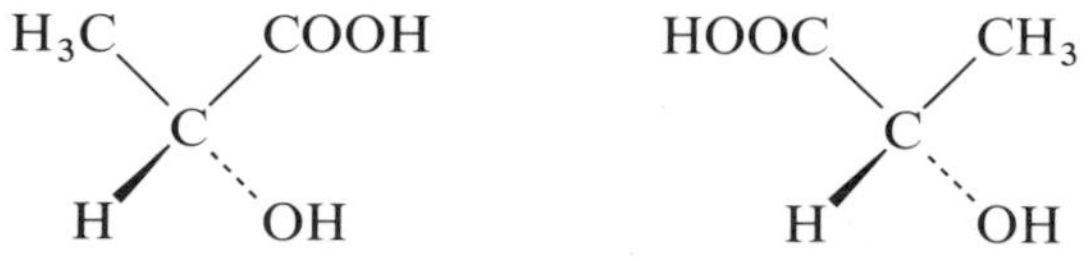

It was one of Werner's accomplishments to recognize that for certain types of octahedral complexes, enantiomorphs should also exist. He prepared and resolved these compounds and used this result to support his hypothesis that the coordination geometry was indeed octahedral. Among the most important enantiomorphous octahedral complexes are those that contain two or three bidentate ligands. The enantiomorphs of a $M(L—L)_2X_2$ complex are shown as 6-XX and 6-XXI. Those of the $M(L—L)_3$ type are 6-XXII and 6-XXIII.

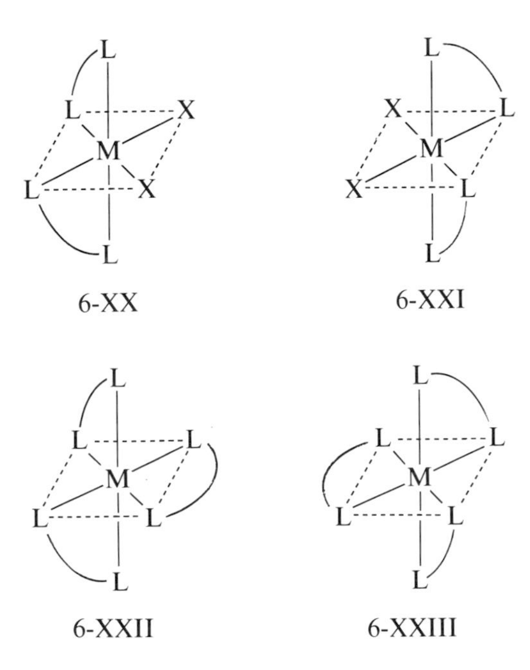

For the latter, which are called trischelate complexes, another useful way to regard them is shown in *Fig. 6-3*, where the view is perpendicular to one pair of opposite triangular faces of the octahedron. Viewed in this way, the molecules have the appearance of helices, like a ship's propellor, with the twist of the helix being opposite in the two cases. *Figure 6-3* also defines a notation for the absolute configurations: Λ (Greek capital lambda) for laevo or left; Δ (Greek capital delta) for dextro or right.

There are a few other sorts of isomerism that are more or less peculiar to coordination compounds, that is, they will not often, if ever, be encountered in organic molecules.

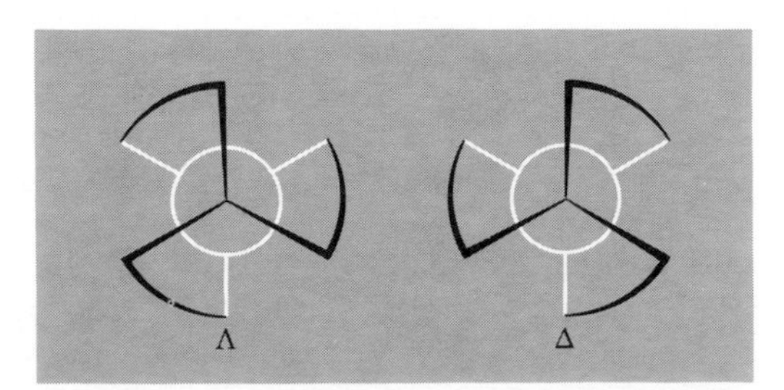

Figure 6-3 *Diagrams of trischelate octahedral complexes showing how the absolute configurations Λ and Δ are defined according to the twist of the helices.*

Ionization Isomerism. Compounds that have the same molecular formula but differ in which anions are coordinated yield different ions when in solution. Examples are

$$[Co(NH_3)_4Cl_2]NO_2 \rightleftharpoons [Co(NH_3)_4Cl_2]^+ + NO_2^-$$

$$[Co(NH_3)_4Cl(NO_2)]Cl \rightleftharpoons [Co(NH_3)_4Cl(NO_2)]^+ + Cl^-$$

and

$$[Coen_2(NO_2)Cl]SCN, \qquad [Coen_2(NO)_2(SCN)]Cl,$$

$$[Coen_2(SCN)Cl]NO_2 \qquad en = H_2NCH_2CH_2NH_2$$

In these illustrations the square brackets are used to enclose the metal atom and all the ligands that are directly bound to it, that is, are in the coordination shell. This use of square brackets is a way of making this distinction in formulas when necessary and will be found in the research literature. They can, however, be omitted when no confusion would arise, as in $Co(NH_3)_3Cl_3$.

The concept of ionization isomerism provides the key to understanding many simple but otherwise puzzling observations. For example, there are three different substances of the composition $CrCl_3 \cdot 6H_2O$. One is violet and is $[Cr(H_2O)_6]Cl_3$; it does not lose water over H_2SO_4 and all Cl^- is immediately precipitated by Ag^+ from a fresh solution. $[Cr(H_2O)_5Cl]Cl_2 \cdot H_2O$ is green; it loses one H_2O over H_2SO_4 and only two thirds of its Cl content is precipitated promptly. $[Cr(H_2O)_4Cl_2]Cl \cdot 2H_2O$ which is also green loses two H_2O over H_2SO_4 and only one third of its Cl content is promptly precipitated.

Linkage Isomerism. Some ligands can bind in more than one way, and often isomeric complexes with the different modes of binding can be isolated. The oldest example is the isomeric pair:

$$\left[(NH_3)Co{-}N\begin{smallmatrix}O\\ \\ O\end{smallmatrix} \right]^{2+}, \qquad [(NH_3)_5Co{-}O{-}N{=}O]^{2+}$$

nitro nitrito

Other ligands prone to give linkage isomers, or at least to bind in different ways in different compounds, are SCN^-, which may use either S or N as the donor atom and sulfoxides, $R_2S{=}O$, which may use either S or O as the donor.

Coordination Isomerism. In compounds where both cation and anion are complex, the distribution of ligands can vary, giving rise to isomers. The following are examples:

$$[Co(NH_3)_6][Cr(CN)_6] \quad \text{and} \quad [Cr(NH_3)_6][Co(CN)_6]$$

$$[Cr(NH_3)_6][Cr(SCN)_6] \quad \text{and} \quad [Cr(NH_3)_4(SCN)_2][Cr(NH_3)_2(SCN)_4]$$

$$[Pt^{II}(NH_3)_4][Pt^{IV}Cl_6] \quad \text{and} \quad [Pt^{IV}(NH_3)_4Cl_2][Pt^{II}Cl_4]$$

Polymerization Isomers. These are not truly isomers, because they do not have the same minimal unit, but they do have the same analytical formula and have traditionally been called isomers.
Examples are

$$(Pt(NH_3)_4Cl_2] \quad \text{and} \quad [Pt(NH_3)_4][PtCl_4]$$

$$[Co(NH_3)_3(NO_2)_3], \quad [Co(NH_3)_6][Co(NO_2)_6]$$

and

$$[Co(NH_3)_5(NO_2)][Co(NH_3)_2(NO_2)_4]_2$$

6-5
Nomenclature for Coordination Complexes

The naming of coordination compounds follows a large body of rules developed and kept up to date by the International Union of Pure and Applied Chemistry. Only a few of the basic ones which find frequent use will be presented here.

A. Designations for Ligands. When they form part of the name of a coordination compound, some ligands have special names:

NH_3	ammine
H_2O	aqua
NO	nitrosyl
CO	carbonyl

Anionic ligands have their usual names but they are modified to end in *o*; for example:

CH_3COO^-	acetato
CN^-	cyano
F^-	fluoro
O^{2-}	oxo
O_2^{2-}	peroxo
OH^-	hydroxo
H^-	hydrido

Organic radicals, even if treated as anions when figuring the formal oxidation number of the metal, are given their usual names as radicals, for example:

CH_3	methyl
C_6H_5	phenyl

Most other ligands receive their ordinary names, but with spaces omitted:

$(CH_3)_2SO$	dimethylsulfoxide
$(NH_2)_2CO$	urea
C_5H_5N	pyridine
$(C_6H_5)_3P$	triphenylphosphine

The ligands N_2 and O_2 are called *dinitrogen* and *dioxygen.*

B. Order of Listing. The ligands are listed first, with the metal last. Neutral ligands are listed first, and then anionic ones.

C. Metal Ions and Oxidation Numbers. For neutral and cationic complexes, the usual name of the metal is used, followed by a Roman numeral in parentheses giving its formal oxidation number. When the complex is an anion, the metal is designated by a word ending in "ate" and sometimes by using a Latin form, for example, ferrate, cuprate, with the formal oxidation number in parentheses.

D. Numerical Prefixes. The occurrence of two or more ligands, or metal atoms, is indicated with the following prefixes.

2	di (bis)
3	tri (tris)
4	tetra (tetrakis)
5	penta (pentakis)
6	hexa (hexakis)
7	hepta
8	octa
9	nona (ennea)
10	deca
11	undeca
12	dodeca
	etc.

The prefixes in parentheses are less used.

At this point, we illustrate with some examples. Note that in formulas the metal atom is mentioned first, followed by the ligands.

$[Co(NH_3)_5CO_3]Cl$	pentamminecarbonatocobalt(III) chloride
$[Cr(H_2O)_4Cl_2]Cl$	tetraquadichlorochromium(III) chloride
$K_2[OsCl_5N]$	potassium pentachloronitridoosmate (VI)
$[(C_6H_5)_4As][PtCl_2HCH_3]$	tetraphenylarsonium dichlorohydridomethylplatinate (II)
$Mo[(C_6H_5)_2PCH_2CH_2P(C_6H_5)_2]_2(N_2)_2$	bis(1,2-diphenylphosphinoethane)-bis(dinitrogen)molybdenum(0)

E. Indicating Isomers. When a particular isomer is to be specified, an italic prefix followed by a hyphen is used. Important ones are: *cis*-, *trans*-, *fac*-, and *mer*-. These last two are used for octahedral complexes of the types 6-XVIII and 6-XIX, respectively, and are abbreviations for *fac*ial and *mer*idional. Thus, if in 6-XIX, the L ligands were $(C_2H_5)_3P$, the X ligands H, and the metal ruthenium, we would have

mer-$[Ru\{(C_2H_5)_3P\}_3H_3]$,	*mer*-tris(triethylphosphine)trihydrido-ruthenium(III)

As we can observe in this and other illustrations, parentheses and curly brackets are sometimes necessary to avoid confusion. Common sense is usually the best guide in using them.

F. Bridging Ligands. A bridging ligand is designated by the prefix μ-. When there are two bridging groups of the same kind di-μ- is used. The bridging ligand(s) is listed in order with the other ligands, set off between hyphens, unless the molecule is symmetrical so that a more compact name results from placing it first. The following examples will illustrate.

$$[(NH_3)_5Co{-}NH_2{-}Co(NH_3)_4(H_2O)]Cl_5$$

pentamminecobalt(III)-μ-amido-tetrammineaquacobalt(III) chloride

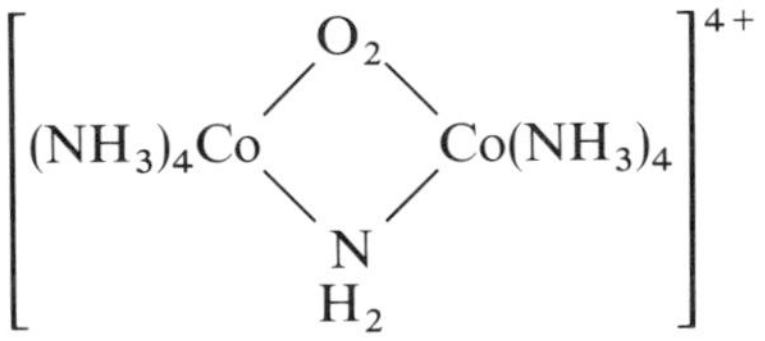

$$\left[(NH_3)_4Co\begin{matrix} O_2 \\ \diagup\quad\diagdown \\ \diagdown\quad\diagup \\ N \\ H_2 \end{matrix}Co(NH_3)_4\right]^{4+}$$

tetramminecobalt(III)-μ-superoxo-μ-amido-tetramminecobalt(III)
(physical data indicate O_2 is O_2^- rather than O_2^{2-}.

$$[(NH_3)_5Cr{-}OH{-}Cr(NH_3)_5]Br_5$$

μ-hydroxobis[pentamminechromium(III)] chloride

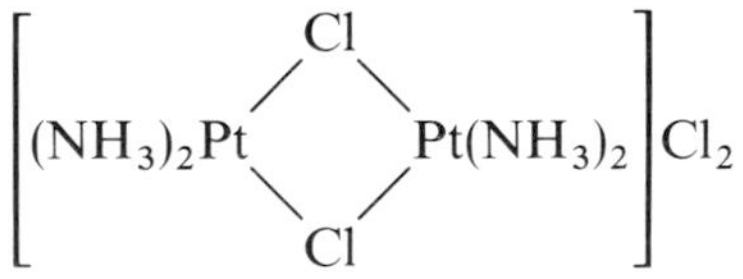

$$\left[(NH_3)_2Pt\begin{matrix} Cl \\ \diagup\quad\diagdown \\ \diagdown\quad\diagup \\ Cl \end{matrix}Pt(NH_3)_2\right]Cl_2$$

di-μ-chlorobis[diammineplatinum(II)] chloride

6-6
Equilibrium Constants for Formation of Complexes in Solution

The formation of complexes in aqueous solution is a matter of great importance not only in inorganic chemistry but also in biochemistry, analytical chemistry, and in a variety of applications. The extent to which an aquo cation combines with ligands to form complex ions is a thermodynamic problem and can be treated in terms of appropriate expressions for equilibrium constants.

Suppose we put a metal ion, M, and some monodentate ligand, L, together in solution. Assuming that no insoluble products are formed, nor any species containing more than one metal ion, equilibrium expressions of the following sort will describe the system:

$$\mathrm{M + L = ML} \qquad K_1 = \frac{[\mathrm{ML}]}{[\mathrm{M}][\mathrm{L}]}$$

$$\mathrm{ML + L = ML_2} \qquad K_2 = \frac{[\mathrm{ML_2}]}{[\mathrm{ML}][\mathrm{L}]}$$

$$\mathrm{ML_2 + L = ML_3} \qquad K_3 = \frac{[\mathrm{ML_3}]}{[\mathrm{ML_2}][\mathrm{L}]}$$

$$\vdots \qquad \vdots \qquad \vdots \qquad \vdots$$

$$\mathrm{ML}_{N-1} + \mathrm{L} = \mathrm{ML}_N \qquad K_N = \frac{[\mathrm{ML}_N]}{[\mathrm{ML}_{N-1}][\mathrm{L}]}$$

There will be N such equilibria, where N represents the maximum coordination number of the metal ion M for the ligand L. N may vary from one ligand to another. For instance, Al^{3+} forms $AlCl_4^-$ and AlF_6^{3-}, and Co^{2+} forms $CoCl_4^{2-}$ and $Co(NH_3)_6^{2+}$, as the highest complexes with the ligands indicated.

Another way of expressing the equilibrium relations is the following:

$$\mathrm{M + L = ML} \qquad \beta_1 = \frac{[\mathrm{ML}]}{[\mathrm{M}][\mathrm{L}]}$$

$$\mathrm{M + 2L = ML_2} \qquad \beta_2 = \frac{[\mathrm{ML_2}]}{[\mathrm{M}][\mathrm{L}]^2}$$

$$\mathrm{M + 3L = ML_3} \qquad \beta_3 = \frac{[\mathrm{ML_3}]}{[\mathrm{M}][\mathrm{L}]^3}$$

$$\vdots \qquad \vdots \qquad \vdots \qquad \vdots$$

$$\mathrm{M} + N\mathrm{L} = \mathrm{ML}_N \qquad \beta_N = \frac{[\mathrm{ML}_N]}{[\mathrm{M}][\mathrm{L}]^N}$$

Since there can be only N independent equilibria in such a system, it is clear that the K_i's and the β_i's must be related. The relationship is indeed, rather obvious. Consider, for example, the expression for β_3. Let us multiply both numerator and denominator by [ML][ML_2] and then rearrange slightly:

$$\beta_3 = \frac{[\mathrm{ML_3}]}{[\mathrm{M}][\mathrm{L}]^3} \cdot \frac{[\mathrm{ML}][\mathrm{ML_2}]}{[\mathrm{ML}][\mathrm{ML_2}]}$$

$$= \frac{[\mathrm{ML}]}{[\mathrm{M}][\mathrm{L}]} \cdot \frac{[\mathrm{ML_2}]}{[\mathrm{ML}][\mathrm{L}]} \cdot \frac{[\mathrm{ML_3}]}{[\mathrm{ML_2}][\mathrm{L}]}$$

$$= K_1 K_2 K_3$$

It is not difficult to see that this kind of relationship is perfectly general, namely:

$$\beta_k = K_1 K_2 K_3 \ldots K_k = \prod_{i=1}^{i=k} K_i$$

The K_i's are called the *stepwise formation constants* (or stepwise stability constants), and the β_i's are called the *overall formation constants* (or overall stability constants); each type has its special convenience in certain cases.

The set of stepwise formation constants, K_i's, provide particular insight into the species present as a function of concentrations. With only a few exceptions, there is generally a slowly descending progression in the values of the K_i's in any particular system. This is illustrated by the data for the $Cd^{2+} - NH_3$ system where the ligands are uncharged and by the $Cd^{2+} - CN^-$ system where the ligands are charged.

$$\begin{aligned} Cd^{2+} + NH_3 &= [Cd(NH_3)]^{2+} & K &= 10^{2.65} \\ [Cd(NH_3)]^{2+} + NH_3 &= [Cd(NH_3)_2]^{2+} & K &= 10^{2.10} \\ [Cd(NH_3)_2]^{2+} + NH_3 &= [Cd(NH_3)_3]^{2+} & K &= 10^{1.44} \\ [Cd(NH_3)_3]^{2+} + NH_3 &= [Cd(NH_3)_4]^{2+} & K &= 10^{0.93} (\beta_4 = 10^{7.12}) \end{aligned}$$

$$\begin{aligned} Cd^{2+} + CN^- &= [Cd(CN)]^+ & K &= 10^{5.48} \\ [Cd(CN)]^+ + CN^- &= [Cd(CN)_2] & K &= 10^{5.12} \\ [Cd(CN)_2] + CN^- &= [Cd(CN)_3]^- & K &= 10^{4.63} \\ [Cd(CN)_3]^- + CN^- &= [Cd(CN)_4]^{2-} & K &= 10^{3.55} (\beta_4 = 10^{18.8}) \end{aligned}$$

Thus, typically, as ligand is added to the solution of metal ion, ML first forms more rapidly than any other complexes in the series. As addition of ligand is continued, the ML_2 concentration rises rapidly, while the ML concentration drops, then ML_3 becomes dominant, ML and ML_2 becoming unimportant, and so forth, until the highest complex. ML_N, is formed to the nearly complete exclusion of all others at very high ligand concentrations. From the formation constants given above for the Cd—CN system, graphical pictures of these relations may be drawn as shown in *Fig. 6-4.*

A steady decrease in K_i as i increases is almost always observed, although occasional exceptions occur because of unusual steric or electronic affects. The principal reason for the decrease is statistical. At any given step, say from ML_n to ML_{n+1}, there is a certain probability for the complexes ML_n to gain another ligand and another for ML_{n+1} to lose a ligand. Obviously, as n increases there are more ligands to be lost and fewer places $(N - n)$ in the coordination shell to accept additional ligands. For a series of steps ML to $ML_2, \ldots ML_5$ to ML_6, the magnitude of $\log K_i$ tends to decrease by about 0.5 at each step for statistical reasons alone.

Many methods of chemical analysis and separation are based on the formation of complexes in solution; accuracy and reliability depend on knowing the values of such constants. For example, different transition metal ions can be selectively determined by adjusting the concentration of EDTA (page 132) and the pH so that one such ion is largely uncomplexed while another is tightly bound. Again, the drastically different abilities of Ni^{2+} and Co^{2+} to form anionic chloro

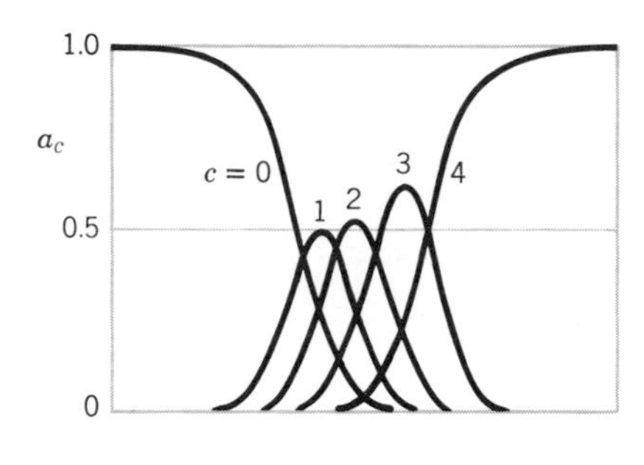

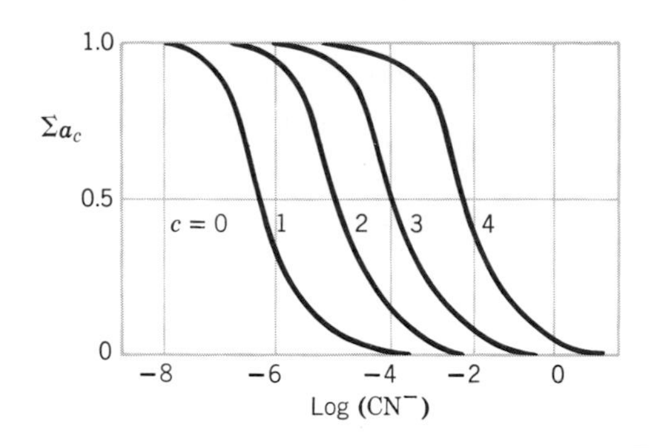

Figure 6-4 *Plots of the proportions of the various complexes* $[Cd(CN)_c]^{(2-c)+}$ *as a function of the ligand concentrations:*

$$\alpha_c = [Cd(CN)_c]/\text{total Cd} \qquad \sum \alpha_c = \sum_{c=0}^{4} [Cd(CN)_c]$$

(*Reproduced by permission from F. J. C. Rossotti in J. Lewis and R. G. Wilkins, eds.,* Modern Coordination Chemistry, *Interscience, New York–London, 1960, p. 10.*)

complexes, MCl_3^- and MCl_4^{2-}, allow a clean separation by using ion exchange columns. As is indicated in Chapter 31, the presence of metal ions in living systems, and their behavior there is bound up with the formation of complexes.

6-7
The Chelate Effect

As a general rule, a complex containing one (or more) 5- or 6-membered chelate rings is more stable (has a higher formation constant) than a complex that is as similar as possible but lacks some or all of the chelate rings. A typical illustration is:

$$Ni^{2+}(aq) + 6NH_3(aq) = [Ni(NH_3)_6]^{2+}(aq) \qquad K = 10^{8.6}$$

$$Ni^{2+}(aq) + 3H_2NCH_2CH_2NH_2(aq) = [Ni(H_2NCH_2CH_2NH_2)_3]^{2+}(aq) \qquad K = 10^{18.3}$$

The complex with three chelate rings is about 10^{10} times more stable. Why should this be true? As with all questions concerning thermodynamic stability, we are dealing with free energy changes, ΔG°, and we first look at the contributions of enthalpy and entropy, to see if one or the other is the main cause of the difference.

We can more directly compare the two reactions above by combining them in the equation:

$$Ni(NH_3)_6{}^{2+}(aq) + 3\,en(aq) = Nien_3{}^{2+}(aq) + 6\,NH_3(aq)$$

(en = ethylenediamine)

for which

$$K = 10^{9.7}$$
$$\Delta G^\circ = -RT \ln K = -67 \text{ kJ mol}^{-1} = \Delta H^\circ - T\,\Delta S^\circ$$
$$\Delta H^\circ = -12 \text{ kJ mol}^{-1}$$
$$-T\,\Delta S^\circ = -55 \text{ kJ mol}^{-1}$$

It is evident that both enthalpy and entropy favor the chelate complex, but the entropy contribution is far more important. Data for a large number of these reactions, with many different metal ions and ligands, show that enthalpy contributions to the chelate effect are sometimes favorable, sometime unfavourable, but always relatively small. The general conclusion is that *the chelate effect is essentially an entropy effect*. The reason for this is as follows.

The nickel ion is coordinated by six water molecules. In each of the first two reactions, these six H_2O are liberated when the nitrogen ligands become coordinated. On that score, the two processes are equivalent. However, in one case *six* NH_3 molecules *lose* their freedom at the same time, and there is no net change in the number of particles. In the other case, only *three* en molecules lose their freedom, and thus there is a net increase of three moles of individual molecules. The reaction with 3 en causes a much greater increase in disorder than does that with $6NH_3$ and, therefore, ΔS° is more positive (more favorable) in the former case than in the latter. It is easy to see that this reasoning is general for all such comparisons of a chelate with a nonchelate process.

Another way to state the matter is to visualize a chelate ligand with one donor atom attached to a metal ion. The other donor atom cannot then get very far away, and the probability of it, too, becoming attached is greater than if it were in an entirely independent molecule, with access to the entire volume of solution. This point of view readily explains why the chelate effect peters out as ring size increases. The effect is greatest for five- and six-membered rings, becomes marginal for seven-membered ones, and is unimportant thereafter. When the ring to be formed is large, the probability of the second donor atom attaching itself promptly to the same metal atom is no longer large as compared with its encountering a different metal atom, or to the dissociation of the first donor atom before the second one makes effective contact.

6-8
Reactions of Complexes

Virtually all of transition metal chemistry and a great deal of the rest of inorganic chemistry could be included under the above title, taken in its broadest sense. Only two aspects will be covered in this and following sections: ligand displacement and exchange reactions in solution, and electron transfer reactions. Additional aspects of the subject, with particular regard to the metal carbonyl and organometallic classes of complex are discussed in Chapters 28 to 30.

The ability of a complex ion to engage in reactions that result in replacing one or more ligands in its coordination sphere by others is called its *lability*. Those complexes for which reactions of this type are very rapid are called *labile*, whereas those for which such reactions proceed only slowly or not at all are called *inert*. We emphasize that these two terms refer to rates of reactions and should not be confused with the terms stable and unstable, which refer to the thermodynamic tendency of species to exist under equilibrium conditions. A simple example of this distinction is provided by the $[Co(NH_3)_6]^{3+}$ ion which will persist for days in an acid medium because of its kinetic inertness or lack of lability despite the fact that it is thermodynamically unstable, as the following equilibrium constant shows:

$$[Co(NH_3)_6]^{3+} + 6H_3O^+ = [Co(H_2O)_6]^{3+} + 6NH_4^+ \qquad K \sim 10^{25}$$

In contrast, the stability of $Ni(CN)_4^{2-}$ is extremely high,

$$[Ni(CN)_4]^{2-} = Ni^{2+} + 4CN^- \qquad K \sim 10^{-22}$$

but the rate of exchange of CN^- ions with isotopically labeled CN^- added to the solution is immeasurably fast by ordinary techniques. Of course, this lack of any necessary relation between thermodynamic stability and kinetic lability is to be found generally in chemistry, but its appreciation here is especially important.

Taube has proposed a practical definition of the terms labile and inert. Inert complexes are those whose ligand-replacement reactions are slow enough (half-times of a minute or more) to be studied by "classical" kinetic experiments, namely, those in which reagents are poured together in a vessel and changes in optical density, pH, gas evolution, and the like, are followed directly by the observer. Labile complexes are those that react so fast that "modern" methods, namely, flow systems, special rapid mixing devices, rapidly responding electronic recorders, and even so-called relaxation methods, must be used to follow them.

In the first transition series, virtually all octahedral complexes save those of Cr^{III} and Co^{III} are normally labile; that is, ordinary complexes come to equilibrium with additional ligands, including H_2O, so rapidly that the reactions appear instantaneous by ordinary techniques of kinetic measurement. Complexes of Cr^{III} and Co^{III} ordinarily undergo ligand replacement reactions with half-times of the order of hours, days, or even weeks at 25°.

Two extreme mechanistic possibilities may be considered for ligand-replacement reactions. First, there is the S_N1 mechanism, in which the complex dissociates, losing the ligand to be replaced, the vacancy in the coordination shell then being taken by the new ligand. This path may be represented as follows:

$$[L_5MX]^{n+} \xrightarrow{\text{Slow}} X^- + \underset{\substack{\text{Five-coordinated} \\ \text{intermediate}}}{[L_5M]^{(n+1)+}} \xrightarrow[\text{Fast}]{+Y^-} [L_5MY]^{n+}$$

The important feature here is that the first step, in which X^- is lost, proceeds *relatively* slowly, and thus determines the rate at which the complete process can proceed. In other words, once it is formed, the intermediate complex, which is only 5-coordinated, will react with the new ligand, Y^-, almost instantly. The rate law for such a process is

$$v = k[L_5MX] \tag{6-1}$$

When this mechanism is operative, the rate of the reaction necessarily is directly proportional to the concentration of $[L_5MX]^{n+}$ but is independent of the concentration of the new ligand, Y^-. The symbol S_N1 stands for *substitution, nucleophilic, unimolecular*.

The other extreme pathway for a ligand exchange is the S_N2 mechanism. In this case the new ligand attacks the original complex directly to form a 7-coordinated activated complex which then ejects the displaced ligand, as is indicated in this scheme:

$$[L_5MX]^{n+} + Y^- \xrightarrow{\text{Slow}} \left[L_5M\begin{matrix} \diagup X \\ \diagdown Y \end{matrix} \right]^{(n-1)+} \xrightarrow{\text{Fast}} [L_5MY]^{n+} + X^-$$

When this mechanism is operative, the rate of the reaction will be proportional to the concentration of $[L_5MX^n]^+$ times that of Y^-, the rate law being

$$v = k[L_5MX][Y^-] \tag{6-2}$$

The symbol S_N2 stands for *substitution, nucleophilic, bimolecular*.

Unfortunately, these two extreme mechanisms are just that—extremes—and real mechanisms are seldom as simple. It is more realistic to recognize that it is likely that some degree of bond formation will occur before bond breaking is complete, that is, the transition state may not be either the truly 5-coordinate species or the one in which the *leaving* and *entering* groups are both strongly bound at once. Subsequently, we shall use the terms S_N1 and S_N2 not to imply necessarily the extremes, but to describe mechanisms that may only approximate to these extremes.

To complicate matters further, a rate law of type (6-1) or (6-2) does not prove that the reaction proceeds by an S_N1 or S_N2 mechanism, even approximately. The three most important cases in illustration of this are (1) solvent intervention, (2) ion-pair formation, and (3) conjugate-base formation.

1. *Solvent Intervention.* Most reactions of complexes have been studied in water, which is itself a ligand and which is present in high and effectively constant concentration ($\sim 55.5\ M$). Thus, the rate law (6-1) might be observed even if the actual course of the reaction were

$$[L_5MX] + H_2O \longrightarrow [L_5MH_2O] + X \qquad \text{Slow} \tag{6-3a}$$

$$[L_5MH_2O] + Y \longrightarrow [L_5MY] + H_2O \qquad \text{Fast} \tag{6-3b}$$

Moreover, we should not be able to tell from the rate law alone whether either (6-3a) or (6-3b) proceeded by S_N1 or S_N2 type processes.

2. *Ion-pair Formation.* When the reacting complex is a cation and the entering group is an anion, especially when one or both have high charges, ion pairs (or *outer sphere complexes*, as they are also called) will form to some extent,

$$[L_5MX]^{n+} + Y^{m-} = \{[L_5MX]Y\}^{n-m} \tag{6-4}$$

with an equilibrium constant K. These equilibrium constants are generally in the range of 0.1 to 20. If the only path by which $[L_5MX]^{n+}$ and Y^{m-} can react with significant velocity involves preliminary formation of the ion pair, then the rate law might be

$$v = k'K[L_5MX][Y] = k''[L_5MX][Y] \tag{6-5}$$

which has the same form as the rate law (6-2). Only by additional experiments can we learn whether the reaction is truly an S_N2 type in the sense of Eq. 6-2 or whether the ion pair is involved. Even if it can be shown that the ion pair is involved (and this is frequently determinable), the question of how the ion pair transforms itself into the products remains unanswered, because S_N1, S_N2 or solvent participation processes might all occur.

3. *Conjugate-base Formation.* Whenever a rate law involving $[OH^-]$ is found, there is the question whether OH^- actually attacks the metal giving an S_N2 reaction in the sense of Eq. 6-2, or whether it appears in the rate law because it first reacts rapidly to remove a proton from a ligand, forming the conjugate base (CB), which then reacts, as in the following sequence:

$$[Co(NH_3)_5Cl]^{2+} + OH^- = [Co(NH_3)_4(NH_2)Cl]^+ + H_2O \qquad \text{Fast} \tag{6-6a}$$

$$[Co(NH_3)_4(NH_2)Cl]^+ \xrightarrow[\text{then } H^+]{+Y^-} [Co(NH_3)_5Y]^{2+} + Cl^- \qquad \text{Slow} \tag{6-6b}$$

In cases where (6-6b) proceeds by an S_N1 mechanism, the overall mechanism represented by (6-6a) and (6-6b) is called an S_N1CB mechanism. Of course, where there is no protonic hydrogen atom available [or if it is known that a process like (6-6a) is too slow] the appearance of $[OH]^-$ in the rate law probably does indicate an authentic S_N2 process.

6-9 Water Exchange and Formation of Complexes from Aquo Ions

Since most reactions in which complexes are formed occur in aqueous solution, one of the most basic reactions to be studied and understood is that in which the water molecules surrounding a cation in aqueous solution are removed from the coordination shell and are replaced by other ligand atoms. Included here is the case in which the new ligand is simply another water molecule, the water-exchange reaction.

With but few exceptions, for example $Cr(H_2O)_6{}^{3+}$, $Rh(H_2O)_6{}^{3+}$, these reactions are very fast and must be studied by relaxation methods. A survey of the results is shown in *Fig.* 6-5. Although all those shown are "fast," a range of 10^{10} in

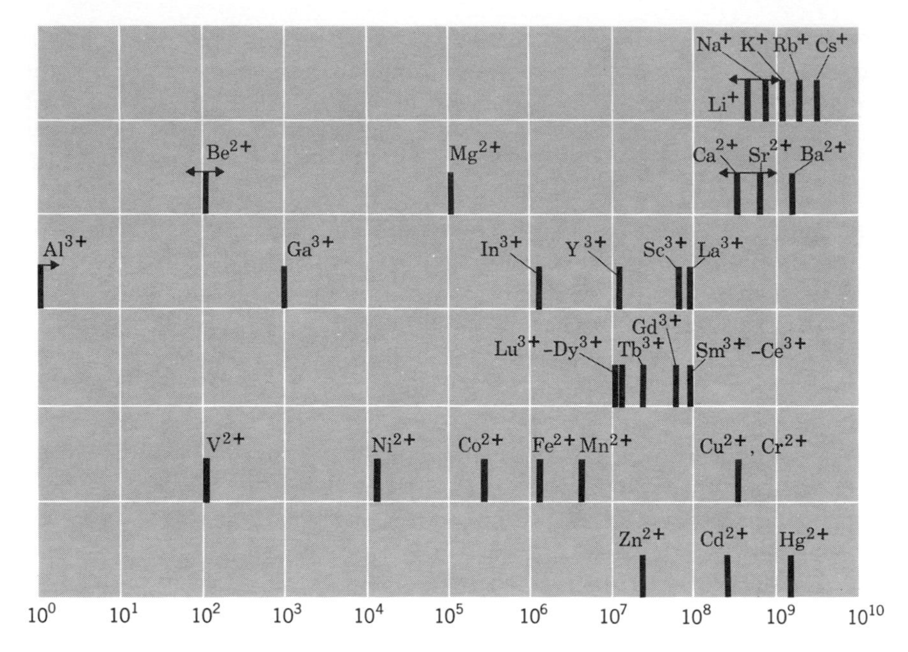

Figure 6-5 *Characteristic rate constants* (sec^{-1}) *for substitutions of inner sphere* H_2O *of various aqua ions. [Adapted from M. Eigen,* Pure Appl. Chem., *1963*, 6, *105, with revised data kindly provided by M. Eigen. See also H. P. Bennetto and E. F. Caldin,* J. Chem. Soc. A, *1971, 2198.]*

velocity is covered. The reasons for the variations are only partly understood. One relationship which is often evident is that in a series such as Zn^{2+}, Cd^{2+}, Hg^{2+}, the alkali ions or the alkaline earth ions, the rate increases with ionic size. This is because, with constant charge and coordination number, the strength of attachment of the H_2O molecules decreases as the ion gets bigger, and the H_2O molecules can more easily dissociate.

Two remarkably general observations have been made concerning the rates at which aquo ions combine with other ligands to form complexes:

1. The rates for a given ion show little or no dependence (less than a factor of 10) on the identity of the ligand.

2. The rates for each ion are practically the same as the rate of water exchange for that ion, usually ~ 10 times slower.

The most reasonable explanation for these observations is that the overall reaction, with an apparent rate constant k_{obs}, occurs in the following three steps:

$$[M(H_2O)_6]^{n+} + X^- \overset{K}{=} \{[M(H_2O)_6]X\}^{(n-1)+}$$

$$\{[M(H_2O)_6]X\}^{(n-1)+} \xrightarrow{k_{aq}} \{[M(H_2O)_5]X\}^{(n-1)+} + H_2O$$

$$\{[M(H_2O)_5]X\}^{(n-1)+} \xrightarrow[\text{Fast}]{\text{Very}} [M(H_2O)_5X]^{(n-1)+}$$

In the first step an ion pair (also called an outer-sphere complex) is formed, with an equilibrium constant K (usually ≈ 1). A coordinated water molecule is then lost from this species with much the same rate constant, k_{aq}, as for the simple hexaaqua ion itself. In the third step, which is very fast, and may not be distinct from the second step, the ligand X^- slips into the coordination sphere. The rate of the complete reaction, v, is then given by:

$$\begin{aligned} v &= k_{obs}[M(H_2O)_6{}^{n+}][X^-] \\ &= k_{aq}[M(H_2O)_6X^{(n-1)+}] \\ &= k_{aq}K[M(H_2O)_6{}^{n+}][X^-] \end{aligned}$$

Thus, the observed rate constant, k_{obs}, is equal to Kk_{aq}. For a given metal ion and the common anions, X^-, the values of K will not vary a great deal, in accord with observation.

6-10
Ligand-displacement Reactions in Octahedral Complexes

A general equation for a ligand-displacement reaction is

$$[L_nMX] + Y = [L_nMY] + X$$

In aqueous solution the special case in which Y is H_2O (or OH^-) is of overwhelming importance. It appears that there are few, if any, reactions in which X is not first replaced by H_2O, and only then does the other ligand, Y, enter the complex by displacing H_2O. Thus our discussion will be restricted almost entirely to the subject of the *aquation or hydrolysis* reaction.

The rates of hydrolyses of cobalt(III) ammine complexes are pH-dependent and generally follow the rate law

$$v = k_A[L_5CoX] + k_B[L_5CoX][OH^-] \tag{6-7}$$

In general, k_B (for *base hydrolysis*) is some 10^4 times k_A (for *acid hydrolysis*).

Acid Hydrolysis. We consider first the term $k_A[L_5CoX]$. Since the entering ligand is H_2O, which is present in high (~ 55.5 *M*) and effectively constant concentration, the rate law tells us *nothing* as to the order in H_2O; the means for deciding whether this is an associative (S_N2) or a dissociative (S_N1) process must be sought elsewhere.

It appears that the majority of substitution reactions in octahedral complexes proceed by an essentially dissociative (S_N1) pathway, but there may be some significant exceptions. The types of experiments that can shed light on this question are both subtle and not entirely conclusive. We do not discuss them explicitly here.

Base Hydrolysis. The interpretation of a term of the type $k_B[ML_5X][OH^-]$ in a rate law for base hydrolysis has long been disputed. It could, of course, be interpreted as representing a genuine S_N2 process, OH^- making a nucleophilic attack. However, the possibility of an S_N1CB mechanism, discussed above, must also be considered. There are arguments on both sides, and it is of course possible that the mechanism may vary for different complexes. Studies of base hydrolysis in octahedral complexes have thus far dealt mainly with those of Co^{III}, and it is now reasonably sure that for these the predominant mechanism is, indeed, S_N1CB.

Base hydrolysis of Co^{III} complexes is generally very much faster than acid hydrolysis, that is, $k_B \gg k_A$ in Eq. 6-7. This, in itself, provides evidence against a simple S_N2 mechanism, and therefore in favor of the S_N1CB mechanism because there is no reason to expect OH^- to be uniquely capable of electrophilic attack on the metal. In the reactions of square complexes it turns out to be a distinctly inferior nucleophile toward Pt^{II}.

The S_N1CB mechanism, of course, requires that the reacting complex have, at least, one protonic hydrogen atom on a nonleaving ligand, and that the rate of reaction of this hydrogen be fast compared with the rate of ligand displacement. It has been found that the rates of proton exchange in many complexes subject to rapid base hydrolysis are, in fact, some 10^5 times faster than the hydrolysis itself [eg., in $Co(NH_3)_5Cl^{2+}$ and $Coen_2NH_3Cl^{2+}$.] Such observations are in keeping with the S_N1CB mechanism but afford no positive proof of it.

If the conjugate base mechanism is indeed correct, there is the question of why the conjugate base so readily dissociates to release the ligand X. In view of the very low acidity of coordinated amines, the concentration of the conjugate base is a very small fraction of the total concentration of the complex. Thus, its reactivity is enormously greater, by a factor far in excess of the mere ratio of k_B/k_A. It can be estimated that the ratio of the rates of aquation of $[Co(NH_3)_4NH_2Cl]^+$ and $[Co(NH_3)_5Cl]^{2+}$ must be greater than 10^6. Two features of the conjugate base have been considered in efforts to account for this reactivity. First, there is the obvious charge effect. The conjugate base has a charge that is one unit less positive than the complex from which it is derived. Although it is difficult to construct a rigorous argument, it seems entirely unlikely that the charge effect, in itself, can account for the enormous rate difference involved. It has been proposed that the amide ligand could labilize the leaving group X by a combination of electron repulsion in the ground state and a π-bonding contribution to the stability of the 5-coordinate intermediate, as is suggested in *Fig. 6-6*.

$$H_2\ddot{N}-Co-X \longrightarrow H_2N{=}Co + :X^-$$

Figure 6-6 *A sketch showing how an amide group could promote the dissociation of another ligand,* X.

However, there are observations that some workers consider to contradict this explanation also and, consequently, the question of why the conjugate base is hyperreactive remains unsettled.

Electrophilic Attack on Ligands. There are some reactions known where ligand exchange does not involve the breaking of metal-ligand bonds, but instead bonds within the ligands themselves are broken and re-formed. One well-known case is the aquation of carbonato complexes. When isotopically labeled water, $H_2{*}O$, is used, it is found that no *O gets into the coordination sphere of the ion during aquation,

$$[Co(NH_3)_5OCO_2]^+ + 2H_3{*}O^+ \qquad [Co(NH_3)_5(H_2O)]^{3+} + 2H_2{*}O + CO_2$$

The most likely path for this reaction involves proton attack on the oxygen atom bonded to Co followed by expulsion of CO_2 and then protonation of the hydroxo complex (Eq. 6-8). Similarly, in the reaction of NO_2^- with

$$\left\{ \begin{array}{c} Co(NH_3)_5{-}O\cdots C(=O){-}O \\ \vdots \\ H^+ \\ \vdots \\ O(H)H \\ \text{Transition state} \end{array} \right\} \longrightarrow [Co(NH_3)_5{-}O(H)]^{2+} \xrightarrow{+H^+} [Co(NH_3)_5(H_2O)]^{3+} \qquad (6\text{-}8)$$

pentaammineaquocobalt(III) ion, isotopic labeling studies show that the oxygen originally in the bound H_2O turns up in the bound NO_2^-. This remarkable result is explained by the reaction sequence (Eq. 6-9):

$$2NO_2^- + 2H^+ = N_2O_3 + H_2O$$

$$[Co(NH_3)_5{*}OH]^{2+} + N_2O_3 \longrightarrow \left\{ \begin{array}{c} (NH_3)_5CO{-}{*}O\cdots H \\ \vdots \qquad \vdots \\ ON\cdots ONO \end{array} \right\}_{\text{Transition state}} \xrightarrow{\text{Fast}} HNO_2 + [Co(NH_3)_5{*}ONO]^{2+} \xrightarrow{\text{Slow}} [Co(NH_3)_5(NO{*}O)]^{2+} \qquad (6\text{-}9)$$

6-11
Ligand-displacement Reactions in Square Complexes

For square complexes, the mechanistic problem is more straightforward and, hence is better understood. One might expect that 4-coordinate complexes would be more likely than octahedral ones to react by an S_N2 mechanism, and extensive studies of Pt^{II} complexes have shown that this is true.

For reactions in aqueous solution of the type in Eq. 6-10 the rate law takes the general form of Eq. 6-11. It is believed that the second term corresponds to

$$PtL_nCl_{4-n} + Y = PtL_nCl_{3-n}Y + Cl^- \qquad (6\text{-}10)$$

$$v = k[PtL_nCl_{4-n}] + k'[PtL_nCl_{4-n}][Y] \qquad (6\text{-}11)$$

a genuine S_N2 reaction of Y with the complex, while the first term represents a two-step path in which one Cl^- is first replaced by H_2O (probably also by an S_N2

mechanism) as the rate-determining step followed by relatively fast replacement of H_2O by Y.

It has been found that the rates of reaction (Eq. 6-10) for the series of four complexes in which L = NH_3 and Y = H_2O vary by only a factor of 2. This is a remarkably small variation, since the charge on the complex changes from -2 to $+1$ as n goes from 0 to 3. Since Pt—Cl bond breaking should become more difficult in this series, while the attraction of Pt for a nucleophile should increase in the same order, the virtual constancy in the rate argues for an S_N2 process in which both Pt—Cl bond breaking and $Pt \cdots OH_2$ bond formation are of comparable importance.

A general representation of the stereochemical course of displacement reactions of square complexes is given in *Fig. 6-7*. It should be carefully noted that this process is entirely stereospecific: *cis* and *trans* starting materials lead, respectively, to *cis* and *trans* products. Whether any of the three intermediate configurations possess enough stability to be regarded as actual intermediates rather than merely phases of the activated complex remains uncertain.

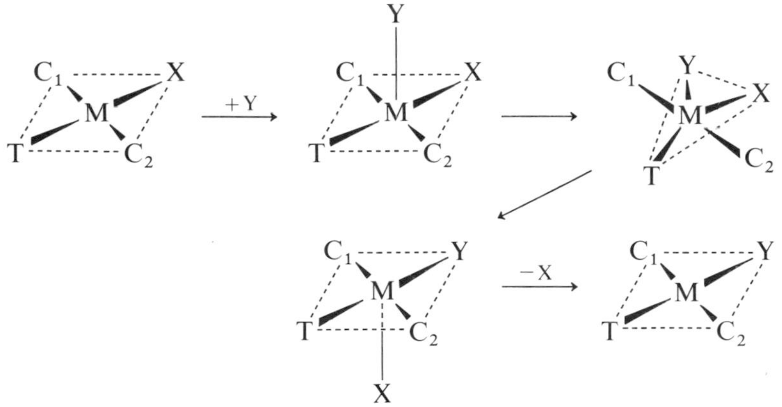

Figure 6-7 *The course of ligand displacement at a planar complex and the trigonal bipyramidal five-coordinate structure of the transition state.*

Although the evidence is less than complete, it appears likely that the S_N2 mechanism is also valid for the reactions of square complexes other than those of Pt^{II}, like those of Ni^{II}, Pd^{II}, Rh^{I}, Ir^{I}, and Au^{III}.

The order of nucleophilic strength of entering ligands, i.e., the order of the rate constants k' in Equation 6-11 for substitution reactions on Pt^{II}, is

$$F^- \sim H_2O \sim OH^- < Cl^- < Br^- \sim NH_3 \sim \text{olefins} < C_6H_5NH_2 < C_5H_5N < NO_2^- < N_3^- < I^- \sim SCN^- \sim R_3P$$

The trans Effect. This is a particular feature of ligand-replacement reactions in square complexes which is of less importance in reactions of octahedral complexes except in some special cases.

Consider the general reaction (6-12):

$$[PtLX_3]^- + Y^- \qquad [PtLX_2Y]^- + X^- \tag{6-12}$$

Sterically, there are two possible reaction products, with *cis* and *trans* orientation of Y with respect to L. It has been observed that the relative proportions of the *cis* and *trans* products vary appreciably with the ligand L. Moreover, in reactions of the type (6-13) either or both of the indicated isomers may be produced. It

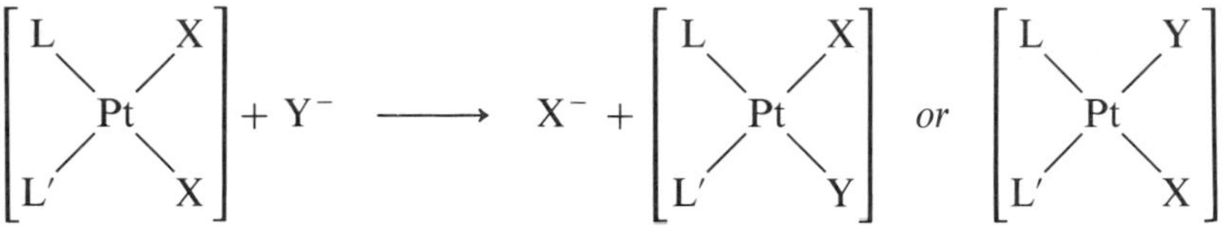

$$\left[\begin{matrix} L & & X \\ & Pt & \\ L' & & X \end{matrix}\right] + Y^- \longrightarrow X^- + \left[\begin{matrix} L & & X \\ & Pt & \\ L' & & Y \end{matrix}\right] \quad or \quad \left[\begin{matrix} L & & Y \\ & Pt & \\ L' & & X \end{matrix}\right] \qquad (6\text{-}13)$$

is found that, both in these types of reaction and in others, a fairly extensive series of ligands may be arranged in the same order with respect to their ability to facilitate substitution in the position *trans* to themselves. This phenomenon is known as the *trans effect*. The approximate order of increasing *trans* influence is:

$$H_2O, OH^-, NH_3, py < Cl^-, Br^- < -SCN^-, I^-, NO_2^-,$$
$$C_6H_5^- < SC(NH_2)_2,\ CH_3^- < H^-, PR_3 < C_2H_4, CN^-, CO$$

It is to be emphasized that the *trans* effect is here defined solely as a kinetic phenomenon. It is the effect of a coordinated group on the rate of substitution at the position *trans* to itself in a square or octahedral complex.

The *trans* effect series has proved very useful in rationalizing known synthetic procedures and in devising new ones. As an example, we consider the synthesis of the *cis* and *trans* isomers of $[Pt(NH_3)_2Cl_2]$. The synthesis of the *cis* isomer is accomplished by treatment of the $[PtCl_4]^{2-}$ ion with ammonia (reaction 6-14). Since Cl^- has a greater *trans* directing influence than does NH_3, substitution of

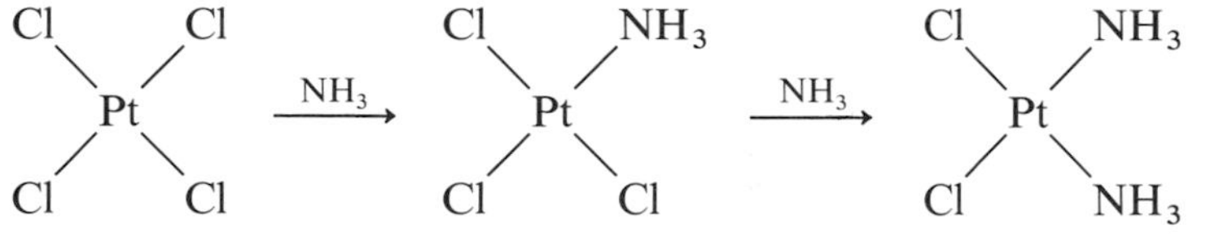

$$\begin{matrix} Cl & & Cl \\ & Pt & \\ Cl & & Cl \end{matrix} \xrightarrow{NH_3} \begin{matrix} Cl & & NH_3 \\ & Pt & \\ Cl & & Cl \end{matrix} \xrightarrow{NH_3} \begin{matrix} Cl & & NH_3 \\ & Pt & \\ Cl & & NH_3 \end{matrix} \qquad (6\text{-}14)$$

NH_3 into $[Pt(NH_3)Cl_3]^-$ is least likely to occur in the position *trans* to the NH_3 already present and, thus, the *cis* isomer is favored. The *trans* isomer is made by treating $[Pt(NH_3)_4]^{2+}$ with Cl^- (reaction 6.15). Here the superior *trans* directing influence of Cl^- causes the second Cl^- to enter *trans* to the first one, producing *trans*-$[Pt(NH_3)_2Cl_2]$.

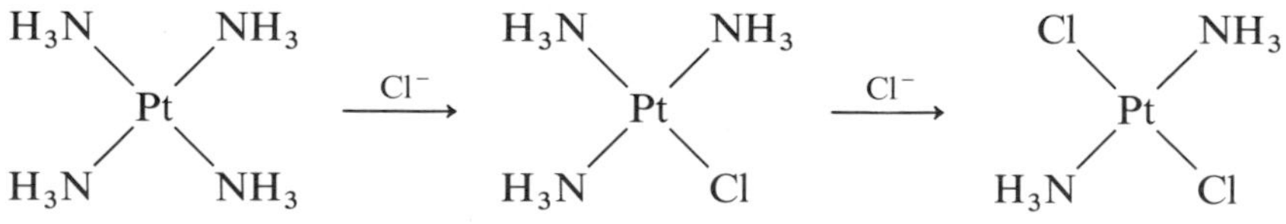

$$\begin{matrix} H_3N & & NH_3 \\ & Pt & \\ H_3N & & NH_3 \end{matrix} \xrightarrow{Cl^-} \begin{matrix} H_3N & & NH_3 \\ & Pt & \\ H_3N & & Cl \end{matrix} \xrightarrow{Cl^-} \begin{matrix} Cl & & NH_3 \\ & Pt & \\ H_3N & & Cl \end{matrix} \qquad (6\text{-}15)$$

All theorizing about the *trans* effect must recognize the fact that since it is a kinetic phenomenon, depending on activation energies, the stabilities of *both* the ground state and the activated complex are relevant. The activation energy can be affected by changes in one or the other of these energies or by changes in both.

The earliest attempt to explain the *trans* effect was the so-called polarization theory of Grinberg, which is primarily concerned with effects in the ground state. This theory deals with a postulated charge distribution as is shown in *Fig. 6-8*. The primary charge on the metal ion induces a dipole in the ligand L which in turn induces a dipole in the metal. The orientation of this dipole on the metal is such as to repel negative charge in the *trans* ligand X. Hence, X is less attracted by the metal atom because of the presence of L. This theory would lead to the expectation that the magnitude of the *trans* effect of L and its polarizability should be monotonically related, and for some ligands in the *trans* effect series, for example, H^-, $I^- > Cl^-$, such a correlation is observed. In effect, this theory says that the *trans* effect is attributable to a ground-state weakening of the bond to the ligand that is to be displaced.

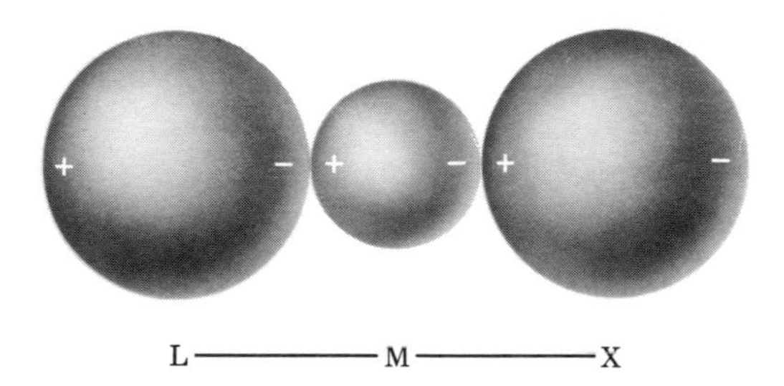

Figure 6-8 *Arrangement of dipoles according to the polarization theory of the trans effect.*

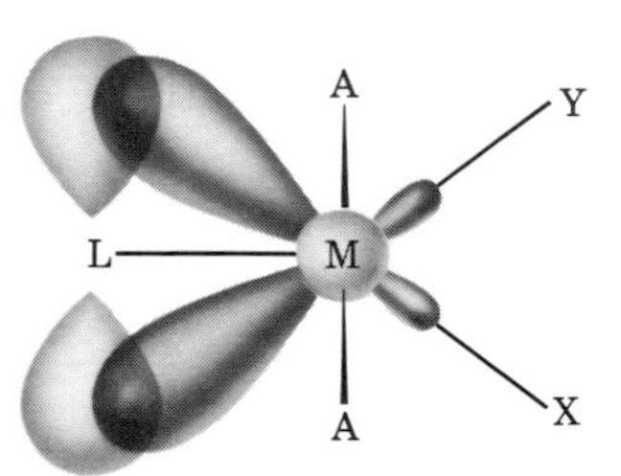

Figure 6-9 *Postulated* **tbp** *5-coordinate activated complex for reaction of* Y *with trans*-MA_2LX, *to displace* X.

An alternative theory of the *trans* effect was developed with special reference to the activity of ligands such as phosphines, CO, and olefins, which are known to be strong π acids (see Chapter 28 for further details). This model attributes their effectiveness primarily to their ability to stabilize a 5-coordinate transition state or intermediate. This model is, of course, only relevant if the reactions are bimolecular; there is good evidence that this is true in the vast majority of, if not all cases. *Figure 6-9* shows how the ability of a ligand to withdraw metal $d\pi$ electron density into its own empty π or π^* orbitals could enhance the stability of a species in which both the incoming ligand, Y, and the outgoing ligand, X, are simultaneously bound to the metal atom.

Very recently, evidence has been presented to show that even in cases where stabilization of a 5-coordinate activated complex may be important, there is still a ground-state effect—a weakening and polarization of the *trans* bond. In the anion

$C_2H_4PtCl_3^-$ the Pt—Cl bond *trans* to ethylene is slightly longer than the *cis* ones, the Pt—*trans*-Cl stretching frequency is lower than the average of the two Pt—*cis*-Cl frequencies, and there is evidence that the *trans*-Cl atom is more ionically bonded.

The present consensus of opinion among workers in the field appears to be that, in each case, over the entire series of ligands whose *trans* effect has been studied both the ground-state bond weakening and the activated-state stabilizing roles may be involved to some extent. For a hydride ion or a methyl group it is probable that we have the extreme of pure, ground-state bond weakening. With the olefins the ground-state effect may play a secondary role compared with activated-state stabilization, although the relative importance of the two effects in such instances remains a subject for speculation, and further studies are needed.

Racemization of Tris-chelate Complexes. As we stated previously, these exist in enantiomeric configurations, Λ and Δ (*Fig. 6-3*). At various rates, depending on the metal ion involved and the experimental conditions, these can interconvert, and a sample consisting entirely of one enantiomer will eventually racemize, that is, become a mixture of both in equal quantities. Possible pathways for racemization fall into two broad classes: (1) those without breaking of metal-ligand bonds, and (2) those with bond rupture.

Two possible pathways without bond rupture that have long been recognized are the trigonal (or Bailar) twist and the rhombic (Ray–Dutt) twist, shown as (*a*) and (*b*) in *Fig. 6-10*. Many dissociative (bond-rupture) type pathways may be imagined; one is shown as (*c*) in *Fig. 6-10*. It appears that racemization most often

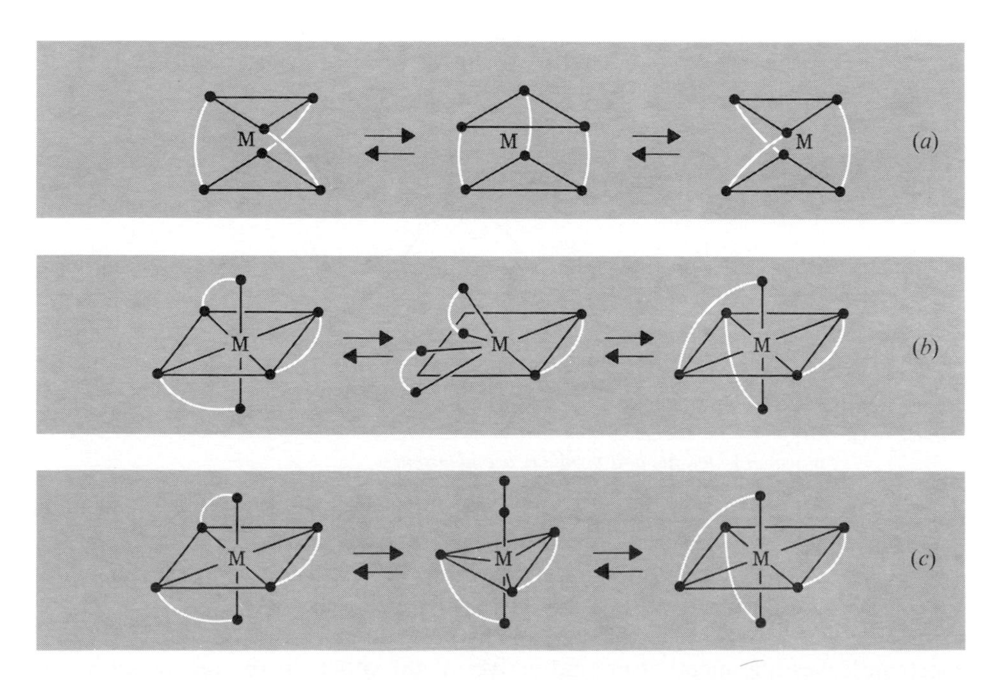

Figure 6-10 *Three possible pathways for racemization of a tris-chelate complex.* (*a*) *The trigonal twist.* (*b*) *The rhombic twist.* (*c*) *One of many pathways involving bond rupture*

occurs via some pathway with bond rupture, although in a few cases there is evidence for the trigonal twist. It appears that the trigonal twist is favored when the bidentate ligands have a short bite (donor atoms close together), since this leads to a distorted structure that can more easily twist. We discuss this further in Section 6-13.

6-12
Electron Transfer Reactions

These are reactions in which two complexes come together and an electron passes from one to the other. In some instances there is an accompanying or immediately following change in the coordination shell of one or both, but not necessarily. Usually, the two complexes are such that the reaction involves net chemical change. Such a reaction is called a redox (oxidation-reduction) reaction, for example

$$Fe^{2+}(aq) + Ce^{4+}(aq) = Fe^{3+}(aq) + Ce^{3+}(aq)$$

However, there are cases in which there is no net chemical change, for example

$$^*Fe(CN)_6^{2-} + Fe(CN)_6^{3-} = {}^*Fe(CN)_6^{3-} + Fe(CN)_6^{2-}$$

These are called *electron-exchange* reactions. They can only be followed, of course, by using isotopic tracers or certain magnetic resonance techniques. They are of interest, however, precisely because the free energy of the system undergoes no net change and the curve of energy versus the reaction coordinate (*Fig. 6-11*) is symmetrical.

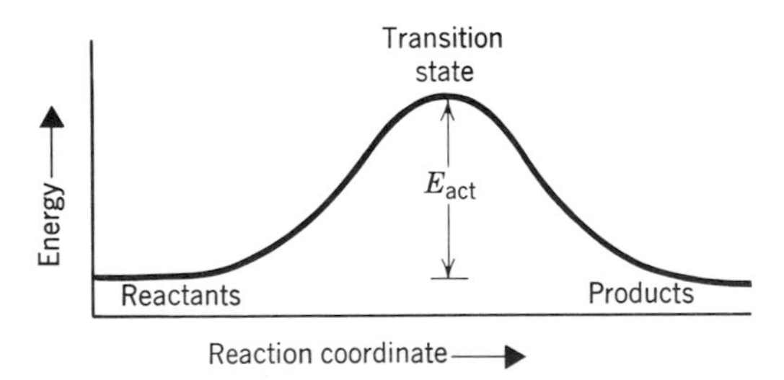

Figure 6-11 *Graph of the energy versus reaction coordinate for an electron-exchange reaction in which reactants and products are identical.*

There are two well-established general mechanisms for electron-transfer processes. In the first, called the *outer-sphere mechanism*, each complex retains its own full coordination shell, and the electron must pass through both. This, of course, is a purely formal statement in that we do not imply that the "same" electron leaves one metal atom and arrives at the other. In the second case, the *inner-sphere mechanism*, the two complexes form an intermediate in which at least one ligand is shared, that is, belongs simultaneously to both coordination shells.

The Outer-Sphere Mechanism. This mechanism is certain to be the correct one when both species participating in the reaction undergo ligand exchange reactions more slowly than they participate in the electron-transfer process. An example is the reaction

$$[Fe^{II}(CN)_6]^{4-} + [Ir^{IV}Cl_6]^{2-} \rightleftharpoons [Fe^{III}(CN)_6]^{3-} + [Ir^{III}Cl_6]^{3-}$$

where both reactants are classified as inert ($t_{1/2}$ for aquation of 0.1 M solution > 1 ms) but the redox reaction has a rate constant of $\sim 10^5$ l mol^{-1} sec^{-1} at 25°.

Table 6-1 lists some electron-exchange reactions believed to proceed by the outer-sphere mechanism. The range covered by the rate constants is very large, extending from 10^{-4} up to perhaps nearly the limit of diffusion control ($\sim 10^9$). It is possible to account qualitatively for the observed variation in rates in terms of the different amounts of energy required to change the metal-ligand bond distances in the reacting species from their initial values to those needed in the transition state. For the case of an electron exchange process, the transition state must be symmetrical; that is, the two halves of the activated complex must be identical. It can be shown that an unsymmetrical transition state would correspond to a higher activation energy and, therefore, would not be part of the preferred reaction path. In the seven fastest reactions in Table 6-1 there is very

Table 6-1 Rates of Some Electron-Exchange Reactions with Outer-Sphere Mechanisms

Reactants	Rate constants (l mole^{-1} sec^{-1})
$[Fe(bipy)_3]^{2+}$, $[Fe(bipy)_3]^{3+}$	$> 10^6$ at 25°
$[Mn(CN)_6]^{3-}$, $[Mn(CN)_6]^{4-}$	
$[Mo(CN)_8]^{3-}$, $[Mo(CN)_8]^{4-}$	
$[W(CN)_8]^{3-}$, $[W(CN)_8]^{4-}$	
$[IrCl_6]^{2-}$, $[IrCl_6]^{3-}$	
$[Os(bipy)_3]^{2+}$, $[Os(bipy)_3]^{3+}$	
$[Fe(CN)_6]^{3-}$, $[Fe(CN)_6]^{4-}$	Second order, $\sim 10^5$ at 25°
$[MnO_4]^-$, $[MnO_4]^{2-}$	Second order, $\sim 10^3$ at 0°
$[Coen_3]^{2+}$, $[Coen_3]^{3+}$	Second order, $\sim 10^{-4}$ at 25°
$[Co(NH_3)_6]^{2+}$, $[Co(NH_3)_6]^{3+}$	
$[Co(C_2O_4)_3]^{3-}$, $[Co(C_2O_4)_3]^{4-}$	

little difference in the metal-ligand bond lengths in the two reacting species and thus very little energy of bond stretching and bond compression is required to achieve the symmetrical transition state. For the MnO_4^-/MnO_4^{2-} pair the difference is greater, and for the last three reactions there is a considerable difference.

In electron-transfer reactions between two dissimilar ions, in which there is a net decrease in free energy, the rates are generally higher than in comparable electron-exchange processes. In other words, one factor favoring rapid electron transfer is the thermodynamic favorability of the overall reaction. This generalization seems to apply not only to the outer-sphere processes now under discussion but also to the inner-sphere mechanism to be discussed shortly.

In several cases the rate constants for reactions in Table 6-1 have been found to depend on the identity and concentration of cations present in the solution. The general effect is an increase in rate with an increase in concentration of the cations, but certain cations are particularly effective. The general effect can be attributed to the formation of ion pairs that then decrease the electrostatic contribution to the activation energy. Certain specific effects found, for example, in the MnO_4^-—MnO_4^{2-} and $[Fe(CN)_6]^{4-}$—$[Fe(CN)_6]^{3-}$ systems are less easily interpreted with certainty. The effect of $[Co(NH_3)_6]^{3+}$ on the former is thought to be due to ion pairing, greatly enhanced by the high charge. There is no evidence that the cations participate in the actual electron transfer, though this may be so in some cases.

Ligand-Bridged or Inner-Sphere Processes. Ligand-bridged transition states have been shown to occur in a number of reactions, mainly through the elegant experiments devised and executed by H. Taube and his school. He has demonstrated that the following general reaction occurs:

$$[Co(NH_3)_5X]^{2+} + Cr^{2+}(aq) + 5H^+ = [Cr(H_2O)_5X]^{2+} + Co^{2+}(aq) + 5NH_4^+$$

$$(X = F^-, Cl^-, Br^-, I^-, SO_4^{2-}, NCS^-, N_3^-, PO_4^{3-}, P_2O_7^{4-}, CH_3COO^-,$$

$$C_3H_7COO^-, \text{crotonate, succinate, oxalate, maleate}) \qquad (6\text{-}16)$$

The significance and success of these experiments rest on the following facts. The Co^{III} complex is not labile while the Cr^{II} aquo ion is, whereas, in the products, the $[Cr(H_2O)_5X]^{2+}$ ion is not labile while the Co^{II} aquo ion is. It is found that the transfer of X from $[Co(NH_3)_5X]^{2+}$ to $[Cr(H_2O)_5X]^{2+}$ is quantitative. The most reasonable explanation for these facts is a mechanism such as that illustrated in (6-17).

$$Cr^{II}(H_2O)_6^{2+} + Co^{III}(NH_3)_5Cl^{2+} = [(H_2O)_5Cr^{II}ClCo^{III}(NH_3)_5]^{4+}$$

$$\Big\updownarrow \text{electron transfer}$$

$$Cr(H_2O)_5Cl^{2+} + Co(NH_3)_5(H_2O)^{2+} = [(H_2O)_5Cr^{III}ClCo^{II}(NH_3)_5]^{4+} \qquad (6\text{-}17)$$

$$\Big\downarrow H^+, H_2O$$

$$Co(H_2O)_6^{2+} + 5NH_4^+$$

Since all Cr^{III} species, including $Cr(H_2O)_6^{3+}$ and $Cr(H_2O)_5Cl^{2+}$, are substitution-inert, the quantitative production of $Cr(H_2O)_5Cl^{2+}$ must imply that electron transfer, $Cr^{II} \rightarrow Co^{III}$, and Cl^- transfer from Co to Cr are mutually interdependent acts, neither possible without the other. Postulation of the binuclear, chloro-bridged intermediate appears to be the only chemically credible way to explain this. As the general Eq. 6-16 above implies, many (although not all) ligands can serve as bridges.

In reactions between Cr^{2+} and CrX^{2+} and between Cr^{2+} and $Co(NH_3)_5X^{2+}$, which are inner sphere, the rates decrease as X is varied in the order $I^- > Br^- > Cl^- > F^-$. This seems a reasonable one if ability to "conduct" the transferred electron is associated with polarizability of the bridging group, and it appeared that this order might even be considered diagnostic of the mechanism. However, the opposite order is found for $Fe^{2+}/Co(NH_3)_5X^{2+}$ and $Eu^{2+}/Co(NH_3)_5X^{2+}$ reactions; the $Eu^{2+}/Cr(H_2O)_5X^{2+}$ reactions give the order first mentioned, thus showing that the order is not simply a function of the reducing ion used. The order must, of course, be determined by the relative stabilities of transition states with different X, and the variation in reactivity order has been rationalized on this basis.

There are now a number of cases, for example, those of $Co(NH_3)_5X^{2+}$ with $Co(CN)_5{}^{3-}$, where X = F^-, CN^-, $NO_3{}^-$, and $NO_2{}^-$, and that of Cr^{2+} with $IrCl_6{}^{2-}$, in which the electron transfer takes place by both inner- and outer-sphere pathways.

6-13
Stereochemically Nonrigid and Fluxional Complexes

No molecule is strictly rigid in the sense that all the interatomic distances and bond angles are fixed at one precise set of values. On the contrary, all molecules, even at the absolute zero constantly execute a set of vibrations, such that all of the atoms oscillate with amplitudes of a few tenths of an angstrom, about their average positions. In this sense, no molecule is rigid, but there are many molecules that undergo rapid deformational rearrangements of a much greater amplitude, in which atoms actually change places with each other. Such rearrangements are found among an enormous variety of compounds, including inorganic molecules such as PF_5, metal carbonyls, organometallic compounds, and organic molecules. Molecules that behave in this way are called stereochemically nonrigid. The recognition of stereochemical nonrigidity and its study is only possible by nuclear magnetic resonance (nmr) spectroscopy. Let us consider one of the earliest inorganic examples, PF_5.

This molecule is known to have a trigonal bipyramidal structure. It would be expected that the fluorine (^{19}F) nmr spectrum would show a complex multiplet of relative intensity 2 for the axial fluorine atoms and another of intensity 3 for the equatorial ones. The multiplets would result from coupling of each type of fluorine to those of the other type, and from coupling of both types to the phosphorus atom which has a spin of 1/2. In fact, only a sharp doublet is seen, indicating that, as far as nmr can tell, all five fluorine atoms are equivalent; the doublet structure results from their coupling to the phosphorus atom.

This result is due to the axial and equatorial fluorine atoms changing places with one another so rapidly ($>$ 10,000 times per second) that the nmr spectrometer cannot sense the two different environments and records all five of them at a frequency that is the weighted average of those which each separate environment would have. The fact that the splitting of the fluorine resonance into a doublet by the phosphorus atom is not lost shows that the exchange of places occurs without breaking the P—F bonds.

The generally accepted explanation for the rapid exchange of axial and equatorial fluorine atoms in PF_3 is shown in *Fig. 6-12*. This rearrangement pathway has two main stages. First, there is a concerted motion of the two axial F atoms and two of the equatorial ones so that these four atoms come into the same plane

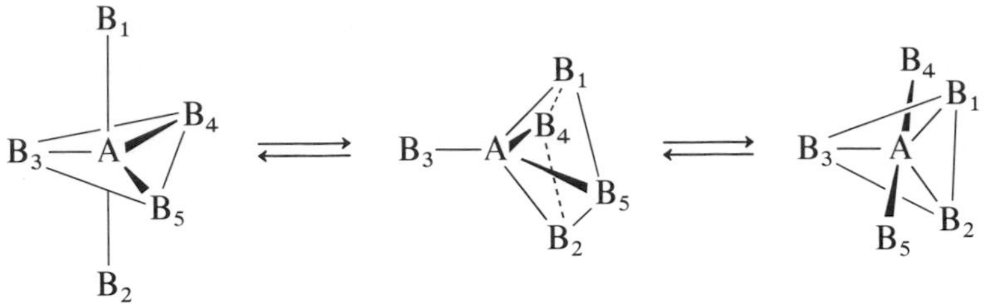

Figure 6-12 *A simple mechanism which interchanges axial and equatorial ligands of a trigonal bipyramid by passage through a square pyramid intermediate.*

and define a square. All of these four are now equivalent to each other, and the entire set of five defines a square pyramid. Second, a trigonal bypyramidal arrangement is now recovered. There are two equally probable ways for this to happen. In one, the same F atoms that were initially axial can return to axial positions. This would do nothing to cause exchange. However, if the other diagonally opposite pair of F atoms that were initially equatorial move to axial positions (while the other two, which were initially axial necessarily become equatorial), an exchange of positions involving all but one of the F atoms is accomplished. The same process can now be repeated in such a way that the equatorial F atom that did not exchange the first time becomes exchanged. If this process is repeated indefinitely, all F atoms will constantly pass back and forth between axial and equatorial positions.

It is to be noted that the molecules that exist immediately before and after the rearrangement steps (or after any number of steps) are chemically identical. They differ only in the interchange of indistinguishable nuclei; the process causes no net chemical change and has $\Delta H^\circ = \Delta S^\circ = \Delta G^\circ = 0$. Molecules of this type are by far the commonest and most important stereochemically nonrigid molecules and are called *fluxional molecules*.

An important fact to note about the process occurring in PF_5 is that it consists of a rearrangement of one of the more symmetrical forms of 5-coordination, the trigonal bipyramid (**tbp**) into the other, the square pyramid (**sp**), and then back to an equivalent version of the first in which some ligands have changed places. This type of process has been called polytopal rearrangment, because the two different arrangements of the ligand set are polytopes.

For coordination number 5, the **tbp** and **sp** arrangements seldom differ greatly in energy, so that whichever one is the preferred arrangement in a given substance, the other one can provide a low energy pathway for averaging the ligand environments. Thus, as a general rule, 5-coordinate species are fluxional, even at very low temperatures.

Polytopal rearrangements are generally facile for complexes with coordination numbers higher than 6 as well. This is because while one symmetrical structure may be somewhat more stable than any other, other arrangements are only a few kilojoules less stable, and with ordinary thermal energies available they provide accessible intermediates for rearrangement. As an example, consider an 8-coordinate complex with dodecahedral structure. The eight ligands are not all equivalent but fall into two sets, the A's and the B's as shown in *Fig. 6-1*. It is easy to see how the dodecahedron could be converted by relatively slight changes in interatomic distances into either a cubic or a square antiprismatic intermediate from which a new dodecahedron with the A's and B's interchanged would be recovered.

Octahedral complexes are generally not fluxional. That is, even when isomers, such as *cis* and *trans* isomers of MX_4Y_2 complexes interconvert, they do so by ligand dissociation and recombination rather than by any intramolecular rearrangement. It has been shown in a few cases that intramolecular rearrangement by way of a twist does occur. These are mostly tris-chelate species where the process studied is racemization. This has been mentioned previously (Section 6-11).

Notice in *Fig. 6-4* that if the top part of the Λ isomer is twisted relative to the bottom half by 120°, the molecule will pass through a trigonal prismatic intermediate structure and then become the Δ isomer. This sort of process, shown in *Fig. 6-10a*, is in general not facile and is rapid only in cases where the chelate ligands have a relatively short distance between their donor atoms (a small "bite"). Since the distance to be spanned is shorter in the eclipsed trigonal prismatic intermediate than in the octahedral structure, such ligands cause the two structures to be closer in stability so that the prism becomes a thermally accessible intermediate or transition state.

Fluxional behavior will be mentioned again later in discussing metal carbonyls (Chapter 28) and organometallic compounds (Chapter 29).

Study Questions

A

1. For each coordination number from 2 to 9, mention the principal geometrical arrangement (or arrangements).
2. What does each of the following abbreviations stand for: **tbp**, **sp**, *fac*, *mer*?
3. What is meant by tetragonal, rhombic, and trigonal distortion of an octahedron?
4. What do the terms *chelate* and *polydentate* mean?
5. What are the structures of the following ligands: acetylacetonate, ethylenediamine, diethylenetriamine, $EDTA^{4-}$?
6. Show with drawings the enantiomorphs of $M(L\text{-}L)_2X_2$ and $M(L\text{-}L)_3$ type complexes.
7. Give one example of each of the following types of isomers: ionization isomers, linkage isomers, coordination isomers, polymerization isomers.
8. Write the names of each of the following: $[Co(NH_3)_4en]Cl_3$, $[CrenCl_4]^-$, $[Pt(acac)NH_3Cl]$, $[Ru(NH_3)_5N_2](NO_3)_2$, $KFeCl_4$, $K_3[Cl_3WCl_3WCl_3]$.
9. What are the two principal sets of equilibrium constants (K_i's and β_i's) for expressing the formation of a series of complexes, ML, ML_2, ML_3, etc? How are they related?
10. Except in rare cases how do the magnitudes of the constants, K_i, vary with increasing i? What is the underlying reason for this, regardless of the charges?
11. What is meant by the *chelate effect*? Give an example.
12. For what ring sizes is it most important? How do you explain it?
13. Explain the difference between kinetic inertness (or lability) and thermodynamic stability (or instability).
14. What are the two limiting mechanisms for ligand exchange? Write equations showing how these rate laws differ.

15. Explain how solvent intervention, ion-pair formation, and conjugate-base formation can affect the observed rate law so that it will not correspond to the actual nature (S_N1 or S_N2) of the rate-limiting step.
16. Why does the rate law tell us nothing as to the true order of an aquation (acid hydrolysis) reaction carried out in aqueous solution?
17. True or false: The high rate of basic hydrolysis of $[Co(NH_3)_5Cl]^{2+}$ is attributable to the exceptional ability of OH^- to attack the cobalt ion nucleophilically. If false, give an alternative explanation of the high rate.
18. Why do many square complexes have two-term rate laws for ligand replacement reactions?
19. What is meant by the term *trans effect*? What two factors contribute to giving high *trans* effects?
20. Discuss the two general mechanisms for electron transfer reactions.
21. Describe the type of reaction and the reasoning used by Taube to prove that certain electron transfer reactions must occur by way of a bridged intermediate.
22. What is meant by a fluxional molecule? What is the experimental evidence that proves PF_5 to be one?

B

1. Show with drawings how axial-equatorial exchange in a square pyramidal complex, AB_5, could occur via a **tbp** intermediate.
2. Draw all the isomers of an octahedral complex having four different monodentate ligands. Indicate optical isomers.
3. Show how the experimental determination of the number of isomers of $[Co(NH_3)_4Cl_2]^+$ would enable you to show that the coordination geometry is octahedral not trigonal prismatic.
4. Why do you think most species, such as $AlCl_3$, $CuCl_3^-$, $Pt(NH_3)_2Cl^+$, are not actually such three-coordinate monomers but, instead, dimerize?
5. Suppose you prepared $[Copn_2Cl_2]^+$ using racemic pn (pn = $CH_3CH(NH_2)CH_2NH_2$). Ignoring possible ring conformation effects, how many isomers, geometric and optical, could be formed?
6. How many other polymerization isomers, besides the three listed in the text can you write for the empirical formula $Co(NH_3)_3(NO_2)_3$?
7. Draw all the possible isomers of the dinuclear complex $L_2X_2M(\mu\text{-}X)_2ML_2X_2$, where L is a ligand which cannot be a bridge. If L = NH_3, X = Cl and M = Ru, write a systematic name for each isomer.
8. Calculate the numerical value of the statistical ratio of K_4/K_3 for a series of octahedral complexes ML, . . . ML_6.
9. For PtX_4^{2-} complexes both ligand exchange rates and thermodynamic stability increase in the order X = Cl < Br < I < CN. Explain why these observations are not inconsistent.
10. Using the *trans* effect sequence given in the text, devise rational procedures for selectively synthesizing each of the three isomers of $[PtpyNH_3NO_2Cl]$.
11. Suppose you had a **tbp** complex in which each of the five ligands was distinguishable, but otherwise equally likely to occupy either an axial or an equatorial position. How many arrangements are there? How many times would the **tbp** → **sp** → **tbp**′ type process have to be repeated to carry any given isomer into every one of the others?
12. Purified *trans*-$[Coen_2Cl_2]Cl$ is converted by excess aqueous KCN to *trans*-$[Coen_2OHCl]Cl$. However, if the impure material that is contaminated with Co(II) is used, considerable $[Co(CN)_5Cl]^{3-}$ is obtained. Suggest an explanation.

Chapter 6
Study Guide

Supplementary Reading

Basolo, F. and Johnson, R. C., *Coordination Chemistry*, Benjamin, 1964.
Basolo, F. and Pearson, R. G., *Mechanisms of Inorganic Reactions*, 2nd ed., Wiley, 1967.
Benson, D., *Mechanisms of Inorganic Reactions in Solution*, McGraw-Hill, 1968.
Edwards, J. O., *Inorganic Reaction Mechanisms*, Parts I and II, Vols. 13 and 17, of *Progress in Inorganic Chemistry*, Wiley, 1970, 1972.
Edwards, J. O., *Inorganic Reaction Mechanisms*, Benjamin, 1964.
Grinberg, A. A., *The Chemistry of Complex Compounds*, Pergamon Press, 1962.
Halpern, J., *J. Chem. Educ.*, 1968, *45*, 372.
Martell, A. E., ed., *Coordination Chemistry*, Vol. I, Van Nostrand-Reinhold, 1971.
Rossotti, F. J. C. and Rossotti, H., *The Determination of Stability Constants*, McGraw-Hill, 1961.
Sykes, A. G., *Kinetics of Inorganic Reactions*, Pergamon Press, 1966.
Taube, H., *Electron Transfer Reactions of Complex Ions in Solution*, Academic Press, 1970.
Wells, A. F., *Structural Inorganic Chemistry*, 3rd ed., Oxford University Press, 1962.

7

solvents, solutions, acids, and bases

The majority of chemical reactions and many measurements of properties are carried out in a solvent. The properties of the solvent are crucial to the success or failure of the study. For the inorganic chemist, water has been the most important solvent, and it will continue to be, but many other solvents have been tried and found useful. A few of them, and the concepts that influence the choice of a solvent, are discussed here. Closely connected with the properties of solvents is the behavior of acids and bases. In this chapter some fundamental concepts concerning acids and bases are also presented.

7-1 Solvent Properties

Properties that chiefly determine the utility of a solvent are:

1. The temperature range over which it is a liquid.
2. Its dielectric constant.
3. Its donor and acceptor (Lewis acid-base) properties.
4. Its protonic acidity or basicity.
5. The nature and extent of autodissociation.

The first two are of rather obvious import and need not detain us long. The others will merit discussion in subsequent sections.

Liquid Range. Solvents that are liquid at room temperature and one atmosphere pressure are most useful because they are easily handled, but it is also desirable that measurements or reactions be feasible at temperatures well above and below room temperature. As Table 7-1 shows, dimethyl formamide, propane-1,2-diol carbonate, and acetonitrile are especially good in this respect.

Dielectric Constant. The ability of a liquid to dissolve ionic solids depends strongly although not exclusively on its dielectric constant, ε. The force, F, of

Table 7-1 Properties of Some Useful Solvents[a]

Name	Abbreviation	Formula	Liq. Range, °C	$\varepsilon/\varepsilon_0$
Water	—	H_2O	0 to 100	82
Acetonitrile	—	CH_3CN	−45 to 82	38
Dimethylformamide	DMF	$HC(O)N(CH_3)_2$	−61 to 153	38
Dimethyl sulfoxide	DMSO	$(CH_3)_2SO$	18 to 189	47
Nitromethane	—	CH_3NO_2	−29 to 101	36
Sulfolane	—	(ring) SO_2	28 to 285	44
Propane--1,2-diol Carbonate	—	(ring) O, C=O, O	−49 to 242	64
Hexamethylphosphoramide	HMP	$OP[N(CH_3)_2]_3$		30
Glycol dimethyl ether	glyme	$CH_3OCH_2CH_2OCH_3$	−58 to 83	3.5
Tetrahydrofuran	THF	(ring) O	−65 to 66	7.6
Dichloromethane	—	CH_2Cl_2	−97 to 40	9
Ammonia	—	NH_3	−78 to −33	23 (−50°)
Hydrogen Cyanide	—	HCN	−14 to 26	107
Sulfuric acid	—	H_2SO_4	−14 to 26	107
Hydrogen Fluoride	—	HF	−83 to 20	84 (0°)

[a] Instead of the absolute value of ε, we give here the ratio of ε to ε_0, the latter being the value for a vacuum. In subsequent sections the term "dielectric constant" refers to this ratio.

attraction between cations and anions immersed in a medium of dielectric constant ε, is inversely proportional to ε.

$$F = \frac{q^+q^-}{4\pi\varepsilon r^2}$$

Thus, water ($\varepsilon = 82\varepsilon_0$ at 25°, where ε_0 is for a vacuum) reduces the attractive force nearly to 1% of its value in absence of a solvent.

7-2 Donor and Acceptor Properties

The ability of a solvent to keep a given solute in solution depends considerably on its ability to solvate the dissolved particles, that is, to interact with them in a quasi-chemical way. For ionic solutes, there are both cations and anions to be solvated. Commonly, the cations are the smaller [e.g., $Ca(NO_3)_2$, $FeCl_3$] and the solvation of the cations is of prime importance. The solvation of simple cations is essentially the process of forming complexes in which the ligands are solvent molecules. The order of coordinating ability toward typical cations for some common solvents is

$$\text{DMSO} > \text{DMF} \approx H_2O > \text{acetone} \approx (CH_3CHCH_2)O_2CO \approx CH_3CN$$
$$> (CH_2)_4SO_2 > CH_3NO_2 > C_6H_5NO_2 \gg CH_2Cl_2$$

Acceptor properties are usually manifested less specifically. The positive ends of the solvent molecule dipoles will orient themselves toward the anions.

It is noteworthy that in general the dielectric constant and the ability to solvate ions are related properties, tending to increase simultaneously, but there is no quantitative correlation. The more polar the molecules of a solvent the higher its dielectric constant tends to be (although the extent of hydrogen bonding plays a very important role also); at the same time, the more polar a molecule the better able it is to use its negative and positive regions to solvate cations and anions, respectively.

7-3
Protic Solvents

These contain ionizable protons and are more or less acidic. Examples are H_2O, HCl, HF, H_2SO_4, and HCN. Even ammonia, which is usually considered a base, is a protic solvent and can supply H^+ to stronger bases. Protic solvents characteristically undergo autodissociation.

Autodissociation of Protic Solvents. For some of the examples just mentioned, the autodissociation reactions can be written in the simplest way as follows:

$$2H_2O = H_3O^+ + OH^-$$
$$2HCl = H_2Cl^+ + Cl^-$$
$$2HF = H_2F^+ + F^-$$
$$2H_2SO_4 = H_3SO_4{}^+ + HSO_4{}^-$$
$$2NH_3 = NH_4{}^+ + NH_2{}^-$$

The significance of autodissociation is that solutes encounter not only the molecules of solvent but the cations and anions that form in the autodissociation process. The autodissociations of several of the acid solvents are discussed in detail in Section 7-11. Here, we give a closer examination of the processes in water and ammonia. The simple equations above do not consider in detail the further solvation of the primary products of autodissociation, and this is of importance.

Water. A more general equation for the autodissociation of water is

$$(n + m + 1)H_2O = [H(H_2O)_n]^+ + [OH(H_2O)_m]^-$$

For the hydrogen ion, $[H(H_2O)_n]^+$, there is strong association of H^+ with one water molecule to give H_3O^+, a flat pyramidal ion (7-I) isoelectronic with NH_3.

This ion is observed in a number of crystalline compounds. In water it is further solvated. Another species actually observed in crystals is (7-II). Probably the

$[H_3O]^+$ (~118°)

7-I

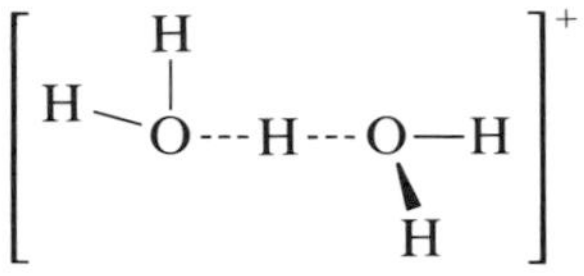

$[H_2O \cdots H \cdots OH_2]^+$

7-II

$[H_2O \cdots H\text{—}O(\text{—}H \cdots OH_2)\text{—}H \cdots OH_2]^+$

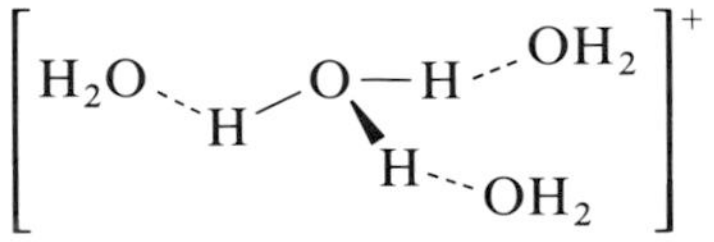

7-III

$H_9O_4^+$ ion (7-III) is the largest well-defined species. The extent of autodissociation is slight,

$$K'_{25^\circ} = \frac{[H^+][OH^-]}{[H_2O]} = (1.0 \times 10^{-14})/55.56$$

In practice, the essentially constant 55.56 *M* concentration of H_2O molecules is omitted (because it *is* constant), and the constant $K_{25^\circ} = [H^+][OH^-] = 1.0 \times 10^{-14}$ is used.

Liquid NH_3. This is very similar to water except that its autodissociation, usually written as shown below is even less:

$$2NH_3 = NH_4^+ + NH_2^-$$

$$K_{-50^\circ} = [NH_4^+][NH_2^-] = 10^{-30}$$

7-4 Aprotic Solvents

There are three broad classes of these:

1. *Nonpolar, or very weakly polar, nondissociated liquids, which do not solvate strongly.* Examples are CCl_4 and hydrocarbons. Because of low polarity, low dielectric constants, and poor donor power, these are not powerful solvents except for other nonpolar substances. Their main value, when they can be used, is that they play a minimal role in the chemistry of reactions carried out therein.

2. *Nonionized but strongly solvating (generally polar) solvents.* Examples of this type are CH_3CN, dimethylformamide (DMF), dimethylsulfoxide (DMSO), tetrahydrofuran (THF), and SO_2. These have in common the facts that they are aprotic, that no autodissociation equilibria are known to occur, and that they strongly solvate ions. In other respects they differ. Some are high boiling (DMSO), others are low boiling (SO_2); some have high dielectric constants (DMSO, 45)

while others are of low polarity (THF, 7.6). For the most part, they solvate cations best by using negatively charged oxygen atoms, but sulfur dioxide has pronounced acceptor ability, and solvates anions and other Lewis bases effectively. For example the molecular compound $(CH_3)_3N \rightarrow SO_2$ can be isolated.

3. *Highly polar, autoionizing solvents.* Some of these are interhalogens:

$$2BrF_3 = BrF_2^+ + BrF_4^-$$

$$2IF_5 = IF_4^+ + IF_6^-$$

although not invariably:

$$2Cl_3PO = Cl_2PO^+ + Cl_4PO^-$$

Solvents in this class are extremely difficult to work with since they are highly reactive. Some react with silica, with noble metals such as gold and platinum, and all are extremely sensitive to moisture. The chemistry that is usually carried out in solvents of this kind is best described in terms of acid-base reactions and is, accordingly, discussed in Section 7-8.

7-5
Molten Salts

These represent a kind of extreme of aprotic, autoionizing solvents. In them ions predominate over neutral molecules which, in some cases, are of negligible concentration. The alkali metal halides and nitrates are among the "totally" ionic molten salts, whereas others such as molten halides of zinc, tin, and mercury contain many molecules as well as ions.

Low melting points are often achieved with either mixtures or by using halides of alkylammonium ions. Thus an appropriate mixture of $LiNO_3$, $NaNO_3$, and KNO_3 has a melting point as low as 160° C and $(C_2H_5)_2H_2NCl$ has a melting point of 215°.

Examples of important reactions carried out in molten salts are the following preparations of low-valent metal salts:

$$CdCl_2 + Cd \xrightarrow{\text{Liq. } AlCl_3} Cd_2^{2+}(AlCl_4^-)_2$$

$$Re_3Cl_9 \xrightarrow{\text{Liq. } (C_2H_5)_2H_2NCl} [(C_2H_5)_2H_2N^+]_2[Re_2Cl_8^{2-}]$$

The industrial production of aluminum is carried out by electrolysis of a solution of Al_2O_3 in molten Na_3AlF_6.

7-6
Solvents for Electrochemical Reactions

A good solvent for electrochemical reactions must meet several criteria. Generally, electrochemical reactions involve ionic substances, so that a dielectric constant of 10 or better is desirable. Second, the solvent must have a wide range of voltage over which it is not oxidized or reduced so its own electrode reactions will not take precedence over those of interest.

Water is a widely useful solvent for electrochemistry. Because of its high dielectric constant and solvating ability, it dissolves many electrolytes. Its intrinsic conductance is suitably low. Its range of redox stability is fairly wide, as shown by the following potentials, although its reduction is often a limitation:

$$O_2 + 4H^+(10^{-7}M) + 4e = 2H_2O \qquad E^\circ = +0.82\ V$$

$$H^+(10^{-7}M) + e = \tfrac{1}{2}H_2 \qquad E^\circ = -0.41\ V$$

Acetonitrile, CH_3CN. This is widely used for solutes such as organometallic compounds or salts containing large alkylammonium ions, which are insufficiently soluble in water. It is stable over a large range of voltages.

Others. Dimethylformamide, $HC(O)N(CH_3)_2$ is similar to CH_3CN but more easily reduced. Dichloromethane is sometimes used for organic solutes as is nitromethane. Molten salts are also useful.

7-7 Purity of Solvents

Although it is obvious that a solvent should be pure if reproducible and interpretable results are to be obtained, it is not always obvious what subtle forms of contamination can occur. Of particular importance are water and oxygen. Oxygen is slightly soluble in virtually all solvents, and saturated solutions are formed on brief exposure to air, for example, when pouring. Oxygen can be partially removed by bubbling nitrogen through the liquid, but only by repeatedly freezing and pumping on a vacuum line can it be completely removed. Certain organic solvents, especially ethers, react with oxygen on long exposure to air, forming peroxides. They can best be purified of these by distillation from reductants (e.g., hydrides) or by passage through "molecular sieves" (page 115).

Water also dissolves readily in solvents exposed to the air or to glass vessels that have not been baked dry. It is important to recognize that even small quantities of H_2O on a weight percentage basis can be important. For example, acetonitrile which contains only 0.1% by weight of water is about 0.04 molar in H_2O, so that the properties of 0.1 *M* solutions can be seriously influenced by the "trace" of water.

7-8 Definitions of Acids and Bases

The concepts of acidity and basicity are so pervasive in chemistry that acids and bases have been defined many times and in various ways. One definition, probably the oldest, is so narrow as to pertain only to water as solvent. According to this, acids and bases are sources of H^+ and OH^-, respectively. A somewhat broader, but closely allied definition that is applicable to all protonic solvents is that of Brønsted and Lowry.

Brønsted–Lowry Definition. An acid is a substance that supplies protons and a base is a proton acceptor. Thus, in water, any substance that increases the concentration of hydrated protons (H_3O^+) above that due to the autodissociation of the water is an acid, and any substance that lowers it is a base. Any solute that supplies hydroxide ions is a base, since these combine with protons to reduce the H_3O^+ concentration. However, other substances such as sulfides, oxides, or anions of weak acids (e.g., F^-, CN^-) are also bases.

Solvent System Definition. This can be applied in all cases where the solvent has a significant autoionization reaction, whether protons are involved or not. Some examples are

$$2H_2O = H_3O^+ + OH^-$$

$$2NH_3 = NH_4^+ + NH_2^-$$

$$2H_2SO_4 = H_3SO_4^+ + HSO_4^-$$

$$2OPCl_3 = OPCl_2^+ + OPCl_4^-$$

$$2BrF_3 = BrF_2^+ + BrF_4^-$$

A solute that increases the cationic species natural to the solvent is an acid; one that increases the anionic species is a base. Thus, for the BrF_3 solvent, a compound such as BrF_2AsF_6, which dissolves to give BrF_2^+ and AsF_6^- ions is an acid, while $KBrF_4$ is a base. If solutions of acid and base are mixed, a neutralization reaction, producing a salt and solvent molecules, takes place:

$$\underbrace{BrF_2^+ + AsF_6^-}_{\text{acid}} + \underbrace{K^+ + BrF_4^-}_{\text{base}} = \underbrace{K^+ + AsF_6^-}_{\text{salt}} + 2BrF_3$$

Even for protonic solvents this is a broader and more useful definition, because it explains why acid or base character is not an absolute property of the solute. Rather, the acid or base character of a substance can only be specified in relation to the solvent used. For example, in water, CH_3COOH (acetic acid) is an acid:

$$CH_3COOH + H_2O = H_3O^+ + CH_3COO^-$$

In the sulfuric acid solvent system, CH_3COOH is a base:

$$H_2SO_4 + CH_3COOH = CH_3CO_2H_2^+ + HSO_4^-$$

As another example, urea, $H_2NC(O)NH_2$, which is essentially neutral in water, is an acid in liquid ammonia:

$$NH_3 + H_2NC(O)NH_2 = NH_4^+ + H_2NC(O)NH^-$$

The Lux and Flood Definition. Consider the following reaction sequences:

$$\left.\begin{array}{l} CaO + H_2O = Ca(OH)_2 \\ CO_2 + H_2O = H_2CO_3 \end{array}\right\} \longrightarrow CaCO_3 + 2H_2O$$

$$CaO + CO_2 \longrightarrow CaCO_3$$

If CaO and CO_2 are first allowed to react with water, the hydration products are readily recognized as a base and an acid, respectively. Their reaction to produce the salt $CaCO_3$ and solvent is then a very conventional neutralization reaction. However, the reaction may be carried out directly, as in the second equation, without the intervention of solvent. It is natural to continue to regard this as an acid-base reaction. Some other examples of direct reactions between acidic and basic oxides are

$$CaO + SiO_2 \longrightarrow CaSiO_3$$

$$3Na_2O + P_2O_5 \longrightarrow 2Na_3PO_4$$

The general principle involved in such processes was recognized by Lux and Flood, who proposed that an acid be defined as an *oxide ion donor* and a base as an *oxide ion acceptor*. Thus, in the above reactions, the acids, CaO and Na_2O provide oxide ions to the bases CO_2, SiO_2, and P_2O_5 so that anions, CO_3^{2-}, SiO_3^{2-}, and PO_4^{3-} can be formed.

The Lux–Flood concept is very useful in dealing with high-temperature, anhydrous systems like those encountered in ceramics and metallurgy. It has an inverse relation to the aqueous chemistry of acids and bases because the acids are oxides that react with water to generate bases, for example,

$$Na_2O + H_2O \longrightarrow 2Na^+ + 2OH^-$$

and the bases are the anhydrides of aqueous acids, for example,

$$P_2O_5 + 3H_2O \longrightarrow 2H_3PO_4$$

The Lewis Definition. One of the most general—and useful—of all definitions was proposed by G. N. Lewis, who defined an acid as an electron-pair acceptor and a base as an electron-pair donor. This definition includes the Brønsted–Lowry definition as a special case, since the proton can be regarded as an electron-pair acceptor and the base, be it OH^-, NH_2^-, HSO_4^-, etc, as an electron-pair donor; for example,

$$H^+ + :OH^- = H:OH$$

The Lewis definition covers a great many systems where protons are not involved at all, however. The reaction between ammonia and BF_3, for example, is an acid-base reaction:

$$\underset{\text{Lewis base}}{H_3N:} + \underset{\text{Lewis acid}}{BF_3} \longrightarrow H_3N:BF_3$$

In the Lewis sense, all of the usual ligands can be regarded as bases and all metal ions can be regarded as acids. The degree of affinity of a metal ion for ligands can be termed its Lewis acidity, and the tendency of a ligand to become bound to a metal ion can be regarded as a measure of its Lewis basicity.

Base and acid strengths in the Lewis sense are not fixed, inherent properties of the species concerned, but vary somewhat with the nature of the partner. That is, the order of base strength of a series of Lewis bases may change when the type of acid with which they are allowed to combine changes. We discuss this in the next section.

Observe that, for a given donor or acceptor atom, basicity or acidity can be influenced greatly by the nature of the substituents. Substituent influence can be either electronic or steric in origin.

Electronic Effects. The electronegativity of substituents exercises an obvious effect. Thus base strength and acid strength are affected oppositely, as the following examples show.

Bases: $(CH_3)_3N > H_3N > F_3N$

Acids: $(CH_3)_3B < H_3B < F_3B$

The more electron-withdrawing (electronegative) the substituent the more it enhances Lewis acidity and diminishes Lewis basicity.

However, more subtle electronic effects can also be important. On simple electronegativity grounds the following order of acid strengths would be predicted: $BF_3 > BCl_3 > BBr_3$. Experimentally, just the opposite is found. This can be understood when the existence of π interactions in the planar molecules is taken into account, and when it is noted that, after the Lewis acid has combined with a base, the BX_3 group becomes pyramidal and the boron atom no longer interacts with the π electrons of the X atoms. Simple calculations indicate that the B—X π interactions will decrease in strength in the order $F \gg Cl > Br$. Therefore, BF_3 is a weaker Lewis acid than BCl_3 because the planar BF_3 molecule is stabilized to a greater extent than BCl_3 by B—X π bonding. Borate esters, $B(OR)_3$, are also surprisingly weak Lewis acids for the same reason.

Steric Effects. These may be of several kinds. For the following three bases (7-IV to 7-VI) base strength toward the proton increases slightly from IV to

7-IV 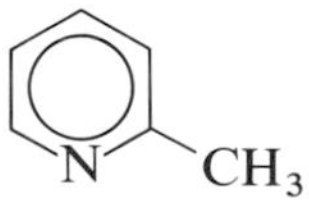　　7-V 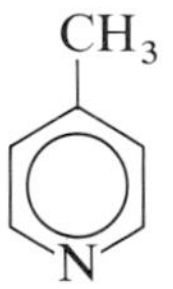　　7-VI

V and is virtually the same for V and VI, as is expected from the ordinary inductive effect of a methyl group. However, with respect to $B(CH_3)_3$, the order of basicity is

$$7\text{-IV} \approx 7\text{-VI} \gg 7\text{-V}$$

This results from the steric hindrance between the ortho methyl group of the base and the methyl groups of $B(CH_3)_3$. For the same reason quinuclidine, (7-VII) is a far stronger base toward $B(CH_3)_3$ than is triethylamine, (7-VIII):

7-VII 7-VIII

A different sort of steric effect results as the bulk on the boron atom in a BR_3 base is increased. Since, as we stated previously the BR_3 molecule goes from planar to pyramidal when it interacts with the acid, the R groups must be squeezed into considerably less space. As the R groups increase in size, this effect strongly opposes the formation of the $A{:}BR_3$ compound, thus effectively decreasing the basicity.

7-9
"Hard" and "Soft" Acids and Bases

It has been known for a long time that metal ions can be sorted into two groups according to their preference for various ligands. Let us consider the ligands formed by the elements of groups V, VI, and VII. For group V we might take a homologous series such as R_3N, R_3P, R_3As, R_3Sb, and for Group VII we take the anions themselves, F^-, Cl^-, Br^-, and I^-. For type (a) metals, complexes are most stable with the lightest ligands and less stable as each group is descended. For the type (b) elements the trend is just the opposite. This is summarized as follows:

Complexes of type (a) metal	Ligands			Complexes of type (b) metal
Strongest ↑	R_3N	R_2O	F^-	Weakest ↓
	R_3P	R_2S	Cl^-	
	R_3As	R_2Se	Br^-	
Weakest	R_3Sb	R_2Te	I^-	Strongest

Type (a) metal ions include principally:

1. Alkali metal ions
2. Alkaline earth ions
3. Lighter and more highly charged ions, for example,

$$Ti^{4+}, Fe^{3+}, Co^{3+}, Al^{3+}$$

Type (b) metal ions include principally:

1. Heavier transition metal ions, such as

$$Hg_2^{2+}, Hg^{2+}, Pt^{2+}, Pt^{4+}, Ag^+, Cu^+$$

2. Low-valent metal ions such as the formally zerovalent metals in metal carbonyls.

This empirical ordering proved very useful in classifying and to some extent predicting relative stabilities of complexes. Later, Pearson observed that it might be possible to generalize the correlation to include a broader range of acid-base interactions. He noted that the type (a) metal ions (acids) were small, compact and not very polarizable and that they preferred ligands (bases) which were also small and less polarizable. He called these acids and bases "hard." Conversely, the type (b) metal ions, and the ligands they prefer, tend to be larger and more polarizable; he described these acids and bases as "soft." The empirical relationship could then be expressed, qualitatively, by the statement that *hard acids prefer hard bases and soft acids prefer soft bases.* Although the point of departure for the "hard and soft" terminology was the concept of polarizability, other factors undoubtedly enter into the problem. There is no unanimity among chemists as to the detailed nature of "hardness" and "softness," but clearly coulombic attraction will be of importance for hard-hard interactions while covalence will be quite significant for soft-soft interactions. The participation of both electrostatic and covalent forces in acid-base interactions will be considered in the next section.

7-10 Covalent and Ionic Components of Lewis Acid-Base Interactions

In an attempt to account quantitatively for the enthalpies, ΔH_{AB}, of combining a Lewis acid, A, with a Lewis base, B, the following type of equation has been proposed.

$$-\Delta H_{AB} = E_A E_B + C_A C_B$$

The form of this equation is based on the notion that for each acid-base interaction there will be both electrostatic and covalent components. It is then further *postulated* that the tendency of an individual acid or base to contribute to electrostatic and covalent interaction with *any* partner is a fixed characteristic that is measured by E_A or E_B for the electrostatic part and C_A or C_B for the covalent part. Thus the electrostatic contribution to the total enthalpy change is given by $E_A E_B$ and the covalent contribution is given by $C_A C_B$. This is a crude idea, and also a somewhat arbitrary one, since there is no mathematically unique set of E and C values, no matter how many $-\Delta H_{AB}$'s are known. It is necessary to assign arbitrarily one each of the E_A, E_B, C_A, and C_B parameters before a unique set can be developed purely by data-fitting procedures.

This scheme, which has been advocated by R. S. Drago and his students, has some advantage over the simple HSAB concept, simply because it (a) has more parameters, and (b) at least attempts to be quantitative. However, the HSAB picture can also be extended by adding the concepts of strong and weak to those of hard and soft. In other words, every acid and base can be classified by its position in the hard-soft scale *and* by its strength. Thus we might find ourselves describing some base as a "moderately weak and fairly soft," "very hard but weak," and so on.

There have been those who have wondered, often to themselves, and occasionally out loud, just where these various attempts to be precise and quantitative about the nature of acidity and basicity become too quixotic to be valuable. It is not impossible that the venerable sport of jousting at windmills is now being practiced by the more zealous defenders of the various acid and base "religions," but perhaps only time will tell for certain.

7-11
Some Common, Aqueous Acids

Sulfuric Acid, H_2SO_4. This acid is of enormous industrial importance and is manufactured in larger quantities than any other. The preparation requires first the burning of sulfur to SO_2. Oxidation of SO_2 to SO_3 must then be catalyzed either homogeneously by oxides of nitrogen (lead chamber process) or heterogeneously by platinum (contact process). Sulfuric acid is ordinarily sold as a 98% mixture with water (18 molar). The pure substance is obtained as a colorless liquid by addition of sufficient SO_3 to react with the remaining H_2O. The solid and liquid are built of SO_4 tetrahedra linked by hydrogen bonds.

Addition of further SO_3 to 100% H_2SO_4 gives *fuming sulfuric acid* or *oleum*, which contains polysulfuric acids such as pyrosulfuric acid, $H_2S_2O_7$ and, with more SO_3, $H_2S_3O_{10}$ and $H_2S_4O_{13}$.

Sulfuric acid is not a very strong oxidizing agent, but it is a powerful dehydrating agent for carbohydrates and other organic substances, often degrading the former to elemental carbon:

$$C_nH_{2n}O_n \xrightarrow{H_2SO_4} nC + H_2SO_4 \cdot nH_2O$$

The equilibria in pure H_2SO_4 are complex. Besides self-ionization

$$2H_2SO_4 = H_3SO_4^+ + HSO_4^- \qquad K_{10^\circ} = 1.7 \times 10^{-4} \text{ mol}^2 \text{ kg}^2$$

there are hydration/dehydration equilibria, such as:

$$2H_2SO_4 = H_3O^+ + HS_2O_7^-$$
$$2H_2SO_4 = H_2O + H_2S_2O_7$$
$$H_2SO_4 + H_2S_2O_7 = H_3SO_4^+ + HS_2O_7^-$$
etc.

Nitric Acid, HNO_3. The normally available, concentrated acid is about 70% by weight HNO_3 in water. It is colorless when pure but is often yellow as a result of photochemical decomposition which gives NO_2:

$$2HNO_3 \xrightarrow{h\nu} 2NO_2 + H_2O + \tfrac{1}{2}O_2$$

Red, "fuming" nitric acid is essentially 100% HNO_3 which contains additional NO_2.

The pure acid is a colorless liquid or solid that must be stored below 0° to avoid thermal decomposition according to the same equation as above for photochemical decomposition. In the pure liquid the following equilibria occur:

$$2HNO_3 = H_2NO_3^+ + NO_3^-$$
$$H_2NO_3^+ = NO_2^+ + H_2O$$

While aqueous nitric acid below 2 M concentration is not strongly oxidizing, the concentrated acid is a very powerful oxidizing agent. It will attack nearly all metals except for Au, Pt, Rh, and Ir and a few others that quickly become passivated (covered with a resistent oxide film) such as Al, Fe, and Cu.

Aqua Regia. Aqua regia (approximately 3 vols HCl to 1 vol HNO_3) contains free Cl_2 and ClNO and attacks even Au and Pt, owing to the ability of Cl^- to stabilize the metal cations as the complexes $AuCl_4^-$ and $PtCl_6^{2-}$

Perchloric Acid, $HClO_4$. This is normally available in concentrations 70 to 72% by weight. The pure substance, which can be obtained by vacuum distillation in presence of the dehydrating agent $Mg(ClO_4)_2$, is stable at 25°C for only a few days, decomposing to give off Cl_2O_7. Both the pure and the concentrated aqueous acid react explosively with organic matter. The ClO_4^- ion is a very weak ligand, and perchloric acid and alkali perchlorates are therefore of use in preparing solutions in which complexing of cations is to be minimized.

The Hydrohalic Acids, HCl, HBr, and HI. These three acids are similar but differ markedly from hydrofluoric acid, which we describe below. The pure compounds are pungent gases at 25°C but are highly soluble in water to give strongly acidic solutions. One molar solutions are virtually 100% dissociated. For aqueous HBr and especially HI, their reactivity as simple acids is complicated by the reducing character of the Br^- and I^- ions.

Only HCl (bp −85°C) has been much studied as a pure liquid. Its self-ionization is small

$$3HCl = H_2Cl^+ + HCl_2^-$$

but many organic and some inorganic compounds dissolve to give conducting solutions. A number of compounds containing the $[Cl—H—Cl]^-$ and $[Br—H—Br]^-$ ions have been isolated.

Hydrofluoric Acid, HF. In aqueous solution HF is a weak acid,

$$HF + H_2O = H_3O^+ + F^- \qquad K_{25^\circ} = 7.2 \times 10^{-5}$$

This is due mainly to the great strength of the H—F bond. The aqueous acid readily attacks glass and silica because the stable SiF_6^{2-} ion can be formed;

$$6HF(aq) + SiO_2 = 2H_3O^+ + SiF_6^{2-}$$

and it is used commercially to etch glass.

In contrast to the aqueous solution, liquid HF (bp 19.5° C) is one of the strongest acids known. The principle self-ionization equilibria are

$$2HF = H_2F^+ + F^-$$
$$F^- + nHF = HF_2^- + H_2F_3^- + H_3F_4^-, \text{etc.}$$

There are a few substances that act as acids toward liquid HF, namely, as fluoride ion acceptors, which can thereby further increase the concentration of H_2F^+. An example is SbF_5:

$$2HF + SbF_5 = H_2F^+ + SbF_6^-$$

Liquid HF has a dielectric constant (84 at 0° C) comparable to that of water and is an excellent solvent for a wide range of both inorganic and organic compounds.

7-12
Some Rules Concerning Strengths of Oxy Acids

Acids consisting of a central atom surrounded by O atoms and OH groups, $XO_n(OH)_m$ are very common, including H_2SO_4, H_3PO_4, HNO_3, etc. For these acids there are two important generalizations:

1. The ratio of successive dissociation constants, K_n/K_{n-1} is 10^{-4} to 10^{-5}. (which is equivalent to $pK_{n-1} - pK_n = 4.5 \pm 0.5$, where $pK = -\log K$).

2. The magnitude of K_1 depends on n, the number of additional oxygen atoms besides those in OH groups. The more of these, the greater the acid strength, according to:

n	K	Acid strength
3	Very, very large	Very strong
2	$\sim 10^2$	Strong
1	10^{-2}–10^{-3}	Medium
0	$10^{-7.5}$–$10^{-9.5}$	Weak

The basis for these rules, and their general validity lies in the delocalization of the charge of the anions. For a given initial dissociation,

$$XO_n(OH)_m = XO_{n+1}(OH)_{m-1}^- + H^+$$

the greater the number of oxygen atoms, $n + 1$, the more the negative charge can be spread out and thus the more stable is the anion. For cases where there are many O atoms and only a single proton, for example $HClO_4$, delocalization is very effective and the dissociation is thus very favorable. When $n = 0$ there is practically no delocalization as in $Te(OH)_5O^-$, and dissociation is not favorable.

The steady decrease in the values of K_1, K_2, K_3, etc., occurs because after each dissociation, there is an increased negative charge that lessens the tendency of the next proton to depart.

Apparent exceptions to rule (2) turn out not to have simple $XO_n(OH)_m$ type structures. For example, phosphorus acid, H_3PO_3 would have $K_1 \approx 10^{-8}$ if it were $P(OH)_3$. In fact, the value of K_1 is about 10^{-2}, which should mean that it has $n = 1$. It actually does belong in that group since its structure is $HPO(OH)_2$, with one hydrogen atom directly attached to P. Similarly, hypophosphorous acid, H_3PO_2 has $K_1 \approx 10^{-2}$ and its actual structure is $H_2PO(OH)$.

Carbonic acid also deviates from expectation, but for a different reason. For $CO(OH)_2$ we expect $K_1 \approx 10^{-2}$ whereas the measured value is $\sim 10^{-6}$. This is because much of the solute in a solution of "carbonic acid" is present as loosely hydrated CO_2 and not as $CO(OH)_2$. When a correction is made for this, the true dissociation constant of $CO(OH)_2$ is found to be $10^{-3.6}$, which is close to the expected range.

7-13 Superacids

There are a number of liquids that are considerably more acidic, by as much as 10^6–10^{10} times than concentrated aqueous solutions of so-called very strong acids such as nitric and sulfuric acids. These are called superacids, and in recent years a great deal of new chemistry has been found to occur in these media. Superacid systems are necessarily nonaqueous, since the acidity of any aqueous system is limited by the fact that the strongest acid that can exist in presence of water is H_3O^+. Any stronger acid simply transfers its protons to H_2O to form H_3O^+.

To measure superacidity it is necessary to define a scale that goes beyond the normal pH scale and that is defined in terms of an experimental measurement. The usual one is the Hammett acidity function, H_0, defined as follows:

$$H_0 = pK_{BH^+} - \log \frac{[BH^+]}{[B]}$$

B is an indicator base, and BH^+ is its protonated form. pK_{BH^+} is $-\log K$ for dissociation of BH^+. The ratio $[BH^+]/[B]$ can be measured spectrophotometrically. By employing bases with very low basicities (very negative pK values), the H_0 scale may be extended to the very negative values appropriate to the superacids. The H_0 scale becomes identical to the pH scale in dilute aqueous solution. Crudely, H_0 values can be thought of as pH values extending below pH = 0.

The first superacid systems to be studied quantitatively were very concentrated solutions of H_2SO_4. Pure H_2SO_4 has $H_0 = -12$; it is thus about 10^{12} times more acidic that 1 molar aqueous H_2SO_4. When SO_3 is added, to produce oleum, H_0 can reach about -15.

Hydrofluoric acid has $-H_0$ of about 11, and this value is increased further on the addition of fluoride ion acceptors such as SbF_5, although numerical values have not been reported.

Superacid media that have found wide application are obtained on addition of AsF_5 or SbF_5 to fluorosulfonic acid, HSO_3F. Pure fluorosulfonic acid has $H_0 = -15$ and is useful because of its wide liquid range, from $-89°$ to $+164°$, its ease of purification, and the fact that it does not attack glass provided it is free of HF. The self-ionization of HSO_3F is

$$2HSO_3F = H_2SO_3F^+ + SO_3F^-$$

and any additive that increases the concentration of $H_2SO_3F^+$ increases the acidity. The addition of about 10 mole % of SbF_5 to HSO_3F increases $-H_0$ to about 19. The highest value of $-H_0$ thus far observed is 19.4 for HSO_3F containing 7% SbF_5. A 1:1 molar mixture of HSO_3F and SbF_5 is colloquially known as "magic acid," although the additional SbF_5 beyond about 10% does little to increase the acidity.

The ability of SbF_5 to increase the acidity of HSO_3F is due mainly to the equilibrium

$$2HSO_3F + SbF_5 = H_2SO_3F^+ + SbF_5(SO_3F)^-$$

Superacid media have been used in many ways. The most obvious is to protonate molecules not normally thought of as bases, for instance, aromatic hydrocarbons. Thus, fluorobenzene in HF/SbF_5 or HSO_3F/SbF_5 produces the ion (7-IX)

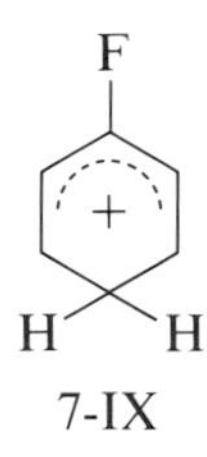

7-IX

Many other cationic species that would be immediately destroyed by even weak bases can be prepared in and isolated from superacid media. These include carbonium ions (Eq. 7-1), and halogen cations (Eq. 7-2), as well as some remarkable polynuclear cations of sulfur, selenium, and tellurium, such as S_4^{2+}, S_8^{2+}, Se_4^{2+}, and Te_4^{2+}.

$$(CH_3)_3COH \xrightarrow{\text{Superacid}} (CH_3)_3C^+ + H_3O^+ \quad (7\text{-}1)$$

$$I_2 \xrightarrow{\text{Superacid}} I_2^+ \text{ and/or } I_3^+ \quad (7\text{-}2)$$

Study Questions

A

1. Name some properties that determine the utility of a solvent.
2. What is the principal effect of dielectric constant?
3. What is the relationship between donor and/or acceptor ability of a solvent and its ability to function as a solvent?

4. Name four protic solvents besides water.
5. Discuss the autodissociation of water and the forms of the hydrated proton.
6. In liquid NH_3 what are the species characteristic of acids? And bases?
7. Describe the three classes of aprotic solvents, mentioning examples of each.
8. Name an important industrial process that employs a molten salt as a solvent.
9. What two properties are generally important in a solvent for electrochemical reactions?
10. Name two common impurities in solvents and indicate how they can be removed.
11. State the Brønsted–Lowry definition of acids and bases.
12. Discuss the solvent system definition and show how it includes the Brønsted–Lowry definition as a special case.
13. Why is acetic acid not an acid in H_2SO_4?
14. To what sort of systems does the Lux–Flood concept apply? Give a representative equation.
15. State the Lewis definition of acids and bases and write three equations that illustrate it, including one that involves a protonic acid.
16. Why is F_3N a much weaker base than H_3N?
17. Why is BBr_3 a stronger acid than BF_3?
18. Describe the origin of the concept of hard and soft acids and bases.
19. Write the type of equation used to account for the combined effect of both electrostatic and covalent forces in acid-base interactions.
20. What are the main properties of each of the following common acids? H_2SO_4,. HNO_3, $HClO_4$, HF.
21. Rank the following acids in order of their strengths: $HClO_2$, $HClO_3$, $HClO_4$; H_2SeO_3, H_3AsO_4, $HMnO_4$, H_2SeO_4. Explain your reasoning.
22. What is the definition of the Hammett acidity function, H_0?
23. Why does the addition of SbF_5 to HSO_3F cause H_0 to become more negative?

B

1. Consider acetic acid as a solvent. Its dielectric constant is ~10. What is its mode of self-ionization likely to be? Name some substances that will be acids and some that will be bases in acetic acid. Will it be a better or poorer solvent than H_2O for ionic compounds?
2. State whether each of the following would act as an acid or a base in liquid HF: BF_3, SbF_5, H_2O, CH_3CO_2H, C_6H_6. In each case write an equation, or equations, to show the basis for your answer.
3. Dimethylsulfoxide is a very good solvent for polar and ionic materials. Why?
4. Why are only superacids good solvents for species such as I_2^+, Se_4^{2+}, S_8^{2+}, etc.? How would they react with less acidic solvents, such as H_2O or HNO_3?
5. Why do you think phosphines, R_3P and phosphine oxides, R_3PO, differ considerably in their base properties?
6. Which member of each pair would you expect to be the more stable: (1) $PtCl_4^{2-}$ or PtF_4^{2-}. (2) $Fe(H_2O)_6^{3+}$ or $Fe(PH_3)_6^{3+}$. (3) F_3B:THF or Cl_3B:THF. (4) $(CH_3)_3B:PCl_3$ or $(CH_3)_3B:P(CH_3)_3$. (5) $(CH_3)_3Al$:pyridine or $(CH_3)_3Ga$:pyridine. (6) $Cl_3B:NCCH_3$ or $(CH_3)_3B:NCCH_3$.
7. In terms of the hard and soft acid and base concept, which end of the SCN^- ion would you expect to coordinate to Cr^{3+}, Pt^{2+}?
8. Estimate pK_1 values for H_2CrO_4, $HBrO_4$, HClO, H_5IO_6, and HSO_3F.
9. Write equations for the probable main self-ionization equilibria in liquid HCN.
10. AlF_3 is insoluble in HF, but dissolves when NaF is present. When BF_3 is passed into the solution, AlF_3 is precipitated. Account for these observations using equations.

Chapter 7
Study Guide

Supplementary Reading

Bell, R. P., *The Proton in Chemistry*, 2nd ed., Chapman Hall, 1973.
Drago, R. S., "A Modern Approach to Acid-Base Chemistry," *J. Chem. Educ.*, *51*, 300 (1974).
Gillespie, R. J., "The Chemistry of Superacid Systems," *Endeavour*, *32*, 541 (1973).
Lagowski, J. J., ed., *The Chemistry of Nonaqueous Solvents*, Vols. 1, 2, Academic Press, 1966, 1967.
Luder, W. F. and Zuffanti, S., *The Electronic Theory of Acids and Bases*, 2nd ed., Dover, 1961.
Pearson, R. G., ed., *Hard and Soft Acids and Bases*, Dowden, Hutchinson and Ross, 1973.

8

the periodic table and the chemistry of the elements

8-1 Introduction

Inorganic chemistry has often been said to comprise a vast collection of unrelatable facts in contrast to organic chemistry, where there appears to be a much greater measure of systematization and order. This is in part true, since the subject matter of inorganic chemistry is far more diverse and complicated and the rules for chemical behavior are often less well established. The subject matter is complicated because even among elements of similar electronic structure, such as the group Li, Na, K, Rb, and Cs, differences arise because of differences in the size of atoms, ionization potentials, hydration, solvation energies, or the like. Some of these differences may be quite subtle—for example, those that enable the human cell and other living systems to discriminate between Li, Na, and K. In short, every element behaves in a different way.

Organic chemistry deals with many compounds that are formed by a *few* elements namely, carbon in sp, sp^2, or sp^3 hybridization states, along with H, O, N, S and the halogens, and less commonly B, Si, Se, P, Hg, etc. The chemistry is mainly one of molecular compounds that are liquids or solids commonly soluble in nonpolar solvents, distillable, or crystallizable and normally stable to, though combustible in, air or oxygen.

Inorganic chemistry by contrast deals with many compounds formed by *many* elements. It involves the study of the chemistry of more than 100 elements that can form compounds as gases, liquids, or solids whose reactions may be—or may have to be—studied at very low or very high temperatures. The compounds may form ionic, extended-covalent, or molecular crystals and their solubility may range from essentially zero in all solvents to high solubility in alkanes; they may react spontaneously and vigorously with water or air. Furthermore, while compounds in organic chemistry almost invariably follow the octet rule (page 194) with a maximum coordination number and a maximum valence of 4 for all elements, inorganic compounds may have coordination numbers up to 14 with those of 4, 5, 6, and 8 being especially common, and valence numbers from -2 to $+8$. Finally, there are types of bonding in inorganic compounds that have no parallel in organic chemistry where σ and $p\pi$—$p\pi$ multiple bonding normally prevail.

Although various concepts help to bring order and system into inorganic chemistry, the oldest and still the most meaningful relies on the periodic table of the elements. This in turn depends on the electron structures of the *gaseous atoms*. As we demonstrated in Chapter 2, by successively adding electrons to the available energy levels we can build up the pattern of electronic structures of the elements to the heaviest one currently known, lawrencium, $Z = 103$. Moreover, on the basis of the electron configurations, the elements can be arranged in a tabular array that has the same form as the conventional long form periodic table. However, the periodic table can also be based entirely on the chemical properties of the elements, and one of its chief uses is to provide a compact mnemonic device for chemical facts.

In this chapter the periodic table is discussed from the chemical instead of the theoretical aspect. In effect, the kinds of chemical observations that originally stimulated chemists such as Mendeleyev to devise the periodic chart are examined. Now, however, we not only can correlate such facts but can interpret them in terms of the electronic structures of the atoms.

THE NATURE AND TYPES OF THE ELEMENTS

It is self-evident that the chemical properties of an element must depend on the electronic structure of the atom. This determines not only how the element can bind to other elements but also to itself. Thus, hydrogen $1s^1$ can clearly only form a diatomic molecule.

8-2
Monatomic Elements: He, Ne, Ar, Kr, Xe, Rn

The noble gases with their closed shell structures are necessarily monatomic. In the vapor mercury, $5d^{10}\ 6s^2$, is also monatomic. However, liquid mercury despite its relatively high vapor pressure and solubility in water and other solvents, has appreciable electrical conductivity and is bright and metallic in appearance. This is because the $6p$ orbitals are able to participate in metallic bonding.

8-3
Diatomic Molecules: H_2, N_2, O_2, F_2, Cl_2, Br_2, I_2

For the halogens and hydrogen, the formation of a single electron-pair bond in a diatomic molecule completes the octet. For nitrogen $2s^2\ 2p^3$ and oxygen $2s^2\ 2p^4$ multiple bonding (page 66) can give a simple diatomic molecule. P_2 and S_2 are also stable at elevated temperatures, but not at 25°.

8-4
Discrete Polyatomic Molecules: P_4, S_n, Se_8

For the second row and heavier elements, $p\pi$—$p\pi$ bonding of the type found in N_2 and O_2 is less effective. The formation by phosphorus and sulfur of the normal number of single electron-pair bonds as expected from their electronic structures, namely three and two respectively, leads either to discrete molecules or to chain structures, which are more stable than the diatomics.

White phosphorus has tetrahedral P_4 molecules (8-I) with the P—P distance 2.21 Å, and the P—P—P angles are, of course, 60°. The small angle implies considerable strain energy, which has been estimated by Pauling to be about 100 kJ per mole of P_4. This means that the total energy of six P—P bonds in the molecule is that much smaller than could be the total energy of six P—P bonds of the same length formed by P atoms with normal bond angles. Thus the structure of the molecule is consistent with its high reactivity. As_4 and Sb_4 molecules are also formed on condensation from vapor but for them the tetrahedral structure is still less stable, readily transforming to the normal form noted below.

P
P—|—P
P

8-I

Sulfur has a profusion of allotropes; these contain multiatom sulfur rings. The largest ring thus far known is S_{20}. The allotropes are referred to as cyclohexa-sulfur, cycloocta-sulfur, and the like. Chains occur in *catenasulfur* S_x. The thermo-dynamically most stable form is orthorhombic sulfur, *Fig. 8-1*.

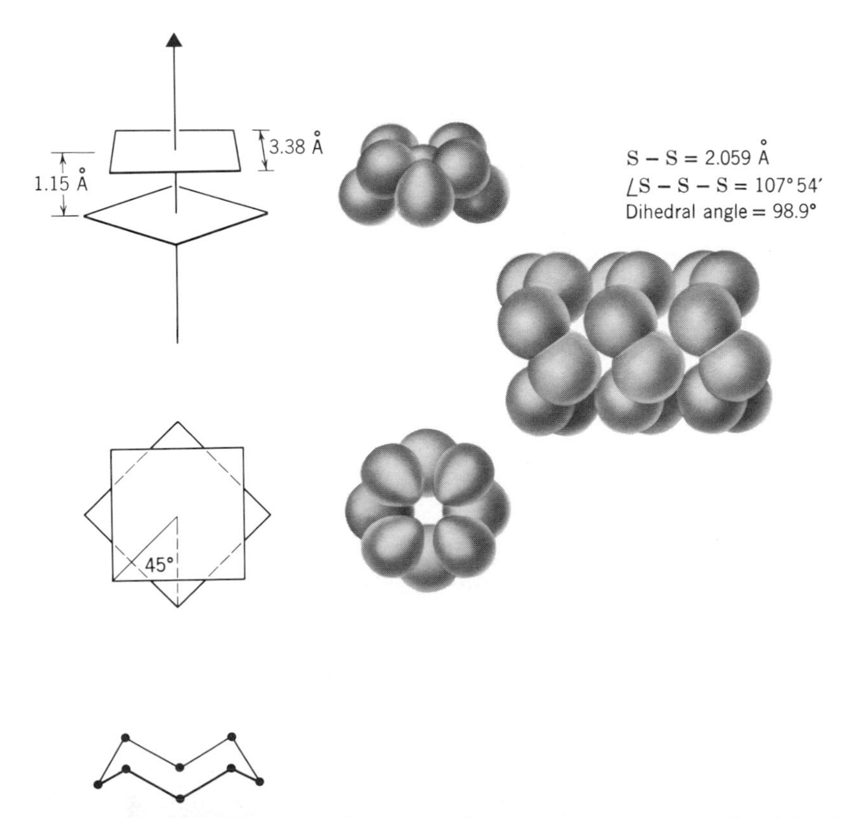

Figure 8-1 *The structure of orthorhombic sulfur in which layers of cyclic S_8 molecules are stacked together.*

8-5
Giant Molecules

Atoms that can form 2, 3, or 4 single covalent bonds to each other give either chains or three-dimensional extended structures. The most important are:

B	C	P[a]	S[a]
	Si	As	Se[a]
	Ge	Sb	Te
	Sn[b]	Bi	

(a) Also molecular (b) Also metallic

Some of these also have allotropes of metallic or molecular types; metal structures are discussed below.

Boron has several allotropes but all are based on B_{12} icosahedra (8-II). α-rhombohedral boron has such B_{12} units packed similar to *ccp* of spheres; the

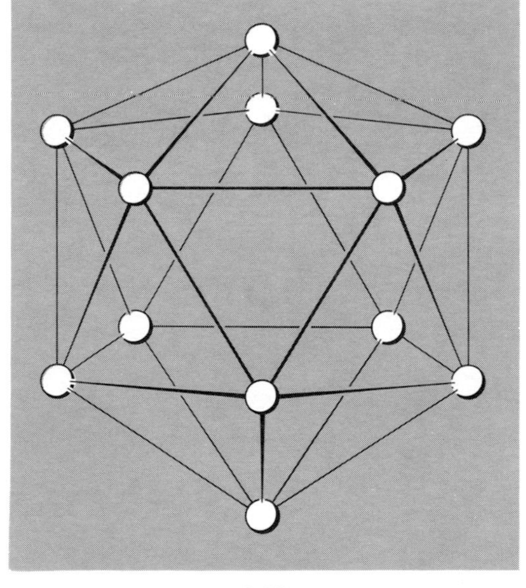

8-II

bonds between the icosahedra are weaker than those within them. A tetragonal form has layers of B_{12} icosahedra connected by single B atoms while β-rhombohedral B also has B_{12} units packed in a complicated way and joined by B—B bonds. The latter, obtained by crystallization of liquid boron is the thermodynamically stable form. The structure clearly accounts for the high melting point (2250 ± 50°C) and for the chemical inertness of boron.

The Group IV elements C, Si, Ge, and Sn all occur with the *diamond* structure shown in *Fig. 8-2*. This has a cubic unit cell but can, for some purposes, be viewed as a stacking of puckered infinite layers. All the atoms are equivalent, each being surrounded by a perfect tetrahedron of four others. Each atom forms a localized 2-electron bond to each of its neighbors. The structure clearly accounts for the extreme hardness of diamond.

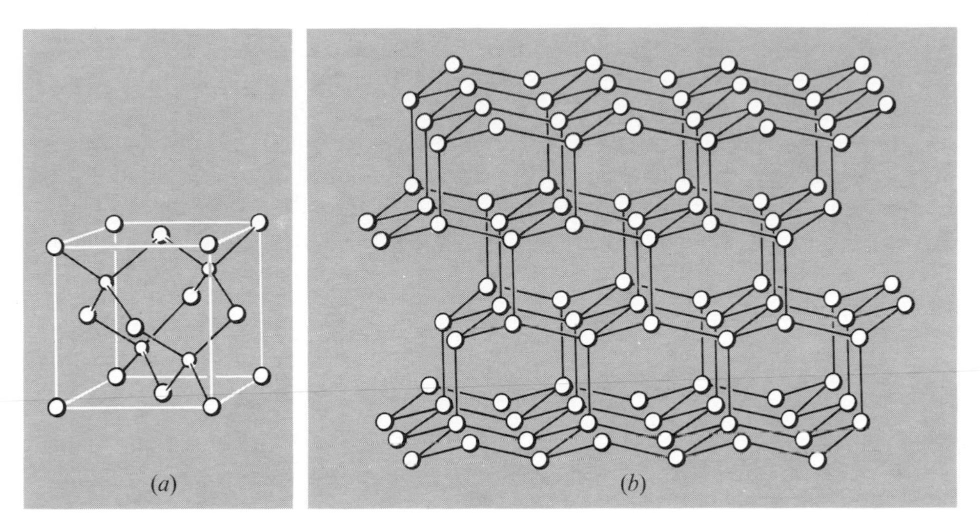

Figure 8-2 The diamond structure seen from two points of view. (a) The conventional unit cell. (b) A view showing how layers are stacked; these layers run perpendicular to the body diagonals of the cube. Remember, however, that this is not a layer structure; its properties are the same in all directions.

Silicon and *germanium* normally have the same structure but *tin* has an equilibrium,

$$\underset{\substack{\text{"grey"} \\ \text{diamond} \\ d^{20} = 5.75}}{\alpha\text{-Sn}} \overset{18^\circ}{\rightleftharpoons} \underset{\substack{\text{"white"} \\ \text{distorted cp.} \\ d^{20} = 7.31}}{\beta\text{-Sn}}$$

where the approach to ideal close packing in white tin compared with the diamond form accounts for the increase in density (d^{20} represents density at 20° C in g cm^{-3}).

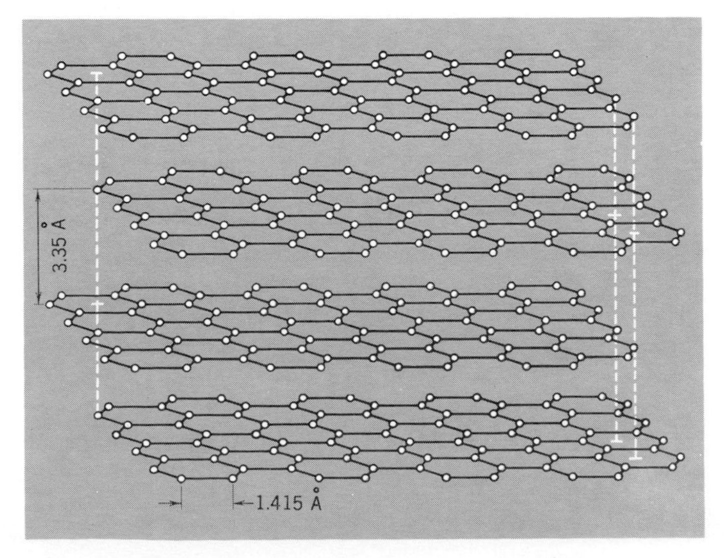

Figure 8-3 The normal structure of graphite.

Carbon also exists as *graphite*, which has the layer structure shown in *Fig. 8-3*. The separation of the layers, 3.35 Å, is about the sum of the van der Waals radii for C and indicates that the forces between the layers should be weak. This accounts for the softness and lubricity of graphite, since the layers can easily slip over one another. Each C atom is surrounded by only three neighbors; after forming one σ bond with each neighbor, each C atom still has one electron and these electrons are paired up into a system of π bonds as in (8-III). Resonance leads to essential equivalence so that the C—C bond distances are all 1.415 Å.

8-III

This is a little longer than the C—C distance in benzene where the bond order is 1.5 and corresponds to a C—C bond order in graphite of $\sim$1.33. Since $p\pi$—$p\pi$ multiple bonding is clearly involved, the other Group IV elements cannot form this type of structure. The π systems in the layers allow of electrical conductivity, and graphite is used for electrode materials. Despite having the diamond structure, both Si and Sn have appreciable conductivity (resistivity 10 and 11 $\mu\Omega$ cm at 0°, respectively) while Ge is semiconducting and has resistivity 5×10^7 $\mu\Omega$ cm at 22°C.

In Group V, *phosphorus* has numerous polymorphs, the red form being structurally uncharacterized. Black phosphorus, obtained by heating white P under pressure has the structure shown in *Fig. 8-4*. Each P atom is bound to three neighbors by single bonds, 2.17 to 2.20 Å. The double layers thus formed are stacked in layers with an interlayer distance of 3.87 Å. As for graphite, the layer

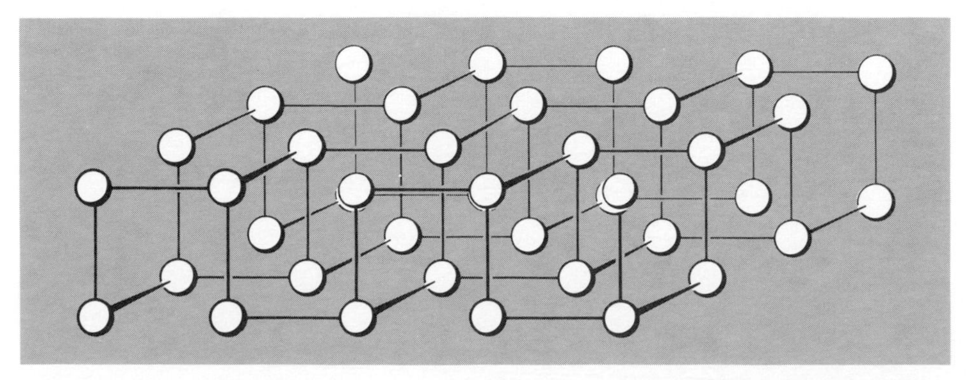

Figure 8-4 *The arrangement of atoms in the double layers found in crystalline black phosphorus.*

structure leads to flakiness of the crystals. It also accounts for the lack of reactivity, for example, to air, compared to P_4.

Arsenic, Sb and Bi all form crystals whose structures are similar to black P. However, they are bright and metallic in appearance and have resistivities of *ca.* 30, 40, and 105 $\mu\Omega$ cm, which are comparable to those of metals such as Ti or Mn (42 and 185 $\mu\Omega$ cm, respectively).

The chain form of *sulfur*, *catena sulfur* is the main component of the so-called plastic sulfur obtained when molten sulfur is poured into water. It can be drawn into long fibers that contain helical chains of sulfur atoms. It slowly transforms to orthorhombic S_8.

The stable form of *selenium*, grey, metal-like crystals obtained from melts, contains infinite spiral chains. There is evidently weak interaction of a metallic nature between neighboring atoms of different chains, but in the dark the electrical conductivity of selenium is not comparable to that of true metals (resistivity 2×10^{11} $\mu\Omega$ cm). However, it is notably photoconductive and is hence used in photoelectric devices.

Tellurium is isomorphous with grey Se, although it is silvery white and semi-metallic (resistivity 2×10^5 $\mu\Omega$ cm). The resistivity of S, Se, and Te has a negative temperature coefficient, usually considered a characteristic of nonmetals.

8-6 Metals

The majority of the elements are metals. These have many physical properties different from those of other solids: notably (1) high reflectivity, (2) high electrical conductance, decreasing with increasing temperature, (3) high thermal conductance, and (4) mechanical properties such as strength and ductility. There are three basic metal structures: *cubic* and *hexagonal close packed* (illustrated page 102) and *body-centered cubic, bcc* (*Fig. 8-5*). In *bcc* packing each atom has only 8 instead of 12 nearest neighbors, although there are 6 next nearest neighbors that are only about 15% further away. It is only 92% as dense an arrangement as the *hcp* and *ccp* structures. The distribution of these three structure types, *hcp*, *ccp*, and *bcc*, in the periodic table is shown in *Fig. 8-6*. The majority of the metals deviate slightly from the ideal structures, especially those with *hcp* structures. For the *hcp* structure the ideal value of c/a, where c and a are the hexagonal unit-cell edges, is 1.633. All metals having this structure have a smaller c/a ratio (usually 1.57 – 1.62) except zinc and cadmium.

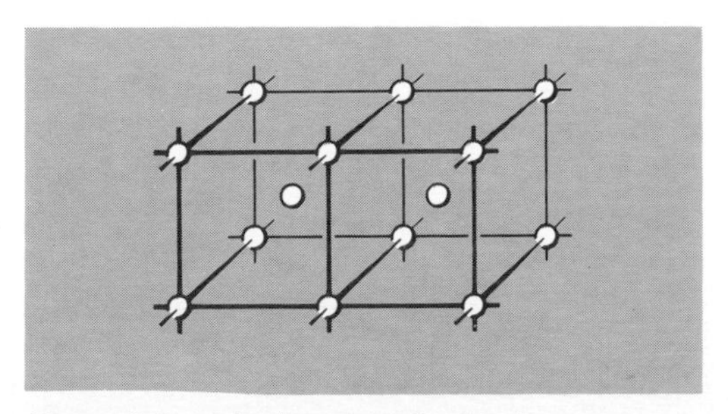

Figure 8-5 A body-centered cubic, bcc, structure.

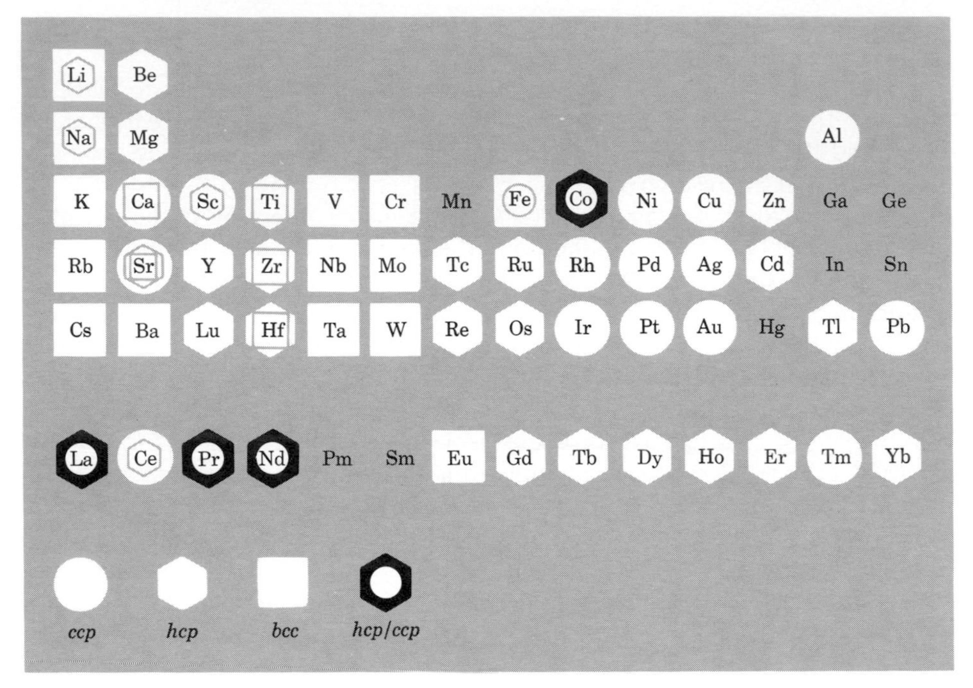

Figure 8-6 *The occurrence of hexagonal close-packed (hcp), cubic close-packed (ccp) and body-centered cubic (bcc) structures among the elements. Where two or more symbols are used, the largest represents the stable form at 25° C. The symbol labeled hcp/ccp signifies a mixed* ...ABCABABCAB... *type of close packing, with overall hexagonal symmetry. (Adapted, with permission, from H. Krebs,* Grundzüge der Anorganishen Kristallchemie, *F. Enke Verlag, 1968.).*

The characteristic physical properties of metals as well as the high coordination numbers (either 12 or 8 nearest neighbors plus 6 more that are not too remote) suggest that the *bonding in metals* is different from that in other substances. There is no ionic contribution, and it is also impossible to have 2-electron covalent bonds between all adjacent pairs of atoms, since there are neither sufficient electrons nor sufficient orbitals. An explanation of the characteristic properties of metals is given by the so-called band theory. This is very mathematical but the principle can be illustrated.

Imagine an array of atoms so far apart that their atomic orbitals do not interact. Now suppose this array contracts. The orbitals of neighboring atoms begin to overlap and interact with each other. So many atoms are involved that at the actual distances in metals, the interaction forms essentially continuous energy bands that spread through the metal (*Fig. 8-7*). The electrons in these bands are *completely delocalized*. Observe also that some bands may overlap: in Na used as an illustration in *Fig. 8-7*, the 3*s* and 3*p* bands overlap.

The energy bands can also be depicted as in *Fig. 8-8*. Here energy is plotted horizontally, and the envelope indicates on the vertical the number of electrons that can be accommodated at each value of the energy. Shading is used to indicate filling of the bands.

Completely filled or completely empty bands, as shown in *Fig. 8-8a*, do not permit net electron flow and the substance is an *insulator*. Covalent solids can be

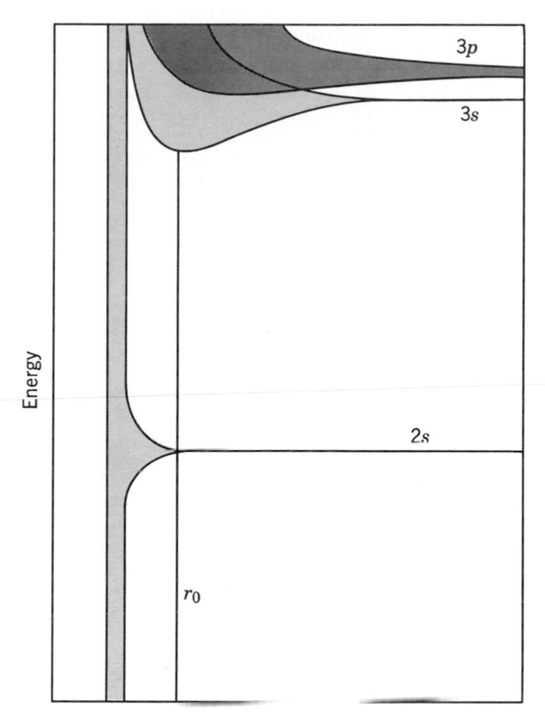

Figure 8-7 *Energy bands of sodium as a function of internuclear distance. r_0 represents the actual equilibrium distance. [Reproduced by permission from J. C. Slater,* Introduction to Chemical Physics, *McGraw-Hill Book Co., 1939.]*

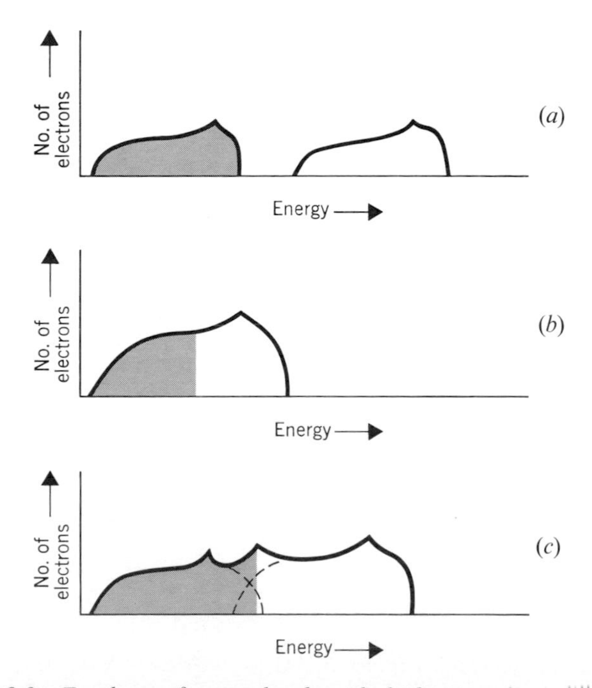

Figure 8-8 *Envelopes of energy bands, with shading to indicate filling.*

discussed from this point of view (though it is unnecessary to do so) by saying that all electrons occupy low-lying bands (equivalent to the bonding orbitals) while the high-lying bands (equivalent to antibonding orbitals) are entirely empty. Metallic conductance occurs when there is a partially filled band, as in *Fig. 8-8b*; the transition metals, with their incomplete sets of *d* electrons, have partially filled *d* bands and this accounts for their high conductances. Overlapping bands, as in Na, are illustrated in *Fig. 8-8c*.

Cohesive Energies of Metals. The strength of binding among the atoms in metals can be measured by the *enthalpies of atomization*, *Fig. 8-9*. Cohesive energy

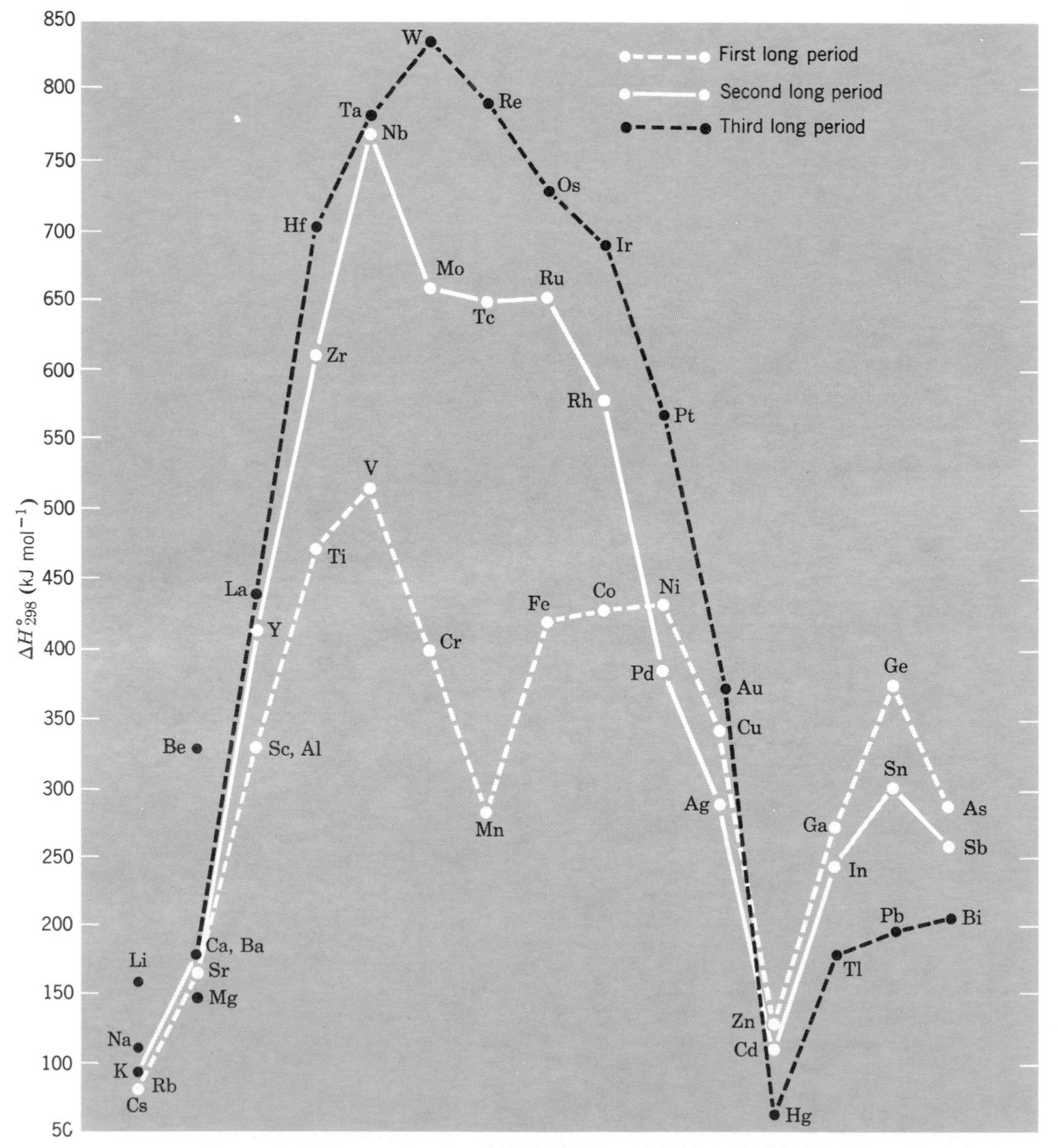

Figure 8-9 *Heats of atomization of metals,* ΔH°_{298} *for* M(s) → M(g). [*Reproduced by permission from W. E. Dasent,* Inorganic Energetics, *Penguin Books, Ltd., 1970.*]

maximizes with elements having partially filled *d* shells, that is, with the transition metals. However, it is particularly with the elements near the middle of the second and third transition series, especially Nb—Ru and Hf—Ir, that the energies are largest, reaching 837 kJ mol^{-1} for tungsten. It is noteworthy that these large cohesive energies are principally due to the structure of the metals where high coordination numbers are involved. For a *hcp* or *ccp* structure, there are 6 bonds per metal atom (since each of the 12 nearest neighbors has a half-share in each of the 12 bonds). Each bond, even when cohesive energy is 800 kJ mol^{-1}, has an energy of only 133 kJ mol^{-1}, roughly half the C—C bond energy in diamond where each carbon atom has only four neighbors, but there are three times as many of them.

CHEMISTRY OF THE ELEMENTS IN RELATION TO THEIR POSITION IN THE PERIODIC TABLE

We can now proceed to more detailed commentary on the chemical reactivity and types of compounds formed by the elements. The periodic table forms the basis for the discussion, starting with the simplest chemistry, namely that of hydrogen, and proceeding to the heaviest elements.

8-7
Hydrogen $1s^1$

The chemistry of hydrogen depends on three electronic processes:

1. *Loss of the* 1s *valence electron.* This forms merely the proton, H^+. Its small size, $r \sim 1.5 \times 10^{-13}$ cm, relative to atomic sizes $r \sim 10^{-8}$ cm, and its small charge result in a unique ability to distort the electron clouds surrounding other atoms. The proton *never* exists as such except in gaseous ion beams. It is invariably associated with other atoms or molecules. Although the hydrogen ion in water is commonly written as H^+, it is actually H_3O^+ or $H(H_2O)_n{}^+$.

2. *Acquisition of an electron.* The H atom can acquire an electron forming the *hydride ion*, H^- with the He $1s^2$ structure. This ion exists only in crystalline hydrides of the most electropositive metals, for example, NaH, CaH_2.

3. *Formation of an electron pair bond.* Nonmetals and even many metals can form covalent bonds to hydrogen.

The chemistry of hydrogen compounds is highly dependent on the nature of the element and on the nature of other groups or ligands also present. The extent to which compounds dissociate in polar solvents and act as acids

$$HX \rightleftharpoons H^+ + X^-$$

is particularly dependent on the nature of X.

Also important is the electronic structure and coordination number of the whole molecule. Consider BH_3, CH_4, NH_3, OH_2, and FH. The first acts as a Lewis acid, and dimerizes instantly to B_2H_6 (page 80); CH_4 is unreactive and neutral; NH_3 has a lone pair and is a base; H_2O with two lone pairs can act as a base or as a very weak acid; HF, a gas, is a much stronger though still weak acid in aqueous solution.

All H—X bonds necessarily have some polar character with the dipole oriented $\overset{\delta+}{H}—\overset{\delta-}{X}$ or $\overset{\delta-}{H}—\overset{\delta+}{X}$. The term "hydride" is usually given to those compounds in which the negative end of the dipole is on hydrogen, for example, in SiH_4, $\overset{\delta+}{Si}—\overset{\delta-}{H}$. However, although HCl as $\overset{\delta+}{H}—\overset{\delta-}{Cl}$ is a strong acid in aqueous solution, nevertheless, it is a gas and is properly termed a covalent hydride.

8-8 Helium, $1s^2$ and the Noble Gases, ns^2np^6

The second element He, Z = 2 has the closed 1*s* shell; its very small size leads to some physical properties that are unique to liquid helium. The physical properties of the other noble gases vary systematically with size. Although the first ionization energies are high, consistent with their chemical inertness, the values fall steadily as the size of the atom increases. The ability to enter into chemical combination with other atoms should increase with decreasing ionization potential and decreasing energy of promotion to states with unpaired electrons—that is, $ns^2np^6 \rightarrow ns^2np^5(n+1)s$. The threshold of chemical activity is reached at Kr but few compounds have been isolated. The reactivity of Xe is much greater, and many compounds of O and F are known (Chapter 21). The reactivity of Rn is presumably greater still but, since the longest lived isotope, ^{222}Rn, has a half-life of only 3.825 days, only tracer studies can be made.

8-9 Elements of the First Short Period

The third element, Li, Z = 3 has the structure $1s^22s$. With increasing *Z*, electrons enter the 2*s* and 2*p* levels until the closed shell configuration, $1s^22s^22p^6$, is reached at neon. The seven elements Li to F constitute the first members of the *groups* of elements.

Although these elements have many properties in common with the heavier elements of their respective groups—which is to be expected in view of the similarity in the outer electronic structures of the *gaseous* atoms—they nevertheless show highly individual behavior in many important respects. We have already seen that O_2 and N_2 form diatomic molecules whereas their congeners, S and P, form polyatomic molecules or chains. Indeed, the differences between the chemistries of B, C, N, and O and Al, Si, P, and S and the heavier members of these groups are sufficiently striking that in many ways it is not useful to regard the first period elements as prototypes for their congeners. The closest analogies between the first row and the heavier elements are for Li and F, followed by Be.

The increase in nuclear charge and consequent changes in the extranuclear structure result in extremes of physical and chemical properties. In *Fig. 8-10* are given the first ionization enthalpies. The low ionization enthalpy for *lithium* is in accord with facile loss of an electron to form the Li^+ ion, which occurs both in solids and in solution. It accords also with the high reactivity of lithium to oxygen, nitrogen, water, and many other elements.

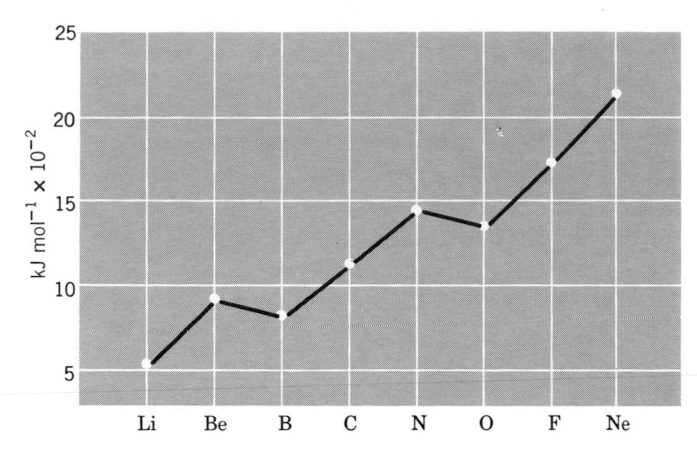

Figure 8-10 *First ionization enthalpies of the elements* Li—Ne.

For *beryllium* the first (899 kJ mol^{-1}) and especially the second (1757 kJ mol^{-1}) ionization enthalpies are sufficiently high that total loss of both electrons to give Be^{2+} does not occur even with the most electronegative elements. Even in BeF_2 the Be—F bonds have appreciable covalent character. The ion in aqueous solution, $[Be(H_2O)_4]^{2+}$, is very strongly aquated and undergoes, hydrolysis quite readily to give species with Be(OH) bonds.

For the succeeding elements, the absence of any simple cations under any conditions is to be expected from the high ionization enthalpies. Note that the values (*Fig. 8-10*) for B, C, and N increase regularly but that they are lower than the values which would be predicted by extrapolation from Li and Be. This arises because p electrons are less penetrating than s electrons—they are therefore shielded by the s electrons and are removed more easily. Another discontinuity occurs between N and O. This one is due to the fact that the $2p$ shell is half-full, that is, $p_x p_y p_z$ at N. The p electrons added in O, F, and Ne thus enter p orbitals that are already single occupied. Hence, they are partly repelled by the p electron already present in the same orbital and are thus less tightly bound.

The electron attachment energies (page 8) rise from Li to F and the electronegativity (page 50) also rises.

Boron, $2s^2 2p$ has no cationic chemistry and is bound covalently in all its compounds, such as oxoanions, organoboron compounds and hydrides.

Anion formation first appears for carbon, which forms $C_2{}^{2-}$ and some other polyatomic ions, although the existence of C^{4-} is uncertain. N^{3-} ions are stable in nitrides of highly electropositive elements. Oxide, O^{2-}, and fluoride, F^-, are common in solids but observe that O^{2-} ions cannot exist in aqueous solutions, compare,

$$O^{2-} + H_2O = 2OH^- \qquad K > 10^{22}$$

$$F^- + H_2O = HF + OH^- \qquad K = 10^{-7}$$

Carbon is a true nonmetal and its chemistry is dominated by single, double and triple bonds to itself or to nitrogen, oxygen, and a few other elements. What distinguishes carbon from other elements is its unique ability to form chains of carbon–carbon bonds (called catenation) in compounds—as distinct from the element itself.

Nitrogen. Nitrogen gas, N_2, is relatively unreactive because of the great strength of the N≡N bond and its electronic structure (page 67). Nitrogen compounds are covalent, usually involving three single bonds, although multiple bonds such as C≡N or Os≡N can exist. With electropositive elements, ionic nitrides containing N^{3-} may be formed.

Oxygen. The diatomic molecule has two unpaired electrons (page 66) and consequently is very reactive. There is an extensive chemistry with covalent bonds as in $(CH_3)_2C{=}O$, $(C_2H_5)_2O$, CO, SO_3, and the like. However, well-defined oxide ions O^{2-}, $O_2^{\ -}$, and $O_2^{\ 2-}$ exist in crystalline solids. Hydroxide ions, OH^-, exist both in solids and in solutions, although in hydroxylic solvents the OH^- ion is doubtless hydrated via hydrogen bonds.

Fluorine is extremely reactive due largely to the low bond energy in F_2. This is a result, in part, of repulsions by nonbonding electrons. Ionic compounds containing F^- ions and covalent compounds containing X—F bonds are well established. Owing to the high electronegativity of fluorine, such covalent bonds are generally quite polar in the sense $X^+{-}F^-$.

Covalent bonds. A few points may be mentioned here.

1. Note that Be, B, and C have fewer electrons in their ground states than the number of electron pair bonds they normally form. This has been explained previously (page 72) in terms of promotion to valence states.

2. The first row elements obey the *octet rule.* Since they have only four orbitals ($2s$, $2p_x$, $2p_y$, $2p_z$) in their valence shell, there are *never* more than eight electrons in their valence shells. This means that the *maximum* number of electron pair bonds is *four*. The octet rule breaks down in the second short period. For example, phosphorus, $3s^23p^33d^0$ can be excited to a valence state $3s^13p^33d^1$ with an expenditure of energy so modest that the heat of formation of the two additional bonds will more than compensate for it. On the other hand, promotion of N, $2s^22p^3$ to any state with five unpaired electrons, such as $2s^12p^33d^1$, would require more promotional energy than could be recovered by the extra bond formation energy.

For C, promotion from $2s^22p^2$ to $2s2p^3$ gives the valence of 4. For *N*, $2s^22p^3$ only three of the five electrons can possibly be unpaired, in O only two and in F, only one. Hence, these elements are limited to valences of 3, 2, and 1. On the other side of C, that is in Li, Be, and B, the valences are less than 4 because of lack of electrons to occupy the orbitals, so that by electron-sharing these can show valences of only 1, 2, and 3, respectively.

3. Where there are fewer electrons than are required to fill the energetically useful orbitals, as in trivalent boron compounds such as BCl_3, BF_3 and $B(CH_3)_3$, there is a strong tendency to utilize these orbitals by combining with compounds that have an excess of electrons. Such compounds are those of trivalent nitrogen such as NH_3, $N(CH_3)_3$, etc., or oxygen, such as H_2O, $(C_2H_5)_2O$, etc., that have unshared electron pairs. The former are thus *acceptors* of electrons and the latter are *donors*. For these compounds the terms *Lewis acids* and *Lewis bases* are commonly used, since the general concept of acids as acceptors and bases as donors of electron pairs was developed by G. N. Lewis. The formation of a dative bond is shown in *Fig. 8-11*.

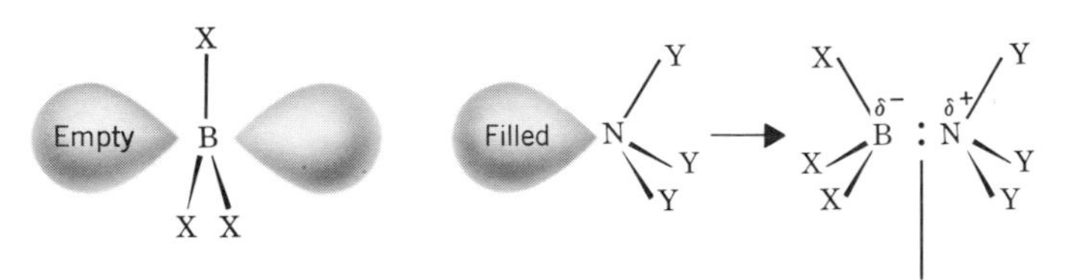

Figure 8-11 *The formation of a dative bond between boron in a* BX_3 *acceptor and nitrogen in an* NY_3 *donor.*

Notice that while nitrogen compounds have only one unshared pair, $:NR_3$, oxygen compounds have two $:\ddot{O}R_2$; normally only one of the electron pairs is used and only in a very few cases does oxygen form four bonds. Beryllium compounds with two empty orbitals usually fill these by forming compounds with two donor molecules, BeX_2L_2.

Note that such donor-acceptor behavior is not confined to first-row elements and is quite general. Adducts may be formed between compounds whenever one has empty orbitals and the other has unshared electron pairs.

Compounds of many elements may act as acceptors, but donors are commonly compounds of trivalent N, P, and As and compounds of divalent O and S. However, a very important class of donors are the halide and pseudohalide ions and ions such as hydride, H^-, and carbanions such as CH_3^- or $C_6H_5^-$. Some representative examples are

$$BF_3 + F^- = BF_4^-$$

$$VF_5 + F^- = VF_6^-$$

$$AlCl_3 + Cl^- = AlCl_4^-$$

$$PtCl_4 + 2Cl^- = PtCl_6^{2-}$$

$$Ni(CN)_2 + 2CN^- = Ni(CN)_4^{2-}$$

$$Co(NCS)_2 + 2SCN^- = Co(NCS)_4^{2-}$$

$$BH_3 + H^- = BH_4^-$$

$$Al(CH_3)_3 + CH_3^- = Al(CH_3)_4^-$$

Lewis-base behavior is also shown by some transition metal compounds, as we discuss later. One example is the compound $(\eta^5\text{-}C_5H_5)_2ReH$, which is as strong a base to protons as NH_3. The reason why some atoms succeed in increasing their coordination numbers from 3 to 4 but seldom from 2 to 4 can be understood if we consider the polar nature of the dative bond. The donor and acceptor molecules are both electrically neutral. When the bond is formed, the donor atom has, in effect, lost negative charge rendering it positive. It has only half ownership of an electron pair that formerly belonged to it entirely. Conversely, the acceptor atom now has extra negative charge. This would be true for complete sharing of the electron pair (8-IV). Lesser polarity is introduced if the electron pair is still more the property of the donor atom than the acceptor (8-V), in which case we indicate only charges $\delta+$ and $\delta-$ on the atoms.

$$\overset{-e}{\mathrm{B}}:\overset{+e}{\mathrm{N}} \qquad \overset{\delta-}{\mathrm{B}}.....:\overset{\delta+}{\mathrm{N}} \qquad \mathrm{R}-\overset{\delta+}{\ddot{\mathrm{O}}}(\mathrm{R}):\overset{\delta-}{\mathrm{BX_3}}$$

8-IV 8-V 8-VI

This charge separation can only be achieved by doing work against coulomb forces, which we must assume is more than compensated by the bond energy when a stable system results. However, if we take a case where one donor bond has been formed (8-VI), then the second unshared pair on oxygen is further restrained by the positive charge on O which arises from the dative bond already formed. There is thus much more coulombic work to be done in forming a second dative bond—enough apparently to make this process energetically unfavorable. Steric hindrance between the first acceptor and a second would also militate against addition of a second. Note that this electrostatic argument is basically the same as that used to explain relative dissociation constants in polyfunctional acids (page 176).

8-10
The Second Short Period

The elements Na, Mg, Al, Si, P, S, and Cl, constitute the second short period. Although their outer electronic structures are similar to those of the first short period, the chemistries differ considerably. In particular, the chemistries of Si, P, S and, to a large extent, chlorine are completely different from their corresponding first-row element. The second-row elements do, however, give a better guide to the chemistries of the heavier elements in their respective groups than do the first-row members. It is the nonmetallic elements in particular that differ substantially. The main reasons for this are as follows.

1. It is not possible to form $p\pi—p\pi$ multiple bonds such as Si=Si, Si=O or P=P. Most likely, this is because, in order to approach close enough to get good overlap of $p\pi$ atomic orbitals, the heavier atoms would encounter large repulsive forces due to overlapping of their filled inner shells, whereas the small compact inner shell of the first-row elements, that is, just $1s^2$, does not produce this repulsion.

The result is that, as we have seen, the nature of the elements themselves is strikingly different. As a dramatic example, silicon has no chemistry comparable to the carbon chemistry of C=C, C≡C, >C=O, and >C=N bonds; that is, no analogs of alkenes, alkynes, ketones, nitriles, and the like. Whereas CO_2 is a gas, SiO_2 forms an infinite polymer.

2. Although in certain types of compounds of P, S, and Cl such as Cl_3PO, Cl_2SO, SO_2, ClO_4^-, ClO_2, etc., there is some multiple bonding, this occurs by an entirely different mechanism involving d orbitals. The low-lying $3d$ orbitals can be utilized not only for $p\pi—d\pi$ multiple bonding, but also for additional

bond formation. The octet rule now no longer holds rigorously and is indeed commonly violated.

3. The possibility of using the 3*d* orbitals then allows promotion to valence states leading to formation of five or six bonds. Hence, we find compounds such as PCl_5 or SF_6 and silicon can form species with 5 and 6 coordination also as in SiF_6^{2-}. For silicon, even where there is some analogy with carbon chemistry, as in compounds with single bonds, the reactions and mechanisms operating in silicon chemistry may be vastly different. A simple example is the unreactivity of CCl_4 toward water, whereas $SiCl_4$ is instantly hydrolyzed.

4. The shapes of molecules as well as the nature of the bonds also differ. This topic is discussed in detail in Section 3-10, page 83, where xenon compounds are also discussed.

5. Even the cation- and anion-forming elements also differ. Thus while beryllium forms only $[Be(H_2O)_4]^{2+}$, the magnesium aquo ion is $[Mg(H_2O)_6]^{2+}$, and there are substantial differences between the chemistries of Li and Na (page 217). Aluminum is an electropositive metal totally different from boron, although in certain covalent compounds there are some similarities.

For Group VII, the Cl—Cl bond strength is actually higher than that of F_2 (page 10) and Cl_2 is much less reactive than F_2, while the electronegativity of the atom is also considerably lower. Solid chlorides commonly have structures quite different from the corresponding fluorides and are much closer to those given by sulfides.

8-11
Remainder of Nontransition Elements

The first and especially the second-row elements just discussed give some guide to the chemistry of the elements in their respective groups. The main features are as follows.

Group I. All the elements (Table 8-1) are highly electropositive giving +1 ions. Of all the groups in the periodic table, these metals show most clearly

Table 8-1 *Some Properties of Group I Elements*

Element	Electronic configuration	mp °C	Ionic radius Å	$E°$V[a]
Li	[He]2*s*	180	0.60	−3.0
Na	[Ne]3*s*	98	0.96	−2.7
K	[Ar]4*s*	64	1.33	−2.9
Rb	[Kr]5*s*	39	1.48	−3.0
Cs	[Xe]6*s*	29	1.69	−3.0
Fr[b]	[Rn]7*s*	—	—	—

[a] For $M^+(aq) + e = M(s)$.
[b] All isotopes are radioactive with short half-lives.

the effect of increasing size and mass on chemical properties. Thus, as examples, the following *decrease* from Li to Cs: (a) melting points and heat of sublimation of the metals; (b) lattice energies of salts except those with very small anions (because of irregular radius ratio effects); (c) effective hydrated radii and hydration energies, and (d) strength of covalent bonds in M_2 molecules.

Group II. Some properties of the elements are given in Table 8-2. Calcium, Sr, Ba, and Ra are also highly electropositive forming +2 ions. Systematic group trends are again shown, for example, by increasing (a) hydration tendencies, (b) insolubilities of sulfates, and (c) thermal stabilities of carbonates or nitrates.

Table 8-2 *Some Properties of Group II Elements*

Element	Electronic configuration	mp °C	Ionic radius M^{2+}, Å	E°V[a]
Be	$[He]2s^2$	1280	0.31	−1.85
Mg	$[Ne]3s^2$	650	0.65	−2.37
Ca	$[Ar]4s^2$	840	0.99	−2.87
Sr	$[Kr]5s^2$	770	1.13	−2.89
Ba	$[Xe]6s^2$	725	1.35	−2.90
Ra	$[Rn]7s^2$	700	1.40	−2.92
Zn	$[Ar]3d^{10}4s^2$	420	0.74	−0.76
Cd	$[Kr]4d^{10}5s^2$	320	0.97	−0.40
Hg	$[Xe]4s^{14}5d^{10}6s^2$	−39	1.10	+0.85

[a] For $M^{2+}(aq) + 2e = M(s)$.

Zinc, Cd, and Hg are also classed in Group II, as IIB. They have two *s* electrons outside filled *d* shells since they follow Cu, Ag, and Au, respectively, after the first, second, and third transition series elements. The chemistries of Zn and Cd are quite similar, but the polarizing power of the M^{2+} ions is larger than would be predicted by comparing the radii with those of the Mg—Ra group. This can be associated with the greater ease of distortion of the filled *d* shell compared with the noble gas shell of the Mg—Ra ions. Zinc and Cd are quite electropositive resembling Mg in their chemistry, although there is a greater tendency to form complexes with NH_3, halide ions, and CN^-.

Mercury is unique. It has a high *positive* potential, and the Hg^{2+} ion does *not* resemble Zn^{2+} or Cd^{2+}. For example, the formation constants for, say, halide ions, are orders of magnitude greater than for Cd^{2+}. Mercury also readily forms the mercurous ion, which has a metal–metal bond, $^{+}Hg—Hg^{+}$.

Group III. Some properties of the elements are given in Table 8-3. This group is quite a numerous one, since it contains the Group IIIA elements, Sc, Y, La, and Ac, and the Group IIIB elements, Al, Ga, In, and Tl. In addition, all of the lanthanide elements could be included, since their chemistry is similar to that of the IIIA elements.

Table 8-3 Some Properties of the Group III Elements

Element	Electronic configuration	mp °C	Ionic radius Å[c]	E° V[a]
Sc	$[Ar]3d^1 4s^2$	1540	0.81	−1.88
Y	$[Kr]4d^1 5s^2$	1500	0.93	−2.37
La	$[Xe]5d^1 6s^2$	920	1.15	−2.52
Ac[b]	$[Rn]6d^1 7s^2$	1050	1.11	~ −2.6
Al	$[Ne]3s^2 3p$	660	0.50	−1.66
Ga	$[Ar]3d^{10} 4s^2 4p$	30	0.62	−0.53
In	$[Kr]4d^{10} 5s^2 5p$	160	0.81	−0.34
Tl	$[Xe]4f^{14} 5d^{10} 6s^2 6p$	300	0.95	+0.72

[a] For $M^{3+}(aq) + 3e = M(s)$.
[b] Isotopes are all radioactive.
[c] For M^{3+}.

However, we consider the lanthanides separately because of their special position in the periodic table. Notice that in the Sc—Ac group the 3-valence electrons are $d^1 s^2$ compared with $s^2 p^1$ for the Al—Tl group. Despite this occupancy of the d levels, the elements show no transition metal-like chemistry. They are highly electropositive metals, and their chemistry is primarily one of the +3 ions which have the noble gas configuration.

Scandium with the smallest ionic radius has chemical behavior intermediate between that of Al, which has a considerable tendency to covalent bond formation, and the mainly ionic natures of the heavier elements.

Gallium, In and Tl, like Al are borderline between ionic and covalent in compounds, even though the metals are quite electropositive and they form M^{3+} ions.

The +1 state becomes progressively more stable as the group is descended, and for Tl the Tl^{I}—Tl^{III} relationship is a dominant factor of the chemistry. The occurrence of an oxidation state two below the group valence is sometimes attributed to the *inert pair effect*, which first makes itself evident here. It could be considered to apply in the low reactivity of mercury, but it is more pronounced still in Groups IV and V. The term refers to the resistance of a pair of s electrons to be lost or to participate in covalent bond formation. Thus Hg is difficult to oxidize allegedly because it contains only an inert pair ($6s^2$), Tl forms Tl^{I} rather than Tl^{III} because of the inert pair in the valence shell ($6s^2 6p$), and so on. The concept of the inert pair tells us little, if anything, about the ultimate reasons for the stability of lower oxidation states. It is a useful label. The true cause of the phenomenon is not intrinsic inertness—that is, unusually high ionization energy of the s^2 pair, but rather the decreasing strength of bonds as a group is descended. Thus the sum of the second and third ionization enthalpies is lower for In (4506 kJ mol^{-1}) than for Ga (4920 kJ mol^{-1}) with Tl (4825 kJ mol^{-1}) intermediate. There is, however, a steady decrease in the mean thermochemical bond energies, for example, in the chlorides, Ga 242, In 206, and Tl 153 kJ mol^{-1}.

Group IV. Some properties of the elements are given in Table 8-4. Note that we are now considering the Group IVB in the periodic table, since Group IVA

Table 8-4 Some Properties of Group IVB Elements

Element	Electronic configuration	mp °C	Covalent radius Å	Self-bond Energy kJ mol^{-1}
C	$[He]2s^22p^2$	>3550	0.77	356
Si	$[Ne]3s^23p^2$	1410	1.17	210–250
Ge	$[Ar]3d^{10}4s^24p^2$	940	1.22	190–210
Sn	$[Kr]4d^{10}5s^25p^2$	232	1.40	105–145
Pb	$[Xe]4f^{14}5d^{10}6s^26p^2$	327	1.44	?

comprises the transition metals Ti, Zr, and Hf, whose chemistry we consider separately. This pattern holds true for the remaining Groups V to VII.

There is no more striking example of the enormous discontinuity in properties between the first- and second-row elements, followed by a relatively smooth change toward metallic character, than in Group IVB. Carbon is nonmetallic; silicon is also nonmetallic but little of the chemistry of Si can be inferred from that of C. Germanium is much like silicon, although it shows more metallic-type behavior in its chemistry. Tin and lead are metals and both have some metal-like chemistry, especially in the divalent state.

The main chemistry in the IV oxidation state for all the elements is essentially one that involves covalent bonds and molecular compounds. Typical examples are $GeCl_4$ and $PbEt_4$. There is a decrease in the tendency to *catenation*, which is such a feature of carbon chemistry in the order $C \gg Si > Ge \simeq Sn \simeq Pb$. This is partly due to the diminishing strength of the C—C, Si—Si, and the like, bonds (Table 8-4). The strengths of covalent bonds to other atoms also generally decrease in going from C to Pb.

The divalent state. Although in CO the *oxidation* state of C is *formally* taken to be 2, this *is* only a formalism and carbon uses more than two valence electrons in bonding. True divalence is found only in *carbenes* such as $:CF_2$, and these species are very reactive due to the accessibility of the sp^2 hybridized lone pair. The divalent compounds of the other Group IV elements can be regarded as carbenelike in the sense that they are angular with a lone pair and can readily undergo an oxidative addition reaction (see also Chapter 30) to give two new bonds to the element, for example,

$$R_2C^{II}: + X\text{—}Y = R_2C^{IV}XY$$

The "inert pair" concept (page 199) is not very useful, especially since the non-bonding electrons are known *not* to be inert in a stereochemical sense. Indeed, the two electrons behave much as a lone pair in every respect.

The increase in stability of the divalent state cannot be attributed to ionization energies as they are very similar in all cases. Factors that doubtless govern the relative stabilities are (i) promotion energies, (ii) bond strengths in covalent compounds, and (iii) lattice energies in ionic compounds.

For CH_4, the factor that stabilizes CH_4 relative to $CH_2 + H_2$ despite the much higher promotional energy required in forming CH_4 is the great strength of the C—H bonds. If we now have a series of reactions

$$MX_2 + X_2 = MX_4$$

in which the M—X bond energies are decreasing, as they do from Si → Pb, then it is possible that bond energy may become too small to compensate for the M^{II}—M^{IV} promotion energy and MX_2 then becomes the more stable.

The change in this group is shown by the reactions:

$$GeCl_2 + Cl_2 = GeCl_4 \quad \text{(very rapid at 25°)}$$
$$SnCl_2 + Cl_2 = SnCl_4 \quad \text{(slow at 25°)}$$
$$PbCl_2 + Cl_2 = PbCl_4 \quad \text{(only under forcing condition)}$$

Also, even $PbCl_4$ decomposes readily while $PbBr_4$ and PbI_4 do not exist, probably because of the reducing power of Br^- and I^-.

It is difficult to give any rigorous argument on lattice energy effects, since there is no evidence for the existence of M^{4+} ions and in only a few compounds are there even Pb^{2+} ions.

Group V. Some properties of these elements are given in Table 8-5.

Like nitrogen, phosphorus is essentially covalent in all its chemistry but arsenic, antimony, and bismuth show increasing tendencies to cationic behavior.

Table 8-5 *Some Properties of Group VB Elements*

Element	Electronic configuration	mp °C	Covalent radius Å	Ionic radius Å
P	$[Ne]3s^23p^3$	44	1.10	2.12 (P^{3-})
As	$[Ar]3d^{10}4s^24p^3$	814 (36 atm)	1.21	
Sb	$[Kr]4d^{10}5s^25p^3$	603	1.41	0.92 (Sb^{3+})
Bi	$[Xe]4f^{14}5d^{10}6s^26p^3$	271	1.52	1.08 (Bi^{3+})

Although electron gain to achieve the electronic structure of the next noble gas is conceivable (as in N^{3-}), considerable energies are involved so that anionic compounds are rare. Similarly, loss of valence electrons is difficult because of high ionization energies. There are no +5 ions and even the +3 ions are not simple, being SbO^+ and BiO^+. BiF_3 seems predominantly ionic.

The increasing metallic character is shown by the oxides that change from acidic for phosphorus to basic for bismuth, and by halides that have increasing ionic character.

Group VI. Table 8-6 gives some properties.

Table 8-6 *Some Properties of Group VIB Elements*

Element	Electronic configuration	mp °C	Covalent radius Å	Ionic (X^{2-}) radius Å
S	$[Ne]3s^23p^4$	119	1.03	1.90
Se	$[Ar]3d^{10}4s^24p^4$	217	1.17	2.02
Te	$[Kr]4d^{10}5s^25p^4$	450	1.37	2.22
Po	$[Xe]4f^{14}5d^{10}6s^26p^4$	254	—	2.30

These atoms can achieve the configuration of the noble gas by forming:

1. The *chalconide* ions, M^{2-}, in salts of highly electropositive elements.
2. Two electron-pair bonds as in H_2S or $SeCl_2$.
3. Anionic species with one bond as in HS^-.
4. Three bonds and one positive charge as in sulfonium ions R_3S^+.

There are also compounds in formal oxidation states IV and VI with 4, 5, or 6 covalent bonds, for example, $SeCl_4$, SeF_5^-, and TeF_6.

Only for polonium is there any evidence for cationic behavior, but there are gradual changes in properties with increasing size and decreasing electronegativity, such as:

(a) Decreasing stability of the hydrides H_2X.

(b) Increasing metallic character of the elements themselves.

(c) Increasing tendency to form anionic complexes such as $SeBr_6^{2-}$, $TeBr_6^{2-}$, PoI_6^{2-}.

Group VII. Some properties are given in Table 8-7.

The halogens atoms are only one electron short of the noble gas configuration, and the elements form the anion X^- or a single covalent bond. Their chemistries

Table 8-7 *Some Properties of Group VIIB Elements*

Element	Electronic configuration	mp °C	bp °C	Radius X^-A	Covalent radius Å
F	$[He]2s^22p^5$	−233	−118	1.19	0.71
Cl	$[Ne]3s^23p^5$	−103	−34.6	1.70	0.97
Br	$[Ar]3d^{10}4s^24p^5$	−7.2	58.8	1.87	1.14
I	$[Kr]4d^{10}5s^25p^5$	113.5	184.3	2.12	1.33
At[a]	$[Xe]4f^{14}5d^{10}6s^26p^5$	—	—	—	—

[a] All isotopes are radioactive with short half-lives.

are completely nonmetallic. The changes in behavior with increasing size are progressive and, with the exception of the Li—Cs group, there are closer similarities within the group than in any other in the periodic table.

The halogens can form compounds in higher formal oxidation states, mainly in halogen fluorides such as ClF_3, ClF_5, BrF_5, and IF_7 and oxo compounds.

No evidence exists for cationic behavior with ions of the type X^+. However, Br_2^+, I_2^+, Cl_3^+, and Br_3^+ and several iodine cations are known. When a halogen forms a bond to another atom more electronegative than itself, for example, ICl, the bond will be polar with a positive charge on the heavier halogen.

8-12 The Transition Elements of the *d* and *f* Blocks

The transition elements may be strictly defined as those that *as elements*, have partly filled *d* or *f* shells. We adopt a broader definition and include also elements that have partly filled *d* or *f* shells *in compounds*. This means that we treat the *coinage metals*, Cu, Ag, and Au, as transition metals, since Cu^{II} has a $3d^9$ configuration, Ag^{II} has a $4d^9$ configuration, and Au^{III} has a $5d^8$ configuration. Appropriately we also consider these elements as transition elements because their chemical behavior is quite similar to that of other transition elements.

There are thus 56 transition elements, counting the heaviest elements through the one of atomic number 104. All these elements have certain common properties:

1. They are all metals.
2. They are practically all hard, strong, high-melting, high-boiling metals that conduct heat and electricity well.
3. They form alloys with one another and with other metallic elements.
4. Many of them are sufficiently electropositive to dissolve in mineral acids, although a few are "noble"—that is, they have such low electrode potentials that they are unaffected by simple acids.
5. With very few exceptions, they exhibit variable valence, and their ions and compounds are colored in one if not all oxidation states.
6. Because of partially filled shells they form at least some paramagnetic compounds.

This large number of transition elements is subdivided into three main groups: (a) the main transition elements or *d*-block elements, (b) the lanthanide elements, and (c) the actinide elements.

The main transition group or *d* block includes those elements that have partially filled *d* shells only. Thus, the element scandium, with the outer electron configuration $4s^2 3d$, is the lightest member. The eight succeeding elements, the *first transition series*, Ti, V, Cr, Mn, Fe, Co, Ni, and Cu, all have partly filled 3*d* shells either in the ground state of the free atom (all except Cu) or in one or more of their chemically important ions (all except Sc). At zinc the configuration is $3d^{10}4s^2$; and this element forms no compound in which the 3*d* shell is ionized, nor does this ionization occur in any of the next nine elements. It is not until we come to yttrium, with ground-state outer electron configuration $5s^2 4d$, that we meet the next transition element. The following eight elements, Zr, Nb, Mo, Tc, Ru, Rh, Pd, and Ag, all have partially filled $4d$ shells whether in the free element (all but Ag) or in one or more of the chemically important ions (all but Y). This group of nine elements constitutes the *second transition series.*

Again there follows a sequence of elements in which there are never d-shell vacancies under chemically significant conditions until we reach the element lanthanum, with an outer electron configuration in the ground state of $6s^2 5d$. Now, if the pattern we have observed twice before were to be repeated, there would follow 8 elements with enlarged, but not complete, sets of $5d$ electrons. This does not happen, however. The $4f$ shell now becomes slightly more stable than the $5d$ shell and, through the next 14 elements, electrons enter the $4f$ shell until at lutetium it becomes filled. Lutetium thus has the outer electron configuration $4f^{14}5d6s^2$. Since both La and Lu have partially filled d shells and no other partially filled shells, it might be argued that both of them should be considered as d-block elements. However, for chemical reasons, it would be unwise to classify them in this way, since all of the 15 elements La ($Z = 57$) through Lu ($Z = 71$) have very similar chemical and physical properties, those of lanthanum being in a sense prototypal; hence, these elements are called the *lanthanides.*

The shielding of one f electron by another from the effects of the nuclear change is quite weak on account of the shapes of the f orbitals. Hence, with increasing atomic number and nuclear charge, the effective nuclear charge experienced by each $4f$ electron increases. This causes a shrinkage in the radii of the atoms or ions as one proceeds from La to Lu (see Table 26-1). This accumulation of successive shrinkages is called the *lanthanide contraction.* It has a profound effect on the radii of subsequent elements, which are smaller than might have been anticipated from the increased mass. Thus Zr^{4+} and Hf^{4+} have almost identical radii despite the atomic numbers of 40 and 72, respectively.

For practical purposes, the *third transition series* begins with hafnium, having the ground-state outer electron configuration $6s^2 5d^2$, and embraces the elements Ta, W, Re, Os, Ir, Pt, and Au, all of which have partially filled $5d$ shells in one or more chemically important oxidation states as well as (excepting Au) in the neutral atom.

Continuing on from mercury, which follows gold, we come via the noble-gas radon and the radioelements Fr and Ra to actinium, with the outer electron configuration $7s^2 6d$. Here we might expect, by analogy to what happened at lanthanum, that in the following elements electrons would enter the $5f$ orbitals, producing a lanthanide-like series of 15 elements. What actually occurs is, unfortunately, not as simple. Although, immediately following lanthanum, the $4f$ orbitals become decisively more favorable than the $5d$ orbitals for the electrons entering in the succeeding elements, there is apparently not so great a difference between the $5f$ and $6d$ orbitals until later. Thus, for the elements immediately following Ac, and their ions, there may be electrons in the $5f$ or $6d$ orbitals, or both. Since it appears that later on, after 4 or 5 more electrons have been added to the Ac configuration, the $5f$ orbitals do become definitely the more stable, and since the elements from about americium on do show moderately homologous chemical behavior, it has become accepted practice to call the 15 elements beginning with Ac the *actinide elements.*

There is an important distinction, based on electronic structures, between the three classes of transition elements. For the *d*-block elements the partially filled shells are *d* shells, 3*d*, 4*d*, or 5*d*. These *d* orbitals project well out to the periphery of the atoms and ions so that the electrons occupying them are strongly influenced by the surroundings of the ion and, in turn, are able to influence the environments very significantly. Thus, many of the properties of an ion with a partly filled *d* shell are quite sensitive to the number and arrangement of the *d* electrons present. In marked contrast to this, the 4*f* orbitals in the lanthanide elements are rather deeply buried in the atoms and ions. The electrons that occupy them are largely screened from the surroundings by the overlying shells (5*s*, 5*p*) of electrons, and therefore reciprocal interactions of the 4*f* electrons and the surroundings of the atom or the ion are of relatively little chemical significance. This is why the chemistry of all the lanthanides is so homologous, whereas there are seemingly erratic and irregular variations in chemical properties as one passes through a series of *d*-block elements. The behavior of the actinide elements lies between those of the two types described above because the 5*f* orbitals are not so well shielded as are the 4*f* orbitals, although they are not so exposed as are the *d* orbitals in the *d*-block elements.

Study Questions

A

1. Which elements are (at 25°C and 1 atm pressure)
   (a) gases, (b) liquids, (c) solids melting below 100°C?
2. Why is white phosphorus much more chemically reactive than black phosphorus?
3. Draw the structure of the most stable form of sulfur.
4. Draw the structure for carbon in (a) diamond, (b) graphite. What is the nature of C—C bonding in the two allotropes?
5. Write down the electronic structures of the first row elements, then answer the following questions.
   (a) What is the first ionization energy of Li (approximately)?
   (b) Why does Be not form a 2+ ion in solids?
   (c) Why is there a discontinuity between the ionization energy of N and O?
   (d) How do the electron attachment energies vary from Li to F?
   (e) Which of the elements can form anions?
6. Why is dinitrogen normally unreactive?
7. What is the octet rule? Why does it apply only to the first row elements?
8. What are Lewis acids and Lewis bases? Give two examples of each.
9. Why is there no silicon analog of graphite?
10. What are the main trends in properties of the alkali *metals*?
11. List the elements of Groups IIA and B. Compare their main chemical features.
12. Give the electronic structures of

    Sc and Ti
    Y and Zr
    La and Hf

    Why are there 14 other elements between La and Hf?
13. How do the following elements attain the noble gas configuration?
    (a) N, (b) S

14. Why are Cu, Ag, and Au considered as transition metals?
15. List the common features of transition metals.
16. What are the main groups of transition metals? Write out their names and give the electronic structures of the first, the middle, and the last.

B

1. Use MO theory to account for the bonding in N_2 and O_2. Why is O_2 paramagnetic?
2. What is an icosahedron? For which element is it the most characteristic structural feature?
3. What are the principal properties and structural types of metals?
4. On what electronic processes does the chemistry of hydrogen depend?
5. Why is carbon unique in forming chains of single bonds in compounds?
6. Why is the bond energy of F_2 much less than that of Cl_2?
7. In the Ga, In, and Tl group why do we observe $+1$ valency?
8. What is the lanthanide contraction and what is its main effect?
9. What are the "actinide" elements, and what relation do they bear to the "lanthanide" elements?

Chapter 8
Study Guide

Supplementary Reading

Donohue, J., *The Structures of the Elements*, Wiley, 1974.
Sanderson, R. T., *Chemical Periodicity*, Van Nostrand-Reinhold, 1960.
Weeks, M. E., *The Discovery of the Elements*, 7th ed., Chemical Education Publishing Co., 1968.

2

The Main Group Elements

9

hydrogen

9-1 Introduction

Hydrogen (not carbon) forms more compounds than any other element. For this and other reasons, many aspects of hydrogen chemistry are treated elsewhere in this book. Protonic acids and the aqueous hydrogen ion have already been discussed in Chapter 7. In this chapter we examine certain topics that most logically should be considered here.

Three isotopes of hydrogen are known: ^1H, ^2H (deuterium or D) and ^3H (tritium or T). Although isotope effects are greatest for hydrogen, justifying the use of distinctive names for the two heavier isotopes, the chemical properties of H, D, and T are essentially identical except in matters such as rates and equilibrium constants of reactions. The normal form of the element is the diatomic molecule; the various possibilities are H_2, D_2, T_2, HD, HT, DT.

Naturally occurring hydrogen contains 0.0156% deuterium, while tritium (formed continuously in the upper atmosphere in nuclear reactions induced by cosmic rays) occurs naturally in only minute amounts that are believed to be of the order of 1 in 10^{17} and is radioactive (β^-, 12.4 yr).

Deuterium as D_2O is separated from water by fractional distillation or electrolysis and is available in ton quantities for use as a moderator in nuclear reactors.

Molecular hydrogen is a colorless, odorless gas (fp 20.28 K) virtually insoluble in water. It is most easily prepared by the action of dilute acids on metals such as Zn or Fe and by electrolysis of water.

Industrially hydrogen is obtained by steam re-forming of methane or light petroleums over a promoted nickel catalyst at *ca.* 750°. The process is complex but the main reaction is

$$CH_4 + H_2O \rightleftharpoons CO + 3H_2 \qquad \Delta H = 205 \text{ kJ mol}^{-1}$$

This is followed by the shift reaction over iron and copper catalysts

$$CO + H_2O \rightleftharpoons CO_2 + H_2 \qquad \Delta H = -42 \text{ kJ mol}^{-1}$$

The CO_2 is removed by scrubbing with K_2CO_3 solution from which it is recovered. For ammonia synthesis (page 278) the small amounts of CO, CO_2 which act as poisons are catalytically converted to methane, which is innocuous.

$$CO + 3H_2 \rightleftharpoons CH_4 + H_2O$$

$$CO_2 + 4H_2 \rightleftharpoons CH_4 + 2H_2O$$

Hydrogen is not exceptionally reactive. It burns in air to form water and will react with oxygen and the halogens explosively under certain conditions. At high temperatures the gas will reduce many oxides either to lower oxides or to the metal. In the presence of suitable catalysts and above room temperature it reacts with N_2 to form NH_3. With electropositive metals and most nonmetals it forms hydrides.

In the presence of suitable catalysts, usually Group VIII metals or their compounds, a great variety of both inorganic and organic substances can be reduced.

The dissociation of hydrogen is highly endothermic, and this accounts in part for its rather low reactivity at low temperatures:

$$H_2 = 2H \qquad \Delta H^\circ = 434.1 \text{ kJ mol}^{-1}$$

9-2
The Bonding of Hydrogen

The chemistry of hydrogen depends mainly on the three electronic processes discussed in Chapter 8, page 191, namely, (1) loss of valence electron to give H^+, (2) acquisition of an electron to give H^-, and (3) formation of a single covalent bond as in CH_4.

However, hydrogen has additional unique bonding features. The nature of the proton and the complete absence of any shielding of the nuclear charge by electron shells allow other forms of chemical activity that are either unique to hydrogen or particularly characteristic of it. Some of these are the following, which we shall discuss in some detail subsequently.

1. The formation of numerous compounds, often nonstoichiometric, with metallic elements. They are generally called hydrides but cannot be regarded as simple saline hydrides (Section 9.6).

2. The formation of hydrogen bridge bonds in electron-deficient compounds such as (9-I) and transition metal complexes such as (9-II).

$H_2B(\mu\text{-}H)_2BH_2$

9-I

$[(OC)_5Cr{-}H{-}Cr(CO)_5]^-$

9-II

The best-studied example of bridge bonds is provided by diborane, 9-I, and related compounds (Chapter 12). The electronic nature of such bridge bonds was discussed in Chapter 3, page 78.

3. The hydrogen bond. This bond is important not only because it is essential to an understanding of much other hydrogen chemistry but also because it is one of the most intensively studied examples of intermolecular attraction. Hydrogen bonds dominate the chemistry of water, aqueous solutions, hydroxylic solvents, and OH-containing species generally, and they are of crucial importance in biological systems, being responsible *inter alia* for the linking of polypeptide chains in proteins and the base pairs of nucleic acids.

9-3
The Hydrogen Bond

When hydrogen is bonded to another atom, X, mainly F, O, N, or Cl such that the X—H bond is quite polar with H bearing a partial positive charge, it can interact with another negative or electron-rich atom, Y, to form what is called a hydrogen bond (H-bond), written as

$$X—H\text{---}Y$$

Although the details are subject to variation, and controversy, it is generally believed that typical hydrogen bonds are due largely to electrostatic attraction of H and Y. The X—H distance becomes slightly longer, but this bond remains essentially a normal 2-electron bond. The H---Y distance is generally much longer than that of a normal covalent H—Y bond.

In the case of the very strongest hydrogen bonds, the X to Y distance becomes quite short and the X—H and Y—H distances come close to being equal. In these cases there are presumably covalent and electrostatic components in both the X—H and Y—H bonds.

Experimental evidence for hydrogen bonding came first from comparisons of the physical properties of hydrogen compounds. Classic examples are the apparently abnormally high boiling points of NH_3, H_2O, and HF (*Fig. 9-1*) which imply association of these molecules in the liquid phase. Other properties such as heats of vaporization provided further evidence for association. Although physical properties reflecting association are still a useful tool in detecting hydrogen bonding, the most satisfactory evidence for solids comes from X-ray and neutron-diffraction crystallographic studies, and for solids, liquids, and solutions from infrared and nuclear magnetic resonance spectra.

Structural evidence for hydrogen bonds is provided by the X to Y distances, which are shorter than the expected van der Waals contact when a hydrogen bond exists. For instance, in crystalline $NaHCO_3$ there are four kinds of O---O distances between HCO_3^- ions with values of 3.12, 3.15, 3.19, and 2.55 Å. The first three are about equal to twice the van der Waals radius of oxygen, but the last one indicates a hydrogen bond, O—H---O. When an X—H group enters into hydrogen bonding, the X—H stretching band in the infrared spectrum is lowered in frequency, broadened, and increased in integrated intensity. These changes afford a very useful means of studying H-bonding in solution.

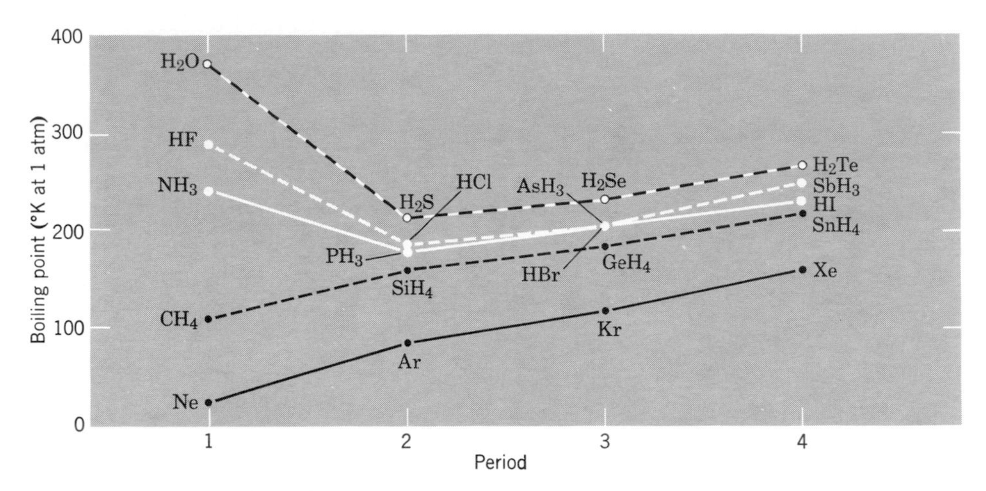

Figure 9-1 *Boiling points of some molecular hydrides.*

The enthalpies of hydrogen bonds are relatively small in most instances: 20–30 kJ mol^{-1}, as compared with covalent bond enthalpies of 200 kJ mol^{-1} and up. Nevertheless, these bonds can have a profound effect on the properties and chemical reactivity of substances in which they occur. This is clear from *Fig. 9-1*, where water, for example, would boil at about $-100°$C instead of $+100°$C if hydrogen bonds did not play their role. Obviously life itself—as we know it—depends on the existence of hydrogen bonds.

9-4
Ice and Water

The structure of water is very important since it is the medium in which so much chemistry, including the chemistry of life, takes place. The structure of ice is of interest for clues about the structure of water. There are nine known modifications of ice, the stability of each depending on temperature and pressure. The ice formed in equilibrium with water at 0°C and 1 atm is called ice I and has the structure shown in *Fig. 9-2*. There is an infinite array of oxygen atoms, each tetrahedrally surrounded by four others with hydrogen bonds linking each pair.

The structural nature of liquid water is still controversial. The structure is not random, as in liquids consisting of more-or-less spherical nonpolar molecules; instead, it is highly structured owing to the persistence of hydrogen bonds; even at 90°C only a few percent of the water molecules appear not to be hydrogen-bonded. Still, there is considerable disorder, or randomness, as befits a liquid.

In an attractive, though not universally accepted, model of liquid water the liquid consists at any instant of an imperfect network, very similar to the network of ice I, but differing in that (a) some interstices contain water molecules that do not belong to the network but, instead, disturb it; (b) the network is patchy and does not extend over long distances without breaks; (c) the short-range ordered regions are constantly disintegrating and re-forming (they are "flickering clusters"); and (d) the network is slightly expanded compared with ice I. The fact that water has a slightly higher density than ice I may be attributed to the presence of enough

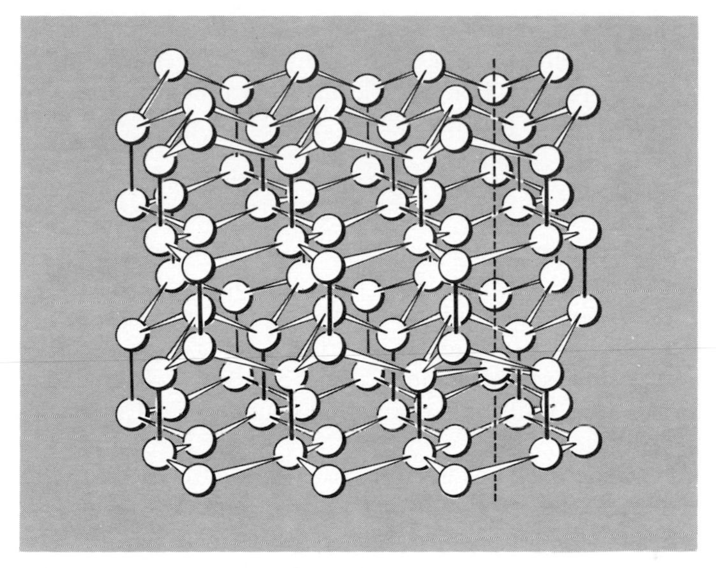

Figure 9-2 *The structure of ice I. Only the oxygen atoms are shown. The* O---O *distances are* 2.75 Å.

interstitial water molecules to more than offset the expansion and disordering of the ice I network. This model of water receives support from X-ray scattering studies.

9-5
Hydrates and Water Clathrates

Solids that consist of molecules of a compound together with water molecules are called *hydrates*. The majority contain discrete water molecules either bound to cations through the oxygen atom or bound to anions or other electron-rich atoms through hydrogen bonds, or both, as is shown in *Fig. 9-3*. In many cases

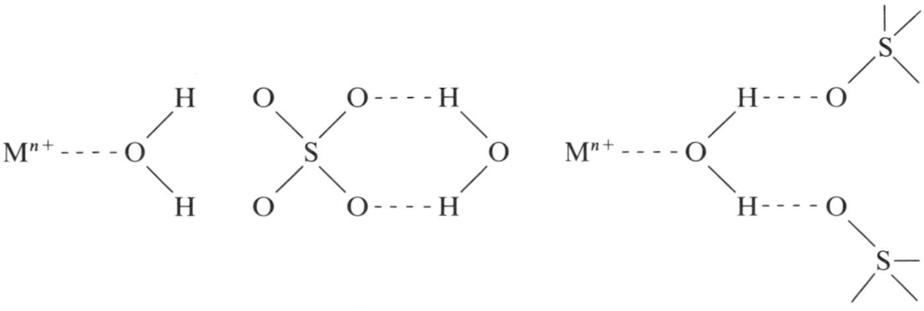

(*a*) (*b*) (*c*)

Figure 9-3 *Three principal ways in which water molecules are bound in hydrates.* (*a*) *Through oxygen to cations;* (*b*) *Through the hydrogen atoms to an anion;* (*c*) *A combination of the preceding two.*

when the hydrate is heated above 100°, the water can be driven off leaving the *anhydrous* compound. However, there are many cases where, instead, something other than water is driven off. For example, many hydrated chlorides give off HCl and a basic or oxo chloride is left:

$$ScCl_3 \cdot 6H_2O \xrightarrow{\text{Heat}} ScOCl + 2HCl(g) + 5H_2O(g)$$

Water also forms compounds called *gas hydrates*, which are actually a type of clathrate compound. A clathrate (from the Latin clathratus, meaning "enclosed or protected by crossbars or gratings") is a substance in which one component crystallizes in a very open structure that contains holes or channels in which atoms or small molecules of the second component can be trapped. There are a number of substances, other than water, for example, *p*-quinol, $[C_6H_4(OH)_2]$, and urea, that can form clathrates.

There are two common gas hydrate structures, both cubic. In one, the unit cell contains 46 molecules of H_2O connected to form six medium-size and two small cages. This structure is adopted when atoms (Ar, Kr, Xe) or relatively small molecules (e.g., Cl_2, SO_2, CH_3Cl) are used, generally at pressures greater than 1 atm for the gases. Complete filling of only the medium cages by atoms or molecules, X, would give a composition $X \cdot 7.67H_2O$, while complete filling of all eight cages would lead to $X \cdot 5.76H_2O$. In practice, complete filling of all cages of one or both types is seldom attained, and these formulas therefore represent limiting rather than observed compositions; for instance, the usual formula for chlorine hydrate is $Cl_2 \cdot 7.30H_2O$. The second structure, often formed in the presence of larger molecules of liquid substances (and thus sometimes called the liquid hydrate structure) such as chloroform and ethyl chloride, has a unit cell containing 136 water molecules with eight large cages and sixteen smaller ones. The anesthetic effect of substances such as chloroform may be due to the formation of liquid hydrate crystals in brain tissue.

A third notable class of clathrate compounds, salt hydrates, is formed when tetraalkylammonium or sulfonium salts crystallize from aqueous solution with high water content, for example, $[(n\text{-}C_4H_9)_4N]C_6H_5CO_2 \cdot 39.5H_2O$ or $[(n\text{-}C_4H_9)_3S]F \cdot 20H_2O$. The structures of these substances are very similar to the gas and liquid hydrate structures in a general way although different in detail. These structures consist of frameworks constructed mainly of hydrogen-bonded water molecules but apparently including also the anions (e.g., F^-) or parts of the anions (e.g., the O atoms of the benzoate ion). The cations and parts of the anions (e.g., the C_6H_5C part of the benzoate ion) occupy cavities in an incomplete and random way.

9-6 **Hydrides**

Compounds of hydrogen may all be called hydrides, although this term is best reserved for those that are neither organic compounds nor acids. As we might expect, they are of various types, as is indicated in *Fig. 9-4*.

The most electropositive elements, the alkali metals and larger alkaline earth metals, form hydrides with considerable ionic character, called *saline* (saltlike)

H																	He
Li	Be											B	C	N	O	F	Ne
Na	Mg											Al	Si	P	S	Cl	Ar
K	Ca	Sc	Ti*	V*	Cr*	Mn*	Fe*	Co*	Ni*	Cu	Zn	Ga	Ge	As	Se	Br	Kr
Rb	Sr	Y	Zr	Nb	Mo*	Tc*	Ru*	Rh*	Pd*	Ag	Cd	In	Sn	Sb	Te	I	Xe
Cs	Ba	La–Lu	Hf	Ta*	W*	Re*	Os*	Ir*	Pt*	Au	Hg	Tl	Pb	Bi	Po	At	Rn
Fr	Ra	Ac			U, Pu												
Saline hydrides		Transition metal hydrides								Borderline hydrides			Covalent hydrides				

Figure 9-4 *A classification of the hydrides. The starred elements are the transition elements for which complex molecules or ions containing* M—H *bonds are known.*

hydrides. They may be thought of as containing metal cations and H^- ions. Their ionic nature is shown by the facts that they conduct electricity just below or at their melting points and, when they are dissolved in molten halides, hydrogen is evolved at the positive electrode. The ionic radius of H^- lies between those of F^- and Cl^-. The alkali metal hydrides, LiH to CsH, all have the NaCl structure.

The saline hydrides are prepared by direct reaction of the metals with hydrogen at 300 to 700° C. They are quite reactive toward water and air (except for LiH). All are powerful reductants or hydrogenation reagents.

Among the covalent hydrides are many molecular compounds, including the H_2X compounds of group VIB, the H_3X compounds of Group VB, and the H_4X compounds of group IVB, $LiAlH_4$, and the many hydrogen compounds formed by boron, all of which are discussed in appropriate subsequent chapters.

The transition metal hydrides are very diverse in their properties. Many are formed by direct action of H_2 on the metals. The best known and most important ones are formed by the lanthanides, the actinides, and the elements of groups IVA and VA. For the most part, these are nonstoichiometric black solids, typical compositions being $LaH_{2.87}$, $YbH_{2.55}$, $TiH_{1.7}$, and $ZrH_{1.9}$. Uranium forms a well-defined stoichiometric hydride, UH_3, which is a useful reactant for preparing other uranium compounds.

These lanthanide and actinide hydrides appear to be mainly ionic in nature, but no entirely satisfactory detailed description of them has yet been developed. The group IVA and VA materials are even less well understood.

Of the remaining transition metals, only palladium and some of its alloys seems to form definite hydride phases.

Study Questions

A

1. What are the three isotopes of hydrogen called? What are their approximate natural abundances? Which one is radioactive?
2. What is the chief large scale use for D_2O?
3. What is one thing that helps to explain the relatively low reactivity of elemental hydrogen?
4. What are the three principal electronic processes that lead to formation of compounds by the hydrogen atom?

5. When a hydrogen bond is symbolized by X—H---Y, what do the solid and dotted lines represent? Which distance is shorter?
6. How does hydrogen bond formation affect the properties of HF, H_2O, and NH_3? Compared with what?
7. What is the usual range of enthalpies of a hydrogen bond?
8. Describe the main features of the structure of ice I. How is the structure of water believed to differ from that?
9. In what two principal ways is water bound in salt hydrates?
10. Can it safely be assumed that whenever a salt hydrate is heated at 100 to 120° the corresponding anhydrous salt will remain?
11. What is the true nature of so-called chlorine hydrate, $Cl_2 \cdot 7.3H_2O$?
12. What is a saline hydride? What elements form them? Why are they believed to contain cations and H^- ions?

B

1. It is often stated (by organic chemists) that carbon forms more compounds than any other element. Explain why this cannot be true.
2. Suggest a means of preparing pure HD.
3. It is believed that the shortest H-bonds become symmetrical. How must the conventional description (X—H---Y) be modified to cover this situation?
4. Which H-bond would you expect to be stronger, and why: S—H---O or O—H---S?
5. Why do only the most electropositive elements form saline hydrides? (Think in terms of a Born–Haber cycle.)

Chapter 9
Study Guide

Supplementary Reading

Evans, E. A., *Tritium and Its Compounds*, Butterworths, 1966.
Franks, F., ed., *Water, A Comprehensive Treatise*, Vol. 1, Plenium, 1972.
Libowitz, G. C., *The Solid State Chemistry of Binary Hydrides*, Benjamin, 1965.
Moore, R. A., ed., *Water and Aqueous Solutions: Structures, Thermodynamics and Transport Processes*, Wiley, 1972.
Pimentel, G. C. and McClellan, A. L., *The Hydrogen Bond*, Freeman, 1960.
Shaw, B. L., *Inorganic Hydrides*, Pergamon, 1967.
Vinogradov, S. N., *Hydrogen Bonding*, Van Nostrand-Reinhold, 1971.
Wiberg, E. and Amberger, E., *Hydrides*, American Elsevier, 1971.

10

the group IA elements: lithium, sodium, potassium, rubidium and cesium

10-1 Introduction

Sodium and potassium are abundant (2.6 and 2.4%, respectively) in the lithosphere. There are vast deposits of rock salt, NaCl, and carnallite, $KCl \cdot MgCl_2 \cdot 6H_2O$, resulting from evaporation of lagoons over geologic time. The Great Salt Lake of Utah and the Dead Sea in Israel are examples of evaporative processes at work now. Lithium, Rb, and Cs have much lower abundances and occur in a few silicate minerals.

The element francium has only very short-lived isotopes which are formed in natural radioactive decay series or in nuclear reactors. Tracer studies show that the ion behaves as expected from the position of Fr in Group I.

Sodium and its compounds are of great importance. The metal, as Na—Pb alloy, is used to make tetraalkylleads (Section 29-9), and there are other industrial uses. The hydroxide, carbonate, sulfate, tripolyphosphate, and silicate are among the top 50 of industrial chemicals with production in the United States for 1972 of one to ten million tons of each. Potassium salts, usually sulfate, are used in fertilizers. The main use for lithium is as metal in the synthesis of lithium alkyls (Section 29-3).

Both Na^+ and K^+ are of physiological importance in animals and plants; cells can differentiate between Na^+ and K^+ probably by some type of complexing mechanism. Lithium salts are used in treatment of certain mental disorders.

Some properties of the elements were given in Table 8-1, page 197. The low ionization enthalpies and the fact that the resulting M^+ ions are spherical and of low polarizability leads to a chemistry of $+1$ ions. The high second ionization enthalpies preclude the formation of $+2$ ions. Despite the essentially ionic nature of Group I compounds, some degree of covalent bonding can occur. The diatomic molecules of the elements, for example, Na_2, are covalent. In some chelate and organo compounds, the M—O, M—N, and M—C bonds have slight covalent nature. The tendency to covalency is greatest for the ion with the greatest polarizing power, that is, Li^+. The charge/radius ratio for Li^+, which is similar to that for

Mg^{2+}, accounts for the similarities in their chemistry where Li^+ differs from the other members (see below).

Some other ions that have +1 charge and radii similar to those of the alkalis may have similar chemistry. The most important are:

1. Ammonium and substituted ammonium ions. The solubilities and crystal structures of salts of NH_4^+ resemble those of K^+.

2. The Tl^+ ion can resemble either Rb^+ or Ag^+; its ionic radius is similar to that of Rb^+ but it is more polarizable.

3. Spherical, +1 complex ions such as $(\eta^5\text{-}C_5H_5)_2Co^+$ (Chapter 29).

10-2
Preparation and Properties of the Elements

Lithium and Na are obtained by electrolysis of fused salts or of low-melting eutectics such as $CaCl_2$ + NaCl. Because of their low melting points and ready vaporization K, Rb, and Cs cannot readily be made by electrolysis, but are obtained by treating molten chlorides with Na vapor. The metals are purified by distillation. Lithium, Na, K, and Rb are silvery but Cs has a golden-yellow cast. Because there is only one valence electron per metal atom, the binding energies in the close-packed metal lattices are relatively weak. The metals are hence very soft with low melting points. Na—K alloy with 77.2% K has a melting point of −12.3°.

Lithium, Na, or K may be dispersed on various solid supports, such as Na_2CO_3, kieselguhr, and the like, by melting. They are used as catalysts for various reactions of alkenes, notably the dimerization of propene to 4-methyl-1-pentene. Dispersions in hydrocarbons result from high-speed stirring of a suspension of the melted metal. These dispersions may be poured in air, and they react with water with effervescence. They may be used where sodium shot or lumps would react too slowly.

The metals are highly electropositive (Table 8-1, page 197) and react directly with most other elements and many compounds on heating. Lithium is usually the least, and Cs the most reactive.

Lithium is only slowly attacked by water at 25° and will not replace the weakly acidic hydrogen in $C_6H_5C{\equiv}CH$, whereas the others will do so. However, Li is uniquely reactive with N_2, slowly at 25° but rapidly at 400°, forming a ruby-red crystalline nitride, Li_3N. Like Mg, which gives Mg_3N_2, lithium can be used to absorb N_2.

With water, Na reacts vigorously, K inflames and Rb and Cs react explosively; large lumps of Na may also react explosively. Lithium, Na and K can be handled in air although they tarnish rapidly. The others must be handled under argon.

A fundamental difference attributable to cation size is shown by the reaction with O_2. In air or O_2 at 1 atm the metals burn. Lithium gives only Li_2O with a trace of Li_2O_2. Sodium normally gives the peroxide, Na_2O_2, but it will take up further O_2 under pressure and heat to give the superoxide, NaO_2. Potassium, Rb, and Cs form the superoxides MO_2. The increasing stability of the per- and superoxides as the size of the alkali ions increases is a typical example of the stabilization of larger anions by larger cations through lattice-energy effects, as is explained in Section 4-6.

The metals react with alcohols to give the alkoxides, and Na or K in C_2H_5OH or *t*-butanol is commonly used in organic chemistry as a reducing agent and a source of the nucleophilic OR^- ions.

Sodium and the other metals dissolve with much vigor in mercury. Sodium amalgam (Na/Hg) is a liquid when low in sodium, but is solid when rich. It is a useful reducing agent and can be used for aqueous solutions.

10-3
Solutions of Metals in Liquid Ammonia and Other Solvents

The Group I metals, and to a less extent Ca, Sr, Ba, Eu, and Yb, are soluble in ammonia giving solutions that are blue when dilute. These solutions conduct electricity and the main current carrier is the solvated electron. While the lifetime of the solvated electron in water is very short, in very pure liquid ammonia it may be quite long (less than 1% decomposition per day).

In *dilute solutions* the main species are metal ions, M^+, and electrons, both solvated. The broad absorption around 15,000 Å which accounts for the common blue color is due to the solvated electrons. Magnetic and electron spin resonance studies show the presence of individual electrons, but the decrease in paramagnetism with increasing concentration suggests that the electrons can associate to form diamagnetic electron pairs. Although there may be other equilibria, the data can be accommodated by the equilibria:

$$\text{Na(s)(dispersed)} \rightleftharpoons \text{Na (in solution)} \rightleftharpoons \text{Na}^+ + e$$

$$2e \rightleftharpoons e_2$$

The most satisfactory models of the solvated electron assume that the electron is not localized but is "smeared out" over a large volume so that the surrounding molecules experience electronic and orientational polarization. The electron is trapped in the resultant polarization field, and repulsion between the electron and the electrons of the solvent molecules leads to the formation of a cavity within which the electron has the highest probability of being found. In ammonia this is estimated to be approximately 3.0 to 3.4 Å in diameter; this cavity concept is based on the fact that solutions are of much lower density than the pure solvent, that is, they occupy far greater volume than that expected from the sum of the volumes of metal and solvent.

As the concentration of metal increases, metal ion clusters are formed. Above 3 *M* concentration, the solutions are copper-colored with a metallic luster. Physical properties, such as their exceedingly high electrical conductivities, resemble those of liquid metals.

The metals are also soluble to varying degrees in other amines, hexamethylphosphoramide, $OP(NMe_2)_3$, and in ethers such as tetrahydrofuran or diglyme giving blue solutions.

The ammonia and amine solutions are widely used in organic and inorganic synthesis. Lithium in methylamine or ethylenediamine can reduce aromatic rings to cyclic monoolefins. Sodium in ammonia is most widely used. This solution is moderately stable, but the reaction

$$Na + NH_3(l) = NaNH_2 + \tfrac{1}{2}H_2$$

can occur photochemically and is catalyzed by transition-metal salts. Sodium amide is prepared by treatment of Na with ammonia in the presence of a trace of ferric chloride. Amines react similarly:

$$Li(s) + CH_3NH_2(l) \xrightarrow{50-60^\circ} LiNHCH_3(s) + \tfrac{1}{2}H_2$$

The *lithium dialkylamides* are used to make compounds with $M—NR_2$ bonds (Section 24-7).

For the amides of K, Rb, and Cs the reaction

$$e^- + NH_3 \rightleftharpoons NH_2^- + \tfrac{1}{2}H_2 \qquad K = 5 \times 10^4$$

is reversible, but for $LiNH_2$ and $NaNH_2$ which are insoluble in ammonia we have, for example,

$$Na^+(am) + e^-(am) + NH_3(l) = NaNH_2(s) + \tfrac{1}{2}H_2 \qquad K = 3 \times 10^9$$

COMPOUNDS OF THE GROUP I ELEMENTS

10-4 Binary Compounds

The metals react directly with most other elements to give binary compounds or alloys. Many of these are described under the appropriate element. The most important are the oxides, obtained by combustion. They are readily hydrolyzed by water:

$$M_2O + H_2O = 2M^+ + 2OH^-$$

$$M_2O_2 + 2H_2O = 2M^+ + 2OH^- + H_2O_2$$

$$2MO_2 + 2H_2O = O_2 + 2M^+ + 2OH^- + H_2O_2$$

10-5 Hydroxides

These are white, very deliquescent crystalline solids; NaOH (mp 318°) and KOH (mp 360°). The solids and their aqueous solutions absorb CO_2 from the atmosphere. They are freely soluble exothermically in water and in alcohols and are used whenever strong alkali bases are required.

10-6 Ionic Salts

Salts of virtually all acids are known; they are usually colorless, crystalline, ionic solids. Color arises from colored anions, except where defects induced in the lattice, for example, by radiation, may cause *color centers*, through electrons being trapped in holes (cf. ammonia solutions above).

The properties of a number of *lithium compounds* differ from those of the other Group I elements but resemble those of Mg^{2+} compounds. Many of these anomalous properties arise from the very small size of Li^+ and its effect on lattice energies, as explained in Section 4-6 (page 101). In addition to examples cited there, we note that LiH is stable to approximately 900° while NaH decomposes at 350°. Li_3N is stable whereas Na_3N does not exist at 25°. Lithium hydroxide decomposes at red heat to Li_2O, whereas the other hydroxides MOH sublime unchanged; LiOH is also considerably less soluble than the other hydroxides. The carbonate, Li_2CO_3, is thermally less stable relative to Li_2O and CO_2 than are other alkali-metal carbonates. The solubilities of Li^+ salts resemble those of Mg^{2+}. Thus LiF is sparingly soluble (0.27 g/100 g H_2O at 18°) and is precipitated from ammoniacal NH_4F solutions; LiCl, LiBr, LiI and, especially $LiClO_4$ are soluble in ethanol, acetone, and ethyl acetate; LiCl is soluble in pyridine.

The alkali metal salts are generally characterized by high melting points, by electrical conductivity of the melts, and by ready solubility in water. They are seldom hydrated when the anions are small, as in the halides, because the hydration energies of the ions are insufficient to compensate for the energy required to expand the lattice. The Li^+ ion has a large hydration energy, and it is often hydrated in its solid salts when the same salts of other alkalis are unhydrated, $LiClO_4 \cdot 3H_2O$. For salts of *strong* acids, the Li salt is usually the *most* soluble in water of the alkali-metal salts, whereas for *weak* acids the Li salts are usually *less* soluble than those of the other elements.

There are few important *precipitation reactions* of the ions. One example is the precipitation by methanolic solutions of 4,4′-diaminodiphenylmethane (L) of Li and Na salts, for example, NaL_3Cl. Generally the larger the M^+ ion the more numerous are its insoluble salts. Thus Na has few insoluble salts; the mixed Na—Zn and Na—Mg uranyl acetates [e.g., $NaZn(UO_2)_3(CH_3COO)_9 \cdot 6H_2O$], which may be precipitated almost quantitatively from dilute acetic acid solutions, are useful for analysis. Salts of the heavier ions, K^+, Rb^+, and Cs^+, with large anions such as ClO_4^-, $PtCl_6^{2-}$, $Co(NO_2)_6^{3-}$, and $B(C_6H_5)_4^-$ are relatively insoluble and form the basis for gravimetric analysis.

10-7 Solvation and Complexation of Alkali Cations

For these cations, as for others, solvation must be considered from two points of view. First there is the primary hydration shell, which is the number of solvent molecules directly coordinated, and second there is the solvation number, which is the total number of solvent molecules on which the ion exercises a substantial restraining influence. Most important is the primary hydration shell.

For Li^+ a primary hydration shell of four tetrahedrally arranged H_2O molecules is observed in various crystalline salts and probably occurs in solution. The ions Na^+ and K^+ may also have fourfold primary hydration; Rb^+ and Cs^+ probably coordinate $6H_2O$. However, electrostatic forces operate beyond the primary hydration sphere and additional layers of water molecules are bound. The extent of this secondary hydration appears to vary *inversely* with the size of the bare cation. Thus, as the *crystal radii increase*, the total hydration numbers, hydrated radii, and hydration energies all *decrease*. As hydrated radii decrease, ionic mobilities increase. These trends can be observed in the data of Table 10-1.

Table 10-1 *Data on Hydration of Aqueous Group I Ions*

	Li^+	Na^+	K^+	Rb^+	Cs^+
Pauling radii,[a] Å	0.60	0.96	1.33	1.48	1.69
Hydrated radii (approx.), Å	3.40	2.76	2.32	2.28	2.28
Approximate hydration numbers[b]	25.3	16.6	10.5	10.0	9.9
Hydration energies, kJ mol^{-1}	519	406	322	293	264
Ionic mobilities (at ∞ dil., 18°)	33.5	43.5	64.6	67.5	68

[a] From Table 4-2, page 98.
[b] From transference data.

The above trends play a role in the behavior of the alkali ions in ion exchange materials and in their passage through cell walls and other biological membranes, although doubtless other factors than size and hydration numbers are also important. In a cation exchange resin, two cations compete for attachment at anionic sites as in the equilibrium:

$$A^+(aq) + [B^+R^-](s) = B^+(aq) + [A^+R^-](s)$$

where R represents the resin and A^+ and B^+ the cations. Such equilibria can be measured quite accurately, and the order of preference of the alkali cations is usually $Li^+ < Na^+ < K^+ < Rb^+ < Cs^+$, although irregular behavior does occur in some cases. The usual order may be explained if we assume that the binding force is essentially electrostatic and that under ordinary conditions the ions within the waterlogged resin are hydrated approximately as they are outside it. Then the ion with the smallest hydrated radius (which is the one with the largest "naked" radius) will be able to approach most closely to the negative site of attachment and will hence be held most strongly according to the coulomb law.

Ethers, polyethers and, especially, cyclic polyethers are particularly suited to solvate Na^+ and other alkali ions. Examples are tetrahydrofuran, the "glyme" solvents, which are linear polyethers, for example, $CH_3O(CH_2CH_2O)_nCH_3$, and the macrocyclic "*crown ethers*" such as cyclohexyl-18-crown-6, (10-I). For this ether, bonding constants increase in the order $Li^+ < Na^+, Cs^+ < Rb < K^+$.

Even more potent and selective agents for binding alkali ions (and others) are the *cryptates*, which differ from the crown ethers in two ways. First, they incorporate nitrogen as well as oxygen atoms. Second, they are polycyclic and, hence, are able to surround the metal ion. One of the cryptates is (10-II) and the structure

10-I 10-II

of a representative complex is shown in *Fig. 10-1*. The cryptates also complex the Group IIA ions very effectively and can, for example, render $BaSO_4$ soluble.

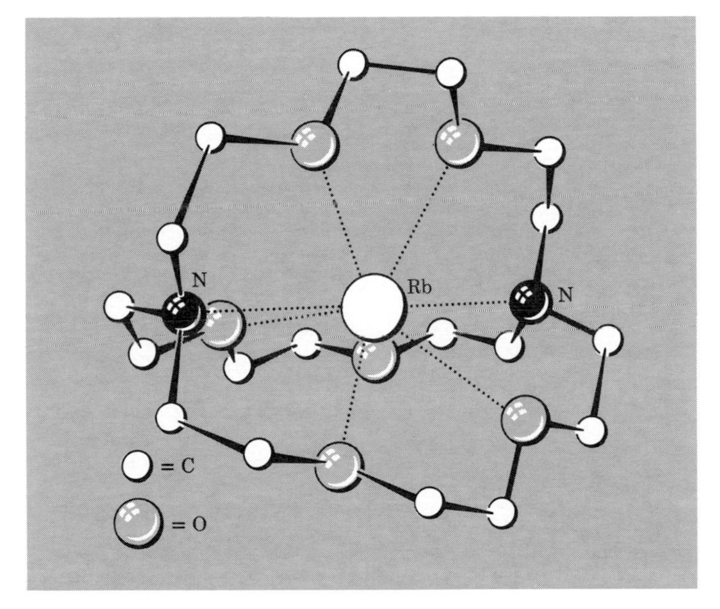

Figure 10-1 *The structure of the cation in the salt* $[RbC_{18}H_{36}N_2O_6]SCN \cdot H_2O$. [*Reproduced by permission from M. R. Truter*, Chem. in Britain, 1971, *203*.]

Study Questions

A

1. Why are the alkali metals soft and volatile?
2. Why are they highly electropositive?
3. Write down the electronic structure of francium.
4. Why are the first ionization energies of the Group IA atoms low?
5. Why does the chemical reactivity of the metals increase from Li to Cs?
6. What other ions have properties similar to the alkali metal ions?
7. How does the charge-radius ratio of Li^+ differ from those of the other Group I ions? List some consequences of this difference.

8. How do the reactivity and the nature of the products vary from Li to Cs when the alkali metals react with oxygen?
9. What is the nature of the solutions of alkali metals in liquid ammonia? What is the chief reaction by which they decompose?
10. How would you make lithium hydride? Why is it more stable than NaH?
11. Draw the crystal structures of NaCl and CsCl. Why do they differ?
12. Why is sodium peroxide a useful oxidizing agent in aqueous solution?
13. In what order are the M^+ ions eluted from a cation exchange resin column?
14. Why is LiF almost insoluble in water whereas LiCl is soluble, not only in water, but in acetone?
15. What is (a) a crown ether, (b) a cryptate?
16. Why are lithium salts commonly hydrated and those of the other alkali ions usually anhydrous?
17. How would you extinguish a sodium fire in the laboratory?
18. Vapors of KOH contain dimers?
19. How are the sizes of the hydrated alkali ions related to the crystallographic radii of the ions? How is this reflected in their behavior on ion exchange columns?

B

1. Vapors of the metals contain $\sim 1\%$ diatomic molecules. Discuss the bonding in such molecules. Why do their dissociation energies decrease with increasing Z?
2. Anhydrous KOH in tetrahydrofuran is one of the strongest known bases and will deprotonate exceedingly weak acids. Why?
3. Sketch a suitable piece of apparatus and give some details as to how you would prepare sodium amide from Na and NH_3.
4. Why is the stability constant for the 1:1 complex by K^+ with

$$N(CH_2CH_2OCH_2CH_2OCH_2CH_2)_3N$$

much larger than the values for other M^+ ions?
5. Why are ethers such as tetrahydrofuran widely used for reactions of sodium with organic substances, metal carbonyls, etc.?
6. Why is there so little variation in the standard potentials for $M^+(aq) + e = M(s)$ for the Group I elements?
7. What would happen if you electrolyzed tetraethylammonium chloride under the same conditions as NaCl in aqueous solution and in a melt?
8. What happens when increasing amounts of sodium are dissolved in liquid ammonia? Explain the phenomena.

Chapter 10
Study Guide

Supplementary Reading

Advances in Chemistry Series, No. 130. *Polyamine Chelated Alkali Metal Compounds*, American Chemical Society, 1974.

The Alkali Metals, *Spec. Publ.* No. 22, The Chemical Society, London, 1967.

Jolly, W. L., *Metal-Ammonia Solutions*, Dowden, Hutchinson, and Ross, 1972.

Jortner, J. and Kestner, N. R., eds., *Electrons in Liquids; The Nature of Metal Ammonia Solutions*, Springer Verlag, 1973.

Kapoor, P. N. and Mehrotra, R. C., "Coordination Compounds of the Alkali and Alkaline Earth Elements with Covalent Characteristics," *Coord. Chem. Revs.*, *14*, 1 (1974).

Mellor's Comprehensive Treatise on Inorganic and Theoretical Chemistry, Vol. II. Supplement 2, Li, Na (1961); Supplement 3, K, Rb, Cs, Fr (1963). Longmans Green.

11

the group IIA elements: beryllium, magnesium, calcium, strontium, barium and radium

11-1
Introduction

Beryllium occurs in the mineral *beryl*, $Be_3Al_2(SiO_3)_6$. Beryllium compounds are exceedingly toxic, especially if inhaled, when they cause degeneration of lung tissue similar to miners' silicosis; they must be handled with great care. The element has only minor technical importance.

Magnesium, Ca, Sr, and Ba are widely distributed in minerals and in the sea. There are substantial deposits of limestone, $CaCO_3$, dolomite, $CaCO_3 \cdot MgCO_3$, and carnallite, $KCl \cdot MgCl_2 \cdot 6H_2O$. Less abundant are strontianite, $SrSO_4$, and barytes, $BaSO_4$. All isotopes of radium are radioactive. ^{226}Ra, α, 1600 yr, which occurs in the ^{238}U decay series was first isolated by Pierre and Marie Curie from the uranium ore pitchblende. It was collected from solutions by co-precipitation with $BaSO_4$ and the nitrates subsequently fractionally crystallized. Its use in cancer therapy has been supplanted by other forms of radiation.

The positions of the Group IIA elements and of the related Group IIB (Zn, Cd, Hg) elements in the periodic table and some of their properties have been given in Chapter 8, page 198.

The atomic radii are smaller than those of the Li—Cs group as a result of the increased nuclear charge (cf. Table 4-2). The number of bonding electrons in the metals is now two, so that these have higher melting and boiling points and densities. The ionization enthalpies are higher than those of Group IA atoms and their enthalpies of vaporization are higher. Nevertheless, the high lattice energies and high hydration energies of M^{2+} ions compensate for these increases. The metals are hence electropositive with high chemical reactivities and standard electrode potentials. Born–Haber cycle calculations show that MX compounds would be unstable, in the sense that the following reactions should have very large negative enthalpies:

$$2MX = M + MX_2$$

There is, however, some evidence in anodic dissolution of Mg for transitory Mg^+ ions.

Because of its exceptionally small atomic radius and high enthalpies of ionization and sublimation, the lattice or hydration energies are insufficient in the case of beryllium to provide complete charge separation. Thus even BeF_2 and BeO show evidence of covalent character, and covalent compounds with bonds to carbon are quite stable. In both these respects Be resembles Zn. Note that to form two covalent bonds promotion from the $2s^2$ to the $2s2p$ configuration is required. Thus BeX_2 molecules should be linear, but since such molecules are coordinatively unsaturated they exist only in the gas phase. In condensed phases fourfold coordination is achieved by:

1. Polymerization to give chains with bridge groups as in $(BeF_2)_n$, $(BeCl_2)_n$, or $(Be(CH_3)_2)_n$, *Fig. 11-1.*

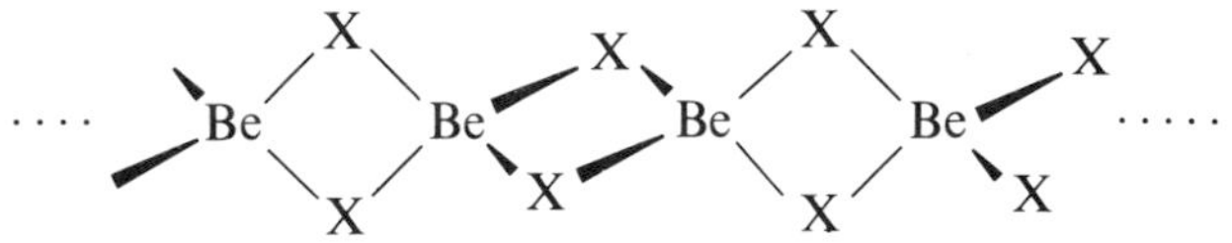

Figure 11-1 *The infinite chain structure of* BeX_2 *compounds,* X = F, Cl, CH_3, *whereby each* Be *atom achieves a coordination number of four.*

2. Formation of a covalent lattice as in BeO or BeS which have the wurtzite (ZnO) or zinc blende (ZnS) structures (see *Fig. 4-1*, page 92).

3. Interaction of BeX_2 as a Lewis acid with solvents to give four-coordinate compounds like $BeCl_2(OEt_2)_2$.

4. Interaction of BeX_2 with anions to give anionic species like BeF_4^{2-}.

The cation $[Be(H_2O)_4]^{2+}$ has its water molecules very firmly bound, but because of the high charge and small size, dissociation of protons occurs readily.

The second member of Group IIA, magnesium, is intermediate in behavior between beryllium and the remainder of the group whose chemistry is almost entirely ionic in nature. The Mg^{2+} ion has high polarizing ability, and there is a decided tendency to nonionic behavior. Magnesium forms bonds to carbon readily (Chapter 29). Like $Be(OH)_2$, $Mg(OH)_2$ is sparingly soluble in water while the other hydroxides are water soluble and highly basic.

Calcium, Sr, Ba, and Ra form a closely related group where the chemical and physical properties change systematically with increasing size. Examples are increases from Ca to Ra in (a) the electropositive nature of the metal—cf. $E°$, Table 8-2, page 198, (b) hydration energies of salts, (c) insolubility of most salts, notably sulfates, and (d) thermal stabilities of carbonates and nitrates. As in Group I, the larger cations can stabilize large anions such as O_2^{2-}, O_2^-, and I_3^- (cf. Section 4-6).

Because of similarity in charge and radius, the +2 ions of the lanthanides (Section 26-5) resemble the Sr—Ra ions. Thus europium, which forms an insoluble sulfate, $EuSO_4$, sometimes occurs in Group II minerals.

BERYLLIUM

11-2
The Element and Its Compounds

The *metal*, obtained by Ca or Mg reduction of $BeCl_2$, is very light and has been used for "windows" in X-ray apparatus. The absorption of electromagnetic radiation depends on the electron density in matter, and Be has the lowest stopping power per unit of mass-thickness of all constructional materials.

The metal, or the hydroxide, dissolve in strong base to give the beryllate ion, $[Be(OH)_4]^{2-}$, behavior comparable to that of Al and $Al(OH)_3$. In strongly acid solutions of noncomplexing acids, the aquo ion is $[Be(OH_2)_4]^{2+}$. Solutions of Be salts are acidic, due to hydrolysis, where the initial reaction

$$[Be(H_2O)_4]^{2+} = [Be(H_2O)_3OH]^+ + H^+$$

is followed by further polymerization reactions. In fluoride solutions the $[BeF_4]^{2-}$ ion is formed. This tetrahedral ion behaves in crystals much like SO_4^{2-}; thus $PbBeF_4$ and $PbSO_4$ have similar structures and solubilities.

Beryllium chloride forms long chains (page 226) in the crystal, and this compound and the similar methyl $[Be(CH_3)_2]_n$ are cleaved by donor molecules to give complexes such as $BeCl_2(OR_2)_2$. This Lewis acid behavior is also typical of Mg, Zn, and Al halides and alkyls.

Inhalation of beryllium or beryllium compounds can cause serious respiratory disease and soluble compounds may produce dermatitis on contact with the skin. Appropriate precautions should be taken in handling the element and its compounds.

MAGNESIUM, CALCIUM, STRONTIUM, BARIUM, AND RADIUM

11-3
The Elements and Their Properties

Magnesium. Magnesium is produced in several ways. The important sources are dolomite rock and seawater, which contains 0.13% Mg. Dolomite is first calcined to give a CaO/MgO mixture from which the calcium can be removed by ion exchange using seawater. The equilibrium is favorable because the solubility of $Mg(OH)_2$ is lower than that of $Ca(OH)_2$:

$$Ca(OH)_2 \cdot Mg(OH)_2 + Mg^{2+} \longrightarrow 2Mg(OH)_2 + Ca^{2+}$$

The most important processes for obtaining the metal are: (a) the electrolysis of fused halide mixtures (e.g., $MgCl_2 + CaCl_2 + NaCl$) from which the least electro positive metal, Mg, is deposited, and (b) the reduction of MgO or of calcined dolomite ($MgO \cdot CaO$). The latter is heated with Ferrosilicon:

$$CaO \cdot MgO + FeSi = Mg + \text{silicates of Ca and Fe}$$

and the Mg is distilled out. MgO can be heated with coke at 2000° and the metal deposited by rapid quenching of the high-temperature equilibrium which lies well to the right:

$$MgO + C \rightleftharpoons Mg + CO$$

Magnesium is greyish white and has a protective surface oxide film. Thus it is not attacked by water despite the favorable potential unless amalgamated. It is, however, readily soluble in dilute acids. It is used in light constructional alloys and for the preparation of Grignard reagents (Chapter 29) by interactions with alkyl or aryl halides in ether solution. It is essential to life because it occurs in chlorophyll (cf. Chapter 31).

Calcium. Calcium, Sr and Ba are made only on a relatively small scale by reduction of the halides with Na. They are soft and silvery resembling Na in their reactivities, although somewhat less reactive. Calcium is used for the reduction to the metal of actinide and lanthanide halides and for the preparation of CaH_2, a useful reducing agent.

11-4 Binary Compounds

The oxides, MO, are obtained by roasting of the carbonates; CaO is made in huge amounts for cement.

Magnesium oxide is relatively inert, especially after ignition at high temperatures, but the other oxides react with H_2O, evolving heat, to form the hydroxides. They absorb CO_2 from the air. Magnesium hydroxide is insoluble in water ($\sim 1 \times 10^{-4}$ g/l at 20°) and can be precipitated from Mg^{2+} solutions; it is a much weaker base than the Ca—Ra hydroxides, although it has no acidic properties and unlike $Be(OH)_2$ is insoluble in excess of hydroxide. The Ca—Ra hydroxides are all soluble in water, increasingly so with increasing atomic number [$Ca(OH)_2$, $\sim$2 g/l; $Ba(OH)_2$, $\sim$60 g/l at $\sim$20°], and all are strong bases.

Halides. The anhydrous halides can be made by dehydration (Section 20-3) of the hydrated salts. Magnesium and Ca halides readily absorb water. The ability to form hydrates, as well as the solubilities in water, decrease with increasing size, and Sr, Ba, and Ra halides are normally anhydrous. This is attributed to the fact that the hydration energies decrease more rapidly than the lattice energies with increasing size of M^{2+}.

The fluorides vary in solubility in the reverse order, that is, $Mg < Ca < Sr < Ba$, because of the small size of the F^- relative to the M^{2+} ion. The lattice energies decrease unusually rapidly because the large cations make contact with one another without at the same time making contact with the F^- ions.

All the halides appear to be essentially ionic. On account of its dispersion and transparency properties, CaF_2 is used for prisms in spectrometers and for cell windows (especially for aqueous solutions). It is also used to provide a stabilizing lattice for trapping lanthanide +2 ions (Section 26-5).

Other compounds. The metals, like the alkalis, react with many other elements. Compounds such as phosphides, silicides, or sulfides are mostly ionic and are hydrolyzed by water.

Calcium *carbide*, obtained by reduction of the oxide with carbon in an electric furnace is an acetylide $Ca^{2+}C_2^{2-}$ (page 259). It used to be employed as a source of acetylene.

11-5
Oxo Salts, Ions, and Complexes

All the elements form *oxo salts*, those of Mg and Ca often being hydrated. The carbonates are all rather insoluble in water and the solubility products decrease with increasing size of M^{2+}; $MgCO_3$ is used in stomach powders to absorb acid. The same solubility order applies to the *sulfates*; magnesium sulfate which, as Epsom salt $MgSO_4 \cdot 7H_2O$, is used as a mild laxative in "health" salts, is readily soluble in water. Calcium sulfate has a hemihydrate $2CaSO_4 \cdot H_2O$ (plaster of Paris) which readily absorbs more water to form the very sparingly soluble $CaSO_4 \cdot 2H_2O$ (gypsum), while Sr, Ba, and Ra sulfates are insoluble and anhydrous. $BaSO_4$ is accordingly used for "barium meals" as it is opaque to X-rays and provides a suitable shadow in the stomach. The *nitrates* of Sr, Ba, and Ra are also anhydrous and the last two can be precipitated from cold aqueous solution by the addition of fuming nitric acid. *Magnesium perchlorate* is used as a drying agent, but contact with organic materials must be avoided because of the hazard of explosions.

For water, acetone and methanol solutions, nuclear magnetic resonance studies have shown that the coordination number of Mg^{2+} is 6, although in ammonia it appears to be 5. The $[Mg(H_2O)_6]^{2+}$ ion is not acidic and in contrast to $[Be(H_2O)_4]^{2+}$ can be dehydrated fairly readily: it occurs in a number of crystalline salts.

Only Mg and Ca show any appreciable tendency to form *complexes* and in solution, with a few exceptions, these are of oxygen ligands. $MgBr_2$, MgI_2, and $CaCl_2$ are soluble in alcohols and polar organic solvents. Adducts such as $MgBr_2(OEt_2)_2$ and $MgBr_2(THF)_4$ can be obtained.

Oxygen chelate complexes, among the most important being those with ethylenediaminetetraacetate (EDTA) type ligands, readily form in alkaline aqueous solution. For example:

$$Ca^{2+} + EDTA^{4-} = [Ca(EDTA)]^{2-}$$

The cyclic polyethers and related nitrogen compounds (page 222) form strong complexes and salts can be isolated. The complexing of calcium by $EDTA^{4-}$ and also by polyphosphates is of importance, not only for removal of Ca^{2+} from water but also for the volumetric estimation of Ca^{2+}.

Both Mg^{2+} and Ca^{2+} have important biological roles (Chapter 31). The tetrapyrrole systems in chlorophyll form an exception to the rule that complexes of Mg (and the other elements) with nitrogen ligands are weak.

Study Questions

A

1. Name the important minerals of the Group IIA elements.
2. Why do the metals have higher melting points than the alkali metals?
3. Why does beryllium tend to form covalent compounds?
4. Why do linear molecules X—Be—X exist only in the gas phase?
5. Which compound, when dissolved in water, would give the most acid solution, $BeCl_2$ or $CaCl_2$?
6. Draw the structures of $BeCl_2$ and $CaCl_2$ in the solid state.
7. How is magnesium made?
8. What are the properties of the hydroxides, $M(OH)_2$?
9. How do the solubilities of (a) hydroxides, (b) chlorides, and (c) sulfates vary in Group II.
10. What and where are the Dolomites from which $MgCO_3 \cdot CaCO_3$ gets its name?
11. What is an important fact about beryllum compounds from a safety point of view?
12. Compare the physical properties of Be, Mg, Ca, and Sr.
13. Do the alkaline earth cations form many complexes? Which cations tend most to do so and what are the best complexing agents?
14. What are the main types of compounds formed by the alkaline earth elements? Are they generally soluble in water?

B

1. Beryllium readily forms a compound of stoichiometry $Be_4O(CO_2CH_3)_6$. What is a likely structure for this compound?
2. From the E° value for Mg, would you expect the metal to react with water? Discuss the actual situation.
3. Barium oxide used to be employed in an old process for making hydrogen peroxide. How did this work?
4. What is the main cause of "hard" water in limestone areas? How do water-softeners work? Why are polyphosphate detergents so useful?
5. How would you use the complexing of Ca^{2+} by $EDTA^{4-}$ as the basis of an analytical procedure for calcium?
6. Explain why CaCl does not exist.
7. Magnesium perchlorate is an excellent drying agent. Why?
8. Why do you think the usual coordination numbers of Be^{2+} and Mg^{2+} are 4 and 6, respectively.

Chapter 11
Study Guide

Supplementary Reading

Bell, N. A., "Beryllium Halides and Complexes," *Adv. Inorg. Chem. Radiochem.*, *14*, 225 (1972).

Everest, D. A., *The Chemistry of Beryllium*, Elsevier, 1964.

Kapoor, P. N. and Mehrotra, R. C., "Coordination Compounds of Alkali and Alkaline Earth Elements with Covalent Characteristics," *Coord. Chem. Rev.*, *14*, 1 (1974).

12

boron

12-1 Introduction

The principal ores are all borates. Borax, $Na_2B_4O_7 \cdot 4H_2O$, occurs in large deposits in the Mojave Desert of California and is the major source of boron.

No boron cations are formed because the ionization enthalpies for boron are so high that lattice or hydration enthalpies cannot supply the required energy. Boron normally forms three covalent bonds using sp^2 hybrid orbitals in a plane at angles of 120° (page 74). All BX_3 compounds are coordinatively unsaturated (cf. BeX_2) and act as strong Lewis acids (page 170). Interaction with neutral molecules or with anions gives tetrahedral species such as $BF_3(OEt_2)$, BF_4^-, or BPh_4^-.

Boron has unique chemistry with only a few features in common with Al and the remaining Group III elements. The main resemblances to silicon and differences from the more metallic Al are as follows.

1. The oxide B_2O_3 and $B(OH)_3$ are acidic. $Al(OH)_3$ is a basic hydroxide although it shows weak amphoteric properties, dissolving in strong NaOH.

2. Borates and silicates are built on similar structural principles (page 114) with sharing of oxygen atoms so that complicated chain, ring, or other structures result.

3. The halides of B and Si (except BF_3) are readily hydrolyzed. The Al halides are solids and only partly hydrolyzed by water. All act as Lewis acids.

4. The hydrides of B and Si are volatile, spontaneously flammable, and readily hydrolyzed. Aluminum hydride is a polymer $(AlH_3)_n$.

The most striking feature of boron chemistry is the existence of great numbers of compounds consisting of boron atoms in closed polyhedra or open, basketlike arrangements. Often the framework of such molecules includes atoms other than boron, for example, carbon, and many of those with carbon, the carboranes, form complexes with transition metals in which there are multicenter attachments of the carborane ligands to the metal atoms.

12-2
Isolation of the Element

Boron is exceedingly difficult to prepare in a pure state because of its high melting point (2250° for β-rhombohedral boron) and the corrosive nature of the liquid. It is made in 95 to 98% purity as an amorphous powder by reduction of B_2O_3 with Mg, followed by washing of the product with aqueous NaOH, HCl, and HF. The several forms of crystalline boron all have structures built up of B_{12} icosahedra (see page 184) for a drawing of an icosahedron, which also occurs in $B_{12}H_{12}^{2-}$). Crystalline boron is very inert and is attacked only by hot concentrated oxidizing agents. Amorphous boron is more reactive. Direct interaction with many other elements gives *borides*, which are hard refractory substances. Interaction with NH_3 at white heat gives BN, a slippery white solid with a layer structure resembling that of graphite but with hexagonal rings of alternate B and N atoms.

12-3
Oxygen Compounds of Boron

These are among the most important boron compounds comprising nearly all the naturally occurring forms of the element. The structures are based on triangular planar BO_3 units, with the occasional occurrence of tetrahedral BO_4 units.

Boric Acid. From borates, or by hydrolysis of boron halides, the acid $B(OH)_3$ can be obtained as white needle crystals. The $B(OH)_3$ units are linked together by hydrogen bonds to form infinite layers of nearly hexagonal symmetry. These layers are 3.18 Å apart and this accounts for the ready cleavage of the crystals.

Some reactions of boric acid are given in *Fig. 12-1*. Boric acid is moderately soluble in water. It is a weak monobasic acid that acts, not as a proton donor, but

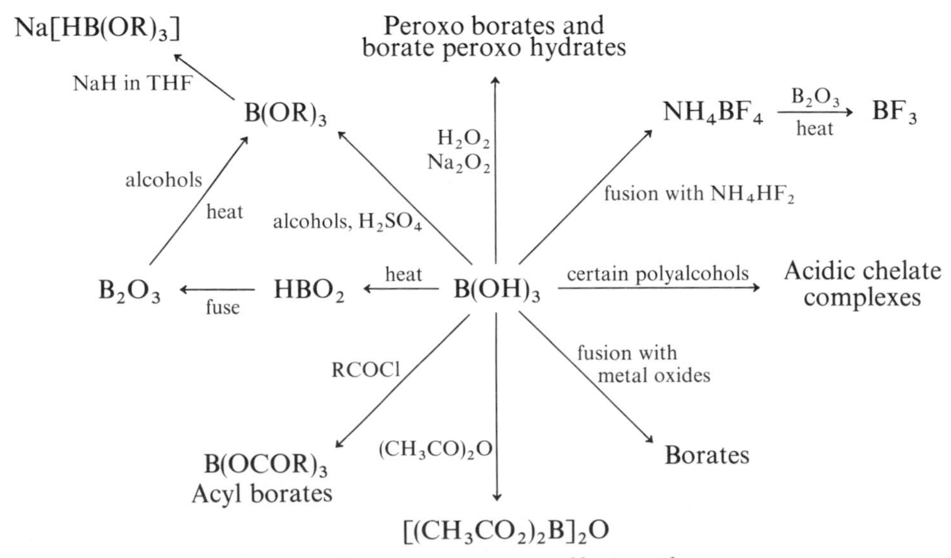

Figure 12-1 *Some reactions of boric acid.*

as a Lewis acid, accepting OH^-:

$$B(OH)_3 + H_2O = B(OH)_4^- + H^+ \qquad pK = 9.0$$

The $B(OH)_4^-$ ion occurs in several minerals, but most borates have more complex structures such as the ring anion 12-I. Boric acid and borates form very stable complexes with diols or polyols of the type 12-II. The acidity of boric acid is increased by complexing and the addition, for example, of glycerol, allows the acid to be titrated directly with NaOH.

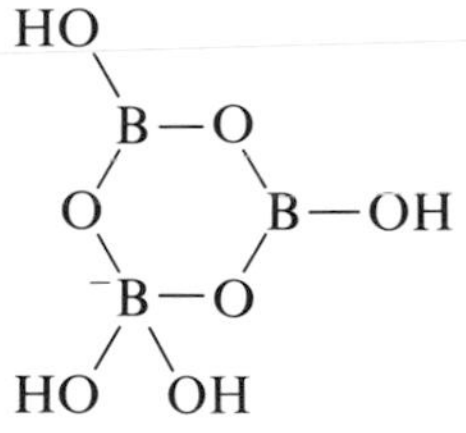

```
  HO
    \
     B—O
    /   \
   O     B—OH
    \   /
    ⁻B—O
    / \
  HO   OH

     12-I
```

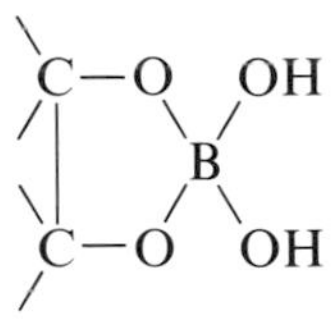

```
  \
   C—O   OH
  /|   \ /
   |    B
  \|   / \
   C—O   OH
  /

     12-II
```

The fusion of boric acid gives the *oxide* B_2O_3 as a glass. The melt readily dissolves metal oxides to give borate glasses. Pyrex and similar glasses are borosilicates.

12-4
The Trihalides of Boron

Boron Trifluoride. This pungent colorless gas (bp -101°) is obtained by heating B_2O_3 with NH_4BF_4 or CaF_2 and concentrated H_2SO_4; it is commercially available in tanks.

Boron trifluoride is one of the strongest Lewis acids known and reacts readily with most Lewis bases such as ethers, alcohols, amines, or water to give adducts, and with F^- to give the tetrafluoroborate ion BF_4^-. The diethyletherate, $(C_2H_5)_2OBF_3$, a viscous liquid, is available commercially. Unlike the other halides, BF_3 is only partially hydrolyzed by water:

$$4BF_3 + 6H_2O = 3H_3O^+ + 3BF_4^- + B(OH)_3$$

$$BF_4^- + H_2O = [BF_3OH]^- + HF$$

Because of this, and its potency as a Lewis acid, BF_3 is widely used to promote various organic reactions. Examples are:

(a) Ethers or alcohols + acids → esters + H_2O or ROH.

(b) Alcohols + benzene → alkylbenzenes + H_2O.

(c) Polymerization of alkenes and alkene oxides such as propylene oxide.

(d) Friedel–Crafts-like acylations and alkylations.

In (a) and (b) the effectiveness of BF_3 must depend on its ability to form an adduct with one or both of the reactants, thus lowering the activation energy of the rate determining step in which H_2O or ROH is eliminated by breaking of C—O bonds. In reactions of type (d), intermediates may be characterized at low temperatures. Thus the interaction of benzene and C_2H_5F proceeds as in Eq. 12-1. It is clear that

$$C_2H_5F + BF_3 \longrightarrow [C_2H_5^{\delta+}\text{---}F\text{---}\overset{\delta-}{B}F_3] \xrightarrow{C_6H_6} \left[C_6H_6(H)(C_2H_5) \right]^+ + BF_4^-$$

$$\longrightarrow C_6H_5C_2H_5 + HBF_4 \qquad (12\text{-}1)$$

BF_3 is not actually a catalyst, since it must be present in stoichiometric amount and is consumed in removing HF as HBF_4.

Fluoroboric acid solutions are formed on dissolving $B(OH)_3$ in aqueous HF:

$$B(OH)_3 + 4HF = H_3O^+ + BF_4^- + 2H_2O$$

The commercial solutions contain 40% acid. Fluoroboric acid is a strong acid and cannot, of course, exist as HBF_4. The ion is tetrahedral and fluoroborates resemble the corresponding perchlorates in their solubilities and crystal structures. Like ClO_4^- and PF_6^- the anion has a low tendency to act as a ligand toward metal ions in *aqueous* solution. In nonaqueous media, there is evidence for complex formation.

Boron trichloride (bp 12°) and the *bromide* (bp 90°) are obtained by direct interaction at elevated temperatures. They fume in moist air and are violently hydrolyzed by water:

$$BCl_3 + 3H_2O = B(OH)_3 + 3HCl$$

The rapid hydrolysis supports other evidence that these halides are stronger Lewis acids than BF_3.

Boron trichloride is the usual source material for synthesis of organoboron compounds, borate esters, and the like.

12-5
The Boron Hydrides (Boranes) and Related Compounds

Boron forms an extensive series of molecular hydrides called boranes. Typical ones are B_2H_6, B_4H_{10}, B_9H_{15}, $B_{10}H_{14}$, and $B_{20}H_{16}$. Boranes were first prepared between 1912 and 1936 by Alfred Stock who developed vacuum line techniques to handle these reactive materials. Stock's original synthesis—the reaction of Mg_3B_2 with acid—is now superseded for all but B_6H_{10}. Most syntheses now involve thermolysis of B_2H_6 under varied conditions, often in the presence of hydrogen.

Diborane. B_2H_6 is a gas (bp $-92.6°$) spontaneously flammable in air and instantly hydrolyzed by water to H_2 and $B(OH)_3$. It is obtained virtually quantitatively by the reaction of sodium borohydride (page 239) with BF_3:

$$3\,NaBH_4 + 4\,BF_3 \xrightarrow{\text{in }(MeOCH_2CH_2)_2O} 3\,NaBF_4 + 2\,B_2H_6$$

Borane, BH_3, has only a transient existence in the thermal decomposition of diborane:

$$2\,B_2H_6 - BH_3 + B_3H_9$$

Diborane has a large number of reactions. Like all BX_3 compounds it can act as a Lewis acid and forms adducts such as Me_3NBH_3; the formation of the BH_4^- ion can be regarded as a reaction with H^-.

One of the principal uses of B_2H_6 is as an extremely versatile reagent for the synthesis of organoboranes, which in turn are very useful intermediates in organic synthesis. B_2H_6 is also a powerful reducing agent for some functional groups, for example, $RCHO \rightarrow RCH_2OH$ and $RCN \rightarrow RCH_2NH_2$.

The reaction of B_2H_6 in ethers with unsaturated hydrocarbons, commonly called *hydroboration*, gives predominantly anti-Markownikoff, *cis*-hydrogenation or hydration (Eq. 12-2):

$$RCH{=}CH_2 \xrightarrow{B_2H_6} (RCH_2CH_2)_3B$$

$$(RCH_2CH_2)_3B \xrightarrow{H_3O^+} RCH_2CH_3$$

$$(RCH_2CH_2)_3B \xrightarrow{H_2O_2} RCH_2CH_2OH$$

$$(RCH_2CH_2)_3B \xrightarrow[100\text{–}125°]{\text{CO diglyme}} [(RCH_2CH_2)_3CBO]_2 \xrightarrow{H_2O_2} (RCH_2CH_2)_3COH \qquad (12\text{-}2)$$

Carbonylations using CO results in the formation of compounds in which carbon is "inserted" between the B and C of the alkyl group.

Higher Boranes. The heavier boranes, for example, B_6H_{10}, are mainly liquids whose flammability in air decreases with increasing molecular weight. One of the most important is decaborane, $B_{10}H_{14}$, a solid (mp 99.7°) that is stable in air and only slowly hydrolyzed by water. It is obtained by heating B_2H_6 at 100° and is an important starting material for the synthesis of the $B_{10}H_{10}^{2-}$ anion and carboranes discussed later.

12-6
Structures and Bonding in the Boranes

The structures of the boranes are unlike those of other hydrides such as those of carbon and are unique. A few of them are shown in *Fig. 12-2*. Observe that in none are there sufficient electrons to allow the formation of conventional 2-electron bonds between all adjacent pairs of atoms (2*c*–2*e* bonds). There is thus the problem of the electron deficiency. It was to rationalize the structures of boranes that the

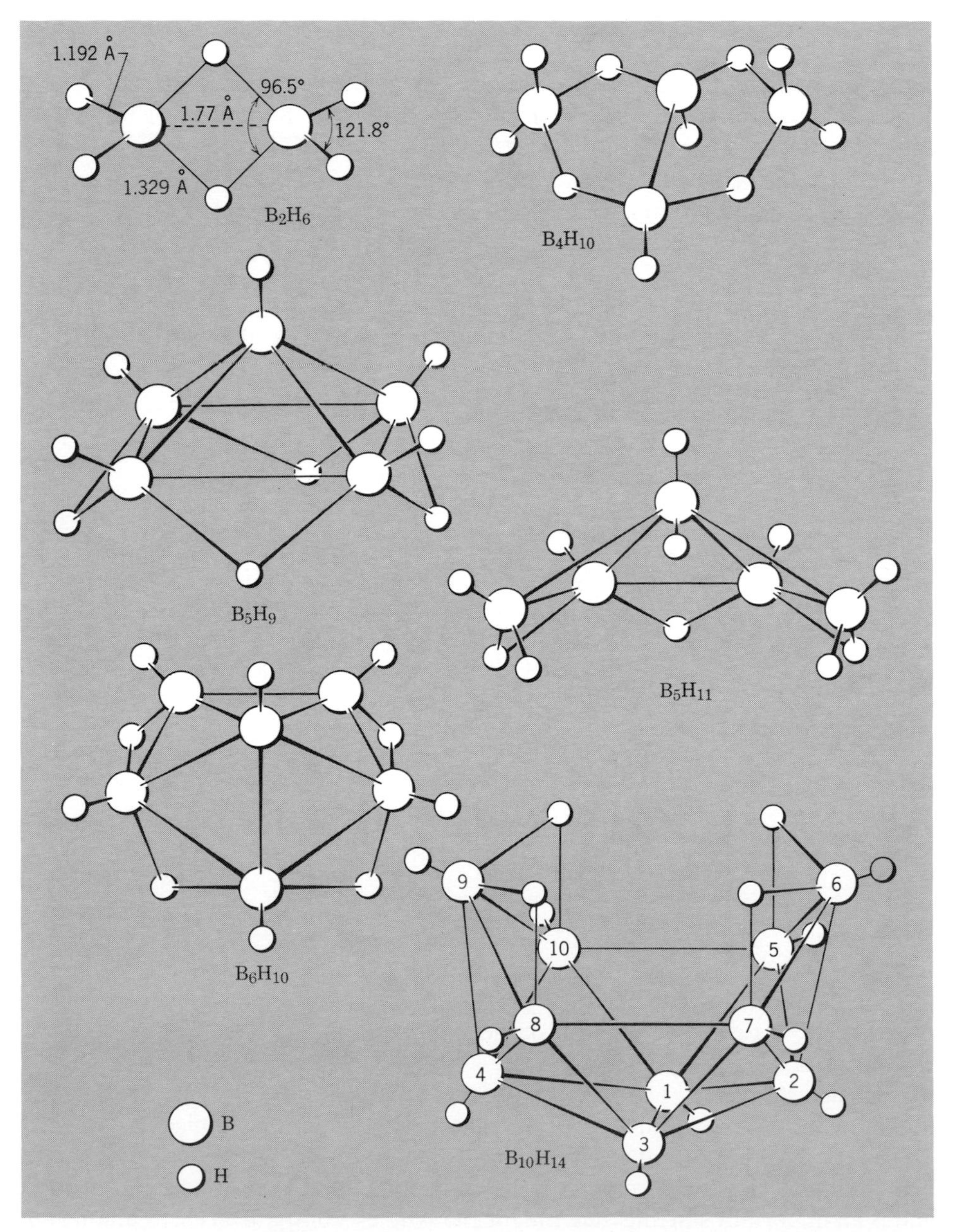

Figure 12-2 *The structures of some boranes.*

earliest of the various concepts of multicenter bonding (Chapter 3) were first developed.

For diborane itself *3c–2e* bonds are required to explain the B—H—B bridges. The terminal B—H bonds may be regarded as conventional *2c–2e* bonds. Thus, each boron atom uses two electrons and two roughly sp^3 orbitals to form *2c–2e* bonds to two hydrogen atoms. The boron atom in each BH_2 group still has one electron and two hybrid orbitals for use in further bonding. The plane of the two remaining orbitals is perpendicular to the BH_2 plane. When two such BH_2 groups approach each other as is shown in *Fig. 12-3*, with hydrogen atoms also lying, as is shown, in the plane of the four empty orbitals, two B—H—B *3c–2e* bonds are formed. The total of four electrons required for these bonds is provided by the one electron carried by each H atom and by each BH_2 group.

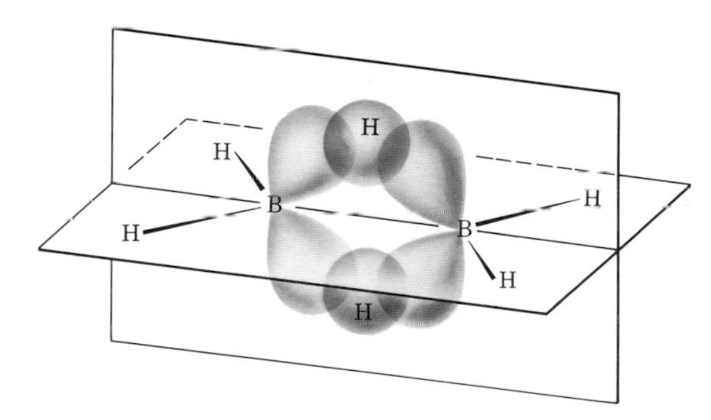

Figure 12-3 *Diagram showing how the approach of two properly oriented* BH_2 *radicals and two* H *atoms leads to the formation of two 3c–2e* B—H—B *bonds.*

We have just seen that two structure/bonding elements are used in B_2H_6, namely, *2c–2e* BH groups and *3c–3e* BHB groups. To account for the structures and bonding of the higher boranes, these elements as well as three others are required. The three others are: *2c–2e* BB groups *3c–2e* open BBB groups and *3c–2e* closed BBB groups. These five structure/bonding elements are conveniently represented in the following way:

Terminal *2c–2e* boron-hydrogen bond	B—H
3c–2e Hydrogen bridge bond	B (H) B, arched
2c–2e Boron–boron bond	B—B
Open *3c–2e* Boron bridge bond	B (B) B, arched
Closed *3c–2e* boron bond	B, B, B joined at a central point

By using these five elements, Lipscomb was able to develop "semitopological" descriptions of the structures and bonding in all of the boranes. The scheme is capable of elaboration into a comprehensive, semipredictive tool for correlating all the structural data. *Figure 12-4* shows a few examples of its use to depict known structures.

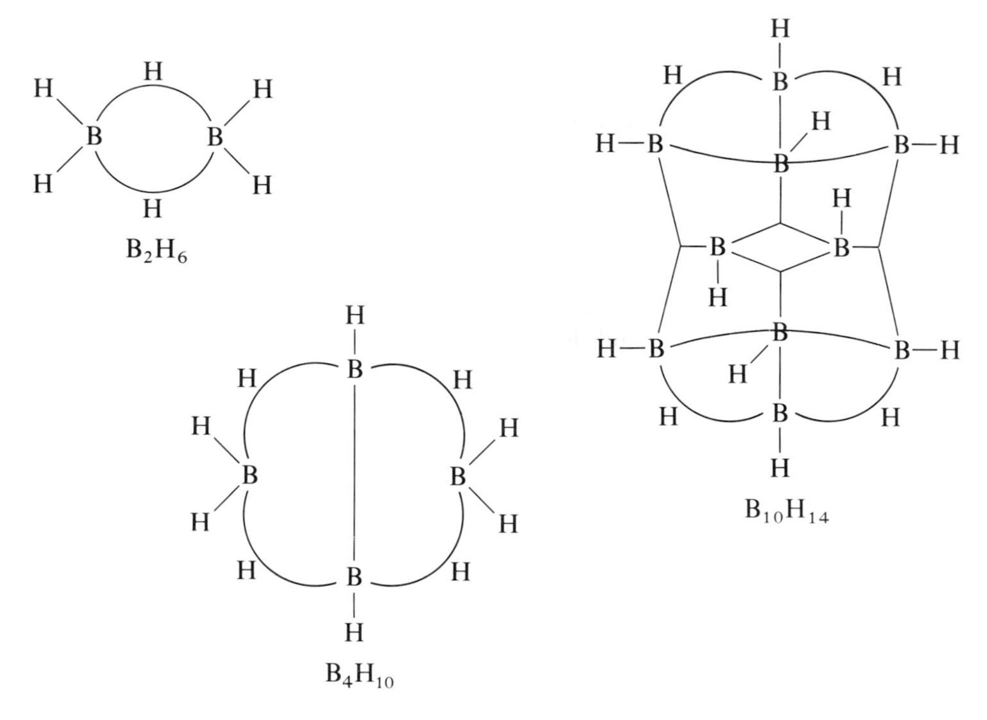

Figure 12-4 *Valence descriptions of boron hydrides in terms of Lipscomb's "semitopological" scheme.*

The semitopological scheme does not always provide the best description of bonding in the boranes and related species such as the polyhedral borane anions and carboranes we discuss below. Where there is symmetry of a high order it is often better to think in terms of a highly delocalized molecular-orbital description of the bonding. For instance, in B_5H_9 (*Fig. 12-2*) where the four basal boron atoms are equivalently related to the apical boron atom, it is *possible* to depict a resonance hybrid involving the localized B B B and B—B elements, namely:

B B B B B ⟷ B B B B B

but it is neater and simpler to formulate a set of seven 5-center MO's with the lowest three occupied by electron pairs. When one approaches the hypersymmetrical species such as $B_{12}H_{12}^{2-}$, use of the full molecular symmetry in an MO treatment becomes the only practical course.

A major motivation for theoretical study of the electronic structures of these molecules is to understand their chemical reactivity. One of the most important types of reaction that the boranes (and also the borane anions and carboranes) undergo is electrophilic substitution. Those boron atoms to which bonding theory assigns the greatest negative charge are those that are preferentially attacked in electrophilic substitution. For example, in $B_{10}H_{14}$, charge-distributions calculated from MO treatments assign considerable ($\sim$0.25e) excess negative charge to boron atoms 2 and 4, approximate neutrality to boron atoms 1 and 3, and positive charge to all others. Experiments show consistently that only positions 1, 2, 3, and 4 can be substituted electrophilically and that positions 2 and 4 are perhaps slightly preferred. Similar agreement between experimental results and calculated charge distributions has been obtained for $B_{10}C_2H_{12}$.

12-7 Derivatives of Boranes

The neutral boranes represent only the beginning of the chemistry of B—H compounds and we now turn to some of their derivatives.

The Tetrahydroborate Ion. This ion, BH_4^-, is the simplest of a number of borohydride anions. It is of great importance as a reducing agent and source of H^- ion both in inorganic and organic chemistry; derivatives such as $[BH(OMe)_3]^-$ and $[BH_3CN]^-$ are also useful, the latter because it can be used in acidic solutions.

Borohydrides of many metals have been made and some representative syntheses are

$$4\,NaH + B(OMe)_3 \xrightarrow{\sim 250^\circ} NaBH_4 + 3\,NaOCH_3$$

$$NaH + B(OMe)_3 \xrightarrow{THF} NaBH(OMe)_3$$

$$2\,LiH + B_2H_6 \xrightarrow{Ether} 2\,LiBH_4$$

$$AlCl_3 + 3\,NaBH_4 \xrightarrow{Heat} Al(BH_4)_3 + 3\,NaCl$$

$$UF_4 + 2\,Al(BH_4)_3 \longrightarrow U(BH_4)_4 + 2\,AlF_2BH_4$$

The most important salt is $NaBH_4$. It is a white crystalline solid, stable in dry air and nonvolatile. It is insoluble in diethyl ether but dissolves in water, tetrahydrofuran and ethyleneglycol ethers from which it can be crystallized.

Many borohydrides are ionic, containing the tetrahedral BH_4^- ion. However BH_4^- can serve as a ligand, interacting more or less covalently with metal ions, by bridging hydrogen atoms. Thus in $(Ph_3P)_2CuBH_4$ there are two Cu—H—B bridges, whereas in $Zr(BH_4)_4$, each BH_4 forms three bridges to Zr. These M—H—B bridges are 3*c*–2*e* bonding systems.

Polyhedral Borane Anions and Carboranes. The polyhedral borane anions have the formula $B_nH_n^{2-}$. The carboranes may be considered to be *formally* derived from $B_nH_n^{2-}$ by replacement of BH^- by the isoelectronic and isostructural CH. Thus two replacements lead to neutral molecules $B_{n-2}C_2H_n$. Carboranes or derivatives with $n = 5$ to $n = 12$ are known, in some of which two or more isomers may be isolated. Sulfur and phosphorus derivatives can also be obtained, PH^+, for example, replacing CH or BH^-.

Geometrically there are two broad classes of compounds:

1. Those in which the boron atom framework closes in on itself to form a polyhedron. These are *closo* (Greek for cage) compounds.

2. Those frameworks that are open or incomplete polyhedra. These are *nido* (nest) compounds.

Some of the *closo* boranes and carboranes are shown in *Fig. 12-5*.

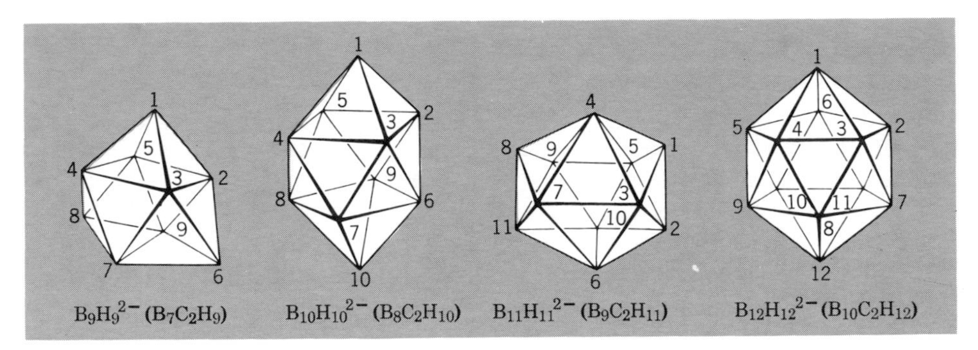

Figure 12-5 *The triangulated-polyhedral structures of some $B_nH_n^{2-}$ and $B_{n-2}C_2H_n$ species with the conventional numbering schemes indicated.*

$B_nH_n^{2-}$ Ions. The most stable and best studied are $B_{10}H_{10}^{2-}$ and $B_{12}H_{12}^{2-}$, which can be synthesized by the reactions

$$B_{10}H_{14} + 2R_3N \xrightarrow{150^\circ} 2(R_3NH)^+ + B_{10}H_{10}^{2-} + H_2$$

$$6B_2H_6 + 2R_3N \xrightarrow{150^\circ} 2(R_3NH)^+ + B_{12}H_{12}^{2-} + 11H_2$$

The most important general reaction of the anions is attack by electrophilic reagents such as Br^+, $C_6H_5N_2^+$, RCO^+ in strongly acid media. $B_{10}H_{10}^{2-}$ is more susceptible to substitution than is $B_{12}H_{12}^{2-}$.

$B_{n-2}C_2H_n$ Carboranes. The most important carboranes are 1,2- and 1,7-dicarba*closo*dodecaborane, $B_{10}C_2H_{12}$, and their *C*-substituted derivatives. The 1,2-isomer may be obtained by the reactions

$$B_{10}H_{14} + 2R_2S = B_{10}H_{12}(R_2S)_2 + H_2$$

$$B_{10}H_{12}(R_2S)_2 + RC{\equiv}CR' = 1,2\text{-}B_{10}H_{10}C_2RR' + 2R_2S + H_2$$

On heating at 450° the 1,2-isomer rearranges to the 1,7-isomer.

Derivatives may be obtained from $B_{10}C_2H_{12}$ by replacement of the hydrogen bound to C by Li. The dilithio derivatives react with many other reagents (Scheme 12-1) where a self-explanatory abbreviation is used for $B_{10}H_{10}C_2$.

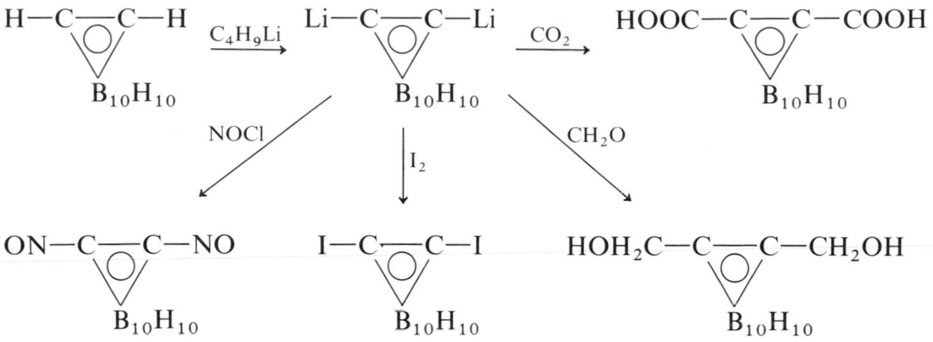

H—C—C—H ($B_{10}H_{10}$) $\xrightarrow{C_4H_9Li}$ Li—C—C—Li ($B_{10}H_{10}$) $\xrightarrow{CO_2}$ HOOC—C—C—COOH ($B_{10}H_{10}$)

Li—C—C—Li ($B_{10}H_{10}$) $\xrightarrow{NOCl}$ ON—C—C—NO ($B_{10}H_{10}$)

Li—C—C—Li ($B_{10}H_{10}$) $\xrightarrow{I_2}$ I—C—C—I ($B_{10}H_{10}$)

Li—C—C—Li ($B_{10}H_{10}$) $\xrightarrow{CH_2O}$ HOH_2C—C—C—CH_2OH ($B_{10}H_{10}$)

Scheme 12-1

An enormous number of compounds have been made, one of the main motives being the incorporation of the thermally stable carborane residues into high polymers such as silicones in order to increase the thermal stability. Chlorinated carboranes can be obtained directly from $B_{10}C_2H_{10}R_2$.

$B_9C_2H_{13-n}{}^{n-}$ Carborane Anions. When the 1,2- and 1,7-dicarbaclosododecaboranes are heated with alkoxide ions, degradation occurs to form isomeric *nido*carborane anions $B_9C_2H_{12}{}^-$:

$$B_{10}C_2H_{12} + EtO^- + 2EtOH = B_9C_2H_{12}{}^- + B(OEt)_3 + H_2$$

This removal of a BH^{2+} unit from $B_{10}C_2H_{12}$ may be interpreted as a nucleophilic attack at the most electron-deficient boron atoms of the carborane. MO calculations show that the C atoms in carboranes have considerable electron-withdrawing power. The most electron-deficient B atoms are those adjacent to carbon. In 1,2-$B_{10}C_2H_{12}$ these will be in positions 3 and 6 while in 1,7-$B_{10}C_2H_{12}$ they will be at positions 2 and 3 (see numbering in *Fig. 12-5*).

While alkoxide ion attack produces only $B_9C_2H_{12}{}^-$, use of the very strong base NaH forms the $B_9C_2H_{11}{}^{2-}$ ions:

$$B_9C_2H_{12}{}^- + NaH = Na^+ + B_9C_2H_{11}{}^{2-} + H_2$$

The structures of the isomeric $B_9C_2H_{11}{}^{2-}$ ions, *Fig. 12-6*, confirm the above expectation. The $B_9C_2H_{11}{}^{2-}$ ions are very strong bases and readily acquire H^+ to give $B_9C_2H_{12}{}^-$. These, in turn, can be protonated to form the neutral *nido*-carboranes $B_9C_2H_{13}$, which are strong acids.

$$B_9C_2H_{11}{}^{2-} \underset{}{\overset{H^+}{\rightleftharpoons}} B_9C_2H_{12}{}^- \overset{H^+}{\rightleftharpoons} B_9C_2H_{13}$$

Heating $B_9C_2H_{13}$ gives yet another *closo*carborane, $B_9C_2H_{11}$, with loss of hydrogen.

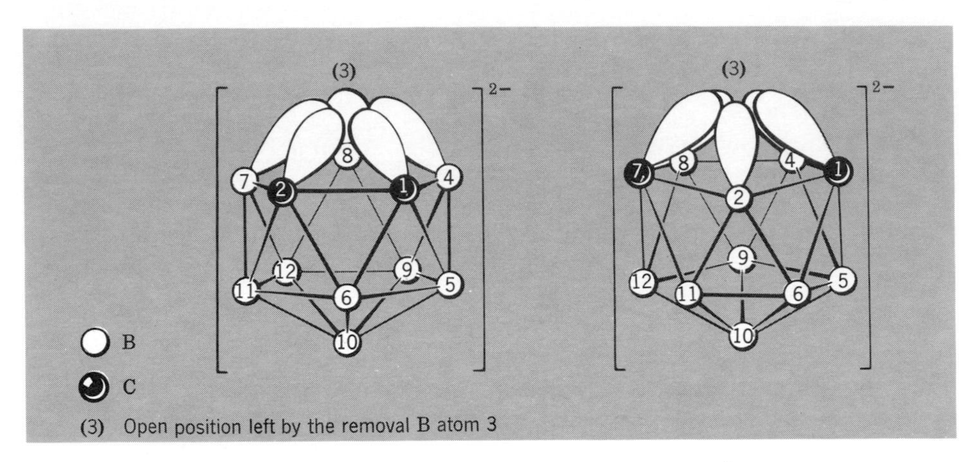

Figure 12-6 *Structures of the isomeric nido* $B_9C_2H_{11}^{2-}$ *ions.*

Metal Complexes of Carborane Anions. The open pentagonal faces of the $B_9C_2H_{11}^{2-}$ ions (*Fig. 12-6*) were recognized by M. F. Hawthorne in 1964 to bear a strong resemblance structurally and electronically to the cyclopentadienyl ion $C_5H_5^-$. The latter forms strong bonds to transition metals, as we discuss in Chapter 29.

Interaction of $Na_2B_9C_2H_{11}$ with metal compounds such as those of Fe^{2+} or Co^{3+} thus leads to species isoelectronic with ferrocene, $(C_5H_5)_2Fe$, or the cobalticinium ion, $(C_5H_5)_2Co^+$, namely, $(B_9C_2H_{11})_2Fe^{2-}$ and $(B_9C_2H_{11})_2Co^-$, respectively. The iron complex undergoes reversible oxidation like ferrocene:

$$[(C_5H_5)_2Fe^{III}]^+ \quad + e = [(C_5H_5)_2Fe^{II}]^0$$

$$[(B_9C_2H_{11})_2Fe]^- + e = [(B_9C_2H_{11})Fe]^{2-}$$

The formal nomenclature for $B_9C_2H_{11}^{2-}$ ion and its complexes is unwieldy and the trivial name "*dicarbollide*" ion was proposed (from the Spanish *olla* for pot, from the potlike shape of the B_9C_2 cage).

The structures of two types of bis(dicarbollide) metal complexes are shown in *Fig. 12-7.* While some complexes have a symmetrical "sandwich" structure (*Fig. 12-7a*) others have the metal disposed asymmetrically.

Finally, comparable with η^5-$C_5H_5Mn(CO)_3$ (Chapter 29), there are mixed complexes with only one dicarbollide unit and other ligands such as CO, Ph_4C_4, C_5H_5, etc., (*Fig. 12-7b*).

Compounds of Boron with Other Elements. There are many other types of compound with bonds to N, P, As, S, and C; the organoboranes were mentioned in Section 12-5. Here we describe only some boron-nitrogen compounds. The —NR′—BR— unit is similar to —CR′═CR— and can replace it in many compounds. We have already referred to the graphitelike structure of BN (page 232).

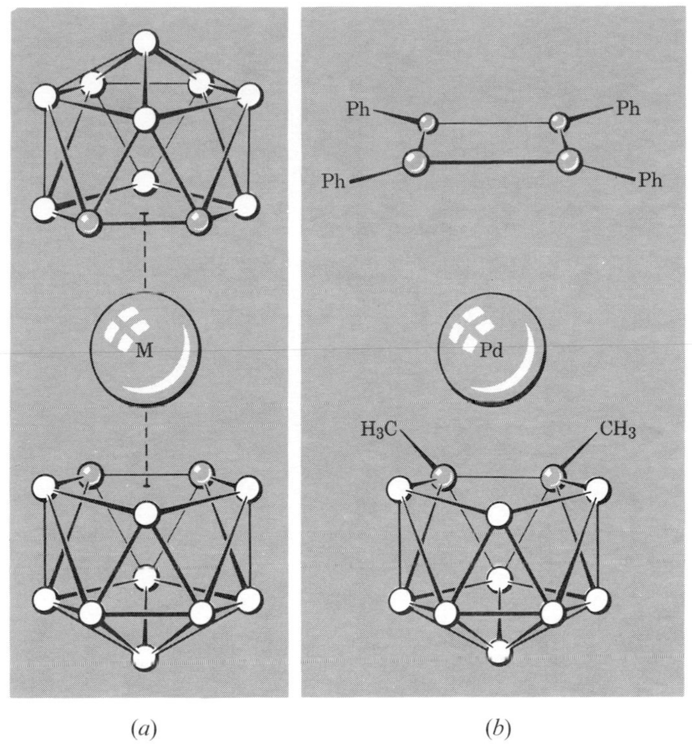

(*a*) (*b*)

Figure 12-7 *(a) The structures of bis(dicarbollide) metal complexes. (b) Structure of a mono-(dicarbollide) complex.*

The analogy has been justified by the assumption that the actual electron distribution in the N—B bond can be described as a resonance hybrid:

$$>\ddot{N}-B< \longleftrightarrow >\overset{+}{N}=\overset{-}{B}<$$

(12-IIIa) (12-IIIb)

where appreciable π-bonding is introduced. Such B—N bonds have appreciable π character but at the same time they lack the polarity expected from 12-IIIb. This apparent paradox is explained by the existence of polarity in the σ-bond in a direction opposite to that in the π-bond. Hence, there is only a small differential polarity.

One of the most interesting B—N compounds is *borazine* (12-IV). It has an

H
B
HN⁺ ⁻ ⁺NH
HB⁻ ⁺ ⁻BH
N
H

12-IV

obvious formal resemblance to benzene and the physical properties of the compounds are similar. However borazine is much more reactive than benzene and readily undergoes addition reactions such as

$$B_3N_3H_6 + 3HX \longrightarrow (-H_2N-BHX-)_3 \qquad X = Cl, OH, OR, \text{etc.}$$

which do not occur with benzene. Borazine also decomposes slowly and may be hydrolyzed to NH_3 and $B(OH)_3$ at elevated temperatures. Like benzene, however, π complexes with transition metals can be obtained (Chapter 29); thus, hexamethylborazine gives $(B_3N_3Me_6)Cr(CO)_3$ (12-V).

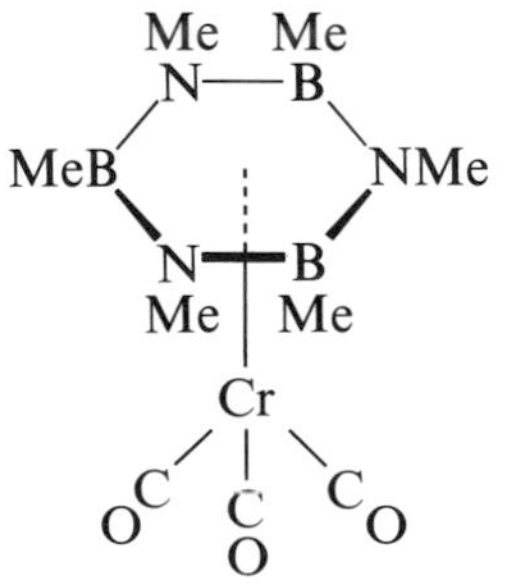

12-V

Borazine and substituted borazines may be synthesized by the reactions in Scheme 12-2, which also illustrates other types of B—N compounds.

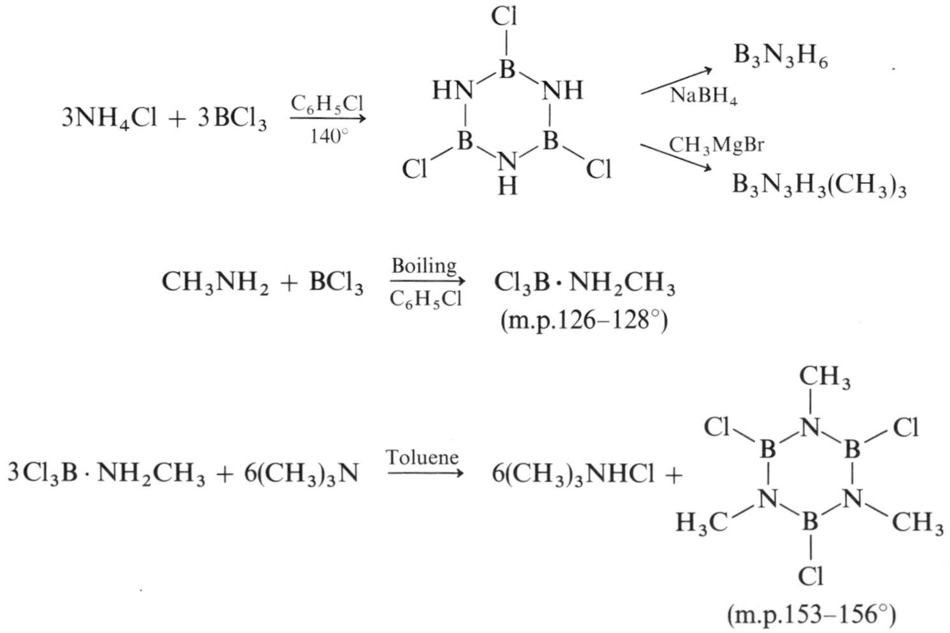

$$3NH_4Cl + 3BCl_3 \xrightarrow[140^\circ]{C_6H_5Cl} (ClBNH)_3 \quad \xrightarrow{NaBH_4} B_3N_3H_6; \quad \xrightarrow{CH_3MgBr} B_3N_3H_3(CH_3)_3$$

$$CH_3NH_2 + BCl_3 \xrightarrow[C_6H_5Cl]{\text{Boiling}} Cl_3B \cdot NH_2CH_3 \ (\text{m.p.}126\text{–}128^\circ)$$

$$3Cl_3B \cdot NH_2CH_3 + 6(CH_3)_3N \xrightarrow{\text{Toluene}} 6(CH_3)_3NHCl + (ClBNCH_3)_3 \ (\text{m.p.}153\text{–}156^\circ)$$

Scheme 12-2

Study Questions

A

1. Why does B not form cationic species?
2. Draw an icosahedron.
3. Draw structures for the cyclic anion in the salt $K_3B_3O_6$ and the chain anion in CaB_2O_4.
4. How does boric acid ionize in water? How strong an acid is it?
5. Why is the acidity of boric acid increased by addition of glycerol?
6. How would you make BF_3 in the laboratory?
7. Why is BBr_3 a better Lewis acid than BF_3?
8. How would you make B_2H_6 in the laboratory?
9. Draw the structure of diborane. Describe the bonding.
10. What is a semitopological description of structure and bonding in boranes? Give examples.
11. How is sodium borohydride made?
12. What is a carborane? Give one example each of a *closo* and a *nido*carborane.
13. Give an example of a transition metal carborane complex.
14. Although BH_3 cannot be isolated, how would you obtain derivatives of BH_3?

B

1. Show the intermediates involved when BF_3 serves as a catalyst in Friedel–Crafts alkylation of benzene.
2. Draw structures for the anions in the salts:

$$KB_5O_8 \cdot 4H_2O,\ Na_2B_4O_7 \cdot 10H_2O,\ CoB_2O_5.$$

3. Methanol solutions of boric acid, when heated, give vapors that will color a flame green. Why?
4. What would you expect the structures of B_2Cl_4 and B_4Cl_4 (obtained by electric discharges in BCl_3) to be?
5. Describe the "hydroboration" reaction of an alkene.
6. Draw the structures of B_5H_9, B_6H_{10}, and $B_{10}H_{14}$.
7. Draw the structure of $Al(BH_4)_3$.
8. Draw the structures of $B_{10}H_{10}^{2-}$ and $B_{10}H_{10}C_2RR'$.
9. Describe the preparation of borazine. How does it compare with benzene in its chemistry?

Chapter 12
Study Guide

Supplementary Reading

Boscke, F., ed., *New Results in Boron Chemistry, Fortschritte der Chemischen Forschung*, Springer Verlag, 1971.
Brown, H. C., *Boranes in Organic Chemistry*, Cornell University Press, 1972.
Grimes, R. N., *Carboranes*, Academic Press, 1971.
Muetterties, E. L., ed., *The Chemistry of Boron and its Compounds*, Wiley, 1967.
Muetterties, E. L. and Knoth, W. H., *Polyhedral Boranes*, Dekker, 1968.

13

the group IIIB elements: aluminum, gallium, indium and thallium

13-1 Introduction

Aluminum is the commonest metallic element in the earth's crust and occurs in rocks such as felspars and micas. More accessible deposits are hydrous oxides such as bauxite, Al_2O_3, nH_2O, and cryolite, Na_3AlF_6. Gallium and In occur only in traces in Al and Zn ores. Thallium, also a rare element, is recovered from flue dusts from the roasting of pyrite and other sulfide ores.

Aluminum metal has many uses and some salts such as the sulfate (*ca.* 10^8 kg, U.S.A., 1972) are made on a large scale. Gallium finds some use in solid state devices as GaAs. Thallium is used mainly as the Tl(III) carboxylates in organic synthesis.

The position of the elements and their relation to the Sc, Y, La group is discussed in Chapter 8, page 199, where Table 8-3 gives some important properties of the elements.

The elements are more metallic than boron, and the chemistry in compounds is more ionic. Nevertheless many of the compounds are on the borderline of ionic-covalent character. All four elements give trivalent compounds, but the univalent state becomes increasingly important for Ga, In, and Tl. For Tl the two states are about equally important and the redox system Tl^I—Tl^{III} dominates the chemistry. The Tl^+ ion is well defined in solutions.

The main reason for the existence of the univalent state is the decreasing strengths of bonds in MX_3; thus, for the chlorides, the mean bond energies are Ga, 242; In, 206; Tl, 153, kJ mol^{-1}. There is hence an increasing drive for the reaction

$$MX_3 = MX + X_2$$

to occur.

The compounds MX_3 or MR_3 resemble similar BX_3 compounds in that they are Lewis acids, with strengths decreasing in the order: B > Al > Ga > In ~ Tl. However, while all BX_3 compounds are planar monomers, the halides of the other elements have crystal structures in which the coordination number is increased to 4 or 6. Also the lower alkyls of Al are dimers, while Lewis acid adducts may also be 5-coordinate like $(Me_3N)_2AlH_3$.

Each of the elements forms an aquo ion, $[M(H_2O)_6]^{3+}$, and gives simple salts and complex compounds in virtually all of which the metals are octahedrally coordinated.

13-2
Occurrence, Isolation, and Properties of the Elements

Aluminum is prepared on a vast scale from bauxite, $Al_2O_3 \cdot nH_2O$ (n = 1-3). This is purified by dissolution in aqueous NaOH and reprecipitation as $Al(OH)_3$ using CO_2. The dehydrated product is dissolved in molten cryolite and the melt at 800 to 1000° is electrolyzed. Aluminum is a hard, strong, white metal. Although highly electropositive, it is nevertheless resistant to corrosion because a hard, tough film of oxide is formed on the surface. Thick oxide films are often electrolytically applied to aluminum, a process called anodizing; the fresh films can be colored by pigments. Aluminum is soluble in dilute mineral acids, but is "passivated" by concentrated HNO_3. If the protective effect of the oxide film is broken, for example, by scratching or by amalgamation, rapid attack even by water can occur. The metal is readily attacked by hot aqueous NaOH, halogens, and various nonmetals.

Gallium, *indium*, and *thallium* are usually obtained by electrolysis of aqueous solutions of their salts; for Ga and In this possibility arises because of large overvoltages for hydrogen evolution of these metals. They are soft, white, comparatively reactive metals, dissolving readily in acids. Thallium dissolves only slowly in H_2SO_4 or HCl, since the Tl^I salts formed are only sparingly soluble. Gallium, like Al, is soluble in aqueous NaOH. The elements react rapidly at room temperature, or on warming, with the halogens and with nonmetals such as sulfur.

CHEMISTRY OF THE TRIVALENT STATE

13-3
Oxides

The only oxide of aluminum is *alumina*, Al_2O_3. However, this simplicity is compensated by the occurrence of polymorphs and hydrated materials whose nature depends on the conditions of preparation. There are two forms of anhydrous Al_2O_3, namely, α-Al_2O_3 and γ-Al_2O_3. Other trivalent metals (e.g., Ga, Fe) form oxides that crystallize in these same two structures. Both have close-packed arrays of oxide ions but differ in the arrangement of the cations.

α-Al_2O_3 is stable at high temperatures and also indefinitely metastable at low temperatures. It occurs in nature as the mineral corundum and may be prepared by heating γ-Al_2O_3 or any hydrous oxide above 1000°. γ-Al_2O_3 is obtained by dehydration of hydrous oxides at low temperatures (~450°). α-Al_2O_3

is hard and is resistant to hydration and to attack by acids. γ-Al_2O_3 readily absorbs water and dissolves in acids; the aluminas used for chromatography and conditioned to different reactivities are γ-Al_2O_3.

There are several hydrated forms of alumina of stoichiometries from $AlO\cdot OH$ to $Al(OH)_3$. Addition of ammonia to a boiling solution of an aluminum salt produces a form of $AlO\cdot OH$ known as *bohmite*. A second form of $AlO\cdot OH$ occurs in nature as the mineral *diaspore*. The true *hydroxide*, $Al(OH)_3$, is obtained as a crystalline white precipitate when CO_2 is passed into alkaline "aluminate" solutions.

Gallium and indium oxides are similar, but Tl gives only brown-black Tl_2O_3, which decomposes to Tl_2O at 100°.

The elements form *mixed oxides* with other metals. Aluminum oxides containing only traces of other metal ions include ruby (Cr^{3+}) and blue sapphire (Fe^{2+}, Fe^{3+}, and Ti^{4+}). Synthetic ruby, blue sapphire, and white sapphire (gem-quality corundum) are manufactured in large quantities. Mixed oxides containing macroscopic proportions of other elements include the minerals *spinel*, $MgAl_2O_4$, and *crysoberyl*, $BeAl_2O_4$. The *spinel structure* (page 104) is important as a prototype for many other $M^{II}M_2^{III}O_4$ compounds. Compounds such as $NaAlO_2$, which can be made by heating Al_2O_3 with sodium oxalate at 1000°, are also ionic mixed oxides.

13-4 Halides

All four halides of each element are known, with one exception. The compound TlI_3, obtained by adding iodine to thallous iodide, is not thallium(III) iodide, but rather thallium(I) triiodide, $Tl^I(I_3)$. This situation may be compared with the nonexistence of iodides of other oxidizing cations such as Cu^{2+} and Fe^{3+}, except that here a lower-valent compound fortuitously has the same stoichiometry as the higher-valent one. The coordination numbers of the halides are shown in Table 13-1. The fluorides of Al, Ga, and In are ionic and high-melting ($>950°$),

Table 13-1 *Coordination Numbers of Metal Atoms in Group III Halides*

	F	Cl	Br	I
Al	6	6	4	4
Ga	6	4	4	4
In	6	6	6	4
Tl	6	6	4	

whereas the chlorides, bromides, and iodides have lower melting points. There is some correlation between melting points and coordination number, since the halides with coordination number 4 consist of discrete dinuclear molecules (*Fig. 13-1*) and the melting points are low. Thus, the three chlorides have the following melting points: $AlCl_3$, 193° (at 1700 mm); $GaCl_3$, 78°; $InCl_3$, 586°. In the vapor, aluminum chloride also is dimeric so that there is a radical change of coordination

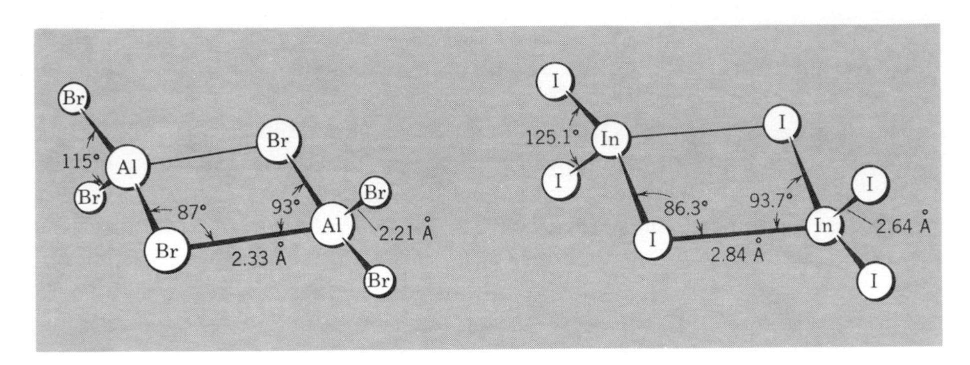

Figure 13-1 *The structures of* Al_2Br_6 *and* In_2I_6.

number on vaporization. The dimer structures persist in the vapor phase at temperatures close to the boiling points but at higher temperatures dissociation occurs, giving triangular monomers analogous to the boron halides.

The covalent halides dissolve readily in nonpolar solvents such as benzene, in which they are dimeric. As *Fig. 13-1* shows, the configuration of halogen atoms about each metal atom is distorted tetrahedral. The formation of such dimers is attributable to the tendency of the metal atoms to complete their octets.

The thallium(III) halides vary considerably in thermal stability. Although TlF_3 is stable to 500°, $TlCl_3$ loses chlorine at about 40° forming TlCl, while $TlBr_3$ loses Br_2 at even lower temperatures to give first "$TlBr_2$", which is actually $Tl^I[Tl^{III}Br_4]$.

The trihalides (fluorides excepted) are strong Lewis acids, and this is one of the most important aspects of their chemistry, as well as that of other MR_3 compounds, such as the alkyls and AlH_3. Adducts are formed quite readily with Lewis bases (including halide ions), the dimeric halides being cleaved thereby to give products such as $Cl_3AlN(CH_3)_3$ and $AlCl_4^-$.

Aluminum chloride and bromide especially are used as catalysts (Friedel–Crafts type) in a variety of reactions. The formation of $AlCl_4^-$ or $AlBr_4^-$ ions is essential to the catalytic action, since in this way carbonium ions are formed simultaneously:

$$RCOCl + AlCl_3 = RCO^+ + AlCl_4^- \text{ (ion pair)}$$

$$RCO^+ + C_6H_6 \longrightarrow [RCOC_6H_6]^+ \longrightarrow RCOC_6H_5 + H^+$$

13-5
The Aquo Ions, Oxo Salts, Aqueous Chemistry

The elements form well-defined octahedral aquo ions, $[M(H_2O)_6]^{3+}$, and many salts are formed including hydrated halides, sulfate, nitrate, and perchlorate. Phosphates are sparingly soluble.

In aqueous solution the octahedral $[M(H_2O)_6]^{3+}$ ions are quite acidic. For the reaction

$$[M(H_2O)_6]^{3+} = [M(H_2O)_5(OH)]^{2+} + H^+$$

the constants are K_a(Al), 1.12×10^{-5}; K_a(Ga), 2.5×10^{-3}; and K_a(In), 2×10^{-4}; K_a(Tl), $\sim 7 \times 10^{-2}$. Although little emphasis can be placed on the exact numbers, the orders of magnitude are important, for they show that aqueous solutions of the M^{III} salts are subject to extensive hydrolysis. Indeed, salts of weak acids—sulfides, carbonates, cyanides, acetates, and the like—cannot exist in contact with water.

In addition to the above hydrolysis reaction there is also a dimerization reaction:

$$2\,AlOH^{2+}(aq) = Al_2(OH)_2{}^{4+}(aq) \qquad K = 600\ mol^{-1}\ (30°)$$

More complex species of general formula $Al[Al_3(OH)_8]_m{}^{m+3}$ have also been postulated and some, such as $[Al_{13}O_4(OH)_{24}(H_2O)_{12}]^{1+}$, have been identified in crystalline basic salts.

An important class of aluminum salts, the *alums*, are structural prototypes and give their name to a large number of analogous salts formed by other elements. They have the general formula $MAl(SO_4)_2 \cdot 12H_2O$ in which M is practically any common univalent, monatomic cation except for Li^+, which is too small to be accommodated without loss of stability of the structure. The crystals are made up of $[M(H_2O)_6]^+$, $[Al(H_2O)_6]^{3+}$, and two $SO_4{}^{2-}$ ions. Salts of the same type, $M^IM^{III}(SO_4)_2 \cdot 12H_2O$, having the same structures are formed by other M^{3+} ions, including those of Ti, V, Cr, Mn, Fe, Co, Ga, In, Rh, and Ir. All such compounds are referred to as alums. The term is used so generally that those alums containing aluminum are redundantly designated aluminum alums.

Thallium carboxylates. Thallium carboxylates, particularly the acetate and trifluoroacetate, which can be obtained by dissolution of the oxide in the acid, are extensively used in organic synthesis. The trifluoroacetate will directly thallate (cf. mercuration, Chapter 29) aromatic compounds to give aryl thallium ditrifluoroacetates, for example, $C_6H_5Tl(OOCCF_3)_2$. It also acts as an oxidant, converting para substituted phenols into *p*-quinones, for example.

Aluminates and gallates. The hydroxides are amphoteric:

$$Al(OH)_3(s) = Al^{3+} + 3OH^- \qquad K \approx 5 \times 10^{-33}$$
$$Al(OH)_3(s) = AlO_2{}^- + H^+ + H_2O \qquad K \approx 4 \times 10^{-13}$$
$$Ga(OH)_3(s) = Ga^{3+} + 3OH^- \qquad K \approx 5 \times 10^{-37}$$
$$Ga(OH)_3(s) = GaO_2{}^- + H^+ + H_2O \qquad K \approx 10^{-15}$$

Like the oxides these also dissolve in bases as well as in acids. By contrast the oxides and hydroxides of In and Tl are purely basic. According to Raman spectra, the main aluminate species from pH 8 to pH 12 appears to be an OH bridged polymer with octahedral Al, but at pH > 13 and concentrations below 1.5 *M* the tetrahedral $Al(OH)_4{}^-$ ion is present. Above 1.5 *M* there is condensation to give the ion $[(HO)_3AlOAl(OH)_3]^{2-}$. This occurs in the crystalline salt $K_2[Al_2O(OH)_6]$ and has an angular Al—O—Al bridge.

13-6
Complex Compounds

The trivalent elements form 4-, 5- and 6-coordinate complexes, which may be cationic, like $[Al(H_2O_6]^{3+}$ or $[Al(OSMe_2)_6]^{3+}$, neutral, for example, $AlCl_3(NMe_3)_2$, or anionic, like AlF_6^{3-} and $In(SO_4)_2^{-}$.

One of the most important salts is *cryolite* whose structure (*Fig. 13-2*) is adopted by many other salts that contain small cations and large octahedral

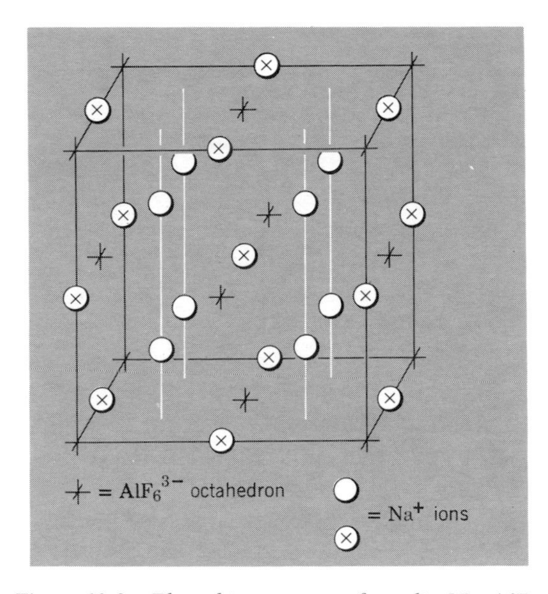

Figure 13-2 *The cubic structure of cryolite* Na_3AlF_6.

anions and, with reversal of cations and anions, by many salts of the same type as $[Co(NH_3)_6]I_3$. It is closely related to the structures adopted by many compounds of the types $M_2{}^+[AB_6]^{2-}$ and $[XY_6]^{2+}Z_2{}^-$. The last two structures are essentially the fluorite (or antifluorite) structures (see *Fig. 4-1*, page 92), except that the anions (or cations) are octahedra whose axes are oriented parallel to the cube edges. The relationship of the two structures can be seen in *Fig. 13-2*, since the Na^+ ions have been indicated by both open ○ and marked ⊗ circles. If all of the latter, which comprise the one at the center, and those on the edges are removed, the cryolite structure becomes the $M_2{}^+[AB_6]^{2-}$ fluorite-type structure.

Many of the important octahedral complexes are those containing chelate rings, typical ones containing β-diketones, pyrocatechol (13-I), dicarboxylic acids (13-II), and 8-quinolinol (13-III). The neutral complexes are soluble in organic

13-I

13-II

13-III

solvents, but insoluble in water. The acetylacetonates have low melting points ($<200°$) and vaporize without decomposition. The anionic complexes are isolated as the salts of large univalent cations. The 8-quinolinolates are used for analytical purposes.

The four elements form *alkoxides*, but only those of aluminum are important. The isopropoxide is widely used in organic chemistry to catalyze the reduction of aldehydes and ketones by alcohols or vice versa (Meerwein–Ponndorf–Oppenauer–Verley reactions). Alkoxides can be made by the reactions

$$Al + 3ROH \xrightarrow[\text{catalyst, warm}]{1\%\ HgCl_2\ \text{as}} (RO)_3Al + \tfrac{3}{2}H_2$$

$$AlCl_3 + 3RONa \xrightarrow{ROH} (RO)_3Al + 3NaCl$$

The alkoxides hydrolyze vigorously in water. The *tert*-butoxide is a cyclic dimer (13-IV) in solvents, whereas the isopropoxide is tetrameric (13-V) at ordinary

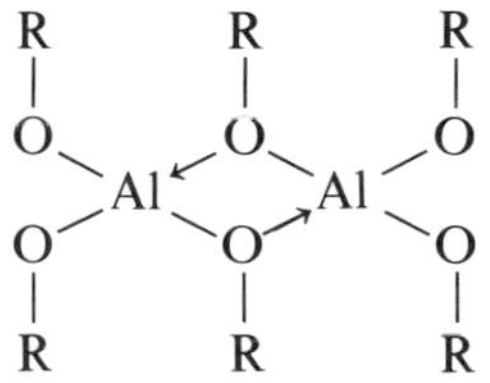

13-IV

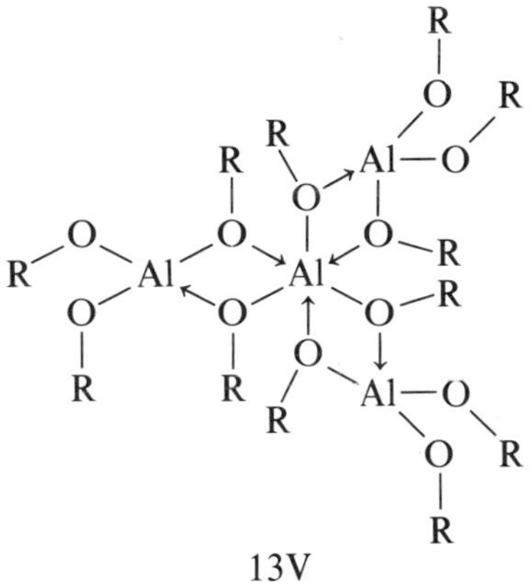

13V

temperatures but trimeric at elevated temperatures. Terminal and bridging alkoxyl groups can be distinguished by nmr spectra. Other alkoxides form dimers and trimers.

13-7 Complex Hydrides

The most important chemistry of Al and Ga is that of the tetrahedral hydride anions, AlH_4^- and GaH_4^-, which are similar to BH_4^- (page 239). The thermal and chemical stabilities vary with the ability of the MH_3 group to act as an acceptor as in the reaction

$$MH_3 + H^- = MH_4^-$$

The order is B > Al > Ga. Thus $LiGaH_4$ decomposes slowly even at 25° to LiH, Ga, and H_2 and is a milder reducing agent than $LiAlH_4$. Similarly, although

BH_4^- is stable in water, the Al and Ga salts are rapidly and often explosively hydrolyzed by water:

$$MH_4^- + 4H_2O = 4H_2 + M(OH)_3 + OH^-$$

The most important compound is *lithium tetrahydridoaluminate* which is widely used in both organic and inorganic chemistry as a reducing agent. It accomplishes many otherwise tedious or difficult reductions, for example, —COOH to —CH_2OH. It is a nonvolatile, crystalline solid, white when pure but usually grey. It is stable below 120° and is soluble in diethyl ether, tetrahydrofuran, and glymes.

Both Al and Ga salts are made by the reaction

$$4LiH + MCl_3 \xrightarrow{Et_2O} LiMH_4 + 3LiCl$$

The sodium salt, $NaAlH_4$, can be obtained by direct interaction:

$$Na + Al + 2H_2 \xrightarrow[150°/2000psi]{THF} NaAlH_4$$

The salt is precipitated by toluene and converted to the lithium salt:

$$NaAlH_4 + LiCl \xrightarrow{Et_2O} NaCl(s) + LiAlH_4$$

The hydrides MH_3 can be obtained as polymeric white powders by action of H_2SO_4 or $LiMH_4$ in tetrahydrofuran. There are numerous adducts with ethers, amines, and the like, for example, $AlH_3(NMe_3)$ and $AlH_3(NMe_3)_2$.

13-8
Lower Valent Compounds

Since the outer electronic configuration is ns^2np, univalent compounds are, in principle, possible. Aluminum forms such species only at high temperature in the gas phase, for example

$$AlCl_3(s) + 2Al(s) \rightleftharpoons 3AlCl(g)$$

Some gallium(I) and indium(I) compounds are known, the so-called dichloride "$GaCl_2$" being actually $Ga^I[Ga^{III}Cl_4]$.

Thallium has a well-defined unipositive state. In aqueous solution it is distinctly more stable than Tl^{III}:

$$Tl^{3+} + 2e = Tl^+ \qquad E° = +1.25\ V$$

The Tl^+ ion is not very sensitive to pH, although the Tl^{3+} ion is extensively hydrolyzed to $TlOH^{2+}$ and the colloidal oxide even at pH 1–2.5. The redox potential is, hence, very dependent on pH as well as on the presence of complexing anions. For example, the presence of Cl^- stabilizes Tl^{3+} more (by formation of complexes) than Tl^+ and the potential is thereby lowered.

The colorless Tl^+ ion has a radius of 1.54 Å, comparable to those of K^+, Rb^+, and Ag^+ (1.44, 1.58, and 1.27 Å). Thus it resembles the alkali ions in some ways and the Ag^+ ion in others. It may replace K^+ in certain enzymes and has potential use as a probe for potassium. In crystalline salts, the Tl^+ ion is usually 6- or 8-coordinate. The yellow hydroxide is unstable, giving the black oxide, Tl_2O, at about 100°. The oxide and hydroxide are soluble in water giving strongly basic solutions. These absorb carbon dioxide from the air, although TlOH is a weaker base than KOH. Many thallous salts, for example, Tl_2SO_4, Tl_2CO_3, or $TlCO_2CH_3$, have solubilities somewhat lower than those of the corresponding K^+ salts, but otherwise they are similar to and quite often isomorphous with them. Thallous fluoride is soluble in water but the other halides are sparingly soluble. Thallous chloride also resembles AgCl in being photosensitive and darkening on exposure to light, but differs in being insoluble in ammonia. Thallium compounds are exceedingly poisonous and in traces cause loss of hair.

Study Questions

A

1. What is bauxite and how is it purified for Al production? Where is Baux from which the mineral gets its name?
2. Why is aluminum resistant to air and water even though it is very electropositive?
3. What is the formula and structure of (a) corundum, (b) the mineral spinel?
4. What is the chemical nature of ruby?
5. What is the structural nature of the trihalide dimers, M_2X_6? Why is this a favorable structure? At higher temperatures what happens to these molecules?
6. Of what ions does the substance TlI_3 consist?
7. Explain why there is no such thing as an aqueous solution of $Al(CN)_3$ or $Al_2(CO_3)_3$.
8. Mention some principal Tl^I compounds and their main properties.
9. What is *an* alum? What species are present in a crystalline alum?
10. For cryolite, give the formula, chief industrial use and structure. What is the relation between this structure and that of fluorite?
11. In what way are the reactivities of BH_4^- and AlH_4^- different?
12. Compare the properties of B_2O_3 and Al_2O_3.
13. Although $AlCl_3$ is a strong Lewis acid like BCl_3, AlF_3 does not resemble BF_3. Explain.
14. What are the *main* areas of resemblance between the chemistries of B and Al?
15. How is $LiAlH_4$ prepared? Why does it explode with H_2O while $LiBH_4$ is soluble in water?

B

1. Discuss the reasons why $Tl^{III}I_3$ is unstable relative to $Tl^{I}I_3$ whereas the opposite is true for Al, Ga, and In. Consider factors such as: Tl^{III}—I bond energy, oxidation-reduction potentials, and lattice energies as a function of ion size.
2. What factors influence the preference of fluorides and chlorides for six-coordinate structures while the bromides and iodides prefer four-coordinate ones?
3. By what experimental tests, other than X-ray crystallography, might one establish the true nature of "$GaCl_2$" as $Ga^I[Ga^{III}Cl_4]$.

4. Show that the removal of the requisite number of cations from the cryolite-type crystal structure to produce the fluorite-type structure is in correct proportion to the formulas. To do this consider how many $[AB_6]^{2-,\,3-}$ ions there are per unit cell.
5. Interaction of Al with alcohols using mercuric chloride as catalyst gives alkoxides; the isopropoxide is used in organic chemistry; it is tetrameric in solvents. Write a structure and say how you could prove this other than by X-ray diffraction.
6. Discuss the electronic structures and position in the periodic table of the Al, Ga, In, Tl, and Sc, Y, La, Ac groups. Compare the main features of their chemistry and discuss the reasons for any similarities and differences.
7. Write an essay on mixed oxides containing M^{3+} ions.
8. Describe what happens when anhydrous $AlCl_3$ is dissolved in water and the solution made progressively more alkaline to say pH 11.
9. Show with equations how $AlCl_3$ functions as a Friedel–Crafts catalyst.
10. Why is the Tl^+/Tl^{3+} couple sensitive to pH and the presence of complexing anions?

Chapter 13
Study Guide

Supplementary Reading

Cucinella, S., Mazzei, A., and Marconi, W., "Synthesis and Reactions of Aluminum Hydride Derivatives," *Inorg. Chim. Acta. Rev.*, *4*, 51 (1970).
Greenwood, N. N., "The Chemistry of Gallium," *Adv. Inorg. Chem. Radiochem.*, *5*, 91 (1963).
Lee, A. G., *The Chemistry of Thallium*, Elsevier, 1971.
Lee, A. G., "Coordination Chemistry of Thallium (I)," *Coord. Chem. Revs.*, *8*, 289 (1972).
Nesmeyanov, A. N. and Sokolik, R. A., *The Organocompounds of B, Al, Ga, In and Tl*, North Holland, 1967.
Sheka, I. A., Chans, I. S., and Mityureva, T. T., *The Chemistry of Gallium*, Elsevier, 1966.

14

carbon

14-1 Introduction

There are more known compounds of carbon than of any other element except hydrogen. Most are best regarded as organic chemicals. In this chapter we consider certain compounds traditionally considered "inorganic" and in Chapter 29 we discuss organometallic or, more precisely, organoelement compounds in which there are bonds to carbon such as Fe—C, P—C, Si—C, Al—C, etc.

The electronic structure of C in its ground state is $1s^2 2s^2 2p^2$ so that to accommodate the normal four-covalence the atom must be promoted to a valence state $2s2p_x2p_y2p_z$ (see page 74). The ion C^{4+} does not arise in any normal chemical process, but C^{4-} may possibly exist in some carbides of the most electropositive metals.

Some cations, anions, and radicals have been detected as transient species in organic reactions, and certain stable species of these types are known. The ions are known as *carbonium* ions, for example, $(C_6H_5)_3C^+$ or *carbanions*, for example, $(NC)_3C^-$. They can be stable only when the charge is extensively delocalized onto the attached groups.

Divalent carbon species or *carbenes*, $:CR_1R_2$ play a role in many reactions, but they are highly reactive. They can be trapped by binding to transition metals and many metal carbene compounds are known (Section 29-17).

The divalent species of some other Group IV elements such as $:SiF_2$ or $:SnCl_2$ can be considered to have carbene-like behavior.

A unique feature of carbon is its propensity for bonding to itself in chains or rings, not only with single bonds, C—C, but also containing multiple bonds, C═C or C≡C. Sulfur and silicon are the elements next most inclined to *catenation*, as this self-binding is called, but they are far inferior to carbon. The reason for the thermal stability of carbon chains is the intrinsic high strength of the C—C single bond, 356 kJ mol^{-1}. The Si—Si (226) bond is weaker but another important factor is that Si—O bonds are much stronger than C—O (368 kJ mol^{-1}; 336 kJ mol^{-1}). Hence, given the necessary activation energy, compounds with Si—Si links are converted very exothermically into ones with Si—O bonds.

14-2
Allotropy of Carbon: Diamond; Graphite

The two best-known forms of carbon, diamond and graphite, differ in their physical and chemical properties because of differences in the arrangement and bonding of the atoms (Chapter 8, page 184). Diamond is denser than graphite (3.51 g cm^{-3}; 2.22 g cm^{-3}), but graphite is the more stable, by 2.9 kJ mol^{-1} at 300 K and 1 atm pressure. From the densities it follows that to transform graphite into diamond, pressure must be applied. From the thermodynamic properties of the allotropes it is estimated that they would be in equilibrium at 300 K under a pressure of ~15,000 atm. Because equilibrium is attained extremely slowly at this temperature, the diamond structure persists under ordinary conditions.

Diamonds can be produced from graphite only by the action of high pressure, and high temperatures are necessary for an appreciable rate of conversion. Naturally occurring diamonds must have been formed when those conditions were provided by geological processes.

Only in 1955 was a successful synthesis of diamonds from graphite reported. Although graphite can be directly converted into diamond at *ca.* 3000° K and pressures above 125 kbar, in order to obtain useful rates of conversion, a transition-metal catalyst such as Cr, Fe, or Pt is used. It appears that a thin film of molten metal forms on the graphite, dissolving some and reprecipitating it as diamond, which is less soluble. Diamonds up to 0·1 carat (20 mg) of high industrial quality can be routinely produced at competitive prices. Some gem quality diamonds have also been made but the cost, thus far, has been prohibitive. Diamond will burn in air at 600 to 800° but its chemical reactivity is much lower than that of graphite or amorphous carbon.

Graphite. The many forms of amorphous carbon, such as charcoals, soot, and lampblack, are all actually microcrystalline forms of graphite. The physical properties of such materials are mainly determined by the nature and extent of their surface areas. The finely divided forms, which present relatively vast surfaces with only partially saturated attractive forces, readily absorb large amounts of gases and solutes from solution. Active carbons impregnated with palladium, platinum, or other metals are widely used as industrial catalysts.

An important aspect of graphite technology is the production of very strong fibers by pyrolysis, at 1500° C or above, or oriented organic polymer fibers, for example, those of polyacrylonitrile, polyacrylate esters, or cellulose. When incorporated into plastics the reinforced materials are light and of great strength. Other forms of graphite such as foams, foils, or whiskers can also be made.

The loose layered structure of graphite allows many molecules and ions to penetrate the layers to form what are called *intercalation* or *lamellar compounds.* Some of these may be formed spontaneously when the reactant and graphite are brought together. Examples of reactants are the alkali metals, halogens, and metal halides and oxides, for example, $FeCl_3$ and MoO_3.

Carbides. The direct interaction of carbon with metals or metal oxides at high temperatures gives compounds generally called carbides. Those of electropositive metals behave as though they contain C^{4-} or C_2^{2-} ions, and react with water to give hydrocarbons, for example,

$$Al_4C_3 + 12H_2O = 4Al(OH)_3 + 3CH_4$$

$$Ca^{2+}C_2^{2-} + 2H_2O = Ca(OH)_2 + HC{\equiv}CH$$

Transition metals give *interstitial carbides* in which carbon atoms occupy octahedral holes in close-packed arrays of metal atoms. Such materials are commonly very hard, electrically conducting, and have very high melting points (3000–4800°). The smaller metals Cr, Mn, Fe, Co, and Ni give carbides that are intermediate between typically ionic and interstitial carbides and these are hydrolyzed by water or dilute acids.

Silicon and boron form SiC and B_4C, which are also extremely hard, infusible, and chemically inert. Silicon carbide has a diamond-like structure in which C and Si atoms are each tetrahedrally surrounded by four of the other kind. Under the name "carborundum" it is used in cutting tools and abrasives.

INORGANIC COMPOUNDS OF CARBON

14-3 Carbon Monoxide

This colorless toxic gas (bp −190°) is formed when carbon is burned in a deficiency of oxygen. At all temperatures there is the equilibrium

$$2CO(s) = C(s) + CO_2(s)$$

but it is rapidly attained only at elevated temperatures. Carbon monoxide is made commercially along with hydrogen (Section 9-1) by steam reforming or partial combustion of hydrocarbons and by the reaction:

$$CO_2 + H_2 = CO + H_2O$$

A mixture of CO and H_2 ("synthesis gas") is very important commercially, being used in the hydroformylation process (Chapter 30) and for the synthesis of methanol. Carbon monoxide is also formed when carbon is used in reduction processes, for example, of phosphate rock to give phosphorus (page 288), in automobile exhausts, and the like. Carbon monoxide is also released by certain marine plants and it occurs naturally in the atmosphere.

Carbon monoxide is formally the anhydride of formic acid (HCOOH), but this is not an important aspect of its chemistry. Although CO is an exceedingly weak base, one of its important properties is the ability to act as a ligand toward transition metals. The metal—CO bond involves a certain type of multiple bonding, $d\pi$—$p\pi$ bonding discussed in Chapter 28. The toxicity of CO arises from this ability to bind to the Fe atom in hemoglobin (Chapter 31) in the blood. Only iron and nickel react directly with CO (Chapter 28) under practical conditions.

14-4
Carbon Dioxide and Carbonic Acid

Carbon dioxide is present in the atmosphere (300 ppm), in volcanic gases, and in supersaturated solution in certain spring waters. It is released on a large scale by fermentation processes, limestone calcination, and all forms of combustion of carbon and carbon compounds. It is involved in geochemical cycles as well as in photosynthesis. In the laboratory it can be made by action of heat or acids on carbonates. Solid carbon dioxide (sublimes $-78.5°$) or "dry ice" is used for refrigeration.

Carbon dioxide is the anhydride of the most important simple acid of carbon, "carbonic acid." For many purposes, the following acid dissociation constants are given for aqueous "carbonic acid":

$$\frac{[H^+][HCO_3^-]}{[H_2CO_3]} = 4.16 \times 10^{-7}$$

$$\frac{[H^+][CO_3^{2-}]}{[HCO_3^-]} = 4.84 \times 10^{-11}$$

The equilibrium quotient in the first equation is incorrect because not all the CO_2 dissolved and undissociated is present as H_2CO_3. The greater part of the dissolved CO_2 is only loosely hydrated, so that the correct first dissociation constant, using the real concentration of H_2CO_3, has the much larger value of about 2×10^{-4}, more in keeping (see page 177) with the structure $(HO)_2CO$.

The rate at which CO_2 comes into equilibrium with H_2CO_3 and its dissociation products when passed into water is measurably slow. This is why we can distinguish analytically between H_2CO_3 and the loosely hydrated $CO_2(aq)$. This slowness is of great importance in biological, analytical, and industrial chemistry.

The slow reaction can be shown by addition of a saturated aqueous solution of CO_2, on the one hand, and of dilute acetic acid, on the other, to solutions of dilute NaOH containing phenolphthalein indicator. The acetic acid neutralization is instantaneous whereas with the CO_2 neutralization it takes several seconds for the color to fade.

The hydration of CO_2 occurs by two paths. For pH < 8 the principal mechanism is direct hydration of CO_2:

$$CO_2 + H_2O = H_2CO_3 \qquad \text{(Slow)}$$

$$H_2CO_3 + OH^- = HCO_3^- + H_2O \qquad \text{(Instantaneous)} \qquad (14\text{-}1)$$

The rate law is pseudo-first order,

$$\frac{-d(CO_2)}{dt} = k_{CO_2}(CO_2); \quad k_{CO_2} = 0.03 \text{ sec}^{-1}$$

At pH > 10, the predominant reaction is direct reaction of CO_2 and OH^-:

$$\begin{aligned} CO_2 + OH^- &= HCO_3^- && \text{(Slow)} \\ HCO_3^- + OH^- &= CO_3^{2-} + H_2O && \text{(Instantaneous)} \end{aligned} \qquad (14\text{-}2)$$

where the rate law is

$$\frac{-d(CO_2)}{dt} = k_{OH}(OH^-)(CO_2); \quad k_{OH^-} = 8500 \text{ sec}^{-1}(\text{mol/l})^{-1}$$

This can be interpreted, of course, merely as the base catalysis of (14-1). In the pH range 8–10 both mechanisms are important. For each hydration reaction (14-1, 14-2) there is a corresponding dehydration reaction:

$$H_2CO_3 \longrightarrow H_2O + CO_2 \qquad k_{H_2CO_3} = k_{CO_2} \times K = 20 \text{ sec}^{-1}$$

$$HCO_3^- \longrightarrow CO_2 + OH^- \qquad k_{HCO_3^-} = k_{OH^-} \times \frac{KK_w}{K_a} = 2 \times 10^{-4} \text{ sec}^{-1}$$

Hence for the equilibrium

$$H_2CO_3 \rightleftharpoons CO_2 + H_2O$$

$$K = \frac{(CO_2)}{(H_2CO_3)} = \frac{k_{H_2CO_3}}{k_{CO_2}} = ca.\ 600 \qquad (14\text{-}3)$$

It follows from (14-3) that the true ionization constant of H_2CO_3, K_a, is greater than the apparent constant as noted above.

14-5
Compounds with C—N Bonds; Cyanides and Related Compounds

An important area of "inorganic" carbon chemistry is that of compounds with C—N bonds. The most important species are the cyanide, cyanate, and thiocyanate ions and their derivatives.

1. *Cyanogen*, $(CN)_2$. This flammable gas (bp $-21°$) is stable despite the fact that it is highly endothermic ($\Delta Hf^\circ_{298} = 297$ kJ mol^{-1}). It can be obtained by catalytic gas phase oxidation of HCN by NO_2

$$2HCN + NO_2 \longrightarrow (CN)_2 + NO + H_2O$$

$$NO + \tfrac{1}{2}O_2 \longrightarrow NO_2$$

Cyanogen can also be obtained from CN^- by aqueous oxidation using Cu^{2+} (cf. the Cu^{2+}—I^- reaction):

$$Cu^{2+} + 2CN^- \longrightarrow CuCN + \tfrac{1}{2}(CN)_2$$

or acidified peroxodisulfate. Dry $(CN)_2$ is made by the reaction:

$$Hg(CN)_2 + HgCl_2 \longrightarrow Hg_2Cl_2 + (CN)_2$$

Although pure $(CN)_2$ is stable, the impure gas may polymerize at 300 to 500°. Cyanogen dissociates into CN radicals and, like halogens, can oxidatively add to lower-valent metal atoms (Chapter 30) giving dicyano complexes, for example,

$$(Ph_3P)_4Pd + (CN)_2 \longrightarrow (Ph_3P)_2Pd(CN)_2 + 2Ph_3P$$

A further resemblance to the halogens is the disproportionation in basic solution:

$$(CN)_2 + 2OH^- \longrightarrow CN^- + OCN^- + H_2O$$

Thermodynamically this reaction can occur in acid solution but is rapid only in base. A stoichiometric mixture of O_2 and $(CN)_2$ burns producing one of the hottest flames (*ca.* 5050° K) known from a chemical reaction.

2. *Hydrogen cyanide.* HCN, like the hydrogen halides, is a covalent, molecular substance, but capable of dissociation in aqueous solution. It is an extremely poisonous (though less so than H_2S), colorless gas and is evolved when cyanides are treated with acids. Liquid HCN (bp 25.6°) has a very high dielectric constant (107 at 25°) that is due (as for H_2O) to association of the polar molecules by hydrogen bonding. Liquid HCN is unstable and can polymerize violently in the absence of stabilizers: in aqueous solutions polymerization is induced by ultraviolet light.

Hydrogen cyanide is thought to have been one of the small molecules in the earth's primeval atmosphere and to have been an important source or intermediate in the formation of biologically important chemicals. For example, under pressure, with traces of water and ammonia, HCN pentamerizes to adenine.

In aqueous solution, HCN is a very weak acid, $pK_{25^\circ} = 9.21$, and solutions of soluble cyanides are extensively hydrolyzed, but the pure liquid is a strong acid.

Hydrogen cyanide is made industrially from CH_4 and NH_3 by the reactions

$$2CH_4 + 3O_2 + 2NH_3 \xrightarrow[> 800^\circ]{\text{Catalyst}} 2HCN + 6H_2O \qquad \Delta H = -475 \text{ kJ mol}^{-1}$$

or

$$CH_4 + NH_3 \xrightarrow[\text{Pt}]{1200^\circ} HCN + 3H_2 \qquad \Delta H = +240 \text{ kJ mol}^{-1}$$

Hydrogen cyanide has many industrial uses. It may be added directly to alkenes; for example; butadiene gives adiponitrile, $NC(CH_2)_4CN$ (for Nylon) in the presence of zerovalent Ni alkylphosphite catalysts which operate by oxidative-addition and transfer reactions (Chapter 30).

3. *Cyanides.* Sodium cyanide is manufactured by fusion of calcium cyanamide with carbon and sodium carbonate:

$$CaCN_2 + C + Na_2CO_3 \longrightarrow CaCO_3 + 2NaCN$$

The cyanide is leached with water. The $CaCN_2$ is made in an impure form contaminated with CaO, CaC_2, C, etc., by the interaction:

$$CaC_2 + N_2 \xrightarrow{ca.\ 1100^\circ} CaNCN + C$$

The linear NCN^{2-} ion is isostructural and isoelectronic with CO_2. Cyanamide itself, H_2NCN, can be made by acidification of CaNCN. The commerical product is the dimer, $H_2NC({=}NH)NHCN$, which also contains much of the tautomer containing the substituted *carbodiimide group*, $H_2N{-}C({=}NH){-}N{=}C{=}NH$. Organocarbodiimides are important synthetic reagents in organic chemistry and $CH_3N{=}C{=}NCH_3$ is stable enough to be isolated.

Sodium cyanide can also be obtained by the reaction

$$NaNH_2 + C \xrightarrow{500-600^\circ} NaCN + H_2$$

Cyanides of electropositive metals are water soluble but those of Ag^I, Hg^I, and Pb^{II} are very insoluble. The cyanide ion is of great importance as a ligand (Chapter 28), and many cyano complexes are known of transition metals, Zn, Cd, Hg, etc.; some, like $Ag(CN)_2^-$ and $Au(CN)_2^-$, are of technical importance and others are employed analytically. The complexes sometimes resemble halogeno complexes, for example, $Hg(CN)_4^{2-}$ and $HgCl_4^{2-}$, but other types exist. Fusion of alkali cyanides with sulfur gives the *thiocyanate* ion, SCN^-.

14-6 Compounds with C—S Bonds

Carbon disulfide, CS_2. This very toxic liquid (bp 46°), usually pale yellow, is prepared on a large scale by interaction of methane and sulfur over silica or alumina catalysts at $\sim 1000^\circ$.

$$CH_4 + 4S = CS_2 + 2H_2S$$

In addition to its high flammability in air, CS_2 is a very reactive molecule and has an extensive chemistry, much of it organic in nature. It is used to prepare carbon tetrachloride industrially:

$$CS_2 + 3Cl_2 \longrightarrow CCl_4 + S_2Cl_2$$

Carbon disulfide is one of the small molecules that readily undergo the "insertion reaction" (Chapter 30) where the $-S-\underset{\underset{S}{\|}}{C}-$ group is inserted between Sn—N, Co—Co, or other bonds. Thus with titanium dialkylamides, dithiocarbamates are obtained:

$$Ti(NR_2)_4 + 4CS_2 \longrightarrow Ti(S_2CNR_2)_4$$

The CS_2 molecule can also serve as a ligand, being either bound as a donor through sulfur or added oxidatively (Chapter 30) to give a three-membered ring as in (14-I):

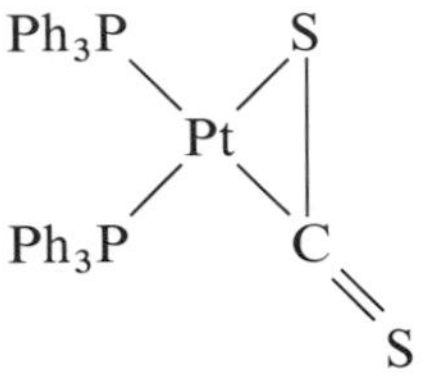

$$\begin{array}{c} Ph_3P \quad\; S \\ \diagdown \;\; \diagup | \\ Pt \;\; | \\ \diagup \;\; \diagdown | \\ Ph_3P \quad\; C \\ \quad\quad\quad \diagdown\!\!\diagdown \\ \quad\quad\quad\quad S \end{array}$$

14-I

Important reactions of CS_2 involve nucleophilic attacks on carbon by the ions OR^- and SH^- and by primary or secondary amines, which lead, in basic solution, respectively to xanthates, thiocarbonates, and dithiocarbamates, for example:

$$\left.\begin{array}{l} RO^- \\ HS^- \\ R_2HN \end{array}\right\} + \begin{array}{c} S \\ \| \\ C \\ \| \\ S \end{array} \longrightarrow \left\{\begin{array}{ll} ROCS_2^- & \text{xanthate} \\ CS_3^{2-} & \text{thiocarbonate} \\ R_2NCS_2^- & \text{dithiocarbamate} \end{array}\right.$$

Dithiocarbamates are normally prepared as Na salts by action of primary or secondary amines on CS_2 in presence of NaOH. The Zn, Mn, and Fe dithiocarbamates are used as agricultural fungicides, and Zn salts are used as accelerators in the vulcanization of rubber.

Dithiocarbamates form many complexes with metals. The CS_2^- group in dithiocarbamates as well as in xanthates, thioxanthates, and thiocarbonates is usually chelated as in (14-II), but monodentate and bridging dithiocarbamates are known.

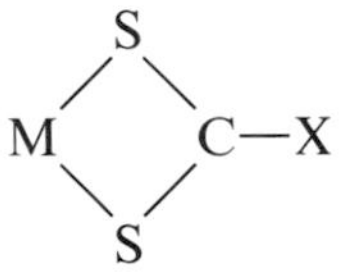

$$\begin{array}{c} S \\ \diagup \;\; \diagdown \\ M \qquad C{-}X \\ \diagdown \;\; \diagup \\ S \end{array} \qquad \begin{array}{l} X = NHR \text{ or } NR_2, \\ OR \text{ or } SR, \\ O, S \text{ or } S{-}S \end{array}$$

14-II

On oxidation of aqueous solutions by H_2O_2, Cl_2, or $S_2O_8^{2-}$, *thiuram disulfides* are obtained, for example,

$$I_2 + 2Me_2NCS_2^- \longrightarrow Me_2N\underset{\underset{S}{\|}}{C}{-}S{-}S{-}\underset{\underset{S}{\|}}{C}NMe_2 + 2I^-$$

Thiuram disulfides, which are strong oxidants, are used as polymerization initiators (for, when heated, they give radicals) and as vulcanization accelerators. Tetraethylthiuram disulfide is "Antabuse," the agent for rendering the body allergic to ethanol.

Study Questions

A

1. The electronic structure of C in its ground state is $1s^22s^22p_x2p_y$. Why does carbon usually form four single bonds and not two?

2. Give examples of a stable carbonium ion, a carbanion, and a free radical. What is a carbene?
3. What is meant by catenation? Why does silicon have much less tendency to catenation than carbon? Could the same be said for nitrogen?
4. Describe the synthesis and main properties of diamond.
5. What is graphite? Draw its structure and explain why its properties differ from those of diamond.
6. List ways in which CO can be made.
7. List ways in which CO_2 can be made.
8. On which side is the equilibrium in the reaction:

$$CO_2(aq) + 2H_2O \rightleftharpoons H_3O^+ + HCO_3^-?$$

9. Why does $CaCO_3$ dissolve to some extent in CO_2 saturated water? Write balanced equations for the reactions involved.
10. How could you make cyanogen in the laboratory? Write balanced equations.
11. List similarities between $(CN)_2$ and CN^- and Cl_2 and Cl^-.
12. Why are solutions of KCN in water alkaline?
13. Give the industrial synthesis and major properties of hydrogen cyanide.
14. How is CS_2 prepared? Write equations for its reaction with C_2H_5ONa in ethanol and with $(C_2H_5)_2NH$ in presence of aqueous NaOH.
15. How would you convert $BaCO_3$ labeled with ^{13}C or ^{14}C, which is the usual source of labeled carbon, to (a) $Ni(^*CO)_4$, (b) $^*C_2H_2$, (c) $^*CH_4$, (d) $^*CS_2$, and (e) $^*CH_3OH$?

B

1. Describe in detail what happens when $CO_2(g)$ dissolves in water.
2. The C—C bond length in graphite is 1.42 Å. How does this compare with that in (a) diamond, (b) ethylene, (c) benzene? How can you explain the bond order of the C—C bond in graphite?
3. Write down molecular orbitals for the isoelectronic molecules CO and N_2. Do the differences suggest major differences in chemistry?
4. Write down the structures and molecular orbitals for the isoelectronic molecules carbon dioxide and allene. What sort of differences in their chemistry do you expect?
5. HCN can give on polymerization, dimers, trimers, tetramers, pentamers, and polymers. Write some plausible structures for these molecules.
6. Explain why HCN is a weak acid in aqueous solution yet in the pure state is a strong acid.
7. What is a likely mechanism for the reaction of $Sn(NEt_2)_4$ with CS_2 to give a dithiocarbamate?
8. Zinc dithiocarbamates are dimeric. Propose a structure.
9. How is cyanamide prepared? What is its structure? It is trimerized industrially to melamine (for plastics), or 2, 4, 6-triamino-1, 3, 5-triazine. How is this likely to occur?

Chapter 14
Study Guide

Scope and Purpose. Most of the chemistry of the element carbon constitutes the field of organic chemistry. The inorganic chemist, however, is legitimately concerned with certain aspects that are very important and that have traditionally not been included in the realm of organic chemistry. These include nearly all of the chemistry of the element itself, of compounds in which carbon is combined with metals and metalloids, and much of the chemistry of the simple, binary compounds with nonmetals (oxides, cyanides, halides). The field of organometallic chemistry, which we examine in Chapters 29 and 30, is a truly interdisciplinary one.

The material in this chapter is diverse and is not, as a whole, well covered in the review literature.

Supplementary Reading

Cotton, F. A. and Wilkinson, G., *Advanced Inorganic Chemistry*, 3rd ed., Wiley, 1972. Chapter 10 gives a more detailed coverage, with many references.

15

the group IVB elements: silicon, germanium, tin and lead

15-1 Introduction

Silicon is second only to oxygen in its natural abundance (*ca.* 28% of earth's crust) and occurs in a great variety of silicate minerals and as quartz, SiO_2.

Germanium, tin, and lead are rare elements (*ca.* 10^{-3}%). Tin and lead have been known since antiquity because of the ease with which they are obtained from their ores.

Cassiterite, SnO_2, occurs mixed in granites, sands, and clays. Lead occurs mainly as *galena*, PbS.

Germanium was discovered in 1886 following the prediction of its existence by Mendeleev. It occurs widely but in small amounts and is recovered from coal and zinc ore concentrates.

The main use of Ge, Sn, and Pb is as the metals, but alkyltin and lead compounds are made on a large scale (Chapter 29).

The position of the elements in the periodic table and some general features, including the reasons for the existence of the lower II oxidation state, were discussed in Chapter 8, page 199. Some properties of the elements were given in Table 8-4, page 200.

Multiple Bonding. Unlike carbon, these elements do not form stable $p\pi$—$p\pi$ multiple bonds. Thus, although stoichiometric similarities may occur, for example, the pairs CO_2, SiO_2; $(CH_3)_2CO$, $(CH_3)_2SiO$, there is no structural or chemical similarity between them. Carbon dioxide is a gas, properly written O=C=O, whereas SiO_2 is a giant molecule with each Si atom singly bound to four oxygen atoms giving linked SiO_4 tetrahedra (page 114). Despite this absence of $p\pi$—$p\pi$ bonding, the elements can use *d* orbitals in multiple bonding. Thus certain structural and chemical features of Si and Ge compounds, particularly those with SiO and SiN bonds, are best explained by some double bond character of the $p\pi$—$d\pi$ type. For example $N(SiH_3)_3$ is planar, whereas $N(CH_3)_3$ is pyramidal; this can be ascribed to $p\pi$—$d\pi$ bonding involving a filled nitrogen $2p_z$ orbital overlapping with empty silicon $3d_{xy}$ orbitals (*Fig. 15-1*).

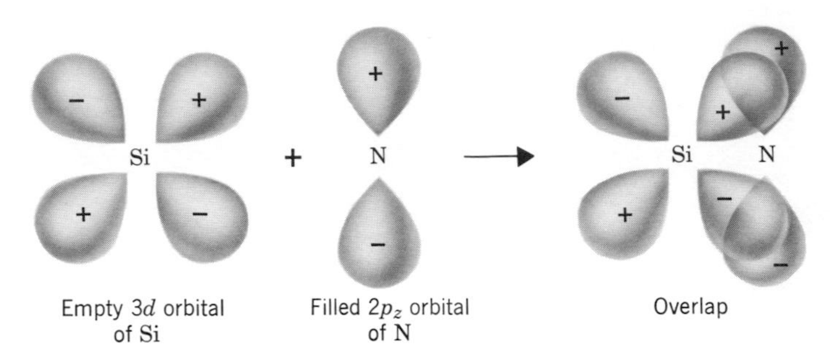

Figure 15-1 *The formation of dπ—pπ bond between* Si *and* N *in trisilylamine.*

Stereochemistry in the IV State. Unlike carbon, the compounds of these elements can have 5, 6 and, in one case, $Sn(NO_3)_4$, 8-coordination. Examples are

5-coordination	$Me_3SnClpy$,	SiF_5^-
6-coordination	$SnCl_4py_2$,	$[Ge(C_2O_4)_3]^{2-}$
	$[Si\ acac_3]^+$,	$PbCl_6^{2-}$

The II-Oxidation State. In many of the compounds of Sn^{II} and to a lesser extent of Ge^{II} and Pb^{II} the lone pair of electrons has important structural and chemical consequences.

First, the structures are such that the lone pairs appear to be occupying a bond position. Thus the $SnCl_3^-$ ion is pyramidal with a lone-pair like NH_3 (page 275). Second, tin(II) compounds, especially $SnCl_2$ and $SnCl_3^-$ can act as donors toward transition metals as in the complex $[Pt^{II}(SnCl_3)_5]^{3-}$.

15-2
Isolation and Properties of the Elements

Silicon is obtained in the ordinary commercial form by reduction of SiO_2 with carbon or CaC_2 in an electric furnace. Similarly, Ge is prepared by reduction of GeO_2 with C or H_2. Silicon and Ge are used as semiconductors, especially in transistors. For this purpose, exceedingly high purity ($<10^{-9}$ atom % of impurities) is essential, and special methods are required to obtain usable materials. The element is first converted to the tetrachloride, which is reduced back to the metal by hydrogen at high temperatures. After casting into rods it is *zone refined.* A rod of metal is heated near one end so that a cross-sectional wafer of molten silicon is produced. Since impurities are more soluble in the melt than they are in the solid they concentrate in the melt, and the melted zone is then caused to move slowly along the rod by moving the heat source. This carries impurities to the end. The process may be repeated. The impure end is then removed. Super-pure Ge is made in a similar way.

Tin and lead are obtained by reduction of the oxide or sulfide with carbon. The metals can be dissolved in acid and deposited electrolytically to effect further purification.

Silicon is ordinarily rather unreactive. It is attacked by halogens giving tetrahalides, and by alkalis giving solutions of silicates. It is not attacked by acids except hydrofluoric; presumably the stability of SiF_6^{2-} provides the driving force here.

Germanium is somewhat more reactive than silicon and dissolves in concentrated H_2SO_4 and HNO_3. Tin and lead dissolve in several acids and are rapidly attacked by halogens. They are slowly attacked by cold alkali, rapidly by hot, to form stannates and plumbites. Lead often appears to be more noble and unreactive than would be indicated by its standard potential of -0.13 V. This low reactivity can be attributed to a high overvoltage for hydrogen and also, in some instances, to insoluble surface coatings. Thus lead is not dissolved by dilute H_2SO_4 and concentrated HCl.

COMPOUNDS OF GROUP IV ELEMENTS

15-3
Hydrides, MH_4

These are colorless gases. Only *monosilane*, SiH_4 is of any importance. This spontaneously flammable gas is prepared by the action of $LiAlH_4$ on SiO_2 at 150–170° or by reduction of $SiCl_4$ with $LiAlH_4$ in an ether. Although stable to water and dilute acids, rapid base hydrolysis gives hydrated SiO_2 and H_2.

Substituted silanes with organic groups are of great importance, as are some closely related tin compounds (Chapter 29). The most important reaction of compounds with Si—H bonds, such as $HSiCl_3$ or $HSiCH_3$, is the Speier or hydrosilation reaction of alkenes:

$$RCH = CH_2 + SiHCl_3 = RCH_2CH_2SiCl_3$$

This reaction, which employs chloroplatinic acid as a catalyst, is commercially important for the synthesis of precursors to silicones.

15-4
Chlorides, MCl_4

Chlorination of the Group IV elements on heating gives colorless or yellow ($PbCl_4$) liquids. They are hydrolyzed by water eventually to the hydrous oxides but partial hydrolysis can give oxochlorides. In presence of aqueous HCl the chlorides of Ge, Sn, and Pb give chloroanions, and $GeCl_4$ differs from $SiCl_4$ in that it can be distilled and separated from concentrated HCl. The principal uses of $SiCl_4$ and $GeCl_4$ are in synthesis of the pure elements and for silicon and tin, in the synthesis of organo compounds (Chapter 29).

15-5
Oxygen Compounds

Silica. Pure SiO_2 occurs in two forms, *quartz* and *cristobalite*. The Si is always tetrahedrally bound to four oxygen atoms but the bonds have considerable ionic character. In cristobalite the silicon atoms are placed as are the carbon atoms in diamond, with the oxygen atoms midway between each pair. In quartz, there are helices so that enantiomorphic crystals occur, and these may be easily recognized and separated mechanically.

Quartz and cristobalite can be interconverted when heated. These processes are slow because the breaking and re-forming of bonds is required and the activation energy is high. However, the rates of conversion are profoundly affected by the presence of impurities, or by the introduction of alkali-metal oxides.

Slow cooling of molten SiO_2 or heating any solid form to the softening temperature gives an amorphous material which is glassy in appearance and is indeed a glass in the general sense, that is, a material with no long-range order but, instead, a disordered array of polymeric chains, sheets, or three-dimensional units.

Silica is relatively unreactive towards Cl_2, H_2, acids, and most metals at 25° or even at slightly elevated temperatures but is attacked by F_2, aqueous HF, alkali hydroxides, and fused carbonates.

Aqueous HF gives solutions containing fluorosilicates, for example, SiF_6^{2-}. The *silicates* have been discussed in Section 5-7. The fusion of excess alkali carbonates with SiO_2 at ~1300° gives water-soluble products commercially sold as syrupy liquid that has many uses. Aqueous sodium silicate solutions appear to contain the ion $[SiO_2(OH)_2]^{2-}$ but, depending on the pH and concentration, polymerized species are also present.

The basicity of the dioxides increases, SiO_2 being purely acidic, GeO_2 less so, SnO_2 amphoteric, and PbO_2 somewhat more basic. When SnO_2 is made at high temperatures or by dissolving Sn in hot concentrated nitric acid, it is, like PbO_2, remarkably inert to attack.

Only lead forms a stable oxide containing both Pb^{II} and Pb^{IV}, namely Pb_3O_4, which is a bright red powder known commercially as red lead. It is made by heating PbO and PbO_2 together at 250°. Although it behaves chemically as a mixture of PbO and PbO_2, the crystal contains $Pb^{IV}O_6$ octahedra linked in chains by sharing opposite edges. The chains are linked by Pb^{II} atoms each bound to three O atoms.

There are no true hydroxides and the products of hydrolysis of the hydrides or halides, and the like, are best regarded as hydrous oxides.

15-6
Complex Compounds

Most of the complex species contain halide ions or donor ligands that are O, N, S, or P compounds.

Anionic Complexes. Silicon forms only fluoroanions, normally SiF_6^{2-}, whose high formation constant accounts for the incomplete hydrolysis of SiF_4 in water:

$$2SiF_4 + 2H_2O = SiO_2 + SiF_6^{2-} + 2H^+ + 2HF$$

The ion is usually made by dissolving SiO_2 in aqueous HF and is stable even in basic solution. Under selected conditions and with ions of the right size, the SiF_5^- ion can be isolated, for example,

$$SiO_2 + HF(aq) + R_4N^+Cl \xrightarrow{CH_3OH} [R_4N][SiF_5]$$

By contrast with SiF_6^{2-}, the GeF_6^{2-} and SnF_6^{2-} ions are hydrolyzed by bases; PbF_6^{2-} ion is hydrolyzed even by water.

Although Si does not, the other elements give chloroanions, and all the elements form oxalato ions $[Mox_3]^{2-}$.

Cationic Complexes. The most important are those of chelating uninegative oxygen ligands such as the acetylacetonates, for example, $[Ge\ acac_3]^+$.

Adducts. The tetrahalides act as Lewis acids; $SnCl_4$ is a good Friedel–Crafts catalyst. The adducts are 1:1 or 1:2 but it is not always clear in the absence of X-ray evidence whether they are neutral, that is, MX_4L_2, or whether they are salts, for example, $[MX_2L_2]X_2$. Some of the best defined are the pyridine adducts, for example, *trans*-py_2SiCl_4.

Alkoxides, Carboxylates and Oxo Salts. All four elements form alkoxides. Those of silicon, for example, $Si(OC_2H_5)_4$, are the most important; the surface of glass or silica can also be alkoxylated. Alkoxides are normally obtained by the standard method:

$$MCl_4 + 4ROH + 4\ \text{amine} \longrightarrow M(OR)_4 + 4\ \text{amine}\cdot HCl$$

Silicon alkoxides are hydrolyzed by water, eventually to hydrous silica. Of the carboxylates, *lead tetraacetate* is the most important as it is used in organic chemistry as a strong but selective oxidizing agent. It is made by dissolving Pb_3O_4 in hot glacial acetic acid or by electrolytic oxidation of Pb^{II} in acetic acid. In oxidations the attacking species is probably $Pb(OOCMe)_3^+$, which is isoelectronic with the similar oxidant, $Tl(OOCMe)_3$, but this is not always so and some oxidations are free radical in nature. The trifluoroacetate is a white solid, which will oxidize even heptane to give $ROOCCF_3$ species whence the alcohol ROH is obtained by hydrolysis; benzene similarly gives phenol.

Tin(IV) sulfate, $Sn(SO_4)_2 \cdot 2H_2O$, can be crystallized from solutions obtained by oxidation of Sn^{II} sulfate; it is extensively hydrolyzed in water.

Tin(IV) nitrate is a colorless volatile solid made by interaction of N_2O_5 and $SnCl_4$; it contains bidentate NO_3^- groups giving dodecahedral coordination. The compound reacts with organic matter.

15-7 The Divalent State

Silicon. Divalent silicon species are thermodynamically unstable under normal conditions. However, several, notably SiF_2, have been identified in high temperature reactions and trapped by chilling to liquid nitrogen temperatures. Thus at *ca.* 1100° and low pressures, the following reaction goes in *ca.* 99.5% yield

$$SiF_4 + Si \rightleftharpoons 2SiF_2$$

SiF_2 is stable for a few minutes at 10^{-4} cm pressure; the molecule is angular and diamagnetic. When the frozen compound warms, it gives fluorosilanes up to $Si_{16}F_{34}$.

Germanium. Germanium dihalides are stable. GeF_2 is a white crystalline solid obtained by action of anhydrous HF on Ge at 200°; it is a fluorine bridged polymer with approximately **tbp** coordination of Ge. $GeCl_2$ gives salts of the $GeCl_3^-$ ion similar to those of Sn noted below.

Tin. The most important compounds are SnF_2 and $SnCl_2$, which are obtained by heating Sn with gaseous HF or HCl. The fluoride is sparingly soluble in water and is used in fluoride-containing toothpastes. Water hydrolyzes $SnCl_2$ to a basic chloride, but from dilute acid solutions $SnCl_2 \cdot 2H_2O$ can be crystallized. Both halides dissolve in solutions containing an excess of halide ion, thus:

$$SnF_2 + F^- = SnF_3^- \qquad pK \approx 1$$

$$SnCl_2 + Cl^- = SnCl_3^- \qquad pK \approx 2$$

In aqueous fluoride solutions SnF_3^- is the major species, but the ions SnF^+ and $Sn_2F_5^-$ can be detected.

The halides dissolve in donor solvents such as acetone, pyridine, or DMSO, to give pyramidal adducts, for example, $SnCl_2OC(CH_3)_2$.

The very air-sensitive stannous ion, Sn^{2+}, occurs in acid perchlorate solutions, which may be obtained by the reaction

$$Cu(ClO_4)_2 + Sn/Hg = Cu + Sn^{2+} + 2ClO_4^-$$

Hydrolysis gives $[Sn_3(OH)_4]^{2+}$, with $SnOH^+$ and $[Sn_2(OH)_2]^{2+}$ in minor amounts:

$$3Sn^{2+} + 4H_2O \rightleftharpoons [Sn_3(OH)_4]^{2+} + 4H^+ \qquad \log K = -6.77$$

The trimeric, probably cyclic, ion appears to provide the nucleus of several basic tin(II) salts obtained from aqueous solutions at fairly low pH. Thus the nitrate appears to be $Sn_3(OH)_4(NO_3)_2$ and the sulfate, $Sn_3(OH)_2OSO_4$. All Sn^{II} solutions are readily oxidized by oxygen and, unless stringently protected from air, normally contain some Sn^{IV}. The chloride solutions are often used as mild reducing agents:

$$SnCl_6^{2-} + 2e = SnCl_3^- + 3Cl^- \qquad E^\circ = \text{ca. } 0.0 \text{ V } (1\ M \text{ HCl}, 4\ M\ Cl^-)$$

Lead. Of the four elements, only lead has a well-defined cationic chemistry.

The plumbous ion, Pb^{2+} is partially hydrolyzed in water. In perchlorate solution

$$Pb^{2+} + H_2O = PbOH^+ + H^+ \qquad \log K \approx -7.9$$

In concentrated solutions and on addition of base, polymeric ions that contain 3, 4, and 6 Pb atoms are formed. The crystalline "basic" salt

$$[Pb_6O(OH)_6]^{4+}(ClO_4^-)_4 \cdot H_2O$$

has the cluster structure in *Fig. 15-2.* The O atom lies at the center of the middle tetrahedron, while the OH groups lie on the faces of the outer tetrahedra.

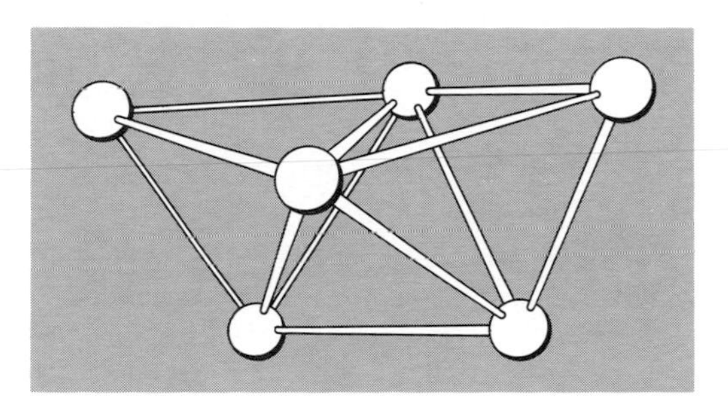

Figure 15-2 *The three face-sharing tetrahedra of* Pb *atoms in the* $Pb_6O(OH)_6^{4+}$ *cluster.*

Addition of more base eventually gives the hydrous oxide which dissolves in excess to give the plumbate ion.

Most lead salts are only sparingly soluble in water and some, for example, $PbSO_4$ or $PbCrO_4$, are insoluble. The common soluble salts are $Pb(NO_3)_2$ and $Pb(CO_2Me)_2 \cdot 2H_2O$, which is incompletely ionized in water. The halides are always anhydrous and in solution form complex species PbX^+, PbX_3^-, etc., except for the fluoride where only PbF^+ occurs.

Study Questions

A

1. Why is CO_2 a gas and SiO_2 a giant molecule?
2. Explain what is meant by $d\pi$—$p\pi$ bonding.
3. Why does tin form divalent compounds more easily than silicon?
4. How is super pure Ge made from GeO_2?
5. Write balanced equations for the synthesis of SiH_4 and its hydrolysis by aqueous NaOH.
6. Why is CCl_4 unreactive to H_2O whereas $SiCl_4$ is rapidly hydrolyzed?
7. Draw structures for (a) a cyclic silicate anion (b) a pyroxene (c) an asbestos.
8. Explain the nature of zeolites and molecular sieves.
9. Why is SiF_4 incompletely hydrolyzed in water?
10. Why does Si have much less tendency to form bonds to itself than does C?
11. How is lead tetraacetate made?
12. What is red lead?
13. What is the nature of Sn^{II} in chloride solution?
14. What is the nature of basic tin and lead salts?
15. How would you recover Ge from a spent petroleum re-forming catalyst comprised of 1% Pt—Ge alloy on alumina?
16. What is the structure of SiC?

B

1. Explain why H_3SiNCS has a linear SiNCS group whereas in H_3CNCS the CNC group is angular.
2. Why are silanols such as $(CH_3)_3SiOH$ stronger acids than their carbon analogs?
3. List the various types of geometries in compounds of the tetravalent Group IV elements and give examples.
4. What methods other than X-ray diffraction could you use to determine the nature of 1:1 and 1:2 adducts of $SnCl_4$ with neutral ligands?
5. Why can Sn^{II} compounds such as $SnCl_3^-$ act as donors to transition metals?
6. The single bond energies for the first- and second-row elements are

$$C > Si, \quad N < P, \quad O < S, \quad F < Cl$$

   Why is the first apparently anomalous?
7. Discuss the main similarities and differences between the chemistries of B and Si.

Chapter 15
Study Guide

Supplementary Reading

Breck, D. W., *Molecular Sieves*, Wiley, 1973.
Burger, H. and Eugen, R. "The Chemistry of Lower-Valent Silicon," Topics in Current Chemistry, No. 5, Springer-Verlag, 1974.
Donaldson, J. D., "The Chemistry of Bivalent Tin," *Prog. Inorg. Chem.*, *8*, 287 (1967).
Drake, J. E. and Riddle, C., "Volatile Compounds of the Hydrides of Silicon and Germanium with Elements of Groups V and VI," *Quart. Rev.*, *24*, 263 (1970).
Eaborn, C., *Organosilicon Compounds*, Butterworth, 1960.
Eitel, W., ed., *Silicate Science*, Vols. I–IV, Academic Press, 1968.
Glocking, F., *The Chemistry of Germanium*, Academic Press, 1969.
Margrave, J. L. and Wilson, P. W., "Silicon Difluoride, Its Reactions and Properties," *Accts. Chem. Research*, *4*, 145 (1971).
Noll. W. *et al.*, *Chemistry and Technology of Silicones*, Academic Press, 1968.
Poller, R. C., *Organotin Chemistry*, Logos Press, 1970.
Zuckerman, J. J., "Mossbauer Spectroscopy of Organotin Compounds," *Adv. Organometallic Chem.*, *9*, 21 (1970).

16

nitrogen

16-1 Introduction

The nitrogen atom, $1s^2 2s^2 2p_x 2p_y 2p_z$, can complete its valence shell in the following ways:

1. Electron gain to form the nitride ion N^{3-}; this ion is found only in saltlike nitrides of the most electropositive metals.

2. Formation of electron-pair bonds: (a) single bonds as in NH_3, or (b) multiple-bond formation as in $:N{\equiv}N:$, $-\ddot{N}{=}\ddot{N}-$, or NO_2.

3. Formation of electron-pair bonds with electron gain as in NH_2^- or NH^{2-}.

4. Formation of electron-pair bonds with electron loss as in the tetrahedral ammonium and substituted ammonium ions $[NR_4]^+$.

There are a few stable species in which, formally, the nitrogen valence shell is incomplete. The best examples are NO, NO_2, and nitroxides $R_2\dot{N}{=}\ddot{\underset{..}{O}}$; these have unpaired electrons and are paramagnetic.

Three-Covalent Nitrogen. The molecules NR_3 are pyramidal; the bonding is best considered as involving sp^3 hybrid orbitals so that the lone-pair occupies the fourth position. There are three points of note:

1. As a result of the nonbonding electron pair, all NR_3 compounds behave as Lewis bases and they give donor-acceptor complexes with Lewis acids, for example, $F_3\overset{-}{B}{-}\overset{+}{N}Me_3$ and they act as ligands toward transition metal ions as in, for example, $[Co(NH_3)_6]^{3+}$.

2. Pyramidal molecules NRR′R″ should be chiral. Optical isomers cannot be isolated, however, because such molecules undergo very rapidly a motion known as *inversion* in which the N atom oscillates through the plane of the three R groups, much as an umbrella can turn inside out (*Fig. 16-1*). The energy barrier for this process is only *ca.* 24 kJ mol^{-1}. In NH_3 the oscillation frequency is 2.387013×10^{10} cps (cycles per second).

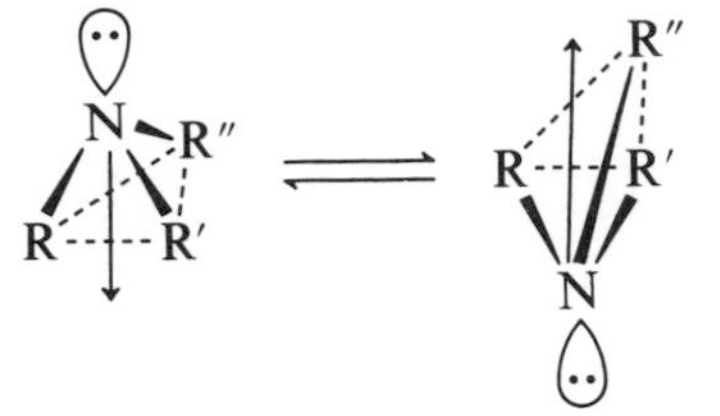

Figure 16-1 *Diagram illustrating the inversion of* NH_3.

3. There are a very few cases where three-covalent nitrogen is planar; in these cases multiple bonding is involved as we discussed for $N(SiMe_3)_3$, page 267. The N-centered triangular metal complexes such as $[NIr_3(SO_4)_6(H_2O)_3]^{4-}$ are similar.

N—N Single-Bond Energy. The N—N single bond is quite weak. If we compare the single-bond energies:

$H_3C—CH_3$	$H_2N—NH_2$	HO—OH	F—F	
350	160	140	150	kJ mol^{-1}

it is clear that there is a profound drop between C and N. This difference is probably attributable to the effects of repulsion between nonbonding lone pairs. The result is that unlike carbon, nitrogen has little tendency to catenation.

Multiple Bonds. The propensity of nitrogen, like carbon, to form $p\pi-p\pi$ multiple bonds is a feature that distinguishes it from phosphorus and the other Group V elements. Thus nitrogen as the element is dinitrogen, N_2, with a very high bond strength and short internuclear distance (1.094 Å), whereas phosphorus forms P_4 molecules or infinite layer structures in which there are only single bonds (page 186).

Where nitrogen forms one single and one double bond, nonlinear molecules, which often have stable stereo isomers, are formed (16-I to 16-IV).

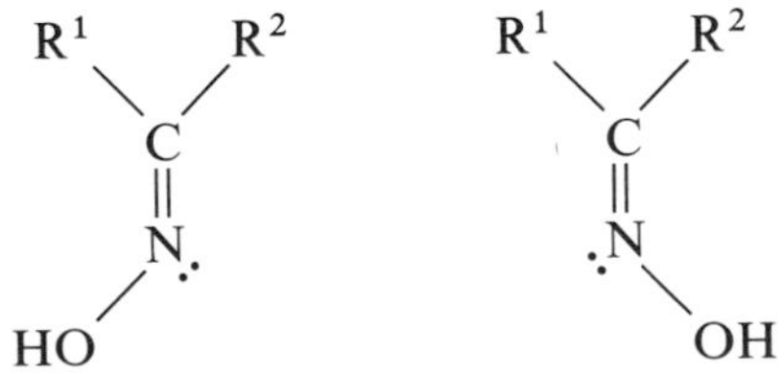

16-I

16-II

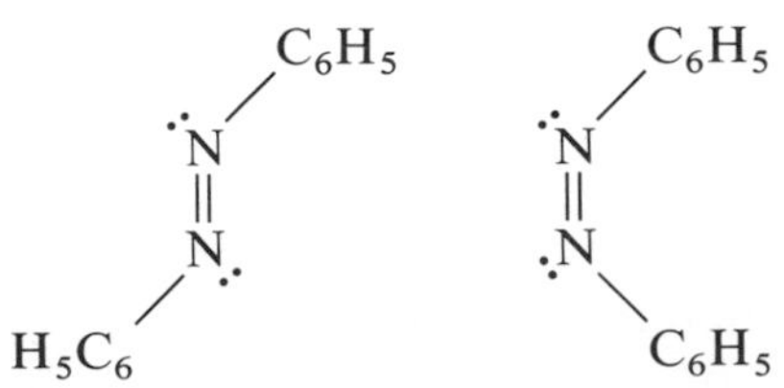

16-III

16-IV

The explanation is that nitrogen uses sp^2 hybrid orbitals, two of which form σ bonds, while the third houses the lone pair. A $p\pi$ bond is then formed using the nitrogen p_z orbital.

In the oxo compounds, NO_2^- and NO_3^-, there are multiple bonds that may be formulated in either resonance or MO terms, as is shown in Chapter 3.

16-2
Occurrence and Properties

Nitrogen occurs in nature mainly as dinitrogen, N_2 (bp 77.3 K), which comprises 78% by volume of the earth's atmosphere. The isotopes ^{14}N and ^{15}N have an absolute ratio $^{14}N/^{15}N = 272.0$. Compounds enriched in ^{15}N are used in tracer studies.

The heat of dissociation of N_2 is extremely large:

$$N_2(g) = 2N(g) \qquad \Delta H = 944.7 \text{ kJ mol}^{-1} \qquad K_{25} = 10^{-120}$$

The great strength of the N≡N bond is principally responsible for the chemical inertness of N_2 and for the fact that most simple nitrogen compounds are endothermic even though they may contain strong bonds. Dinitrogen is notably unreactive in comparison with isoelectronic, triply bonded systems such as X—C≡C—X, :C≡O:, X—C≡N:, and X—N≡C:. Both —C≡C— and —C≡N groups can act as donors by using their π electrons, whereas N_2 does not. It can, however, form complexes similar to those formed by CO, although to a much more limited extent, in which there are M ← N≡N: and M ← C≡O: configurations (Chapter 28).

Nitrogen is obtained by liquefaction and fractionation of air. It usually contains some argon and, depending on the quality, upward of ~30 ppm of oxygen. Spectroscopically pure N_2 is made by thermal decomposition of sodium or barium azide:

$$2NaN_3 \longrightarrow 2Na + 3N_2$$

The only reactions of N_2 at room temperature are with metallic Li to give Li_3N, with certain transition-metal complexes, and with nitrogen-fixing bacteria, either free-living or symbiotic on root nodules of clover, peas, beans, and the like. The mechanism by which these bacteria fix N_2 is unknown.

At elevated temperatures nitrogen becomes more reactive, especially when catalyzed, typical reactions being:

$$N_2(g) + 3H_2(g) = 2NH_3(g) \qquad K_{25^\circ} = 10^3 \text{ atm}^{-2}$$

$$N_2(g) + O_2(g) = 2NO(g) \qquad K_{25^\circ} = 5 \times 10^{-31}$$

$$N_2(g) + 3Mg(s) = Mg_3N_2(s)$$

$$N_2(g) + CaC_2(s) = C(s) + CaNCN(s)$$

NITROGEN COMPOUNDS

16-3
Nitrides

Nitrides of electropositive metals have structures with discrete nitrogen atoms and can be regarded as ionic, for example, $(Ca^{2+})_3(N^{3-})_2$, $(Li^+)_3N^{3-}$, etc. Their ready hydrolysis to ammonia and the metal hydroxides is consistent with this. Such nitrides are prepared by direct interaction or by loss of ammonia from amides on heating, for example:

$$3\,Ba(NH_2)_2 \longrightarrow Ba_3N_2 + 4\,NH_3$$

Many transition metals "nitrides" are often nonstoichiometric and have nitrogen atoms in the interstices of close-packed arrays of metal atoms. Like the similar carbides or borides they are hard, chemically inert, high-melting, and electrically conducting.

16-4
Nitrogen Hydrides

Ammonia. NH_3, is formed by action of base on an ammonium salt:

$$NH_4X + OH^- \longrightarrow NH_3 + H_2O + X^-$$

Industrially ammonia is made by the Haber process in which the reaction

$$N_2(g) + 3\,H_2(g) = 2\,NH_3(g) \qquad \Delta H = -46\ \text{kJ mol}^{-1}; \qquad K_{25^\circ} = 10^3\ \text{atm}^{-2}$$

is carried out at 400 to 500°C and pressures of 10^2 to 10^3 atm in the presence of a catalyst. Although the equilibrium is most favorable at low temperature, even with the best catalysts elevated temperatures are required to obtain a satisfactory rate. The best catalyst is α-iron containing some oxide to widen the lattice and enlarge the active interface.

Ammonia is a colorless pungent gas (bp −33.35°C). The liquid has a large heat of evaporation ($1.37\ \text{kJ g}^{-1}$ at the boiling point) and can be handled in ordinary laboratory equipment. Liquid NH_3 resembles water in its physical behavior, being highly associated via strong hydrogen bonding. Its dielectric constant (~22 at −34°; cf. 81 for H_2O at 25°) is sufficiently high to make it a fair ionizing solvent. Its self-ionization has been discussed previously (page 166).

Liquid NH_3 has lower reactivity than H_2O toward electropositive metals and dissolves many of them (page 219).

Because NH_3(l) has a much lower dielectric constant than water, it is a better solvent for organic compounds but generally a poorer one for ionic inorganic compounds. Exceptions occur when complexing by NH_3 is superior to that by water. Thus AgI is exceedingly insoluble in water but very soluble in NH_3. Primary solvation numbers of cations in NH_3 appear similar to those in H_2O, for example, 5.0 ± 0.2 and 6.0 ± 0.5 for Mg^{2+} and Al^{3+}, respectively.

Ammonia burns in air

$$4NH_3(g) + 3O_2(g) = 2N_2(g) + 6H_2O(g) \qquad K_{25^\circ} = 10^{228}$$

However, despite the fact that this process is thermodynamically favorable, at 750 to 900° in the presence of a platinum or platinum-rhodium catalyst reaction with oxygen can be made to proceed according to the equation

$$4NH_3 + 5O_2 = 4NO + 6H_2O \qquad K_{25^\circ} = 10^{168}$$

thus affording a useful synthesis of NO. The latter reacts with excess of O_2 to produce NO_2, and the mixed oxides can be absorbed in water to form nitric acid:

$$2NO + O_2 \longrightarrow 2NO_2$$

$$3NO_2 + H_2O \longrightarrow 2HNO_3 + NO, \text{ etc.}$$

Thus the sequence in industrial utilization of atmospheric nitrogen is as follows:

$$N_2 \xrightarrow[\text{Haber process}]{H_2} NH_3 \xrightarrow[\text{Ostwald process}]{O_2} NO \xrightarrow{O_2 + H_2O} HNO_3(aq)$$

Ammonia is extremely soluble in water. Although aqueous solutions are generally referred to as solutions of the weak base NH_4OH, called "ammonium hydroxide," undissociated NH_4OH probably does not exist. The solutions are best described as $NH_3(aq)$, with the equilibrium written as

$$NH_3(aq) + H_2O = NH_4^+ + OH^- \qquad K_{25^\circ} = \frac{[NH_4^+][OH^-]}{[NH_3]}$$

$$= 1.81 \times 10^{-5} (pK_b = 4.75)$$

Ammonium Salts. Stable crystalline salts of the tetrahedral NH_4^+ ion are mostly water-soluble. Ammonium salts generally resemble those of potassium and rubidium in solubility and structure, since the three ions are of comparable (Pauling) radii: $NH_4^+ = 1.48$ Å, $K^+ = 1.33$ Å, $Rb^+ = 1.48$ Å. Salts of strong acids are fully ionized, and the solutions are slightly acidic:

$$NH_4Cl = NH_4^+ + Cl^- \qquad K \approx \infty$$

$$NH_4^+ + H_2O = NH_3 + H_3O^+ \qquad K_{25^\circ} = 5.5 \times 10^{-10}$$

Thus, a 1 *M* solution will have a pH of ~4.7. The constant for the second reaction is sometimes called the hydrolysis constant; however, it may equally well be considered as the acidity constant of the cationic acid NH_4^+, and the system regarded as an acid-base system in the following sense:

$$\underset{\text{Acid}}{NH_4^+} + \underset{\text{Base}}{H_2O} = \underset{\text{Acid}}{H_3O^+} + \underset{\text{Base}}{NH_3(aq)}$$

Many ammonium salts volatilize with dissociation around 300°, for example:

$$NH_4Cl(s) = NH_3(g) + HCl(g) \qquad \Delta H = 177\ \text{kJ mol}^{-1}; \quad K_{25^\circ} = 10^{-16}$$

$$NH_4NO_3(s) = NH_3(g) + HNO_3(g) \qquad \Delta H = 171\ \text{kJ mol}^{-1}$$

Salts that contain oxidizing anions may decompose when heated, with oxidation of the ammonia to N_2O or N_2 or both. For example:

$$(NH_4)_2Cr_2O_7(s) = N_2(g) + 4H_2O(g) + Cr_2O_3(s) \qquad \Delta H = -315\ \text{kJ mol}^{-1}$$

$$NH_4NO_3(l) = N_2O(g) + 2H_2O(g) \qquad \Delta H = -23\ \text{kJ mol}^{-1}$$

Hydrazine. Hydrazine, N_2H_4, may be thought of as derived from ammonia by replacement of a hydrogen atom by the $—NH_2$ group. It is a bifunctional base:

$$N_2H_4(aq) + H_2O = N_2H_5^+ + OH^- \qquad K_{25^\circ} = 8.5 \times 10^{-7}$$

$$N_2H_5^+(aq) + H_2O = N_2H_6^{2+} + OH^- \qquad K_{25^\circ} = 8.9 \times 10^{-15}$$

and two series of hydrazinium salts are obtainable. Those of $N_2H_5^+$ are stable in water while those of $N_2H_6^{2+}$ are extensively hydrolyzed. Salts of $N_2H_6^{2+}$ can be obtained by crystallization from aqueous solution containing a large excess of the acid, since they are usually less soluble than the monoacid salts.

Anhydrous N_2H_4 is a fuming colorless liquid (bp 114°). It is surprisingly stable in view of its endothermic nature ($\Delta H_f^\circ = 50\ \text{kJ mol}^{-1}$). It burns in air with considerable evolution of heat.

$$N_2H_4(l) + O_2(g) = N_2(g) + 2H_2O(l) \qquad \Delta H^\circ = -622\ \text{kJ mol}^{-1}$$

Aqueous hydrazine in a powerful reducing agent in basic solution, normally being oxidized to nitrogen. Hydrazine is made by the interaction of aqueous ammonia with sodium hypochlorite:

$$NH_3 + NaOCl \longrightarrow NaOH + NH_2Cl \qquad \text{(Fast)}$$

$$NH_3 + NH_2Cl + NaOH \longrightarrow N_2H_4 + NaCl + H_2O$$

However, there is a competing reaction that is rather fast once some hydrazine has been formed:

$$2NH_2Cl + N_2H_4 \longrightarrow 2NH_4Cl + N_2$$

To obtain appreciable yields, it is necessary to add gelatine. This sequesters heavy-metal ions that catalyze the parasitic reaction: even the part per million or so of Cu^{2+} in ordinary water will almost completely prevent the formation of hydrazine if no gelatine is used. Since simple sequestering agents such as EDTA are not as beneficial as gelatine, the latter is assumed to have a catalytic effect as well.

Hydroxylamine. Hydroxylamine, NH_2OH, is a weaker base than NH_3:

$$NH_2OH(aq) + H_2O = NH_3OH^+ + OH^- \qquad K_{25^\circ} = 6.6 \times 10^{-9}$$

It is prepared by reduction of nitrates or nitrites either electrolytically or with SO_2, under controlled conditions. Hydroxylamine is a white unstable solid. In aqueous solution or as its salts $[NH_3OH]Cl$ or $[NH_3OH]_2SO_4$ it is used as a reducing agent.

Azides. Sodium azide can be obtained by the reaction

$$3NaNH_2 + NaNO_3 \xrightarrow{175^\circ} NaN_3 + 3NaOH + NH_3$$

Heavy metal azides are explosive and lead or mercury azide have been used in detonation caps. The azide ion which is linear and symmetrical behaves rather like a halide ion and can act as a ligand in metal complexes. The pure acid, HN_3, is a dangerously explosive liquid.

16-5 Nitrogen Oxides

Nitrous Oxide. Nitrous oxide, N_2O, is obtained by thermal decomposition of molten ammonium nitrate:

$$NH_4NO_3 \xrightarrow{250^\circ} N_2O + 2H_2O$$

The contaminants are NO, which can be removed by passage through ferrous sulfate solution and 1 to 2% of nitrogen.

Nitrous oxide has a linear structure NNO. It is relatively unreactive, being inert to the halogens, alkali metals and ozone at room temperature, but on heating it decomposes to N_2 and O_2, reacts with alkali metals, and many organic compounds, and supports combustion. It is used as an anaesthetic.

Nitric Oxide. Nitric oxide, NO, is formed in many reactions involving reduction of nitric acid and solutions of nitrates and nitrites. For example, with 8 *N* nitric acid:

$$8HNO_3 + 3Cu \longrightarrow 3Cu(NO_3)_2 + 4H_2O + 2NO$$

Reasonably pure NO is obtained by the aqueous reactions:

$$2NaNO_2 + 2NaI + 4H_2SO_4 \longrightarrow I_2 + 4NaHSO_4 + 2H_2O + 2NO$$

$$2NaNO_2 + 2FeSO_4 + 3H_2SO_4 \longrightarrow Fe_2(SO_4)_3 + 2NaHSO_4 + 2H_2O + 2NO$$

or, dry,

$$3KNO_2(l) + KNO_3(l) + Cr_2O_3(s) \longrightarrow 2K_2CrO_4(s, l) + 4NO$$

Nitric oxide reacts instantly with O_2:

$$2NO + O_2 \longrightarrow 2NO_2$$

It is oxidized to nitric acid by strong oxidizing agents; the reaction with permanganate is quantitative and provides a method of analysis. It is reduced to N_2O by SO_2 and to NH_2OH by Cr^{2+}, in acid solution in both cases.

Nitric oxide is thermodynamically unstable and at high pressures it readily decomposes in the range 30 to 50°:

$$3NO \longrightarrow N_2O + NO_2$$

The NO molecule is paramagnetic with the electron configuration

$$(\sigma_1)^2(\sigma_2)^2(\sigma_3)^2(\pi)^4(\pi^*).$$

The electron in the π^* orbital is relatively easily lost to give the *nitrosonium ion*, NO^+, which forms many salts. Because the electron removed comes out of an antibonding orbital, the bond is stronger in NO^+ than in NO; the bond length decreases by 0.09 Å and the vibration frequency rises from 1840 cm^{-1} in NO to 2150–2400 cm^{-1} (depending on environment) in NO^+.

The ion is formed when N_2O_3 or N_2O_4 is dissolved in concentrated sulfuric acid:

$$N_2O_3 + 3H_2SO_4 = 2NO^+ + 3HSO_4^- + H_3O^+$$

$$N_2O_4 + 3H_2SO_4 = NO^+ + NO_2^+ + 3HSO_4^- + H_3O^+$$

The compound $NO^+HSO_4^-$, nitrosonium hydrogen sulfate, is an important intermediate in the lead-chamber process for manufacture of sulfuric acid.

Nitric oxide forms many complexes with transition metals (Chapter 28) some of which can be considered to arise from NO^+.

Nitrogen Dioxide, NO_2, and Dinitrogen Tetroxide, N_2O_4. The two oxides, NO_2 and N_2O_4, exist in a strongly temperature-dependent equilibrium

$$\underset{\substack{\text{Brown} \\ \text{paramagnetic}}}{2NO_2} \rightleftharpoons \underset{\substack{\text{Colorless} \\ \text{diamagnetic}}}{N_2O_4}$$

both in solution and in the gas phase. In the solid state, the oxide is wholly N_2O_4. In the liquid, partial dissociation occurs; it is pale yellow at the freezing point (−11.2°) and contains 0.01% of NO_2, which increases to 0.1% in the deep red-brown liquid at the boiling point, 21.15°. Dissociation is complete in the vapor above 140°. NO_2 has an unpaired electron. The other "free radical" molecules, NO and ClO_2 (page 326), have little tendency to dimerize, and the difference may be that in NO_2 the electron is localized mainly on the N atom. The dimer has three isomeric forms of which the most stable and normal form has the planar structure $O_2N—NO_2$. The N—N bond is rather long, 1.75 Å, as would be expected from its weakness. The dissociation energy of N_2O_4 is only 57 kJ mol^{-1}.

The mixed oxides, i.e., NO_2 plus N_2O_4, are obtained by heating metal nitrates, by oxidation of NO, and by reduction of nitric acid and nitrates by metals and other reducing agents. The gases are highly toxic and attack metals rapidly. They react with water:

$$2NO_2 + H_2O = HNO_3 + HNO_2$$

the nitrous acid decomposing, particularly when warmed:

$$3HNO_2 = HNO_3 + 2NO + H_2O$$

The thermal decomposition

$$2NO_2 \rightleftharpoons 2NO + O_2$$

begins at 150° and is complete at 600°.

The oxides are fairly strong oxidizing agents in aqueous solution, comparable in strength to bromine:

$$N_2O_4(g) + 2H^+_{(aq)} + 2e = 2HNO_{2(aq)} \qquad E^\circ = +1.07\ V$$

The mixed oxides, "nitrous fumes," are used in organic chemistry as selective oxidizing agents; the first step is hydrogen abstraction:

$$RH + NO_2 = R^{\bullet} + HONO$$

Interaction of stoichiometric quantities of NO_2 and NO gives rise to the oxide N_2O_3, which exists pure only at low temperatures as it readily dissociates

$$NO + NO_2 \rightleftharpoons N_2O_3$$

Liquid N_2O_4 can be used as a solvent and has been utilized to make anhydrous nitrates and nitrate complexes. Thus Cu dissolves in N_2O_4 in ethyl acetate to give $Cu(NO_3)_2 \cdot N_2O_4$, which loses N_2O_4 on heating to give $Cu(NO_3)_2$.

N_2O_4 dissociates ionically in anhydrous HNO_3

$$N_2O_4 = NO^+ + NO_3^-$$

Dinitrogen Pentoxide. This oxide, N_2O_5, forms unstable colorless crystals. It is made by the reaction

$$2HNO_3 + P_2O_5 = 2HPO_3 + N_2O_5$$

N_2O_5 is the anhydride of nitric acid. In the solid state it is nitronium nitrate, $NO_2^+NO_3^-$.

16-6 The Nitronium Ion

Just as NO readily loses its odd electron, so does NO_2. The *nitronium* ion NO_2^+ is involved in the dissociation of HNO_3, in solutions of nitrogen oxides in acids, and in nitration reactions of aromatic compounds. Indeed, it was studies on nitration reactions that lead to proper recognition of the importance of NO_2^+ as the attacking species.

The nitronium ion is formed in ionizing solvents such as H_2SO_4, CH_3NO_2, or CH_3COOH by ionizations such as

$$2HNO_3 = NO_2^+ + NO_3^- + H_2O$$
$$HNO_3 + H_2SO_4 = NO_2^+ + HSO_4^- + H_2O$$

The actual nitration process can then be formulated

$$C_6H_6 + NO_2^+ \xrightarrow{\text{Slow}} [C_6H_6(H^+)(NO_2)] \xrightarrow{\text{Fast}} C_6H_5{-}NO_2 + H^+$$

Nitronium salts can be readily isolated. They are stable but rapidly hydrolyzed. Typical preparations are

$$N_2O_5 + HClO_4 = NO_2^+ClO_4^- + HNO_3$$
$$HNO_3 + 2SO_3 = NO_2^+HS_2O_7^-$$

16-7 Nitrous Acid

Solutions of the weak acid, HONO ($pK_a = 3.3$), are made by acidifying cold solutions of nitrites. The aqueous solution can be obtained free of salts by the reaction

$$Ba(NO_2)_2 + H_2SO_4 \longrightarrow 2HNO_2 + BaSO_4(s)$$

The pure liquid acid is unknown, but it can be obtained in the vapor phase. Even aqueous solutions of nitrous acid are unstable and decompose rapidly when heated:

$$3HNO_2 \rightleftharpoons H_3O^+ + NO_3^- + 2NO$$

Nitrites of the alkali metals are prepared by heating the nitrates with a reducing agent such as carbon, lead, iron, or the like. They are very soluble in water. Nitrites are very toxic but have been used for preservation of ham and other meat products; there is evidence that they can react with proteins to give carcinogenic nitrosoamines.

The main use of nitrites is to generate nitrous acid for the synthesis of organic diazonium compounds from primary aromatic amines. Organic derivatives of the NO_2 group are of two types: nitrites, R—ONO, and nitro compounds, R—NO_2. Similar tautomerism occurs in inorganic complexes, in which either oxygen or nitrogen is the actual donor atom when NO_2^- is a ligand (page 135).

Study Questions

A

1. Give the electronic structure of the N atom and list the ways by which the octet can be completed.
2. Formulate the multiple bonding using (a) valence bond, and (b) MO theory for NO_2^-.
3. How would you make spectroscopically pure nitrogen?
4. Compare the properties of H_2O and NH_3 that make them good solvents.
5. Sodium dissolves in liquid NH_3 to give a blue solution but reacts with water to give hydrogen. Explain this difference.
6. Write balanced equations for the synthesis of nitric acid from NH_3 and O_2.
7. Why is it wrong to refer to ammonia solutions as ammonium hydroxide?
8. Write equations for the action of heat on (a) $NaNO_3$, (b) NH_4NO_3, and (c) $Cu(NO_3)_2 \cdot nH_2O$.
9. How is hydrazine prepared?
10. Write balanced equations for three different preparations of nitric oxide.
11. How is the nitronium ion obtained? Explain its significance in nitration of aromatic hydrocarbons.
12. In acid solution we have

$$HNO_2 + H^+ + e = NO + H_2O \qquad E^\circ = 1.0 \text{ V}$$

Write balanced equations for the interaction of nitrous acid with I^-, Fe^{2+}, and $C_2O_4^{2-}$.
13. How can NO_2^- and NO_3^- be bound to transition metals in complexes.
14. List the main properties of nitric acid.
15. Write electronic structures for the NO^+ and NO_2^+ ions.

B

1. Compare the electronic structures of CO and N_2. Why does N_2 form complexes with metals less readily than CO?
2. NR_3 compounds cannot be resolved but PR_3 compounds can. What is the difference?
3. Why does nitrogen form only a diatomic molecule unlike phosphorus and other members of Group V?
4. Why is a 1 *M* solution of NH_4Cl acidic (pH ~ 4.7)?
5. Nitrogen trichloride is an extremely dangerous explosive oil but NF_3 is a very stable gas that reacts only above *ca.* 250°. Explain this difference.
6. Three isomers of N_2O_4 are known. Draw likely structures for them.
7. Although NO_2 readily dimerizes, NO does not. Why this difference?

Chapter 16
Study Guide

Supplementary Reading

Colburn, C. B., ed., *Developments in Inorganic Nitrogen Chemistry*, Vols. 1 and 2, Elsevier, 1966, 1973.
Jolly, W. L., *The Inorganic Chemistry of Nitrogen*, Benjamin, 1964.
Mellor's Comprehensive Treatise on Inorganic and Theoretical Chemistry, Vol. VIII, Supplements I and II, Longman's Green, 1967.
Wright, A. N. and Winkler, C. A., *Active Nitrogen*, Academic Press, 1968.

17

the group VB elements: phosphorus, arsenic, antimony and bismuth

17-1 Introduction

Phosphorus occurs mainly in minerals of the *apatite* family, $Ca_9(PO_4)_6 \cdot CaX_2$; X=F, Cl or OH, which are the main component of amorphous phosphate rock, millions of tons of which are processed annually. Arsenic, Sb and Bi, occur mainly as sulfide minerals such as *mispickel*, FeAsS, or *stibnite*, Sb_2S_3.

Some properties of the elements are given in Table 8-5 page 201, and some general features and trends are noted on page 201.

The valence shells of the atoms, ns^2np^3, are similar to the electron configuration of N, but beyond the similarity in stoichiometries of compounds such as NH_3, PH_3, etc., there is little resemblance between even P and N in their chemistries. Phosphorus is a true nonmetal in its chemistry but As, Sb, and Bi show an increasing trend to metallic character and cationic behavior.

The principal factors responsible for the differences between nitrogen and phosphorus group chemistry are those responsible for the C—Si differences, namely, (a) the inability of the second-row element to form $p\pi—p\pi$ multiple bonds, and (b) the possibility of utilizing the lower-lying 3*d* orbitals.

The first explains features such as the fact that nitrogen forms esters O=NOR, whereas phosphorus gives $P(OR)_3$. Nitrogen oxides and oxoacids all involve multiple bonds (page 276) whereas the phosphorus oxides have single P—O bonds as in P_4O_6 and phosphoric acid is $PO(OH)_3$ in contrast to $NO_2(OH)$.

The utilization of *d* orbitals has three effects. First, it allows some $p\pi—d\pi$ bonding as in $R_3P{=}O$ or $R_3P{=}CH_2$. Thus amine oxides R_3NO have only a single canonical structure $R_3\overset{+}{N}—\overset{-}{O}$ and are chemically reactive, while P—O bonds are shorter than expected for the sum of single bond radii, indicating multiple bonding, and are very strong, *ca.* 500 kJ mol^{-1}. Second, there is the possibility of expansion of the valence shell, whereas nitrogen has a covalency maximum of four. Thus we have compounds such as PF_5, PPh_5, $P(OMe)_6^-$, and PF_6^-.

Notice that for many of the 5-coordinate species, especially of phosphorus, the energy difference between the trigonal bipyramidal and square pyramidal configurations is small and that such species are usually stereochemically nonrigid (page 157).

When higher coordination occurs in the III state, as in SbF_5^{2-}, there are five bonding pairs plus a lone pair which occupies a bond position, so that the ion can be regarded as octahedral, with an electron pair serving as one ligand.

Finally, while trivalent nitrogen and the other elements in compounds such as NEt_3, PEt_3, $AsPh_3$, etc., have lone pairs and act as donors, there is a profound difference in their donor ability toward transition metals. This follows from the fact that although NR_3 has no low-lying acceptor orbitals, the others do have such orbitals, namely, the empty *d* orbitals. These can accept electron density from filled metal *d* orbitals to form $d\pi—d\pi$ bonds, as we shall discuss in detail later (page 489).

17-2
The Elements

Phosphorus is obtained by reduction of phosphate rock with coke and sand in an electric furnace. Phosphorus distills and is condensed under water as P_4. Phosphorus allotropes have been discussed (pages 183, 186).

$$2Ca_3(PO_4)_2 + 6SiO_2 + 10C = P_4 + 6CaSiO_3 + 10CO$$

P_4 is stored under water to protect it from air in which it will inflame. Red and black P are stable in air but will burn on heating. P_4 is soluble in CS_2, benzene and similar organic solvents; it is very poisonous.

Arsenic. Sb and Bi are obtained as the metals (page 187) by reduction of their oxides with carbon or hydrogen. The metals burn on heating in oxygen to give the oxides.

All the elements react readily with halogens but are unaffected by nonoxidizing acids. Nitric acid gives, respectively, phosphoric acid, arsenic acid, antimony trioxide, and bismuth nitrate, which well illustrates the increasing metallic character as the group is descended.

Interaction with various metals and nonmetals gives phosphides, arsenides, and the like, which may be ionic, covalent polymers or metal-like solids. Gallium arsenide, GaAs—one of the so-called III-V compounds of a group III and a group V element—has semiconductor properties similar to those of Si and Ge.

COMPOUNDS OF GROUP V ELEMENTS

17-3
Hydrides, MH_3

The stability of these gases falls rapidly and SbH_3 and BiH_3 are very unstable thermally. The average bond energies are N—H, 391; P—H, 322; As—H, 247; and Sb—H, 255 kJ mol^{-1}.

Phosphine, PH_3, is made by action of acids on zinc phosphide. When pure, it is not spontaneously flammable, but often inflames owing to traces of P_2H_4 or P_4 vapor. It is exceedingly poisonous. Unlike NH_3, it is not associated in the liquid state. It is only sparingly soluble in water and is a very weak base. The proton affinities of PH_3 and NH_3 (Eq. 17-1) differ considerably:

$$EH_3(g) + H^+(g) = EH_4{}^+(g) \tag{17-1}$$

$$\Delta H^\circ = -770 \text{ kJ mol}^{-1} \qquad \text{for E} = \text{P}$$

$$\Delta H^\circ = -866 \text{ kJ mol}^{-1} \qquad \text{for E} = \text{N}$$

but PH_3 and gaseous HI react to give PH_4I as unstable colorless crystals. In view of the low basicity, complete hydrolysis occurs in water

$$PH_4I(s) + H_2O = H_3O^+ + I^- + PH_3(g)$$

PH_3 dissolves in very strong acids such as $BF_3 \cdot H_2O$ to give $PH_4{}^+$. PH_3 is used industrially to make organophosphorus compounds (Chapter 29).

17-4 Halides, MX_3, MX_5, and Oxohalides

The trihalides, except PF_3, are obtained by direct halogenation, keeping the element in excess. An excess of the halogen gives MX_5. The trihalides are rapidly hydrolyzed by water and are rather volatile; the gaseous molecules have pyramidal structures. The chlorides and bromides as well as PF_3 and PI_3 have molecular lattices. AsI_3, SbI_3, and BiI_3 have layer structures based on hexagonal close packing of iodine atoms with the group III atoms in octahedral holes. BiF_3 is known in two forms, in both of which it has coordination number 8, while SbF_3 has an intermediate structure in which SbF_3 molecules are linked through F bridges to give each Sb^{III} a very distorted octahedral environment.

Phosphorus trifluoride. This is a colorless, toxic gas, made by fluorination of PCl_3. It forms complexes with transition metals similar to those formed by carbon monoxide (page 490). Unlike the other trihalides, PF_3 is hydrolyzed only slowly by water, but it is attacked rapidly by alkalis. It has no Lewis acid properties.

Phosphorus trichloride. This is a low-boiling liquid that is violently hydrolyzed by water to give phosphorous acid. It reacts readily with oxygen to give $OPCl_3$. *Figure 17-1* illustrates some of the important reactions of PCl_3. Many of these reactions are typical of other MX_3 compounds and also, with obvious changes in formulas, of $OPCl_3$ and other oxo halides.

Arsenic trihalides. Arsenic trihalides are similar to those of phosphorus. $SbCl_3$ differs in that it dissolves in a limited amount of water to give a clear solution that, on dilution, gives insoluble oxo chlorides such as SbOCl and $Sb_4O_5Cl_2$. No simple Sb^{3+} ions exist in the solutions. $BiCl_3$, a white, crystalline solid, is hydrolyzed by water to BiOCl but this reaction is reversible;

$$BiCl_3 + H_2O \rightleftharpoons BiOCl + 2HCl$$

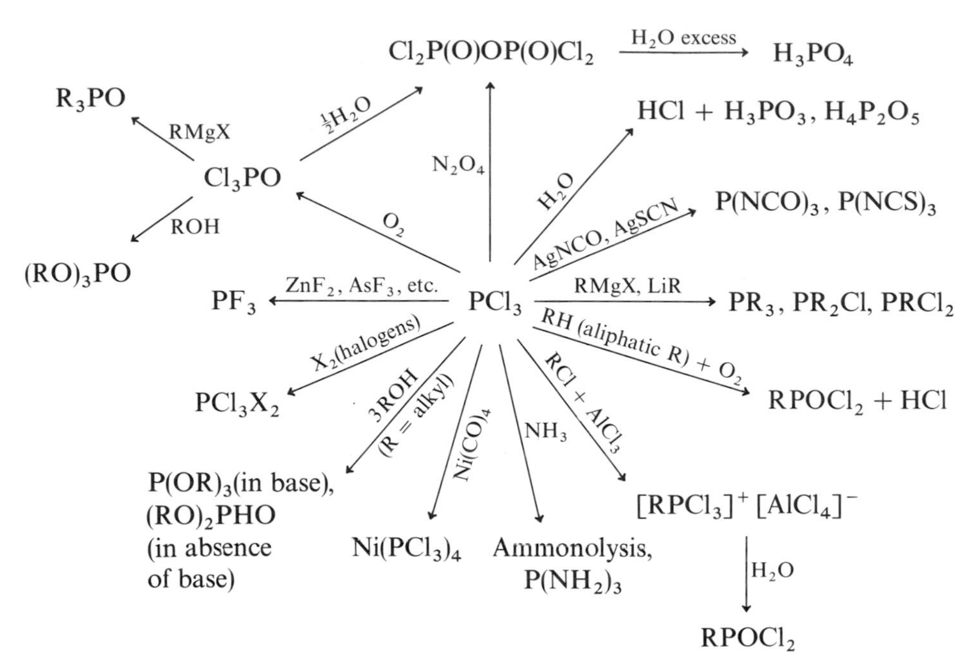

Figure 17-1 *Some important reactions of* PCl_3. *Many of these are typical for other* MX_3 *compounds as well as for* MOX_3 *compounds.*

Phosphorus pentafluoride. Phosphorus pentafluoride, PF_5 is prepared by the interaction of PCl_5 with CaF_2 at 300 to 400°. It is a very strong Lewis acid and forms complexes with amines, ethers, and other bases as well as with F^-, in which phosphorus becomes 6-coordinate. However, these organic complexes are less stable than those of BF_3 and are rapidly decomposed by water and alcohols. Like BF_3, PF_5 is a good catalyst, especially for ionic polymerization. AsF_5 is similar.

Antimony pentafluoride. Antimony pentafluoride is a viscous liquid (bp 150°). Its association is due to polymerization through fluorine bridging. The crystal has cyclic tetramers. Its main use is in "superacids" (page 177).

AsF_5, SbF_5, and PF_5 are potent fluoride ion acceptors, forming MF_6^- ions. The PF_6^- ion is a common and convenient "noncomplexing" anion.

Phosphorus(V) chloride has a trigonal bipyramidal structure in the gas, melt, and solution in nonpolar solvents, but the solid is $[PCl_4]^+[PCl_6]^-$ and in polar solvents like CH_3NO_2 it is ionized. The tetrahedral PCl_4^+ ion can be considered to arise here by transfer of Cl^- to the Cl^- acceptor, PCl_5. It is not therefore surprising that many salts of the PCl_4^+ ion are obtained when PCl_5 reacts with other Cl^- acceptors, namely,

$$PCl_5 + TiCl_4 \longrightarrow [PCl_4]_2^+[Ti_2Cl_{10}]^{2-} \quad \text{and} \quad [PCl_4]^+[Ti_2Cl_9]^-$$

$$PCl_5 + NbCl_5 \longrightarrow [PCl_4]^+[NbCl_6]^-$$

Solid phosphorus pentabromide is also ionic, but differs, being $PBr_4^+Br^-$ Antimony, but not arsenic, forms a pentachloride, which is a fuming liquid, colorless when pure, usually yellow. It is a powerful chlorinating agent.

Phosphoryl halides. Phosphoryl halides are X_3PO, in which X may be F, Cl, or Br. The most important one is Cl_3PO, obtainable by the reactions

$$2\,PCl_3 + O_2 \longrightarrow 2\,Cl_3PO$$

$$P_4O_{10} + 6\,PCl_5 \longrightarrow 10\,Cl_3PO$$

The reactions of Cl_3PO are much like those of PCl_3 (*Fig. 17-1*). Hydrolysis by water yields phosphoric acid. Cl_3PO also has donor properties and many complexes are known, in which oxygen is the ligating atom.

The oxohalides SbOCl and BiOCl are precipitated when solutions of Sb^{III} and Bi^{III} in concentrated HCl are diluted.

17-5 Oxides

The oxides of the Group V elements clearly exemplify two important trends that are manifest to some extent in all groups of the periodic table: (1) the stability of the higher oxidation state decreases with increasing atomic number, and (2) in a given oxidation state the metallic character of the elements, and therefore the basicity of the oxides, increase with increasing atomic number. Thus, P^{III} and As^{III} oxides are acidic, Sb^{III} oxide is amphoteric and Bi^{III} oxide is strictly basic.

Phosphorus Oxides. Phosphorus pentoxide is so termed for historical reasons but the correct molecular formula is P_4O_{10} (*Fig. 17-2a*). It is made by burning phosphorus in excess oxygen. It has at least three solid forms. Two are polymeric but one is a white, crystalline material that sublimes at 360° and 1 atm. Sublimation is an excellent method of purification, since the products of incipient hydrolysis, which are the commonest impurities, are comparatively nonvolatile. This form and the vapor consist of molecules in which the P atoms are at the corners of a tetrahedron with six oxygen atoms along the edges. The remaining four O atoms lie along extended threefold axes of the tetrahedron. The P—O—P

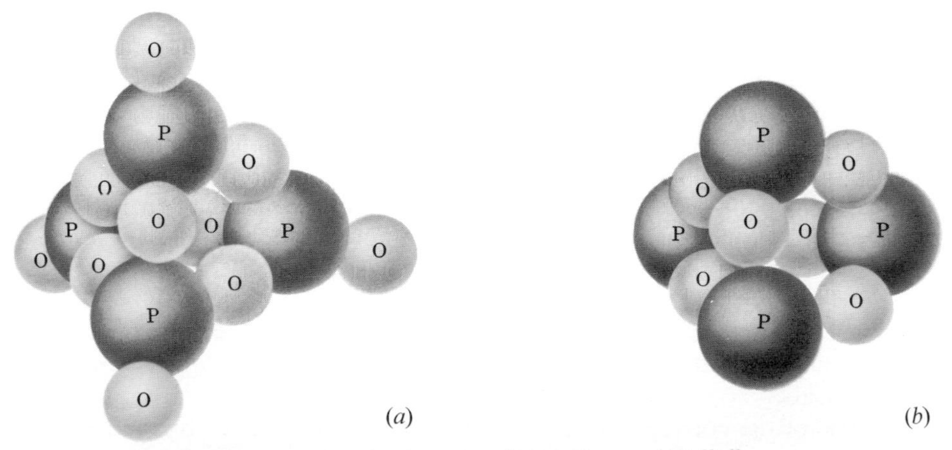

Figure 17-2 *The structure of (a)* P_4O_{10}*, and (b)* P_4O_6.

bonds are single but the lengths of the four apical P—O bonds indicates $p\pi$—$d\pi$ bonding, that is, P=O.

P_4O_{10} is one of the most effective drying agents known at temperatures below 100°. It reacts with water to form a mixture of phosphoric acids (see below) whose composition depends on the quantity of water and other conditions. It will even extract the elements of water from many other substances themselves considered good dehydrating agents; for example, it converts pure HNO_3 into N_2O_5 and H_2SO_4 into SO_3. It also dehydrates many organic compounds, for example, converting amides into nitriles.

The *trioxide* is also polymorphous: one form contains discrete molecules, P_4O_6. The structure (*Fig. 17-2b*) is similar to that of P_4O_{10} except that the four nonbridging apical oxygens present in the latter are missing. P_4O_6 is a colorless, volatile compound that is formed in about 50% yield when P_4 is burned in a deficit of oxygen. As_4O_6 and Sb_4O_6 are similar to P_4O_6 both structurally and in their acidic nature. Bi_2O_3 and the hydroxide $Bi(OH)_3$ precipitated from bismuth(III) solution have no acidic properties.

17-6 Sulfides

Phosphorus and sulfur combine directly above 100° to give several sulfides, the most important being P_4S_3, P_4S_5, P_4S_7, amd P_4S_{10}. Each compound is obtained by heating stoichiometric quantities of red P and sulfur. P_4S_3 is used in matches. It is soluble in organic solvents such as carbon disulfide and benzene. P_4S_{10} has the same structure as P_4O_{10}. The others also have structures based on a tetrahedral group of phosphorus atoms with P—S—P bridges or apical P=S groups. P_4S_{10} reacts with alcohols:

$$P_4S_{10} + 8\,ROH \longrightarrow 4(RO)_2P(S)SH + 2\,H_2S$$

to give dialkyl and diaryl dithiophosphates which form the basis of many extreme-pressure lubricants, of oil additives, and of flotation agents.

Arsenic. Arsenic forms As_4S_3, As_4S_4, As_2S_3, and As_2S_5 by direct interaction. The last two can also be precipitated from hydrochloric acid solutions of As^{III} and As^{V} by H_2S. As_2S_3 is insoluble in water and acids but is acidic dissolving in alkali sulfide solutions to give thio anions. As_2S_5 behaves similarly. As_4S_4, which occurs as the mineral *realgar*, has a structure with an As_4 tetrahedron.

Antimony forms Sb_2S_3 either by direct interaction or by precipitation with H_2S from Sb^{III} solutions; it dissolves in an excess of sulfide to give anionic thio complexes, probably mainly SbS_3^{3-}. Sb_2S_3, as well as Bi_2S_3, has a ribbon like polymeric structure in which each Sb atom and each S atom is bound to three atoms of the opposite kind, forming interlocking SbS_3 and SSb_3 pyramids.

Bismuth gives dark brown Bi_2S_3 on treatment of Bi^{III} solutions with H_2S; it is not acidic.

Some of the corresponding selenides and tellurides of As, Sb, and Bi have been studied intensively as semiconductors.

17-7
The Oxo Acids

The nature and properties of the oxoanions of the Group V elements have been discussed in Chapter 5. Here we discuss only the important acids and some of their derivatives.

Phosphorous acid. Phosphorous acid is obtained when PCl_3 or P_4O_6 are hydrolyzed by water. It is a deliquescent colorless solid (mp 70°, pK = 1.8). The acid and its mono and diesters differ from PCl_3 in that there are *four* bonds to P, one being P—H. The presence of hydrogen bound to P can be demonstrated by nmr or other spectroscopic techniques. Phosphorous acid is, hence, best written $HP(O)(OH)_2$ as in (17-I). Hypophosphorous acid H_3PO_2, also has P—H bonds (17-II). By contrast the triesters have only three bonds to phosphorus, thus being analogous to PCl_3. The trialkyl and aryl phosphites, $P(OR)_3$ have excellent donor properties toward transition metals and many complexes are known.

17-I

17-II

Phosphorous acid may be oxidized by chlorine or other agents to phosphoric acid, but the reactions are slow and complex. However, the triesters are quite readily oxidized and must be protected from air:

$$2(RO)_3P + O_2 = 2(RO)_3PO$$

They also undergo the Michaelis–Arbusov reaction with alkyl halides, forming dialkyl phosphonates:

$$P(OR)_3 + R'X \longrightarrow \underset{\text{Phosphonium intermediate}}{[(RO)_3PR']X} \longrightarrow RO{-}\overset{O}{\overset{\|}{P}}{-}R' + RX$$

(with OR on P)

Trimethylphosphite easily undergoes spontaneous isomerization to the dimethyl ester of methylphosphonic acid:

$$P(OCH_3)_3 \longrightarrow CH_3PO(OCH_3)_2$$

Orthophosphoric acid. Orthophosphoric acid, H_3PO_4, commonly called phosphoric acid, is one of the oldest known and most important phosphorus compounds. It is made in vast quantities, usually as 85% syrupy acid, by the direct reaction of ground phosphate rock with sulfuric acid and also by the direct burning of phosphorus and subsequent hydration of P_4O_{10}. The pure acid is a colorless crystalline solid (mp 42.35°). It is very stable and has essentially no oxidizing properties below 350 to 400°. At elevated temperatures it is fairly reactive toward metals, which reduce it, and it will attack quartz. *Pyrophosphoric acid* is also produced:

$$2H_3PO_4 \longrightarrow H_2O + H_4P_2O_7$$

but this conversion is slow at room temperature.

The acid is tribasic: at 25°, $pK_1 = 2.15$, $pK_2 = 7.1$, $pK_3 \approx 12.4$. The pure acid and its crystalline hydrates have tetrahedral PO_4 groups connected by hydrogen bonds. Hydrogen bonding persists in the concentrated solutions and is responsible for the sirupy nature. For solutions of concentration less than ~50%, the phosphate anions are hydrogen-bonded to the liquid water rather than to other phosphate anions.

Phosphates and the polymerized phosphate anions (for which the free acids are unknown) are discussed on page 117. Large numbers of *phosphate esters* can be made by the reaction

$$OPCl_3 + 3ROH = OP(OR)_3 + 3HCl$$

or by oxidation of trialkylphosphites. Phosphate esters such as tributylphosphate are used in the extraction of certain +4 metal ions (see page 450) from aqueous solutions.

Phosphate esters are also of fundamental importance in living systems. It is because of this that their hydrolysis has been much studied. Triesters are attacked by OH^- at P and by H_2O at C, depending on pH.

$$OP(OR)_3 \xrightarrow{^{18}OH^-} OP(OR)_2(^{18}OH) + RO^-$$

$$OP(OR)_3 \xrightarrow{H_2{}^{18}O} OP(OR)_2(OH) + R^{18}OH$$

Diesters, which are strongly acidic, are completely in the anionic form at normal (and physiological) pH's:

$$RO{-}\overset{\overset{\displaystyle O}{\|}}{\underset{\underset{\displaystyle OH}{|}}{P}}{-}OR' \rightleftharpoons R'OPO_2OR^- + H^+ \qquad K \approx 10^{-1.5}$$

They are thus relatively resistant to nucleophilic attack by either OH^- or H_2O, and this is why enzymic catalysis is indispensible to achieve useful rates of reaction.

Relatively little has been firmly established as yet concerning the mechanisms of most phosphate ester hydrolyses, especially the many enzymic ones. Two important possibilities are the following.

1. One-step nucleophilic displacement (S_N2) with inversion:

$$H_2O(\text{or } OH^-) + (O{=})({}^-O)(R'O)P{-}OR \longrightarrow HO{-}P({=}O)(O^-)(OR') + HOR$$

2. Release of a short-lived "metaphosphate" group, which rapidly recovers the 4-connected orthophosphate structure:

```
                         O
      O       O          ||
 —O—P—O—P—O—P—OH   ——→
      O       O   ↗  |    ↘
               H⁺  O⁻    H⁺
```

```
   O  O
 —OPOPOH + "PO3⁻"  —H2O→  H2PO4⁻
   O  O
```

$$\begin{array}{c} \mathrm{O\;\;O} \\ -\mathrm{OPOPOH} \\ \mathrm{O\;\;O} \end{array} + \text{"}PO_3^{-}\text{"} \xrightarrow{H_2O} H_2PO_4^{-}$$

17-8
Complex Chemistry of Group VB Elements

The main aqueous chemistry of Sb^{III} is in oxalato, tartrato, and similar hydroxyacid complexes.

The $Sb(C_2O_4)_3^{3-}$ ion forms isolable salts and has been shown to have the incomplete pentagonal-bipyramid structure (*Fig. 17-3*) with a lone pair at one axial position. The tartrate complexes of antimony(III) have been much studied, and have been used medicinally as "tartar emetic" for more than 300 years. The structure of the anion in this salt, $K_2[Sb_2(d\text{-}C_4H_2O_6)_2]\cdot 3H_2O$, is shown in *Fig. 17-4*.

Only for Bi is there a true cationic chemistry. Aqueous solutions contain well-defined hydrated cations, but there is no evidence for a simple aquo ion $[Bi(H_2O)_n]^{3+}$. In neutral perchlorate solutions the main species is $[Bi_6O_6]^{6+}$ or its hydrated form, $[Bi_6(OH)_{12}]^{6+}$, and at higher pH $[Bi_6O_6(OH)_3]^{3+}$ is formed. The $[Bi_6(OH)_{12}]^{6+}$ species contains an octahedron of Bi^{3+} ions with an OH^- bridging each edge.

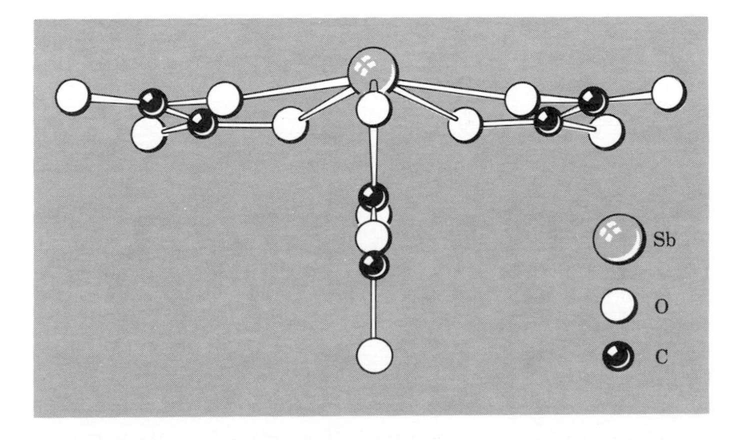

Figure 17-3 *The $[Sb(C_2O_4)_3]^{3-}$ ion projected on a plane approximately perpendicular to the basal plane of the pentagonal pyramid*

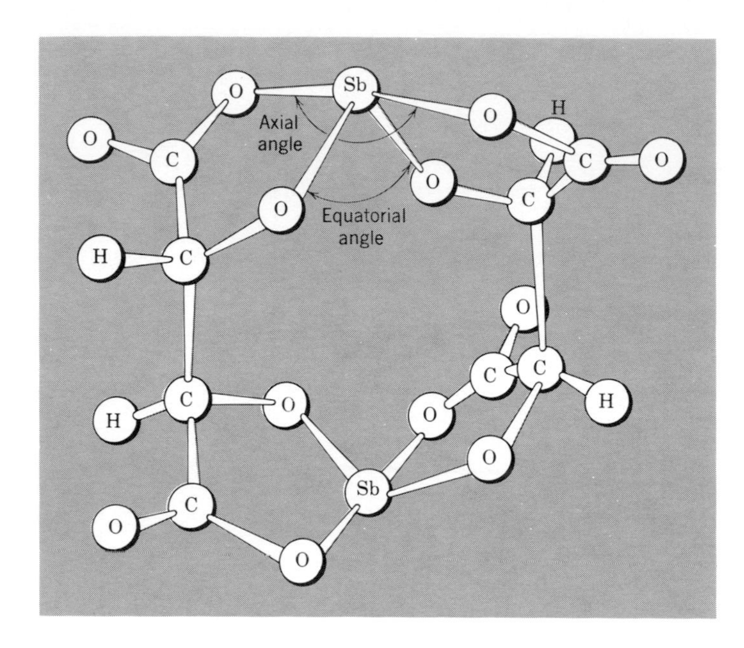

Figure 17-4 *Geometry of the anion* $[Sb_2(C_4H_2O_6)_2]^{2-}$*; water molecules link the anions into sheets by hydrogen bonding to carboxylate oxygen.* (*Reproduced by permission from* Coord. Chem. Rev., *1969*, 4, *323*.)

17-9
Phosphorus–Nitrogen Compounds

Many compounds are known with P—N and P=N bonds. R_2N—P bonds are particularly stable and occur widely in combination with bonds to other univalent groups, such as P—R, P—Ar, and P—halogen.

Phosphazenes. These are cyclic or chain compounds that contain alternating phosphorus and nitrogen atoms with two substituents on each phosphorus atom. The three main structural types are the cyclic trimer (17-III), cyclic tetramer (17-IV), and the oligomer or high polymer (17-V). The alternating sets of single and double bonds in (17-III) to (17-V) are written for convenience but, in general,

$(NPR_2)_3$ (cyclic) — 17-III $(NPR_2)_4$ (cyclic) — 17-IV $[N{=}PR_2]_x$ — 17-V

all P—N distances are found to be equal. It appears that they are of order ≈1.5, since their lengths, 1.56 to 1.61 Å are appreciably shorter than expected (~1.80) for P—N single bonds. Hexachlorocyclotriphosphazene, $(NPCl_2)_3$, is a key intermediate in the synthesis of many other phosphazenes and is manufactured by the reaction

$$nPCl_5 + nNH_4Cl \xrightarrow{\text{in } C_2H_2Cl_4 \text{ or } C_6H_5Cl} (NPCl_2)_n + 4nHCl$$

This reaction produces a mixture of $[NPCl_2]_n$ species with $n = 3, 4, 5, \ldots$ and low-polymeric linear species. Favorable conditions give 90% yields of the $n = 3$ or 4 species, which can be separated by extraction, crystallization or sublimation.

The majority of phosphazene reactions involve replacement of halogen atoms by other groups (OH, OR, NR_2, NHR, or R) to give partially or fully substituted derivatives:

$$(NPCl_2)_3 + 6NaOR \longrightarrow [NP(OR)_2]_3 + 6NaCl$$

$$(NPCl_2)_3 + 6NaSCN \longrightarrow [NP(NCS)_2]_3 + 6NaCl$$

$$(NPF_2)_3 + 6PhLi \longrightarrow (NPPh_2)_3 + 6LiF$$

In partly substituted molecules many isomers are possible.

The rings in $(NPF_2)_x$ where $x = 3$ or 4 are planar, and those when $x = 5$ or 6 approach planarity. For other $(NPX_2)_n$ compounds the six-rings are planar or nearly so, but larger rings are generally nonplanar. *Figure 17-5* shows the structures of $(NPCl_2)_3$ and $(NPClPh)_4$.

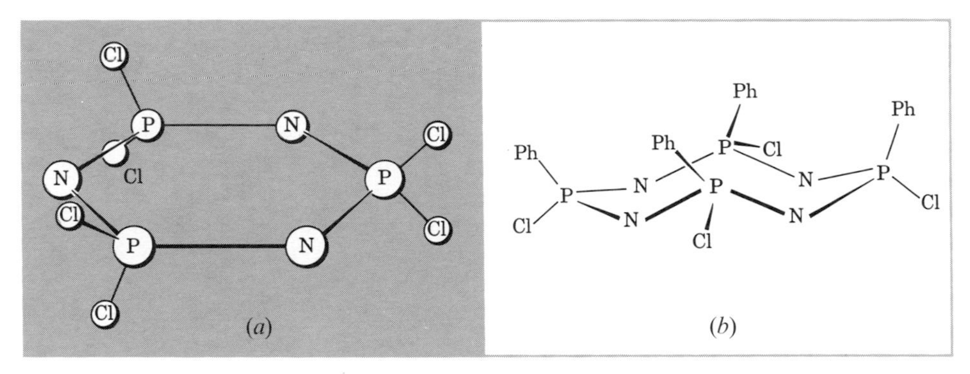

Figure 17-5 *The structures of two representative cyclic phosphazenes. (a)* $(NPCl_2)_3$, *(b) all-cis-*$(NPClPh)_4$.

The high-polymeric linear phosphazenes are potentially useful materials as far as physical and mechanical properties are concerned, but they have been generally useless because of hydrolytic instability. Useful polymers have perfluoroalkoxy and other side groups, for example, $[NP(OCH_2CF_3)_2]_n$ which resembles polyethylene and is crystalline and water repellant.

Study Questions

A

1. Why does phosphorus form P_4 molecules while nitrogen is N_2?
2. How are white and red phosphorus obtained from phosphate rock?

3. What are the principal factors responsible for the differences between the chemistry of nitrogen and the chemistry of phosphorus?
4. Explain the differences in (a) basisity and (b) donor ability toward transition metals of NMe_3 and PMe_3.
5. Write balanced equations for the reactions:
   (i) $P_4 + HNO_3$ (ii) $AsCl_3 + H_2O$ (iii) $POCl_3 + H_2O$
   (iv) $P_4O_{10} + HNO_3$ (v) $P_4O_6 + H_2O$ (vi) Zn_3P + dilute HCl
6. How is PCl_5 made? What is its structure in solutions and in the solid state?
7. Draw the structures of P_4O_{10} and As_4O_6.
8. What happens when H_2S is passed into acidic (HCl) solution of trivalent P, As, Sb, and Bi?
9. What are the structures of (a) phosphorous acid, and (b) triethylphosphite?
10. What is the Michaelis–Arbusov reaction?
11. Why is pure phosphoric acid syrupy?
12. What is the structure of "tartar emetic"?
13. What are phosphazenes and how are they made?
14. Describe the interaction of water with $SbCl_3$ and $BiCl_3$.
15. How is PF_5 prepared? Give its main chemical properties.
16. Compare the structure and properties of nitric and phosphoric acids.

B

1. Discuss $d\pi$—$p\pi$ bonding and its effects. Give examples, with explanations for differences between the chemistries of N and P.
2. What are stereochemically nonrigid molecules? What is their role in phosphorus chemistry?
3. NF_3 has no donor properties at all, but PF_3 forms numerous complexes, for example, $Ni(PF_3)_4$, with transition metals and is very toxic. Explain.
4. P and Sb both form pentachlorides but As does not. Why?
5. What is the importance and the mechanism of hydrolysis of phosphate esters?
6. Nitrogen forms heterocyclic compounds like pyridine. Would you expect P to do the same. What would the properties be like?
7. Write an essay comparing the oxides of nitrogen with those of phosphorus.
8. Why is NCl_3 unstable and highly explosive whereas PCl_3 is not? How would you expect them to react with water or dilute NaOH?

Chapter 17
Study Guide

Supplementary Reading

Alcock, H. R., *Phosphorus-Nitrogen Compounds*, Academic Press, 1972.
Corbridge, D. E., *The Structural Chemistry of Phosphorus*. Elsevier, 1974.
Doak, G. O. and Freedman, L. D., *Organometallic Compounds of Arsenic, Antimony and Bismuth*, Wiley-Interscience, 1970.
Fluck, E., "The Chemistry of Phosphine," *Topics in Current Chemistry*, Springer-Verlag, 1973.
Grayson, M. and Griffiths, E. J., eds., *Topics in Phosphorus Chemistry*, Vols. 1–7, Wiley, 1964–1973.
Mellor's Comprehensive Treatise on Inorganic and Theoretical Chemistry, Vol. III, Supplement III, Longmans Green, 1971.
Van Wazer, J. R., *Phosphorus and Its Compounds*, Interscience, 1958.

18

oxygen

18-1
Introduction

Oxygen compounds of all the elements except He, Ne, and possibly Ar are known. Molecular oxygen (dioxygen, O_2) reacts with all other elements except the halogens, a few noble metals, and the noble gases either at room temperature or on heating.

The chemistry of oxygen involves the completion of the neon configuration by one of the following:

1. Electron gain to form O^{2-}.

2. Formation of two single covalent bonds —O— or a double bond =O, as in $(CH_3)_2C{=}O$ or $Cl_5Re{=}O$.

3. Formation of one single bond and electron gain as in OH^- or OEt^-.

4. Formation of three, or less commonly, four covalent bonds as in oxonium ions H_3O^+, R_3O^+, and $Be_4O(CO_2CH_3)_6$.

Oxides. The range of physical properties shown by binary oxides of the elements is attributable to the range of bond types from essentially ionic to essentially covalent (see also Chapter 4, page 104 and Chapter 5, page 108).

The formation of the oxide ion from molecular oxygen requires much energy, *ca.* 1000 kJ mol^{-1}:

$$\tfrac{1}{2}O_2(g) = O(g) \qquad \Delta H = 248 \text{ kJ mol}^{-1}$$

$$O(g) + 2e = O^{2-}(g) \qquad \Delta H = 752 \text{ kJ mol}^{-1}$$

In forming an ionic metal oxide, energy must also be expended in vaporizing and ionizing the metal atoms. The existence of the many ionic oxides is a result of the high lattice energies of oxides that contain the small (1.40 Å) double charged O^{2-} ion.

Where the lattice energy is insufficient to provide the necessary energy for complete ionization, oxides with substantial covalent character are formed, examples being BeO, SiO_2, B_2O_3 etc.

Essentially covalent molecular oxides are compounds such as CO_2, SO_2, NO_2, etc., in which multiple bonding is important. Covalent oxides with single bonds are also formed; P_4O_{10} is one example. The chemical properties of oxide and hydroxide ions are discussed in Chapter 5, page 108.

Two-coordinate covalent oxygen. The compounds of formula R_2O are invariably angular. The bonding can be considered as involving sp^3 hybrid orbitals with two 2-electron covalent bonds and two lone-pairs. There are wide variations, depending on the nature of R from the tetrahedral angle, for example, H_2O, 104.5°, Me_2O, 111°. Where the atoms bound to oxygen have d orbitals available some $d\pi$—$p\pi$ character is often present in the bonds to oxygen and the X—O—X angles may be even larger, for example, H_3Si—O—SiH_3, >150°, Si—O—Si in quartz, 142°. The extreme of a linear X——O—X bond occurs in some transition metal complexes, for example, $[Cl_5Ru—O—RuCl_5]^{4-}$. The σ bonds are formed by sp hybrids on O, thus leaving two pairs of π electrons in pure p orbitals. These can interact with empty $d\pi$ orbitals on the metal atom. Oxygen compounds R_2O behave as Lewis bases and, when R_2O functions as a base, the oxygen atom becomes 3-coordinate, as for example,

$$Et_2O + BF_3 = Et_2OBF_3$$

The formation of oxonium ions is analogous to the formation of ammonium ions;

$$:NH_3 + H^+ = NH_4{}^+$$
$$\ddot{\underset{\cdot\cdot}{O}}H_2 + H^+ = :OH_3{}^+$$

Oxygen is less basic than nitrogen, and the oxonium ions are therefore less stable. Notice that ions of the type $OH_4{}^{2+}$ are unlikely to be obtained even though $:OH_3{}^+$ still has a lone pair, because of the electrostatic repulsion of the charged ion to another incoming proton. Like NR_3, the $OR_3{}^+$ species undergo rapid inversion (page 275).

Multiply-bonded oxygen. There are many compounds with unicoordinate, multiply-bonded oxygen. Some examples are $OSCl_2$, $OPMe_3$, $OV(acac)_2$, OUO^{2+}. The order of the X—O bond may vary from essentially unity in amine oxides, for example, $R_3\overset{+}{N}:\ddot{\underset{\cdot\cdot}{O}}:^-$ through varying degrees of π bonding up to 2. The simplest π-bonding example is that in the ketones RR′C=O where there is one π bond perpendicular to the molecule plane.

18-2
Occurrence, Properties, and Allotropes

Oxygen has three isotopes, ^{16}O (99.759%), ^{17}O (0.0374%), and ^{18}O (0.2039%). Fractional distillation of water allows concentrates containing up to 97 atom % ^{18}O or up to 4 atom % ^{17}O to be prepared. ^{18}O is used as a tracer in studying reaction mechanisms of oxygen compounds. Although ^{17}O has a nuclear spin (5/2), its low abundance means that, even when enriched samples are used, spectrum accumulation and/or the Fourier transform method are required. An example of

^{17}O resonance studies is the distinction between H_2O in a complex, for example, $[Co(NH_3)_5H_2O]^{3+}$, and solvent water.

Oxygen has two allotropes; dioxygen, O_2 and trioxygen or ozone, O_3. O_2 is paramagnetic in all states and has the rather high dissociation energy of 496 kJ mol^{-1}. Simple valence-bond theory predicts the electronic structure $:\ddot{O}=\ddot{O}:$ which, though accounting for the strong bond, fails to account for the paramagnetism. However, simple MO theory (page 66) readily accounts for the triplet ground state having a double bond. There are several low-lying singlet states that are important in photochemical oxidations (page 302). Like NO, which has one unpaired electron in an antibonding (π^*) MO, oxygen molecules associate only weakly, and true electron pairing to form a symmetrical O_4 species does not occur even in the solid. Both liquid and solid O_2 are pale blue.

Ozone. The action of a silent electric discharge on O_2 produces O_3 in concentrations up to 10%. Ozone gas is perceptibly blue and is diamagnetic. Pure ozone obtained by fractional liquefaction of O_2—O_3 mixtures gives a deep blue, explosive liquid. The action of ultraviolet light on O_2 produces traces of O_3 in the upper atmosphere. The maximum concentration is at an altitude of ~25 km. It is of vital importance in protecting the earth's surface from excessive exposure to ultraviolet light. Ozone is very endothermic:

$$O_3 = \tfrac{3}{2}O_2 \qquad \Delta H = -142 \text{ kJ mol}^{-1}$$

but it decomposes only slowly at 250° in absence of catalysts and ultraviolet light.

The O_3 molecule is symmetrical and bent; ∠ O—O—O, 117°; O—O, 1.28 Å. Since the O—O bond distances are 1.49 Å in HOOH (single bond) and 1.21 Å in O_2 (~ double bond), it is apparent that the O—O bonds in O_3 must have considerable double-bond character. In terms of a reasonance description, this can be accounted for as follows:

$$\ddot{O}::\ddot{O}:\ddot{O}: \longleftrightarrow :\ddot{O}:\ddot{O}::\ddot{O}$$

Chemical Properties of O_2 and O_3. Ozone is a much more powerful oxidizing agent than O_2 and reacts with many substances under conditions where O_2 will not. The reaction

$$O_3 + 2KI + H_2O \longrightarrow I_2 + 2KOH + O_2$$

is quantitative and can be used for analysis. Ozone is used for oxidations of organic compounds and in water purification. Oxidation mechanisms probably involve free radical chain processes as well as intermediates with —OOH groups. In acid solution, O_3 is exceeded in oxidizing power only by F_2, the perxenate ion, atomic oxygen, OH radicals, and a few other such species.

The following potentials indicate the oxidizing strengths of O_2 and O_3 in ordinary aqueous solution:

$$O_2 + 4H^+(10^{-7}M) + 4e = 2H_2O \qquad E^\circ = +0.815 \text{ V}$$

$$O_3 + 2H^+(10^{-7}M) + 2e = O_2 + H_2O \qquad E^\circ = +1.65 \text{ V}$$

The first step in the reduction of O_2 in aprotic solvents such as dimethyl sulfoxide and pyridine appears to be a one-electron step to give the superoxide anion:

$$O_2 + e = O_2^-$$

whereas in aqueous solution a two-electron step occurs to give HO_2^-:

$$O_2 + 2e + H_2O = HO_2^- + OH^-$$

It can also be seen from the potential given above that neutral water saturated with O_2 is a fairly good oxidizing agent. For example, although Cr^{2+} is just stable toward oxidation in pure water, in air-saturated water it is rapidly oxidized; Fe^{2+} is oxidized (only slowly in acid, but rapidly in base) to Fe^{3+} in presence of air, although in air-free water Fe^{2+} is quite stable:

$$Fe^{3+} + e = Fe^{2+} \qquad E^\circ = +0.77 \text{ V}$$

Many oxidations by oxygen in acid solution are slow, but the rates of oxidation may be vastly increased by catalytic amounts of transition-metal ions, especially Cu^{2+}, where a Cu^I–Cu^{II} redox cycle is involved.

O_2 is readily soluble in organic solvents, and merely pouring these liquids in air serves to saturate them with O_2. This should be kept in mind when determining the reactivity of air-sensitive materials in solution in organic solvents.

Measurements of electronic spectra of alcohols, ethers, benzene, and even saturated hydrocarbons show that there is reaction of the charge-transfer type with the oxygen molecule. However, there is no true complex formation, since the heats of formation are negligible and the spectral changes are due to contact between the molecules at van der Waals distances. The classic example is that of N,N-dimethylaniline which becomes yellow in air or oxygen but colorless again when the oxygen is removed. Such weak charge-transfer complexes make certain electronic transitions in molecules more intense; they are also a plausible first stage in photo-oxidations.

With certain transition-metal complexes, O_2 adducts may be formed, sometimes reversibly (page 307). Although the O_2 entity remains intact, the complexes may be described as having coordinated O_2^- or O_2^{2-} ions, bound to the metal in a three-membered ring or as a bridging group. Coordinated O_2 is more reactive than free O_2, and substances not directly oxidized under mild conditions can be attacked in presence of metal complexes.

Singlet O_2 and Photochemical Oxidations. The lowest-energy electron configuration of the O_2 molecule that contains two electrons in π^* orbitals, gives rise to three states, as is shown below. Oxygen molecules in excited singlet states, especially the ${}^1\Delta_g$ state, which has a much longer lifetime than the ${}^1\Sigma_g^+$ state, react with a variety of unsaturated organic substrates to cause limited, specific

State	π_a^*	π_b^*	Energy
${}^1\Sigma_g^+$	↑	↓	155 kJ (~13,000 cm^{-1})
${}^1\Delta_g$	↑↓	—	92 kJ (~8,000 cm^{-1})
${}^3\Sigma_g^-$	↑	↑	0 (ground state)

oxidations, a very typical reaction being a Diels–Alder-like 1,4-addition to a 1,3-diene:

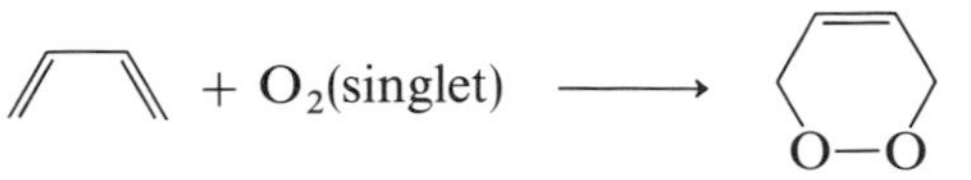

$$+ \; O_2(\text{singlet}) \longrightarrow \quad \text{O—O}$$

Singlet oxygen molecules are generated photochemically by irradiation in presence of a sensitizer or "sens" (typically a fluorescein derivative, methylene blue, certain porphyrins, or certain polycyclic aromatic hydrocarbons) probably as follows:

$$^1\text{Sens} \xrightarrow{h\nu} {}^1\text{Sens}^*$$

$$^1\text{Sens}^* \longrightarrow {}^3\text{Sens}^*$$

$$^3\text{Sens}^* + {}^3O_2 \longrightarrow {}^1\text{Sens} + {}^1O_2$$

$$^1O_2 + \text{Substrate} \longrightarrow \text{Products}$$

Energy transfer from triplet excited sensitizer, $^3\text{Sens}^*$, to $^3O_2$ to give $^1O_2$ is a spin-allowed process. Singlet oxygen (mainly $^1\Delta_g$) is also generated chemically in the reaction

$$H_2O_2 + ClO^- \longrightarrow Cl^- + H_2O + O_2(^1\Delta_g)$$

and by carrying out this reaction in alcohol in presence of substrates, useful amounts of products may be produced. Singlet oxygen may be involved in many biological and other oxidations using O_2, especially in presence of light.

OXYGEN COMPOUNDS

Most oxygen compounds are described under the chemistry of other elements. We mention a few important compounds and classes of compounds here.

18-3
Hydrogen Peroxide

Pure H_2O_2 is a colorless liquid (bp 152.1°, fp −0.41°). It resembles water in many of its physical properties and is even more highly associated via hydrogen bonding and 40% denser than is H_2O. It has a high dielectric constant, but its utility as an ionizing solvent is limited by its strong oxidizing nature and its ready decomposition in the presence of even traces of many heavy-metal ions according to the equation:

$$2H_2O_2 = 2H_2O + O_2 \qquad \Delta H = -99 \text{ kJ mol}^{-1}$$

In dilute aqueous solution it is more acidic than water:

$$H_2O_2 = H^+ + HO_2^- \qquad K_{20^\circ} = 1.5 \times 10^{-12}$$

The molecule H_2O_2 has a skew, chain structure (*Fig. 18-1*).

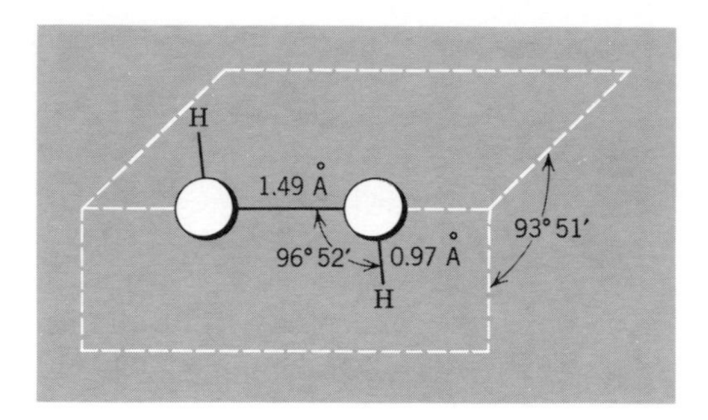

Figure 18-1 *The structure of hydrogen peroxide.*

There are two methods for large-scale production of hydrogen peroxide. One is by autoxidation of an anthraquinol, such as 2-ethylanthraquinol:

$$\text{(OH)}_2\text{C}_{14}\text{H}_7\text{—C}_2\text{H}_5 \underset{\text{H}_2/\text{Pd}}{\overset{\text{O}_2}{\rightleftharpoons}} \text{(O=)}_2\text{C}_{14}\text{H}_7\text{—C}_2\text{H}_5 + \text{H}_2\text{O}_2$$

The resulting quinone is reduced with H_2 gas. The H_2O_2 is obtained as a 20% aqueous solution. Only O_2 and H_2 and H_2O are required as raw materials.

An older and more expensive method is electrolytic oxidation of sulfuric acid or ammonium sulfate–sulfuric acid solutions to give peroxodisulfuric acid, which is then hydrolyzed to yield H_2O_2:

$$2HSO_4^- \longrightarrow HO_3S{-}O{-}O{-}SO_3H + 2e^-$$

$$H_2S_2O_8 + H_2O \longrightarrow H_2SO_5 + H_2SO_4 \text{ (Rapid)}$$

$$H_2SO_5 + H_2O \longrightarrow H_2O_2 + H_2SO_4 \text{ (Slow)}$$

Fractional distillation can then give 90 to 98% H_2O_2.

The redox chemistry of H_2O_2 in aqueous solution is summarized by the potentials:

$$H_2O_2 + 2H^+ + 2e = 2H_2O \qquad E^\circ = 1.77\text{ V}$$

$$O_2 + 2H^+ + 2e = H_2O_2 \qquad E^\circ = 0.68\text{ V}$$

$$HO_2^- + H_2O + 2e = 3OH^- \qquad E^\circ = 0.87\text{ V}$$

These show that hydrogen peroxide is a strong oxidizing agent in either acid or basic solution. It behaves as a reducing agent only toward very strong oxidizing agents such as MnO_4^-.

Dilute or 30% hydrogen peroxide solutions are widely used as oxidants. In acid solution oxidations with hydrogen peroxide are slow, whereas in basic solution they are usually fast. Decomposition to H_2O and O_2, which may be considered a self-oxidation, occurs most rapidly in basic solution; hence, an excess of H_2O_2 may best be destroyed by heating in basic solution.

Many reactions involving H_2O_2 (and also O_2) in solutions involve free-radicals. Metal-ion-catalyzed decomposition of H_2O_2 and other reactions form radicals of which HO_2 and OH are most important. HO_2 has been detected in aqueous solutions where H_2O_2 interacts with Ti^{3+}, Fe^{2+}, or Ce^{4+} ions.

18-4 Peroxides and Superoxides

These are substances derived formally from O_2^{2-} and O_2^-, respectively. *Ionic peroxides* are formed by alkali metals, Ca, Sr, and Ba. Sodium peroxide is made commercially by air oxidation of Na, first to Na_2O, then to Na_2O_2; it is a yellowish powder, very hygroscopic though thermally stable to 500°, which contains also, according to esr studies, about 10% of the superoxide.

The ionic peroxides give H_2O_2 with water or dilute acids. All are powerful oxidizing agents. They convert organic materials to carbonate even at moderate temperatures. Na_2O_2 also vigorously oxidizes some metals. For example, Fe violently gives FeO_4^{2-}. Na_2O_2 is used for oxidizing fusions. The alkali peroxides also react with CO_2:

$$2CO_2(g) + 2M_2O_2 \longrightarrow 2M_2CO_3 + O_2$$

Peroxides can also act as reducing agents for such strongly oxidizing substances as permanganate.

Other electropositive metals such as Mg, the lanthanides, or uranyl ion also give peroxides; these are intermediate in character between the ionic ones and the essentially covalent peroxides of metals such as Zn, Cd, and Hg.

Many ionic peroxides form well-crystallized hydrates such as $Na_2O_2 \cdot 8H_2O$ and $M^{II}O_2 \cdot 8H_2O$. These contain discrete O_2^{2-} ions to which the water molecules are hydrogen-bonded, giving chains of the type

$$\text{---}O_2^{2-}\text{---}(H_2O)_8\text{---}O_2^{2-}\text{---}(H_2O)_8\text{---}.$$

Ionic Superoxides. Ionic superoxides, MO_2, are formed by the interaction of O_2 with K, Rb, or Cs as yellow to orange crystalline solids. NaO_2 can be obtained only by reaction of Na_2O_2 with O_2 at 300 atm and 500°. LiO_2 cannot be isolated. Alkaline-earth, Mg, Zn, and Cd superoxides occur only in small concentrations as solid solutions in the peroxides. The O_2^- ion has one unpaired electron. Superoxides are very powerful oxidizing agents. They react vigorously with water:

$$2O_2^- + H_2O = O_2 + HO_2^- + OH^-$$

$$2HO_2^- = 2OH^- + O_2 \text{ (Slow)}$$

The reaction with CO_2, which involves peroxocarbonate intermediates, is used for removal of CO_2 and regeneration of O_2 in closed systems (e.g. submarines). The over-all-reaction is

$$4MO_2(s) + 2CO_2(g) = 2M_2CO_3(s) + 3O_2(g)$$

18-5
Other Peroxo Compounds

There are many *organic peroxides* and *hydroperoxides*. Peroxo carboxylic acids, for example, peracetic acid, $CH_3C(O)OOH$, can be obtained by action of H_2O_2 on acid anhydrides. The peroxo acids are useful oxidants and sources of free radicals, for example, by treatment with Fe^{2+} (aq). Benzoyl peroxide and cumyl hydroperoxide are moderately stable and widely used where free-radical initiation is required, as in polymerization reactions.

Organic peroxo compounds are also obtained by *autoxidation* of ethers, alkenes, and the like, on exposure to air. Autoxidation is a free-radical chain reaction initiated by radicals generated by interaction of oxygen and traces of metals such as Cu, Co, or Fe. The attack on specific reactive C—H bonds by a radical, $X^{\cdot}$, gives first $R^{\cdot}$ and then hydroperoxides which can react further:

$$RH + X^{\cdot} \longrightarrow R^{\cdot} + HX$$
$$R^{\cdot} + O_2 \longrightarrow RO_2^{\cdot}$$
$$RO_2^{\cdot} + RH \longrightarrow ROOH + R^{\cdot}$$

Explosions can occur on distillation of oxidized solvents, and they should be washed with acidified $FeSO_4$ solution or, for ethers and hydrocarbons, passed through a column of activated alumina. Peroxides are absent when $Fe^{2+} + SCN^-$ reagent gives no red color, indicative of the $Fe(SCN)^{2+}$ ion.

There are also many inorganic peroxo compounds where —O— is replaced by —O—O— groups, such as peroxodisulfuric acid, $(HO)_2S(O)OOS(O)(OH)_2$, mentioned above. Potassium and ammonium peroxodisulfates (page 318) are commonly used as a strong oxidizing agent in acid solution, for example, to convert C into CO_2, Mn^{2+} into MnO_4^-, or Ce^{3+} into Ce^{4+}. The last two reactions are slow and normally incomplete in the absence of silver ion as a catalyst.

It is important to make the distinction between true peroxo compounds, which contain —O—O— groups, and compounds that contain hydrogen peroxide of crystallization such as $2Na_2CO_3 \cdot 3H_2O_2$ or $Na_4P_2O_7 \cdot nH_2O_2$.

18-6
The Dioxygenyl Cation

The interaction of PtF_6 with O_2 gives an orange solid O_2PtF_6, isomorphous with $KPtF_6$, which contains the paramagnetic O_2^+ ion. This reaction was of importance in that it lead Bartlett to treat PtF_6 with xenon (page 338). A number of other salts of the O_2^+ ion are known.

It is instructive to compare the various $O_2^{n\pm}$ species, since they provide an interesting illustration of the effect of varying the number of antibonding electrons on the length and stretching frequency of a bond, as the data in Table 18-1 show.

Table 18-1 *Bond Values for Oxygen Species*

Species	O—O dist. (Å)	Number of π^* electrons	ν_{O-O}(cm^{-1})
O_2^+	1.12	1	1860
O_2	1.21	2	1556
O_2^-	1.33	3	1145
O_2^{2-}	1.49	4	~770

18-7
Complexes of Dioxygen

Although the most common mode of reaction of molecular oxygen with transition-metal complexes is oxidation, that is, extraction of electrons from the metal (or, on occasion, from the ligand system), in appropriate circumstances the oxygen molecule or *dioxygen* can become a ligand. The reaction of dioxygen with a complex so as to incorporate the dioxygen intact is called *oxygenation*, as contrasted to oxidation, in which O_2 loses its identity.

Oxygenation reactions are generally although not invariably reversible. That is, on increasing temperature and/or reducing the partial pressure of O_2, the dioxygen ligand is lost by dissociation or transferred to another acceptor (which may become oxidized). The process of reversible oxygenation plays an essential role in life processes. The best-known examples involve the hemoglobin and myoglobin molecules of higher animals; they are discussed in Chapter 31. There are several synthetic oxygen-carrying cobalt complexes (Section 24-33), though it is not always firmly established how the oxygen is bound to the cobalt.

Our principal concern here is with a class of compounds that first came to light in 1963, with Vaska's discovery of the reaction

$$\text{Ir(Ph}_3\text{P)(Cl)(OC)(PPh}_3) + O_2 \rightleftharpoons \text{Ir(Ph}_3\text{P)(O}_2\text{)(OC)(PPh}_3\text{)(Cl)}$$

This reaction is reversible. Diamagnetic dioxygen complexes of Fe, Ru, Rh, Ir, Ni, Pd, and Pt are now known. In all those complexes that have been studied by X-ray diffraction the metal atom and the dioxygen ligand form an isosceles triangle. However, the O—O distances vary greatly, from 1.31 to 1.63 Å, as is

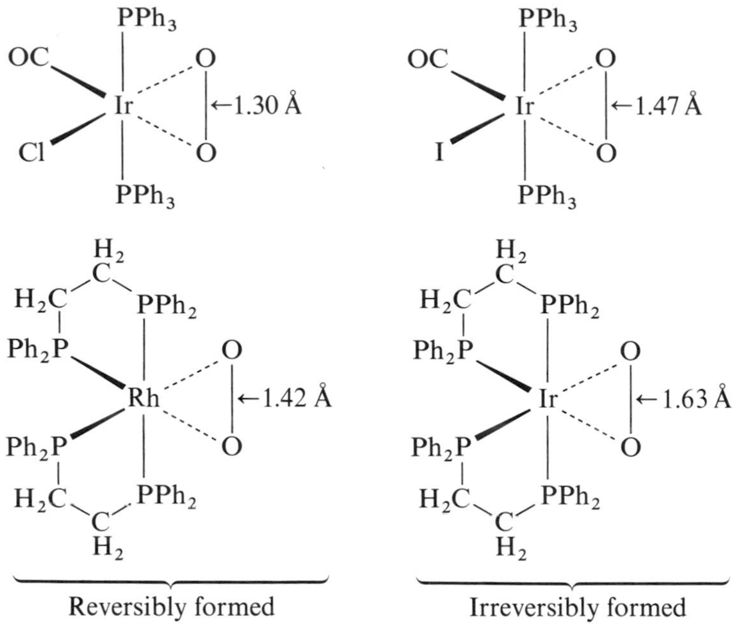

Reversibly formed

Irreversibly formed

Figure 18-2 *Four representative dioxygen complexes, illustrating the correlation between* O—O *distances and the reversibility of their formation reactions.*

illustrated by structures shown in *Fig. 18-2*. This variation seems to depend on the electron density at the metal atom which, in turn, depends markedly on the other ligands present. In addition, there is a close correlation between the O—O bond length and the degree of reversibility of the reaction: the compounds with the longest O—O bonds are formed irreversibly.

The nature of the metal-to-dioxygen bonding is not well understood. Both σ and π orbitals of the oxygen atoms play some role. In the most irreversibly formed complexes, that is, those with the longest O—O bonds, the electronic structure can be described fairly accurately by a set of three single bonds, two M—O and one O—O. However, this may be too simple a picture, since electron-spectroscopic results imply the transfer of *ca.* 1.4 electrons to O_2 in $(Ph_3P)_2PtO_2$, which is formed irreversibly. At least in a *formal* sense, the formation of an O_2 compound can be viewed as an oxidative-addition reaction (Chapter 30). We discuss later the catalytic use of O_2 complexes for oxidations.

Study Questions

A

1. Give the electronic configuration of the oxygen atom. How can oxygen complete its octet?
2. Give two examples of oxonium ions. What is their structure?
3. Describe the C—O bond in acetone.
4. Describe the interaction of acidic, basic, and neutral oxides with water. Give two examples in each case.

5. What is a nonstoichiometric oxide? Give an example.
6. Explain why the oxygen molecule is paramagnetic.
7. Why is ozone of importance? How is it prepared in the laboratory?
8. How is hydrogen peroxide made? What are its main properties?
9. Write balanced equations for the reactions:
   (a) H_2O_2 and $KMnO_4$ in acid solution,
   (b) $Fe(OH)_2$ and O_2 in basic solution,
   (c) sodium peroxide and CO_2,
   (d) the electrolytic oxidation of H_2SO_4 solution,
   (e) potassium superoxide and water,
   (f) $K_2S_2O_8$ and Mn^{2+} in HNO_3 solution.
10. What is autoxidation? How would you purify diethylether before distilling it?
11. What is the difference between oxygenation and oxidation?
12. In dioxygen complexes, what is the relationship between the O—O distance and the reversibility of the complex formation reaction?

B

1. What is the likely path for reduction of O_2 in solvents?
2. What is meant by the terms singlet and triplet oxygen? How can singlet oxygen be generated?
3. Write electronic structures for all ions $O_2^{n\pm}$ noting which are paramagnetic.
4. Why is the aquo ion $Fe(H_2O)_6^{3+}$ more acidic than $Fe(H_2O)_6^{2+}$?
5. If you had water enriched in ^{17}O and wanted to make the following oxygen compounds incorporating the label, how would you proceed: CH_3COOH; NO_3^-; CO_3^-; and SO_2.
6. How could you remove traces of oxygen from nitrogen or argon (a) if you wanted dry gas, (b) if wet gas, or an additional drying stage, would suffice?

Chapter 18
Study Guide

Supplementary Reading

Advances in Chemistry Series, No. 21, *Ozone Chemistry and Technology*, American Chemical Society, 1959.
Ardon, M., *Oxygen, Elementary Forms and Hydrogen Peroxide*, Benjamin, 1965.
Patai, S., ed., *The Chemistry of the Hydroxyl Group*, Wiley-Interscience, 1971.
Severn, D., *Organic Peroxides*, 3 Vols., Wiley, 1972.

19

the group VIB elements: sulfur, selenium, tellurium and polonium

19-1 Introduction

The position of these elements in the periodic table has been discussed in Chapter 8, page 187, and some properties are listed in Table 8-6, page 202. There is very little resemblance to the chemistry of oxygen, the main reasons being the following:

1. Sulfur, Se, Te, and Po have lower electronegativities than oxygen, which means that their compounds have less ionic character. The relative stabilities of bonds to other elements are also different, and in particular the importance of hydrogen bonding is drastically lowered. Only very weak S---H—S bonds exist and, for example, H_2S is totally different from H_2O (page 212).

2. For sulfur particularly, as in other second-row elements, there is multiple $d\pi$—$p\pi$ bonding but no $p\pi$—$p\pi$ bonding. The short S—O distances in the sulfate ion, where s and p orbitals are used in σ bonding is a result of multiple $d\pi$—$p\pi$ bond character through flow of electrons from filled $p\pi$ orbitals on O to empty $d\pi$ orbitals on S.

3. The valence is not confined to 2, and d orbitals can be utilized to form more than four bonds to other elements. Examples are SF_6 and $Te(OH)_6$.

4. Sulfur has a strong tendency to catenation and forms compounds with no O, Se, or Te analogs. Examples are polysufide ions, S_n^{2-}, polythionate anions, $O_3SS_nSO_3^{2-}$ and compounds XS_nX, where X = H, Cl, CN, or NR_2. The changes in properties of compounds on going from S to Po can be associated with increasing size of the atoms and decreasing electronegativity. Some examples are:
(a) The decreasing thermal stability of H_2X.
(b) The increasing tendency to form complex ions like $SeBr_6^{2-}$.
(c) The appearance of even some metal-like properties in Te and Po. Thus the oxides MO_2 are ionic and react with HCl to give the chlorides.

19-2
Occurrence and Reactions of the Elements

Sulfur occurs widely in nature as the element, as H_2S and SO_2, in metal sulfide ores, and as sulfates such as gypsum and anhydrite ($CaSO_4$), magnesium sulfate, etc. Sulfur is obtained on a vast scale from natural hydrocarbon gases such as those in Alberta, Canada, which contain up to 30% H_2S; this is removed by interaction with SO_2, obtained from burning sulfur in air,

$$S + O_2 = SO_2$$

$$2H_2S + SO_2 = 3S + 2H_2O$$

Selenium and *tellurium* are less abundant but frequently occur as selenide and telluride minerals in sulfide ores, particularly those of Ag and Au. They are recovered from flue dusts from combustion chambers for sulfide ores.

Polonium occurs in U and Th minerals as a product of radioactive decay series. The most accessible isotope, ^{210}Po. (α, 138.4d) can be made in gram quantities by irradiation of Bi in nuclear reactors:

$$^{209}Bi(n, \gamma)^{210}Bi \xrightarrow{\beta^-} {}^{210}Po$$

The Po can be separated by sublimation on heating. It is intensely radioactive and special handling techniques are required. The chemistry resembles that of Te but is somewhat more "metallic."

The physical properties and structures of the elements have been described (pages 183, 187). On melting, S_8 first gives a yellow, transparent, mobile liquid that becomes dark and increasingly viscous above *ca.* 160°. The maximum viscosity occurs *ca.* 200° but on further heating the mobility increases until the boiling point, 444.6°, where the liquid is dark red. The "melting point" of S_8 is actually a decomposition point. Just after melting, rings with an average of 13.8 sulfur atoms are formed and at higher temperature, still larger rings. Then in the high viscosity region there are giant macro molecules that are probably chains with radical ends. At higher temperatures, highly colored S_3 and S_4 molecules are present to the extent of 1 to 3% at the boiling point. The nature of the physical changes and of the species involved are by no means fully understood.

Sulfur vapor contains S_8 and at higher temperatures S_2 molecules. The latter, like O_2, are paramagnetic with two unpaired electrons, and account for the blue color of the hot vapor.

Cyclosulfurs, S_n, $n = 6 - 12$, soluble in CS_2, benzene, and cyclohexane are light sensitive and thermally unstable at 25°.

Sulfur, Se, and Te burn in air on heating to form the dioxides; they also react on heating with halogens, most metals, and nonmetals. They are attacked by hot oxidizing acids like H_2SO_4 or HNO_3.

In oleums (page 174), S, Se, and Te dissolve to give highly colored solutions that contain cations in which the element is in a fractional oxidation state. Salts of these cations that have stoichiometries $M_4{}^{2+}$, $M_8{}^{2+}$, and $M_{16}{}^{2+}$ have been obtained by selective oxidation of the elements with SbF_5 or AsF_5 in liquid HF. For example,

$$S_8 + 3SbF_5 = S_8{}^{2+} + 2SbF_6{}^- + SbF_3$$

or by reactions in molten $AlCl_3$, for example,

$$7Te + TeCl_4 + 4AlCl_3 = 2Te_4^{2+} + 4AlCl_4^-$$

The S_4^{2+}, Se_4^{2+}, and Te_4^{2+} ions are square (19-I) and there is probably a six π-electron quasi-aromatic system. The green Se_8^{2+} ion has a ring structure (19-II). The S_{16}^{2+} and Se_{16}^{2+} ions have two M_8 rings joined together.

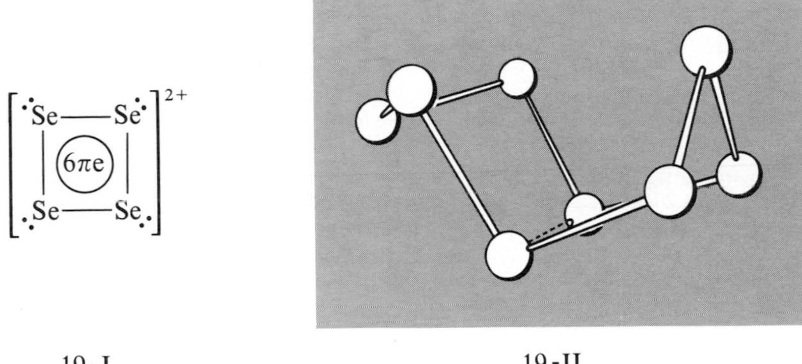

19-I 19-II

The reaction of sulfur with the double bonds of natural and synthetic rubbers, a process called vulcanization, is of great technical importance: it leads to formation of S bridges between carbon chains and, hence, to strengthening of rubber.

All reactions of S_8 must involve initial ring opening to give sulfur chains or chain compounds. Many reactions involve nucleophilic reactants, for example,

$$S_8 + 8CN^- \longrightarrow 8SCN^-$$

$$S_8 + 8Na_2SO_3 \longrightarrow 8Na_2S_2O_3$$

$$S_8 + 8Ph_3P \longrightarrow 8Ph_3PS$$

Such reactions proceed by a series of steps such as:

$$S_8 + CN^- \longrightarrow SSSSSSSSCN^-$$

$$S_6{-}S{-}SCN^- + CN^- \longrightarrow S_6SCN^- + SCN^-, \text{etc.}$$

Sulfur–sulfur bonds occur in a variety of compounds, and —S—S— bridges are especially important in certain enzymes and other proteins.

COMPOUNDS OF GROUP VI ELEMENTS

19-3 Hydrides, MH_2

These are obtained by action of acids on metal sulfides, selenides, or tellurides. They are extremely poisonous gases with revolting odors. The toxicity of H_2S far exceeds that of HCN. The thermal stability and bond strengths decrease down the series, whereas the acidity in water increases.

Hydrogen sulfide dissolves in water to give a solution *ca.* 0.1 *M* at 1 atm. Its dissociation constants are

$$H_2S + H_2O = H_3O^+ + HS^- \qquad K = 1 \times 10^{-7}$$

$$HS^- + H_2O = H_3O^+ + S^{2-} \qquad K = \sim 10^{-14}$$

Owing to this small second dissociation constant, essentially only SH^- ions are present in solutions of ionic sulfides, and S^{2-} occurs only in very alkaline solutions (>8 *M* NaOH) as

$$S^{2-} + H_2O = SH^- + OH^- \qquad K = \sim 1$$

The compounds H_2S_2 to H_2S_6 are generally known as sulfanes; they contain —S—S— to —SSSSSS— chains. They can be obtained by reactions such as

$$2H_2S(l) + S_nCl_2 = 2HCl(g) + H_2S_{n+2}(l)$$

19-4
Halides and Oxohalides of Sulfur

Sulfur fluorides. Direct fluorination of S_8 yields mainly SF_6 and traces of S_2F_{10} and SF_4.

The *tetrafluoride*, SF_4 (bp −30°) is evolved as a gas when SCl_2 is refluxed with NaF in acetonitrile at 78 to 80°.

$$3SCl_2 + 4NaF = SF_4 + S_2Cl_2 + 4NaCl$$

SF_4 is extremely reactive, and instantly hydrolyzed by water to SO_2 and HF. It is a very selective fluorinating agent converting C=O and P=O groups smoothly into CF_2 and PF_2, and COOH and P(O)OH groups into CF_3 and PF_3 groups.

Sulfur hexafluoride is very resistant to chemical attack. Because of its inertness, high dielectric strength, and molecular weight, it is used as a gaseous insulator in high-voltage generators and other electrical equipment. The low reactivity is presumably due to a combination of factors including high S—F bond strength and the fact that sulfur is both coordinately saturated and sterically hindered. It is due to kinetic factors and not to thermodynamic stability, since the reaction of SF_6 with H_2O to give SO_3 and HF would be decidedly favorable ($\Delta F = -460$ kJ mol^{-1}).

Sulfur Chlorides. The chlorination of molten sulfur gives S_2Cl_2, an orange liquid of revolting smell. By using an excess of Cl_2 and traces of $FeCl_3$ or I_2, as catalyst, at room temperature, an equilibrium mixture containing *ca.* 85% of SCl_2 is obtained. The dichloride dissociates within a few hours:

$$2SCl_2 \rightleftharpoons S_2Cl_2 + Cl_2$$

but it can be obtained pure as a dark-red liquid by fractional distillation in presence of some PCl_5, which stabilizes SCl_2.

Sulfur chlorides are solvents for sulfur, giving dichlorosulfanes up to about $S_{100}Cl_2$, which are used in the vulcanization of rubber. They are also used as mild chlorinating agents.

Thionyl chloride, $SOCl_2$, is obtained by the reaction

$$SO_2 + PCl_5 = SOCl_2 + POCl_3$$

It is a colorless fuming liquid (bp 80°) readily hydrolyzed by water

$$SOCl_2 + H_2O = SO_2 + 2HCl$$

Because of this reaction in which the volatile products are easily removed, $SOCl_2$ can often be used to prepare anhydrous chlorides. Examples are:

$$Fe(OH)_3 + 3SOCl_2 \longrightarrow 3SO_2 + 3HCl + FeCl_3$$

$$FeCl_3 \cdot 6H_2O + 6SOCl_2 \longrightarrow 6SO_2 + 12HCl + FeCl_3$$

Thionyl chloride has a pyramidal structure with sulfur using a set of roughly sp^3 hybrid orbitals, one of which holds the lone pair. $SOCl_2$ can hence act as a weak Lewis base.

Sulfuryl chloride, SO_2Cl_2, is obtained by the reaction

$$SO_2 + Cl_2 = SO_2Cl_2$$

in presence of a catalyst such as $FeCl_3$. It is a colorless liquid fuming in moist air and is used for chlorinating organic compounds. It is readily hydrolyzed by water.

19-5 Oxides and Oxo Acids

The *dioxides* are obtained by burning the elements in air. Sulfur dioxide is produced when many sulfides are heated in air. Selenium and tellurium dioxides are also obtained by treating the metals with hot nitric acid to form H_2SeO_3 and $2TeO_2 \cdot HNO_3$, respectively, and then heating these to drive off water or nitric acid.

Sulfur dioxide is a gas with a pungent smell. The molecule is angular. Liquid SO_2 dissolves many organic and inorganic substances and is used as a solvent for nmr studies as well as in preparative reactions. The liquid does not undergo self-ionization and its conductivity is mainly a reflection of the purity.

Sulfur dioxide has lone pairs and can act as a Lewis base. However, it also acts as a Lewis acid giving complexes, for example, with amines as in Me_3NSO_2 and with electron-rich transition metal complexes. In the crystalline compound $SbF_5 \cdot SO_2$, which is of interest because of the use of SO_2 as a solvent for superacid systems (page 177), the SO_2 is bound as in (19-III). The bonding in

(19-IV), differs in that the S atom is bound to the metal. Metal-sulfur bonding appears to be general in transition-metal species. Sulfur dioxide also undergoes "insertion" reactions (Chapter 30) with metal-carbon bonds, for example,

$$RCH_2HgOAc + SO_2 \longrightarrow RCH_2SO_2HgOAc$$
$$(CH_3)_4Sn + SO_2 \longrightarrow (CH_3)_3SnSO_2CH_3$$

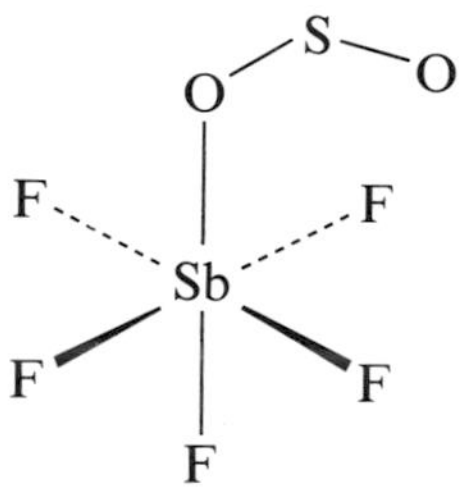

19-III

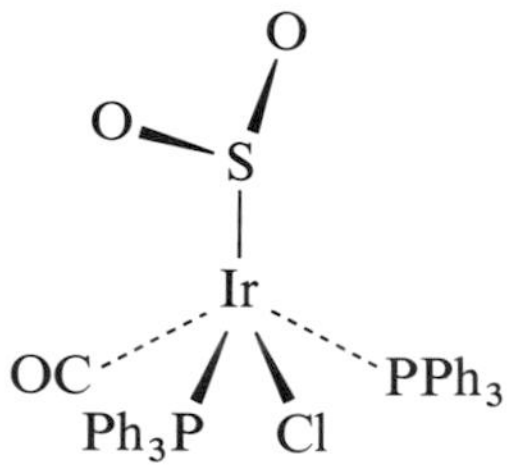

19-IV

SO_2 is quite soluble in water; such solutions, which possess acidic properties, have long been referred to as solutions of *sulfurous acid*, H_2SO_3. However, H_2SO_3 is either not present or present only in infinitesimal quantities in such solutions. The so-called hydrate, $H_2SO_3 \cdot \sim 6H_2O$, is the gas hydrate (Section 9-5), $SO_2 \cdot \sim 7H_2O$. The equilibria in aqueous solutions of SO_2 are best represented as

$$SO_2 + xH_2O = SO_2 \cdot xH_2O \quad \text{(hydrated } SO_2\text{)}$$
$$[SO_2 \cdot xH_2O = H_2SO_3 \qquad K \ll 1]$$
$$SO_2 \cdot xH_2O = HSO_3^-(aq) + H_3O^+ + (x-2)H_2O$$

and the first acid dissociation constant for "sulfurous acid" is properly defined as follows:

$$K_1 = \frac{[HSO_3^-][H^+]}{[\text{Total dissolved } SO_2] - [HSO_3^-] - [SO_3^{2-}]} = 1.3 \times 10^{-2}$$

Although sulfurous acid does not exist, two series of salts, the *bisulfites*, containing HSO_3^-, and the *sulfites*, containing SO_3^{2-}, are well known. The SO_3^{2-} ion in crystals is pyramidal. Only the water-soluble alkali sulfites and bisulfites are commonly encountered.

Heating solid bisulfites or passing SO_2 into their aqueous solutions affords *pyrosulfites*:

$$2MHSO_3 \xrightleftharpoons{\text{Heat}} M_2S_2O_5 + H_2O$$

$$HSO_3^-(aq) + SO_2 = HS_2O_5^-(aq)$$

Whereas pyro acids, for example, pyrosulfuric, $H_2S_2O_7$, (page 174) usually have oxygen bridges, the pyrosulfite ion has an unsymmetrical structure, $O_2S{-}SO_3$. Some important reactions of sulfites are shown in *Fig. 19-1*.

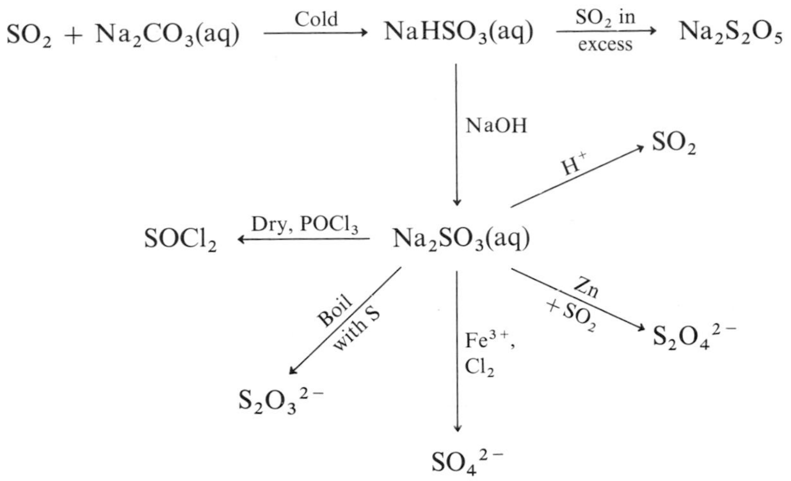

$SO_2 + Na_2CO_3(aq) \xrightarrow{\text{Cold}} NaHSO_3(aq) \xrightarrow[\text{excess}]{SO_2 \text{ in}} Na_2S_2O_5$

$NaHSO_3(aq) \xrightarrow{NaOH} Na_2SO_3(aq)$

$Na_2SO_3(aq) \xrightarrow{H^+} SO_2$

$Na_2SO_3(aq) \xrightarrow{\text{Dry, } POCl_3} SOCl_2$

$Na_2SO_3(aq) \xrightarrow{\text{Boil with S}} S_2O_3^{2-}$

$Na_2SO_3(aq) \xrightarrow{Fe^{3+}, Cl_2} SO_4^{2-}$

$Na_2SO_3(aq) \xrightarrow{Zn + SO_2} S_2O_4^{2-}$

Figure 19-1 *Some reactions of sulfites.*

Solutions of SO_2 and of sulfites possess reducing properties and are often used as reducing agents:

$$SO_4^{2-} + 4H^+ + (x - 2)H_2O + 2e = SO_2 \cdot xH_2O \qquad E^\circ = 0.17 \text{ V}$$

$$SO_4^{2-} + H_2O + 2e = SO_3^{2-} + 2OH^- \qquad E^\circ = -0.93 \text{ V}$$

Sulfur trioxide is obtained by reaction of SO_2 with O_2, a reaction that is thermodynamically very favorable but extremely slow in the absence of a catalyst such as platinum sponge, V_2O_5, or NO. SO_3 reacts vigorously with water to form sulfuric acid. Industrially, SO_3 is absorbed in concentrated H_2SO_4 to give oleum (page 174), which is then diluted. SO_3 is used as such for preparing sulfonated oils and alkyl arenesulfonate detergents. It is also a powerful but generally indiscriminate oxidizing agent.

The SO_3 molecule, in the gas phase, has a planar, triangular structure involving both $p\pi$—$p\pi$ and $p\pi$—$d\pi$, S—O bonding. It forms polymers in the solid state.

Sulfuric Acid. This has been discussed on page 174.

Selenic and Telluric Acids. Selenic acid is similar to H_2SO_4, including the isomorphism of the hydrates and salts. It differs in being less stable, evolving oxygen about 200°, and being a strong but usually not kinetically fast oxidizing agent,

$$SeO_4^{2-} + 4H^+ + 2e = H_2SeO_3 + H_2O \qquad E^\circ = 1.15 \text{ V}$$

Telluric acid, obtained by oxidation of Te or TeO_2 with H_2O_2 or other powerful oxidants, is very different in structure, being $Te(OH)_6$ in the crystal. It is a very weak dibasic acid, $K_1 \approx 10^{-7}$, and is also an oxidant. Most tellurates contain TeO_6 octahedra as in $K[TeO(OH)_5]$ or Hg_3TeO_6.

Thiosulfates. Thiosulfates are readily obtained by boiling solutions of sulfites with sulfur. The free acid is unstable at ordinary temperatures. The alkali thiosulfates are manufactured for use in photography where they are used to dissolve unreacted silver bromide from emulsion by formation of the complexes $[Ag(S_2O_3)]^-$ and $[Ag(S_2O_3)_2]^{3-}$; the thiosulfate ion also forms complexes with other metal ions.

The thiosulfate ion has the structure $S{-}SO_3^{2-}$.

Dithionates. The reduction of sulfites in aqueous solutions containing an excess of SO_2, by zinc dust, gives ZnS_2O_4. The Zn^{2+} and Na^+ salts are commonly used as powerful and rapid reducing agents in alkaline solution:

$$2SO_3^{2-} + 2H_2O + 2e = 4OH^- + S_2O_4^{2-} \qquad E^\circ = -1.12\ V$$

In the presence of 2-anthraquinonesulfonate as catalyst aqueous $Na_2S_2O_4$ efficiently removes oxygen from inert gases. The ion has the structure $O_2S{-}SO_2^{2-}$ with a long weak S—S bond.

Polythionates. These anions have the general formula $[O_3SS_nSO_3]^{2-}$. The corresponding acids are not stable, decomposing rapidly into S, SO_2, and sometimes SO_4^{2-}. The well-established polythionate anions are those with $n = 1$–4. They are named according to the total number of sulfur atoms and are thus called: trithionate $S_3O_6^{2-}$, tetrathionate $S_4O_6^{2-}$, and so on. There is evidence for anions having chains with up to 20 sulfur atoms.

Tetrathionates are obtained by treatment of thiosulfates with iodine in the reaction used in the volumetric determination of iodine:

$$2S_2O_3^{2-} + I_2 \longrightarrow 2I^- + S_4O_6^{2-}$$

Peroxodisulfates. The NH_4^+ or Na^+ salts are obtained by electrolysis of the corresponding sulfates at low temperatures and high current densities. The $S_2O_8^{2-}$ ion has the structure $O_3S{-}O{-}O{-}SO_3$, with approximately tetrahedral angles about each S atom.

The ion is one of the most powerful and useful of oxidizing agents:

$$S_2O_8^{2-} + 2e = 2SO_4^{2-} \qquad E^\circ = 2.01\ V$$

However, the reactions are complicated mechanistically. Oxidations by $S_2O_8^{2-}$ are slow and are usually catalyzed by addition of Ag^+ which is converted to Ag^{2+}, the actual oxidant.

Study Questions

A

1. What are the two principal forms in which sulfur occurs in nature?
2. Ordinary solid sulfur consists of what species? Summarize briefly what is observed when sulfur is heated from below its melting point to above its boiling point and explain the reasons for these changes.

3. What types of species are formed on dissolving S, Se, and Te in oleums or other superacids?
4. Discuss the aqueous chemistry of H_2S, HS^-, and S^{2-}.
5. What are the principal fluorides of sulfur?
6. Write equations for the preparations and for the reactions with water of thionyl chloride and sulfuryl chloride.
7. Write equations for the two most important reactions, or types of reaction, of SO_3.
8. Of what use(s) is SO_2?
9. Mention the chief similarities and differences among sulfuric, selenic, and telluric acids.
10. Give general formulas for three series of compounds that contain chains of more than two S atoms.

B

1. How can SO_2 be removed from natural gas streams? Suggest other ways of removing SO_2, for example, from stack gases.
2. Why is oxygen O_2 and sulfur S_8?
3. How and why do the boiling points and acid strengths of H_2X vary from O to Te?
4. Although SF_6 is unreactive, TeF_6 is hydrolyzed by water. Explain.
5. Why should compounds with S—S bonds be more numerous and stable than those with O—O, Se—Se or Te—Te bonds?
6. For SF_4 and SF_6 describe their: (a) preparation from S_8, (b) structure, and (c) chemical reactions and uses.
7. Why is it that $SOCl_2$ can act both as a Lewis acid and a Lewis base? What would be the structure of $SeOCl_2py_2$?
8. Unlike SO_2, SeO_2 is a solid that has a chain structure. Draw a reasonable diagram for such a structure.
9. Describe the interrelationships of SO_3, sulfuric acid, and oleum. What is the structure of SO_3? Why is H_2SO_4 a syrupy liquid?
10. Does sulfurous acid exist? If not, why not?
11. Name and draw the structures of the following anions: $S_2O_3^{2-}$, $S_2O_4^{2-}$, $S_2O_6^{2-}$, and $S_2O_8^{2-}$.
12. What is a likely structure for red salts containing the ion $Pt^{IV}S_{15}^{2-}$ obtained by boiling H_2PtCl_6 with NH_4S_x?
13. The bond order of the S—O bond *decreases* in the series $OSF_2 > OSCl_2 > OSBr_2$. Explain.
14. What would you expect the stoichiometry of a potassium polonate to be?

Chapter 19
Study Guide

Supplementary Reading

Bagnall, K. W., *The Chemistry of Se, Te and Po*, Elsevier, 1966.
Cooper, W. C., *Tellurium*, Van Nostrand-Reinhold, 1972.
Heal, H. G., "Sulfur-Nitrogen Compounds," *Adv. Inorg. Chem. Radiochem.*, *15*, 375 (1972).
Janickis, J., "Polythionates and Selenopolythionates," *Accts. Chem. Research*, *2*, 316 (1969).
Nickless, G., ed., *Inorganic Sulfur Chemistry*, Elsevier, 1968.
Rahman, R., Safe, S., and Taylor, A., "The Stereochemistry of Polysulfides," *Quart. Rev.*, *24*, 208 (1970).
Schmidt, M., "Elemental Sulfur," *Angew. Chem. Internat. Ed.* (English), *12*, 445 (1973).

20

the halogens: fluorine, chlorine, bromine, iodine and astatine

20-1 Introduction

With the exception of He, Ne, and Ar, all of the elements in the periodic table form halides. Ionic or covalent halides are among the most important and common compounds. They are often the easiest to prepare and are widely used as source materials for the synthesis of other compounds. Where an element has more than one valence, the halides are often the best known and most accessible compounds in all of the oxidation states. There is also an extensive and varied chemistry of organic halogen compounds; the fluorine compounds, especially where F completely replaces H, have unique properties.

The position of the elements in the periodic table is outlined, page 44, and some properties are listed in Table 8-7, page 202. For the element *astatine*, named from the Greek for "unstable," the longest lived isotope has a half-life of only 8.3 hours. As far as can be ascertained by tracer studies, At behaves like I but is perhaps somewhat less electronegative. It is made by the reaction $^{209}Bi(\alpha, 2n)^{211}At$.

20-2 Occurrence, Isolation, and Properties

Fluorine occurs widely, for example, as *fluorspar*, CaF_2, *cryolite*, Na_3AlF_6, and *fluorapatite*, $3Ca_3(PO_4)_2Ca(F, Cl)_2$. It is more abundant than chlorine. Fluorine was first isolated in 1886 by Moissan. The greenish gas is obtained by electrolysis of molten fluorides. The most commonly used electrolyte is $KF \cdot 2\text{–}3\ HF$ (mp 70–100°). As the electrolysis proceeds the melting point increases but the electrolyte is readily regenerated by resaturation with HF from a storage tank. Fluorine cells are constructed of steel, Cu, or Ni—Cu alloy, which become coated with an unreactive layer of fluoride. The cathodes are steel or Cu, the anodes ungraphitized carbon. Although F_2 is often handled in metal apparatus, it can be handled in glass provided traces of HF, which attacks glass rapidly, are removed by passing the gas through anhydrous NaF or KF with which HF forms the bifluorides, MHF_2.

Fluorine is the most chemically reactive of all the elements and combines directly at ordinary or elevated temperatures with all the elements other than O_2, He, Ne, and Kr, often with extreme vigor. It also attacks many other compounds, breaking them down to fluorides; organic materials often inflame and burn in F_2.

The great reactivity of F_2 is in part attributable to the low dissociation energy (Table 1-1) of the F—F bond, and the fact that reactions of atomic fluorine are strongly exothermic. The low F—F bond energy is probably due to repulsion between nonbonding electrons. A similar effect may account for the low bond energies in H_2O_2 and N_2H_4.

Chlorine occurs as NaCl, KCl, $MgCl_2$, etc., in seawater, salt lakes, and as deposits originating from the prehistoric evaporation of salt lakes. Chlorine is obtained by electrolysis of brine using a mercury anode in which sodium dissolves:

$$Na^+ + e = Na$$

$$Cl^- = \tfrac{1}{2}Cl_2 + e$$

The sodium is then separately removed by washing the amalgam with water to give very pure NaOH. A liability of this procedure is that losses of Hg constitute a major pollution hazard and some plants have been closed. Use of other electrodes gives less pure NaOH.

Chlorine is a greenish gas. It is moderately soluble in water with which it reacts (see page 327).

Bromine occurs as bromides, in much smaller amounts along with chlorides. Bromine is obtained from brines by the reaction

$$2Br^- + Cl_2 \xrightarrow{pH \sim 3.5} 2Cl^- + Br_2$$

It is swept out in a current of air. Bromine is a dense, mobile, dark-red liquid at room temperature. It is moderately soluble in water and miscible with nonpolar solvents such as CS_2 and CCl_4.

Iodine occurs as iodide in brines and as iodate in Chile salt peter (guano). Various forms of marine life concentrate iodine. Production of I_2 involves either oxidizing I^- or reducing iodates to I^- followed by oxidation. MnO_2 in acid solutions is commonly used as the oxidant.

Iodine is a black solid with a slight metallic luster. At atmospheric pressure it sublimes without melting. It is readily soluble in nonpolar solvents such as CS_2 and CCl_4. Such solutions are purple, as in the vapor. In polar solvents, unsaturated hydrocarbons, and liquid SO_2, brown or pinkish-brown solutions are formed. These colors indicate the formation of weak complexes $I_2 \ldots S$ known as *charge-transfer complexes*. The bonding energy results from partial transfer of charge in the sense $I_2^-S^+$. The complexes of I_2 and also of Br_2, Cl_2, and ICl can sometimes be isolated as crystalline solids at low temperatures.

Iodine forms a blue complex with starch, where the iodine atoms are aligned in channels in the polysaccharide amylose.

COMPOUNDS OF THE HALOGENS

20-3 Halides

There are almost as many ways of classifying halides as there are types of halides. Binary halides may form simple molecules, or complex, infinite arrays. For ionic compounds some common types of lattices are given in Chapter 4 and some general points on halides are discussed in Chapter 5, page 119. Other types of halide compounds include oxide halides such as $VOCl_3$, hydroxy halides, organo halides, etc. The covalent and ionic radii are given in Table 8-7, page 202.

Preparation of Anhydrous Halides

1. *Direct interaction of elements with halogens.* The halogens are normally used for most elements. HF, HCl, and HBr may also be used for metals.

Direct fluorination normally gives fluorides in the higher oxidation states. Most metals and nonmetals react very vigorously with F_2; with nonmetals such as P_4, the reaction may be explosive. For rapid formation in dry reactions of *chlorides*, *bromides*, and *iodides* elevated temperatures are usually necessary. For metals, the reaction with Cl_2 and Br_2 may be more rapid when tetrahydrofuran or some other ether is used as reaction medium; the halide is then obtained as a solvate.

2. *Dehydration of hydrated halides.* The dissolution of metals, oxides, or carbonates in aqueous halogen acids followed by evaporation or crystallization gives hydrated halides. These can sometimes be dehydrated by heating in vacuum, but this often leads to impure products or oxohalides. Dehydration of chlorides can be effected by thionyl chloride, and halides in general can be treated with 2,2-dimethoxypropane:

$$CrCl_3 \cdot 6H_2O + 6SOCl_2 \xrightarrow{\text{Reflux}} CrCl_3 + 12HCl + 6SO_2$$

$$MX_n \cdot mH_2O \text{ in } CH_3C(OCH_3)_2CH_3 \longrightarrow MX_n + m(CH_3)_2CO + 2mCH_3OH$$

The acetone and/or methanol may give a solvated halide, but these are generally removed easily by gentle heating or pumping.

3. *Treatment of oxides with other halogen compounds.* Compounds such as ClF_3, BrF_3, CCl_4, $CCl_3CClCCl_2$, NH_4Cl, $SOCl_2$, and SO_2Cl_2 at elevated temperatures are used in reactions such as

$$NiO + ClF_3 \longrightarrow NiF_2$$

$$UO_3 + CCl_2{=}CCl{-}CCl{=}CCl_2 \xrightarrow{\text{Reflux}} UCl_4$$

$$Pr_2O_3 + 6NH_4Cl(s) \xrightarrow{300^\circ} 3PrCl_3 + 3H_2O + 6NH_3$$

$$Sc_2O_3 + CCl_4 \xrightarrow{600^\circ} ScCl_3$$

4. *Halogen exchange.* Many halides react with either elemental halogens, the acids or soluble halides, or an excess of another halide so that one halogen is replaced by another. Chlorides can often be converted to bromides and especially iodides by KBr or KI in acetone in which KCl is less soluble.

Halogen exchange is especially important for the synthesis of fluorides from chlorides by use of various metal fluorides such as CoF_3 or AsF_5. This type of replacement is much used for organic fluorine compounds (see page 331).

Other fluorinating agents each having special advantages under certain conditions are AgF_2, SbF_3 ($+SbCl_5$ as a catalyst), HgF_2, KHF_2, ZnF_2, AsF_3, etc.

Examples are

$$PCl_3 + ZnF_2 \longrightarrow PF_3$$

$$PhCCl_3 + SbF_3 \longrightarrow PhCF_3 + SbCl_3$$

Molecular Halides. Most of the electronegative elements, and the metals in high oxidation states, form molecular halides. These are gases, liquids, or volatile solids with molecules held together only by van der Waals forces. There is probably a rough correlation between increasing metal-to-halogen covalence and increasing tendency to the formation of molecular compounds. Thus the molecular halides are sometimes also called the covalent halides. The designation molecular is preferable, since it states a fact.

The formation of halide bridges between two or, less often, three other atoms is an important structural feature. Between two metal atoms, the most common situation involves two halogen atoms, but examples with one and three bridge atoms are known. Such bridges used to be depicted as involving a covalent bond to one metal atom and donation of electron pair to the other as in 20-I, but structural data show that the two bonds to each bridging halogen atom are equivalent as in 20-II. Molecular-orbital theory provides a simple, flexible formulation in which the M—X—M group is treated as a 3-center, 4-electron group.

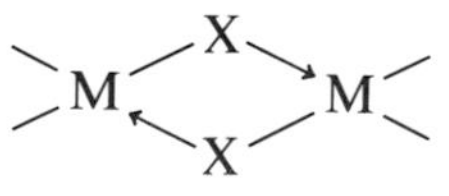

20-I

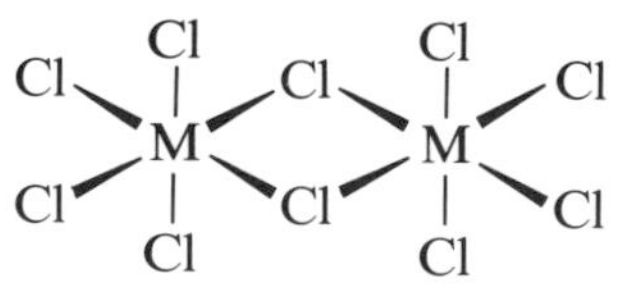

20-II

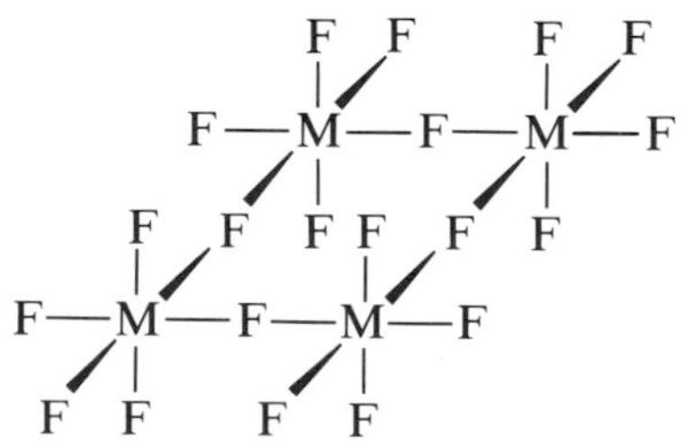

20-III

With Cl^- and Br^-, bridges are characteristically bent, whereas fluoride bridges may be either bent or linear. Thus, in BeF_2 there are infinite chains, ---BeF_2BeF_2---, with bent bridges, similar to the situation in $BeCl_2$. On the other hand, transition-metal pentahalides afford a notable contrast. While the pentachlorides dimerize with bent M—Cl—M bridges (20-II), the pentafluorides form cyclic tetramers with linear M—F—M bridges (20-III). The fluorides probably adopt the tetrameric structures with linear bridges, in part because the smaller size of F than of Cl would introduce excessive metal–metal repulsion in a bent bridge.

Molecular fluorides of both metals and nonmetals are usually gases or volatile liquids. Their volatility is due to the absence of intermolecular bonding other than van der Waals forces, since the polarizability of fluorine is very low and no suitable outer orbitals exist for other types of attraction. Where the central atom has suitable vacant orbitals available, and especially if the polarity of the single bonds M—F would be such as to leave a considerable charge on M, as in, say, SF_6, multiple bonding can occur using filled *p* orbitals of fluorine for overlap with vacant orbitals of the central atom. This multiple bonding is a major factor in the shortness and high strength of many bonds of fluorine. Because of the high electronegativity of fluorine the bonds in these compounds tend to be very polar. Because of the low dissociation energy of F_2 and the relatively high energy of many bonds to F (e.g., C—F, 486; N—F, 272; P—F, 490 kJ mol^{-1}), molecular fluorides are often formed very exothermically.

The high electronegativity of fluorine often has a profound effect on the properties of molecules in which several F atoms occur. Representative are facts such as (a) CF_3COOH is a strong acid; (b) $(CF_3)_3N$ and NF_3 have no basicity; and (c) CF_3 derivatives in general are attacked much less readily by electrophilic reagents in anionic substitutions than are CH_3 compounds. The CF_3 group may be considered as a kind of large pseudohalogen with an electronegativity about comparable to that of Cl.

Reactivity. A fairly general property of molecular halides is their easy hydrolysis, for example,

$$BCl_3 + 3H_2O \longrightarrow B(OH)_3 + 3H^+ + 3Cl^-$$

$$PBr_3 + 3H_2O \longrightarrow HPO(OH)_2 + 3H^+ + 3Br^-$$

$$SiCl_4 + 4H_2O \longrightarrow Si(OH)_4 + 4H^+ + 4Cl^-$$

Where the maximum covalency is attained, as in CCl_4 or SF_6, the halides may be quite inert toward water. However, this is a result of *kinetic* and not thermodynamic factors. Thus for CF_4 the equilibrium constant for the reaction

$$CF_4(g) + 2H_2O(l) = CO_2(g) + 4HF(g)$$

is *ca.* 10^{23}. The necessity for a means of attack is also illustrated by the fact that SF_6 is not hydrolyzed, whereas SeF_6 and TeF_6 are hydrolyzed at 25°. Expansion of the coordination sphere is possible only for Se and Te.

20-4 Halogen Oxides

Oxygen fluorides have been studied as potential rocket fuel oxidizers. F_2O is obtained as a pale yellow gas on passing F_2 gas rapidly through 2% NaOH solution. O_2F_2 is an unstable orange-yellow solid made by action of electric discharges on F_2—O_2 mixtures; O_2F_2 is an extremely potent oxidizing and fluorinating agent.

Chlorine oxides are reactive, unstable, and tend to explode. The *dioxide*, ClO_2, is a powerful oxidant and is used diluted with air commercially, for example for bleaching wood pulp. It is always made where required by the reaction

$$2NaClO_3 + SO_2 + H_2SO_4 = 2ClO_2 + 2NaHSO_4$$

or by reduction of $KClO_3$ with moist oxalic acid at 90°, which reaction also produces CO_2 as a diluent.

Iodine pentoxide is made by heating iodic acid whose anhydride it is.

$$2HIO_3 \underset{H_2O,\ \text{Fast}}{\overset{240^\circ}{\rightleftharpoons}} I_2O_5 + H_2O$$

Iodine pentoxide is an oxidizing agent, one use being in the determination of CO where the liberated iodine is determined by iodometry

$$5CO + I_2O_5 = I_2 + 5CO_2$$

I_2O_5 has a 3-dimensional network structure with O_2IOIO_2 units linked by strong intermolecular I---O interactions.

20-5 The Oxo Acids

The chemistry of the halogen oxo acids is complicated. Solutions of the acids and several of the anions may be obtained by interaction of the free halogens with water or aqueous bases. In this section the term halogen refers to Cl, Br, and I only; fluorine forms only FOH as discussed below.

Reaction of Halogens with H_2O and OH^-. The potentials and equilibrium constants necessary to understand these systems can be derived from data given in Table 20-1.

Table 20-1 *Standard Potentials (in Volts) for Reactions of the Halogens*

	Reaction	Cl	Br	I
(1)	$H^+ + HOX + e = \frac{1}{2}X_2(g, l, s) + H_2O$	1.63	1.59	1.45
(2)	$3H^+ + HXO_2 + 3e = \frac{1}{2}X_2(g, l, s) + 2H_2O$	1.64	—	—
(3)	$6H^+ + XO_3^- + 5e = \frac{1}{2}X_2(g, l, s) + 3H_2O$	1.47	1.52	1.20
(4)	$8H^+ + XO_4^- + 7e = \frac{1}{2}X_2(g, l, s) + 4H_2O$	1.42	1.59	1.34
(5)	$\frac{1}{2}X_2(g, l, s) + e = X^-$	1.36	1.07	0.54[a]
(6)	$XO^- + H_2O + 2e = X^- + 2OH^-$	0.89	0.76	0.49
(7)	$XO_2^- + 2H_2O + 4e = X^- + 4OH^-$	0.78	—	—
(8)	$XO_3^- + 3H_2O + 6e = X^- + 6OH^-$	0.63	0.61	0.26
(9)	$XO_4^- + 4H_2O + 8e = X^- + 8OH^-$	0.56	0.69	0.39

[a] Indicates that I^- can be oxidized by oxygen in aqueous solution.

The halogens are all soluble in water to some extent. However, in such solutions there are species other than solvated halogen molecules, since a *disproportionation* reaction occurs *rapidly*:

$$X_2(g, l, s) = X_2(aq) \qquad K_1$$

$$X_2(aq) = H^+ + X^- + HOX \qquad K_2$$

The values of K_1 are: Cl_2, 0.062; Br_2, 0.21; I_2, 0.0013. The values of K_2 computed from the potentials in Table 20-1 are 4.2×10^{-4} for Cl_2, 7.2×10^{-9} for Br_2, and 2.0×10^{-13} for I_2. We can also estimate from

$$\tfrac{1}{2}X_2 + e = X^-$$

and

$$O_2 + 4H^+ + 4e = 2H_2O \qquad E^\circ = 1.23 \text{ V}$$

that the potentials for the reactions

$$2H^+ + 2X^- + \tfrac{1}{2}O_2 = X_2 + H_2O$$

are −1.62 V for fluorine, −0.13 V for chlorine, 0.16 V for bromine, and 0.69 V for iodine.

Thus for saturated solutions of the halogens in water at 25° we have the results shown in Table 20-2. There is an appreciable concentration of HOCl in a saturated aqueous solution of Cl_2, a smaller concentration of HOBr in a saturated solution of Br_2, but only a negligible concentration of HOI in a saturated solution of I_2.

Table 20-2 *Equilibrium Concentrations in Aqueous Solutions of the Halogens, 25°, mol l^{-1}*

	Cl_2	Br_2	I_2
Total solubility	0.091	0.21	0.0013
Concentration X_2(aq), mol l^{-1}	0.061	0.21	0.0013
$[H^+] = [X^-] = [HOX]$	0.030	1.15×10^{-3}	6.4×10^{-6}

Hypohalous Acids. The colorless, very unstable gas, FOH, is made by passing F_2 over ice and collecting the gas in a trap. It reacts rapidly with water. The other XOH compounds are also unstable. They are known only in solution from the interaction of the halogen and mercuric oxide:

$$2X_2 + 2HgO + H_2O \longrightarrow HgO\cdot HgX_2 + 2HOX$$

The hypohalous acids are very weak acids but good oxidizing agents, especially in acid solution (see Table 20-1).

The *hypohalite ions* can be produced in principle by dissolving the halogens in base according to the general reaction

$$X_2 + 2OH^- \longrightarrow XO^- + X^- + H_2O$$

and for these *rapid* reactions the equilibrium constants are all favorable: 7.5×10^{15} for Cl_2, 2×10^8 for Br_2, and 30 for I_2.

However, the hypohalite ions tend to disproportionate in basic solution to produce the *halate ions*:

$$3XO^- = 2X^- + XO_3^-$$

For these reactions, the equilibrium constants are very favorable: 10^{27} for ClO^-, 10^{15} for BrO^-, and 10^{20} for IO^-. Thus the *actual* products obtained on dissolving the halogens in base depend on the rates at which the hypohalite ions initially produced undergo disproportionation. These rates vary with temperature.

The disproportionation of ClO^- is slow at and below room temperature. Thus, when Cl_2 reacts with base "in the cold," reasonably pure solutions of Cl^- and ClO^- are obtained. In hot solutions, $\sim 75°$, the rate of disproportionation is fairly rapid and good yields of ClO_3^- can be secured.

The disproportionation of BrO^- is moderately fast even at room temperature. Solutions of BrO^- can only be made and/or kept at around 0°. At temperatures of 50 to 80° quantitative yields of BrO_3^- are obtained:

$$3Br_2 + 6OH^- \longrightarrow 5Br^- + BrO_3^- + 3H_2O$$

The rate of disproportionation of IO^- is so fast that it is unknown in solution. Reaction of I_2 with base hence gives IO_3^- quantitatively according to an equation analogous to that for Br_2.

Halous Acids. The only certain acid is *chlorous acid*, $HClO_2$. This is obtained in aqueous solution by treating a suspension of barium chlorite with H_2SO_4, filtering off the $BaSO_4$. It is a relatively weak acid ($K_a \approx 10^{-2}$) and cannot be isolated. *Chlorites*, $MClO_2$, are obtained by reaction of ClO_2 with solutions of bases:

$$2ClO_2 + 2OH^- \longrightarrow ClO_2^- + ClO_3^- + H_2O$$

Chlorites are used as bleaching agents. In alkaline solution ClO_2^- is quite stable even on boiling. In acid solutions, the decomposition is rapid and is catalysed by Cl^-:

$$5HClO_2 \longrightarrow 4ClO_2 + Cl^- + H^+ + 2H_2O$$

Halic Acids. *Iodic acid* is HIO_3, a stable white solid obtained by oxidizing I_2 with concentrated HNO_3, H_2O_2, O_3 etc. *Chloric* and *bromic acids* are obtained in solution by treating the barium halates with H_2SO_4.

The halic acids are strong acids and are powerful oxidizing agents. The ions, XO_3^-, are pyramidal, as is to be expected from the presence of an octet, with one unshared pair, in the halogen valence shell.

Iodates of the +4 ions of Ce, Zr, Hf, and Th can be precipitated from 6 *M* nitric acid to provide a useful means of separation.

Halates. Although disproportionation of ClO_3^- is thermodynamically very favorable,

$$4ClO_3^- = Cl^- + 3ClO_4^- \qquad K \sim 10^{29}$$

the reaction occurs very slowly in solution and is not a useful preparative procedure. *Perchlorates* are prepared by electrolytic oxidation of chlorates. The properties of *perchloric acid* are discussed on page 175 and perchlorates are discussed on page 113.

The disproportionation of BrO_3^- to BrO_4^- and Br^- is extremely unfavorable ($K \sim 10^{-33}$). *Perbromates* can be obtained only by oxidation of BrO_3^-, preferably by F_2, in basic solution

$$BrO_3^- + F_2 + 2OH^- = BrO_4^- + 2F^- + H_2O$$

They are exceedingly powerful oxidants

$$BrO_4^- + 2H^+ + 2e = BrO_3^- + H_2O \qquad E^\circ = +1.76 \text{ V}$$

Solutions of $HBrO_4$ up to 6 *M* are stable, but decompose when stronger.

Periodates resemble tellurates in their stoichiometries. The main equilibria in acid solutions are

$$H_5IO_6 = H^+ + H_4IO_6^- \qquad K = 1 \times 10^{-3}$$

$$H_4IO_6^- = IO_4^- + 2H_2O \qquad K = 29$$

$$H_4IO_6^- = H^+ + H_3IO_6^{2-} \qquad K = 2 \times 10^{-7}$$

In aqueous solutions at 25° the main ion is IO_4^-. The pH-dependent equilibria are established rapidly. Kinetic studies of the hydration of IO_4^- suggest either one-step or two-step paths (*Fig. 20-1*), the latter being more likely. Periodic acid and its salts are used in organic chemistry as oxidants that usually react smoothly and rapidly. They are useful analytical oxidants; for example, they oxidize Mn^{2+} to MnO_4^-.

20-6 Interhalogen Compounds

The halogens form many compounds among themselves in binary combinations that may be neutral or ionic, for example, BrCl, IF_5, Br_3^+, I_3^-. *Ternary* combinations occur only in polyhalide ions, for example, $IBrCl^-$.

Neutral interhalogen compounds are of the type XX'_n where *n* is an *odd* number and X′ is always the lighter halogen when $n > 1$. Because *n* is odd, the compounds are diamagnetic; their valence electrons are present either as bonding pairs or as unshared pairs. The principles involved in the bonding are similar to those in xenon fluorides and have been discussed in Section 3-10.

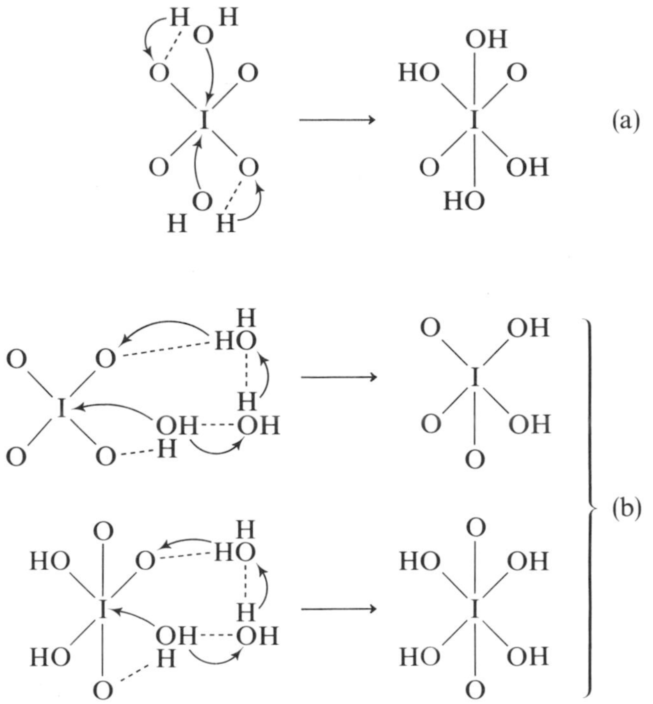

Figure 20-1 *Schematic representation of (a) the one-step, and (b) the two-step mechanism for the aquation of* IO_4^- *to* $IO_2(OH)_4^-$*. Dotted lines represent hydrogen bonds.*

Chlorine trifluoride is a liquid (bp 11.8°) that is commercially available in tanks. It is made by direct combination at 200 to 300°. Reaction of ClF_3 with excess Cl_2 gives *chlorine monofluoride*, which is a gas (bp −100°).

Bromine trifluoride, a red liquid (bp 126°), is also made by direct interaction.

These three substances, typical of all halogen fluorides, are very reactive. They react explosively with H_2O and organic substances. They are powerful fluorinating agents for inorganic compounds, and when diluted with N_2, for organic compounds.

Interhalogen ions. There are both cations and anions. Halogen fluorides react with fluoride ion acceptors, for example,

$$2ClF + AsF_5 = FCl_2^+AsF_6^-$$

or with fluoride ion donors,

$$IF_5 + CsF = Cs^+IF_6^-$$

It is not always clear that such products contain *discrete* ions. For instance in "$ClF_2^+SbF_6^-$" each Cl atom has two close and two distant (belonging to SbF_6^-) fluorine neighbours in a much distorted square.

The pale yellow *triiodide* ion is formed on dissolving I_2 in aqueous KI. There are numerous salts of I_3^-. Other ions are not usually stable in aqueous solution although they can be obtained in CH_3OH or CH_3CN and as crystalline salts of large cations such as Cs^+ or R_4N^+. For chlorine, the ion is formed only in concentrated solution:

$$Cl^-(aq) + Cl_2 \rightleftharpoons Cl_3^-(aq) \qquad K \approx 0.2$$

The electrical conductance of molten I_2 is ascribed to self-ionization

$$3I_2 \rightleftharpoons I_3^+ + I_3^-$$

20-7 Organic Compounds of Fluorine

Although the halogens form innumerable organic compounds, the methods of making organic fluorine compounds and some of their unusual properties are of inorganic interest. Fluorination of other halogen compounds by treatment with metal fluorides has been discussed, page 324. These methods are expensive so that alternative cheaper methods suitable for industrial procedures have been developed.

1. *Replacement of chlorine using hydrogen fluoride.* Anhydrous HF is cheap and can be used to replace Cl in chloro compounds. Catalysts such as $SbCl_5$ or CrF_4 and moderate temperature and pressure are required. Examples are

$$2CCl_4 + 3HF \longrightarrow CCl_2F_2 + CCl_3F + 3HCl$$

$$CCl_3COCCl_3 \longrightarrow CF_3COCF_3$$

2. *Electrolytic replacement of hydrogen by fluorine.* One of the most important laboratory and industrial methods is the electrolysis of organic compounds in liquid HF at voltages (~4.5–6) below that required for the liberation of F_2. Steel cells with Ni anodes and steel cathodes are used. Fluorination occurs at the anode. Although many organic compounds give conducting solutions in liquid HF, a conductivity additive may be required. Examples of such fluorinations are

$$(C_2H_5)_2O \longrightarrow (C_2F_5)_2O$$

$$C_8H_{18} \longrightarrow C_8F_{18}$$

$$(CH_3)_2S \longrightarrow CF_3SF_5 + (CF_3)_2SF_4$$

$$(C_4H_9)_3N \longrightarrow (C_4F_9)_3N$$

$$CH_3COOH \longrightarrow CF_3COOF \xrightarrow{H_2O} CF_3COOH$$

3. *Direct replacement of hydrogen by fluorine.* Although most organic compounds normally inflame or explode with fluorine, direct fluorination of many compounds is possible as follows.

(a) Catalytic fluorination where the reacting compound and F_2 diluted with N_2 are mixed *in the presence* of copper gauze, or cesium fluoride catalyst.

An example is

$$C_6H_6 + 9F_2 \xrightarrow{\text{Cu, } 265^\circ} C_6F_{12} + 6HF$$

(b) The reaction of the substrate in the *solid* state with F_2 diluted with He over a rather long period (12–36 hours) at a low temperature in presence of a heat sink in the form of the reactor and containers. The purpose is to allow the heat generated in the exothermic reaction (overall for replacement of H by F, *ca.* 420 kJ mol^{-1}), which could lead to C—C bond breaking, to be efficiently dissipated. The replacement reaction proceeds by several steps, each less exothermic than the C—C average bond strength, so that, provided the reaction time allows separate completion of individual steps, fluorination without degradation is possible. Examples of materials that can be fluorinated in this way are polystyrene, anthracene, phthalocyanine, carboranes, etc.

(c) Inorganic fluorides such as cobaltic fluoride are used for the vapor-phase fluorination of organic compounds, for example,

$$(CH_3)_3N \xrightarrow{CoF_3} (CF_3)_3N + (CF_3)_2NF + CF_3NF_2 + NF_3$$

4. *Other methods.* A useful and selective fluorinating agent for oxygen compounds is SF_4 (page 314); for example, ketones RR′CO may be converted to $RR'CF_2$, and carboxylate groups, —COOH to —CF_3.

Cesium fluoride acts as a catalyst in various fluorination reactions for example,

$$R_FCN + F_2 \xrightarrow{\text{CsF, } -78^\circ} R_FCF_2NF_2 \qquad (R_F = \text{perfluoralkyl})$$

The F^- ion is very nucleophilic toward unsaturated fluorocarbons and adds to the positive center of a polarized multiple bond. The carbanion so produced may then undergo double-bond migration or may act as a nucleophile leading to the elimination of F^- or another ion by an S_N2 mechanism. Fluoride-initiated reactions of these types have wide scope. The reactions can be carried out in DMF or diglyme by using either the sparingly soluble CsF or the more soluble Et_4NF.

An example is

$$CF_2{=}CFCF_3 \xrightarrow{F^-} (CF_3)_2CF^- \xrightarrow{I_2} (CF_3)_2CFI + I^-$$

Thermal decomposition of aromatic diazonium fluoroborates gives fluoro-aromatic compounds:

$$C_6H_5N_2Cl \xrightarrow{NaBF_4} C_6H_5N_2BF_4 \xrightarrow{\text{Heat}} C_6H_5F + N_2 + BF_3$$

Properties of Organofluorine Compounds. The C—F bond energy is very high (486 kJ mol^{-1}; cf. C—H 415, and C—Cl 332 kJ mol^{-1}), but organic fluorides are not necessarily particularly stable thermodynamically. The low reactivities of fluorine derivatives can be attributed to the impossibility of expansion of the octet of fluorine and the inability of, say, water to coordinate to fluorine or carbon as the first step in hydrolysis. With chlorine this may be possible using outer d orbitals. Because of the small size of the F atom, H can be replaced by F and with least introduction of strain or distortion, as compared with replacement by other halogen atoms. The F atoms also effectively shield the C atoms from attack. Finally, since C bonded to F can be considered to be effectively oxidized (whereas in C—H it is reduced), there is no tendency for oxidation by oxygen. Fluorocarbons are attacked only by hot metals, for example, molten Na. When pyrolyzed, they split at C—C rather than C—F bonds.

The replacement of H by F leads to increased density, but less than by other halogens. Completely fluorinated (called perfluoro) derivatives, C_nF_{2n+2}, have very low boiling points for their molecular weights and low intermolecular forces; the weakness of these forces is also shown by the very low coefficient of friction for polytetrafluoroethylene, $(CF_2{-}CF_2)_n$.

Chlorofluorocarbons are used as nontoxic, inert refrigerants, aerosol bomb propellants, and heat transfer agents. Fluoroolefins are used as monomers for free-radical-initiated polymerizations to give oils, greases, and the like, and also as chemical intermediates. $CF_3CHBrCl$ is a safe anaesthetic. $CHClF_2$ is used for making tetrafluoroethylene:

$$2CHClF_2 \xrightarrow{500-1000^\circ} CF_2{=}CF_2 + 2HCl$$

Tetrafluoroethylene (bp -76.6°) can be polymerized thermally or in aqueous emulsion; the polymer is used for coating frying pans, resistant gaskets, and the like.

Fluorinated carboxylic acids are strong acids, for example, for CF_3COOH, $K_a = 5.9 \times 10^{-1}$, whereas CH_3COOH has $K_a = 1.8 \times 10^{-5}$. Many reactions of carboxylic acids leave the fluoroalkyl group intact, for example,

$$C_3F_7COOH \xrightarrow[C_2H_5OH]{H_2SO_4} C_3F_7COOC_2H_5 \xrightarrow{NH_3} C_3F_7CONH_2$$

$$C_3F_7CONH_2 \xrightarrow{P_2O_5} C_3F_7CN$$

$$C_3F_7CONH_2 \xrightarrow{LiAlH_4} C_3F_7CH_2NH_2$$

Perfluoroalkylhalides are made by the reaction

$$R_FCOOAg + I_2 \xrightarrow{\text{Heat}} R_FI + CO_2 + AgI$$

These halides are relatively reactive, undergoing free-radical reactions when heated or irradiated. Because of the very strong electron-attracting nature of the perfluoroalkyl groups, they do not undergo most nucleophilic reactions of alkyl halides.

Trifluoromethyl iodide is readily cleaved

$$CF_3I = CF_3^{\cdot} + I^{\cdot} \qquad \Delta H = 115\ \text{kJ mol}^{-1}$$

Radical reactions of CF_3I with metals and nonmetals gives CF_3 derivatives, for example,

$$CF_3I + P \xrightarrow{\text{Heat}} (CF_3)_nPI_{3-n}$$

Study Questions

A

1. Where and in what chemical form are the halogens found in nature?
2. How are the free halogens prepared from their ionic halides?
3. Why are solutions of I_2 in alcohols and ketones brown, whereas those in nonpolar solvents have the same purple color as the vapor?
4. List the main methods for preparation of anhydrous chlorides.
5. By what reaction would you obtain (give balanced equations)

$CrCl_3$	from	$[Cr(H_2O)_6]Cl_3$	$FeCl_3$	from	Fe
PBr_3	from	red P	CuI	from	aqueous $CuSO_4$
$FeCl_2$	from	Fe	$GdCl_3$	from	Gd_2O_3

6. Why is it impossible to make iodides of elements in high oxidation states when the corresponding bromides or chlorides are known?
7. What would you expect the trend in properties to be for the fluorides CrF_2, CrF_3, CrF_4, CrF_5, and CrF_6?
8. Which elements give (a) fluorides, (b) chlorides that are essentially insoluble in water or dilute HNO_3?
9. The halides as ligands can act as bridge groups. List the various ways.
10. Why is $SiCl_4$ hydrolyzed readily but CCl_4 is not?
11. How are the following oxo compounds made (a) ClO_2, (b) I_2O_5, (c) NaOCl(aq), (d) $NaClO_2$, (e) $NaClO_3$, (f) $NaClO_4$? Give balanced equations.
12. What are the general formulas and names of the four types of oxo acids of the halogens and their anions. In the case of iodine there is one of unique stoichiometry. What is its formula?
13. When Cl_2 reacts with base the products vary with temperature. Explain
14. Name at least one cationic, one neutral, and one anionic interhalogen compound. In those consisting of three or more atoms, what rule predicts which will be the central atom?
15. I_2 is virtually insoluble in H_2O. It dissolves readily in a solution of KI. Why?
16. Describe at least two methods for making fluoroorganic compounds, using different starting materials.
17. What are the typical physical and chemical properties of $C_nF_{2n+1}X$(X=F, I, COOH, MgBr) compared with those of their $C_nH_{2n+1}X$ analogs?

B

1. F_2O_2 has a very short O—O bond, 1.217 Å compared with those in H_2O_2, 1.48 Å and O_2^{2-}, 1.49 and also relatively long O—F bonds (1.575) compared with those in OF_2. Why?
2. ClO_2 is a free radical with one unpaired electron but has no tendency to dimerize like NO_2. Why is this?
3. Suggest a reason why perfluoroalkylhalides scarcely form Grignard reagents or lithium alkyls.

4. List some differences between the chemistry of fluorine and those of the other halogens.
5. Draw the shapes of the following molecules showing also the lone pairs:

$$ClF,\ BrF_3,\ IF_5,\ IF_7,\ ClF_4^-,\ I_3^-,\ BrF_4^+,\ ICl_2^+$$

6. Why is fluorine more reactive than the other halogens?
7. Describe how and why HF differs from HCl.
8. What is the order of acid strength for the following: $HClO$, $HClO_2$, $HClO_3$, $HClO_4$? Why?
9. Why cannot F_2 be obtained by electrolysis of aqueous solutions of NaF?
10. Why are solutions of I_2 in CCl_4 violet and those in benzene brown-pink?
11. What would you expect for the reaction of astatine with
(a) H_2O (b) aqueous sulfite (c) conc HNO_3?
Would HAt be stronger or weaker than HI?
12. What is the structure of O_2F_2, ClO_2, BrO_3^-, $H_4IO_6^-$?
13. How do you calculate from data in Table 20-1 the values of K for

$$X_2(aq) = H^+ + X^- + HOX?$$

14. Write the complete equation for the oxidation of aqueous HCl by (i) MnO_2, (ii) $KMnO_4$.
15. Describe the differences in the reactions of Cl_2, Br_2, and I_2 with aqueous 2N NaOH at (a) 0°, (b) 70°C.
16. How might you make CF_3NO from CF_3I?
17. 1.86 g of a metal carbonyl were heated with excess iodine dissolved in pyridine, the gas passed over I_2O_5, and the resulting I_2 dissolved in CCl_4. This was determined by sodium thiosulfate 20 ml of 1 M solution being required. Calculate the formula of the metal carbonyl and explain why it should react with iodine in pyridine.
18. Why is HF a weak acid in *water*?
19. Fluorine, F_2, has a low bond energy and F has a low electron affinity. Discuss the possible relation between these facts.

Chapter 20
Study Guide

Supplementary Reading

Emeléus, H. J., *The Chemistry of Fluorine and its Compounds*, Academic Press, 1969.
Gillespie, R. J. and Morton, M. J., "Halogen and Interhalogen Cations," *Quart. Rev.*, *25*, 553 (1971).
Gutmann, V., ed., *Halogen Chemistry*, Vols. 1–3, Academic Press, 1967.
Jolles, Z. E., ed., *Bromine and its Compounds*, Academic Press, 1966.
Neumark, H. R., ed., *The Chemistry and Chemical Technology of Fluorine*, Interscience-Wiley, 1967.
Sheppard, W. A. and Sharts, C. M., *Organic Fluorine Chemistry*, Benjamin, 1970.
Tatlow, J. C. *et al.*, eds., *Advances in Fluorine Chemistry*, Vols. 1–7, 1966–1973.

21

the noble gases

21-1
Occurrence, Isolation and Applications

The noble gases (Table 21-1) are minor constituents of the atmosphere, from which Ne, Ar, Kr, and Xe were first isolated by Sir William Ramsey. He also found that a gas, isolated by Hillebrand from uranium minerals, had the same spectrum as the element identified spectroscopically in the sun by Lockyer and Frankland in 1868 and called helium. Helium occurs in radioactive minerals and notably in some natural gases in the United States. Its origin is entirely from the decay of uranium or thorium isotopes that emit α-particles. These helium nuclei acquire electrons from surrounding elements, oxidizing them, and if the rock is sufficiently impermeable, the helium remains trapped. The gas radon, all of whose isotopes are radioactive with short half-lives, was characterized in the decay series from uranium and thorium.

Table 21-1 *Some Properties of the Noble Gases*

Element	Outer Configuration	1st Ionization enthalpy kJ mol^{-1}	Normal bp K	Vol. % in atmosphere ($\times 10^4$)
He	$1s^2$	2369	4.2	5.2
Ne	$2s^2\ 2p^6$	2078	27.1	18.2
Ar	$3s^2\ 3p^6$	1519	87.3	9340.0
Kr	$4s^2\ 4p^6$	1349	120.3	11.4
Xe	$5s^2\ 5p^6$	1169	166.1	0.08
Rn	$6s^2\ 6p^6$	1036	208.2	—

Ne, Ar, Kr, and Xe are obtained by fractionation of liquid air. The gases were originally termed inert, and because of their apparent low chemical reactivity they provided the key to the problem of valency, the interpretation of the periodic table, and the concept of the closed electron shell configuration.

The main use of He is as the liquid in cryoscopy. Argon may be used to provide an inert atmosphere in laboratory apparatus, in welding, and in gas-filled electric light bulbs. Neon is used for discharge lighting tubes.

21-2
The Chemistry of Xenon

During studies with the very reactive gas PtF_6, N. Bartlett found that with oxygen, a crystalline solid $[O_2^+][PtF_6^-]$ was formed. He noted that since the ionization enthalpy of Xe is almost identical with that of O_2, an analogous reaction might be expected and, indeed, in 1962 he reported the first compound containing à noble gas, a red crystalline solid first believed to be $Xe^+PtF_6^-$ but now known to be more complex.

There is now an extensive chemistry of xenon with bonds to F and O; one compound with a Xe—N bond is known, but compounds with bonds to other elements are highly unstable. A few krypton compounds exist but while there should be an extensive chemistry of Rn, the short lifetimes of the isotopes make study impossible. Xenon reacts directly only with fluorine, but oxygen compounds can be obtained from the fluorides. Certain compounds are very stable and can be made in large quantities. Table 21-2 lists some of the more important compounds and their properties.

Table 21-2 *Some Xenon Compounds*

Oxidation state	Compound	Form	Mp (°C)	Structure	Remarks
II	XeF_2	Colorless crystals	129	Linear	Hydrolyzed to Xe + O_2; v. soluble in HF(l)
IV	XeF_4	Colorless crystals	117	Square	Stable
VI	XeF_6	Colorless crystals	49.6	Complex, see text	Stable
	Cs_2XeF_8	Yellow solid		Archim. antiprism	Stable to 400°
	$XeOF_4$	Colorless liquid	−46	Square pyramid	Stable
	XeO_3	Colorless crystals		Pyramidal	Explosive, hygroscopic; stable in solution
VIII	XeO_4	Colorless gas		Tetrahedral	Explosive
	XeO_6^{4-}	Colorless salts		Octahedral	Anions $HXeO_6^{3-}$, $H_2XeO_6^{2-}$, $H_3XeO_6^-$ also exist

Fluorides. Thermodynamic studies of the reactions

$$Xe + F_2 = XeF_2$$

$$XeF_2 + F_2 = XeF_4$$

$$XeF_4 + F_2 = XeF_6$$

show that only these three fluorides exist. The equilibria are established rapidly only above 250°, and the synthesis must be performed above this temperature.

The three fluorides are volatile substances, subliming readily at 25°. They can be stored in nickel vessels, but XeF_4 and XeF_6 are exceptionally readily hydrolyzed and even traces of water must be excluded.

Xenon difluoride is best made by interaction of Xe with a deficiency of F_2 at high pressure. It dissolves in water to give solutions with a pungent odor of XeF_2. The hydrolysis is slow in acid solution, but rapid in the presence of bases:

$$XeF_2 + 2OH^- = Xe + \tfrac{1}{2}O_2 + 2F^- + H_2O$$

The solutions are strong oxidizers converting HCl to Cl_2 and Ce^{III} to Ce^{IV}. XeF_2 is also a mild fluorinating agent for organic compounds; for example, benzene forms C_6H_5F.

Xenon tetrafluoride is the easiest of the three to prepare. On heating a 1:5 mixture of Xe and F_2 at 400° and *ca.* 6 atm pressure for a few hours, XeF_4 is formed quantitatively. It resembles XeF_2 except for its behavior on hydrolysis, which is discussed below. XeF_4 will specifically fluorinate aromatic rings in compounds like toluene.

Xenon hexafluoride is obtained by interaction of XeF_4 and F_2 under pressure or directly from Xe and fluorine at temperatures above 250° and pressures >50 atm. XeF_6 is extremely reactive, even with quartz:

$$SiO_2 + 2XeF_6 = 2XeOF_4 + SiF_4$$

The colorless crystals contain both tetramers and hexamers each made up of XeF_5^+ units linked by unsymmetrical, bent F^- bridges, as is shown in *Fig. 21-1.* A solution in $(F_5S)_2O$, below $-45°$, appears to contain only Xe_4F_{24} units; all F atoms appear equivalent in the nmr, because of rapid scrambling over various positions in the molecule. Monomeric XeF_6 in the liquid or vapor is yellowish green and has a distorted octahedral structure because of the lone pair.

Xenon Fluoride Complexes. The fluorides will react with strong fluoro-Lewis acids such as SbF_5 or AsF_5 to give adducts. Although $XeF_2 \cdot IF_5$ has a molecular lattice, in other cases, fluoride ion transfer occurs to give solids that contain XeF^+, XeF_5^+, and $Xe_2F_3^+$ ions as in $(Xe_2F_3^+)(AsF_6^-)$ or $XeF_5^+PtF_6^-$. The $Xe_2F_3^+$ ion is planar (21-I).

F 2.14 Å
Xe Xe 1.90 Å
F 151° F

21-I

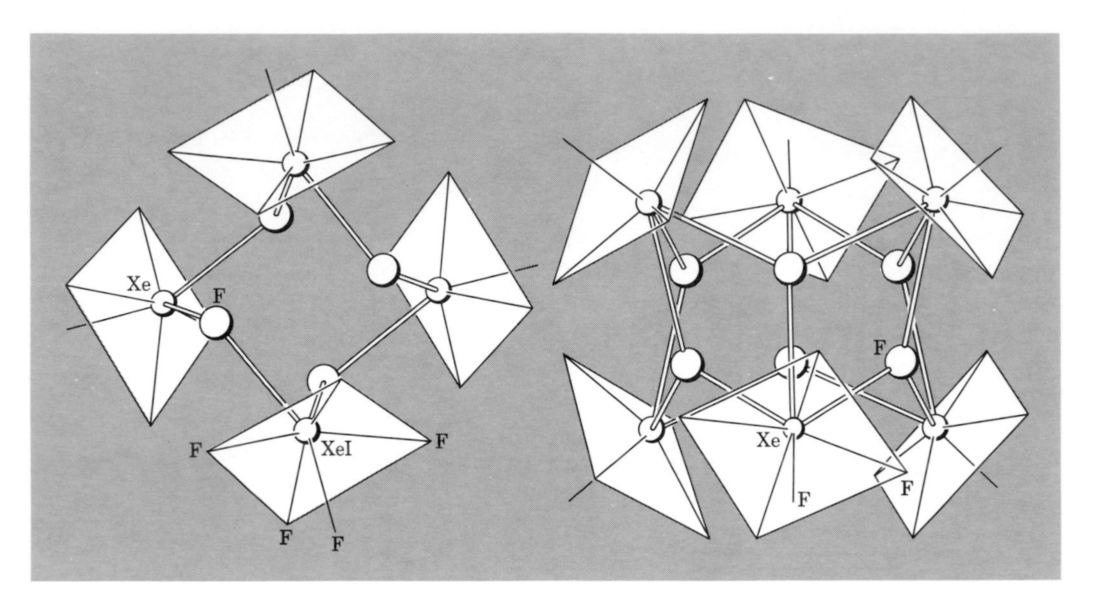

Figure 21-1 *The tetrameric (left) and hexameric (right)* $[XeF_5{}^+F^-]_{4,6}$ *units which make up the crystal structure of* XeF_6.

These studies suggested that Bartlett's original reaction is best written

$$Xe + PtF_6 \xrightarrow{25^\circ} XePtF_6$$

$$XePtF_6 + PtF_6 \xrightarrow{25^\circ} XeF^+PtF_6{}^- + PtF_5 \xrightarrow{60^\circ} XeF^+Pt_2F_{11}{}^-$$

The hexafluoride can act as a Lewis acid toward F^- and can be converted to heptafluoro or octafluoro xenates:

$$XeF_6 + RbF = RbXeF_7$$

The Rb and Cs salts are the most stable xenon compounds known and decompose only above 400°. The sodium salt is less stable and can be used to purify XeF_6 as it is decomposed below 100°.

Xenon-Oxygen Compounds. *Xenon trioxide* is formed in the hydrolysis of XeF_4 and XeF_6

$$3XeF_4 + 6H_2O = XeO_3 + 2Xe + \tfrac{3}{2}O_2 + 12HF$$

$$XeF_6 + 3H_2O = XeO_3 + 6HF$$

The colorless, odorless, and stable aqueous solutions of XeO_3 appear to contain XeO_3 molecules. On evaporation XeO_3 is obtained as a white deliquescent solid that is dangerously explosive. In basic solution a xenate(VI) ion is formed:

$$XeO_3 + OH^- = HXeO_4{}^-$$

but the $HXeO_4^-$ ion slowly disproportionates to give a xenate(VIII) or perxenate ion:

$$2HXeO_4^- + 2OH^- = XeO_6^{4-} + Xe + O_2 + 2H_2O$$

Perxenates are formed not only by disproportionation of $HXeO_4^-$ but when this ion is oxidized by ozone. The perxenate solutions are yellow and are powerful and rapid oxidants. The salts such as $Na_4XeO_6 \cdot 8H_2O$ are stable and sparingly soluble in water.

In alkaline solution, the main form is the ion $HXeO_6^{3-}$, and perxenates are only slowly reduced by water. However, in acid solution, reduction is almost instantaneous:

$$H_2XeO_6^{2-} + H^+ = HXeO_4^- + H_2O + \tfrac{1}{2}O_2$$

and the hydroxyl radical is involved as an intermediate.

When barium perxenate is heated with concentrated H_2SO_4, *xenon tetroxide* is formed as an explosive, unstable gas.

The aqueous chemistry of xenon is summarized by the potentials:

$$\text{Acid solution:}\quad H_4XeO_6 \xrightarrow{2.36\text{ V}} XeO_3 \xrightarrow{2.12\text{ V}} Xe$$

$$XeF_2 \xrightarrow{2.64\text{ V}} Xe$$

$$\text{Alkaline solution:}\quad HXeO_6^{3-} \xrightarrow{0.94\text{ V}} HXeO_4^- \xrightarrow{1.26\text{ V}} Xe$$

Study Questions

1. What is the origin of helium?
2. Why do the boiling points of the noble gases vary systematically with atomic number?
3. How are XeF_2, XeF_4, and XeF_6 prepared from Xe?
4. Write balanced equations for the reactions of water with XeF_2, XeF_4, and XeF_6.
5. How are xenates and perxenates made?
6. Write an equation for the reduction of XeO_3 by I^- in acid solutions to give Xe.
7. How is the interaction of PtF_6 and Xe best interpreted?
8. What are the structures of XeF_4, XeO_3, XeO_6^{4-}?

Chapter 21
Study Guide

Supplementary Reading

Bartlett, N., *The Chemistry of the Noble Gases*, Elsevier, 1971.
Holloway, J. H., *Noble Gas Chemistry*, Methuen, 1968.
Moody, G. J., "A Decade of Xenon Chemistry," *J. Chem. Educ.*, *51*, 628 (1974).

22

zinc, cadmium, and mercury

22-1
Introduction

The position of Zn, Cd, and Hg in the periodic table is discussed in Chapter 2, page 44, and some properties are given in Table 8-2, page 198.

Although these elements characteristically form +2 cations, they do not have much in common with the Be, Mg, Ca—Ra group except for some resemblances between Zn, Be and Mg. Thus BeO, $Be(OH)_2$, and BeS have the same structures as ZnO, $Zn(OH)_2$, and ZnS, and there is some similarity in the solution and complex chemistry of Zn^{2+} and Mg^{2+}. The main cause of the differences between the IIA and the IIB ions arises from the ease of distortion of the filled *d* shell compared with the noble gas-like ions of the IIA elements.

Mercury shows such unique behavior that it cannot be considered as homologous to Zn and Cd.

22-2
Occurrence, Isolation, and Properties of the Elements

The elements have relatively low abundance in nature (of the order 10^{-6} of the earth's crust for Zn and Cd), but have long been known because they are easily obtained from their ores.

Zinc occurs widely but the main source is *sphalerite*, (ZnFe)S, which commonly occurs with galena, PbS; cadmium minerals are scarce but, as a result of its similarity to Zn, Cd occurs by isomorphous replacement in almost all zinc ores. Methods of isolation involve flotation and roasting; Zn and Pb are recovered simultaneously by a blast furnace method. Cadmium is invariably a by-product and is usually separated from Zn by distillation or by precipitation from sulfate solutions by Zn dust:

$$Zn + Cd^{2+} = Zn^{2+} + Cd \qquad E = +0.36\ V$$

The only important ore of mercury is *cinnabar*, HgS; this is roasted to give the oxide which, in turn, decomposes at *ca.* 500°, the mercury vaporizing.

Zinc and cadmium are white, lustrous, but tarnishable metals. Their structures deviate from perfect hexagonal close packing by elongation along the sixfold axis. Mercury is a shiny liquid at ordinary temperatures. All are remarkably volatile for heavy metals, mercury uniquely so. Mercury gives a monatomic vapor and has an appreciable vapor pressure (1.3×10^{-3} mm) at 20°. It is also surprisingly soluble in both polar and nonpolar liquids: a saturated solution in water at 25° has 6×10^{-8} g/g. Because of its high volatility and toxicity, mercury should always be kept in stoppered containers and handled in well-ventilated areas. In the biosphere it is exceptionally toxic because of its conversion by bacteria to CH_3Hg^+ (Section 31-5). Mercury is readily lost from aqueous solutions of mercuric salts owing to reduction by traces of reducing materials and by disproportionation of Hg_2^{2+}.

Both Zn and Cd react readily with nonoxidizing acids, releasing H_2 and giving the divalent ions; Hg is inert to nonoxidizing acids. Zinc also dissolves in strong bases because of its ability to form zincate ions (see below), commonly written ZnO_2^{2-}:

$$Zn + 2OH^- \longrightarrow ZnO_2^{2-} + H_2$$

Cadmium does not dissolve in bases.

Zinc and Cd react readily when heated in O_2, to give the oxides. Although Hg and O_2 are unstable with respect to HgO at 25°, their rate of combination is exceedingly slow; the reaction proceeds at a useful rate at 300 to 350°, but above *ca.* 400° the ΔG becomes positive and HgO decomposes rapidly into the elements:

$$HgO(s) = Hg(l) + \tfrac{1}{2}O_2 \qquad \Delta H_{diss} = 90.4 \text{ kJ mol}^{-1}$$

This ability of Hg to absorb O_2 from air and regenerate it as O_2 was of considerable importance in the earliest studies of oxygen by Lavoisier and Priestley.

All three elements react with halogens and with nonmetals such as S, Se, P, etc.

Zinc and Cd form many alloys. Some, such as brass, which is a copper-zinc alloy, are of technical importance. Mercury combines with many other metals, sometimes with difficulty but sometimes, as with Na or K very vigorously, giving *amalgams*. Many amalgams are of continuously variable compositions, while others are compounds, such as Hg_2Na. Some of the transition metals do not form amalgams, and iron is commonly used for containers of Hg. Sodium amalgams and amalgamated Zn are frequently used as reducing agents for aqueous solutions.

22-3 The Univalent State

Zinc, Cd, and Hg form the ions M_2^{2+}. The Zn_2^{2+} and Cd_2^{2+} ions are unstable, especially Zn_2^{2+}, and are known only in melts or solids. Thus addition of Zn to fused $ZnCl_2$ gives a yellow solution and, on cooling, a yellow glass that contains Zn_2^{2+}.

The ions have a metal–metal bond, $^+M{-}M^+$; Raman spectra allow the estimation of force constants, and they show that the order of bond strength is $Zn_2^{2+} < Cd_2^{2+} < Hg_2^{2+}$.

The *mercurous ion*, Hg_2^{2+}, is formed on reduction of mercuric salts in aqueous solution. X-ray diffraction studies on many compounds such as Hg_2Cl_2, Hg_2SO_4, and $Hg_2(NO_3)_2 \cdot 2H_2O$, show that the Hg—Hg distances range from 2.50 to 2.70 Å, depending on the associated anions. The shortest distances are found with the least covalently bound anions, for example, NO_3^-.

Hg^I—Hg^{II} Equilibria. An understanding of the thermodynamics of these equilibria is essential to an understanding of the chemistry of the mercurous state. The important values are the potentials

$$Hg_2^{2+} + 2e = 2Hg \qquad E^\circ = 0.789 \text{ V} \qquad (22\text{-}1)$$

$$2Hg^{2+} + 2e = Hg_2^{2+} \qquad E^\circ = 0.920 \text{ V} \qquad (22\text{-}2)$$

$$Hg^{2+} + 2e = Hg \qquad E^\circ = 0.854 \text{ V} \qquad (22\text{-}3)$$

For the disproportionation equilibrium

$$Hg_2^{2+} = Hg + Hg^{2+} \qquad E^\circ = -0.131 \text{ V} \qquad (22\text{-}4)$$

From which it follows that

$$K = \frac{[Hg^{2+}]}{[Hg_2^{2+}]} = 6.0 \times 10^{-3}$$

The implication of the standard potentials is clearly that only oxidizing agents with potentials in the range −0.79 to −0.85 V can oxidize mercury to Hg^I but not to Hg^{II}. Since no common oxidizing agent meets this requirement, it is found that when mercury is treated with an excess of oxidizing agent it is entirely converted into Hg^{II}. However, when mercury is in at least 50% excess, only Hg^I is obtained since, according to Eq. 22-4, Hg readily reduces Hg^{2+} to Hg_2^{2+}.

The equilibrium constant for reaction 22-4 shows that Hg_2^{2+} is stable with respect to disproportionation, but by only a small margin. Thus any reagents that reduce the activity (by precipitation or complexation) of Hg^{2+}, to a significantly greater extent than they lower the activity of Hg_2^{2+}, will cause *disproportionation* of Hg_2^{2+}. There are many such reagents, so that the number of stable Hg^I compounds is quite restricted.

Thus, when OH^- is added to a solution of Hg_2^{2+}, a dark precipitate consisting of Hg and HgO is formed; evidently mercurous hydroxide, if it could be isolated, would be a stronger base than HgO. Similarly, addition of sulfide ions to a solution of Hg_2^{2+} gives a mixture of Hg and the extremely insoluble HgS. Mercurous cyanide does not exist because $Hg(CN)_2$, although soluble, is so slightly dissociated. The reactions in these cited cases are

$$Hg_2^{2+} + 2OH^- \longrightarrow Hg + HgO(s) + H_2O$$

$$Hg_2^{2+} + S^{2-} \longrightarrow Hg + HgS(s)$$

$$Hg_2^{2+} + 2CN^- \longrightarrow Hg + Hg(CN)_2(aq)$$

Mercurous Compounds. As we indicate above, no hydroxide, oxide, or sulfide can be obtained by addition of the appropriate anion to aqueous Hg_2^{2+}, nor have these compounds been otherwise made.

Among the best known mercurous compounds are the *halides*. The fluoride is unstable toward water, being hydrolyzed to hydrofluoric acid and unisolable mercurous hydroxide (which disproportionates as above). The other halides are insoluble, which thus precludes the possibilities of hydrolysis or disproportionation to give Hg^{II} halide complexes. Mercurous nitrate, and perchlorate are soluble in water, but Hg_2SO_4 is sparingly soluble.

22-4
Divalent Zinc and Cadmium Compounds

Binary Compounds. The *oxides*, ZnO and CdO, are formed on burning the metals in air or by pyrolysis of the carbonates or nitrates; oxide smokes can be obtained by combustion of the alkyls, cadmium oxide smokes being exceedingly toxic. Zinc oxide is normally white but turns yellow on heating. CdO varies in color from greenish-yellow through brown to nearly black, depending on its thermal history. These colors are the result of various kinds of lattice defects. Both oxides sublime at very high temperatures.

The hydroxides are precipitated from solutions of salts by addition of bases. $Zn(OH)_2$ readily dissolves in an excess of alkali bases to give "zincate" ions, and solid zincates such as $NaZn(OH)_3$ and $Na_2[Zn(OH)_4]$ can be crystallized from concentrated solutions. $Cd(OH)_2$ is insoluble in bases. Both Zn and Cd hydroxide readily dissolve in an excess of strong ammonia to form the ammine complexes, for example, $[Zn(NH_3)_4]^{2+}$. The complete set of formation constants for the cadmium system was presented in Section 6-6.

Sulfides. These are obtained by direct interaction or by precipitation by H_2S from aqueous solutions, acidic for CdS, neutral or basic for ZnS. The sulfides as well as the selenides and tellurides all have the wurtzite or zinc blende structures shown in Chapter 4, page 92.

Halides. The fluorides are essentially ionic, high melting solids, whereas the other halides are more covalent in nature. The fluorides are sparingly soluble in water, a reflection of the high lattice energies of the ZnF_2 (rutile) and CdF_2 (fluorite) structures. The other halides are much more soluble, not only in water but in alcohols, ketones, and similar donor solvents. Aqueous solutions of cadmium halides contain all the species Cd^{2+}, CdX^+, CdX_2, and CdX_3^- in equilibrium.

Oxo Salts and Aquo Ions. Salts of oxo acids such as the nitrate, sulfate, sulfite, perchlorate, and acetate are soluble in water. The Zn^{2+} and Cd^{2+} ions are rather similar to Mg^{2+}, and many of their salts are isomorphous with magnesium salts, for example, $Zn(Mg)SO_4 \cdot 7H_2O$. The aquo ions are acidic, and aqueous solutions of salts are hydrolyzed. In perchlorate solution the only species for Zn, Cd (and Hg) below 0.1 *M* are the MOH^+ ions, for example,

$$Zn^{2+}(aq) + H_2O \rightleftharpoons ZnOH^+(aq) + H^+$$

For more concentrated cadmium solutions, the principal species is Cd_2OH^{3+}:

$$2Cd^{2+}(aq) + H_2O \rightleftharpoons Cd_2OH^{3+}(aq) + H^+$$

In presence of complexing anions, for example, halide, species such as Cd(OH)Cl or $CdNO_3^+$ may be obtained.

Complexes. All of the halide ions except F^- form complex halogeno anions when present in excess, but for Zn^{2+} and Cd^{2+} the formation constants are many orders of magnitude smaller than those for Hg^{2+}. The same applies to complex cations with NH_3 and amines, many of which can be isolated as crystalline salts.

Zinc dithiocarbamates (page 264) are industrially important as accelerators in the vulcanization of rubber by sulfur. Zinc complexes are also of great importance biologically (page 564). Zinc compounds, especially $ZnCO_3$ and ZnO, are used in ointments, since zinc apparently promotes healing processes.

By contrast, cadmium compounds are extremely poisonous, possibly because of substitution of Cd for Zn in an enzyme system, and consequently they constitute a serious environmental hazard, for example, in the neighborhood of Zn smelters.

22-5 Divalent Mercury Compounds

Binary Compounds. Red HgO is formed on gentle pyrolysis of mercurous or mercuric nitrate, by direct interaction at 300 to 350°, or as red crystals by heating of an alkaline solution of K_2HgI_4. Addition of OH^- to aqueous Hg^{2+} gives a yellow precipitate of HgO; the yellow form differs from the red only in particle size.

No hydroxide has been obtained, but the oxide is soluble (10^{-3} to 10^{-4} mol l^{-1}) in water, the exact solubility depending on particle size, to give a solution of what is commonly assumed to be the hydroxide, although there is no proof for such a species. This "hydroxide" is an extremely weak base:

$$K = \frac{[Hg^{2+}][OH^-]^2}{[Hg(OH)_2]} = 1.8 \times 10^{-22}$$

and is somewhat amphoteric, though more basic than acidic.

Mercuric sulfide, HgS, is precipitated from aqueous solutions as a black, highly insoluble compound. The solubility product is 10^{-54}, but the sulfide is somewhat more soluble than this figure would imply because of hydrolysis of Hg^{2+} and S^{2-} ions. The black sulfide is unstable with respect to a red form identical with the mineral cinnabar and changes into it when heated or digested with alkali polysulfides or mercurous chloride.

Mercuric fluoride is essentially ionic and crystallizes in the fluorite structure; it is almost completely decomposed even by cold water, as would be expected for an ionic compound that is the salt of a weak acid and an extremely weak base.

In sharp contrast to the fluoride, the other halides show marked covalent character. *Mercuric chloride* crystallizes in an essentially molecular lattice. Relative to ionic HgF_2, the other halides have very low melting and boiling points, for example, $HgCl_2$, mp 280°. They also show marked solubility in many organic solvents. In aqueous solution they exist almost exclusively (~99%) as HgX_2 molecules, but some hydrolysis occurs, the principal equilibrium being, for example,

$$HgCl_2 + H_2O \rightleftharpoons Hg(OH)Cl + H^+ + Cl^-$$

Mercuric Oxo Salts. Among the mercuric salts that are essentially ionic and, hence, highly dissociated in aqueous solution are the nitrate, sulfate, and perchlorate. Because of the great weakness of mercuric hydroxide, aqueous solutions of these salts tend to hydrolyze extensively and must be acidified to be stable.

In aqueous solutions of $Hg(NO_3)_2$ the main species are $Hg(NO_3)_2$, $HgNO_3^+$, and Hg^{2+}, but at high concentrations of NO_3^- the complex anions $[Hg(NO_3)_{3,4}]^{-,2-}$ are formed.

Mercuric carboxylates, especially the acetate and the trifluoroacetate, are of considerable importance because of their utility in attacking unsaturated hydrocarbons (Section 29-6). They are made by dissolving HgO in the hot acid and crystallizing. The trifluoroacetate is also soluble in benzene, acetone, and tetrahydrofuran, which increases its utility, while the acetate is soluble in water and alcohols.

Mercuric ions catalyze a number of reactions of complex compounds such as the aquation of $[Cr(NH_3)_5X]^{2+}$. Bridged transition states, for example,

$$[(H_2O)_5CrCl]^{2+} + Hg^{2+} = [(H_2O)_5Cr—Cl—Hg]^{4+}$$

are believed to be involved.

Mercuric Complexes. The Hg^{2+} ion forms many strong complexes. The characteristic coordination numbers and stereochemical arrangements are two-coordinate, *linear*, and four-coordinate, *tetrahedral*. Octahedral coordination is less common; a few 3- and 5-coordinate complexes are also known. There appears to be considerable covalent character in the mercury-ligand bonds, especially in the 2-coordinate compounds.

In addition to halide or pseudo halide complex ions, such as $[HgCl_4]^{2-}$ or $[Hg(CN)_4]^{2-}$, there are cationic species such as $[Hg(NH_3)_4]^{2+}$ and $[Hg\ en_3]^{2+}$.

There are also a number of novel compounds in which —Hg— or —HgX is bound to a transition metal. Some of these compounds may be obtained by reaction of $HgCl_2$ with carbonylate anions (Section 28-9), for example,

$$2Na^+Co(CO)_4^- + HgCl_2 = 2NaCl + (CO)_4Co—Hg—Co(CO)_4$$

Study Questions

A

1. Give the electronic structures of Zn, Cd, and Hg and explain their position in the periodic table.
2. Compare the main features of the chemistry of Zn and Cd with that of Mg.

3. Describe the properties of elemental mercury.
4. Write equations for the action of (a) 3 M HCl, and (b) 3 M KOH on Zn.
5. Describe the interaction of Hg and O_2 and the properties of HgO.
6. What is an amalgam? Give two examples. What is their use?
7. What is the structure of the ions of univalent Zn, Cd, and Hg?
8. What is meant by the term disproportionation? What factors alter the ease of disproportionation of Hg^I?
9. Why does no hydroxide, oxide, or sulfide of Hg^I exist?
10. Draw the structures of (a) rutile, (b) fluorite, and (c) zinc blende.
11. What is the nature of $HgCl_2$ in the solid state and in aqueous solution?

B

1. Suggest the reason in thermodynamic terms why the sign of ΔG for the reaction $\frac{1}{2}O_2(g) + Hg(l) = HgO(s)$ changes from $-$ to $+$ at about 400°.
2. Why is it that when Hg is oxidized with an excess of oxidant only Hg^{II} is formed, yet when Hg is in excess only Hg^I is formed?
3. By what methods can it be proved that the mercurous ion is Hg_2^{2+} *in solution*?
4. What is the likely structure of $Zn_4O(CO_2CH_3)_6$? Why is it rapidly hydrolyzed unlike its Be analog.
5. The Zn and Cd dithiocarbamates are dimeric $[M(S_2CNR_2)_2]_2$. Draw a plausible structure.

Chapter 22
Study Guide

Supplementary Reading

Roberts, H. L., "The Chemistry of Mercury," *Adv. Inorg. Chem. Radiochem.*, *11*, 309 (1968).

3

Transition Elements

23

introduction to transition elements. ligand field theory

23-1 Introduction

As we noted in Section 8-12, the transition elements are often defined as those which, *as elements*, have partly filled *d* or *f* shells. For practical purposes, however, we shall consider as transition elements all those which have partly filled *d* or *f* shells in any of their important compounds as well. Thus we include the coinage metals, Cu, Ag, and Au.

The transition elements are all metals, mostly hard strong ones which conduct heat and electricity well. They form many colored and paramagnetic compounds because of their partially filled shells.

In this part of the book we treat them in detail, beginning here with an account of their electronic structures, spectra, magnetic properties, and some other related matters. We then deal with the *d*-block elements, followed by the lanthanides and then the actinides. It would be well for the student to review Section 8-12 before proceeding.

23-2 Ligand Field Theory

The term "ligand field theory" refers to the entire body of theoretical apparatus used to understand the bonding and associated electronic (magnetic, spectroscopic, etc.) properties of complexes and other compounds formed by the transition elements.

There is nothing fundamentally different about the bonding in transition metal compounds as compared with that in compounds of the main group elements. All the usual forms of valence theory that are applied to the main group elements can be applied, successfully, to the transition elements. In general, the molecular orbital method applied to the transition metal compounds gives valid and useful results, the more so as the level of approximation is raised, just as in all other cases.

There are, however, two things that set the study of the electronic structures of transition metal compounds apart from the remaining body of valence theory.

One is the presence of partly filled *d* and *f* shells. This leads to experimental observations not possible in most other cases: paramagnetism, visible absorption spectra, and apparently irregular variations in thermodynamic and structural properties. The second is that there is a crude but effective approximation, called *crystal field theory*, that provides a powerful yet simple method of understanding and correlating all of those properties which arise primarily from the presence of the partly filled shells.

The crystal field theory provides a way of determining, by simple electrostatic considerations, how the energies of the metal ion orbitals will be affected by the set of surrounding atoms or ligands. It works best when the symmetry is high but, with additional effort, can be applied more generally. Crystal field theory is a *model* and not a realistic description of the forces actually at work. However, its simplicity and convenience have earned it a place in the coordination chemist's "toolbox."

In the immediately following sections the crystal field theory is described and illustrated. Then the more complete molecular orbital method is outlined. After that, the electronic properties of transition metal complexes are discussed in terms of the "orbital splittings," which the crystal field theory enables us to work out relatively easily.

Our attention will be confined entirely to the *d*-block elements, and will be focused primarily on those of the 3*d* series. This is where the crystal field theory works best. The splittings of *f* orbitals are generally so small that they are not chemically important.

23-3
The Crystal Field Theory

Let us consider a metal ion, M^{m+}, lying at the center of an octahedral set of point charges, as is shown in *Fig. 23-1*. Let us suppose that this metal ion has a single *d* electron outside of closed shells; such an ion might be Ti^{3+}, V^{4+}, etc. In the free ion, this *d* electron would have had equal probability of being in any one of the five *d* orbitals, since all are equivalent. Now, however, the *d* orbitals are not all equivalent. Some are concentrated in regions of space closer to the negative ions than

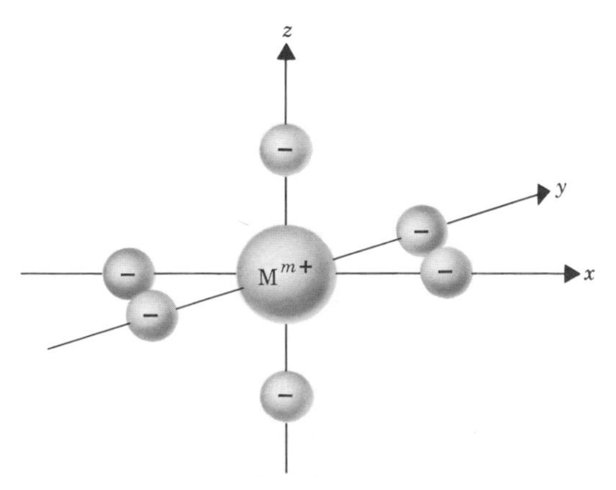

Figure 23-1 *Sketch showing six negative charges arranged octahedrally around a central* M^{m+} *ion with a set of Cartesian axes for reference.*

are others, and the electron will obviously prefer to occupy the orbital(s) in which it can get as far as possible from the negative charges. Recalling the shapes of the *d* orbitals (*Fig. 2-3*) and comparing them with *Fig. 23-1*, we see that both the d_{z^2} and $d_{x^2-y^2}$ orbitals have lobes that are heavily concentrated in the vicinity of the charges, whereas the d_{xy}, d_{yz} and d_{zx} orbitals have lobes that project between the charges. This is illustrated in *Fig. 23-2*. It can also be seen that each of the three orbitals in the latter group, namely, d_{xy}, d_{yz}, d_{zx}, is equally favorable for the electron; these three orbitals have entirely equivalent environments in the octahedral complex. The two relatively unfavorable orbitals, d_{z^2} and $d_{x^2-y^2}$, are also equivalent; this is not obvious from inspection of *Fig. 23-2*, but *Fig. 23-3* shows why it is so. As indicated, the d_{z^2} orbital can be resolved into a linear combination

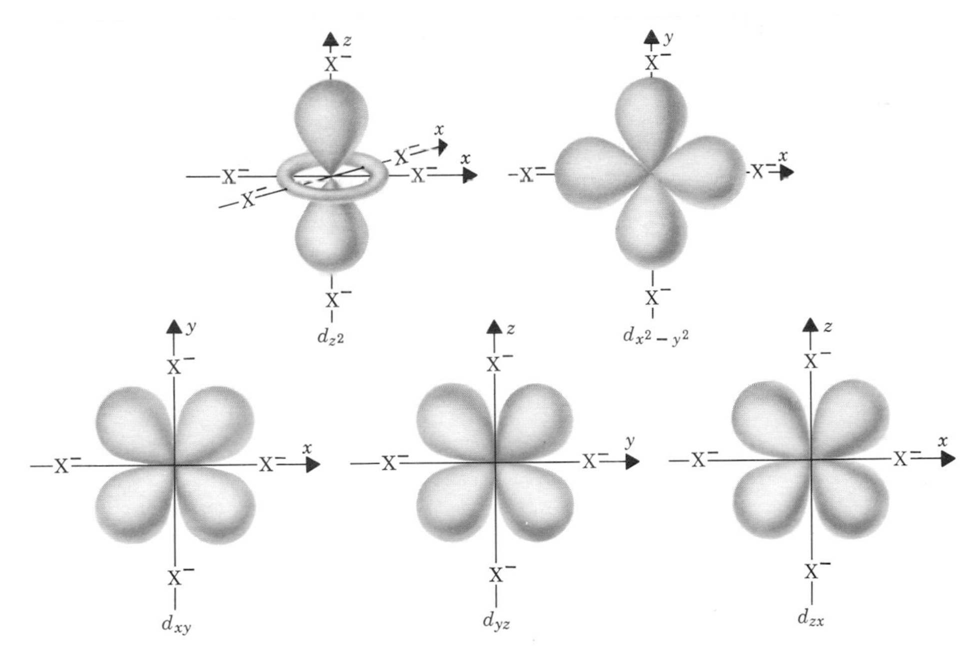

Figure 23-2 *Sketches showing the distribution of electron density in the five d orbitals with respect to a set of six octahedrally arranged negative charges (cf. Fig. 23-1).*

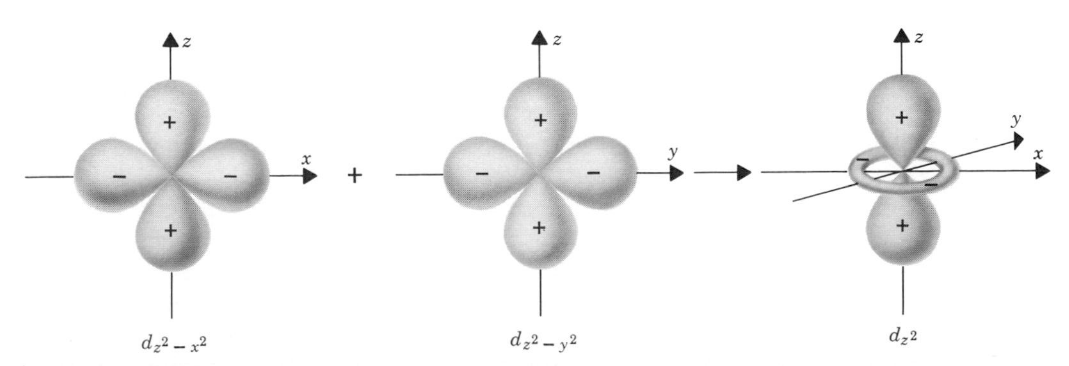

Figure 23-3 *Sketches of $d_{z^2-x^2}$ and $d_{z^2-y^2}$ orbitals which are usually combined to make the d_{z^2} orbital.*

of two orbitals, $d_{z^2-x^2}$ and $d_{z^2-y^2}$, each of which is obviously equivalent to the $d_{x^2-y^2}$ orbital. It is to be stressed, however, that these two orbitals do not have separate existences, and the resolution of the d_{z^2} orbital in this way is only a device to persuade the reader *pictorially* that d_{z^2} is equivalent to $d_{x^2-y^2}$ in relation to the octahedral distribution of charges.

Thus, in the octahedral environment of six negative charges, the metal ion now has two kinds of d orbitals: three of one kind, equivalent to one another and labeled t_{2g}, and two of another kind, equivalent to each other, labeled e_g; furthermore, the e_g orbitals are of higher energy than the t_{2g} orbitals. These results may be expressed in an energy level diagram as shown in *Fig. 23-4a.*

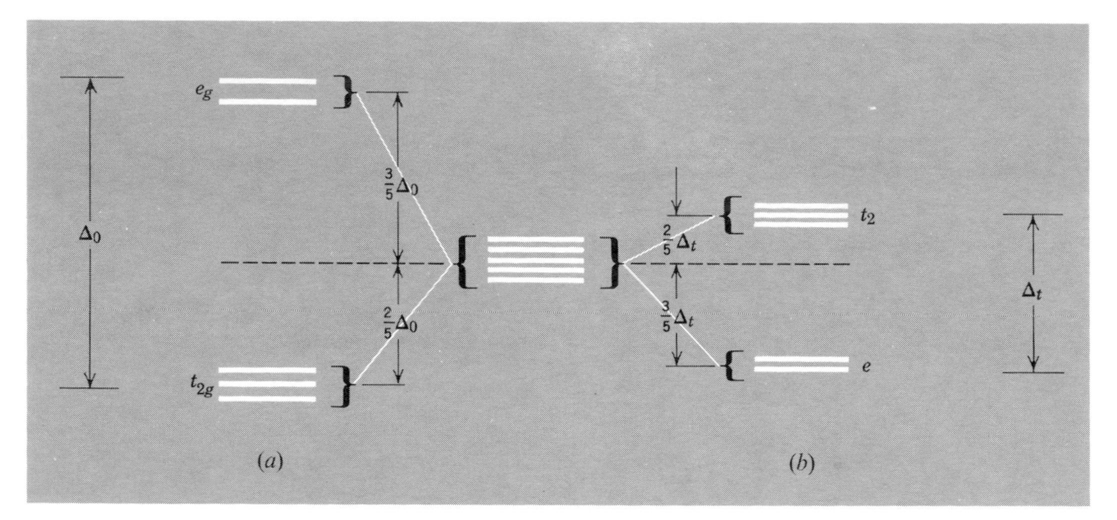

Figure 23-4 *Energy-level diagrams showing the splitting of a set of d orbitals by octahedral and tetrahedral electrostatic crystal fields.* (*a*) *Octahedral complex.* (*b*) *Tetrahedral complex.*

In *Fig. 23-4a* it will be seen that we have designated the energy difference between the e_g and the t_{2g} orbitals as Δ_o, where the subscript o stands for octahedral. The additional feature of *Fig. 23-4a*—the indication that the e_g levels lie $\frac{3}{5}\Delta_o$ above and the t_{2g} levels lie $\frac{2}{5}\Delta_o$ below the energy of the unsplit d orbitals—will now be explained. Let us suppose that a cation containing ten d electrons, two in each of the d orbitals, is first placed at the center of a hollow sphere whose radius is equal to the M—X internuclear distance and that charge of total quantity $6e$ is spread uniformly over the sphere. In this spherically symmetric environment the d orbitals are still fivefold degenerate.* The entire energy of the system, that is, the metal ion and the charged sphere, has a definite value. Now suppose the total charge on the sphere is caused to collect into six discrete point charges, each of magnitude e, and each lying at a vertex of an octahedron but still on the surface of the sphere. Merely redistributing the negative charge over the surface of the sphere in this manner cannot alter the total energy of the system when the metal ion consists entirely of spherically symmetrical electron shells, and yet we have already seen that, as a result of this redistribution, electrons in e_g orbitals now have higher energies than those in t_{2g} orbitals. It must therefore be that the

* The energy of all orbitals is, of course, greatly raised when the charged sphere encloses the ion.

total increase in energy of the four e_g electrons equals the total decrease in energy of the six t_{2g} electrons. This then implies that the rise in the energy of the e_g orbitals is 6/4 times the drop in energy of the t_{2g} orbitals, which is equivalent to the 3/5 : 2/5 ratio shown.

This pattern of splitting, in which the algebraic sum of all energy shifts of all orbitals is zero, is said to "preserve the center of gravity" of the set of levels. This center of gravity rule is quite general for any splitting pattern when the forces are purely electrostatic and where the set of levels being split is well removed in energy from all other sets with which they might be able to interact.

By an analogous line of reasoning it can be shown that the electrostatic field of four charges surrounding an ion at the vertices of a tetrahedron causes the d shell to split up as shown in *Fig. 23-4b*. In this case the d_{xy}, d_{yz}, and d_{zx} orbitals are less stable than the d_{z^2} and $d_{x^2-y^2}$ orbitals. This may be appreciated qualitatively if the spatial properties of the d orbitals are considered with regard to the tetrahedral array of four negative charges as is depicted in *Fig. 23-5*. If the cation, the

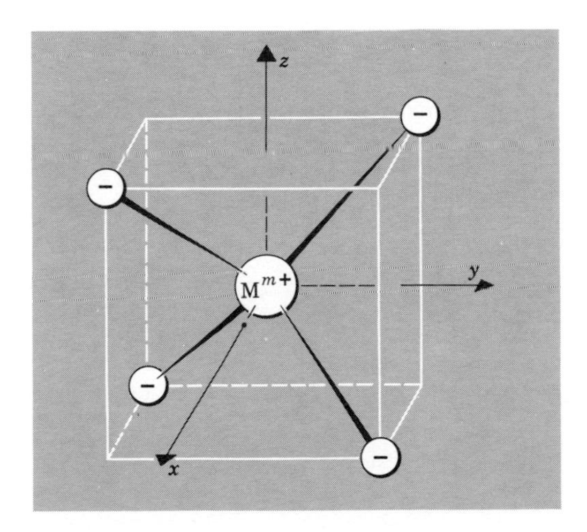

Figure 23-5 *Sketch showing the tetrahedral arrangement of four negative charges around a cation, M^{m+}, with respect to coordinate axes which may be used in identifying the d orbitals.*

anions, and the cation-anion distance are the same in both the octahedral and tetrahedral cases, it can be shown that

$$\Delta_t = \frac{4}{9}\,\Delta_0$$

In other words, other things being about equal, the crystal field splitting in a tetrahedral complex will be about one half the magnitude of that in an octahedral complex.

The above results have been derived on the assumption that ionic ligands, such as F^-, Cl^-, or CN^-, may be represented by point negative charges. Ligands that are neutral, however, are dipolar (e.g., 23-I and 23-II), and they approach

$$\delta^-\!:\!N\!\left(H^{\delta+}\right)_3 \qquad \delta^-\!:\!\ddot{O}\!\left(H^{\delta+}\right)_2$$

23-I 23-II

the metal ion with their negative poles. Actually, in the field of the positive metal ion such ligands are further polarized. Thus, in a complex such as a hexammine, the metal ion is surrounded by six dipoles with their negative ends closest; this array has the same general effects on the d orbitals as an array of six anions, so that all of the above results are valid for complexes containing neutral, dipolar ligands.

We next consider the pattern of splitting of the d orbitals in tetragonally distorted octahedral complexes and in planar complexes. We begin with an octahedral complex, MX_6, from which we slowly withdraw two *trans*-ligands. Let these be the two on the z axis. As soon as the distance from M^{m+} to these two ligands becomes greater than the distance to the other four, new energy differences among the d orbitals arise. First, the degeneracy of the e_g orbitals is lifted, the z^2 orbital becoming more stable than the $(x^2 - y^2)$ orbital. This happens because the ligands on the z axis exert a much more direct repulsive effect on a d_{z^2} electron than upon a $d_{x^2-y^2}$ electron. At the same time the threefold degeneracy of the t_{2g} orbitals is lifted. As the ligands on the z axis move away, the yz and zx orbitals remain equivalent to one another, but they become more stable than the xy orbital because their spatial distribution makes them more sensitive to the charges along the z axis than is the xy orbital. Thus for a small tetragonal distortion of the type considered, we may draw the energy-level diagram shown in *Fig. 23-6*. It should be obvious

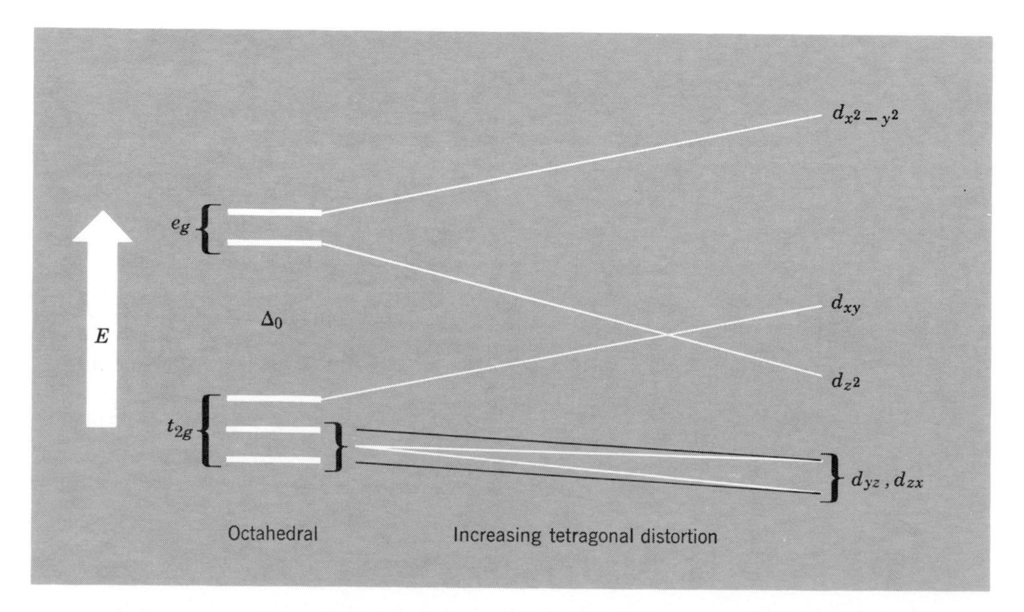

Figure 23-6 *Energy-level diagram showing the further splitting of the d orbitals as an octahedral array of ligands becomes progressively distorted by the withdrawal of two trans ligands, specifically those lying on the z axis.*

that for the opposite type of tetragonal distortion, that is, one in which two *trans*-ligands lie closer to the metal ion than do the other four, the relative energies of the split components will be inverted.

As *Fig. 23-6* shows, it is in general *possible* for the tetragonal distortion to become so large that the z^2 orbital eventually drops below the xy orbital. Whether this will actually happen for any particular case, even when the two *trans*-ligands are completely removed so that we have the limiting case of a square, 4-coordinated complex, depends on quantitative properties of the metal ion and the ligands concerned. Semiquantitative calculations with parameters appropriate for square complexes of Co^{II}, Ni^{II}, and Cu^{II} lead to the energy-level diagram shown in *Fig. 23-7*, in which the z^2 orbital has dropped so far below the xy orbital that it is nearly as stable as the (yz, zx) pair. As *Fig. 23-6* indicates, the d_{z^2} level might even drop below the (d_{xz}, d_{yz}) levels and, in fact, experimental results suggest that in some cases (e.g., $PtCl_4^{2-}$) it does.

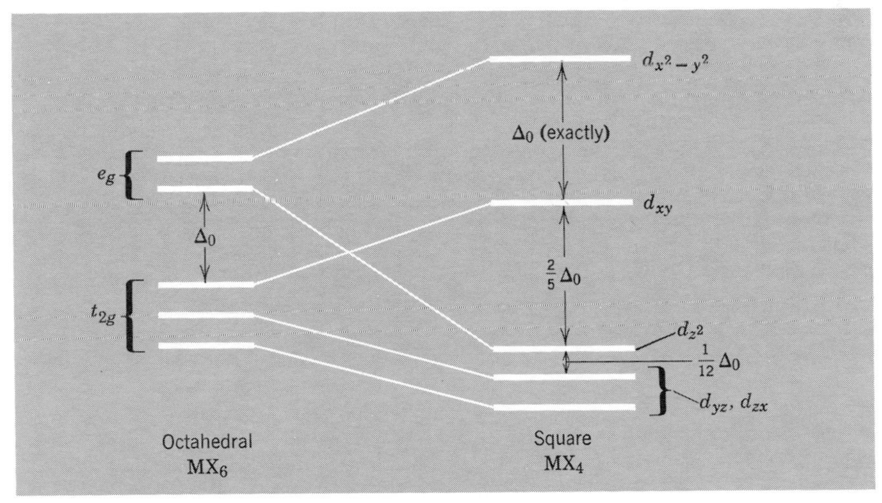

M = Co^{II}, Ni^{II}, Cu^{II}

Figure 23-7 *Approximate energy-level diagram for corresponding octahedral and square complexes of some metal ions in the first transition series.*

23-4
Other Forms of Ligand Field Theory

The electrostatic crystal field theory is the simplest model which can account for the fact that the d orbitals split up into subsets in ligand environments. It is, of course, a physically unrealistic model in certain ways, and it is also incomplete as a treatment of metal-ligand bonding, since it deals only with the d orbitals. It is possible to treat the electronic structures of complexes from a molecular orbital point of view. This is more general, more complete, and potentially more accurate. It includes the crystal field model as a special case.

Let us consider first an octahedral complex MX_6 in which the ligand X has only a sigma orbital, directed toward the metal atom, and no π orbitals. The six σ

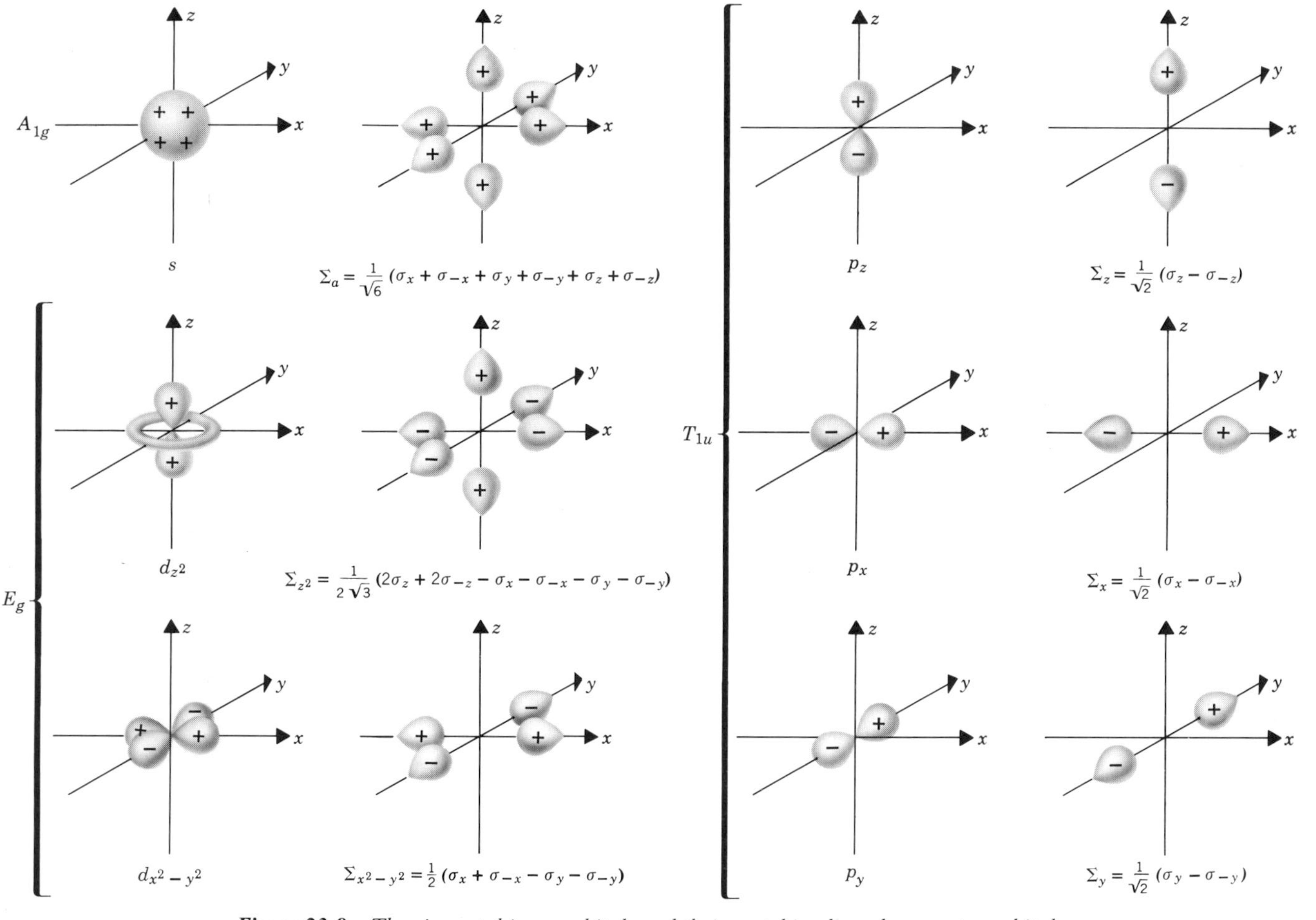

Figure 23-8 *The six metal ion σ orbitals and their matching ligand symmetry orbitals.*

orbitals, $\sigma_x, \sigma_{-x}, \sigma_y, \ldots, \sigma_{-z}$, can combine, by suitable sign combinations, in six ways, each of which is able to overlap with one, and only one, of the six metal orbitals, d_{z^2}, $d_{x^2-y^2}$, s, p_x, p_y, and p_z, as is shown in *Fig. 23-8*. Each such overlap results in the formation of one bonding and one antibonding molecular orbital, according to the general principles of MO theory described in Chapter 3. Since the three d orbitals d_{xy}, d_{yz}, and d_{zx} have no net overlap with any of the ligand σ orbitals, they remain unaffected. They are designated, as a set, t_{2g}. *Figure 23-9* gives an energy-level diagram that shows the results of these σ interactions.

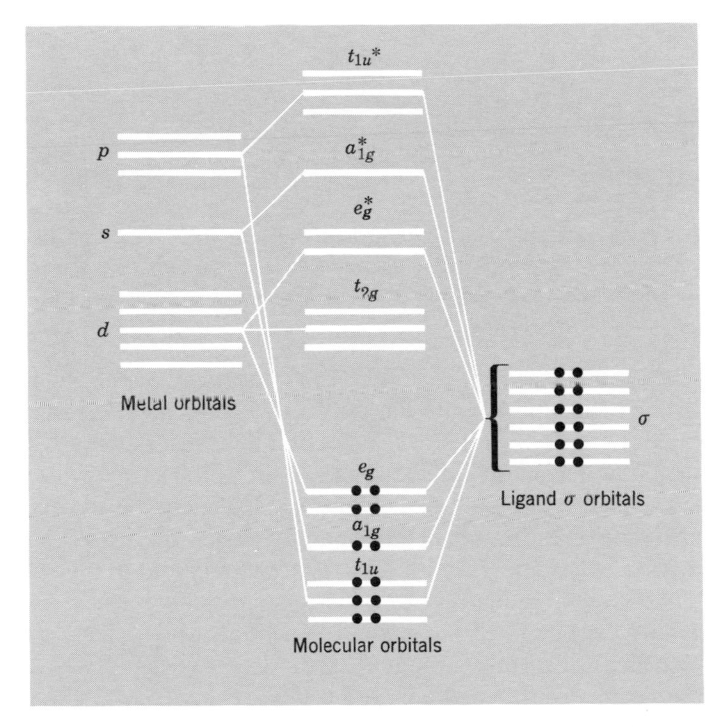

Figure 23-9 *A molecular orbital energy-level diagram for an octahedral complex,* MX_6, *where the ligand atom,* X, *has no* π *orbitals.*

The three MO's (bonding or antibonding) derived from the p orbitals have the same energy (they are degenerate), and are denoted t_{1u} (or t_{1u}^*). Similarly, the two MO's derived from the d_{z^2} and $d_{x^2-y^2}$ orbitals are degenerate and are denoted e_g (or e_g^*). The s orbital forms MO's denoted a_{1g} or a_{1g}^*. If each of the ligand σ orbitals originally contained an electron pair (which is the only situation of practical interest), these six electron pairs will then be found in the six ($3t_{1u}$, $2e_g$, a_{1g}) σ-bonding orbitals of the complex, as also is shown in *Fig. 23-9*.

It is evident that the MO discussion has lead to a result qualitatively the same as that from the crystal field theory with regard to the metal d orbitals: they are split into a set of two, e_g^*, and a set of three, t_{2g}, with the former having a higher energy than the latter, The MO picture also shows explicitly how the main binding energy of the complex arises, namely, by the formation of six 2-electron bonds. The main difference between the MO and the crystal field results is that the e_g^* orbitals as they are obtained in the MO treatment *are not pure metal d orbitals.*

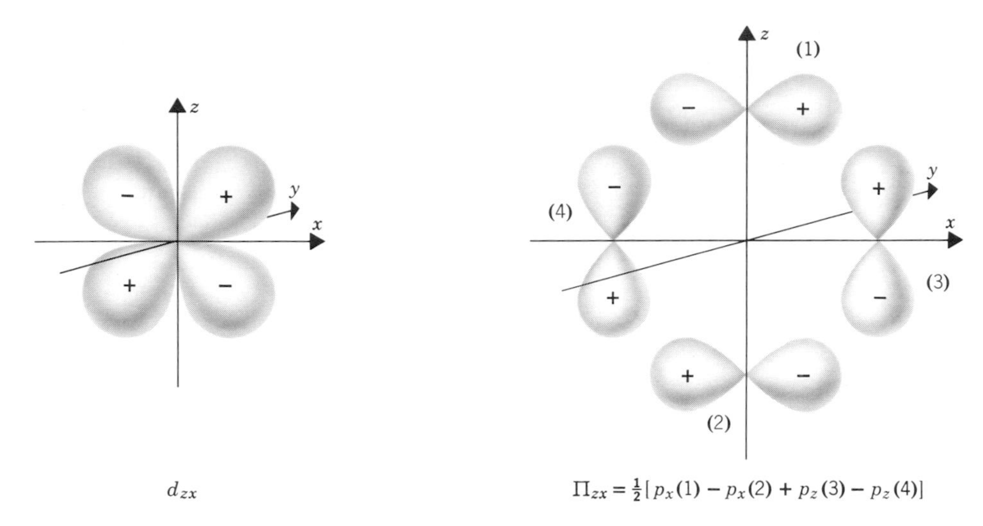

Figure 23-10 *At the right is the symmetry orbital made up of ligand p orbitals which has the proper symmetry to give optimum interaction with the metal ion d_{zx} orbital shown at the left. There are quite analogous symmetry orbitals, π_{xy} and π_{yz}, which are similarly related to the metal ion d_{xy} and d_{yz} orbitals.*

We can generalize the MO treatment by supposing that the ligand atoms also possess π orbitals. Such π orbitals can overlap with the d_{xy}, d_{yz}, and d_{zx} orbitals, as is illustrated for the d_{zx} orbital in *Fig. 23-10*. Thus, instead of only one set of t_{2g} molecular orbitals, which are pure *d* orbitals, there will now be two sets. The positions of these sets of t_{2g} and t_{2g}^* orbitals in the MO energy-level diagram is quite variable depending on the nature of the ligand π orbitals. One case of rather general importance arises when the ligand π orbitals are empty and of higher energy than the metal *d* orbitals. Ligands that provide this situation include (1) phosphines, where the empty π orbitals are phosphorus 3*d* orbitals, and (2) CN^- and CO where the empty π orbitals are antibonding $p\pi^*$ orbitals.

The interaction of the high-energy ligand π orbitals with the metal t_{2g} orbitals results in depressing the latter and thus increasing the separation between the t_{2g} and e_g^* orbitals, as shown in *Fig. 23-11*.

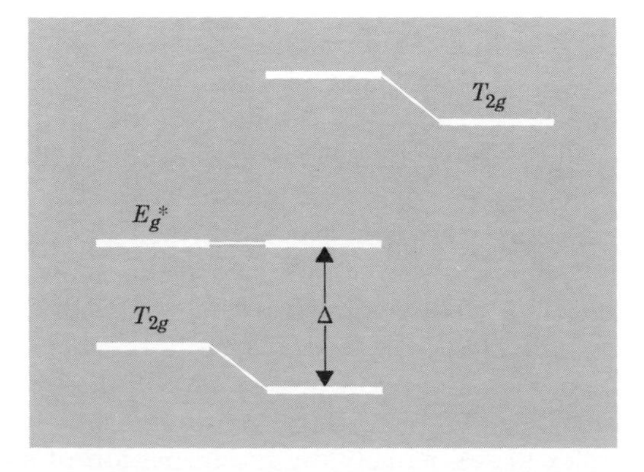

Figure 23-11 *Energy-level diagrams showing how π interactions can affect the value of Δ, when the ligands have π orbitals of higher energy than the metal t_{2g} orbitals.*

From the molecular orbital point of view, we see that a number of factors influence the ligand field splitting of the metal "*d* orbitals" and, further, that the "*d* orbitals" of crystal field theory are actually not pure *d* orbitals. It is remarkable, however, that the simple crystal field model is nevertheless a useful, qualitative working tool. In practice we do not try to use it to make quantitative predictions; that is, we do not try to calculate Δ_o (or Δ_t or any other *d* orbital splitting) from theory. Instead we derive these splittings from electronic spectra and only use the qualitative features of the *d*-orbital splitting patterns as given by crystal field theory.

23-5 Magnetic Properties of Transition Metal Complexes

One of the most useful applications of ligand field theory—whether in the simple electrostatic (crystal field) form or in a more sophisticated form—is to understand and correlate the magnetic properties of transition metal complexes. This is important because, when properly interpreted, the magnetic properties of these compounds are very useful in identifying and characterizing them.

The most basic questions to ask concerning any paramagnetic ion is: How many unpaired electrons are present? We now see how this question may be handled in terms of the orbital splittings described in the preceding sections. We have already noted (Section 2-6) that according to Hund's first rule, if a group of n or less electrons (say n') occupy a set of n degenerate orbitals, they will spread themselves out among the orbitals and give n' unpaired spins. This is true because pairing of electrons is an unfavorable process; energy must be expended to make it occur. If two electrons are not only to have their spins paired but also to be placed in the same orbital, there is a further unfavorable energy contribution because of the increased electrostatic repulsion between electrons that are compelled to occupy the same regions of space. Let us suppose now that in some hypothetical molecule we have two orbitals separated by an energy ΔE and that two electrons are to occupy these orbitals. By referring to *Fig. 23-12*, we see that when we place one electron in each orbital, their spins will remain uncoupled and their combined energy will be $(2E_0 + \Delta E)$. If we place both of them in the lower orbital, their spins will have to be coupled to satisfy the exclusion principle, and the total energy will

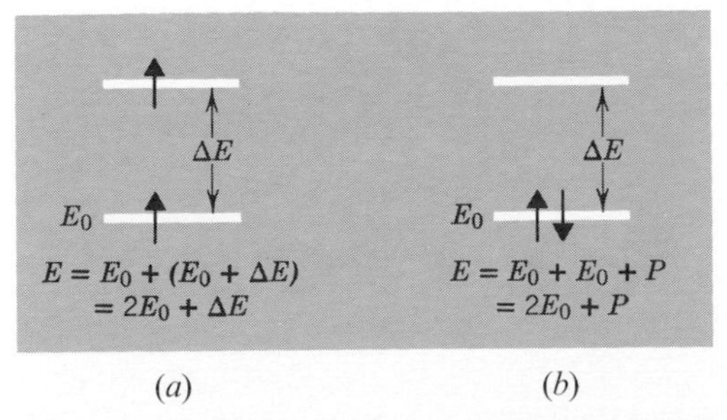

(*a*) (*b*)

Figure 23-12 *A hypothetical two-orbital system in which two possible distributions of two electrons and the resulting total energies are as shown.*

be $(2E_0 + P)$, where P stands for the energy required to cause pairing of two electrons in the same orbital. Thus, whether this system will have distribution (a) or (b) for its ground state depends on whether ΔE is greater or less than P.

Octahedral Complexes. An argument of the above type can be applied to octahedral complexes, using the *d*-orbital splitting diagram previously deduced. As is indicated in *Fig. 23-13*, we may place one, two, or three electrons in the *d* orbitals without any possible uncertainty about how they will occupy the orbitals. They will naturally enter the more stable t_{2g} orbitals with their spins all parallel, and this will be true irrespective of the strength of the crystal field as measured by the magnitude of Δ_o. Furthermore, for ions with eight, nine, and ten *d* electrons, there is only one possible way in which the orbitals may be occupied to give the lowest energy (see *Fig. 23-13*). For each of the remaining configurations, d^4, d^5, d^6, and d^7, two possibilities exist, and the question of which one represents the ground state can only be answered by comparing the values of Δ_o and P, an average pairing energy. The two configurations for each case, together with simple expressions for their energies, are set out in *Fig. 23-14*. The configurations with the maximum possible number of unpaired electrons are called the *high-spin* configurations, and those with the minimum number of unpaired spins are called the

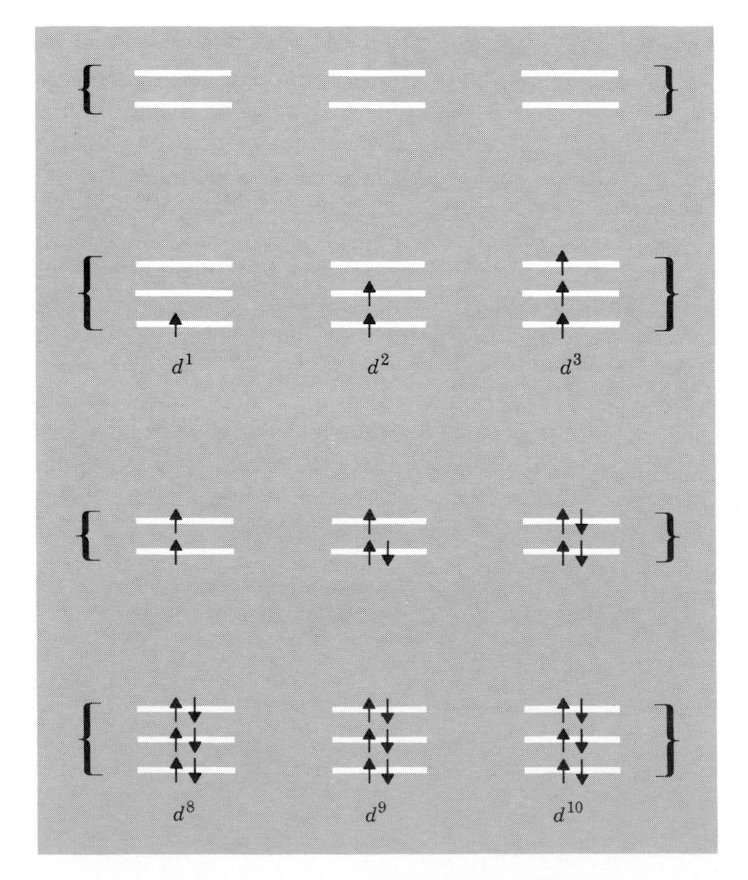

Figure 23-13 *Sketches showing the unique ground state occupancy schemes for d orbitals in octahedral complexes with d configurations d^1, d^2, d^3, d^8, d^9, d^{10}.*

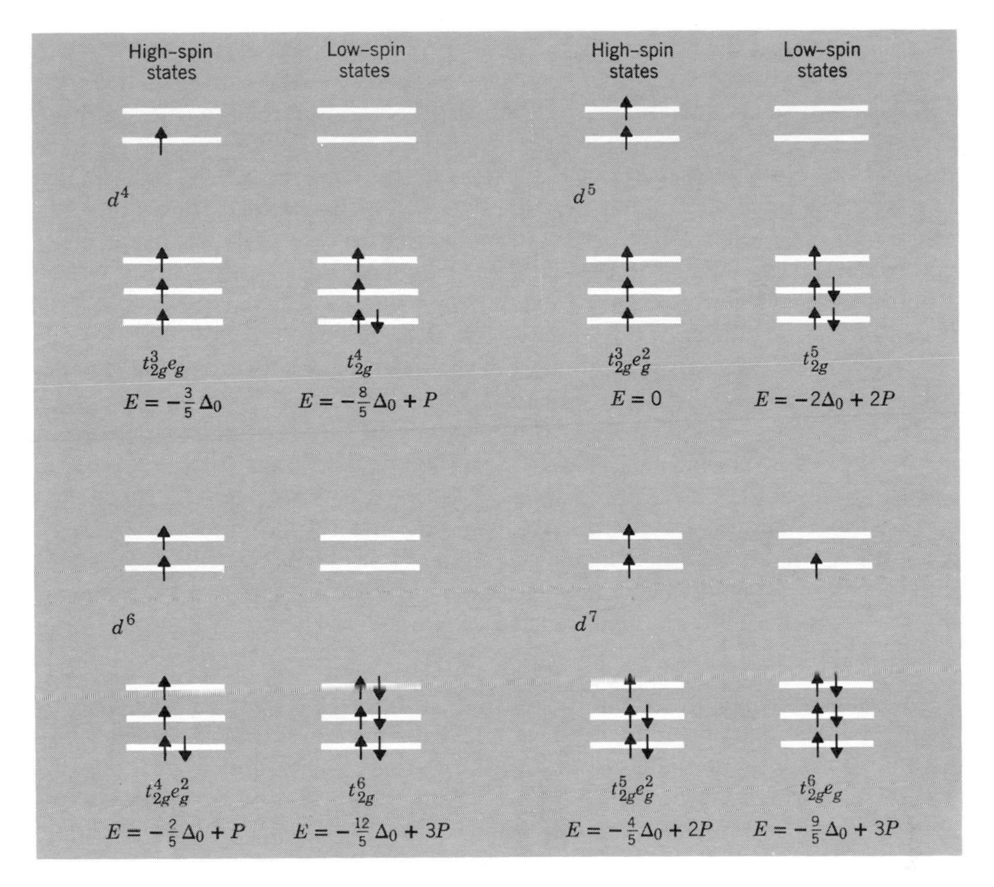

Figure 23-14 *Diagrams showing the possible high-spin and low-spin ground states for d^4, d^5, d^6 and d^7 ions in octahedral crystal fields including the notation for writing out the configurations and expressions for their energies, derived as explained in the text.*

low-spin or spin-paired configurations. These configurations can be written out in a notation similar to that used for electron configurations of free atoms, whereby we list each occupied orbital or set of orbitals, using a right superscript to show the number of electrons present. For example, the ground state for a d^3 ion in an octahedral field is t_{2g}^3; the two possible states for a d^5 ion in an octahedral field are t_{2g}^5 and $t_{2g}^3e_g^2$. This notation is further illustrated in *Fig. 23-14*. The energies are referred to the energy of the unsplit configuration (the energy of the ion in a spherical shell of the same total charge) and are simply the sums of $-2/5\Delta_o$ for each t_{2g} electron, $+3/5\Delta_o$ for each e_g electron and P for every pair of electrons occupying the same orbital.

For each of the four cases where high- and low-spin states are possible, we may obtain from the equations for the energies which are given in *Fig. 23-13* the following expression for the relation between Δ_o and P at which the high- and low-spin states have equal energies:

$$\Delta_o = P$$

The relationship is the same in all cases, and means that the spin state of any ion in an octahedral electrostatic field depends simply on whether the magnitude of the field as measured by the splitting energy, Δ_o, is greater or less than the mean pairing energy, P, for the particular ion. For a particular ion of the d^4, d^5, d^6, or d^7 type, the stronger the crystal field, the more likely it is that the electrons will crowd as much as possible into the more stable t_{2g} orbitals, whereas in the weaker crystal fields, where $P > \Delta_o$, the electrons will remain spread out over the entire set of d orbitals as they do in the free ion. For ions of the other types, d^1, d^2, d^3, d^8, d^9, and d^{10}, the number of unpaired electrons is fixed at the same number as in the free ion irrespective of how strong the crystal field may become.

Approximate theoretical estimates of the mean pairing energies for the relevant ions of the first transition series have been made from spectroscopic data. In Table 23-1 these energies, along with Δ_o values for some complexes (derived by methods described in the next section), are listed. It is seen that the theory developed above affords correct predictions in all cases. We note further that the mean pairing energies vary irregularly from one metal ion to another as do the values of Δ_o for a given set of ligands. Thus, as Table 23-1 shows, the d^5 systems should be exceptionally stable in their high-spin states, whereas the d^6 systems should be exceptionally stable in their low-spin states. These expectations are in excellent agreement with the experimental facts.

Tetrahedral Complexes. Metal ions in tetrahedral electrostatic fields may be treated by the same procedure outlined above for the octahedral cases. For tetrahedral fields it is found that for the d^1, d^2, d^7, d^8, and d^9 cases only high-spin states are possible, whereas for d^3, d^4, d^5, and d^6 configurations both high-spin and low-spin states are in principle possible. Once again the existence of low-spin states requires that $\Delta_t > P$. Since Δ_t values are only about half as great as Δ_o values, it is to be expected that low-spin tetrahedral complexes of first transition series ions with d^3, d^4, d^5, and d^6 configurations would be scarce or even unknown. None have been found.

Square and Tetragonally-Distorted Octahedral Complexes. These two cases must be considered together because, as we noted previously, they merge into one another.

Table 23-1 *Crystal Field Splittings, Δ_o, and Mean Electron Pairing Energies, P, for Several Transition Metal Ions (Energies in cm^{-1})*

Configuration	Ion	P	Ligands	Δ_o	Predicted	Observed
d^4	Cr^{2+}	23,500	$6H_2O$	13,900	High	High
	Mn^{3+}	28,000	$6H_2O$	21,000	High	High
d^5	Mn^{2+}	25,500	$6H_2O$	7,800	High	High
	Fe^{3+}	30,000	$6H_2O$	13,700	High	High
d^6	Fe^{2+}	17,600	$6H_2O$	10,400	High	High
			$6CN^-$	33,000	Low	Low
	Co^{3+}	21,000	$6F^-$	13,000	High	High
			$6NH_3$	23,000	Low	Low
d^7	Co^{2+}	22,500	$6H_2O$	9,300	High	High

Even when the strictly octahedral environment does not permit the existence of a low-spin state, as in the d^8 case, distortions of the octahedron will cause further splitting of degenerate orbitals which may become great enough to overcome pairing energies and cause electron pairing. Let us consider as an example the d^8 system in an octahedral environment which is then subjected to a tetragonal distortion. We have already seen (*Fig. 23-6*) how a decrease in the electrostatic field along the z axis may arise, either by moving the two z-axis ligands out to a greater distance than are their otherwise identical neighbors in the xy plane, or by having two different ligands on the z axis which make an intrinsically smaller contribution to the electrostatic potential than do the four in the xy plane. Irrespective of its origin, the result of a tetragonal distortion of an initially octahedral field is to split apart the $(x^2 - y^2)$ and z^2 orbitals. We have also seen that if the tetragonal distortion, that is, the disparity between the contributions to the electrostatic potential of the two z axis ligands and the other four, becomes sufficiently great, the z^2 orbital may fall below the xy orbital. In either case, the two least stable d orbitals are now no longer degenerate but are separated by some energy, Q. Now the question of whether the tetragonally distorted d^8 complex will have high- or low-spin depends on whether the pairing energy, P, is greater or less than the energy Q. *Figure 23-15a* shows the situation for the case of a "weak" tetragonal distortion, that is, for one in which the second highest d orbital is still d_{z^2}.

Figure 23-15b shows a possible arrangement of levels for a strongly tetragonally distorted octahedron, or for the extreme case of a square, four-coordinate complex (compare with *Fig. 23-7*), and the low-spin form of occupancy of these levels for a d^8 ion. In this case, due to the large separation between the highest and second highest orbitals, the high-spin configuration is impossible of attainment with the pairing energies of the real d^8 ions, for example, Ni^{II}, Pd^{II}, Pt^{II}, Rh^{I}, Ir^{I}, and

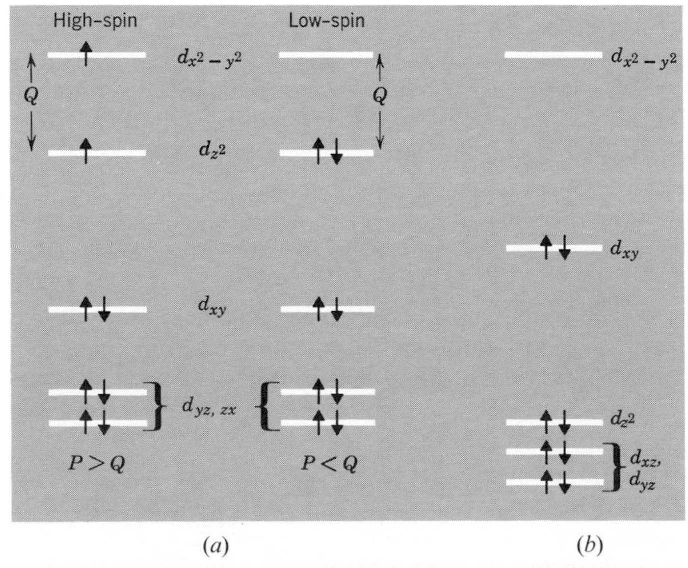

Figure 23-15 *Energy-level diagrams showing the possible high-spin and low-spin ground states for a d^8 system (e.g., Ni^{2+}) in a tetragonally distorted octahedral field. (a) Weak tetragonal distortion. (b) Strong distortion or square field.*

Au^{III}, which normally occur, and all square complexes of these species are diamagnetic. Similarly, for a d^7 ion in a square complex, as exemplified by certain Co^{II} complexes, only the low-spin state with one unpaired electron should occur, and this is in accord with observation.

Other Forms of Magnetic Behavior. In this section, we have just indicated how the number of unpaired electrons on a transition metal ion in a complex, or other compound, can be understood in terms of the *d*-orbital splitting. The experimental method for determining the number of unpaired electrons has been discussed in Section 2-8; it is based on measuring the magnetic susceptibility of the substance. Here we must point out that certain additional factors must be considered in attempting to relate the magnetic moments of individual ions with the measured susceptibilities of bulk compounds.

Diamagnetism (which was briefly mentioned in Section 2-8) is a property of all forms of matter. All substances contain at least some if not all electrons in closed shells. In closed shells there is no net angular momentum since the spin momenta cancel each other and so do the orbital momenta, and no net magnetic moment can result. However, when a substance is placed in a magnetic field, the closed shells are affected in such a way that the orbitals are all tipped and a small, net magnetic moment is set up in opposition to the applied field. This is called diamagnetism, and because the small induced moment is opposed to the applied field, the substance is repelled. In a substance that has no unpaired electrons, this will be the only response to the field. The substance will tend to move away from the strongest part of the field, and it is said to be diamagnetic. The susceptibility of a diamagnetic substance is negative and is independent of field strength and of temperature.

It is important to realize that even a substance that does have unpaired electrons also has diamagnetism because of whatever closed shells of electrons are also present. Thus the positive susceptibility measured is less than that expected for the unpaired electrons alone, because the diamagnetism partially cancels the paramagnetism. This is a small effect, typically amounting to less than 10% of the true paramagnetism, but in accurate work a correction for it must be applied.

Paramagnetism has already been discussed in Section 2-8. Simple paramagnetism occurs when the individual ions having the unpaired electrons are far enough apart to behave independently of one another. Curie's law (Eq. 2-8-1) is thus followed. The magnetic moment thus obtained can be directly, with allowance for small contributions (positive or negative) from orbital motion, interpreted in terms of the number of unpaired electrons.

Ferromagnetism and antiferromagnetism occur in substances where the individual paramagnetic atoms or ions are close together and each one is strongly influenced by the orientation of the magnetic moments of its neighbors. In ferromagnetism (so called because it is very conspicuous in metallic iron) the interaction is such as to cause all moments to tend to point in the same direction. This enormously enhances the magnitude of the susceptibility of the substance as compared with what it would be if all the individual moments behaved independently. Ferromagnetism is generally found in the transition metals, and also in some of their compounds.

Antiferromagnetism occurs when the nature of the interaction between neighboring paramagnetic ions is such as to favor opposite orientations of their magnetic moments, thus causing partial concellation. Antiferromagnetic substances thus have magnetic susceptibilities less than those expected for an array of independent magnetic ions. It occurs quite often among simple salts of ions such as Fe^{3+}, Mn^{2+}, and Gd^{3+}, which have large intrinsic magnetic moments. The antiferromagnetic coupling involves interaction through the anions lying between the metal atoms in the crystal, and disappears in dilute solutions.

Ferro- or antiferromagnetic behavior causes deviations from the Curie law, as is shown in *Fig. 23-16*. In each case there is a temperature at which the temperature dependence of the susceptibility changes abruptly. This is the Curie temperature, T_c, which is a characteristic property of the substance. Above T_c, the behavior is similar to that of the Curie law. Below T_c, the susceptibility either rises (ferromagnetism) or falls (antiferromagnetism) in a manner quite different from that implied by the Curie law. At the Curie temperature the effect of thermal energy in tending to randomize the individual spin orientations begins to get the upper hand over the ferro- or antiferromagnetic coupling interactions.

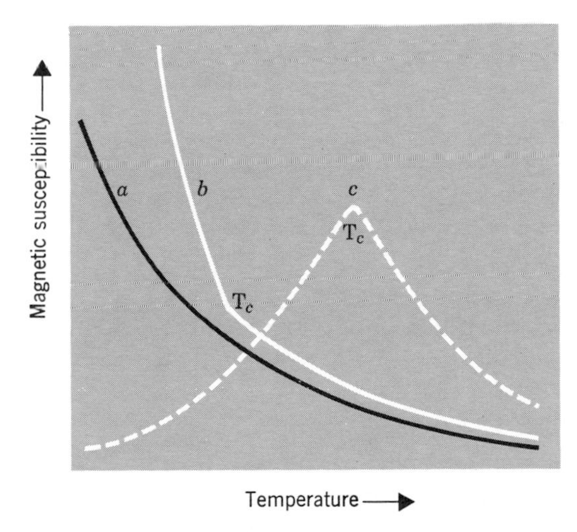

Figure 23-16 *Susceptibility versus temperature plots for (a) a simple paramagnetic (Curie law), (b) a ferromagnetic, and (c) an antiferromagnetic. T_c denotes the Curie temperatures in (b) and (c). For antiferromagnetics the Curie point is also often called the Néel temperature.*

23-6
Absorption Spectra

The simplest possible case is an ion with a d^1 configuration, lying at the center of an octahedral field, for example, the Ti^{III} ion in $[Ti(H_2O)_6]^{3+}$. The d electron will occupy a t_{2g} orbital. On irradiation with light of frequency ν, equal to Δ_o/h, where h is Planck's constant and Δ_o is the energy difference between the t_{2g} and the e_g orbitals, it should be possible for such an ion to capture a quantum of radiation and convert that energy into energy of excitation of the electron from the t_{2g} to the e_g orbitals. The absorption band that results from this process is found in the visible spectrum of the hexaquotitanium(III) ion, shown in *Fig. 23-17*, and is responsible

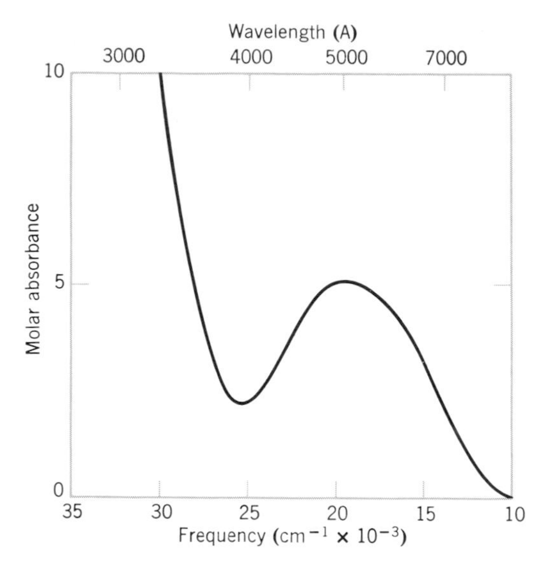

Figure 23-17 *The visible absorption spectrum of* $[Ti(H_2O)_6]^{3+}$.

for its violet color. Two features of this absorption band are of importance here: its position, and its intensity.

In discussing the positions of absorption bands in relation to the splittings of the d orbitals, it is convenient and common practice to use the same unit, the reciprocal centimeter or wave number, abbreviated cm^{-1}, for both the unit of frequency in the spectra and the unit of energy for the orbitals. With this convention, we see that the spectrum of *Fig. 23-17* tells us that Δ_o in $[Ti(H_2O)_6]^{3+}$ is 20,000 cm^{-1}.

We note in *Fig. 23-17* that the absorption band is very weak. Its molar absorbance at the maximum is 5 whereas one-electron transitions which are theoretically "allowed," usually have absorbances of 10^4–10^5. This suggests that the transition in question is not "allowed" but is instead "forbidden" according to quantum theory. That is indeed the case. An electronic transition in which there is no change in the value of the quantum number l is forbidden under the ordinary stimulus of the oscillating electric field of light. In the present case an electron is moving from one $3d$ orbital to another $3d$ orbital, both of which have $l = 2$ (by definition, since they are d orbitals). Thus, to a first approximation, the transition should have zero intensity.

The transition gives rise to a band of low, but not zero, intensity, because the orbitals involved do not actually have *pure* $3d$ character, as implied by the electrostatic crystal field theory. That simple picture is always perturbed slightly by various things not yet considered. In the case of $Ti(H_2O)_6{}^{3+}$ some of the vibrations of the ligands allow a little bit of p character to mix into the d orbitals. The transition is not, therefore, a pure d—d transition and thus gains a little of the intensity characteristic of an allowed $p \rightarrow d$ or $d \rightarrow p$ type transition.

In general, the absorption bands responsible for the colors of transition metal ions in their complexes are these so-called d—d transitions. They are always weak because they are basically forbidden, but they gain slight intensity because of deviations from pure d—d character.

For ions with more than one d electron, the d—d spectra have more than one band and the interpretation requires a more elaborate development of the theory than can be given here. Suffice it to say, however, that d—d spectra for ions in octahedral or tetrahedral complexes can always be analyzed to give a value for Δ_o or Δ_t, the d-orbital splitting.

Certain generalizations may be made about the dependence of the magnitudes of Δ values on the valence and atomic number of the metal ion, the symmetry of the coordination shell and the nature of the ligands. For octahedral complexes containing high-spin metal ions, it may be inferred from the accumulated data for a large number of systems that:

1. Δ_o values for complexes of the first transition series are 7500–12,500 cm^{-1} for divalent ions and 14,000–25,000 cm^{-1} for trivalent ions.

2. Δ_o values for corresponding complexes of metal ions in the same group and with the same valence increases by 30 to 50% on going from the first transition series to the second and by about this amount again from the second to the third. This is well illustrated by the Δ_o values for the complexes $[Co(NH_3)_6]^{3+}$, $[Rh(NH_3)_6]^{3+}$, and $[Ir(NH_3)_6]^{3+}$, which are, respectively, 23,000, 34,000, and 41,000 cm^{-1}.

3. Δ_t values are about 40 to 50% of Δ_o values for complexes differing as little as possible except in the geometry of the coordination shell, in agreement with theoretical expectation.

4. The dependence of Δ values on the identity of the ligands follows a regular order known as the spectrochemical series which will now be explained.

23-7 The Spectrochemical Series

It has been found by experimental study of the spectra of a large number of complexes containing various metal ions and various ligands, that ligands may be arranged in a series according to their capacity to cause d-orbital splittings. This series, for the more common ligands, is: $I^- < Br^- < Cl^- < F^- < OH^- < C_2O_4^{2-} < H_2O <$ —$NCS^- <$ py $< NH_3 <$ en $<$ bipy $<$ o-phen $< NO_2^- < CN^-$. The idea of this series is that the d-orbital splittings and, hence, the relative frequencies of visible absorption bands for two complexes containing the same metal ion but different ligands can be predicted from the above series whatever the particular metal ion may be. Naturally, one cannot expect such a simple and useful rule to be universally applicable. The following qualifications must be remembered in applying it.

1. The series is based on data for metal ions in common oxidation states. Because the nature of the metal-ligand interaction in an unusually high or unusually low oxidation state of the metal may be in certain respects qualitatively different from that for the metal in a normal oxidation state, striking violations of the order shown may occur for complexes in unusual oxidation states.

2. Even for metal ions in their normal oxidation states inversions of the order of adjacent or nearly adjacent members of the series are sometimes found.

23-8
Structural and Thermodynamic Effects of *d*-Orbital Splittings

Regardless of what type or level of theory is used to account for the existence of the *d*-orbital splittings, the fact that they do exist is of major importance. Their existence affects both structural and thermodynamic properties of the ions and their complexes.

Ionic Radii. *Figure 23-18* shows a plot of the octahedral radii of the divalent ions of the first transition series. The points for Cr^{2+} and Cu^{2+} are indicated with open circles because the Jahn–Teller effect, to be discussed below, makes it difficult to obtain these ions in truly octahedral environments, thus rendering the assessment of their "octahedral" radii somewhat uncertain. A smooth curve has also been drawn through the points for Ca^{2+}, Mn^{2+}, and Zn^{2+} ions, which have the electron configurations $t_{2g}^0e_g^0$, $t_{2g}^3e_g^2$, and $t_{2g}^6e_g^4$, respectively. In these three cases the distribution of *d*-electron density around the metal ion is spherical because all *d* orbitals are either unoccupied or equally occupied. Because the shielding of one *d* electron by another from the nuclear charge is imperfect, there is a steady contraction in the ionic radii. It is seen that the radii of the other ions are all below the values expected from the curve passing through Ca^{2+}, Mn^{2+}, and Zn^{2+}. This is because the *d* electrons in these ions are not distributed uniformly (i.e., spherically) about the nuclei as we shall now explain.

The Ti^{2+} ion has the configuration t_{2g}^2. This means that the negative charge of two *d* electrons is concentrated in those regions of space away from the metal-ligand bond axes. Thus, compared with the effect that they would have if distributed spherically around the metal nucleus, these two electrons provide abnormally little shielding between the positive metal ion and the negative ligands; therefore, the ligand atoms are drawn in closer than they would be if the *d* electrons were spherically distributed. Thus, in effect, the radius of the metal ion is smaller than that for the hypothetical, isoelectronic spherical ion. In V^{2+} this same effect is found in even greater degree because there are now three t_{2g} electrons providing much less shielding between metal ion and ligands than would three spherically

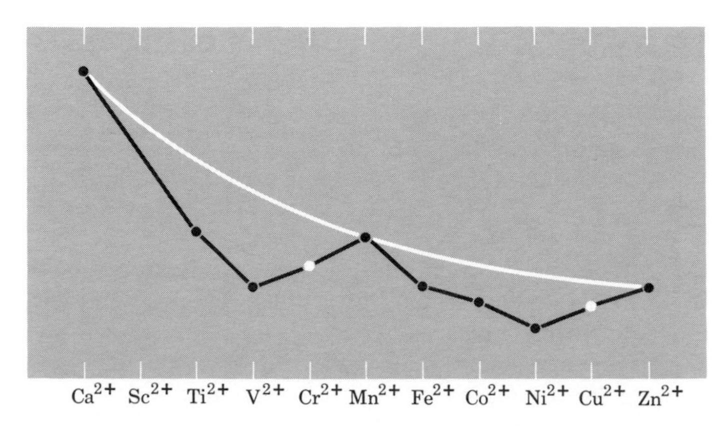

Figure 23-18 *The relative ionic radii of divalent ions of the first transition series. The dashed line is a theoretical curve explained in the text.*

distributed d electrons. For Cr^{2+} and Mn^{2+}, however, we have the configurations $t_{2g}^3e_g$ and $t_{2g}^3e_g^2$, in which the electrons added to the t_{2g}^3 configuration of V^{2+} go into orbitals that concentrate them mainly between the metal ion and the ligands. These e_g electrons thus provide a great deal more screening than would be provided by spherically distributed electrons, and indeed the effect is so great that the radii actually increase. The same sequence of events is repeated in the second half of the series. The first three electrons added to the spherical $t_{2g}^3e_g^2$ configuration of Mn^{2+} go into the t_{2g} orbitals where the screening power is abnormally low, and the radii therefore decrease abnormally rapidly. On going from Ni^{2+}, with the configuration $t_{2g}^6e_g^2$, to Cu^{2+} and Zn^{2+}, electrons are added to the e_g orbitals where their screening power is abnormally high, and the radii again cease to decrease and actually show small increases. Similar effects are found with trivalent ions, with ions of other transition series, and in tetrahedral complexes.

The Jahn–Teller Effect. In 1937 Jahn and Teller showed that in general no nonlinear molecule can be stable in a degenerate electronic state. The molecule must become distorted in such a way as to break down the degeneracy. It develops that one of the most important areas of application of this Jahn–Teller theorem is the stereochemistry of the complexes of certain transition metal ions.

To illustrate, we consider an octahedrally coordinated Cu^{2+} ion. There is one vacancy in the e_g orbitals, in either the $d_{x^2-y^2}$ or the d_{z^2} orbital. If the coordination is strictly octahedral, the two configurations $d_{x^2-y^2}^2d_{z^2}$ and $d_{x^2-y^2}d_{z^2}^2$, are of equal energy. This is the sense in which the electronic state of the Cu^{2+} ion is doubly degenerate. However, this is a state which, according to the Jahn–Teller theorem, cannot be stable, and the octahedron must distort so that the two configurations just mentioned are no longer of equal energy.

Actually, it is easy to see why this will happen. Suppose the actual configuration in the e_g orbitals is $d_{x^2-y^2}d_{z^2}^2$, The ligands along the z axis are much more screened from the charge of the Cu^{2+} ion than are the four ligands along the x and y axes. The z-axis ligands will therefore tend to move further away. As they do so, however, the d_{z^2} orbital will become more stable than the $d_{x^2-y^2}$ orbital, thus removing the degeneracy, as is shown in *Fig. 23-6*. Of course, if we begin with a $d_{x^2-y^2}^2d_{z^2}$ configuration, a distortion of the opposite kind would be expected. The question of which situation will actually occur is very difficult to predict, and there are, in fact, still other possibilities. However, it is the former type of distortion, the elongation on one axis, which is actually observed in a large number of Cu^{2+} complexes.

This is well illustrated by the copper(II) halides. In each case the Cu^{2+} ion has a coordination number of six, with four near neighbors in a plane and two more remote ones. The actual distances are shown in *Fig. 23-19*.

It is not difficult to see that the reasoning involved in the Cu^{2+} case will apply in all cases where an odd number (1 or 3) of electrons would occupy the e_g orbitals in an octahedral complex. In the case of a single electron, either the $d_{x^2-y^2}$ or the d_{z^2} orbital could be occupied, and the occupied orbital should "push away" the ligands toward which it is directed. The important cases in which this may be expected are

$t_{2g}^3e_g$: high-spin Cr^{2+} and Mn^{3+}

$t_{2g}^6e_g$: low-spin Co^{2+} and Ni^{3+}

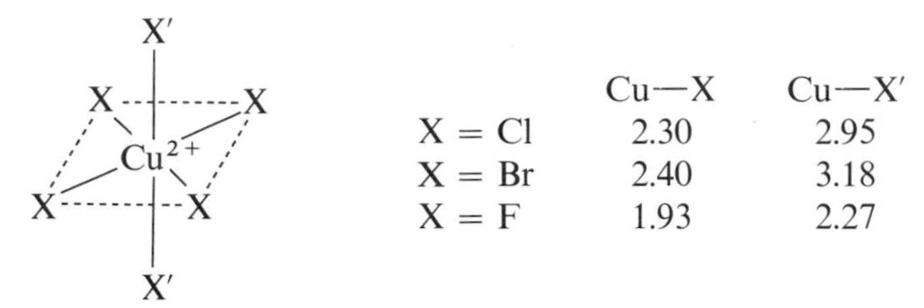

Figure 23-19 *The distorted six-coordination found in the copper*(II) *halides, distances in Ångstroms. An example of the Jahn–Teller effect.*

Distortions similar to those for Cu^{2+} are, indeed, found for the "octahedral" complexes of these ions.

Thermodynamic Effects. We have learned in Section 23-2 that the d orbitals of an ion in an octahedral field are split so that three of them become more stable (by $2\Delta_o/5$) and two of them less stable (by $3\Delta_o/5$) than they would be in the absence of the splitting. Thus, for example, a d^2 ion will have each of its two d electrons stabilized by $2\Delta_o/5$, giving a total stabilization of $4\Delta_o/5$. Recalling from Section 23-6 that Δ_o values run about 10,000 and 20,000 cm^{-1} for di- and trivalent ions of the first transition series, we can see that these "extra" stabilization energies—extra in the sense that they would not exist if the d shells of the metal ions were symmetrical as are the other electron shells of the ions—will amount to ~100 and ~200 kJ mol^{-1}, respectively, for di- and trivalent d^2 ions. These *ligand field stabilization energies*, LFSE's, are of course of the same order of magnitude as the energies of most chemical changes, and they will therefore play an important role in the thermodynamic properties of transition metal compounds.

Let us first consider high-spin octahedral complexes. Every t_{2g} electron represents a stability increase (i.e., energy lowering) of $2\Delta_o/5$, whereas every e_g electron represents a stability decrease of $3\Delta_o/5$. Thus, for any configuration $t_{2g}^p e_g^q$, the net stabilization will be given by $(2p/5-3q/5)\Delta_o$.

The results obtained for all of the ions, that is, d^0 to d^{10}, using this formula are collected in Table 23-2. Since the magnitude of Δ_o for any particular complex can be obtained from the spectrum, it is possible to determine the magnitudes of these

Table 23-2 *Ligand Field Stabilization Energies, LFSE's, for Octahedrally and Tetrahedrally Coordinated High-Spin Ions*

Number of d electrons	Stabilization energies		Difference,
	Oct.	Tetra.	Oct.–Tetra.[b]
1, 6	$2\Delta_o/5$	$3\Delta_t/5$	$\Delta_o/10$
2, 7[a]	$4\Delta_o/5$	$6\Delta_t/5$	$2\Delta_o/10$
3, 8	$6\Delta_o/5$	$4\Delta_t/5$	$8\Delta_o/10$
4, 9	$3\Delta_o/5$	$2\Delta_t/5$	$4\Delta_o/10$
0, 5, 10	0	0	0

[a] For the d^2 and d^7 ions, the figure obtained in this way and given above is not exactly correct because of the effect of configuration interaction.

[b] Assuming $\Delta_o = 2\Delta_t$.

crystal field stabilization energies independently of thermodynamic measurements and, thus, to determine what part they play in the thermodynamics of the transition metal compounds.

The enthalpies of hydration of the divalent ions of the first transition series are the energies of the processes:

$$M^{2+}(\text{gas}) + \infty H_2O = [M(H_2O)_6]^{2+}(\text{aq})$$

They can be estimated by using thermodynamic cycles. The energies calculated are shown by the filled circles in *Fig. 23-20*. It will be seen that a smooth curve, which is nearly a straight line, passes through the points for the three ions, Ca^{2+} (d^0), Mn^{2+}(d^5), and Zn^{2+} (d^{10}), which have no LFSE while the points for all other ions lie above this line. If we subtract the LFSE from each of the actual hydration energies, the values shown by open circles are obtained, and these fall on the smooth curve. It may be noted that, alternatively, LFSE's could have been estimated from *Fig. 23-20* and used to calculate Δ_o values. Either way, the agreement between the spectrally and thermodynamically assessed Δ_o values provides evidence for the fundamental correctness of the idea of *d*-orbital splitting.

Another important example of the thermodynamic consequences of ligand field splittings is shown in *Fig. 23-21*, where the lattice energies of the dichlorides of the metals from calcium to zinc are plotted versus atomic number. Once again they define a curve with two maxima and a minimum at Mn^{2+}. As previously, the energies for all the ions having LFSE's lie above the curve passing through the energies of the three ions which do not have any ligand field stabilization energy. Similar plots are obtained for the lattice energies of other halides and of the chalconides of di- and trivalent metals.

It is important to note that the LFSE's, critical as they may be in explaining the *difference* in energies between various ions in the series, make up only a small fraction, 5 to 10%, of the *total* energies of combination of the metal ions with the ligands. In other words, the LFSE's though crucially important in many ways, are not by any means major sources of the binding energies in complexes.

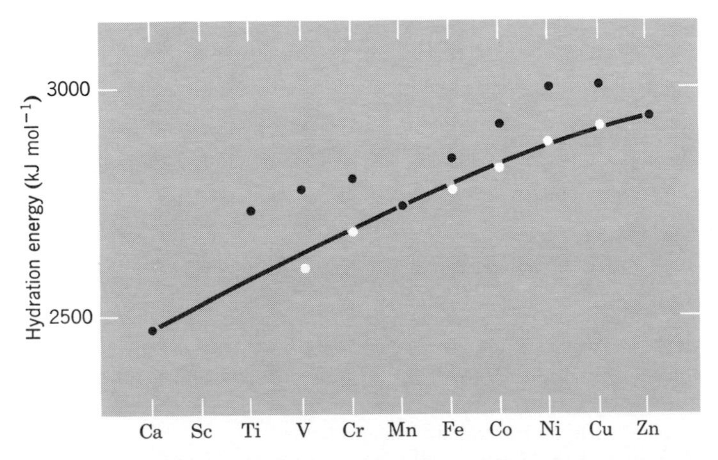

Figure 23-20 *Hydration energies of some divalent ions. Solid circles are the experimentally derived hydration energies. Open circles are energies corrected for* LFSE.

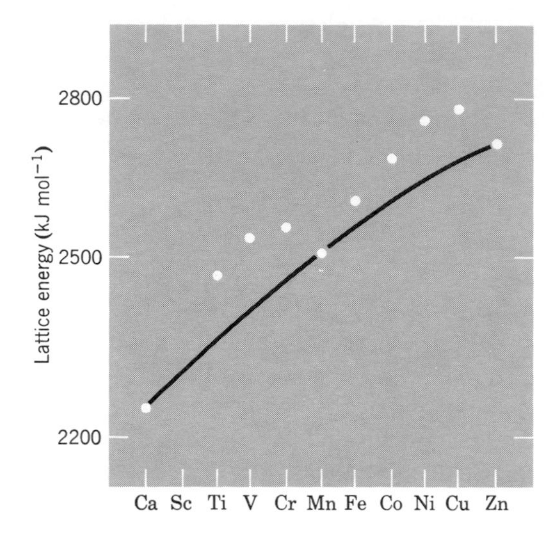

Figure 23-21 *The lattice energies of the dichlorides of the elements from* Ca *to* Zn.

Formation Constants of Complexes. It is a fairly general observation that the equilibrium constants for the formation of analogous complexes of the divalent metal ions of Mn through Zn with ligands which contain nitrogen as the donor atom fall in the following order of the metal ions: $Mn^{2+} < Fe^{2+} < Co^{2+} < Ni^{2+} < Cu^{2+} > Zn^{2+}$. LFSE's are responsible for this general trend. If it is assumed that ΔS° values in the formation of a particular complex by the different metal ions will be essentially constant, then the above order of formation constants is also the order of $-\Delta H^\circ$ values for complex formation. *Figure 23-20* shows that the above order is the same as the order of hydration energies of gaseous ions. When an aqueous aquo ion, $M(H_2O)_6{}^{2+}$, reacts with a set of ligands to form a complex, LFSE in the complex is usually greater than that in the aquo ion. In each case, it will be greater by about the same fraction, say 20%. Thus in each case, the replacement of water molecules by the new ligands will have a $-\Delta H^\circ$ value which is proportional to the LFSE and the magnitudes of these $-\Delta H^\circ$ values are in the same order as the LFSE's themselves.

Study Questions

A

1. What is a practical definition of the transition elements? What fraction of all the *ca.* 106 known elements are of this type?
2. List at least five characteristic properties of transition elements.
3. Make drawings of and give the correct designations (e.g., d_{xy}, etc.) to each of the e_g and t_{2g} orbitals arising from the d shell of a cation in an octahedral field.
4. What is the center of gravity rule and how does it apply to the splitting of d orbitals in octahedral and tetrahedral ligand fields?
5. The ratio of the d-orbital splittings in a tetrahedral and an octahedral field, each provided by the same kind of ligand atoms with the same metal-ligand distances, has what value?

6. Draw a diagram showing how the *d*-orbital pattern changes as we go from a regular octahedral set of ligands through intermediates of increasingly greater tetragonal distortion (by elongation of the metal-ligand distances along the *z* axis) to the limit of a square, 4-coordinate complex.
7. According to crystal field theory, the e_g and t_{2g} orbitals are pure *d* orbitals. How does this description change in MO theory?
8. By using orbital splitting diagrams, show which d^n configurations are capable of giving both high-spin and low-spin configurations in an octahedral ligand field.
9. Explain the nature and causes of (a) diamagnetism, (b) paramagnetism, (c) ferromagnetism, and (d) antiferromagnetism. In which case is Curie's law followed?
10. Why are *d—d* transitions weak? Why are they not absent altogether?
11. How does Δ_o change in going from one octahedral complex to another with the same ligand set but (a) M^{3+} in place of M^{2+}; (b) a second or third transition series element (e.g., Ru^{n+} or Os^{n+}) in place of the first series element (e.g., Fe^{n+})?
12. What is the spectrochemical series, and what limitations must be remembered in using it?
13. Draw a carefully labeled diagram showing how the radii of the +2 ions of the elements from Ti to Zn vary. Then explain why this particular variation is observed.
14. What is the Jahn–Teller effect as observed in the ground state structures of certain transition metal complexes? Use the Cr^{2+} ion as an illustration.
15. Calculate, in units of Δ_o, the LFSE's of the following high-spin ions in their octahedral complexes: Fe^{2+}, Mn^{2+}, Mn^{3+}, Co^{2+}.

B

1. What happens to the orbital splitting pattern in a tetrahedral field if the tetrahedron is flattened? Elongated?
2. What *d*-orbital splitting pattern would you expect in each of the following cases: (a) a linear L—M—L complex; (b) a planar ML_3 complex with the ligands defining an equilateral triangle; (c) a pyramidal ML_3 complex; (d) a trigonal bipyramidal ML_5 complex; and (e) a square pyramidal ML_5 complex?
3. What *d*-orbital splitting pattern would you expect for an ML_8 complex with the ligands at the vertices of a cube? How would the magnitude of the splitting compare with that in an analogous tetrahedral ML_4 complex?
4. Hausmannite, a mineral of composition Mn_3O_4, has the spinel structure (cf. page 104, but in a distorted form, such that the crystals are tetragonal (one long axis and two, equal short ones) instead of cubic. Suggest an explanation.
5. On the basis of LFSE's, which of the following ions would have the greatest tendency to form tetrahedral rather than octahedral complexes? Which one would have the least: Fe^{2+}, Co^{2+}, or Ni^{2+}? In actual fact, the order appears to be $Co^{2+} > Fe^{2+} > Ni^{2+}$. Discuss the reason(s) for any discrepancies between prediction and observation.
6. The complex $NiCl_4{}^{2-}$ is paramagnetic, with two unpaired electrons, while $Ni(CN)_4{}^{2-}$ is diamagnetic. By using ligand field theory explain these observations in terms of the structures of these complexes (which you must deduce from the data given).
7. Beginning with your answer to question 3 in this section, derive an orbital splitting diagram for the *d* orbitals in a cubic antiprism, derived from the cube by twisting one of the faces perpendicular to the *z* axis through 45° but keeping all other parameters the same. How will the separation between the highest and lowest orbitals compare in the two cases?

Chapter 23
Study Guide

Supplementary Reading

Ballhausen, C. J., *Introduction to Ligand Field Theory*, McGraw-Hill, 1962.

Ballhausen, C. J. and Gray, H. B., "Electronic Structures of Metal Complexes," *Coordination Chemistry*, A. E. Martell, ed., Van Nostrand-Reinhold, 1971.

Fackler, J. P., *Symmetry in Coordination Chemistry*, Academic Press, 1971.

Figgis, B. N., *Introduction to Ligand Fields*, Wiley, 1966.

Lever, A. B. P., *Inorganic Electronic Spectroscopy*, Elsevier, 1968.

Mabbs, F. E. and Machin, D. J., *Magnetism and Transition Metal Complexes*, Wiley/Halsted, 1973.

McClure, D. S. and Stephens, P. J., "Electronic Spectra of Coordination Compounds," *Coordination Chemistry*, A. E. Martell, ed., Van Nostrand-Reinhold, 1971.

Schläfer, H. L. and Glieman, G., *Basic Principles of Ligand Field Theory*, Wiley-Interscience, 1969.

24

the first series of d-block transition elements

As we have seen from their position in the periodic table (page 44) these metals show variable valency. In this chapter we first discuss some of their common features and then consider the chemistry of individual elements.

24-1
The Metals

These are hard, refractory, electropositive, and good conductors of heat and electricity. The exception is copper, a soft and ductile metal, relatively noble, but second only to Ag as a conductor of heat and electricity. Some properties are given in Table 24-1. Manganese and iron are attacked fairly readily but the others are generally unreactive at room temperature. All react on heating with halogens, sulfur, and other nonmetals. The carbides, nitrides, and borides are commonly nonstoichiometric, interstitial, hard, and refractory.

24-2
The Lower Oxidation States

The oxidation states are given in Table 24-2, the most common and important (especially in aqueous chemistry) are in bold type. Table 24-2 also gives the d electronic configurations. The chemistries can be classified on this basis, for example, the d^6 series is V^{-I}, Cr^{O}, Mn^{I}, Fe^{II}, Co^{III}, Ni^{IV}. Comparisons of this kind can occasionally emphasize similarities in spectra and magnetic properties. However, the differences in properties of the d^n species due to differences in the nature of the metal, its energy levels, and especially the charge on the ion, often exceed the similarities.

1. *The oxidation states less than II* With the exception of copper, where cuprous binary compounds and complexes, and the Cu^+ ion are known, the

Table 24-1 *Some Properties of the First Transition Series Metals*

	Ti	V	Cr	Mn	Fe	Co	Ni	Cu
mp °C	1668	1890	1875	1244	1537	1493	1453	1083
Properties	Hard, corrosion resistant	Hard, corrosion resistant	Brittle, corrosion resistant	White, brittle, reactive	Lustrous, reactive	Hard, bluish color	Quite corrosion resistant	Soft and ductile, reddish color
Density, gcm^{-3}	4.51	6.11	7.19	7.18	7.87	8.90	8.91	8.94
E° volts[a]	—[b]	−1.19	−0.91	−1.18	−0.44	−0.28	−0.24	+0.34
Solubility in acids	Hot HCl HF	HNO_3, HF conc. H_2SO_4	dil. HCl, H_2SO_4	dil. HCl H_2SO_4 etc.	dil. HCl H_2SO_4 etc.	Slowly in dil. HCl etc.	dil. HCl, H_2SO_4	HNO_3; hot conc. H_2SO_4

[a] For $M_{aq}^{2+} + 2e = M(s)$.
[b] No +2 ion in aqueous solution.

Table 24-2 *Oxidation States of First Series Transition Element*[a]

Ti	V	Cr	Mn	Fe	Co	Ni	Cu
	0 d^5	0 d^6	0 d^7	**0** d^8	0 d^9	**0** d^{10}	
	1 d^4	1 d^5	**1** d^6		**1** d^8	1 d^9	**1** d^{10}
2 d^2	2 d^3	**2** d^4	**2** d^5	**2** d^6	**2** d^7	**2** d^8	**2** d^9
3 d^1	**3** d^2	**3** d^3	**3** d^4	**3** d^5	**3** d^6	3 d^7	3 d^8
4 d^0	**4** d^1	4 d^2	4 d^3	4 d^4	4 d^5	4 d^6	
	5 d^0	5 d^1	5 d^2		5 d^4		
		6 d^0	6 d^1	6 d^2			
			7 d^0				

[a] Formal negative oxidation states are known in compounds of π-acid ligands, for example, Fe^{-II} in $[Fe(CO)_4]^{2-}$, Mn^{-I} in $[Mn(CO)_5]^-$, etc.

chemistry of the I, O, −I, and −II formal oxidation states is entirely concerned with:

(a) π-Acid ligands such as CO, NO, PR_3, CN^-, 2,2′-bipyridine, etc.

(b) Organo chemistry in which olefins, acetylenes, or aromatic systems such as benzene are bound to the metal.

There is an extensive chemistry of mixed compounds such as $(\eta^6\text{-}C_6H_6)Cr(CO)_3$ or $(\eta^4\text{-}C_4H_6)Fe(CO)_3$. These topics are described in Chapters 28 and 29. Some organo compounds in higher oxidation states are known, however, mainly for the cyclopentadienyl ligand as in $(\eta^5\text{-}C_5H_5)_2Ti^{IV}Cl_2$, $(\eta^5\text{-}C_5H_5)_2Fe^{II}$, and $[(\eta^5\text{-}C_5H_5)_2Co^{III}]^+$. With π-acid or organo ligands, transition metals also form many compounds with bonds to hydrogen, for example, $H_2Fe(PF_3)_4$. Compounds with M—H bonds are very important in certain catalytic reactions (Chapter 30).

2. *The II oxidation state.* The binary compounds in this state are usually ionic. The oxides, MO, are basic; they have the NaCl structure but are often nonstoichiometric, particularly for Ti, V, and Fe. The *aquo ions*, $[M(H_2O)_6]^{2+}$, except for the unknown Ti^{2+} ion, are well characterized in solution and in crystalline solids. The potentials and colors are given in Table 24-3. Note that the V^{2+}, Cr^{2+}, and Fe^{2+} ions are oxidized by air in acidic solution.

The aquo ions may be obtained by dissolution of the metals, oxides, carbonates, etc., in acids and by electrolytic reduction of M^{3+} salts. Hydrated salts with noncomplexing anions usually contain $[M(H_2O)_6]^{2+}$; typical ones are

$$Cr(ClO_4)_2 \cdot 6H_2O,\ Mn(ClO_4)_2 \cdot 6H_2O,\ FeF_2 \cdot 8H_2O,\ FeSO_4 \cdot 7H_2O.$$

However, certain *halide hydrates* do *not* contain the aquo ion. Thus $VCl_2 \cdot 4H_2O$ is *trans*-$VCl_2(H_2O)_4$, $MnCl_2 \cdot 4H_2O$ is a polymer with *cis*-$MnCl_2(H_2O)_4$ units; the diaquo species of Mn, Fe, Co, Ni, and Cu have a linear polymeric edge-shared chain structure with *trans*-$[MCl_4(H_2O)_2]$ octahedra. $FeCl_2 \cdot 6H_2O$ contains *trans*-$FeCl_2(H_2O)_4$ units.

The water molecules of $[M(H_2O)_6]^{2+}$ can be displaced by ligands such as NH_3, ethylenediamine, $EDTA^{4-}$, CN^-, acetylacetonate, etc. The resulting

Table 24-3 *Standard Potentials*[a] *(volts, acid solution) and Colors for* $[M(H_2O)_6]^{2+}$ *and* $[M(H_2O)_6]^{3+}$

	Ti	V	Cr	Mn	Fe	Co	Ni	Cu[b]
$M^{2+} + 2e = M$	—	−1.19	−0.91	−1.18	−0.44	−0.28	−0.24	+0.34
$M^{3+} + e = M^{2+}$	−0.37	−0.25	−0.41	+1.59	+0.77	+1.84	—	—
Color M^{2+}aq	—	Violet	Sky blue	Pale pink	Pale green	Pink	Green	Blue green
Color M^{3+}aq	Violet	Blue	Violet	Brown	v. Pale purple	Blue	—	—

[a] Some potentials depend on acidity and complexing anions, for example, for $Fe^{3+} - Fe^{2+}$ in 1 *M* acids: HCl, +0.70, $HClO_4$, +0.75; H_3PO_4, +0.44; 0.5 M H_2SO_4, +0.68 V.

[b] $Cu^{2+} + e = Cu^+$, $E_0 = +0.15$ V; $Cu^+ + e = Cu$, $E_0 = +0.52$ V.

complexes may be cationic, neutral, or anionic depending on the charge of the ligands. For Mn^{2+}, the complex constants in aqueous solution are low compared with those of the other ions, because of the absence of ligand field stabilization energy in the d^5 ion (page 376). In complexes the ions are normally *octahedral*, but for the Cu^{2+} and Cr^{2+} ions two H_2O molecules in trans positions are much further from the metal than the other four equatorial ones, because of the Jahn–Teller effect (page 373). For Mn, the complex $[Mn(EDTA)H_2O]$ is 7-coordinate. With halide ions, SCN^- and some other ligands, *tetrahedral* species MX_4^{2-} and MX_2L_2, may be formed, the tendency being greatest for Co, Ni, and Cu.

Addition of OH^- to the M^{2+} solutions gives *hydroxides*, some of which can be obtained as crystals. $Fe(OH)_2$ and $Ni(OH)_2$ have the brucite, $Mg(OH)_2$, structure. On addition of HCO_3^- the carbonates of Mn, Fe, Co, Ni, and Cu are precipitated.

24-3
The +3 Oxidation State

All of the elements form at least some compounds in this state but for Cu only a few complexes, not stable toward water, are known.

The fluorides, MF_3, and oxides, M_2O_3, are generally ionic but the chlorides, bromides, and iodides (where known) as well as sulfides and similar compounds may have considerable covalent character.

The elements Ti—Co form octahedral *ions*, $[M(H_2O)_6]^{3+}$. The Co^{3+} and Mn^{3+} ions are very readily reduced by water (Table 24-3). The Ti^{3+} and V^{3+} ions are oxidized by air. In aqueous solution high acidities are required to prevent hydrolysis, for example,

$$[Ti(H_2O)_6]^{3+} = [Ti(H_2O)_5OH]^{2+} + H^+ \qquad K = 1.3 \times 10^{-4}$$

Addition of OH^- to the solutions gives *hydrous oxides* rather than true hydroxides. In fairly concentrated halide solutions, complexes of the type $[MCl(H_2O)_5]^{2+}$, $[MCl_2(H_2O)_4]^+$, etc., are commonly formed, and crystalline chlorides of V, Fe, and Cr are of the type *trans*-$[VCl_2(H_2O)_4]^+Cl^- \cdot 2H_2O$. The alums, such as $CsTi(SO_4)_2 \cdot 12H_2O$, or $KV(SO_4)_2 \cdot 12H_2O$ contain the hexaaquo ion as do certain hydrates like $Fe(ClO_4)_3 \cdot 10H_2O$.

There are many M^{III} complexes, anionic, cationic, or neutral, mostly *octahedral*. For Cr^{III} and Co^{III} especially, hundreds of octahedral complexes that are substitutionally inert are known. Representative octahedral complexes are $[TiF_6]^{3-}$, $[V(CN)_6]^{3-}$, $Cr(acac)_3$, and $[Co(NH_3)_6]^{3+}$.

The halides, MX_3, act as Lewis acids and form adducts such as $VX_3(NMe_3)_2$, and $CrCl_3(THF)_3$, as well as the ionic species $[VCl_4]^-$, $[CrCl_4]^-$, etc.

A special feature of the M^{3+} ions is the formation of basic carboxylates in which an O atom is in the center of a triangle of metal atoms (24-I). The latter are linked by carboxylate bridge groups, and the sixth coordination position is occupied by a water molecule or other ligand. This oxo-centered unit has been proved for carboxylates of Cr, Mn, Fe, Ru, Rh, and Ir; and a similar structure has been proved for Co^{III}.

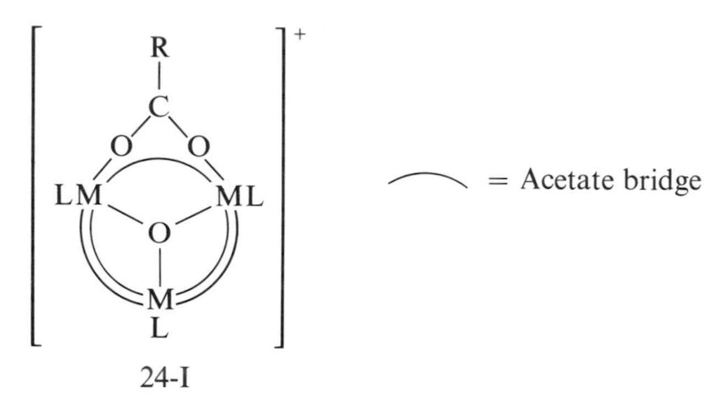

24-I

24-4
The IV and Higher Oxidation States

This is the most important state for Ti where the main chemistry is that of TiO_2 and $TiCl_4$ and derivatives. Although there are compounds like VCl_4, the main V^{IV} chemistry is that of the oxovanadium(IV) or vanadyl ion VO^{2+}. This ion can behave like an M^{2+} ion, and it forms many complexes that may be cationic, neutral, or anionic, depending on the ligand.

For the remaining elements, the IV oxidation is not very common or well established except in fluorides, fluorocomplex ions, oxo anions, and a few complexes. Some tetrahedral compounds with —OR, $—NR_2$, or $—CR_3$ groups are known for a few elements, notably Cr; examples are $Cr(OCMe_3)_4$ and $Cr(1\text{-norbornyl})_4$.

The oxidation states V, and above, are known for V, Cr, Mn, and Fe in fluorides, fluorocomplexes or oxo anions, for example, CrF_5, $KMnO_4$, and K_2FeO_4. All are powerful oxidizing agents.

TITANIUM

24-5
General Remarks; the Element

Titanium has the electronic structure, $3d^24s^2$. The energy of removal of four electrons is so high that Ti^{4+} ion does not exist and titanium(IV) compounds are covalent. There are some resemblances between Ti^{IV} and Sn^{IV} and their radii are similar. Thus TiO_2 (rutile) is isomorphous with SnO_2 (cassiterite) and is similarly yellow when hot. Titanium tetrachloride, like $SnCl_4$, is a distillable liquid readily hydrolyzed by water, behaving as a Lewis acid and giving adducts with donor molecules. The bromide and iodide, which form crystalline molecular lattices, are also isomorphous with the corresponding Group IVB halides.

Titanium is relatively abundant in the earth's crust (0.6%). The main ores are *ilmenite*, $FeTiO_3$, and *rutile*, one of the several crystalline varieties of TiO_2. The metal cannot be made by reduction of TiO_2 with C because a very stable carbide is produced. The rather expensive Kroll process is used. Ilmenite or rutile is treated at red heat with C and Cl_2 to give $TiCl_4$, which is fractionated to free it from impurities such as $FeCl_3$. The $TiCl_4$ is then reduced with molten Mg at $\sim 800°$ in an atmosphere of argon. This gives Ti as a spongy mass from which the excess of Mg and $MgCl_2$ is removed by volatilization at 1000°. The sponge may then be fused in an electric arc and cast into ingots; an atmosphere of Ar or He must be used as titanium readily reacts with N_2 and O_2 when hot.

Titanium is lighter than other metals of similar mechanical and thermal properties and is unusually resistant to corrosion. It is used in turbine engines and industrial chemical, aircraft, and marine equipment. It is unattacked by dilute acids and bases. It dissolves in hot HCl giving Ti^{III} chloro complexes and in HF or HNO_3 + HF to give fluoro complexes. Hot HNO_3 gives a hydrous oxide.

TITANIUM COMPOUNDS

The most important stereochemistries in titanium compounds are the following.

Ti^{II}	Octahedral	in most compounds and in solution
Ti^{III}	Octahedral	
Ti^{IV}	Tetrahedral	in $TiCl_4$, $Ti(CH_2Ph)_4$, etc.
	Octahedral	in TiO_2 and Ti^{IV} complexes

24-6
Binary Compounds

Titanium tetrachloride, a colorless liquid (bp 136°), has a pungent odor, fumes strongly in moist air, and is vigorously, though not violently, hydrolyzed by water:

$$TiCl_4 + 2H_2O = TiO_2 + 4HCl$$

With a deficit of water or on addition of $TiCl_4$ to aqueous HCl, partially hydrolyzed species are formed.

Titanium oxide has three crystal forms—rutile (see page 92), anatase, and brookite—all of which occur in nature. The dioxide that is used in large quantities as a white pigment in paints is made by vapor phase oxidation of $TiCl_4$ with oxygen. The precipitates obtained by addition of OH^- to Ti^{IV} solutions are best regarded as hydrous TiO_2, not a true hydroxide. This material is amphoteric and dissolves in concentrated NaOH.

Materials called "titanates" are of technical importance, for example, as ferroelectrics. Nearly all of them have one of the three major mixed metal oxide structures (page 104). Indeed, the names of two of the structures are those of the titanium compounds that were the first found to possess them, namely, $FeTiO_3$, *ilmenite*, and $CaTiO_3$, *perovskite*.

24-7
Titanium(IV) Complexes

Aqueous Chemistry; Oxo Salts. There is no Ti^{4+} aquo ion. In aqueous solutions of Ti^{IV} there are only oxo species; basic oxo salts, or hydrated oxides may be precipitated. Although these oxo salts have formulas such as $TiOSO_4 \cdot H_2O$ and $(NH_4)_2TiO(C_2O_4)_2 \cdot H_2O$, no discrete TiO^{2+} ion is known. Instead, chains or rings, $(Ti—O—Ti—O—)_x$ occur.

Anionic Complexes. The solutions obtained by dissolving the metal or hydrous oxide in aqueous HF contain fluoro complex ions, mainly TiF_6^{2-}, which can be isolated as crystalline salts. In aqueous HCl, $TiCl_4$ gives yellow oxo complex anions but from solutions saturated with gaseous HCl, salts of the $[TiCl_6]^{2-}$ ion may be obtained.

Adducts of TiX_4. The halides form adducts, TiX_4L or TiX_4L_2, which are crystalline solids often soluble in organic solvents. These adducts are invariably *octahedral.* Thus $[TiCl_4(OPCl_3)]_2$ and $[TiCl_4(MeCOOEt)]_2$ are dimeric, with two chlorine bridges, while $TiCl_4(OPCl_3)_2$ has octahedral coordination with *cis*-$OPCl_3$ groups.

Peroxo Complexes. One of the most characteristic reactions of aqueous Ti solutions is the development of an intense orange color on addition of H_2O_2. This reaction can be used for the colorimetric determination of either Ti or of H_2O_2. Below pH 1, the main species is $[Ti(O_2)(OH)aq]^+$.

Solvolytic Reactions of $TiCl_4$; Alkoxides and Related Compounds. Titanium tetrachloride reacts with compounds containing active hydrogen atoms with loss of HCl. The replacement of chloride is usually incomplete in absence of an HCl acceptor such as an amine or alkoxide ion. The *alkoxides* are typical of other transition-metal alkoxides, which we shall not discuss. They can be obtained by reactions such as

$$TiCl_4 + 4ROH + 4R'NH_2 \longrightarrow Ti(OR)_4 + 4R'NH_3Cl$$

The alkoxides are liquids or solids that can be distilled or sublimed. They are soluble in organic solvents such as benzene, but are exceedingly readily hydrolyzed by even traces of water, to give polymeric species with —OH— or —O— bridges. The initial hydrolytic step probably involves coordination of water to the metal; a proton on H_2O could then interact with the oxygen of an OR group through hydrogen bonding, leading to hydrolysis:

$$H_2O—M(OR)_x \longrightarrow H(H)O—M(OR)_{x-1}(\cdots H \cdots :O—R) \longrightarrow M(OH)(OR)_{x-1} + ROH$$

Although monomeric species can exist, for example, when made from secondary and tertiary alcohols, and in dilute solution, alkoxides are usually polymers. Solid $Ti(OC_2H_5)_4$ is a tetramer, with the structure shown in *Fig. 24-1*. The alkoxides

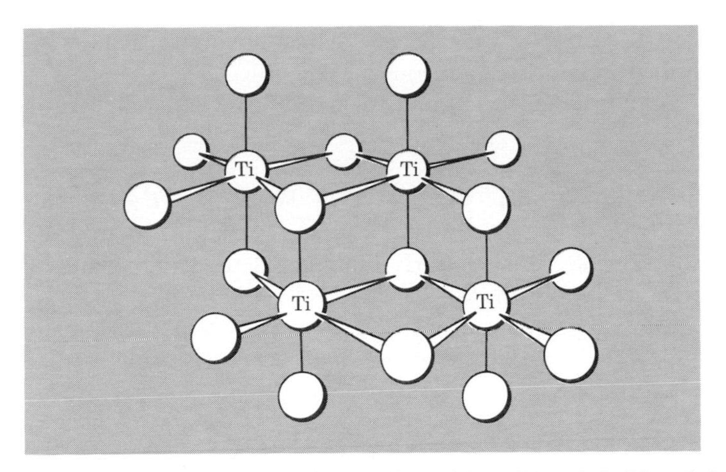

Figure 24-1 *The tetrameric structure of crystalline* $Ti(OC_2H_5)_4$. *Only* Ti *and* O *atoms are shown; each* Ti *is octahedrally coordinated.*

are often referred to as "alkyl titanates" and under this name they are used in heat-resisting paints, where eventual hydrolysis to TiO_2 occurs.

Another class of titanium compounds, the *dialkylamides*, are also representative of similar compounds of other transition metals. These are liquids or volatile solids readily hydrolyzed by water. Unlike the alkoxides they are not polymeric. They are made by reaction of the metal halide with lithium dialkylamides

$$TiCl_4 + 4LiNEt_2 = Ti(NEt_2)_4 + 4LiCl$$

Such amides can undergo a wide range of "insertion" reactions (page 534); thus with CS_2, the dithiocarbamates are obtained

$$Ti(NEt_2)_4 + 4CS_2 = Ti(S_2CNEt_2)_4$$

24-8
The Chemistry of Titanium(III), d^1, and Titanium(II), d^2

Binary Compounds. *Titanium trichloride*, $TiCl_3$, has several crystalline forms. The violet α-form is made by H_2 reduction of $TiCl_4$ vapor at 500 to 1200°. The reduction of $TiCl_4$ by aluminum alkyls (Section 30-10) in inert solvents gives a brown β-form which is converted into the α-form at 250 to 300°. The α-form has a layer lattice containing $TiCl_6$ groups. β-$TiCl_3$ is fibrous with single chains of $TiCl_6$ octahedra sharing edges. This structure is of particular importance for the stereospecific polymerization of propene using $TiCl_3$ as catalyst (Ziegler-Natta process) (Section 30-9).

The *dichloride* is obtained by high temperature syntheses.

$$TiCl_4 + Ti = 2TiCl_2$$

or

$$2TiCl_3 = TiCl_2 + TiCl_4$$

Aqueous Chemistry and Complexes. Aqueous solutions of the $[Ti(H_2O)_6]^{3+}$ ion are obtained by reducing aqueous Ti(IV) either electrolytically or with zinc. The violet solutions reduce O_2 and, hence, must be handled in a N_2 or H_2 atmosphere:

$$\text{"}TiO^{2+}(aq)\text{"} + 2H^+ + e = Ti^{3+} + H_2O \qquad E^\circ = ca.\ 0.1\ V$$

The Ti^{3+} solutions are used as fairly rapid, mild reducing agents in volumetric analysis. In HCl solutions the main species is $[TiCl(H_2O)_5]^{2+}$.

There is no aqueous chemistry of Ti^{II} because of its ready oxidation but a few Ti^{II} complexes such as $TiCl_4{}^{2-}$ can be made in nonaqueous media.

VANADIUM

24-9 The Element

Although V could be regarded as a Group V element just as Ti can be regarded as a Group IV element, there is little similarity to the elements of the P group other than in stoichiometry. The oxo anions, vanadates, are not similar to phosphates.

Vanadium is widely spread but there are few concentrated deposits. Vanadium occurs in petroleum from Venezuela, and is recovered as V_2O_5 from flue dusts after combustion.

Very pure vanadium is rare because, like titanium, it is quite reactive toward O_2, N_2, and C at the elevated temperatures used in metallurgical processes. Since its chief commercial use is in alloy steels and cast iron, to which it lends ductility and shock resistance, commercial production is mainly as an iron alloy, *ferrovanadium.*

Vanadium metal is not attacked by air, alkalis, or nonoxidizing acids other than HF at room temperature. It dissolves in HNO_3, concentrated H_2SO_4, and aqua regia.

VANADIUM COMPOUNDS

The stereochemistries for the most important classes of vanadium compounds are the following

V^{II}, V^{III} — Octahedral as in $[V(H_2O)_6]^{2+}$, $VF_3(s)$ or $[Vox_3]^{3-}$

V^{IV} — Tetrahedral as in VCl_4 or $V(CH_2SiMe_3)_4$.
Square pyramidal in O=V $acac_2$.
Octahedral in VO_2, K_2VCl_6, O=$Vacac_2py$, etc.

V^V — Octahedral as in $[VO_2ox_2]^{3-}$, $VF_5(s)$.

24-10
Binary Compounds

Halides. In the highest oxidation state only VF_5 is known. This colorless liquid (bp 48°), has a high viscosity (cf. SbF_5, page 290) and has chains of VF_6 octahedra linked by *cis*-V—F—V bridges; it is monomeric in the vapor.

The *tetrachloride* is obtained from V + Cl_2 or from CCl_4 on red-hot V_2O_5. It is a dark red oil (bp 154°), which is violently hydrolyzed by water to give solutions of oxovanadium(IV) chloride. It has a high dissociation pressure and loses chlorine slowly when kept, but rapidly on boiling, leaving violet VCl_3. The latter may be decomposed to pale green VCl_2 which is then stable:

$$2VCl_3(s) \longrightarrow VCl_2(s) + VCl_4(g)$$

$$VCl_3(s) \longrightarrow VCl_2(s) + \tfrac{1}{2}Cl_2(g)$$

Vanadium(V) Oxide. Addition of dilute H_2SO_4 to solutions of ammonium vanadate gives a brick-red precipitate of V_2O_5. This oxide is acidic and dissolves in NaOH to give colorless solutions containing the *vanadate* ion VO_4^{3-}. On acidification, a complicated series of reactions occurs involving the formation of hydroxo anions and poly anions (cf. page 118). In very strong acid solutions, the *dioxovanadium(V) ion*, VO_2^+ is formed.

24-11
Oxovanadium Ions and Complexes

The two oxo anions, VO_2^+ and VO^{2+}, have an extensive chemistry and form numerous complex compounds. All of the compounds show infrared and Raman bands that are characteristic for M=O groups. The VO_2^+ group is angular. Examples of complexes are *cis*-$[VO_2Cl_4]^{3-}$, *cis*-$[VO_2EDTA]^{3-}$, and *cis*-$[VO_2ox_2]^{3-}$. The *cis* arrangement for dioxo compounds of metals with no *d* electrons is preferred over the *trans* arrangement that is found in some other metal dioxo systems, for example, RuO_2^{2+}, because the strongly π-donating O ligands then have exclusive use of one $d\pi$ orbital each (d_{xz}, d_{yz}) and share a third one (d_{xy}), whereas in the *trans* configuration they would have to share two $d\pi$ orbitals and leave one unused.

The oxovanadium(IV) or vanadyl compounds are among the most stable and important of vanadium species, and the VO unit persists through a variety of chemical reactions. Solutions of V^{3+} are oxidized in air, while V^V is readily reduced by mild reducing agents to form the blue oxovanadium(IV) ion, $[VO(H_2O)_5]^{2+}$:

$$VO^{2+} + 2H^+ + e = V^{3+} + H_2O \qquad E° = 0.34 \text{ V}$$

$$VO_2^+ + 2H^+ + e = VO^{2+} + H_2O \qquad E° = 1.0 \text{ V}$$

Addition of base to $[VO(H_2O)_5]^{2+}$ gives the yellow hydrous oxide $VO(OH)_2$ which redissolves in acids giving the cation.

Oxovanadium(IV) compounds are usually blue-green. They may be either 5-coordinate square pyramidal (24-II) or 6 coordinate with a distorted octahedron.

Examples are $[VObipy_2Cl]^+$, $VO(acac)_2$, and $[VO(NCS)_4]^{2-}$. The VO bond is short (1.56–1.59 Å), so that VO bonds can properly be regarded as multiple ones, the π component arising from electron flow $O(p\pi) \rightarrow V(d\pi)$. Even in VO_2, which has a distorted rutile structure, one bond (1.76 Å) is conspicuously shorter than the others in the VO_6 unit (note that in TiO_2 all Ti—O bonds are substantially equal).

All of the 5-coordinate complexes such as (24-II) take up a sixth ligand quite readily, becoming octahedral.

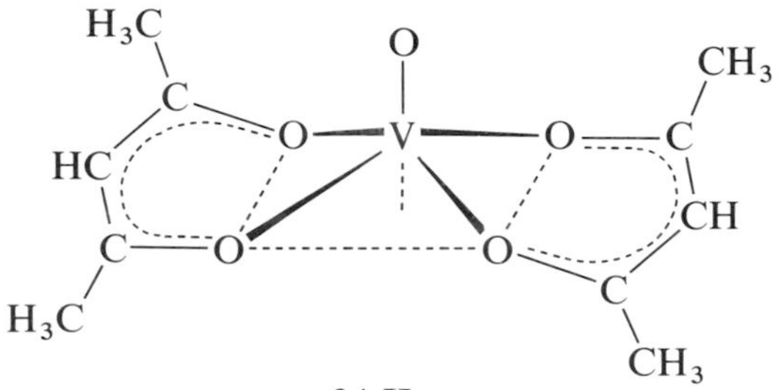

24-II

24-12 The Vanadium(III) Aquo-ion and Complexes

The electrolytic or chemical reduction of acid solutions of vanadates or V^{IV} solutions gives solutions of V^{III} that are quite readily reoxidized to VO^{2+}. Crystalline salts can be obtained. Addition of OH^- precipitates the hydrous oxide V_2O_3.

24-13 Vanadium(II)

When V^{III} solutions are reduced by Zn in acid, violet air-sensitive solutions of $[V(H_2O)_6]^{2+}$ are obtained. These are oxidized by water with evolution of hydrogen despite the fact that the V^{3+}/V^{2+} potential (Table 24-3) suggests otherwise. Vanadous solutions are often used to remove traces of O_2 from inert gases.

The salt $VSO_4 \cdot 6H_2O$ is obtained as violet crystals on addition of ethanol to reduced sulfate solutions. Because of its d^3 configuration the $[V(H_2O)_6]^{2+}$ ion like $[Cr(H_2O)_6]^{3+}$ is kinetically inert, and its substitution reactions are relatively slow.

CHROMIUM

24-14 The Element

Apart from stoichiometric similarities, chromium resembles the Group VIB elements of the sulfur group only in the acidity of CrO_3 and the covalent nature and ready hydrolysis of CrO_2Cl_2 (cf. SO_3, SO_2Cl_2).

The chief ore is *chromite*, $FeCr_2O_4$, which is a spinel with Cr^{III} on octahedral sites and Fe^{II} on the tetrahedral ones. It is reduced by C to the carbon-containing alloy ferrochromium:

$$FeCr_2O_4 + 4C \xrightarrow{\text{Heat}} Fe + 2Cr + 4CO$$

When pure Cr is required, the chromite is first treated with molten NaOH and O_2 to convert the Cr^{III} to CrO_4^{2-}. The melt is dissolved in water and sodium dichromate precipitated. This is then reduced:

$$Na_2Cr_2O_7 + 2C \xrightarrow{\text{Heat}} Cr_2O_3 + Na_2CO_3 + CO$$

The oxide is then reduced:

$$Cr_2O_3 + 2Al \xrightarrow{\text{Heat}} Al_2O_3 + 2Cr$$

Chromium is resistant to corrosion, hence its use as an electroplated protective coating. It dissolves fairly readily in HCl, H_2SO_4, and $HClO_4$, but it is passivated by HNO_3.

CHROMIUM COMPOUNDS

The most common stereochemistries for chromium compounds are the following:

Cr^{II}, Cr^{III} } Octahedral as in $[Cr(H_2O)_6]^{2+}$ or $[Cr(NH_3)_6]^{3+}$

Cr^{IV} Tetrahedral as in $Cr(O^tBu)_4$.

Cr^{V}, Cr^{VI} } Tetrahedral as in CrO_4^{3-}, CrO_4^{2-}, CrO_3.

24-15 Binary Compounds

Halides. The anhydrous Cr^{II} halides are obtained by action of HCl, HBr, or I_2 on the metal at 600 to 700° or by reduction of the trihalides with H_2 at 500 to 600°. $CrCl_2$ dissolves in water to give a blue solution of Cr^{2+} ion.

The red-violet trichloride $CrCl_3$ is made by the action of $SOCl_2$ on the hydrated chloride. The flaky form of $CrCl_3$ is due to its layer structure.

Chromic chloride forms adducts with donor ligands. The violet tetrahydrofuranate, $CrCl_3 \cdot 3THF$, which crystallizes from solutions formed by action of a little zinc on $CrCl_3$ in THF, is a particularly useful material for the preparation of other chromium compounds such as carbonyls or organo compounds.

Oxides. The green α-Cr_2O_3 (corundum structure) is formed on burning Cr in O_2, on thermal decomposition of CrO_3, or on roasting the hydrous oxide, $Cr_2O_3 \cdot nH_2O$. The latter, commonly called "chromic hydroxide," although its water content is variable, is precipitated on addition of OH^- to solutions of Cr^{III} salts. The hydrous oxide is amphoteric, dissolving readily in acid to give $[Cr(H_2O)_6]^{3+}$, and in concentrated alkali to form "chromites."

Chromium oxide and chromium supported on other oxides such as Al_2O_3 are important catalysts for a wide variety of reactions.

Chromium(VI) oxide, CrO_3, is obtained as an orange-red precipitate on adding sulfuric acid to solutions of $Na_2Cr_2O_7$. It is thermally unstable above its melting point (197°), losing O_2 to give Cr_2O_3. The structure consists of infinite chains of CrO_4 tetrahedra sharing corners. It is soluble in water and highly poisonous.

Interaction of CrO_3 and organic substances is vigorous and may be explosive, but CrO_3 is used in organic chemistry as an oxidant, usually in acetic acid as solvent.

24-16
The Chemistry of Chromium(II), d^4

Aqueous solutions of the blue *chromous ion* are best prepared by dissolving electrolytic Cr metal in dilute mineral acids. The solutions must be protected from air (Table 24-3, page 382)—even then, they decompose at rates varying with the acidity and the anions present, by reducing water with liberation of H_2.

The mechanisms of reductions of other ions by Cr^{2+} have been extensively studied, since the resulting Cr^{3+} complex ions are substitution-inert. Much information regarding ligand-bridged transition states (page 156) has been obtained in this way.

Chromium(II) *acetate*, $Cr_2(O_2CCH_3)_4(H_2O)_2$, is precipitated as a red solid when a Cr^{2+} solution is added to a solution of sodium acetate. Its structure is typical of carboxylate-bridged complexes with water end groups. The short Cr—Cr bond, 2.36 Å, and diamagnetism are accounted for by the existence of a quadruple Cr—Cr bond, consisting of a σ, two π, and a δ component typical of carboxylate bridged complexes with water end groups, cf. 24-X, page 415. This was the first compound containing a quadruple bond to be discovered (1844).

24-17
The Chemistry of Chromium(III), d^3

Chromium(III) Complexes. There are thousands of chromium(III) complexes which, with a few exceptions, are all 6-coordinate. The principal characteristic is their relative kinetic inertness in aqueous solutions. It is because of this that so many complex species can be isolated, and why much of the classical complex chemistry studied by early workers, notably S. M. Jørgensen and A. Werner, involved chromium. They persist in solution, even where they are thermodynamically unstable.

The hexaquo ion, $[Cr(H_2O)_6]^{3+}$, occurs in numerous salts such as the violet hydrate, $[Cr(H_2O)_6]Cl_3$, and alums, $M^ICr(SO_4)_2 \cdot 12H_2O$. The chloride has three isomers, the others being the dark green *trans*-$[CrCl_2(H_2O)_4]Cl \cdot 2H_2O$, which is the usual form, and pale green $[CrCl(H_2O)_5]Cl_2 \cdot H_2O$. The ion is acidic and the hydroxo ion condenses to give a dimeric hydroxo bridged species:

$$[Cr(H_2O)_6]^{3+} \underset{H^+}{\overset{-H^+}{\rightleftharpoons}} [Cr(H_2O)_5OH]^{2+}$$

$$\updownarrow$$

$$\left[(H_2O)_5Cr \underset{\underset{H}{O}}{\overset{\overset{H}{O}}{\diamond}} Cr(H_2O)_5 \right]^{4+}$$

On addition of further base, soluble polymeric species of high molecular weight and eventually dark green gels of the hydrous oxide are formed.

The most numerous complexes are those of amine ligands. These provide examples of virtually all the kinds of isomerism possible in octahedral complexes. In addition to the mononuclear species, for example $[Cr(NH_3)_5Cl]^{2+}$, there are many polynuclear complexes in which two or sometimes more metal atoms are bridged by hydroxo groups or, less commonly, oxygen in a linear Cr—O—Cr group. A representative example is $[(NH_3)_5Cr(OH)Cr(NH_3)_5]^{5+}$.

24-18 The Chemistry of Chromium(IV), d^2 and Chromium(V) d^1

The most readily accessible of these rare oxidation states are those with bonds to C, N, and O. A representative synthesis is

$$CrCl_3 \cdot (THF)_3 + 4LiCH_2SiMe_3 \overset{\text{ether}}{=} Li^+[Cr^{III}(CH_2SiMe_3)_4]^- + 3LiCl$$

$$[Cr(CH_2SiMe_3)_4]^- = Cr(CH_2SiMe_3)_4 + e$$

The oxidation of the green Cr^{III} anion to the purple, petroleum-soluble Cr^{IV} compound can be made by air. The alkoxides and dialkylamides are similarly made from $CrCl_3 \cdot (THF)_3$; one example is the dark blue $Cr(OCMe_3)_4$.

For Cr^V some chromites containing $CrO_4{}^{3-}$ are known. Reduction of CrO_3 with concentrated HCl in presence of alkali ions at 0° gives salts $M_2[Cr^VOCl_5]$.

24-19 The Chemistry of Chromium(VI), d^0

Chromate and Dichromate Ions. In basic solutions above pH 6, CrO_3, forms the tetrahedral yellow *chromate* ion, $CrO_4{}^{2-}$. Between pH 2 and pH 6, $HCrO_4{}^-$

and the orange-red *dichromate* ion $Cr_2O_7^{2-}$ are in equilibrium. At pH's below 1 the main species is H_2CrO_4. The equilibria are

$$HCrO_4^- \rightleftharpoons CrO_4^{2-} + H^+ \qquad K = 10^{-5.9}$$

$$H_2CrO_4 \rightleftharpoons HCrO_4^- + H^+ \qquad K = 4.1$$

$$Cr_2O_7^{2-} + H_2O \rightleftharpoons 2HCrO_4^- \qquad K = 10^{-2.2}$$

In addition there are the base-hydrolysis equilibria:

$$Cr_2O_7^{2-} + OH^- \rightleftharpoons HCrO_4^- + CrO_4^{2-}$$

$$HCrO_4^- + OH^- \rightleftharpoons CrO_4^{2-} + H_2O$$

The CrO_4^- ion is tetrahedral; $Cr_2O_7^{2-}$ has the structure shown in *Fig. 24-2*.

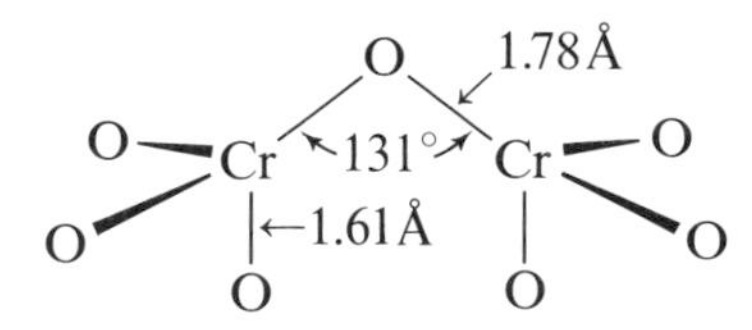

Figure 24-2 *The structure of the dichromate ion as found in* $Na_2Cr_2O_7$.

The pH-dependent equilibria are quite labile and on addition of cations that form insoluble chromates, for example, Ba^{2+}, Pb^{2+}, Ag^+, the chromates and not the dichromates are precipitated. Only for HNO_3 and $HClO_4$ are the equilibria as given above. When HCl is used, there is essentially quantitative conversion into the *chlorochromate* ion, while with sulfuric acid a sulfato complex results:

$$CrO_3(OH)^- + H^+ + Cl^- \longrightarrow CrO_3Cl^- + H_2O$$

$$CrO_3(OH)^- + HSO_4^- \longrightarrow CrO_3(OSO_3)^{2-} + H_2O$$

Acid solutions of dichromate are strong oxidants:

$$Cr_2O_7^{2-} + 14H^+ + 6e = 2Cr^{3+} + 7H_2O \qquad E^\circ = 1.33 \text{ V}$$

In alkaline solution, the chromate ion is much less oxidizing

$$CrO_4^- + 4H_2O + 3e = Cr(OH)_3(s) + 5OH^- \qquad E^\circ = 0.13 \text{ V}$$

Chromium(VI) does not give rise to the extensive and complex series of polyacids and polyanions characteristic of the somewhat less acidic oxides of V^V, Mo^{VI}, and W^{VI}. The reason for this is perhaps the greater extent of multiple bonding, Cr═O, for the smaller chromium ion.

Chromyl chloride, CrO_2Cl_2, a deep red liquid is formed by the action of HCl on chromium(VI) oxide:

$$CrO_3 + 2HCl \longrightarrow CrO_2Cl_2 + H_2O$$

or on warming dichromate with an alkali-metal chloride in concentrated sulfuric acid:

$$K_2Cr_2O_7 + 4KCl + 3H_2SO_4 \longrightarrow 2CrO_2Cl_2 + 3K_2SO_4 + 3H_2O$$

It is photosensitive but otherwise rather stable. It vigorously oxidizes organic matter. It is hydrolyzed by water to CrO_4^{2-} and HCl.

24-20
Peroxo Complexes of Chromium(IV), (V), and (VI)

Like other transition metals, notably Ti, V, Nb, Ta, Mo, and W, chromium forms peroxo compounds in the higher oxidation states. They are all more or less unstable both in and out of solution, decomposing slowly with the evolution of O_2. Some are explosive or flammable in air.

When acid dichromate solutions are treated with H_2O_2, a deep blue color rapidly appears:

$$HCrO_4^- + 2H_2O_2 + H^+ \longrightarrow CrO(O_2)_2 + 3H_2O$$

The blue species decomposes fairly readily, giving Cr^{3+}, but it may be extracted into ether where it is more stable. On addition of pyridine to the ether solution, the compound $pyCrO_5$ is obtained. The pyridine complex has the bisperoxo structure shown in *Fig. 24-3a*.

Treatment of alkaline chromate solutions with 30% H_2O_2 gives the red-brown peroxochromates, $M_3^ICrO_8$ (*Fig. 24-3b*). They are paramagnetic with one unpaired electron.

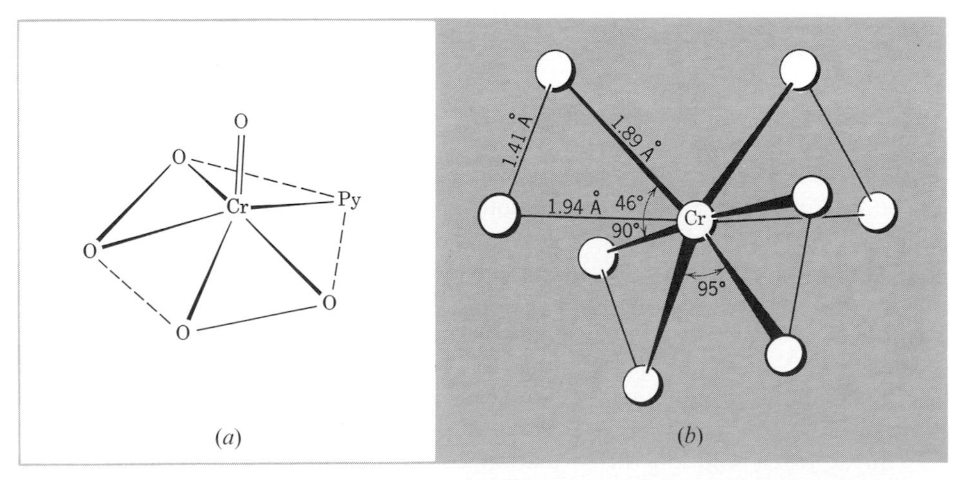

Figure 24-3 *(a) The structure of* $CrO(O_2)_2 \cdot py$. *The coordination polyhedron is approximately a pentagonal pyramid with the oxide oxygen at the apex. (b) The dodecahedral structure of the* CrO_8^{3-} *ion, a tetraperoxide complex.*

MANGANESE

24-21
The Element

The highest oxidation state of manganese corresponds to the total number of 3*d* and 4*s* electrons, but it occurs only in the oxo compounds MnO_4^-, Mn_2O_7, and MnO_3F. These compounds show some similarity to corresponding compounds of the halogens.

Manganese is relatively abundant, and occurs in substantial deposits, mainly oxides, hydrous oxides, or the carbonate. From them, or the Mn_3O_4 obtained by roasting them, the metal can be obtained by reduction with Al.

Manganese is quite electropositive, and readily dissolves in dilute, non-oxidizing acids.

MANGANESE COMPOUNDS

The most common stereochemistries of manganese compounds are the following:

$$\left.\begin{matrix} Mn^{II} \\ Mn^{III} \\ Mn^{IV} \end{matrix}\right\} \text{ Octahedral as in } Mn(H_2O)_6^{2+}, [Mnox_3]^{3-} \text{ and } MnCl_6^{2-}$$

$$\left.\begin{matrix} Mn^{VI} \\ Mn^{VII} \end{matrix}\right\} \text{ Tetrahedral as in } MnO_4^{2-}, MnO_4^-$$

24-22
The Chemistry of Manganese(II), d^5

Manganous salts are mostly water soluble. Addition of OH^- to Mn^{2+} solutions produces the gelatinous white *hydroxide*. This rapidly darkens in air due to oxidation, as shown by the base potentials

$$MnO_2 \xrightarrow{-0.1\text{ V}} Mn_2O_3 \xrightarrow{-0.2\text{ V}} Mn(OH)_2$$

Addition of SH^- gives hydrous MnS, which also oxidizes becoming brown in air; on boiling in absence of air the salmon pink material changes into green crystalline MnS. The sulfate, $MnSO_4$, is very stable and may be used for Mn analysis as it can be obtained on fuming down sulfuric acid solutions to dryness. The phosphate and carbonate are sparingly soluble. The equilibrium constants for the formation of manganous complexes are relatively low as the Mn^{2+} ion has no ligand field stabilization energy (page 376). However, chelating ligands such as ethylenediamine, oxalate, or $EDTA^{4-}$ form complexes isolable from aqueous solution.

In aqueous solution the formation constants for halogeno complexes are very low, for example,

$$Mn_{aq}^{2+} + Cl^- \rightleftharpoons MnCl_{aq}^+ \qquad K \approx 4$$

but in ethanol or acetic acid, salts of complex anions of varying types may be isolated, such as

MnX_3^-:	Octahedral with perovskite structure
MnX_4^{2-}:	Tetrahedral (green-yellow) or polymeric octahedral with halide bridges (pink)
$MnCl_6^{4-}$:	Only Na and K salts known; octahedral

Mn^{2+} ions may occupy tetrahedral holes in certain glasses and substitute for Zn^{II} in ZnO. Tetrahedral Mn^{II} has a green-yellow color, far more intense than the pink of the octahedrally coordinated ion, and it very often exhibits intense yellow-green fluorescence. Most commercial phosphors are manganese-activated zinc compounds, wherein Mn^{II} ions are substituted for some of the Zn^{II} ions in tetrahedral surroundings, as for example in Zn_2SiO_4.

Only the very strong ligand fields give rise to spin-pairing as in the ions $[Mn(CN)_6]^{4-}$ and $[Mn(CNR)_6]^{2+}$, which have only one unpaired electron.

24-23 The Chemistry of Manganese(III), d^4

Oxides. When any manganese oxide or hydroxide is heated at 1000°, black crystals of Mn_3O_4, *haussmannite*, are formed. This is a spinel, $Mn^{II}Mn^{III}{}_2O_4$. When $Mn(OH)_2$ is allowed to oxidize in air, a hydrous oxide is formed which gives MnO(OH) on drying.

The Manganese(III) Aquo Ion. The manganic ion can be obtained by electrolytic or persulfate oxidation of Mn^{2+} solutions or by reduction of MnO_4^-. It cannot be obtained in high concentrations as it is reduced by water (Table 24-3). It also has a strong tendency to hydrolyze and to disproportionate in weakly acid solution:

$$2Mn^{3+} + 2H_2O = Mn^{2+} + MnO_2(s) + 4H^+ \qquad K \approx 10^9$$

The dark brown crystalline *acetylacetonate*, $Mn(acac)_3$ is readily obtained by oxidation of basic solutions of Mn^{2+} by O_2 or Cl_2 in the presence of acetylacetone.

The basic oxo centered acetate (page 384), which is obtained by action of $KMnO_4$ on Mn^{II} acetate in acetic acid, will oxidize olefins to lactones. It is used industrially for oxidation of toluene to phenol.

Manganese(III) and manganese(IV) complexes are probably important in photosynthesis, where oxygen evolution depends on manganese.

24-24 The Chemistry of Manganese(IV), d^3 and Manganese(V), d^2

The only really important compound of Mn^{IV} is *manganese dioxide*, a grey to black solid found in nature as *pyrolusite*. When made by the action of oxygen on manganese at a high temperature it has the rutile structure found for many other oxides, MO_2, for example, those of Ru, Mo, W, Re, Os, Ir, and Rh. However, as normally made, for example by heating $Mn(NO_3)_2 \cdot 6H_2O$ in air, it is nonstoichiometric. A hydrated form is obtained by reduction of aqueous $KMnO_4$ in basic solution.

Manganese dioxide is inert to most acids except when heated, but it does not dissolve to give Mn^{IV} in solution; instead, it functions as an oxidizing agent, the exact manner of this depending on the acid. With HCl, chlorine is evolved:

$$MnO_2(s) + 4H^+ + 4Cl^- = Mn^{2+} + 2Cl^- + Cl_2 + 2H_2O$$

With sulfuric acid at 110°, oxygen is evolved and an Mn^{III} acid sulfate is formed. Hydrated manganese dioxide is used in organic chemistry for the oxidation of alcohols and other compounds.

Mn^V is little known except in bright blue "hypomanganates" that are formed by reduction of permanganate with an excess of sulfite.

24-25 The Chemistry of Manganese(VI), d^1, and -(VII), d^0

Manganese(VI) is known only as the deep green *manganate* ion, MnO_4^{2-}. This is formed on oxidizing MnO_2 in fused KOH with KNO_3 or air.

The manganate ion is stable only in very basic solutions. In acid, neutral, or slightly basic solutions it readily disproportionates according to the equation

$$3MnO_4^{2-} + 4H^+ = 2MnO_4^- + MnO_2(s) + 2H_2O \qquad K \approx 10^{58}$$

Manganese(VII) is best known in the form of salts of the *permanganate ion.* $KMnO_4$ is manufactured by electrolytic oxidation of a basic solution of K_2MnO_4. Aqueous solutions of MnO_4^- may be prepared by oxidation of solutions of Mn^{II} ion with very powerful oxidizing agents such as PbO_2 or $NaBiO_3$. The ion has an intense purple color, and crystalline salts appear almost black.

Solutions of permanganate are intrinsically unstable, decomposing slowly but observably in acid solution:

$$4MnO_4^- + 4H^+ \longrightarrow 3O_2(g) + 2H_2O + 4MnO_2(s)$$

In neutral or slightly alkaline solutions in the dark, decomposition is immeasurably slow. It is catalyzed by light so that standard permanganate solutions should be stored in dark bottles.

In *basic* solution, permanganate is a powerful oxidant:

$$MnO_4^- + 2H_2O + 3e = MnO_2(s) + 4OH^- \qquad E^\circ = +1.23 \text{ V}$$

In very strong base and with an excess of MnO_4^-, however, manganate ion is produced:

$$MnO_4^- + e = MnO_4^{2-} \qquad E^\circ = +0.56 \text{ V}$$

In *acid* solution permanganate is reduced to Mn^{2+} by an excess of reducing agent:

$$MnO_4^- + 8H^+ + 5e = Mn^{2+} + 4H_2O \qquad E^\circ = +1.51 \text{ V}$$

but because MnO_4^- oxidizes Mn^{2+}:

$$2MnO_4^- + 3Mn^{2+} + 2H_2O = 5MnO_2(s) + 4H^+ \qquad E^\circ = +0.46 \text{ V}$$

the product in presence of an excess of permanganate is MnO_2. The addition of small amounts of $KMnO_4$ to concentrated H_2SO_4 gives a clear green solution believed to contain the planar ion MnO_3^+:

$$KMnO_4 + 3H_2SO_4 \rightleftharpoons K^+ + MnO_3^+ + H_3O^+ + 3HSO_4^-$$

With larger amounts of $KMnO_4$, the dangerous explosive oil, Mn_2O_7, separates. This can be extracted into CCl_4 or chlorofluorocarbons in which it is reasonably stable and safe.

IRON

24-26 The Element

Beginning with this element, there is no oxidation state equal to the total number of valence-shell electrons, which in this case is eight. The highest oxidation state is VI, and it is rare. Even the trivalent state, which rose to a peak of importance at chromium, now loses ground to the divalent state.

Iron is the second most abundant metal, after Al, and the fourth most abundant element in the earth's crust. The core of the earth is believed to consist mainly of Fe and Ni. The major ores are *hematite*, Fe_2O_3 *magnetite*, Fe_3O_4, *limonite*, FeO(OH), and *siderite*, $FeCO_3$.

Pure iron is quite reactive. In moist air it is rather rapidly oxidized to give a hydrous ferric oxide (rust) which affords no protection, since it flakes off, exposing fresh metal surfaces. Finely divided iron is pyrophoric.

The metal dissolves readily in dilute mineral acids. With nonoxidizing acids and in absence of air, Fe^{II} is obtained. With air present or when warm dilute HNO_3 is used, some of the iron goes to Fe^{III}. Very strongly oxidizing media such as concentrated HNO_3 or acids containing dichromate passivate iron. Air-free water and dilute air-free solutions of OH^- have little effect, but iron is attacked by hot concentrated NaOH (see below).

IRON COMPOUNDS

The main stereochemistries of iron compounds are as follows:

Fe^{II}, Fe^{III} } Octahedral as in $Fe(OH)_2$, $Fe(H_2O)_6^{2+}$, $Fe(CN)_6^{4-}$, $Fe(H_2O)_6^{3+}$, $Fe(acac)_3$

24.27
Binary Compounds

Oxides and Hydroxides. The addition of OH^- to Fe^{2+} solutions gives the pale green *hydroxide*, which is very readily oxidized by air to give red-brown hydrous ferric oxide.

$Fe(OH)_2$, a true hydroxide with the $Mg(OH)_2$ structure, is somewhat amphoteric. Like Fe, it dissolves in hot concentrated NaOH, from which solutions blue crystals of $Na_4[Fe^{II}(OH)_6]$ can be obtained.

The *oxide*, FeO, may be obtained as a black pyrophoric powder by ignition of Fe^{II} oxalate: it is usually nonstoichiometric $Fe_{0.95}O$, which means that some Fe^{III} is present. The addition of OH^- to ferric solutions gives a red-brown gelatinous mass commonly called ferric hydroxide, but it is best described as a *hydrous oxide* $Fe_2O_3 \cdot nH_2O$. This has several forms; one, FeO(OH), occurs in the mineral *lepidocrocite* and can be made by high temperature hydrolysis of ferric chloride. On heating at 200° the hydrous oxides form red-brown, α-Fe_2O_3, which occurs as the mineral *hematite*. This has the corundum structure with an *hcp* array of O and Fe^{3+} in the octahedral interstices.

The black crystalline oxide, Fe_3O_4, a mixed Fe^{II}–Fe^{III} oxide, occurs in nature as *magnetite*. It can be made by ignition of Fe_2O_3 above 1400°. It has the inverse spinel structure (page 104).

Chlorides. These are used as source materials for the synthesis of other iron compounds. Anhydrous *ferrous chloride* can be made by passing HCl gas over heated iron powder, by reducing $FeCl_3$ with Fe in tetrahydrofuran, or by refluxing $FeCl_3$ in chlorobenzene. It is a very pale green, almost white, solid.

Ferric chloride. This is obtained by action of chlorine on heated iron as almost black, red-brown crystals. Although in the gas phase there are dimers, Fe_2Cl_6, in the crystal the structure is nonmolecular and there are Fe^{III} ions occupying 2/3 of the octahedral holes in alternate layers of Cl ions.

Ferric chloride quite readily hydrolyzes in moist air. It is soluble in ethers and other polar solvents.

24-28
Chemistry of Iron(II), d^6

The ferrous ion, $[Fe(H_2O)_6]^{2+}$, gives many crystalline salts. Mohr's salt, $(NH_4)_2SO_4 \cdot [Fe(H_2O)_6] \cdot SO_4$, is reasonably stable toward air and loss of water, and is commonly used to prepare standard solutions of Fe^{2+} for volumetric analysis and as a calibration substance in magnetic measurements. By contrast, $FeSO_4 \cdot 7H_2O$ slowly effloresces and turns brown-yellow when kept in air.

Addition of HCO_3^- or SH^- to aqueous solutions of Fe^{2+} precipitates $FeCO_3$ and FeS, respectively. The Fe^{2+} ion is oxidized in acid solution by air to Fe^{3+}. With ligands other than water present, substantial changes in the potentials may occur, and the Fe^{II}–Fe^{III} system provides an excellent example of the effect of ligands on the relative stabilities of oxidation states:

$$[Fe(CN)_6]^{3-} + e = [Fe(CN)_6]^{4-} \qquad E^\circ = 0.36 \text{ V}$$

$$[Fe(H_2O)_6]^{3+} + e = [Fe(H_2O)_6]^{2+} \qquad E^\circ = 0.77 \text{ V}$$

$$[Fe\ phen_3]^{3+} + e = [Fe\ phen_3]^{2+} \qquad E^\circ = 1.12 \text{ V}$$

Complexes. Octahedral complexes are generally paramagnetic, and quite strong ligand fields are required to cause spin pairing. Diamagnetic complex ions are $[Fe(CN)_6]^{4-}$ and $[Fe\ dipy_3]^{2+}$. Formation of the red 2,2′-bipyridine and 1,10-phenanthroline complexes is used as a test for Fe^{2+}.

Some tetrahedral complexes like $FeCl_4{}^{2-}$ are known. Among the most important complexes are those involved in biological systems (Chapter 31) or models for them. An important iron(II) compound is ferrocene (Section 29-14).

24-29 The Chemistry of Iron(III), d^5

Iron(III) occurs in crystalline salts with most anions other than those such as iodide that are incompatible because of their reducing properties:

$$Fe^{3+} + I^- = Fe^{2+} + \tfrac{1}{2}I_2$$

The salts containing the aquo ferric ion, $[Fe(H_2O)_6]^{3+}$, such as $Fe(ClO_4)_3 \cdot 10H_2O$ are pale pink to nearly white and the aquo ion is pale purple. Unless Fe^{3+} solutions are quite strongly acid, hydrolysis occurs and the solutions are commonly yellow because of the formation of hydroxo species that have charge-transfer bands in the ultraviolet region tailing into the visible region.

The initial hydrolysis equilibria are

$$[Fe(H_2O)_6]^{3+} = [Fe(H_2O)_5(OH)]^{2+} + H^+ \qquad K = 10^{-3.05}$$

$$[Fe(H_2O)_5(OH)]^{2+} = [Fe(H_2O)_4(OH)_2]^+ + H^+ \qquad K = 10^{-3.26}$$

$$2[Fe(H_2O)_6]^{3+} = [Fe(H_2O)_4(OH)_2Fe(H_2O)_4]^{4+} + 2H^+ \qquad K = 10^{-2.91}$$

The binuclear species in the third equation probably has the structure (24-III).

$$\begin{array}{c} (H_2O)_4Fe(\mu\text{-}OH)_2Fe(H_2O)_4 \end{array}$$

24-III

From the constants it is clear that, even at pH's of 2-3, hydrolysis is extensive. In order to have solutions containing, say, ~ 99% $[Fe(H_2O)_6]^{3+}$ the pH must be around zero. As the pH is raised to about 2-3, more highly condensed species than the dinuclear one are formed, attainment of equilibrium becomes sluggish, and soon colloidal gels are formed. Ultimately, hydrous Fe_2O_3 is precipitated.

Ferric ion has a strong affinity for F^-:

$$Fe^{3+} + F^- = FeF^{2+} \qquad K_1 \approx 10^5$$

$$FeF^{2+} + F^- = FeF_2^{\ +} \qquad K_2 \approx 10^5$$

$$FeF_2^{\ +} + F^- = FeF_3 \qquad K_3 \approx 10^3$$

The corresponding constants for chloro complexes are only ~10, ~3, and ~0.1, respectively. In very concentrated HCl the tetrahedral $FeCl_4^{\ -}$ ion is formed, and its salts with large cations may be isolated. Complexes with SCN^- are an intense red, and this serves as a sensitive qualitative and quantitative test for ferric ion; $Fe(SCN)_3$ and/or $Fe(SCN)_4^{\ -}$ may be extracted into ether. Fluoride ion, however, will discharge this color. In the solid state, $FeF_6^{\ 3-}$ ions are known but in solutions only species with fewer F atoms occur.

The hexacyanoferrate ion, $[Fe(CN)_6]^{3-}$, in contrast to $[Fe(CN)_6]^{4-}$, is quite poisonous because the CN^- ions rapidly dissociate, whereas $[Fe(CN)_6]^{4-}$ is not labile.

The affinity of Fe^{III} for NH_3 and amines is low except for chelates such as $EDTA^{4-}$; 2,2′-bipyridine and 1,10-phenanthroline produce ligand fields strong enough to cause spin-pairing (cf. Fe^{II}) and form quite stable ions that can be isolated with large anions.

A number of hydroxo and oxygen-bridged species, one of which has been mentioned above, are of interest in that they may show unusual magnetic properties due to coupling between the iron atoms via the bridges. One example is the bissalicylaldehydeethylenediiminato complex $[Fesalen]_2O$ which has a nonlinear $Fe\overset{O}{\diagup\ \diagdown}Fe$ group, while Fe(salen)Cl can form both mononuclear and binuclear complexes (24-IV). The latter has marked antiferromagnetic coupling between the Fe atoms.

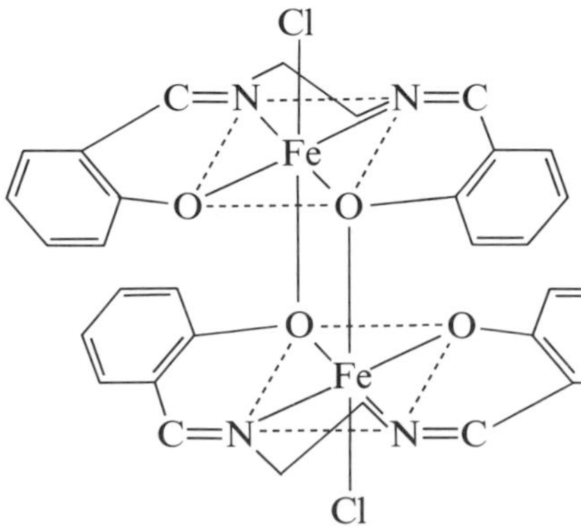

24-IV

24-30
The Chemistry of Iron(IV) and (VI)

For Fe^{IV} there are only a few complexes such as $[Fe(S_2CNR_2)_3]^+$ and [Fe diars$_2$ Cl$_2$]$^{2+}$, and the unusual hydrocarbon soluble alkyl, Fe(1-norbornyl)$_4$ (Section 29-11). No stable Fe^V compounds are known.

The best known compound of iron(VI) is the oxo anion, FeO_4^{2-}, obtained by chlorine oxidation of suspensions of $Fe_2O_3 \cdot nH_2O$ in concentrated NaOH or by fusing Fe powder with KNO_3. The red-purple ion is paramagnetic with two unpaired electrons. The Na and K salts are quite soluble but the Ba salt can be precipitated. The ion is relatively stable in basic solution but decomposes in neutral or acid solution according to the equation

$$2FeO_4^{2-} + 10H^+ = 2Fe^{3+} + \tfrac{3}{2}O_2 + 5H_2O$$

It is an even stronger oxidizing agent than MnO_4^- and can oxidize NH_3 to N_2, Cr^{II} to CrO_4^{2-}, and also primary amines and alcohols to aldehydes.

COBALT

24-31
The Element

The trends toward decreased stability of the very high oxidation states and the increased stability of the II state relative to the III state, which occur through the series Ti, V, Cr, Mn, and Fe, persist with Co. The highest oxidation state is now IV, and only few such compounds are known. Cobalt(III) is relatively unstable in simple compounds, but the low-spin complexes are exceedingly numerous and stable, especially where the donor atoms (usually N) make strong contributions to the ligand field. There are also numerous complexes of Co^I. This oxidation state is better known for cobalt than for any other element of the first transition series except copper. All Co^I complexes have π-acid ligands (Chapter 28).

Cobalt always occurs in association with Ni and usually also with As. The chief sources of cobalt are "speisses," which are residues in the smelting of arsenical ores of Ni, Cu, and Pb.

Cobalt is relatively unreactive, although it dissolves slowly in dilute mineral acids.

COBALT COMPOUNDS

The main stereochemistries found in cobalt compounds are the following:

Co^{II} { Tetrahedral as in $CoCl_4^{2-}$, $CoCl_2(PEt_3)_2$
Octahedral as in $Co(H_2O)_6^{2+}$

Co^{III} Octahedral as in $Co(NH_3)_6^{3+}$

24-32
Chemistry of Cobalt(II), d^7

The dissolution of Co, or the hydroxide or carbonate, in dilute acids gives the pink aquo ion $[Co(H_2O)_6]^{2+}$ which forms many hydrated salts.

Addition of OH^- to Co^{2+} gives the *hydroxide* which may be blue or pink depending on the conditions; it is weakly amphoteric dissolving in very concentrated OH^- to give a blue solution containing the $[Co(OH)_4]^{2-}$ ion.

Complexes. The most common Co^{II} complexes may be either octahedral or tetrahedral. There is only a small difference in stability and both types, with the same ligand, may be in equilibrium. Thus for water there is a very small but finite concentration of the tetrahedral ion

$$[Co(H_2O)_6]^{2+} \rightleftharpoons [Co(H_2O)_4]^{2+} + 2H_2O$$

Addition of excess Cl^- to pink solutions of the aquo ion readily gives the blue tetrahedral species

$$[Co(H_2O)_6]^{2+} + 4Cl^- = [CoCl_4]^{2-} + 6H_2O$$

Tetrahedral complexes $CoX_4{}^{2-}$ are formed by halide, pseudohalide, and OH^- ions. Cobalt(II) forms tetrahedral complexes more readily than any other transition-metal ion. Co^{2+} is the only d^7 ion of common occurrence. For a d^7 ion, ligand-field stabilization energies disfavor the tetrahedral configuration relative to the octahedral one to a smaller extent (see Table 23-2, page 374) than for any other $d^n (1 \leq n \leq 9)$ configuration. This argument is valid only in comparing the behavior of one metal ion with another and not for assessing the absolute stabilities of the configurations for any particular ion.

24-33
The Chemistry of Cobalt(III)

In the absence of complexing agents, the oxidation of $Co(H_2O)_6{}^{2+}$ is very unfavorable (Table 24-3, page 382) and Co^{3+} is reduced by water. However, electrolytic or O_3 oxidation of cold acidic solutions of $Co(ClO_4)_2$ gives the aquo ion, $[Co(H_2O)_6]^{3+}$, in equilibrium with $[Co(OH)(H_2O)_5]^{2+}$. At 0°, the half-life of these diamagnetic ions is about a month. In the presence of complexing agents, such as NH_3, the stability of Co^{III} is greatly improved:

$$[Co(NH_3)_6]^{3+} + e = [Co(NH_3)_6]^{2+} \qquad E^\circ = 0.1\ V$$

In presence of OH^- ion, cobaltous hydroxide is readily oxidized by air to a black hydrous oxide:

$$CoO(OH)(s) + H_2O + e = Co(OH)_2(s) + OH^- \qquad E^\circ = 0.17\ V$$

The Co^{3+} ion shows a particular affinity for N donors such as NH_3, en, EDTA, —NCS, etc., and complexes are exceedingly numerous. They generally undergo ligand-exchange reactions relatively slowly, like Cr^{3+} and Rh^{3+}. Hence, from the days of Werner and Jørgensen, they have been extensively studied. A large part of our knowledge of the isomerism, modes of reaction, and general properties of octahedral complexes as a class is based on studies of Co^{III} complexes. All known Co^{III} complexes are octahedral.

Cobalt(III) complexes are synthesized by oxidation of Co^{2+} in solution in the presence of the ligands. Oxygen or hydrogen peroxide and a catalyst, such as activated charcoal, are used. For example,

$$4Co^{2+} + 4NH_4^+ + 20NH_3 + O_2 \longrightarrow 4[Co(NH_3)_6]^{3+} + 2H_2O$$

$$4Co^{2+} + 8en + 4en\cdot H^+ + O_2 = 4[Co\ en_3]^{3+} + 2H_2O$$

From a reaction similar to the above, in presence of HCl the green salt, *trans*-$[Co\ en_2Cl_2][H_5O_2]Cl_2$, is obtained. This may be isomerized to the purple racemic *cis*-isomer on evaporation of a neutral aqueous solution at 90 to 100°. Both the *cis*- and the *trans*-isomer are aquated when heated in water:

$$[Co\ en_2Cl_2]^+ + H_2O \longrightarrow [Co\ en_2Cl(H_2O)]^{2+} + Cl^-$$

$$[Co\ en_2Cl(H_2O)]^{2+} + H_2O \longrightarrow [Co\ en_2(H_2O)_2]^{3+} + Cl^-$$

and on treatment with solutions of other anions are converted into other $[Co\ en_2X_2]^+$ species, for example,

$$[Co\ en_2Cl_2]^+ + 2NCS^- \longrightarrow [Co\ en_2(NCS)_2]^+ + 2Cl^-$$

The initial reaction of Co^{II} with oxygen may involve oxidative addition (Section 30-2) of O_2 to Co^{II} to give a transient Co^{IV} species which then reacts with another Co^{II} to produce a binuclear *peroxo-bridged* species:

$$L_nCo^{2+} + O_2 \longrightarrow L_nCo^{IV}\begin{matrix}O^{2+}\\ |\\ O\end{matrix}$$

$$L_nCo^{IV}\begin{matrix}O^{2+}\\ |\\ O\end{matrix} + L_nCo^{2+} \longrightarrow L_nCo{-}O{-}O{-}CoL_n^{4+}$$

Complexes such as $[(NH_3)_5CoOOCo(NH_3)_5]^{4+}$ or $[(NC)_5CoOOCo(CN)_5]^{6-}$ have been isolated, although these ions decompose fairly readily in water or acids. The open-chain species $[(NH_3)_5CoO_2Co(NH_3)_5]^{4+}$ can be cyclized in presence

of base to

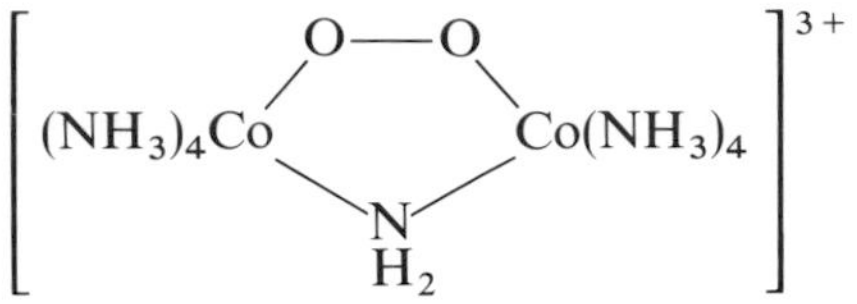

$$\left[(NH_3)_4Co \begin{matrix} O-O \\ \diagup \quad \diagdown \\ \diagdown \quad \diagup \\ \underset{H_2}{N} \end{matrix} Co(NH_3)_4 \right]^{3+}$$

Such peroxo species, open-chain or cyclic, contain low-spin Co^{III} and bridging peroxide, O_2^{2-}, ions.

The O_2-bridged binuclear complexes can often be oxidized in a one-electron step. The resulting ions were first prepared by Werner, who formulated them as peroxo-bridged complexes of Co^{III} and Co^{IV}. However, esr data have shown that the single unpaired electron is distributed equally over *both* cobalt ions, and is best regarded as belonging formally to a superoxide, O_2^- ion but delocalized over the planar $Co^{III}—O—O—Co^{III}$ group. The structures (*Fig. 24-4*) show that the O—O distance is close to that for the O_2^- ion (1.28) and much shorter than the distance (1.47 Å) in the peroxo complexes.

Although no cobalt-containing complex is known to be involved in oxygen metabolism, there are several that provide models for metal-to-oxygen binding in biological systems. Of greatest interest are those that undergo *reversible* oxygenation and deoxygenation in solution. The Schiff base complexes such as Co(acacen) (24-V) in dimethylformamide or pyridine take up O_2 reversibly below 0° C, for example,

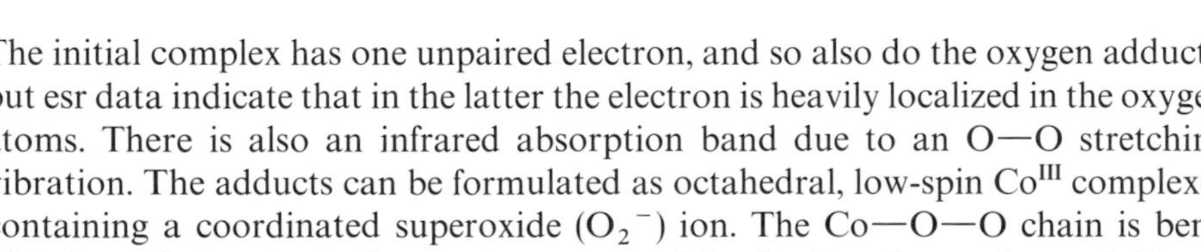

$$\text{Co(acacen)} + \text{DMF} + O_2 \longrightarrow \text{Co(acacen)(DMF)}O_2 \qquad K = 2.1 \text{ at } -10^\circ$$

24-V

The initial complex has one unpaired electron, and so also do the oxygen adducts, but esr data indicate that in the latter the electron is heavily localized in the oxygen atoms. There is also an infrared absorption band due to an O—O stretching vibration. The adducts can be formulated as octahedral, low-spin Co^{III} complexes containing a coordinated superoxide (O_2^-) ion. The Co—O—O chain is bent. This formulation is similar to that proposed by Pauling for oxyhemoglobin. A second type of complex involves the reversible formation of oxygen bridges, Co—O—O—Co, which are similar to those discussed above.

Finally, we note in connection with oxidation that in acid solutions, cobalt(III) carboxylates catalyze not only the oxidation of alkyl side chains in aromatic hydrocarbons, but even alkanes. A cobalt catalyzed process is used commercially for oxidation of toluene to phenol. The actual nature of "cobaltic acetate," a green material made by ozone oxidation of Co^{2+} acetate in acetic acid is uncertain; it can, however, be converted by pyridine to an oxo centered species similar to those known for other M^{III} carboxylates (page 384).

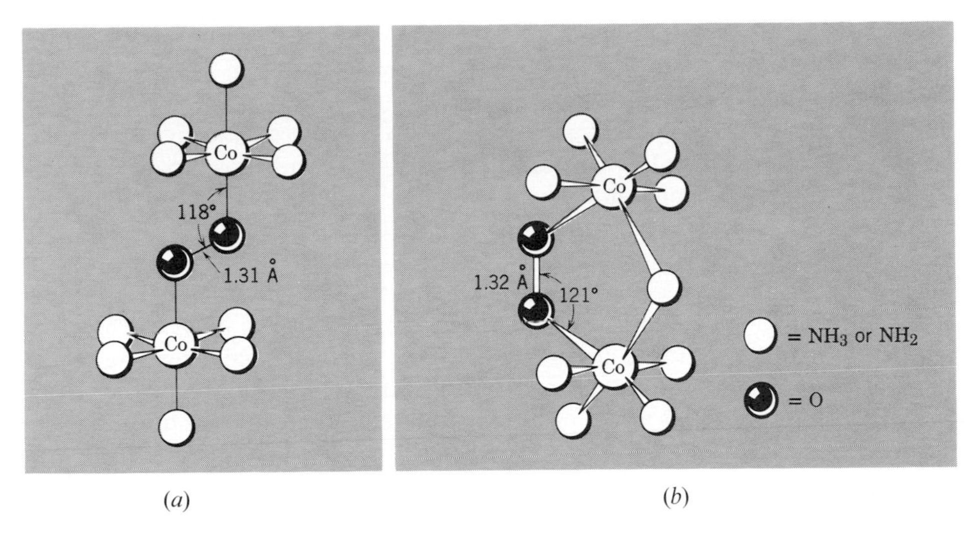

Figure 24-4 *The structures of (a)* $[(NH_3)_5CoO_2Co(NH_3)_5]^{5+}$,

and (b) $\left[(HN_3)_4Co\begin{matrix} O_2 \\ \\ NH_2 \end{matrix}Co(NH_3)_4\right]^{4+}$. *There is octahedral coordination about each cobalt ion and the angles and distances shown are consistent with the assumption that there are bridging superoxo groups. The five-membered ring in (b) is essentially planar.*

24-34
Complexes of Cobalt(I), d^8

With the exception of reduced vitamin B_{12} and models for this system (Section 31-11) which appear to be Co^I species, all Co^I compounds involve ligands of the π-acid type (Chapter 28). The coordination is trigonal bipyramidal or tetrahedral. The compounds are usually made by reducing $CoCl_2$ in presence of the ligand by agents such as N_2H_4, Zn, $S_2O_4{}^{2-}$, or Al alkyls.

Representative examples are $CoH(N_2)(PPh_3)_3$, $[Co(CNR)_5]^+$, $CoCl(PR_3)_3$.

NICKEL

24-35
The Element

The trend toward decreased stability of higher oxidation states continues, so that only Ni^{II} normally occurs with a few compounds *formally* containing Ni^{III} and Ni^{IV}. The relative simplicity of nickel chemistry in terms of oxidation number is balanced by considerable complexity in coordination numbers and geometries.

Nickel occurs in combination with arsenic, antimony, and sulfur as in *millerite*, NiS, and in *garnierite*, a magnesium-nickel silicate of variable composition. Nickel is also found alloyed with iron in meteors; the interior of the earth is believed to contain considerable quantities. In general, the ore is roasted in air to give NiO, which is reduced to Ni with C. Nickel is usually purified by electrodeposition but some high purity nickel is still made by the carbonyl process. Carbon monoxide reacts with impure nickel at 50° and ordinary pressure or with nickel-copper matte under more strenuous conditions, giving volatile $Ni(CO)_4$, from which metal of 99.90 to 99.99% purity is obtained on thermal decomposition at 200°.

Nickel is quite resistant to attack by air or water at ordinary temperatures when compact and is, therefore, often electroplated as a protective coating. It dissolves readily in dilute mineral acids. The metal or high Ni alloys are used to handle F_2 and other corrosive fluorides. The finely divided metal is reactive to air and may be pyrophoric. Nickel absorbs considerable amounts of hydrogen when finely divided and special forms of Ni, for example, Raney nickel, are used for catalytic reductions.

NICKEL COMPOUNDS

24-36
The Chemistry of Nickel(II), d^8

The binary compounds such as NiO, $NiCl_2$, etc., need no special comment.

Nickel(II) forms a large number of *complexes* with coordination numbers 6, 5, and 4, having all the main structural types, namely, octahedral, trigonal bipyramidal, square-pyramidal, tetrahedral, and square. It is characteristic that complicated equilibria, which are generally temperature-dependent and sometimes concentration-dependent, often exist between these structural types.

Six-Coordinate Species. The commonest is the green aquo ion $[Ni(H_2O)_6]^{2+}$ which is formed on dissolution of Ni, $NiCO_3$ etc., in acids and gives salts like $NiSO_4 \cdot 7H_2O$.

The water molecules in the aquo ion can be readily displaced especially by amines to give complexes such as *trans*-$[Ni(H_2O)_2(NH_3)_4]^{2+}$, $[Ni(NH_3)_6]^{2+}$ or $[Ni\,en_3]^{2+}$. These amine complexes are usually blue or purple because of shifts in absorption bands when H_2O is replaced by a stronger field ligand (cf. page 371).

Four-Coordinate Species. Most of these complexes are square. This is a consequence of the d^8 configuration, since the planar ligand set causes one of the d orbitals ($d_{x^2-y^2}$) to be uniquely high in energy, and the eight electrons can occupy the other four d orbitals but leave this strongly antibonding one vacant. In tetrahedral coordination, on the other hand, occupation of antibonding orbitals is unavoidable. With the congeneric d^8 systems Pd^{II} and Pt^{II} this factor becomes so important that no tetrahedral complex is formed.

Planar complexes of Ni^{II} are thus invariably diamagnetic. They are frequently red, yellow, or brown owing to the presence of an absorption band of medium intensity ($\varepsilon \approx 60$) in the range 450–600 nm.

Probably the best known example is the red *bis(dimethylglyoximato) nickel(II)*, $Ni(dmgH)_2$, which is used for the gravimetric determination of nickel; it is precipitated on addition of ethanolic $dmgH_2$ to ammonical nickel(II) solutions. It has the structure (24-VI) where the H-bond is symmetrical, but these units are *stacked* one on top of the other in the crystal. Here, and in similar square compounds of Pd^{II} and Pt^{II} (Section 25-28), there is evidence of metal–metal interaction, even though the distance in the stack is too long for true bonding.

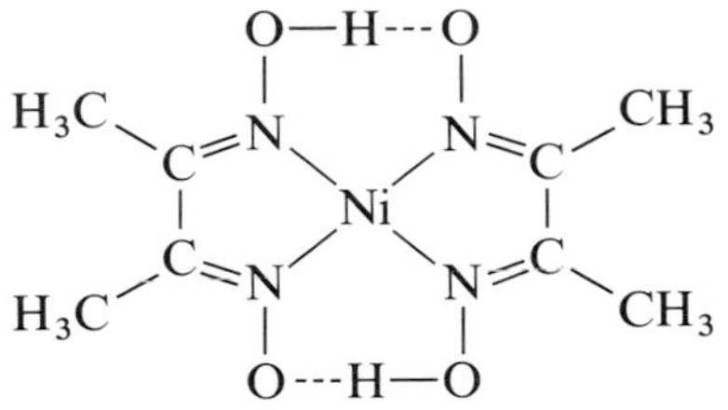

24-VI

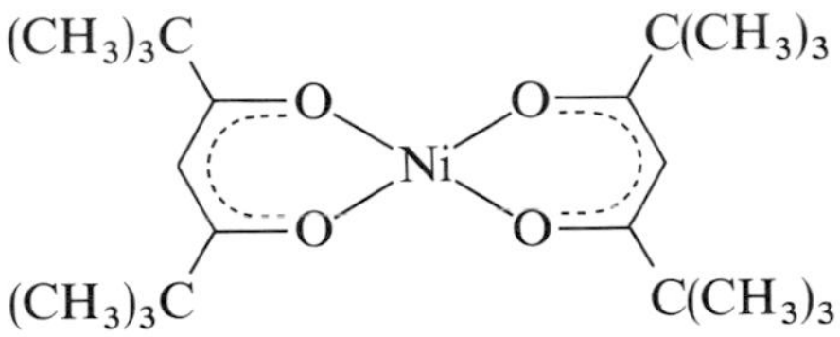

24-VII

Similar square complexes are given by certain β-ketoenolates, for example, (24-VII) as well as by unidentate π-acid ligands, for example, $NiBr_2(PEt_3)_2$, and by CN^- and SCN^-. The cyano complex, $Ni(CN)_4^{2-}$, is readily formed on addition of CN^- to $Ni^{2+}(aq)$. The green $Ni(CN)_2$ first precipitated redissolves to give the yellow ion which can be isolated as, for example, $Na_2[Ni(CN)_4]\cdot 3H_2O$. On addition of an excess of CN^-, the red ion $[Ni(CN)_5]^{3-}$ is formed which can be precipitated only by use of large cations.

Tetrahedral Complexes. They are less common than planar ones, and are all paramagnetic. They are of the types NiX_4^{2-}, NiX_3L^-, NiL_2X_2, and $Ni(L—L)_2$ where X is halogen, L is a neutral ligand such as R_3P, R_3PO, and L—L is a bidentate uninegative ligand; NiL_4^{2+} is known where L = hexamethylphosphoramide.

Five-Coordinate Complexes. These usually have trigonal bipyramidal geometry but some are square pyramidal. Many contain the tetradentate "tripod" ligands such as $N[CH_2CH_2NMe_2]_3$ (see page 132).

24-37
Conformational Properties of Nickel(II) Complexes

The main structural and conformational changes which nickel(II) complexes undergo are the following.

1. *Formation of 5- and 6-Coordinate Complexes by Addition of Ligands to Square Ones.* For *any* square complex, NiL_4, the following equilibria with additional ligands, L′ must in principle exist:

$$ML_4 + L' = ML_4L'$$

$$ML_4 + 2L' = ML_4L'_2$$

Where L = L′ = CN, only the 5-coordinate species is formed, but in most systems in which L′ is a good donor such as pyridine, H_2O, C_2H_5OH, etc., the equilibria lie far in favor of the 6-coordinate species. These have a *trans*-structure and a high-spin electron configuration; many may be isolated as pure compounds. Thus, the β-diketone complex (24-VII) is normally prepared in the presence of water and/or alcohol and is first isolated as the green, paramagnetic dihydrate or dialcoholate, from which the red, square complex is then obtained by heating to drive off the solvent.

Another type of square complex that picks up H_2O, anions or solvent is (24-VIII).

$$\left[\begin{array}{ccccc} & H_2 & & H_2 & \\ PhCH- & N & & N & -CHPh \\ | & & \diagdown \quad \diagup & & | \\ | & & Ni & & | \\ | & & \diagup \quad \diagdown & & | \\ PhCH- & N & & N & -CHPh \\ & H_2 & & H_2 & \end{array}\right]^{2+}$$

24-VIII

2. *Monomer-Polymer Equilibria.* Four-coordinate complexes may associate or polymerize, to give 5- or 6-coordinate species. In some cases, the association is very strong and the 4-coordinate monomers are observed only at high temperatures. In others the position of the equilibrium is such that both red, diamagnetic monomers and green or blue, paramagnetic polymers are present in a temperature- and concentration-dependent equilibrium around room temperature. A clear example of this situation is provided by the *acetylacetonate* (*Fig. 24-5*). As a result of the sharing of some oxygen atoms, each nickel atom achieves octahedral coordination. This trimer is very stable, and detectable quantities of monomer appear only at temperatures around 200° in a noncoordinating solvent. It is, however, readily cleaved by donors such as H_2O or pyridine, to give 6-coordinate monomers.

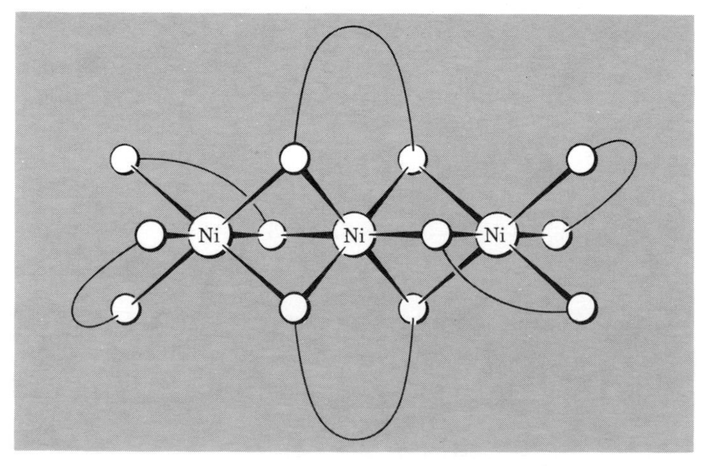

Figure 24-5 *Sketch indicating the trimeric structure of nickel acetylacetonate. The unlabeled circles represent oxygen atoms, and the curved lines connecting them in pairs represent the remaining portions of the acetylacetonate rings. (Reproduced by permission from J. C. Bullen, R. Mason and P. Pauling,* Inorg. Chem., *1965,* 4, *456.)*

When the methyl groups of the acetylacetonate ligand are replaced by the very bulky $C(CH_3)_3$ group trimerization is completely prevented and the planar monomer (24-VII above) results. When groups sterically intermediate between CH_3 and $C(CH_3)_3$ are used, temperature- and concentration-dependent monomer-trimer equilibria are observed in noncoordinating solvents.

3. *Square-Tetrahedral Equilibria and Isomerism.* Complexes, NiL_2X_2 in which L represents a mixed alkylarylphosphine exist in solution in an equilibrium distribution between the tetrahedral and square forms. In some cases it is possible to isolate two crystalline forms of the compound, one yellow to red and diamagnetic, the other green or blue with two unpaired electrons. There is even a case, $Ni[(C_6H_5CH_2)(C_6H_5)_2P]_2Br_2$, in which *both* tetrahedral and square complexes are found together in the same crystalline substance.

24-38 Higher Oxidation States of Nickel

Oxides and Hydroxides. The action of Br_2 on alkaline solutions of Ni^{2+} gives a black hydrous oxide NiO(OH). Other black substances can be obtained by electrolytic oxidation; some of them contain alkali metal ions.

The Edison or nickel-iron battery, which uses KOH as the electrolyte, is based on the reaction

$$Fe + 2NiO(OH) + 2H_2O \underset{\text{charge}}{\overset{\text{Discharge}}{\rightleftarrows}} Fe(OH)_2 + 2Ni(OH)_2 \qquad (\sim 1.3\ V)$$

but the mechanism and the true nature of the oxidized nickel species are not fully understood.

Complexes. There are several authentic complexes of *nickel(III)*. Oxidation of $NiX_2(PR_3)_2$ with the appropriate halogen gives $NiX_3(PR_3)_2$.

Nickel(IV) complexes are even rarer, and the dithiolene complexes (Section 28-18) which could formally be regarded as containing Ni^{4+} and $S_2CR_2{}^{2-}$ ligands are best regarded as Ni^{II} complexes.

COPPER

24-39 The Element

Copper has a single *s* electron outside the filled 3*d* shell. It has little in common with the alkalis except formal stoichiometries in the +I oxidation state. The filled *d* shell is much less effective than is a noble-gas shell in shielding the *s* electron from the nuclear charge, so that the first ionization potential of Cu is higher than those of the alkalis. Since the electrons of the *d* shell are also involved in metallic bonding, the heat of sublimation and the melting point of copper are also much higher than those of the alkalis. These factors are responsible for the more noble character of copper. The effect is to make compounds more covalent and to give them higher lattice energies, which are not offset by the somewhat smaller radius of Cu^+, 0.93 Å compared with Na^+, 0.95; and K^+, 1.33 Å.

The second and third ionization potentials of Cu are very much lower than those of the alkalis and account in part for the transition-metal character.

Copper is not abundant (55 ppm) but is widely distributed as metal, in sulfides, arsenides, chlorides, and carbonates. The commonest mineral is chalcopyrite $CuFeS_2$. Copper is extracted by oxidative roasting and smelting, or by microbial-assisted leaching, followed by electrodeposition from sulfate solutions.

Copper is used in alloys such as brass and is completely miscible with gold. It is very slowly superficially oxidized in moist air, sometimes giving a green coating of hydroxo carbonate and hydroxo sulfate (from SO_2 in the atmosphere).

Copper readily dissolves in nitric acid and in sulfuric acid in presence of oxygen. It is also soluble in KCN or ammonia solutions in the presence of oxygen, as indicated by the potentials

$$Cu + 2NH_3 \xrightarrow{-0.12\text{ V}} [Cu(NH_3)_2]^+ \xrightarrow{-0.01\text{ V}} [Cu(NH_3)_4]^{2+}$$

COPPER COMPOUNDS

The stereochemistry of the more important copper compounds is as follows:

- Cu^I — Tetrahedral as in CuI(s) or $Cu(CN)_4{}^{3-}$
- Cu^{II} — Square as in CuO(s), $Cupy_4{}^{2+}$, or $CuCl_4{}^{2-}$; Distorted octahedral with two longer *trans* bonds, for example, $Cu(H_2O)_6{}^{2+}$, $CuCl_2$(s)

24-40 The Chemistry of Copper(I), d^{10}

Cuprous compounds are diamagnetic and, except where color results from the anion or charge-transfer bands, colorless.

The relative stabilities of the cuprous and cupric states are indicated by the potentials

$$Cu^+ + e = Cu \qquad E^\circ = 0.52\text{ V}$$

$$Cu^{2+} + e = Cu^+ \qquad E^\circ = 0.153\text{ V}$$

From these we have

$$Cu + Cu^{2+} = 2Cu^+ \qquad E^\circ = -0.37\text{ V}; \qquad K = \frac{[Cu^{2+}]}{[Cu^+]^2} = \sim 10^6$$

The relative stabilities depend very strongly on the nature of anions or other ligands present, and vary considerably with solvent or the nature of neighboring atoms in a crystal.

In *aqueous* solution only low equilibrium concentrations of Cu^+ ($<10^{-2}$ M) can exist (see below). The only cuprous compounds that are stable to water are the highly insoluble ones such as CuCl or CuCN. This instability toward water is due partly to the greater lattice and solvation energies and higher formation constants for complexes of the Cu^{2+} ion so that ionic Cu^I derivatives are unstable.

The equilibrium $2Cu^I \rightleftharpoons Cu + Cu^{II}$ can readily be displaced in either direction. Thus with CN^-, I^-, and Me_2S, Cu^{II} reacts to give the Cu^I compound. The Cu^{II} state is favored by anions that cannot give covalent bonds or bridging groups, for example, ClO_4^- and SO_4^{2-}, or by complexing agents that have their greater affinity for Cu^{II}. Thus ethylenediamine reacts with cuprous chloride in aqueous potassium chloride solution:

$$2CuCl + 2en = [Cu\,en_2]^{2+} + 2Cl^- + Cu^0$$

The latter reaction also depends on the chelate nature of the ligand. Thus for ethylenediamine, K is $\sim 10^7$, for pentamethylenediamine (which does not chelate 3×10^{-2}, and for ammonia 2×10^{-2}. Hence, in the last case the reaction is

$$[Cu(NH_3)_4]^{2+} + Cu^0 = 2[Cu(NH_3)_2]^+$$

The lifetime of the Cu^+ ion in water is usually very short (<1 sec), but dilute solutions from reduction of Cu^{2+} with V^{2+} or Cr^{2+} may last for several hours in absence of air.

An excellent illustration of how the stability of the cuprous ion relative to that of the cupric ion may be affected by solvent is the case of acetonitrile. Cu^I is very effectively solvated by CH_3CN, and the halides have relative high solubilities (e.g., CuI, 35 g/kg CH_3CN) vs. negligible solubilities in H_2O. Cu^I is more stable than Cu^{II} in CH_3CN, and Cu^{II} acts as a comparatively powerful oxidizing agent.

Cuprous Binary Compounds. The *oxide* and *sulfide* are more stable than the corresponding Cu^{II} compounds at high temperatures. Cu_2O is made as a yellow powder by controlled reduction of an alkaline solution of a cupric salt with hydrazine or, as red crystals, by thermal decomposition of CuO. Cu_2S is a black crystalline solid prepared by heating copper and sulfur in absence of air; it is, however, markedly nonstoichiometric.

Cuprous chloride and *bromide* are made by boiling an acidic solution of the cupric salt with an excess of copper; on dilution, white CuCl or pale yellow CuBr is precipitated. Addition of I^- to a solution of Cu^{2+} forms a precipitate that rapidly and quantitatively decomposes to CuI and iodine. CuF is unknown. The halides have the zinc blende structure (tetrahedrally coordinated Cu^+). They are insoluble in water, but the solubility is enhanced by an excess of halide ions owing to formation of, for example, $CuCl_2^-$, $CuCl_3^{2-}$, and $CuCl_4^{3-}$.

Cuprous Complexes. The most common types of Cu^I complexes are those of simple halide or amine ligands that are almost invariably *tetrahedral.* Even those with stoichiometries such as K_2CuCl_3 or K_2CuCl_3 still have tetrahedral coordination as there are chains sharing halide ions.

Copper(I) forms several kinds of polynuclear complex in which four Cu atoms lie at the vertices of a tetrahedron. In $Cu_4I_4L_4$ (L = R_3P, R_3As) species, there is a triply bridging I atom on each face of the Cu_4 tetrahedron and one ligand, L, is coordinated to a Cu atom at each vertex (24-IX).

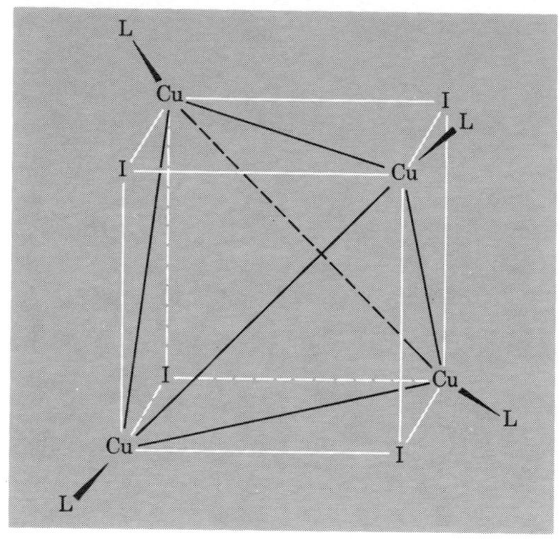

(24-IX)

24-41
The Chemistry of Copper(II), d^9

Most Cu^I compounds are fairly readily oxidized to Cu^{II}, but further oxidation to Cu^{III} is difficult. There is a well-defined aqueous chemistry of Cu^{2+}, and a large number of salts of various anions, many of which are water-soluble, exist in addition to a wealth of complexes.

Before we discuss copper(II) chemistry, it is pertinent to note the stereochemical consequences of the d^9 configuration of Cu^{II}. This makes Cu^{II} subject to distortions (page 373) if placed in an environment of cubic (i.e., regular octahedral or tetrahedral) symmetry. The result is that Cu^{II} is nearly always found in environments appreciably distorted from these regular symmetries. The characteristic distortion of the octahedron is such that there are *four short* Cu—L *bonds in the plane and two trans long ones.* In the limit, this elongation leads to a situation indistinguishable from square coordination as found in CuO and many discrete complexes of Cu^{II}. Thus the cases of tetragonally distorted "octahedral" coordination and square coordination cannot be sharply differentiated.

Some distorted tetrahedral complexes, such as $M_2^ICuX_4$, are also known provided M is large like Cs. $(NH_4)_2CuCl_4$ has a planar anion.

Binary Compounds. Black crystalline CuO is obtained by pyrolysis of the nitrate or other oxo salts; above 800° it decomposes to Cu_2O. The *hydroxide* is obtained as a blue bulky precipitate on addition of NaOH to Cu^{2+} solutions; warming an aqueous slurry dehydrates this to the oxide. The hydroxide is readily soluble in strong acids and also in concentrated NaOH, to give deep blue anions, probably of the type $[Cu_n(OH)_{2n-2}]^{2+}$. In ammoniacal solutions the deep blue tetraammine complex is formed.

The common *halides* are the yellow chloride and the almost black bromide, having structures with infinite parallel bands of square CuX_4 units sharing edges. The bands are arranged so that a tetragonally elongated octahedron is completed about each copper atom by halogen atoms of neighboring chains. $CuCl_2$ and $CuBr_2$ are readily soluble in water, from which hydrates may be crystallized, and also in donor solvents such as acetone, alcohol, and pyridine.

The Aquo Ion and Aqueous Chemistry. Dissolution of copper, the hydroxide, carbonate, etc., in acids gives the blue-green aquo ion which may be written $[Cu(H_2O)_6]^{2+}$. Two of the H_2O molecules are further from the metal than the other four. Of the numerous crystalline hydrates the blue sulfate, $CuSO_4 \cdot 5H_2O$, is best known. It may be dehydrated to the virtually white anhydrous substance. Addition of ligands to aqueous solutions leads to the formation of complexes by successive displacement of water molecules. With NH_3, for example, the species $[Cu(NH_3)(H_2O)_5]^{2+} \cdots [Cu(NH_3)_4(H_2O)_2]^{2+}$ are formed in the normal way, but addition of the fifth and sixth molecules of NH_3 is difficult. The sixth can be added only in liquid ammonia. The reason for this unusual behavior is connected with the Jahn–Teller effect. Because of it, the Cu^{II} ion does not bind the fifth and sixth ligands strongly (even the H_2O). When this intrinsic weak binding of the fifth and sixth ligands is added to the normally expected decrease in the stepwise formation constants (page 140), the formation constants, K_5 and K_6, are very small indeed. Similarly, it is found with ethylenediamine that $[Cu\,en(H_2O)_4]^{2+}$ and $[Cu\,en_2(H_2O)_2]^{2+}$ form readily, but $[Cu\,en_3]^{2+}$ is formed only at extremely high concentrations of en.

Multidentate ligands which coordinate through O or N, such as amino acids, form cupric complexes of considerable stability. The blue solutions formed by addition of tartrate to Cu^{2+} solutions (known as *Fehling's solution* when basic and when *meso*-tartrate is used) may contain monomeric, dimeric, or polymeric species at different pH values. The dimer, $Na_2[Cu\{(\pm)C_4O_6H_2\}] \cdot 5H_2O$, has square Cu^{II} coordination, two tartrate bridges, and a Cu—Cu distance of 2.99 Å.

Polynuclear Compounds with Magnetic Anomalies. Copper forms many compounds in which the Cu—Cu distances are short enough to indicate significant M—M interaction, but in no case is there an actual bond. Particular examples are the bridged *carboxylates* and the related 1,3-triazinato complexes (24-X, 24-XI). Although in other cases of carboxylates with the same structure (Cr_2^{II},

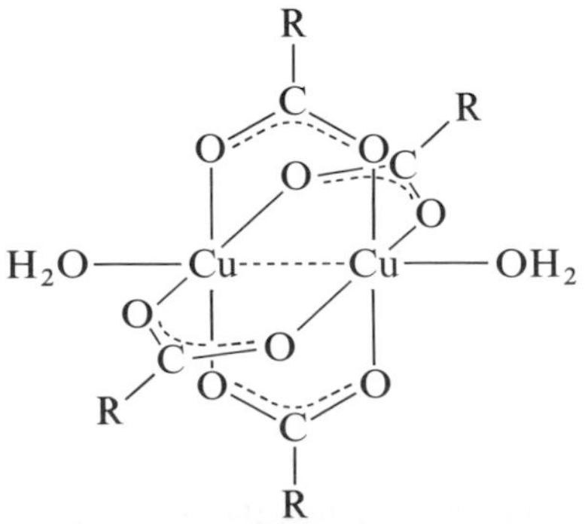

24-X

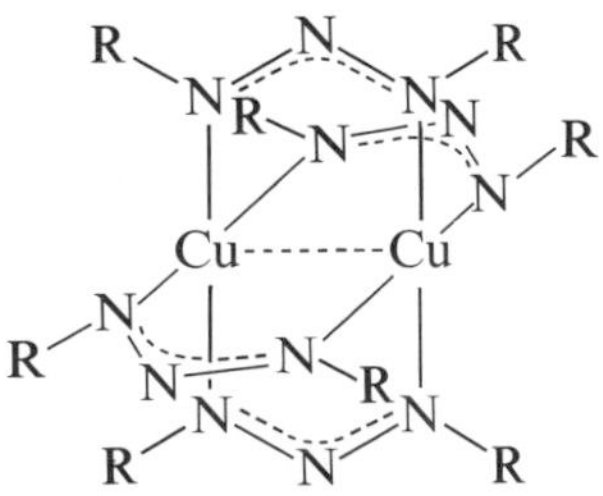

24-XI

Mo_2^{II}, Rh_2^{II}, $Ru_2^{II,III}$) there is a definite M—M bond, this is *not* so for Cu. However, there is weak coupling of the unpaired electrons, one on each Cu^{II} ion, giving rise to a singlet ground state with a triplet state lying only a few kJ mol^{-1} above it; the latter state is thus appreciably populated at normal temperatures and the compounds are paramagnetic. At 25° μ_{eff} is typically about 1.4 BM per Cu atom and the temperature-dependence is very pronounced. The interaction involves either the $d_{x^2-y^2}$ orbitals of the two metal atoms directly or transmission through the π orbitals of the bridge group, or both.

Catalytic Properties of Copper Compounds. Copper compounds catalyze an exceedingly varied array of reactions, heterogeneously, homogeneously, in the vapor phase, in organic solvents, and in aqueous solutions. Many of these reactions, particularly if in aqueous solutions, involve oxidation-reduction systems and a Cu^{I}—Cu^{II} redox cycle. Molecular oxygen can often be utilized as oxidant, for example, in copper-catalyzed oxidations of ascorbic acid and in the Wacker process (Section 30-10).

The oxidation probably involves an initial oxidative-addition (Section 30-2):

$$Cu^+ + O_2 = CuO_2^+$$

$$CuO_2^+ + H^+ = Cu^{2+} + HO_2$$

$$Cu^+ + HO_2 = Cu^{2+} + HO_2^-$$

$$H^+ + HO_2^- = H_2O_2$$

Copper compounds have many uses in organic chemistry for oxidations, for example, of phenols by Cu^{2+}-amine complexes, halogenations, coupling reactions, and the like. Copper(II) has considerable biochemical importance (see Chapter 31).

Study Questions

A

1. Write down the electronic structures of: Ti^{IV}, V^{II}, Cr^{V}, Mn^{VI}, Fe^{O}, Co^{I}, Ni^{II}, Cu^{III}.
2. Which of the +2 and +3 aquo ions are (a) reduced by water, (b) oxidized by water, (c) oxidized by O_2?
3. Which of the +2 hydroxides are oxidized by air? Write complete equations.
4. What is the chief structural difference between $TiOSO_4 \cdot H_2O$ and $VOSO_4 \cdot 5H_2O$?
5. Give two examples each of the Lewis acid behavior of MCl_4 and MCl_5 where M is a transition metal.
6. Why is the $[Ti(H_2O)_6]^{3+}$ ion violet?
7. Explain why VF_5 is an extremely viscous liquid.
8. What is meant by the term disproportionation? Give two examples.
9. How does the V—O stretching frequency alter when bis(acetylacetonato)-oxovanadium(IV) is dissolved in pyridine?
10. What happens when a solution of $K_2Cr_2O_7$ is added to solutions of (a) F^-, Cl^-, Br^-, I^-, (b) OH^-, (c) NO_2^-, (d) SO_4^{2-}, (e) H_2O_2?
11. What are the structural features of the oxides TiO_2, CrO_2, CrO_3, MnO_2, Mn_3O_4, Mn_2O_7, $FeO_{0.95}$?

12. What are (a) spinels, (b) perovskites, and (c) alums? Give two examples of each.
13. Describe why trivalent aquo ions give acid solutions in water. What happens to them as the pH is increased?
14. Draw structures of $Cr_2(CO_2Me)_4(H_2O)_2$ and $Cr_3O(CO_2Me)_6(H_2O)_3$.
15. What differences are there between Cr^{II} and Cu^{II} acetates?
16. What are the structures of $[Ni(acac)_2]_3$, $CrCl_3(THF)_3$, VO_2^+, CrO_5py?
17. Why is it that the freshly precipitated hydroxide
    (a) of Mn^{II} is white but turns dark brown in air?
    (b) of Co^{II} is blue but turns pink on warming?
    (c) of Cu^{II} is blue but turns black on warming?
18. What are thc number of unpaired electrons in complexes of (a) spin-paired Mn^{II}, (b) Cr^{IV} tetrahedral, (b) Co^{III} octahedral, (d) V^{III}octahedral and tetrahedral?
19. Compare the chemistry of metal species having the electronic configurations (a) d^0, (b) d^4, and (c) d^8.
20. How is oxygen bound in the complexes
    (a) $Cs[TiO_2F_5]$
    (b) $K_3[CrO_8]$
    (c) $[Co_2O_2(NH_3)_{10}](SO_4)_2$?
21. Write electronic structures for the per- and superoxidc ions and explain how they may form complexes.
22. Enumerate the possible isomers of $[Co\ en_2(SCN)_2]^+$ and name each one according to proper nomenclature.

B

1. Describe the preparations and main chemical and sterochemical properties of alkoxides and dialkylamides.
2. What is meant by the term "substitution inert?" Which ions show this behavior? What is the reason?
3. Most M—O—M bonds are angular but some are linear, for example, Cr—O—Cr in $[(NH_3)_5\text{-}CrOCr(NH_3)_5]^{4+}$. Is there any explanation?
4. Describe the types of configurational changes in Ni^{II} complexes.
5. The densities of Ca, Sc, and Zn are, respectively, 1.54, 3.00, and 7.13 g cm^{-3}. Make a plot of these along with those of the first transition series as given in Table 24-1 and suggest explanations for the various features of the graph obtained.
6. Dimethyl sulfoxide (DMSO) reacts with $Co(ClO_4)_2$ in absolute ethanol to form a pink product which is a 1:2 electrolyte, and has a magnetic moment of 4.9 BM. $CoCl_2$ reacts with DMSO to form a dark blue product with a magnetic moment (per Co) of 4.6 BM. It is a 1:1 electrolyte and has an empirical formula of $Co(DMSO)_3Cl_2$. What do you think these two compounds are?

Chapter 24
Study Guide

Supplementary Reading

Basolo, F., Bunnett, J. F., and Halpern, J., eds., *Collected Accounts of Transition Metal Chemistry*, Vol. 1, American Chemical Society, 1973.

Clark, R. J. H., *The Chemistry of Titanium and Vanadium*, Elsevier, 1968.

Colton, R. and Canterford, J. H., *Halides of the First Row Transition Metals*, Wiley, 1969.

Hatfield, W. E. and Whyman, R., "Coordination Chemistry of Copper," *Transition Metal Chemistry*, 5, 47 (1969).

Kepert, D. L., *The Early Transition Elements*, Academic Press, 1972.

Toth, L. E., *Transition Metal Carbides and Nitrides*, Academic Press, 1971.

25

the elements of the second and third transition series

25-1 General Remarks

Some important features of these elements compared with those of the first series are as follows.

1. *Radii* The radii of the heavier metals and ions are larger than those of the first series. Because of the lanthanide contraction (pages 204 and 448) the radii of the third series are very little different from those of the second series, despite the increased atomic number and total number of electrons.

2. *Oxidation States*

(a) The higher oxidation states are much more stable than those of the first series. The oxo anions MO_4^{n-} of Mo, W, Tc, Re, Ru, and Os are less readily reduced than those of Cr, Mn, and Fe. Some compounds such as WCl_6, ReF_7, RuO_4, and PtF_6 have no analogues in the first series. The elements in groups IVA to VIA prefer their highest oxidation state.

(b) The II oxidation state is of relatively little importance except for Ru. For molybdenum it is important, but is quite different (Mo_2^{4+}) from chromium (Cr^{2+}), Pd, Pt. Similarly the III state is relatively unimportant except for Rh, Ir, Ru, and Re.

3. *Metal—Metal Bonding.* For the first-row elements, M—M bonding occurs only in metal carbonyls and related compounds (Chapter 28) and in a few binuclear complexes, notably the divalent carboxylates, for example, $Cr_2(CO_2Me)_4(H_2O)_2$ (page 392). The heavier elements are more prone to metal—metal bonding.

(a) There are binuclear species such as $M_2(CO_2Me)_4$, M = Mo, Rh and Ru similar to that of Cr. However, there are also binuclear halides of Mo, Tc, and Re, for example, $Re_2Cl_8^{2-}$, that have strong multiple M—M bonds.

(b) There are lower halides of Nb, Ta, Mo, W, and Re that are cluster compounds, for example, $Ta_6Cl_{12}^{2+}$, and $Re_3Cl_{12}^{3-}$. Some Au_{11}^{3+} clusters are also known.

4. *Magnetic Properties.* The heavier elements tend to give *low-spin* compounds. Ions with an even number of electrons are often diamagnetic. Even where there is an odd number of *d* electrons, there is frequently only one unpaired electron. The simple interpretation of magnetic moments that is usually possible for first-row paramagnetic species can seldom be made because of complications due to spin-orbit coupling. The spin-pairing can be attributed to the greater spatial extension of the 4*d* and 5*d* orbitals. Double occupancy of an orbital produces less interelectronic repulsion than in the smaller 3*d* orbitals. The electronic absorption spectra are also more difficult to interpret in general. A given set of ligands produces splittings in the order $5d > 4d > 3d$.

5. *Stereochemistries.* For the early members of the second and third rows especially, higher coordination numbers of 7 and 8 are more common than in the first-row elements. However, for the platinum group metals, the maximum coordination number, with two exceptions, is 6.

ZIRCONIUM AND HAFNIUM

25-2
General Remarks: The Elements

The atomic and ionic radii of Zr and Hf are virtually identical. Their chemistry is hence remarkably similar. There are usually only small differences, for example, in solubilities or volatilities of compounds.

The significant differences from Ti are:

(I) There are few compounds in oxidation states below IV.

(II) The +4 ions have a high charge, there is no partly filled *d* shell that might give stereochemical preferences, and they are relatively large (0.74, 0.75 Å). Thus they usually have coordination numbers in compounds and complexes of 7 and 8 rather than 6. There are also a variety of polyhedra, especially with O and F. For example, we have

ZrF_6^{2-}	Octahedron in Li_2ZrF_6 and $CuZrF_6 \cdot 4H_2O$
ZrF_7^{3-}	Pentagonal bipyramid, in Na_3ZrF_7
ZrF_7^{3-}	Capped trigonal prism in $(NH_4)_3ZrF_7$
$Zr_2F_{12}^{4-}$	Pentagonal bipyramids sharing an edge in $K_2CuZr_2F_{12} \cdot 6H_2O$
ZrF_8^{4-}	Square antiprism, in $Cu_2ZrF_8 \cdot 12H_2O$
$Zr_2F_{14}^{6-}$	Square antiprisms sharing an edge in $Cu_3Zr_2F_{14} \cdot 16H_2O$

(III) Though there are a few compounds, particularly the halides, of Zr^{III}, this is not nearly so important an oxidation state as for titanium.

Zirconium occurs as *baddeleyite*, a form of ZrO_2, and *zircon*, $ZrSiO_4$. Hafnium always accompanies Zr to the extent of fractions of a percent Zr. Separations are difficult, but solvent extraction or ion-exchange procedures are effective. The metals are made by the Kroll process (page 384). They are similar to Ti both physically, being hard, resistant, and stainless-steel-like in appearance, and chemically, being readily attacked only by HF, to give fluoro complexes.

COMPOUNDS OF Zr^{IV}

Since the chemistries are so similar we shall refer only to Zr. All the compounds are white unless the anion is colored.

25-3 Binary Compounds

The *oxide*, ZrO_2, is made by heating the *hydrous oxide* which is precipitated on addition of OH^- to Zr^{IV} solutions. ZrO_2 is very refractory (mp 2700°) and exceptionally resistant to attack. It is used for crucibles and furnace linings, etc.

The *tetrachloride* is made by direct interaction or by action of Cl_2 on a mixture of ZrO_2 and C. It is tetrahedral in the vapor but in the crystal there are chains of $ZrCl_6$ octahedra.

Like $TiCl_4$, $ZrCl_4$ is a Lewis acid and forms adducts with donors such as Cl^-, $POCl_3$, ethers, and the like. It is only partially hydrolyzed by water at room temperature to give the stable oxide chloride, $ZrOCl_2$.

25-4 Aqueous Chemistry and Complexes

ZrO_2 is more basic than TiO_2 and is virtually insoluble in an excess of base. There is a more extensive aqueous chemistry of zirconium because of a lower tendency toward complete hydrolysis. Nevertheless it is doubtful whether Zr^{4+} aquo ions exist even in strongly acid solutions. The hydrolyzed ion is often referred to as the "zirconyl" ion, ZrO^{2+}, but Zr=O bonds do *not* exist. The complex $ZrOCl_2 \cdot 8H_2O$, which crystallizes from dilute HCl solutions, contains the ion $[Zr_4(OH)_8(H_2O)_{16}]^{8+}$. Here the Zr atoms lie in a distorted square, linked by pairs of hydroxo bridges, and also bound to four H_2O molecules so that the Zr atom is coordinated by eight oxygen atoms in a distorted dodecahedron.

Like Ce^{4+} and other +4 ions, Zr has an *iodate* insoluble in 6 *M* HNO_3. In fairly concentrated HF solutions, *only* $ZrF_6{}^{2-}$ is present in solution. The salts that crystallize *from* these solutions may contain $ZrF_7{}^{3-}$, $ZrF_8{}^{4-}$, and binuclear anions as discussed on page 420.

Other 8-coordinate complexes are the carboxylate, $Zr(CO_2R)_4$, the acetylacetonate, $Zr(acac)_4$, the oxalate, $Na_4[Zr\ ox_4]$, and the nitrate, $Zr(NO_3)_4$.

NIOBIUM AND TANTALUM

25-5 The Elements

Niobium and Ta, though metallic, have chemistries in the V oxidation state that are similar to those of typical nonmetals. They have virtually no cationic chemistry but form numerous anionic complexes most of which have coordination numbers of 7 and 8. In their lower oxidation states they form many metal-atom cluster compounds. Only niobium forms lower states in aqueous solution.

Niobium is 10 to 12 times more abundant than Ta. The *columbite-tantalite* series of minerals have the general composition $(Fe/Mn)(Nb/Ta)_2O_6$, with variable ratios Fe/Mn and Nb/Ta. Niobium is also obtained from *pyrochlore*, a mixed calcium–sodium niobate. Separation and production of the metals is complex. Both metals are bright, high-melting, and resistent to acids. They dissolve with vigor in an HNO_3—HF mixture, and, very slowly, in fused NaOH.

NIOBIUM AND TANTALUM COMPOUNDS

25-6 The Chemistry of Niobium and Tantalum(V), d^0

Binary Compounds. The *oxides*, M_2O_5, are dense white powders, commonly made by ignition of other Nb or Ta compounds in air. Addition of OH^- to halide solutions gives the gelatinous hydrous oxides. The oxides are scarcely attacked by acids other than HF but are dissolved by fused $NaHSO_4$ or NaOH. Alkali fusion gives oxo anions that are stable in aqueous solution only at high pH.

Halides and their Complexes. The pentafluorides are volatile white solids obtained by direct interaction. They have a tetrameric structure (*Fig. 25-1a*) that is characteristic for other pentafluorides. The *pentachlorides* are yellow solids obtained by direct interaction. They are hydrolyzed to the hydrous oxide. In the crystal and in solvents like CCl_4 they are dimeric (*Fig. 25-1b*). Both halides abstract oxygen from donors like Me_2SO or ether on heating to give oxochlorides $MOCl_3$. Nb_2Cl_{10} is reduced by amines to give Nb^{IV} complexes, for example, $NbCl_4py_2$.

The halides are Lewis acids and give adducts with neutral donors and complex anions with halide ions.

The *fluoride solutions* contain $NbOF_5^{2-}$, NbF_6^-, and TaF_6^- plus TaF_7^{2-}. However, from these solutions salts of different stoichiometry can be obtained: $NbOF_6^{3-}$, NbF_7^{2-}, TaF_8^{3-}.

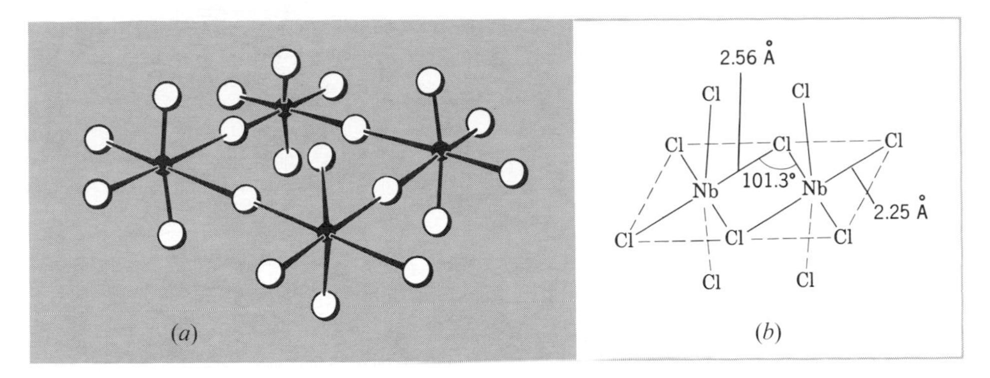

Figure 25-1 (*a*) *The tetrameric structures of* NbF_5 *and* TaF_5 (*also* MoF_5 *and, with slight distortion,* RuF_5 *and* OsF_5). Nb—F *bond lengths:* 2.06 Å (*bridging*), 1.77 Å (*nonbridging*). (*Adapted by permission from A. J. Edwards,* J. Chem. Soc., 1964, *3714.*) (*b*) *The dinuclear structure of crystalline* Nb_2Cl_{10}. *The octahedra are distorted.*

25-7
Lower Oxidation States of Niobium and Tantalum

Niobium sulfate or chloride solutions are reduced by zinc to blue solutions of uncertain nature. The pentahalides of both metals are reduced by H_2, Al, Zn, etc., either to the tetrahalides or to more reduced cluster compounds.

The tetrachlorides are Lewis acids. $TaCl_4$ readily abstracts oxygen to give oxohalides but $NbCl_4$ can react with oxo ligands to give adducts or with β-diketonates to give 8-coordinate complexes, for example, $Nb(acac)_4$.

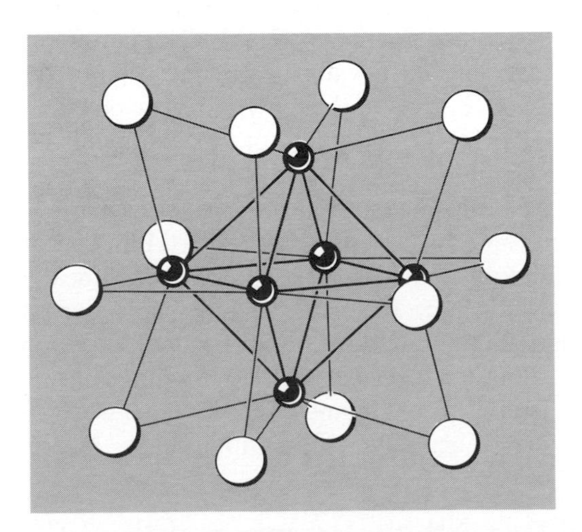

Figure 25-2 *The structure of the* $[M_6X_{12}]^{n+}$ *units found in many halogen compounds of lower-valent niobium and tantalum.* [*Reproduced by permission from L. Pauling.* The Nature of the Chemical Bond, *3rd edn., Cornell Univ. Press, 1960.*]

The *cluster halides* have stoichiometries $MX_{2.33}$ or $MX_{2.5}$. They contain the $[M_6X_{12}]^{n+}$ unit shown in *Fig. 25-2*. This has an octahedron of metal atoms with halogen bridges. In aqueous solution, redox reactions occur:

$$\underset{\text{diamagnetic}}{[M_6Cl_{12}]^{2+}} \rightleftharpoons \underset{\text{1 unpaired electron}}{[M_6Cl_{12}]^{3+}} \rightleftharpoons \underset{\text{diamagnetic}}{[M_6Cl_{12}]^{4+}}$$

Salts of these cations can be isolated.

MOLYBDENUM AND TUNGSTEN

25-8 The Elements

Molybdenum and W do not resemble Cr except in compounds with π-acid ligands. Thus the +2 state is not well known except in compounds with quadruply bonded ${Mo_2}^{4+}$ units. The high stability of Cr^{III} in complexes has no counterpart in Mo or W chemistry. For Mo and W, the higher oxidation states are more common and more stable against reduction.

Both Mo and W have a great range of stereochemistries in addition to the variety of oxidation states, and their chemistry is among the most complex of the transition elements.

Molybdenum occurs chiefly as molybdenite, MoS_2. Tungsten is found almost exclusively in tungstates such as *scheelite*, $CaWO_4$, or *wolframite*, $Fe(Mn)WO_4$.

Molybdenum is roasted to the oxide MoO_3. Tungsten is recovered after alkaline fusion and dissolution in water by precipitation of WO_3 with acids. The oxides are reduced with H_2 to give the metals as grey powders. These are readily attacked only by $HF—HNO_3$ mixtures or by oxidizing alkaline fusions with Na_2O_2 or $KNO_3—NaOH$.

The chief use of both metals is in alloy steels; even small amounts cause tremendous increases in hardness and strength. "High-speed" steels, which are used to make cutting tools that remain hard even at red heat, contain W and Cr. Tungsten is also used for lamp filaments. The elements give hard, refractory, and chemically inert interstitial compounds with B, C, N, or Si on direct reaction at high temperatures. Tungsten carbide is used for tipping cutting tools, and the like.

Molybdenum is used in oxide and other systems as a catalyst for a variety of reactions, one example being the "ammonoxidation" synthesis of acrylonitrile

$$CH_2{=}CHCH_3 + NH_3 + \tfrac{3}{2}O_2 \longrightarrow CH_2{=}CHCN + 3H_2O$$

Molybdenum is present in some enzymes, notably those that reduce N_2.

MOLYBDENUM AND TUNGSTEN COMPOUNDS

25-9
Oxides and Oxyanions

The *trioxides*, are obtained on heating the metal or other compounds in air. MoO_3 is white and WO_3 yellow. They are not attacked by acids other than HF but dissolve in bases to form molybdates or tungstates. Alkali metal or NH_4^+ salts which are water-soluble contain the tetrahedral ions MoO_4^{2-} and WO_4^{2-}. Most other cations give insoluble salts; $PbMoO_4$ can be used for the gravimetric determination of Mo.

When solutions of molybdates or tungstates are made weakly acid, condensation occurs giving complicated polyanions. In more strongly acid solutions the *hydrated oxides* $MoO_3 \cdot 2H_2O$ (yellow) and $WO_3 \cdot 2H_2O$ (white) are formed. These contain MO_6 octahedra sharing corners.

Unlike chromates (page 393) the Mo and W anions are weak oxidants.

25-10
Halides

Interaction of Mo or W with F_2 gives the colorless *hexafluorides*, MoF_6 (bp 35°), WF_6 (bp 17°). Both are readily hydrolyzed.

Chlorination of hot Mo gives only the pentachloride Mo_2Cl_{10} which is a dark red solid with a structure in the crystal very similar to that of Nb_2Cl_{10} (*Fig. 25-1b*).

Mo_2Cl_{10} is soluble in benzene and in polar organic solvents. It is monomeric in solution and is presumably solvated. It readily abstracts oxygen from oxygenated solvents to give oxo species and is rapidly hydrolyzed by water. The preparation of other Mo chlorides is shown in *Fig. 25-3*.

Chlorination of hot W gives the dark blue-black monomeric *hexachloride*, WCl_6. It is soluble in CS_2, CCl_4, alcohol, and ether. It reacts slowly with cold water, rapidly with hot, to give tungstic acid. Mo_2Cl_{10} and WCl_6 are the usual starting materials for synthesis of a variety of compounds such as dialkylamides, alkoxides, organometallics, and carbonyls.

The so-called "dihalides," M_6Cl_{12}, contain $M_6Cl_8^{4+}$ clusters (*Fig. 25-4*) similar to those of Nb and Ta, but with 8 face-bridging rather than 12 edge-bridging

$$MoCl_3 \xleftarrow[400^\circ]{H_2} Mo_2Cl_{10} \xrightarrow[\text{benzene}]{\text{Reflux}} MoCl_4 \xleftarrow[250^\circ]{CCl_4} MoO_2$$

$$Mo_2Cl_{10} \xrightarrow{\text{HCl aq}} [MoOCl_5]^{2-} \qquad Mo \xrightarrow{Cl_2,\ 500^\circ} Mo_2Cl_{10}$$

$$Mo \xrightarrow[750^\circ]{COCl_2} Mo_6Cl_{12} \longrightarrow [Mo_6Cl_8]^{4+}\ \text{salts}$$

$$MoCl_4 \xrightarrow{Mo,\ 600^\circ} Mo_6Cl_{12} \qquad Mo_6Cl_{12} \xrightarrow{Cl_2\ \text{in}\ CCl_4,\ 250^\circ} MoCl_4$$

Figure 25-3 Preparation of molybdenum chlorides.

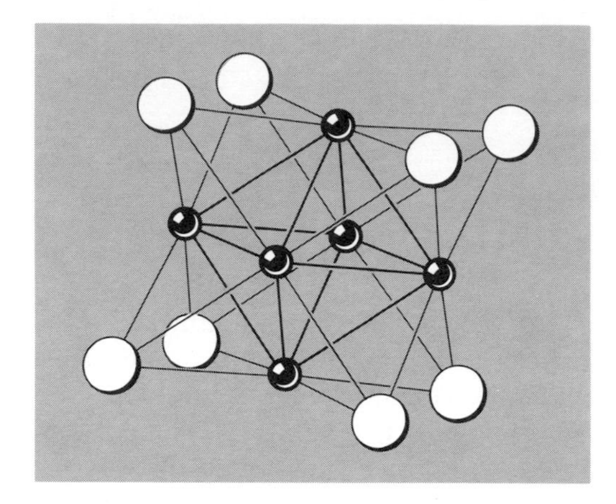

Figure 25-4 *The key structural unit* $M_6X_8^{4+}$, *found in halide cluster compounds.*

chlorine atoms. The Mo clusters differ in that they do not undergo reversible oxidation, but W_6Cl_{12} can be oxidized by Cl_2 at high temperatures. The $(M_6X_8)^{4+}$ units can coordinate six electron-pair donors, one to each metal atom along a fourfold axis of the octahedron. Thus, in molybdenum dichloride, the $(Mo_6Cl_8)^{4+}$ units are connected by bridging Cl atoms (four per unit) and there are nonbridging Cl atoms in the remaining two coordination positions.

The bridging groups in the $M_6X_8^{4+}$ units can undergo replacement reaction only slowly, whereas the six outer ligands are labile. Thus mixed halides such as $Mo_6Cl_8Br_4$ and complexes such as $[Mo_6Cl_8(Me_2SO)_6]^{4+}$ and $Mo_6Cl_8Cl_4(PPh_3)_2$ can be made.

In aqueous solution $M_6X_8^{4+}$ units are unstable to strongly nucleophilic groups such as OH^-, CN^-, or SH^-.

25-11
Complexes of Molybdenum and Tungsten

There are very many complexes of all types in oxidation states from II to VI.

Mo^{II} *species with* Mo—Mo *quadruple bonds.* Interaction of $Mo(CO)_6$ (Section 28-8) with carboxylic acids gives dimers $Mo_2(CO_2R)_4$ that have the same tetrabridged structure as $Cr_2(CO_2R)_4$, page 392. Although $Cr_2(CO_2Me)_4$ with HCl gives only Cr^{2+}, the Mo—Mo bond is *much* stronger and persists giving chloro complexes with quadruple Mo—Mo bonds

$$Mo_2^{II}(CO_2Me)_4 \xrightarrow{\text{HCl aq}} [Mo_2^{II}Cl_8]^{4-} \xrightarrow{4L} Mo_2Cl_4L_4$$

where L is almost any neutral ligand.

Oxo Complexes. The most accessible complex used for the synthesis of other complexes is the emerald green pentachlorooxomolybdate(V) ion, $[MoOCl_5]^{2-}$. This is obtained by reduction of MoO_4^{2-} in HCl solutions or by dissolving Mo_2Cl_{10} in concentrated aqueous HCl. Paramagnetic salts such as

$K_2[MoOCl_5]$ can be isolated. On addition of NaOH to acid solutions, equilibria involving dimeric species occur:

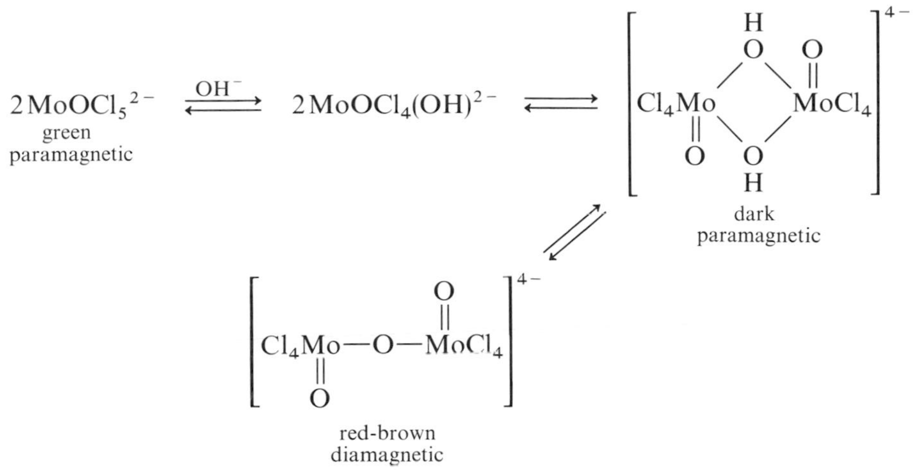

$$2MoOCl_5^{2-} \xrightleftharpoons{OH^-} 2MoOCl_4(OH)^{2-} \rightleftharpoons [Cl_4Mo(O)(\mu\text{-OH})_2Mo(O)Cl_4]^{4-}$$

$2MoOCl_5^{2-}$: green, paramagnetic

$[Cl_4Mo(O)(\mu\text{-OH})_2Mo(O)Cl_4]^{4-}$: dark, paramagnetic

$$\rightleftharpoons [Cl_4Mo(O)\text{—O—}Mo(O)Cl_4]^{4-}$$

red-brown, diamagnetic

The red-brown species represents a common type of Mo^V oxo species. These have an Mo_2O_3 unit with either a linear or a bent Mo—O—Mo bridge. Other types have dioxo or disulfur bridges.

In view of the interest in models for enzyme systems such as xanthine oxidase, complexes with amino acids, cysteine, and organic sulfur compounds have been much studied. Two examples are the xanthate $Mo^V{}_2O_3(S_2COEt)_4$ (25-I) and the histidine complex, $[Mo^V{}_2O_4(\text{L-histidine})_2]\cdot 3H_2O$ (25-II).

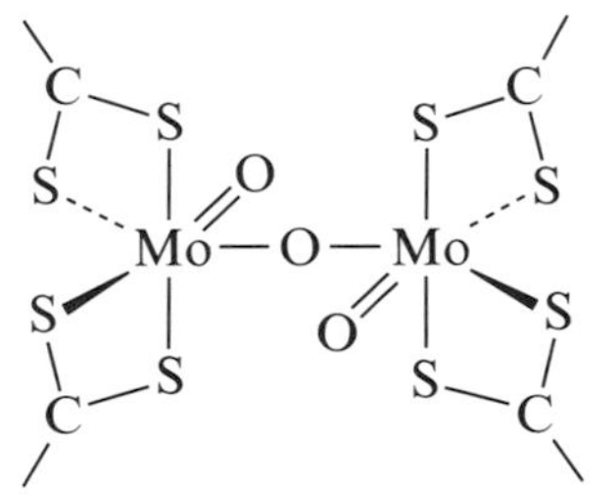

25-I

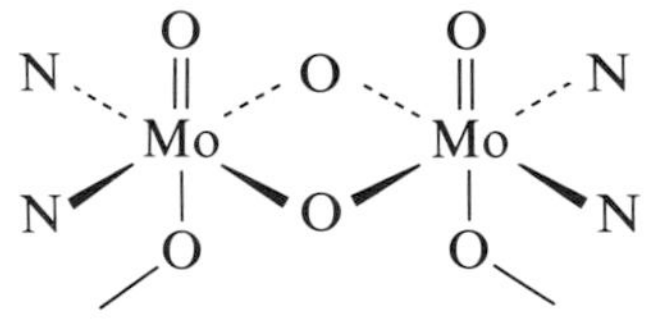

25-II

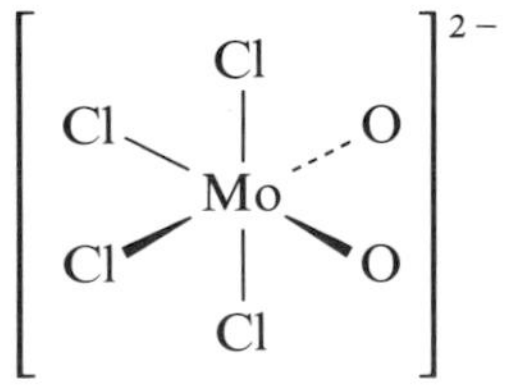

$[MoO_2Cl_4]^{2-}$

25-III

Molybdenum(VI) commonly forms *dioxo* species in which the two Mo=O bonds are *cis*. Thus MoO_3 in 12 M HCl gives $[MoO_2Cl_4]^{2-}$ (25-III).

Tungsten does not form a comparable variety of oxo complexes, although a few are known.

TECHNETIUM AND RHENIUM

25-12
The Elements

These elements differ considerably from Mn the first-row element.

1. There is no analog of Mn^{2+}(aq) and only a very few complexes are known in the II state.

2. There is little cationic chemistry in any oxidation state even in complexes.

3. Both elements have an extensive chemistry in the IV and V states, and especially as oxo compounds in the V state.

4. The oxo anions MO_4^- are much weaker oxidants than permanganate.

5. The formation of clusters and metal—metal bonds is a feature of the chemistry in the II to IV states.

Rhenium is recovered from the flue dusts in the roasting of MoS_2 ores and from residues in the smelting of some Cu ores. It is usually left in solutions as perrhenate ion, ReO_4^-. After concentration, the addition of KCl precipitates the sparingly soluble salt, $KReO_4$.

All isotopes of technetium are radioactive. ^{99}Tc (2.12×10^5 yr) is recovered in kilogram quantities from fission product wastes. There may be more ^{99}Tc in existence on the earth than Re.

The metals are obtained by H_2 reduction of the oxides or the $(NH_4)MO_4$ compounds. They are very high-melting and unreactive at room temperature. They burn in O_2 at 400° to give the volatile oxides M_2O_7. They dissolve to give the oxo acid in warm aqueous Br_2 or hot HNO_3. Re dissolves in 30% H_2O_2.

Rhenium is used mainly in a Pt—Re alloy supported on alumina for catalytic reforming of petroleums. Technetium because of its radioactivity is used for radiographic scanning of the liver and other internal organs.

RHENIUM COMPOUNDS

For present purposes, technetium compounds can be assumed to be similar to those of Re.

25-13
Binary Compounds

The yellow volatile oxide Re_2O_7 is very hygroscopic and dissolves in water, from which the oxide hydrate $O_3ReOReO_3(H_2O)_2$ can be obtained by evaporation. In NaOH, the perrhenate ion, ReO_4^-, is formed.

Saturation of HCl or H_2SO_4 solution of ReO_4^- with H_2S gives the black *sulfide* Re_2S_7. This procedure is used for recovery of Re from residues.

The only *halides* in the VI and VII states are the volatile ReF_6 and ReF_7. Chlorination of Re at *ca.* 550° gives dark red brown Re_2Cl_{10}. It is a dimer like Mo_2Cl_{10} or Ta_2Cl_{10} (page 423). On heating, the liquid decomposes to give the *trichloride*. This is a cluster compound whose structure is shown in *Fig. 25-5*. The Re_3Cl_9 units are linked into a polymer by sharing of Cl atoms. This unit is extremely stable, persists in the vapor at 600° and forms the structural basis for much of Re^{III} chemistry.

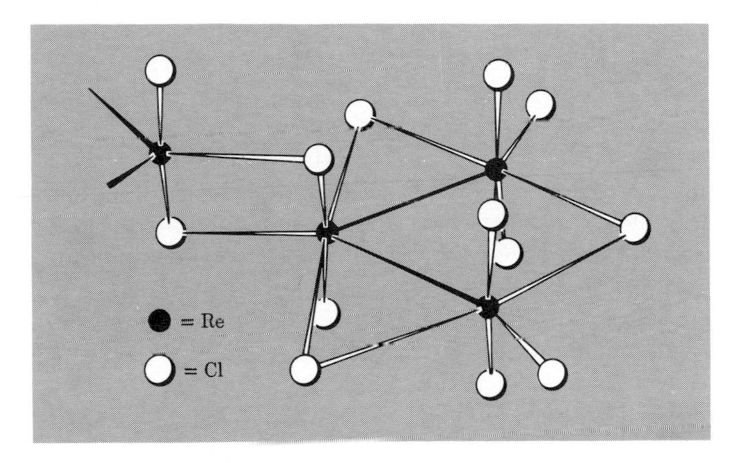

Figure 25-5 *The cluster structure of* Re_3Cl_9.

25-14
Oxo Compounds and Complexes

As with Mo and W, oxo compounds are important especially in the V and VII oxidation states.

The salts of the *perrhenate* ion, ReO_4^-, have solubilities similar to perchlorates, but salts of TcO_4^- are more soluble than either. An insoluble perrhenate suitable for gravimetric analysis is given by tetraphenylarsonium Ph_4As^+.

The ions are stable in water. They are weak oxidants. In HCl solution ReO_4^- is reduced by hypophosphite, partially to the chlorocomplex ion $[Re^{IV}Cl_6]^{2-}$, which forms stable salts such as K_2ReCl_6, and partially to the $Re_2Cl_8^{2-}$ ion, which is isoelectronic with the $Mo_2Cl_8^{4-}$ ion and contains a quadruple bond between the metal atoms.

Oxo Complexes. These are numerous, as with Mo. Re_2Cl_{10} dissolves in aqueous strong HCl to give $[ReOCl_5]^{2-}$. Oxo species may have the groups Re=O, Re—O—Re and *trans* O=Re=O (Mo^{VI} has *cis* dioxo groups) and linear O=Re—O—Re=O.

There is an extensive chemistry of oxorhenium(V) compounds containing phosphine ligands. The complexes $ReOCl_3(PR_3)_2$ are obtained by interaction of ReO_4^- with PR_3 in ethanol containing HCl. The halide ion (or other ligand) opposite to the Re=O bond is labile; in ethanol, for example, it is rapidly replaced, giving the $ReOCl_2(OEt)(PR_3)_2$.

THE PLATINUM METALS

25-15
General Remarks

Ruthenium, osmium, rhodium, iridium, palladium, and platinum are the six heaviest members of Group VIII. They are rare elements; Platinum is the commonest with an abundance of about $10^{-6}\%$ whereas the others have abundances of the order of $10^{-7}\%$. They occur as metals, often as alloys such as osmiridium, and in arsenide, sulfide, and other ores. The elements are usually associated not only with one another but also with nickel, copper, silver, and gold.

The compositions of the ores and the extraction methods vary considerably. An important source is South African Ni-Cu sulfide. The ore is concentrated by gravitation and flotation, after which it is smelted with lime, coke, and sand and is bessemerized in a convertor. The resulting Ni-Cu sulfide "matte" is cast into anodes. On electrolysis in sulfuric acid solution, Cu is deposited at the cathode, and Ni remains in solution, from which it is subsequently recovered by electrodeposition, while the platinum metals, silver and gold collect in the anode slimes. The subsequent procedures for separation of the elements are very complicated. Although most of the separations involve classical precipitations or crystallizations, ion-exchange and solvent-extraction procedures are also feasible.

The metals are greyish-white and are obtained initially as powders by ignition of salts such as $(NH_4)_2PtCl_6$. Almost all compounds of these elements give the metal when heated. However, Os is readily oxidized by air to the very volatile oxide OsO_4, and Ru gives RuO_2 so that reduction by hydrogen is necessary.

The metals can also be thrown out from acid solutions by action of Zn—a common recovery procedure known as "footing."

The metals are chemically inert especially when massive. Ru and Os are best attacked by an alkaline oxidizing fusion, Rh and Ir by HCl + $NaClO_3$ at 125° to 150°, and Pd and Pt by concentrated HCl + Cl_2 or aqua regia.

The metals, as gauze or foil and especially on supports such as alumina or charcoal, on to which the metal salts are absorbed and reduced *in situ*, are extensively used as catalysts in industry. One of the biggest uses of Pt is as Pt-Re or Pt-Ge on alumina catalysts in the reforming or "platforming" of crude petroleum. Pd and Rh compounds are used in homogeneous catalytic syntheses (Chapter 30). The catalytic "after burners" going into use on automobile exhausts use a platinum catalyst.

Platinum or its alloys are used in electrical contacts.

Both Pd and Pt are capable of absorbing large volumes of molecular hydrogen, and Pd is used for the purification of H_2 by diffusion since Pd metal is permeable to hydrogen uniquely.

25-16
General Remarks on the Chemistry of the Platinum Metals

The chemistries of these elements have some common features, but there are wide variations depending on differing stabilities of oxidation states, stereochemistries, and the like. There is little similarity to Fe, Co, and Ni except in some compounds of π-acid ligands such as CO and in stoichiometries of compounds. The important oxidation states are listed in Table 25-1. Some general points are as follows.

1. *Binary compounds.* The halides, oxides, sulfides, phosphides, and the like, are not of great importance.

2. *Aqueous chemistry.* This is almost exclusively that of complex compounds. Aquo ions of Ru^{II}, Ru^{III}, Rh^{III}, and Pd^{II} exist, in solutions of noncomplexing anions, namely, ClO_4^-, BF_4^-, $CF_3SO_3^-$ or *p*-toluenesulfonate, but are not ordinarily of importance.

A vast array of complex ions, predominantly with halide or nitrogen donor ligands, are water-soluble. Exchange and kinetic studies have been made with many of these because of interest in (a) *trans*-effects, especially with square Pt^{II}, (b) differences in substitution mechanisms between the ions of the three transition-metal series, and (c) the unusually rapid electron-transfer processes with heavy-metal complex ions.

3. *Compounds with π-acid ligands.* (a) Binary carbonyls are formed by all but Pd and Pt, the majority of them polynuclear. Substituted polynuclear carbonyls are known for Pd and Pt, and all six elements give carbonyl halides and a variety of carbonyl complexes containing other ligands.

(b) For Ru, nitrosyl (NO) complexes are an essential feature of the chemistry.

(c) There is an extensive chemistry of complexes with tertiary phosphines and phosphites, and to a lesser extent with R_3As and R_2S. Some of these are useful homogeneous catalysts (Chapter 30).

Mixed complexes of PR_3 with CO, alkenes, halides, and hydride ligands in at least one oxidation state are common for all of the elements.

(d) All the elements have a strong tendency to form bonds to carbon, especially with alkenes and alkynes; Pt^{II}, Pt^{IV}, and to a lesser extent Pd^{II} have a strong tendency to form σ-bonds, while Pd^{II} very readily forms π-allyl species (Section 29-16).

(e) A characteristic feature is the formation of complexes with M—H bonds when the metal halides in higher oxidation states are reduced, especially in presence of tertiary phosphines or other ligands. Hydrogen abstraction from reaction media such as alcohols or dimethylformamide is common.

4. *Oxidation states.* The main *oxidation states* are given in Table 25-1.

Table 25-1 *Oxidation States of Platinum Metals* (Bold Type Shows Main States)

Ru	Os	Rh	Ir	Pd	Pt
0	**0**	0	0	**0**	**0**
—	—	**1**	**1**	—	—
2	2	2	2	**2**	**2**
3	**3**	**3**	**3**	—	—
4	**4**	4	**4**	4	**4**
5[ab]	5[ab]	5[a]	5[a]	5[a]	5[a]
6[ab]	6[ab]	6[a]	6[a]	6[a]	6[a]
7[ab]	7[ab]				
8[ab]	8[ab]				

[a] In fluorides or fluoro complexes.
[b] In oxides or oxo anions.

5. *Stereochemistry.* In only a few compounds does the coordination number exceed six, for example, $OsH_4(PR_3)_3$ and $IrH_5(PR_3)_2$. Most complexes in the +3 and +4 states are octahedral. The d^8 species Rh^I, Ir^I, Pd^{II}, and Pt^{II} normally are square or 5-coordinate.

The +2 states for Ru and Os are 5- or 6-coordinate.

RUTHENIUM AND OSMIUM

25-17
Oxo Compounds of Ruthenium and Osmium

One of the most characteristic features of the chemistry of Ru and Os is the oxidation by aqueous oxidizing agents to give the volatile *tetraoxides.*

Orange-yellow RuO_4 (mp 25°) is formed when acid solutions containing Ru are oxidized by MnO_4^-, Cl_2, or hot $HClO_4$. It can be distilled from the solutions or swept out by a gas stream.

Colorless OsO_4 (mp 40°) is more easily obtained and HNO_3 is a sufficiently powerful oxidant. The distillation first of OsO_4 and then of RuO_4 is used in their separation from other platinum metals. The RuO_4 is collected in strong HCl solutions where it is reduced to a mixture of Ru^{III} and Ru^{IV} chloro complexes. The evaporated product is sold as "$RuCl_3 \cdot 3H_2O$," the commonest starting material for syntheses of Ru compounds.

The tetraoxides consist of tetrahedral molecules. They are extracted from aqueous solutions by CCl_4. Both are powerful oxidants. OsO_4 is used in organic chemistry as it oxidizes olefins to *cis*-diols. It is also used for biological staining as organic matter reduces it. It presents an especial hazard to the eyes and must be handled carefully. RuO_4 is much more reactive and can react vigorously with organic matter; it is very toxic.

Dissolution of OsO_4 in base gives a colorless *oxo anion*

$$OsO_4 + 2OH^- = [OsO_4(OH)_2]^{2-}$$

which can be reduced to $[OsO_2(OH)_4]^{2-}$.

A much less stable orange oxo ruthenate RuO_4^{2-} is obtained by fusing Ru compounds with Na_2O_2 and dissolving the melt in water. The difference in stoichiometry may be due to the greater ability of the 5*d* anion to increase its coordination shell.

Reduction of RuO_4 by HCl in presence of KCl gives red crystals of $K_4[Ru_2OCl_{10}]$. This oxo species of $Ru^{IV}(d^4)$ is diamagnetic because the electrons become paired in a molecular orbital extending over the *linear* Ru—O—Ru bridge.

25-18
Ruthenium Chloro Complexes and Aquo Ions

The commercial product "$RuCl_3 \cdot 3H_2O$," on evaporation with HCl, is reduced to ruthenium(III) chloro complexes. With high concentrations of Cl^-, the ion

$[RuCl_6]^{3-}$ may be obtained. The rate of replacement of Cl^- by H_2O decreases as the number of Cl^- ions decreases so that while the aquation of $[RuCl_6]^{3-}$ to $[RuCl_5(H_2O)]^{2-}$ occurs within seconds in water, the half reaction time for conversion of $[RuCl(H_2O)_5]^{2-}$ to $[Ru(H_2O)_6]^{3+}$ is about 1 year. The intermediate species such as *trans*-$[RuCl_2(H_2O)_4]^+$ can be isolated by ion exchange procedures.

The Cl^- can be removed by $AgBF_4$ and the +3 ion electrolytically reduced to the easily oxidized +2 aquo ion.

$$[Ru(H_2O)_6]^{3+} + e = [Ru(H_2O)_6]^{2+} \qquad E^\circ = 0.23 \text{ V}$$

25-19
Ruthenium Ammine Complexes

There is an extensive chemistry of Ru with N-ligands. Some of the chemistry is summarized in *Fig. 25-6*. The $Ru(NH_3)_5{}^{2+}$ group has remarkable π-bonding properties. It forms complexes with CO, RNC, N_2O, SO_2, and its dinitrogen complex was the first N_2 complex to be made.

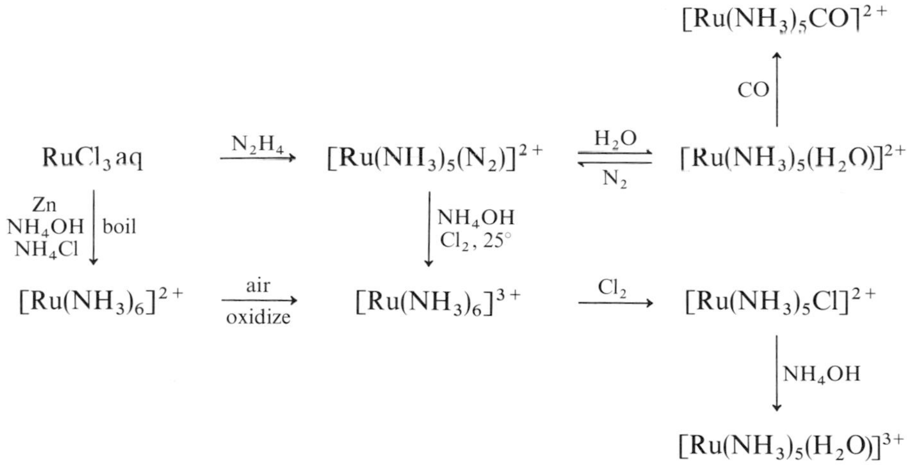

$$\begin{array}{ccccc}
 & & & & [Ru(NH_3)_5CO]^{2+} \\
 & & & & \uparrow \text{CO} \\
RuCl_3\,aq & \xrightarrow{N_2H_4} & [Ru(NH_3)_5(N_2)]^{2+} & \underset{N_2}{\overset{H_2O}{\rightleftharpoons}} & [Ru(NH_3)_5(H_2O)]^{2+} \\
\text{Zn, } NH_4OH, NH_4Cl \downarrow \text{boil} & & \downarrow NH_4OH,\ Cl_2,\ 25^\circ & & \\
[Ru(NH_3)_6]^{2+} & \underset{\text{oxidize}}{\xrightarrow{\text{air}}} & [Ru(NH_3)_6]^{3+} & \xrightarrow{Cl_2} & [Ru(NH_3)_5Cl]^{2+} \\
 & & & & \downarrow NH_4OH \\
 & & & & [Ru(NH_3)_5(H_2O)]^{3+}
\end{array}$$

Figure 25-6 *Some reactions of ruthenium ammines*

25-20
Nitric Oxide Complexes

Both elements form octahedral complexes ML_5NO that have an M—NO group. Depending on the nature of L, they may be cationic, anionic, or neutral. The MNO group can survive many chemical transformations of such complexes.

Ruthenium solutions that have at any time been treated with HNO_3 can, and usually do, contain nitrosyl species that are then difficult to remove. They are readily detected by their infrared absorption in the region 1930 to 1845 cm^{-1} (Section 28-14).

25-21
Tertiary Phosphine Complexes

Both elements have an extensive chemistry with these π-acid ligands. Some representative reactions are shown for Ru in *Fig. 25-7*. The complexes $RuHCl(PPh_3)_3$ and $RuH_2(PPh_3)_3$ are of interest in that they are highly active catalysts for the selective homogeneous hydrogenation of alk-l-enes (Section 30-7).

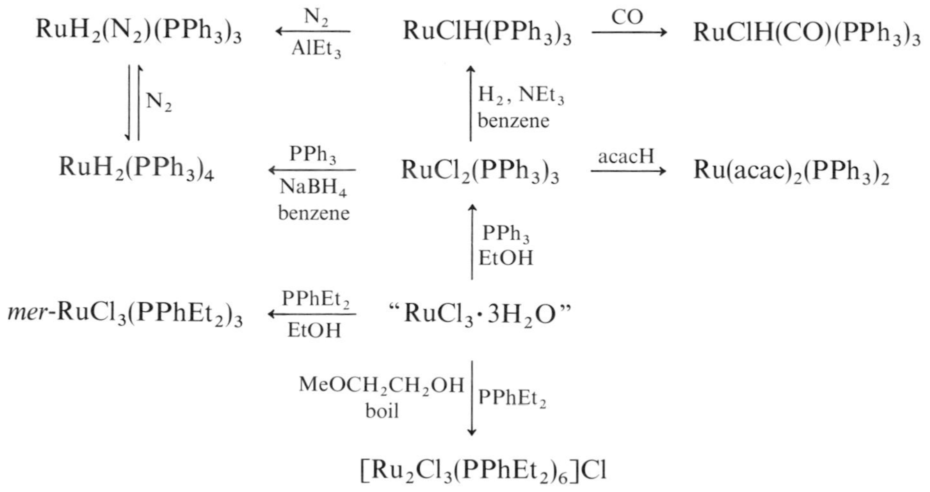

$$RuH_2(N_2)(PPh_3)_3 \xleftarrow[AlEt_3]{N_2} RuClH(PPh_3)_3 \xrightarrow{CO} RuClH(CO)(PPh_3)_3$$

$$RuH_2(N_2)(PPh_3)_3 \underset{}{\overset{N_2}{\rightleftharpoons}} RuH_2(PPh_3)_4$$

$$RuCl_2(PPh_3)_3 \xrightarrow{H_2,\ NEt_3,\ benzene} RuClH(PPh_3)_3$$

$$RuH_2(PPh_3)_4 \xleftarrow[NaBH_4,\ benzene]{PPh_3} RuCl_2(PPh_3)_3 \xrightarrow{acacH} Ru(acac)_2(PPh_3)_2$$

$$\text{"}RuCl_3\cdot 3H_2O\text{"} \xrightarrow{PPh_3,\ EtOH} RuCl_2(PPh_3)_3$$

$$mer\text{-}RuCl_3(PPhEt_2)_3 \xleftarrow[EtOH]{PPhEt_2} \text{"}RuCl_3\cdot 3H_2O\text{"}$$

$$\text{"}RuCl_3\cdot 3H_2O\text{"} \xrightarrow[MeOCH_2CH_2OH,\ boil]{PPhEt_2} [Ru_2Cl_3(PPhEt_2)_6]Cl$$

Figure 25-7 *Some reactions of tertiary phosphine complexes of ruthenium. Note that the use of different phosphines may give different products.*

RHODIUM AND IRIDIUM

25-22
Complexes of Rhodium(III) and Iridium(III), d^6

There are many diamagnetic, kinetically inert octahedral complexes similar to those of Co^{III}. They differ from Co^{III}, first, in that octahedral halogeno complexes are readily formed, for example, $[RhCl_5(H_2O)]^{2-}$, and $[IrCl_6]^{3-}$. Second, on reduction of the trivalent complexes the divalent complex is not obtained, except under special circumstances for Rh. When the ligands are halogens, ammines, or water, reduction gives the metal, or under controlled conditions, a *hydride* complex like $[Rh(NH_3)_5H]SO_4$; when π-acid ligands are present, reduction to Rh^I or Ir^I, or to iridium(III) hydrido complexes occurs.

Chlororhodates; The Rh^{III} Aquo Ion. Fusion of Rh with NaCl in Cl_2 followed by dissolution in water and crystallization gives $Na_3[RhCl_6]$. Addition of OH^- to this pink ion gives the *hydrous oxide* Rh_2O_3. Dissolution of this in dilute $HClO_4$, gives $[Rh(H_2O)_6]^{3+}$, yellow salts of which can be crystallized.

When Rh_2O_3 is dissolved in HCl and the solutions are evaporated, a dark red deliquescent material, $RhCl_3\cdot nH_2O$, is obtained. This is the usual starting material for synthesis of Rh compounds. It is soluble in alcohols as well as water. Fresh solutions do not give AgCl with Ag^+ ion, but on boiling, the red-brown solutions do turn to the yellow of $[Rh(H_2O)_6]^{3+}$. Some of its reactions are shown in *Fig. 25-8*.

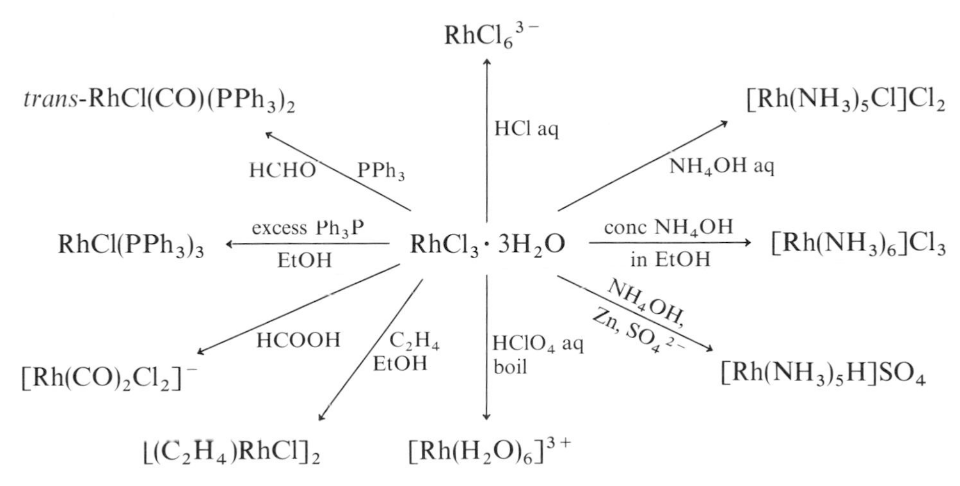

Figure 25-8 *Some reactions of rhodium trichloride.*

25-23
Complexes of Rhodium(IV) and Iridium(IV), d^5

It is very difficult to oxidize Rh^{III} and only a few unstable compounds of Rh^{IV} are known. Octahedral complexes of Ir^{IV} are stable; they have an unpaired electron (t_{2g}^5).

Hexachloroiridates. Hexachloroiridates are made by heating Ir + NaCl in Cl_2. The black salt Na_2IrCl_6 is very soluble in water; the so-called "chloroiridic acid" (cf. chloroplatinic acid page 440) is an oxonium salt $(H_3O)_2IrCl_6{\cdot}4H_2O$. These materials are used to prepare other Ir complexes.

The dark red brown $Ir^{IV}Cl_6^{2-}$ ion is rapidly and quantitatively reduced in strong OH^- solution to give yellow-green $Ir^{III}Cl_6^{3-}$:

$$2IrCl_6^{2-} + 2OH^- \rightleftharpoons 2IrCl_6^{3-} + \tfrac{1}{2}O_2 + H_2O$$

The $IrCl_6^{2-}$ ion will oxidize many organic compounds, and it is also quantitatively reduced by KI and $C_2O_4^{2-}$.

In *acid* solution we have

$$2IrCl_6^{2-} + H_2O = 2IrCl_6^{3-} + \tfrac{1}{2}O_2 + 2H^+$$
$$K = 7 \times 10^{-8}\ \text{atm}^{1/2}\ \text{mol}^3 \text{l}^{-2} (25^\circ)$$

so that in 12 M HCl, oxidation of $Ir^{III}Cl_6^{3-}$ occurs partially at 25° and completely on boiling.

25-24 Complexes of Rhodium(II), d^7

Only a few of these are known, the major ones being the diamagnetic binuclear *carboxylates* that have the common tetra-bridged structure. The end positions may be occupied by solvent molecules; with oxygen donors the complexes are green or blue, but with π-acids, such as PPh_3, they are orange-red. The carboxylates are made by boiling $RhCl_3 \cdot aq$ with $NaCO_2R$ in methanol. Action of very strong noncomplexing acids gives the Rh_2^{4+} aquo ion which also has a Rh—Rh bond.

25-25 Complexes of Rhodium(I) and Iridium(I), d^8

These square or 5-coordinate, diamagnetic complexes all have π-acid ligands. They are formed by reduction of Rh^{III} or Ir^{III} in presence of the ligand. They have been much studied because they provide the best systems for study of the oxidative-addition reaction (page 531) that is a characteristic feature of square d^8 complexes. For *trans*-$IrX(CO)(PR_3)_2$ the equilibria, for example,

$$trans\text{-}Ir^{I}Cl(CO)(PPh_3)_2 + HCl \rightleftharpoons Ir^{III}Cl_2H(CO)(PPh_3)_2$$

lie well to the Ir^{III} side and the Ir^{III} complexes can be readily characterized. For Rh, the Rh^{III} complexes are much less stable.

trans-Chlorocarbonylbis(triphenylphosphine)rhodium and -iridium, *trans*-MCl(CO)$(PPh_3)_2$, are yellow compounds obtained by reducing the halides in alcohols containing PPh_3 by HCHO, which acts as a reductant and source of CO.

Bis(dicarbonyl)dichlorodirhodium, $[Rh(CO)_2Cl]_2$, is obtained by passing CO saturated with ethanol over $RhCl_3 \cdot 3H_2O$ at *ca.* 100°, when it sublimes as red needles. It has the structure shown in *Fig. 25-9* where the coordination around each Rh atom is planar, and there are bridging chlorides with a marked dihedral angle, along the Cl—Cl line. There appears to be some direct interaction between electrons in rhodium orbitals perpendicular to the planes of coordination.

This carbonyl chloride is a useful source of other rhodium(I) species, and it is cleaved by donor ligands to give *cis*-dicarbonyl complexes, for example,

$$[Rh(CO)_2Cl]_2 + 2L \longrightarrow 2RhCl(CO)_2L$$

$$[Rh(CO)_2Cl]_2 + 2Cl^- \longrightarrow 2[Rh(CO)_2Cl_2]^-$$

$$[Rh(CO)_2Cl]_2 + acac^- \longrightarrow 2Rh(CO)_2(acac) + 2Cl^-$$

Hydridocarbonyltris(triphenylphosphine)rhodium, is a yellow crystalline solid with a **tbp** structure with equatorial phosphine groups. It is prepared by the reaction

$$trans\text{-}RhCl(CO)(PPh_3)_2 + PPh_3 \xrightarrow[EtOH]{NaBH_4} RhH(CO)(PPh_3)_3$$

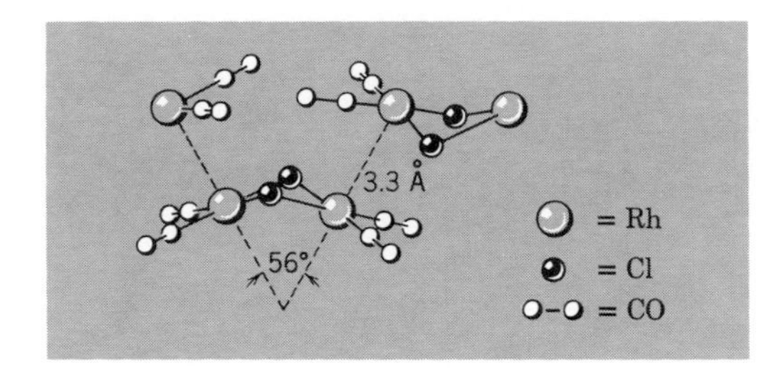

Figure 25-9 *The structure of crystalline* $[Rh(CO)_2Cl]_2$.

but it is also formed by action of CO + H_2 under pressure with virtually any rhodium compound in presence of an excess of PPh_3. Its main importance is as a hydroformylation catalyst for alkenes (Section 30-8).

Chlorotris(triphenylphosphine)rhodium, $RhCl(PPh_3)_3$. This red-violet crystalline solid is formed by reduction of ethanolic solutions of $RhCl_3 \cdot 3H_2O$ with an excess of PPh_3. It is a catalyst for hydrogenation of olefins and other unsaturated substances (Section 30-7). It undergoes many oxidative-addition reactions (Section 30-2), and it abstracts CO readily from metal carbonyl complexes and from organic compounds such as acyl chlorides and aldehydes, often at room temperature to give $RhCl(CO)(PPh_3)_2$.

PALLADIUM AND PLATINUM

25-26 Chlorides

Palladous chloride, $PdCl_2$, is obtained by chlorination of Pd. Above 550° an unstable α-form is produced, while below 550°, a β-form. There are α- and β-forms also of $PtCl_2$. The β-forms have a molecular structure with M_6Cl_{12} units (25-IV); the stabilization is due to halogen bridges rather than metal—metal bonds. Although the structure of α-$PtCl_2$ is not certain, it differs from that of α-$PdCl_2$ which has a flat chain (25-V). In both structures the metal atom has the square coordination characteristic Pd^{II} and Pt^{II}.

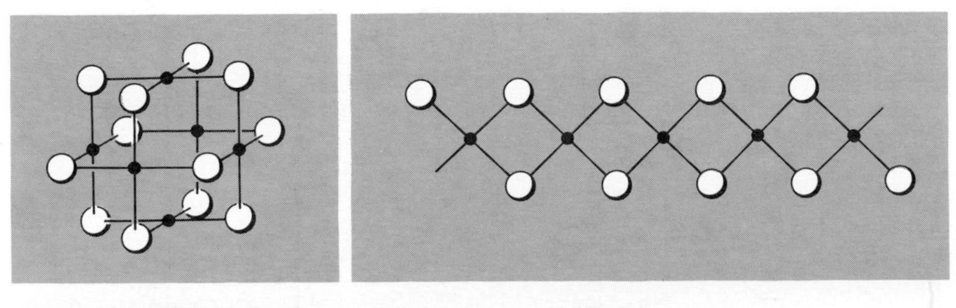

25-IV 25-V

Platinic chloride, $PtCl_4$, is obtained as red-brown crystals by heating chloroplatinic acid, $(H_3O)_2PtCl_6$, in chlorine. It is soluble in water and in HCl. The analogous chloride of Pd^{IV} does not exist.

25-27
Complexes of Palladium(II) and Platinum(II), d^{10}

The palladous ion, Pd^{2+}, occurs in PdF_2 and is paramagnetic. However the *aquo ion* $Pd(H_2O)_4{}^{2+}$ is spin-paired and all Pd and Pt complexes are diamagnetic. Brown deliquescent salts like $[Pd(H_2O)_4](ClO_4)_2$ can be obtained when Pd is dissolved in HNO_3, or PdO in $HClO_4$.

Palladous acetate is obtained as brown crystals when Pd sponge is dissolved in acetic acid containing HNO_3. It is a trimer, $[Pd(CO_2Me)_2]_3$. The metal atoms form a triangle with bridging acetate groups. The acetate acts like Pb^{IV} and Hg^{II} acetates (page 271) in attacking aromatic hydrocarbons; such "palladation" reactions are involved in many catalytic processes (cf. Chapter 30).

Palladium(II) and platinum(II) complexes are *square* or *5-coordinate* with the formulas $ML_4{}^{2+}$, $ML_5{}^{2+}$, ML_3X^+, *cis*- and *trans*-ML_2X_2, $MX_4{}^-$, ML_3X_2, where L is a neutral ligand and X a uninegative ion. The palladium ones are thermodynamically and kinetically less stable than their Pt^{II} analogs. Otherwise the two series of complexes are similar. The kinetic inertness of the Pt^{II} (and also Pt^{IV}) complexes has allowed them to play a very important role in the development of coordination chemistry. Many studies of geometrical isomerism and reaction mechanisms using platinum complexes have had a profound influence on our understanding of complex chemistry (cf. also Cr^{III}, Co^{III}, Rh^{III}).

There is a preference for amine ligands, halogens, CN^-, tertiary phosphines, and sulfides R_2S but little affinity for oxygen ligands and F^-. The concepts of hard and soft acids and bases, or class A and class B metals, are clearly shown here (cf. page 172). The strong binding of heavy donor atoms such as P is due in part to π bonding.

Many complexes have halide or other bridges, for example,

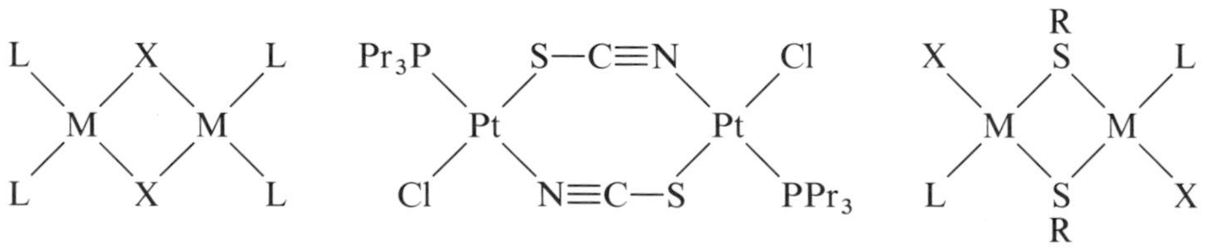

```
L     X     L      Pr3P      S—C≡N      Cl       X     R     L
                                                         S
  M     M              Pt          Pt                M       M
                                                         S
L     X     L      Cl        N≡C—S      PPr3     L     R     X
```

Bridged complexes can be cleaved by donors to give mononuclear species, for example,

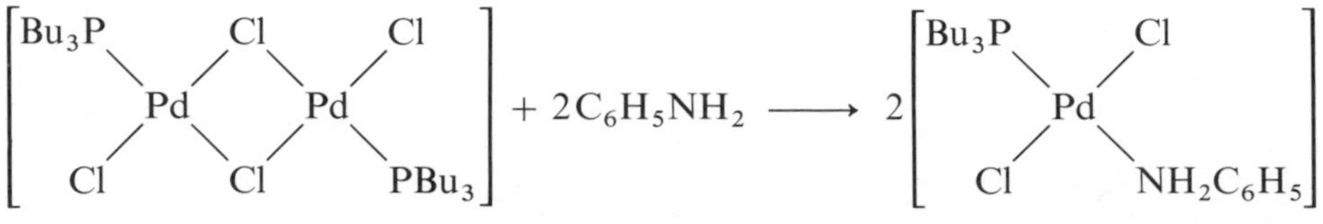

```
[Bu3P    Cl    Cl  ]                          [Bu3P     Cl       ]
[    Pd      Pd    ] + 2C6H5NH2  ———→  2 [      Pd           ]
[Cl      Cl    PBu3]                          [Cl       NH2C6H5  ]
```

Halogeno anions, MCl_4^-. Salts of the halogeno anions, MCl_4^{2-}, are common source materials. The yellowish $PdCl_4^{2-}$ ion is obtained when $PdCl_2$ is dissolved in HCl. The red $PtCl_4^{2-}$ ion is made by reduction of $PtCl_6^{2-}$ with oxalic acid or N_2H_5Cl.

25-28
Metal—Metal Interactions in Square Complexes

In crystals the square complexes are often stacked one above the other. Even though the metal—metal distances may be too long for true bonding, weak interactions can occur between *d* orbitals on adjacent metal atoms. An example is $PtenCl_2$ shown in *Fig. 25-10a*; others are Ni and Pd dimethylglyoximates. Salts such as $[Pt(NH_3)_4][PtCl_4]$, $[Pd(NH_3)_4][Pd(SCN)_4]$, or $[Cu(NH_3)_4][PtCl_4]$ also have stacked cations and anions so that there are chains of metal atoms. When *both* metal atoms are Pt^{II}, the crystal is green, although the constituent cations are colorless or pale yellow and the anions red. There is, also (a) marked dichroism with high absorption of light polarized in the direction of the metal chains, and (b) increased electrical conductivity along the chain. If steric hindrance is too large as in $[Pt(EtNH_2)_4][PtCl_4]$ the structure is different and the crystal has a pink color, the sum of the colors of the constituents.

A related class of compounds with chainlike structures differ in that the metals are linked by halide bridges (*Fig. 25-10b*). Again there is high electrical conductivity along the $—X—M^{II}—X—M^{IV}—X—$chain.

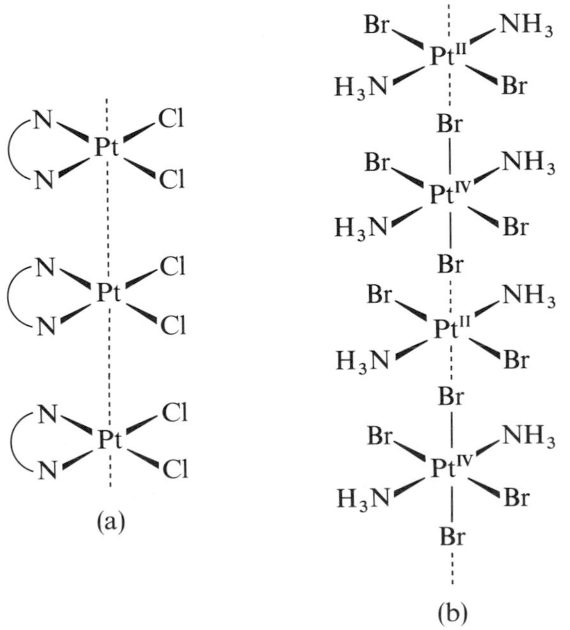

(a)

(b)

Pt^{II}----Br(chain) = ~3.1 Å

Figure 25-10 *(a) Linear stacks of planar* $PtenCl_2$ *molecules. (b) Chains of alternating* Pt^{II} *and* Pt^{IV} *atoms with bridging bromide ions in* $Pt(NH_3)_2Br_3$.

Five-coordinate complexes. Substitution and isomerization of square Pd^{II} and Pt^{II} complexes proceed by an associative pathway involving 5-coordinate intermediates. Some stable complexes have multifunctional ligands such as tris[o-(diphenylarsino)phenylarsine], (QAS) which gives salts, for example, $[Pd(QAS)I]^+$. Platinum gives the salts $(R_4N)_3[Pt(SnCl_3)_5]$.

25-29
Complexes of Platinum(IV), d^6

There are few complexes of Pd^{IV}, a nitrato complex being formed when Pd is dissolved in concentrated HNO_3. However, platinum(IV) forms many thermally stable and kinetically inert octahedral complexes, ranging from cationic ones such as $[Pt(NH_3)_6]Cl_4$ to anionic ones like $K_2[PtCl_6]$.

The most important are the sodium or potassium *hexachloroplatinates*, which are starting materials for synthesis of other compounds. The "acid" called "chloroplatinic acid" is an oxonium salt, $(H_3O)_2PtCl_6$. It is formed as orange crystals when the solution of Pt in aqua regia or in HCl saturated with chlorine is evaporated.

25-30
Complexes of Palladium(O) and Platinum(O)

All of these involve π-acid ligands, mainly tertiary phosphines. The complex $M(PPh_3)_4$ is obtained when K_2PdCl_4 or K_2PtCl_4 is reduced by N_2H_4 in ethanol containing PPh_3. These complexes readily undergo oxidative-addition reactions (page 531) in which two PPh_3 molecules are lost, for example,

$$Pt(PPh_3)_4 + CH_3I = PtI(CH_3)(PPh_3)_2 + 2PPh_3$$

They also give complexes with O_2, alkenes and alkynes (Chapter 29).

SILVER AND GOLD

25-31
General Remarks

In spite of the similarity in electronic structures, with an *s* electron outside a completed *d* shell and high ionization potentials, there are only limited resemblances between Ag, Au, and Cu. These are as follows.

1. The metals crystallize with the same face-centered cubic (*ccp*) lattice.

2. Cu_2O and Ag_2O have the same body-centered cubic structure where the metal atom has two close O neighbors and every O is tetrahedrally surrounded by four metal atoms.

3. Although the stability constant sequence for halogeno complexes of many metals is F > Cl > Br > I, Cu^I and Ag^I belong to the group of ions of the more noble metals for which it is the reverse.

4. Cu^I and Ag^I (and to a lesser extent Au^I) form similar types of complexes, such as $[MCl_2]^-$, $[Et_3AsMI]_4$, and K_2MCl_3.

5. Certain complexes of Cu^{II} and Ag^{II} are isomorphous, and Ag^{III}, Au^{III}, and Cu^{III} also give similar complexes.

The only stable cation, apart from complex ions, is Ag^+. The Au^+ ion is exceedingly unstable with respect to the disproportionation

$$3\,Au^+(aq) = Au^{3+}(aq) + 2\,Au(s) \qquad K \approx 10^{10}$$

Gold(III) is *invariably* complexed in all solutions, usually as anionic species such as $[AuCl_3OH]^-$. The other oxidation states, Ag^{II}, Ag^{III}, and Au^I, are either unstable to water or exist only in insoluble compounds or complex species. Intercomparisons of the standard potentials are of limited utility, particularly since these strongly depend on the nature of the anion; some useful ones are:

$$Ag^{2+} \xrightarrow{2.0\ V} Ag^+ \xrightarrow{0.799\ V} Ag$$

$$Ag(CN)_2^- \xrightarrow{-0.31\ V} Ag + 2CN^-$$

$$AuCl_4^- \xrightarrow{1.00\ V} Au + 4Cl^-$$

$$Au(CN)_2^- \xrightarrow{-0.6\ V} Au + 2CN^-$$

25-32 The Elements

The elements are widely distributed as metals, in sulfides and arsenides, and as AgCl. Silver is usually recovered from the work-up of other ores, for example, of lead, the platinum metals and particularly, copper. The elements are extracted by treatment with cyanide solutions in presence of air, whereby the cyano complexes, $M(CN)_2^-$ are formed, and are recovered from them by addition of zinc. They are purified by electrodeposition.

Silver is white, lustrous, soft, and malleable (mp 961°) with the highest known electrical and thermal conductivities. It is less reactive than copper, except toward sulfur and hydrogen sulfide, which rapidly blacken silver surfaces. Silver dissolves in oxidizing acids and in cyanide solutions in presence of oxygen or peroxide.

Gold is soft and yellow (mp 1063°) with the highest ductility and malleability of any element. It is unreactive and is not attacked by oxygen or sulfur but reacts readily with halogens or with solutions containing or generating chlorine such as aqua regia. It dissolves in cyanide solutions in presence of air or hydrogen peroxide to form $[Au(CN)_2]^-$.

Both silver and gold form many useful alloys.

SILVER AND GOLD COMPOUNDS

25-33
Silver(I), d^{10}, Compounds

The argentous ion, Ag^+, is evidently solvated in aqueous solution but an aquo ion does *not* occur in salts, practically all of which are anhydrous. $AgNO_3$, $AgClO_3$, and $AgClO_4$ are water-soluble but Ag_2SO_4 and $AgOOCCH_3$ are sparingly so. The salts of oxo anions are ionic. Although the water-insoluble halides AgCl and AgBr have the NaCl structure, there appears to be appreciable covalent character in the Ag $\cdots$ X interactions. The addition of NaOH to Ag^+ solutions produces a dark brown *oxide* that is difficult to free from alkali ions. It is basic, and its aqueous suspensions are alkaline:

$$\tfrac{1}{2}Ag_2O(s) + \tfrac{1}{2}H_2O = Ag^+ + OH^- \qquad \log K = -7.42$$

$$\tfrac{1}{2}Ag_2O(s) + \tfrac{1}{2}H_2O = AgOH \qquad \log K = -5.75$$

they absorb CO_2 from the air to give Ag_2CO_3. The oxide decomposes above ~160° and is reduced to the metal by hydrogen. The treatment of water-soluble halides with a suspension of silver oxide is a useful way of preparing hydroxides, since the silver halides are insoluble.

The action of hydrogen sulfide on argentous solutions gives the black sulfide Ag_2S. The coating often found on silver articles is Ag_2S; this can be readily reduced by contact with aluminum in dilute Na_2CO_3 solution.

Silver fluoride is unique in forming hydrates such as $AgF \cdot 4H_2O$. The other halides are precipitated by the addition of X^- to Ag^+ solutions; the color and insolubility in water increase Cl < Br < I. *Silver chloride* can be obtained as rather tough sheets that are transparent over much of the infrared region and have been used for cell materials. Silver chloride and bromide are light-sensitive and have been intensively studied because of their importance in photography. For monodentate ligands, the *complex ions*, AgL^+, $AgL_2{}^+$, $AgL_3{}^+$, and $AgL_4{}^+$ exist. The constants K_1 and K_2 are usually high whereas K_3 and K_4 are relatively small. The main species are, hence, $AgL_2{}^+$, which are linear. Because of this, chelating ligands cannot form simple ions, and they give polynuclear complexes instead. The commonest complexes are those such as $[Ag(NH_3)_2]^+$ formed by dissolving silver chloride in NH_3, $[AgCN)_2]^-$ and $[Ag(S_2O_3)_2]^{3-}$. Silver halides also dissolve in solutions with excess halide ion and excess Ag^+, for example,

$$AgI + nI^- \rightleftharpoons AgI_{n+1}{}^{n-}$$

$$AgI + nAg^+ \rightleftharpoons Ag_{n+1}I^{n+}$$

25-34
Silver(II), d^9 and Silver(III), d^8 Compounds

Silver(II) fluoride, is a brown solid formed on heating Ag in F_2; it is a useful fluorinating agent. A black oxide obtained by oxidation of Ag_2O in alkaline solution is $Ag^IAg^{III}O_2$.

Both Ag^{II} and Ag^{III} occur in complexes with appropriate ligands; the usual procedure is to oxidize Ag^+ in presence of the ligand. Thus oxidation by $S_2O_8^{2-}$ in presence of pyridine gives the red ion $[Agpy_4]^{2+}$ while in alkaline periodate solution the ion $[Ag(IO_6)_2]^{7-}$ is obtained.

25-35
Gold Compounds

The *oxide* Au_2O_3 decomposes to Au and O_2 at about 150°. Chlorination of gold at 200° gives *auric chloride*, Au_2Cl_6, as red crystals; on heating at 160° this in turn gives *aurous chloride*, AuCl.

Complexes. The *dicyanoaurate ion*, $[Au(CN)_2]^-$, is readily formed by dissolving gold in cyanide solutions in presence of air or H_2O_2.

The interaction of Au_2Cl_6 in ether with tertiary phosphines gives gold(I) complexes R_3PAuCl; Cl^- can be replaced by I^-, or SCN^-. On reduction with $NaBH_4$, these complexes give *gold cluster compounds* of stoichiometry $Au_{11}X_3(PR_3)_7$. The cluster is an incomplete icosahedron with a central Au atom.

Gold alkylsulfides, $[Au(SR)]_n$, and similar compounds made from sulfurized terpenes are very soluble in organic solvents and are also probably cluster compounds. They are used as "liquid gold" for decorating ceramic and glass articles, which are then fired leaving a gold film.

Gold(III) d^8 is isoelectronic with Pt(II), and its compounds are hence *square*. Dissolution of Au in aqua regia or of Au_2Cl_6 in HCl gives a solution that on evaporation deposits yellow crystals of $[H_3O][AuCl_4]\cdot 3H_2O$. The tetrachloroaurate(III) ion quite readily hydrolyzes to $[AuCl_3OH]^-$.

From dilute HCl solutions Au^{III} can be extracted with a very high partition coefficient into ethyl acetate or diethylether. The yellow species in the organic layer is probably $[H_3O][AuCl_3OH]$.

Study Questions

A

1. State the chief differences between the second- and third-row transition elements on the one hand and those of the first series on the other with respect to (a) atomic and ionic radii, (b) oxidation states, (c) formation of metal—metal bonds, (d) stereochemistry, and (e) magnetic properties.
2. Why are the chemical and physical properties of hafnium and zirconium compounds so similar?
3. What elements characteristically form cluster compounds in their lower oxidation states? Gives examples of the three major types, two of which have six metal atoms, and the other three.
4. Draw the structures of the following: $Mo_2(O_2CCF_3)_4$, $Re_2Cl_8^{2-}$, "$TaCl_5$," "NbF_5," $Mo_2O_3(S_2COEt)_4$, and $Rh_2Cl_2(CO)_4$.
5. Describe the chemical and physical properties of RuO_4 and OsO_4, including preparations and toxicology.

6. List all the elements in the group called the "platinum metals" and show how and where they are arranged in the periodic table. Indicate the relative importance of oxidation states I to VI for each.
7. What is the true nature of the so-called "dihalides" of molybdenum and tungsten?
8. Discuss the terrestrial abundance and commercial availability of technetium.
9. What evidence is there for metal—metal interactions in compounds containing square complexes of Ni^{II}, Pd^{II}, and Pt^{II} stacked so the metal atoms form chains perpendicular to the parallel planes of the complexes?
10. Show with sketches the structures of the α- and β-forms of $PdCl_2$. What role is direct metal—metal bonding thought to play in each?
11. What is the structure of Pd^{II} acetate?
12. How is $Pt(Ph_3P)_4$ prepared? What product is formed when it reacts with methyl iodide?
13. Contrast the chemistry of Cu with that of Ag and Au. First mention the important similarities and then several important differences.
14. Compare the chemistries of Ag^I and Au^I.
15. Write balanced equations for the following processes: (a) leaching of metallic gold by CN^- in presence of oxygen. (b) The reaction of AgI with a solution of thiosulfate (photographer's "hypo"). (c) The reaction of aqueous $AgNO_3$ with $S_2O_8^{2-}$ in presence of excess pyridine.
16. Name the most important silver salts that are (a) soluble in water, and (b) insoluble in water.
17. Starting with a Ni—Cu sulfide ore which contains significant amounts of the platinum metals, what are the main steps by which the latter, as a group, are isolated?
18. Complete and balance the following:

$$Mo + F_2 \longrightarrow$$

$$W + F_2 \longrightarrow$$

$$Mo + Cl_2 \xrightarrow{\text{heat}}$$

$$W + Cl_2 \xrightarrow{\text{heat}}$$

B

1. What is the lanthanide contraction and what effect does it have on the chemistry of the heavier transition elements?
2. How would you most easily: (a) dissolve tantalum metal; (b) precipitate zirconium from aqueous solution in presence of aluminum; (c) prepare 50 g of molybdenum (V) chloride from MoO_3; (d) prepare rhenium(III) chloride; (e) dissolve WO_3; (f) prepare $Rh(CO)H(PPh_3)_3$; (g) make $K_2[MoOCl_5]$ from MoO_3.
3. Why is it that compounds such as K_3ZrF_7, $K_4Zr_2F_{12}$, and K_4ZrF_8 can be crystallized from solutions which contain all the zirconium as ZrF_6^{2-}? What is the structure of $Zr_2F_{12}^{4-}$?
4. State the most important differences between each of the following elements and its first-row congener: Ta, Re, Rh.
5. How would you dissolve a Au-Ag alloy and obtain the two metals separately?
6. What is meant by the term "bridge-cleaving reaction"? Give two examples using Rh or Pt complexes. What is a likely mechanism for such reactions?
7. How is commercial "$RuCl_3 \cdot 3H_2O$" made, and what does it actually contain? Suggest the products when it is (a) dissolved in concentrated HCl and evaporated carefully to dryness; (b) heated with aqueous hydrazine; (c) boiled in aqueous

$NH_4Cl + NH_4OH$ with zinc powder; (d) heated with triphenylphosphine in ethanol.

8. What happens when commercial "$RhCl_3 \cdot 3H_2O$" is (a) boiled with aqueous HCl; (b) warmed with excess Ph_3P in ethanol; (c) heated with ammonia in ethanol; (d) boiled with sodium acetate in methanol.
9. How would you separate the long-lived isotope of Tc from fission products (isotopes of elements from Se to Pm)?
10. Suggest explanations for the following: (a) the aquo nickel(II) ion is paramagnetic while the aquo palladium(II) ion is diamagnetic, although both NiF_2 and PdF_2, which are isostructural, contain paramagnetic metal ions; (b) the $[Ru_2OCl_{10}]^{4-}$ ion has no unpaired electrons; (c) there is important metal—metal bonding in the $M_6Cl_{12}{}^{n+}$ cluster species when M = Nb or Ta, but not when M = Pd or Pt.

Chapter 25
Study Guide

Supplementary Reading

Canterford, J. H. and Colton, R., *Halides of the Second and Third Row Transition Series*, Interscience-Wiley, 1968.

Colton, R., *The Chemistry of Rhenium and Technetium*, Wiley, 1965.

Cotton, F. A., "Compounds with Multiple Metal to Metal Bonds," *Chem. Soc. Rev.*, *4*, 27 (1975).

Fairbrother, F., *The Chemistry of Niobium and Tantalum*, Elsevier, 1967.

Griffith, W. P., *The Chemistry of the Platinum Metals*, Interscience-Wiley, 1967.

Hartley, F. R., *The Chemistry of Palladium and Platinum*, Applied Science Publishers, 1973.

Kepert, D. L., *The Early Transition Elements*, Academic Press, 1972.

Larsen, E. M., "The Chemistry of Zirconium and Hafnium," *Adv. Inorg. Chem. Radiochem.*, *13*, 1 (1970).

Mitchell, P. C. H., ed., *Chemistry and Uses of Molybdenum*, *J. Less-Common Metals*, Vol. 36, 1974.

Rouschias, G., "Recent Advances in the Chemistry of Rhenium," *Chem. Rev.*, *73*, 531 (1973).

26

scandium, yttrium, lanthanum and the lanthanides

26-1
General Features

The position of these elements in the periodic table is discussed on page 44. Note that actinium, although the first member of the actinide elements (Chapter 27), is a true member of the Group IIIA series, Sc, Y, La, Ac. Except for some similarities in the chemistries of Sc and Al, little resemblance exists between these elements and the Group IIIB elements, Al—Tl.

The elements and some of their properties are given in Table 26-1. Strictly, the elements are the fourteen that follow La and in which the $4f$ electrons are successively added to the La configuration. The term lanthanide is taken to include lanthanum itself, as indeed, this element is the prototype for the succeeding fourteen. The progressive decrease in radii of the atoms and ions of these elements which when summed is called the *lanthanide contraction* has been discussed, page 204.

The elements are all highly electropositive with the M^{3+}/M potential varying from -2.25 V (Lu) to -2.52 V (La). The chemistry is predominantly ionic and of the M^{3+} ions.

Yttrium, which lies above La in Group IIIA has a similar $+3$ ion with a noble gas core; due to the effect of the lanthanide contraction the Y^{3+} radius is close to the values for Tb^{3+} and Dy^{3+}. Consequently, Y occurs in lanthanide minerals. The lighter element in Group IIIA, *scandium*, is also considered here, although it has a smaller ionic radius and shows chemical behavior intermediate between that of Al and that of Y and the lanthanides.

Variable Valency. Certain lanthanides (Table 26-1) form $+2$ or $+4$ ions. The $+2$ ions are readily oxidized and the $+4$ ions are readily reduced to the $+3$ ion. A simplified explanation for the occurrence of these valences is that empty, half-filled or filled f shells are especially stable. A similar phenomenon has been noted concerning the ionization enthalpies of the elements of the first transition series (page 48), and half-filling of the $3d$ shell accounts for the stability of manganese(II). For the lanthanides, the oxidation state IV for cerium gives Ce^{4+} with the empty f shell configuration of La^{3+}. Similarly the formation of Yb^{2+} gives this

Table 26-1 *Some Properties of Scandium, Yttrium and the Lanthanides*

Z	Name	Symbol	Electronic Configuration	Valences	Radius M^{3+} (Å)	Color M^{3+}
21	Scandium	Sc	[Ar] $3d^1 4s^2$	3	0.68	Colorless
39	Yttrium	Y	[Kr] $4d^1 5s^2$	3	0.88	Colorless
57	Lanthanum	La	[Xe] $5d^1 6s^2$	3	1.06	Colorless
58	Cerium	Ce	[Xe] $4f^1 5d^1 6s^2$	3, 4	1.03	Colorless
59	Praseodymium	Pr	[Xe] $4f^3 6s^2$	3, 4	1.01	Green
60	Neodymium	Nd	[Xe] $4f^4 6s^2$	3	0.99	Lilac
61	Promethium	Pm	[Xe] $4f^5 6s^2$	3	0.98	Pink
62	Samarium	Sm	[Xe] $4f^6 6s^2$	2, 3	0.96	Yellow
63	Europium	Eu	[Xe] $4f^7 6s^2$	2, 3	0.95	Pale pink
64	Gadolinium	Gd	[Xe] $4f^7 5d\, 6s^2$	3	0.94	Colorless
65	Terbium	Tb	[Xe] $4f^9 6s^2$	3, 4	0.92	Pale pink
66	Dysprosium	Dy	[Xe] $4f^{10} 6s^2$	3	0.91	Yellow
67	Holmium	Ho	[Xe] $4f^{11} 6s^2$	3	0.89	Yellow
68	Erbium	Er	[Xe] $4f^{12} 6s^2$	3	0.88	Lilac
69	Thulium	Tm	[Xe] $4f^{13} 6s^2$	3	0.87	Green
70	Ytterbium	Yb	[Xe] $4f^{14} 6s^2$	2, 3	0.86	Colorless
71	Lutetium	Lu	[Xe] $4f^{14} 5d\, 6s^2$	3	0.85	Colorless

ion an f^{14} configuration. The half-filled f^7 configuration of Gd^{3+} is formed by reduction to give Eu^{2+} or oxidation to give Tb^{4+}. That other factors are involved, however, is shown by the existence of many +2 ions stabilized in CaF_2 lattices and of Pr^{4+} and Nd^{4+} fluoride complexes.

Magnetic and Spectral Properties. The lanthanide ions that have unpaired electrons are colored and are paramagnetic. There is a fundamental difference from the *d*-block elements in that the 4*f* electrons are inner electrons and are very effectively shielded from the influence of external forces by the overlying $5s^2$ and $5p^6$ shells. Hence, there are essentially only very weak effects of ligand fields. As a result, electronic transitions between *f* orbitals give rise to extremely narrow absorption bands, quite unlike the broad bands resulting from *d*—*d* transitions, and the magnetic properties of the ions are little affected by their chemical surroundings.

Coordination Numbers and Stereochemistry. It is characteristic of the M^{3+} ions that *coordination numbers exceeding six are common.* Very few 6-coordinate species are known but coordination numbers of 7, 8, and 9 are important. In the ion $[Ce(NO_3)_6]^{2-}$, the Ce is surrounded by 12 oxygen atoms of chelate NO_3 groups.

The decrease in radii from La to Lu also has the effect that different crystal structures and coordination numbers may occur for different parts of the lanthanide group. For example, the metal atoms in the trichlorides La—Gd are 9-coordinate, whereas the chlorides of Tb—Lu have an $AlCl_3$ type structure with the metal being octahedrally coordinated. Similar differences in coordination numbers occur for ions in solution.

26-2
Occurrence and Isolation

Scandium is quite a common element being as abundant as As and twice as abundant as B. However, it is not readily available, partly owing to a lack of rich ores, and partly due to the difficulty of separation. It may be separated from Y and the lanthanides, which may be associated with Sc minerals, by cation exchange procedures using oxalic acid as elutant.

The lanthanide elements, including La and Y, were originally known as the Rare Earths—from their occurrence in oxide (or in old usage, earth) mixtures. They are *not* rare elements and their absolute abundances are relatively high. Thus even the scarcest, Tm, is as common as Bi and more common than As, Cd, Hg, or Se. The major source is *monazite*, a heavy dark sand of variable composition. Monazite is essentially a lanthanide orthophosphate, but may contain up to 30% thorium. La, Ce, Pr, and Nd usually account for *ca.* 90% of the lanthanide content of minerals, with Y and the heavier elements accounting for the rest. Minerals carrying lanthanides in the +3 oxidation state are usually poor in Eu which, because of its tendency to give the +2 state, is often concentrated in minerals of the Ca group.

Promethium occurs only in traces in U ores as a spontaneous fission fragment of ^{238}U. Milligram quantities of pink $^{147}Pm^{3+}$ salts can be isolated by ion exchange methods from fission products in spent fuel of nuclear reactors where ^{147}Pm (β^-, 2.64 yr) is formed.

The lanthanides are separated from most other elements by precipitation of oxalates or fluorides from HNO_3 solution, and from each other by ion-exchange on resins. Cerium and europium are normally first removed. Cerium is oxidized to Ce^{IV} and is then precipitated from 6*M* HNO_3 as ceric iodate or separated by solvent extraction. Europium is reduced to Eu^{2+} and is removed by precipitation as $EuSO_4$.

The ion-exchange behavior depends primarily on the hydrated ionic radius. As with the alkalis (page 227), the smallest ion crystallographically (Lu) has the largest hydrated radius, while La has the smallest hydrated radius. Hence, La is the most tightly bound and Lu is the least, and the elution order is Lu → La (*Fig. 27-3*, page 467). This trend is accentuated by use of complexing agents at an appropriate pH; the ion of smallest radius also forms the strongest complexes and, hence, the preference for the aqueous phase is enhanced. Typical complexing ligands are α-hydroxyisobutyric acid, $(CH_3)_2CH(OH)COOH$, $EDTAH_4$, and other hydroxo or amino carboxylic acids. From the eluates the M^{3+} ions are recovered by acidification with dilute HNO_3 and addition of oxalate ion, which precipitates the oxalates essentially quantitatively. These are then ignited to the oxides.

Cerium(IV) (also Zr^{IV}, Th^{IV} and Pu^{IV}) is readily extracted from HNO_3 solutions by tributyl phosphate dissolved in kerosene or other inert solvent and can be

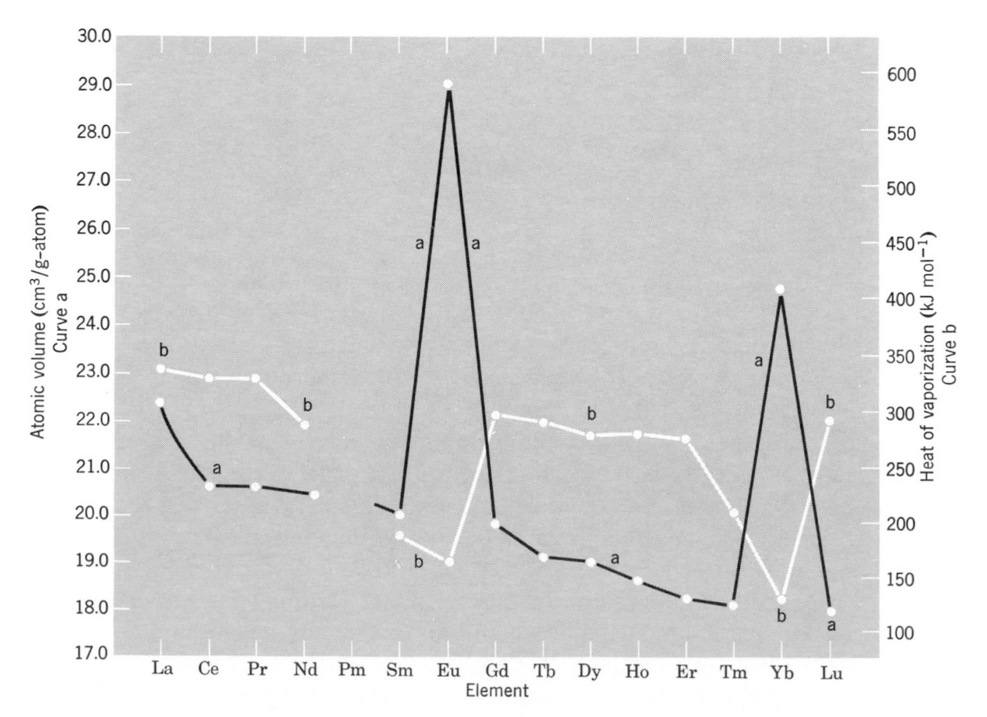

Figure 26-1 *The atomic volumes (curve a) and heats of vaporization (curve b) of the lanthanide metals.*

separated from the +3 lanthanide ions. The +3 lanthanide nitrates can also be extracted under suitable conditions with various phosphate esters or acids. Extractability under given conditions increases with increasing atomic number; it is higher in strong acid or high NO_3^- concentrations.

The Metals. The lighter metals (La—Gd) are obtained by reduction of the trichlorides with Ca at 1000° or more. For Tb, Dy, Ho, Er, Tm, and also Y the trifluorides are used because the chlorides are too volatile. Pm is made by reduction of PmF_3 with Li. Eu, Sm, and Yb trichlorides are reduced only to the dihalides by Ca. Reduction of the +3 oxides with La at high temperatures gives the metals.

The metals are silvery-white and highly electropositive. They react with water, slowly in the cold, rapidly on heating, to liberate hydrogen. They tarnish in air and burn easily to give the oxides M_2O_3; cerium is the exception giving CeO_2. Lighter "flints" are mixed metals mostly cerium. Yttrium is resistant to air even up to 1000° owing to formation of a protective oxide coating. The metals react with H_2, C, N_2, Si, P, S, halogens, and other nonmetals at elevated temperatures.

Many physical properties of the metals change smoothly along the series, except for Eu and Yb and occasionally Sm and Tm (cf. *Fig. 26-1*). The deviations occur with those lanthanides that have the greatest tendency to exist in the +2 state; presumably these elements tend to donate only two electrons to the conduction bands of the metal, thus leaving larger cores and affording lower binding forces. Note, too, that Eu and Yb dissolve in ammonia (see page 219).

LANTHANIDE COMPOUNDS

26-3
The Trivalent State

Oxides and Hydroxides. The oxide Sc_2O_3 is less basic than the other oxides and closely resembles Al_2O_3; it is similarly amphoteric dissolving in NaOH to give a "scandate" ion $[Sc(OH)_6]^{3-}$.

The oxides of the remaining elements resemble CaO and absorb CO_2 and H_2O from the air to form carbonates and hydroxides, respectively. The *hydroxides*, $M(OH)_3$, are true compounds whose basicities decrease with increasing Z, as would be expected from the decrease in ionic radius. They are precipitated from aqueous solutions by bases as gelatinous masses. They are not amphoteric.

Halides. *Scandium* is again exceptional. Its fluoride resembles AlF_3, being soluble in excess HF to give the ScF_6^{3-} ion; Na_3ScF_6 is like cryolite (page 252). However, $ScCl_3$ is not a Friedel–Crafts catalyst like $AlCl_3$ and does not behave as a Lewis acid; its structure is like that of $FeCl_3$, page 400.

Lanthanide fluorides are of importance because of their insolubility. Addition of HF or F^- precipitates MF_3 from solutions even 3 *M* in HNO_3 and is a characteristic test for lanthanide ions. The fluorides of the heavier lanthanides are slightly soluble in an excess of HF owing to complex formation. Fluorides may be redissolved in $3M$ HNO_3 saturated with H_3BO_3, which removes F^- as BF_4^-.

The *chlorides* are soluble in water, from which they crystallize as hydrates. The anhydrous chlorides are best made by the reaction

$$M_2O_3 + 6NH_4Cl \xrightarrow{\sim 300^\circ} 2MCl_3 + 3H_2O + 6NH_3$$

Aquo Ions, Oxo Salts and Complexes. Scandium forms a hexa-aquo ion $[Sc(H_2O)_6]^{3+}$ that is readily hydrolyzed. Scandium β-diketonates are also octahedral like those of Al and unlike those of the lanthanides.

For the lanthanides and yttrium, the aquo ions, $[M(H_2O)_n]^{3+}$, have coordination numbers exceeding 6 as in $[Nd(H_2O)_9]^{3+}$. These are hydrolyzed in water:

$$[M(H_2O)_n]^{3+} + H_2O \rightleftharpoons [M(OH)(H_2O)_{n-1}]^{2+} + H_3O^+$$

The tendency to hydrolyze increases from La to Lu, which is consistent with the decrease in the ionic radii. Yttrium also gives predominantly $Y(OH)^{2+}$. For Ce^{3+}, however, only about 1% of the metal ion is hydrolyzed without forming a precipitate and the main equilibrium appears to be

$$3Ce^{3+} + 5H_2O \rightleftharpoons [Ce_3(OH)_5]^{4+} + 5H^+$$

In aqueous solutions, rather weak *fluoride* complexes, MF_{aq}^{2+} are formed. Complex anions are *not* formed, a feature that distinguishes the +3 lanthanides as a group from the +3 actinide elements that *do* form anionic complexes in strong HCl solutions.

The most stable and common complexes are those with *chelating oxygen ligands.* The formation of water-soluble complexes by citric and other hydroxo acids is utilized in ion-exchange separations, as we note above. The complexes usually have coordination numbers greater than 6.

β-diketone ligands such as acetylacetone are especially important, since some of the fluorinated β-diketones give complexes that are volatile and suitable for gas-chromatographic separation. The preparation if β-diketonates by conventional methods invariably gives hydrated or solvated species such as $M(acac)_3 \cdot C_2H_5OH \cdot 3H_2O$ that have coordination numbers >6. Prolonged drying over $MgClO_4$ gives the very hygroscopic $M(\beta\text{-dik})_3$.

An important use of Eu and Pr β-diketonate complexes that are soluble in organic solvents, such as those derived from 1,1,1,2,2,3,3-heptafluoro-7,7-dimethyl-4,6-octanedione, is as shift reagents in nmr spectrometry. The paramagnetic complex deshields the protons of complicated molecules, and vastly improved separation of the resonance lines may be obtained.

Other uses for lanthanide compounds depend on their spectroscopic properties. Y and Eu in oxide or silicate lattices have fluorescent or luminescent behavior and the phosphors are used in color television tubes. In CaF_2 lattices the +2 ions show laser activity as do salts of $Eu(\beta\text{-dik})_4^-$.

26-4
The Tetravalent State

Cerium(IV). This is the only +4 lanthanide that exists in aqueous solution as well as in solids. The dioxide, CeO_2, is obtained by heating $Ce(OH)_3$, or oxo salts in air. It is unreactive and is dissolved by acids only in the presence of reducing agents (H_2O_2, Sn^{II}, etc.) to give Ce^{3+} solutions. Hydrous ceric oxide, $CeO_2 \cdot nH_2O$, is a yellow, gelatinous precipitate obtained on treating Ce^{IV} solutions with OH^-; it redissolves in acids.

The *ceric ion* in solution is obtained by oxidation of Ce^{3+} in HNO_3 or H_2SO_4 with $S_2O_8^{2-}$ or bismuthate. Its chemistry is similar to that of Zr^{4+}, and +4 actinides. Thus Ce^{4+} gives phosphates insoluble in $4M$ HNO_3 and iodates insoluble in $6M\,HNO_3$, as well as an insoluble oxalate. The phosphate and iodate precipitations can be used to separate Ce^{4+} from the trivalent lanthanides.

The yellow-orange hydrated ion, $[Ce(H_2O)_n]^{4+}$, is fairly strong acidic, hydrolyzes readily, and probably exists only in strong $HClO_4$ solution. In other acids complex formation accounts for the acid-dependence of the potential:

$$Ce^{IV} + e = Ce^{III} \qquad E^\circ = +1.28\text{ V }(2M\text{ HCl}),\ +1.44\text{ V }(1M\ H_2SO_4),\ +1.61\text{ V }(1M\ HNO_3),\ +1.70\text{ V }(1M\ HClO_4)$$

Comparison of the potential in H_2SO_4, where at high SO_4^{2-} concentrations the major species is $[Ce(SO_4)_3]^{2-}$, with that for the oxidation of water:

$$O_2 + 4H^+ + 4e = 2H_2O \qquad E^\circ = +1.229\text{ V}$$

shows that the acid Ce^{IV} solutions commonly used in analysis are metastable.

Cerium(IV) is used as an oxidant in analysis and in organic chemistry, where it is commonly used in acetic acid. The solutions oxidize aldehydes and ketones at the α-carbon atom. Benzaldehyde gives benzoin.

Complex anions are formed quite readily. The analytical standard "ceric ammonium nitrate," which can be crystallized from HNO_3, contains the hexanitratocerate anion, $[Ce(NO_3)_6]^{2-}$.

Praseodymium(IV) and Terbium(IV). These exist only in oxides and fluorides. The oxide systems are very complex and nonstoichiometric. The potential Pr^{IV}/Pr^{III} is estimated to be +2.9 V so that it is not surprising that Pr^{IV} does not exist in aqueous solution.

26-5
The Divalent State

The +2 state is known in both solutions and solid compounds of Sm, Eu, and Yb (Table 26-2). Less well-established are Tm^{2+} and Nd^{2+}, but the +2 ions of all the lanthanides can be prepared and stabilized in CaF_2 or BaF_2 lattices by reduction of, for example, MF_3 in CaF_2 with Ca.

Table 26-2 Properties of the Lanthanide +2 Ions

Ion	Color	$E°(V)^a$	Crystal radius (Å)[b]
Sm^{2+}	Blood-red	−1.55	1.11
Eu^{2+}	Colorless	−0.43	1.10
Yb^{2+}	Yellow	−1.15	0.93

[a] For $M^{3+} + e = M^{2+}$.
[b] Pauling radii, Ca^{2+}, 0.99, Sr^{2+}, 1.13, Ba^{2+}, 1.35.

The *europous* ion can be made by reducing aqueous Eu^{3+} solutions with Zn or Mg. The other ions require the use of Na amalgam. All three can be prepared by electrolytic reduction in aqueous solution or in halide melts.

The ions Sm^{2+} and Yb^{2+} are quite rapidly oxidized by water. Eu^{2+} is oxidized by air.

The Eu^{2+} ion resembles Ba^{2+}. Thus the sulfate and carbonate are insoluble whereas the hydroxide is soluble. The stability of the Eu^{2+} complex with $EDTA^{4-}$ is intermediate between those of Ca^{2+} and Sr^{2+}.

Crystalline compounds of Sm, Eu, and Yb are usually isostructural with the Sr^{2+} or Ba^{2+} analogs.

Study Questions

A

1. Name the lanthanide elements and give their electronic configurations.
2. Explain the position of the lanthanides in the periodic table and their relation to the Al, Ga, In, and Tl group.
3. What is the "lanthanide contraction"? What effect does it have on the chemistry of later elements?
4. Compare the main features of the chemistry of ions of highly electropositive elements with charges +1, +2, and +3.
5. Why are scandium and yttrium usually considered along with the lanthanide elements?
6. Which lanthanide elements show departure from usual +3 oxidation state? Give the electronic configuration of these ions.
7. What is characteristic about the coordination numbers of lanthanide ions? Give examples.
8. How are the lanthanide ions separated from each other?
9. What are the characteristic precipitation reactions of lanthanide +2, +3, and +4 ions?
10. How are anhydrous lanthanide chlorides made?
11. What are the most interesting features of lanthanide β-diketonates?

B

1. Work out the number of unpaired electrons in the ions Pr^{3+}, Pm^{3+}, Sm^{2+}, Gd^{3+}, Tb^{4+}, Tm^{3+}, Lu^{2+}.

2. Why do the electronic absorption spectra of lanthanide ions have sharp bands unlike the broad bands in the spectra of 3*d* elements?
3. Describe and explain the differences between the properties of Eu and Yb metals and those of the other lanthanides.
4. Discuss the pH and anion dependence of the Ce^{III}—Ce^{IV} potential.
5. Explain why sulfides M_2S_3 have three structural types: La—Dy; Ho—Tm and Yb—Lu.

Chapter 26
Study Guide

Supplementary Reading

Bagnall, K. W., ed., *Lanthanides and Actinides*, Butterworth, 1972.
Topp, N. E., *The Chemistry of the Rare Earth Elements*, Elsevier, 1965.

27

the actinide elements

27-1 General Features

The actinide elements and the electronic structures of the atoms are given in Table 27-1. Their position in the periodic table and their relation to the lanthanide elements are discussed in Chapter 8. It will be evident in the following pages that the term *actinides* is not so apt for these elements as is the term lanthanides for elements 59 to 72. The elements immediately following Ac, which is similar to La and has only the +3 state, do not resemble it very closely at all. Thorium, protactinium and, to a lesser extent, uranium are homologous with their vertical groups in the periodic table, that is, Hf, Ta, and W. However, beginning with americium, there is

Table 27-1 *The Actinide Elements and Some of Their Properties*

Z	Name	Symbol	Electronic Structure[a] of Atom	Radii Å M^{3+}	Radii Å M^{4+}
89	Actinium	Ac	$6d\,7s^2$	1.11	—
90	Thorium	Th	$6d^27s^2$	—	0.90
91	Protactinium	Pa	$5f^26d7s^2$ or $5f^16d^27s^2$	—	0.96
92	Uranium	U	$5f^36d7s^2$	1.03	0.93
93	Neptunium	Np	$5f^57s^2$	1.01	0.92
94	Plutonium	Pu	$5f^67s^2$	1.00	0.90
95	Americium	Am	$5f^77s^2$	0.99	0.89
96	Curium	Cm	$5f^76d7s^2$	0.985	0.88
97	Berkelium	Bk	$5f^86d7s^2$ or $5f^97s^2$	0.98	
98	Californium	Cf	$5f^{10}7s^2$	0.977	
99	Einsteinium	Es	$5f^{11}7s^2$		
100	Fermium	Fm	$5f^{12}7s^2$		
101	Mendelevium	Md	$5f^{13}7s^2$		
102	Nobelium	No	$5f^{14}7s^2$		
103	Lawrencium	Lr	$5f^{14}6d7s^2$		
104	Rutherfordium	Rf			

[a] Outside Rn structure.

pronounced lanthanide-like behavior. This, coupled with the existence of the +3 state for all the elements, justifies the term actinide.

The atomic spectra of these heavy elements are very complex, and it is difficult to identify levels in terms of quantum numbers and configurations. The energies of the 5*f*, 6*d*, 7*s*, and 7*p* levels are comparable, and the energies involved in an electron moving from one level to another may lie within the range of chemical binding energies. Thus the electronic structure of an ion in a given oxidation state may be different in different compounds, and in solution may be dependent on the nature of the ligands. It is thus often impossible to say which orbitals are being used in bonding or to decide whether the bonding is covalent or ionic.

A difference from the 4*f* group is that the 5*f* orbitals have a greater spatial extension relative to the 6*s* and 6*p* orbitals than the 4*f* orbitals have relative to the 5*s* and 5*p*. Thus 5*f* orbitals can, and do, participate in bonding to a far greater extent than the 4*f* orbitals. A reflection of this potential for covalent bonding is shown by the formation of organometallic compounds similar to those formed by the *d*-block elements. Examples are di-η^8-cyclooctatetraenyl uranium, $(\eta^8\text{-}C_8H_5)_2U$, and tri-$\eta^5$-cyclopentadienyl uranium benzyl, $(\eta^5\text{-}C_5H_5)_3UCH_2C_6H_5$.

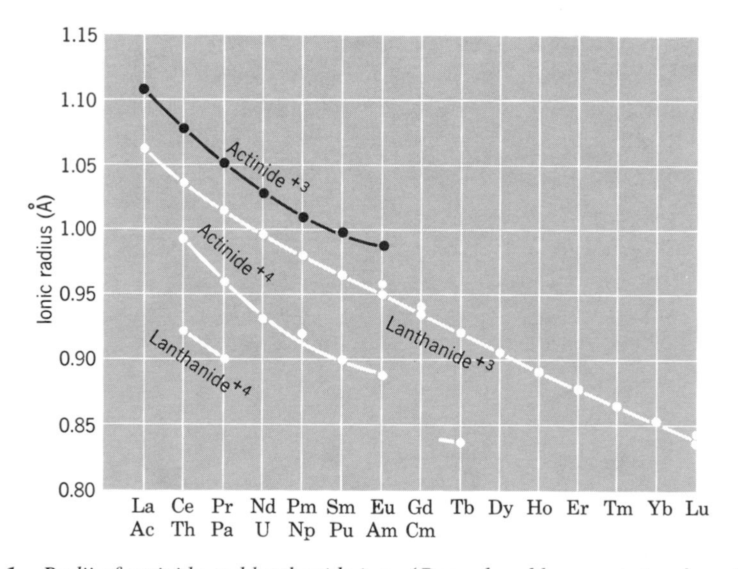

Figure 27-1 *Radii of actinide and lanthanide ions.* (*Reproduced by permission from D. Brown,* Halides of Lanthanides and Actinides, *Interscience–Wiley, 1968.*)

Ionic Radii. The ionic radii of actinide and lanthanide ions are compared in *Fig. 27-1*. Notice that there is an "actinide contraction" similar to the lanthanide contraction.

Magnetic and spectroscopic properties. The magnetic properties of the actinide ions are complicated and difficult to interpret. The electronic absorption spectra which result from *f*—*f* transitions consist, like those of the lanthanides, of quite narrow bands.

Oxidation States. There is a far greater range of oxidation states compared with the lanthanides, which is in part attributable to the fact that the 5*f*, 6*d*, and 7*s* levels are of comparable energies. The known states are given in Table 27-2.

Table 27-2 *Oxidation States of Actinides with Corresponding Members of Lanthanides*

Ac	Th	Pa	U	Np	Pu	Am	Cm	Bk	Cf	Es	Fm	Md	No	Lr
						2			2	2	2	2	**2**	
3	3	3	3	3	3	**3**	**3**	**3**	**3**	**3**	**3**	**3**	3	3
	4	4	4	4	**4**	4	4	4						
		5	5	**5**	5	5			5?					
			6	6	6	6								
				7	7									
La	Ce	Pr	Nd	Pm	Sm	Eu	Gd	Tb	Dy	Ho	Er	Tm	Yb	Lu
f^0							f^7							f^{14}

27-2
Occurrence and Properties of the Elements

All of the elements are radioactive. The terrestrial occurrence of U and Th is due to the half-lives of ^{235}U, ^{238}U, and ^{232}Th, which are sufficiently long to have enabled them to persist since genesis. These isotopes are the ones formed in the radioactive series and found in uranium and thorium minerals. The half-lives of even the most stable of the elements following uranium are so short that any amounts formed at genesis would have disappeared quite quickly.

The first new elements, neptunium and plutonium, named like uranium after the planets, were made in 1940 by McMillan and Abelson and by Seaborg, McMillan, Kennedy, and Wahl, respectively, by bombardments of uranium using particles from the cyclotron in Berkeley. Both are now obtained from spent uranium fuel elements of nuclear reactors where they are formed by capture of neutrons produced in the fission of ^{235}U fuel.

$$^{238}\text{U} \xrightarrow{n\gamma} {}^{239}\text{U} \xrightarrow[23.5m]{\beta^-} {}^{239}\text{Np} \xrightarrow[2.35d]{\beta^-} {}^{239}\text{Pu}(24{,}360\ \text{yr})$$

$$\left.\begin{matrix} ^{235}\text{U} \xrightarrow{2n\gamma} \\ ^{238}\text{U} \xrightarrow{n,\,2n} \end{matrix}\right. {}^{237}\text{U} \xrightarrow[6.75d]{\beta^-} {}^{237}\text{Np}(2.2 \times 10^6\ \text{yr})$$

Only Pu is normally recovered since ^{239}Pu has fission properties similar to ^{235}U and can be used as a fuel or in nuclear weapons. Some ^{237}Np is used to prepare ^{238}Pu (86.4 yr), which is used as a power source for satellites.

Isotopes of elements following Pu are made by successive neutron capture in ^{239}Pu in nuclear reactors. Examples are

$$^{239}\text{Pu} \xrightarrow{n\gamma} {}^{240}\text{Pu} \xrightarrow{n\gamma} {}^{241}\text{Pu} \xrightarrow[13.2\text{ yr}]{\beta^-} {}^{241}\text{Am}(433\text{ yr})$$

$$^{239}\text{Pu} \xrightarrow{4n\gamma} {}^{243}\text{Pu} \xrightarrow[5\text{ hr}]{\beta^-} {}^{243}\text{Am} \xrightarrow{n\gamma} {}^{244}\text{Am} \xrightarrow[26\text{ min}]{\beta^-} {}^{244}\text{Cm}(7.6\text{ yr})$$

The elements 100–104 are made by bombardment of Pu, Am, or Cm with accelerated ions of B, C, or N.

The isotopes ^{237}Np and ^{239}Pu can be obtained in multikilograms, Am and Cm in >100 g amounts, Bk, Cf, and Es in milligrams, and Fm in 10^{-6} g quantities. The isotopes of elements above Fm are short-lived and only tracer quantities are yet accessible. The metals are all chemically very reactive. The intense radiation from the elements with short half-lives can cause rapid decomposition of compounds. Ac and Cm glow in the dark.

27-3
General Chemistry of the Actinides

The chemistry is very complicated, especially in solutions. It has been studied in great detail because of its relevance to nuclear energy, and the chemistry of plutonium is better known than that of many natural elements.

The principal features of the actinides, all of which are electropositive metals, are the following.

1. *Actinium* has only the +3 state and is entirely lanthanide-like.

2. *Thorium* and *protactinium* show limited resemblance to the other elements. They can perhaps best be regarded as the heaviest members of the Ti, Zr, Hf and V, Nb and Ta groups, respectively.

3. *Uranium, Np, Pu, and Am* are all quite similar differing mainly in the relative stabilities of their oxidation states, which range from +3 to +6.

4. *Curium* is lanthanide-like and corresponds to gadolinium in that at Cm the $5f$ shell is half-full. It differs from Gd in having +4 compounds. By comparison with the lanthanides the previous element *americium* should show the +2 state, like Eu, and the succeeding element, berkelium the +4 state, like Tb. This is the case.

5. The elements Cm and Lr are lanthanide-like. Lawrencium, like Lu, has a filled f shell so that element 104 should and, as far as is yet known does, have hafnium-like behavior. The elements from 104 onward should be analogs of Hf, Ta, W, etc. For example, element 112, for which an unsubstantiated claim was made, should resemble Hg. It is uncertain how many more elements can be synthesized. The observation of element 106 was recently reported.

6. A characteristic feature of the compounds and complexes of actinides, like the lanthanides, is the occurrence of *high coordination numbers* up to 12 as in $[Th(NO_3)_6]^{2-}$. Coordination geometries in solids are especially complicated.

7. The various cations of U, Np, Pu, and Am have a very complex solution chemistry. The free energies of various oxidation states differ little, and for Pu the +3, +4, +5, and +6 states can actually coexist. The chemistry is complicated by hydrolysis, polymerization, complexing, and disproportionation reactions. Also, for the most radioactive species, chemical reactions are induced by the intense radiation.

The Metals. These are prepared by the reduction of anhydrous fluorides, chlorides, or oxides by Li, Mg, or Ca at 1100 to 1400°. They are silvery-white and reactive, tarnishing in air, being pyrophoric when finely divided. They are soluble in common acids; HNO_3 or HCl are the best solvents.

Uranium normally has a black oxidized film. When enriched in ^{235}U, the metal can initiate a nuclear explosion above a a certain critical mass, and this is true also for Pu. U, Np, and Pu are similar and all are the densest of metals.

Americium and Cm are much lighter metals with higher melting points than U, Np, and Pu and resemble the lanthanides. The metallic radius of *californium* indicates that it is divalent like Eu and Yb.

Oxidation States. The oxidation states have been summarized in Table 27-2, page 459.

The +3 state is the one common to all actinides except for Th and Pa. It is the preferred state for Ac, Am, and all the elements following Am. The most readily oxidized +3 ion is U^{3+}, which is oxidized by air or more slowly by water.

The chemistry is similar to that of the lanthanides. For example, the fluorides are precipitated from dilute HNO_3 solutions. Since the ionic sizes of both series are comparable, there is considerable similarity in the formation of complex ions, such as citrates, and in the magnitude of the formation constants. The separation of +3 lanthanides and actinides into groups and from each other requires ion-exchange methods (page 450).

The +4 state. This is the principal state for Th. For Pa, U, Np, Pu, and Bk, +4 cations are known in solution but for Am and Cm in solution there are only complex fluoroanions. All form solid +4 compounds. Element 104 has been found only in the +4 state.

The +4 cations in acid solution can be precipitated by iodate, oxalate, phosphate, and fluoride. The *dioxides*, MO_2, from Th to Bk have the fluorite structure. The tetrafluorides, MF_4, for both actinides and lanthanides are isostructural.

The +5 state. For Pa this is the preferred state, in which it resembles Ta. For U—Am only a few solid compounds are known. For these elements the *dioxo ions*, MO_2^+(aq), are of importance (see below).

The +6 state. The only simple compounds are the hexafluorides, MF_6, of U, Np, and Pu. The principal chemistry is that of the *dioxo ions* MO_2^{2+} of U, Np, Pu, and Am (see below).

The +2 and +7 states. These are quite rare. The +2 state is confined to Am (the 5*f* analog of Eu) where the +2 ion can occur in CaF_2 lattices, and to Cf, Es, Fm, Md, and No, which have +2 ions in solution. These are chemically similar to Ba^{2+}. The Md^{2+} ion is less readily oxidized than Eu^{2+}($E^\circ = -0.15$ V *vs* -0.43 V).

The +7 state is known only in oxoanions of Np and Pu when alkaline solutions are oxidized by O_3 or PuO_2 and Li_2O are heated in oxygen. Representative oxo anions, are $NpO_4(OH)_2^{3-}$, PuO_6^{5-}.

The Dioxo ions MO_2^+, MO_2^{2+}. The stabilities of the MO_2^+ ions are determined by the ease of disproportionation, for example,

$$2UO_2^+ + 4H^+ = U^{4+} + UO_2^{2+} + 2H_2O$$

The stability order is Np > Am > Pu > U but, of course, there is dependence on the acid concentration. The UO_2^+ ion has only a transient existence in solution but is most stable in the pH range 2–4.

The MO_2^{2+} ions are quite stable; AmO_2^{2+} is most easily reduced, the stability order being U > Pu > Np > Am.

The AmO_2^+ and AmO_2^{2+} ions undergo reduction at a few percent per hour by the products of their own α-radiation.

The linear dioxo ions can persist through a variety of chemical changes. They also appear as structural units in crystalline higher oxides. The ions are normally coordinated by solvent molecules or anions with 4, or most often, 5 or 6 ligand atoms in or near the equatorial plane of the linear O—M—O group. These equatorial ligands are often not exactly coplanar. An example is the anion in sodium uranyl acetate shown in *Fig. 27-2*. Similar structures occur in $UO_2(NO_3)_2(H_2O)_2$, $Rb[UO_2(NO_3)_3]$ etc.

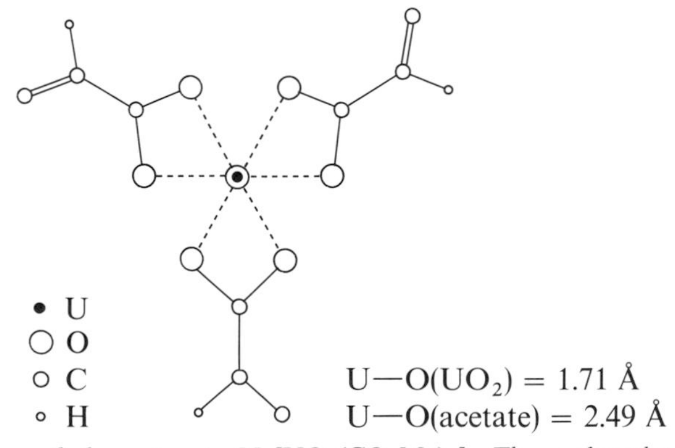

Figure 27-2 *The structure of the anion in* $Na[UO_2(CO_2Me)_3]$. *The carboxylate groups are bidentate and equivalent. The* U—O *distance in* UO_2 *is much shorter than the* U—O *distance in the equatorial plane.*

27-4
Actinium

Actinium occurs in traces in U minerals but now can be made on a milligram scale by the neutron reaction

$$^{226}Ra(n\gamma)^{227}Ra \xrightarrow{\beta^-} {}^{227}Ac(\alpha, 21.7 \text{ yr})$$

It is lanthanum-like in its chemistry, which is difficult to study because of the intense radiation of the decay products.

27-5 Thorium

Thorium is widely distributed, but the chief mineral is *monazite* sand, a complex phosphate that contains also lanthanides. The sand is digested with sodium hydroxide and the insoluble hydroxides are dissolved in hydrochloric acid. When the pH of the solution is adjusted to 5.8, the thorium, uranium and about 3% of the lanthanides are precipitated as hydroxides. The thorium is recovered by extraction from $> 6M$ hydrochloric acid solution by tributylphosphate in kerosene.

The commonest thorium compound is the *nitrate* $Th(NO_3)_4 \cdot 5H_2O$. This is soluble in water and alcohols, ketones, and esters. In aqueous solution the Th^{4+} ion is hydrolyzed at pH higher than ~ 3. It forms complex salts such as $K_4[Th\ ox_4] \cdot 4H_2O$, and $M^{II}[Th(NO_3)_6]$. On heating, the nitrate gives the white refractory dioxide ThO_2. Action of CCl_4 on this at 600° gives the white crystalline $ThCl_4$, which acts as a Lewis acid.

27-6 Protactinium

Protactinium can be isolated from residues after the extraction of uranium from pitchblende. It is exceedingly difficult to handle, except in fluoride solutions where it forms complexes (cf., Ta). In most other acid solutions it hydrolyzes to give polymeric species and colloids that are adsorbed on vessels and precipitates. Only a few compounds, some of Pa^{IV} but mostly Pa^{V} are known; they generally resemble those of Ta. For example, the chloride is Pa_2Cl_{10}, the oxide is Pa_2O_5, and the fluoroanions PaF_6^-, PaF_7^{2-}, and PaF_8^{3-} are formed.

27-7 Uranium

Until the discovery of nuclear fission by Hahn and Strassman in 1939, uranium was used only for coloring glass and ceramics, and the main reason for working its ores was to recover radium for use in cancer therapy. The isotope ^{235}U(0.72% abundance) is the prime nuclear fuel; although natural uranium can be used in nuclear reactors moderated by D_2O, most reactors and nuclear weapons use enriched uranium. Large-scale separation of ^{235}U employs gaseous diffusion of UF_6, but a gas centrifuge method now appears more economical.

Uranium is widely distributed and is more abundant than Ag, Hg, Cd, or Bi. It has few economic ores, the main one being *uraninite* (one form is *pitchblende*) an oxide of approximate composition UO_2. Uranium is recovered from nitric acid solutions by

1. Extraction of uranyl nitrate into diethylether or isobutylmethylketone; a salt such as NH_4^+, Ca^{2+} or Al^{3+} nitrate is added as a "salting out" agent to increase the extraction ratio to technically usable values. If tributylphosphate in kerosene is used, no salting out agent is necessary.

2. Removal from organic solvent by washing with dilute HNO_3,
3. Recovery as U_3O_8 or UO_3 (see below) by precipitation with ammonia.

Oxides. The U-O system is extremely complex. The main oxides are orange-yellow UO_3, black U_3O_8, and brown UO_2. UO_3 is made by heating the hydrous oxide, mainly $UO_2(OH)_2 \cdot H_2O$, obtained by adding NH_4OH to $UO_2{}^{2+}$ solutions. The other oxides are obtained by the reactions

$$3UO_3 \xrightarrow{700^\circ} U_3O_8 + \tfrac{1}{2}O_2$$

$$UO_3 + CO \xrightarrow{350^\circ} UO_2 + CO_2$$

All oxides dissolve in HNO_3 to give uranyl nitrate, $UO_2(NO_3)_2 \cdot nH_2O$.

Halides. The *hexafluoride*, UF_6, is obtained as colorless volatile crystals, (mp 64°) by fluorination at 400° of UF_3 or UF_4. It is a very powerful oxidizing and fluorinating agent and is vigorously hydrolyzed by water.

The green *tetrachloride* is obtained on refluxing UO_3 with hexachloropropene. It is soluble in polar organic solvents and in water. Action of Cl_2 on UCl_4 gives U_2Cl_{10} and, under controlled conditions, the rather unstable UCl_6.

Hydride. Uranium reacts with hydrogen even at 25° to give a pyrophoric black powder:

$$U + \tfrac{3}{2}H_2 \underset{\text{heat}}{\overset{250^\circ}{\rightleftharpoons}} UH_3$$

This hydride is often more suitable for the preparation of uranium compounds than is the massive metal. Some typical reactions are

$$UH_3 + \left\{\begin{array}{ll} H_2O & 350^\circ \\ Cl_2 & 200^\circ \\ H_2S & 450^\circ \\ HF & 400^\circ \\ HCl & 250^\circ \end{array}\right\} = \left\{\begin{array}{l} UO_2 \\ UCl_4 \\ US_2 \\ UF_4 \\ UCl_3 \end{array}\right.$$

Uranyl Salts. The most common uranium salt is the yellow uranyl nitrate which may have 2, 3, or 6 molecules of water depending on whether it is crystallized from fuming, concentrated or dilute nitric acid. When extracted from aqueous solution into organic solvents uranyl nitrate is accompanied by $4H_2O$ and the $NO_3{}^-$ ions and water are coordinated in the equatorial plane (see *Fig. 27-2*).

On addition of an excess of sodium acetate to $UO_2{}^{2+}$ solutions in dilute acetic acid, the insoluble salt $Na[UO_2(CO_2Me)_3]$ is precipitated. The uranyl ion is reduced to red-brown U^{3+} by Na/Hg or zinc, and U^{3+} is oxidizable by air to green U^{4+}. The potentials (1 *M* $HClO_4$) are:

$$UO_2{}^{2+} \xrightarrow{0.06\text{ V}} UO_2{}^{+} \xrightarrow{0.58\text{ V}} U^{4+} \xrightarrow{-0.63\text{ V}} U^{3+} \xrightarrow{-1.8\text{ V}} U$$

$$UO_2{}^{2+} \xrightarrow{0.32\text{ V}} U^{4+}$$

27-8
Neptunium, Plutonium, and Americium

The extraction of plutonium from uranium fuel elements involves (a) removal of the highly radioactive fission products that are produced simultaneously in comparable amounts, (b) recovery of the uranium for reprocessing, (c) remote control of all the chemical operations because of the radiation hazard. An additional hazard is the extreme toxicity of Pu, 10^{-6} g of which is potentially lethal; a particle of $^{239}PuO_2$ only 1 micrometer diameter can give a very high dose of radiation, enough to be strongly carcinogenic.

The separation methods of Np, Pu, and Am from U are based on the following chemistry.

1. *Stabilities of oxidation states.* The stabilities of the major ions involved are: $UO_2^{2+} > NpO_2^{2+} > PuO_2^{2+} > AmO_2^{2+}$; $Am^{3+} > Pu^{3+} \gg Np^{3+}, U^{4+}$. It is thus possible by choice of suitable oxidizing or reducing agents to obtain a solution containing the elements in different oxidation states; they can then be separated by precipitation or solvent-extraction. For example, Pu can be oxidized to PuO_2^{2+} while Am remains as Am^{3+}. The former can then be removed by solvent extraction or the latter by precipitation of AmF_3.

2. *Extractability into organic solvents.* The MO_2^{2+} ions are extracted from nitrate solutions into ethers. The M^{4+} ions are extracted into tributyl phosphate in kerosene from 6 M nitric acid solutions; the M^{3+} ions are similarly extracted from 10–16 M nitric acid, and neighboring actinides can be separated by a choice of conditions.

3. *Precipitation reactions.* Only M^{3+} and M^{4+} give insoluble fluorides or phosphates from acid solutions. The higher oxidation states give either no precipitate or can be prevented from precipitation by complex formation with sulfate or other ions.

4. *Ion exchange methods.* These are used mainly for small amounts of material as in the separation of Am and the following elements, as discussed later.

The following are examples of the separation of Pu from a nitric acid solution of the U fuel (plus its aluminum or other protective jacket).

The combination of oxidation-reduction cycles coupled with solvent extraction and/or precipitation methods removes the bulk of fission products (FP's). Certain elements—notably Ru, which forms cationic, neutral, and anionic nitrosyl complexes—may require special elimination steps. The initial uranyl nitrate solution contains Pu^{4+}, since nitric acid cannot oxidize this to Pu^{V} or Pu^{VI}.

Lanthanum fluoride cycle. This classical procedure was first developed by McMillan and Abelson for the isolation of neptunium, but it is of great utility. For the U—Pu separation, the cycle in Scheme 27-1, is repeated with progressively smaller amounts of lanthanum carrier and smaller volumes of solution until plutonium becomes the bulk phase.

Tributylphosphate solvent extraction cycle. The extraction coefficients from 6 N nitric acid solutions into 30% tributylphosphate (TBP) in kerosene are $Pu^{4+} > PuO_2^{2+}$; $Np^{4+} \sim NpO_2^{+} \gg Pu^{3+}$; $UO_2^{2+} > NpO_2^{+} > PuO_2^{2+}$; the

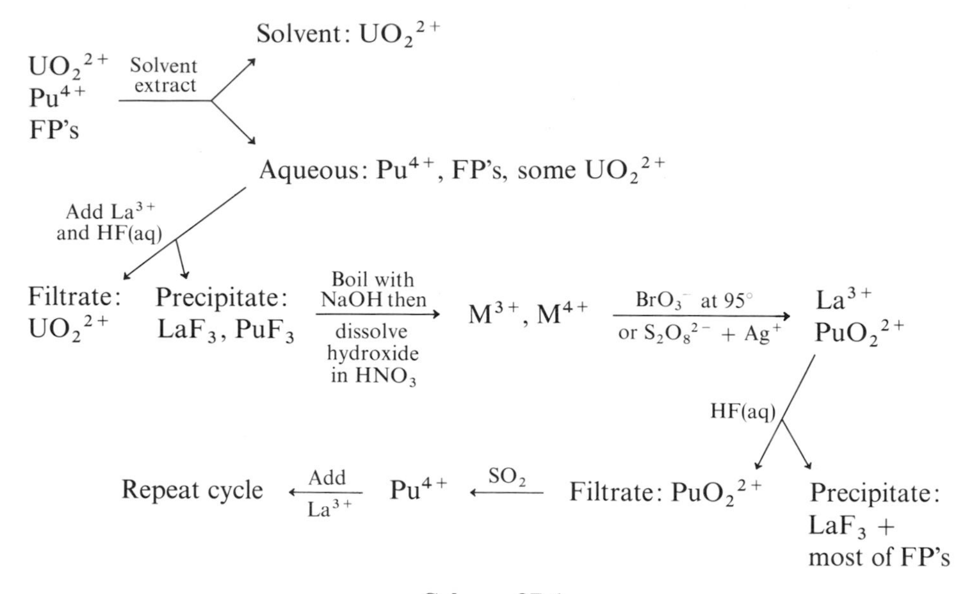

Scheme 27-1

M^{3+} ions have very low extraction coefficients in $6M$ acid, but from $12M$ hydrochloric acid or $16M$ nitric acid the extraction increases and the order is Np < Pu < Am < Cm < Bk.

Thus in the U—Pu separation, after addition of NO_2^- to adjust all of the plutonium to Pu^{4+}, we have Scheme 27-2.

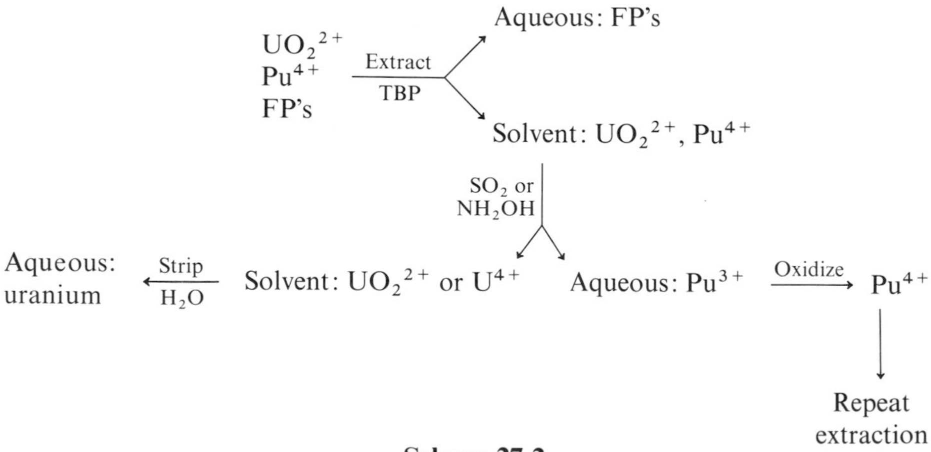

UO_2^{2+}
Pu^{4+}
FP's
Extract TBP
Aqueous: FP's
Solvent: UO_2^{2+}, Pu^{4+}
SO_2 or NH_2OH
Aqueous: uranium
Strip H_2O
Solvent: UO_2^{2+} or U^{4+}
Aqueous: Pu^{3+}
Oxidize
Pu^{4+}
Repeat extraction

Scheme 27-2

The extraction of ^{237}Np involves similar principles of adjustment of oxidation state and solvent extraction; Pu is reduced by ferrous sulfamate plus hydrazine to unextractable Pu^{III}, while Np^{IV} remains in the solvent from which it is differentially stripped by water to separate it from U.

The chemistries of U, Np, Pu, and Am are quite similar and solid compounds are usually isomorphous. The main differences are in stabilities of oxidation states in solution.

For Np, the oxidation states are well separated, but by contrast to UO_2^+, NpO_2^+ is reasonably stable. Plutonium chemistry is complicated because the potentials are not well separated and, indeed, in 1 *M* $HClO_4$ all four oxidation states can coexist.

For Am, the normal state is Am^{3+} and powerful oxidants are required to reach the higher states.

The cations all tend to hydrolyze in water, the ease of hydrolysis being Am > Pu > Np > U and $M^{4+} > MO_2^{2+} > M^{3+} > MO_2^+$. The tendency to complexing also decreases Am > Pu > Np > U.

27-9
The Elements Following Americium

The isotope ^{242}Cm was first isolated among the products of α-bombardment of ^{239}Pu, and its discovery actually preceded that of americium. Isotopes of the other elements were first identified in products from the first hydrogen bomb explosion (1952) or in cyclotron bombardments.

Ion-exchange methods have been indispensible in the separation of the elements following americium (often called the trans-americium elements) and also for tracer quantities of Np, Pu, and Am. By comparison with the elution of lanthanide ions, where La is eluted first and Lu last (page 450), and by extrapolating data for Np^{3+} and Pu^{3+} the order of elution of the ions can be predicted accurately. Even a few *atoms* of an element can be identified because of the characteristic nuclear radiation.

The actinides as a group may be separated from lanthanides (always present as fission products from irradiations that produce the actinides) by use of concentrated HCl or 10 *M* LiCl. This is because the actinide ions more readily form chloro

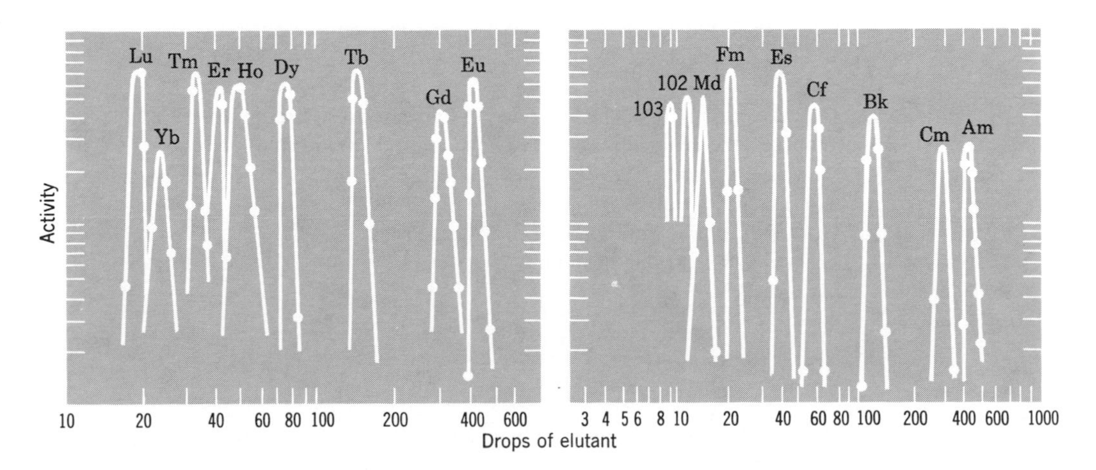

Figure 27-3 *Elution of lanthanide +3 ions (left) and actinide +3 ions (right) from Dowex 50 cation-exchange resin. Buffered ammonium 2-hydroxybutyrate was the elutant. The predicted positions of elements 102 and 103 (unobserved here) are also shown here. (Reproduced by permission from J. J. Katz and G. T. Seaborg,* The Chemistry of the Actinide Elements, *Methuen, London, 1975.)*

complex anions than lanthanides. Hence, actinides can be removed from cation exchange resins, or conversely, adsorbed on anion exchange resins. There is also, in addition to the group separation, some separation of Pu, Am, Cm, Bk, and Cf—Es.

The actinide ions are usually separated from each other by elution with citrate or similar elutant; some typical elution curves in which the relative positions of the corresponding lanthanides are also given are shown in *Fig. 27-3*. Observe that a striking similarity occurs in the spacings of corresponding elements in the two series. There is a distinct break between Gd and Tb and between Cm and Bk, which can be attributed to the small change in ionic radius occasioned by the half-filling of the 4*f* and 5*f* shells, respectively. The elution order is not always as regular as that in *Fig. 27-3*.

After separation by ion exchange, macro amounts of the actinides can be precipitated by F^- or oxalate; tracer quantities can be collected by using La^{3+} carrier.

Solid compounds of Cm, Bk, Cf, and Es, mainly oxides and halides, have been characterized.

Study Questions

A

1. Name the actinide elements and list their electronic configurations.
2. List the oxidation states for actinide elements.
3. Which actinide isotopes can be obtained in macroscopic amounts?
4. What are characteristic reactions of actinide +3 and +4 ions?
5. Which +3 ion has its 6*f* shell half-full? What oxidation states do the preceding and succeeding elements show?
6. Which actinide element corresponds to Lu?
7. How are actinide metals made? What are their main features?
8. What is the structure of the dioxo ions MO_2^{2+} in, for example, uranyl nitrate hydrate?
9. How is actinium made? Which element does it most resemble?
10. What are the main sources of (a) thorium, and (b) uranium?
11. Uranium is usually recovered as uranyl nitrate. How is this converted to the metal?
12. What are the properties and main use of UF_6?
13. How is uranium hydride obtained? What are its uses?
14. What elements would the unknown elements 105, 107, 112, and 118 be expected to resemble?

B

1. What are the main principles upon which the separations of Np, Pu, and Am from U are made?
2. Describe the lanthanum fluoride cycle for separation of Np or Pu from U.
3. Describe the tributylphosphate extraction separation of Np and Pu from U.
4. How are the elements Am-Lw usually separated? Why is it first necessary to separate lanthanides as a group from the actinides as a group and how is this done?
5. Compare and contrast the chemistries of the dioxo ions of U, Np, Pu, and Am.

Chapter 27
Study Guide

Supplementary Reading

Bagnall, K. W., ed., *Lanthanides and Actinides*, Butterworth, 1972.
Bagnall, K. W., *The Actinide Elements*, Elsevier, 1972.
Cleveland, J. M., *The Chemistry of Plutonium*, Gordon and Breach, 1970.
Cordfunke, E. H. P., *The Chemistry of Uranium*, Elsevier, 1969.
Katz, J. J. and Seaborg, G. T., *The Chemistry of the Actinide Elements*, Methuen, 1957.
Seaborg, G. T., *Man-made Transuranium Elements*, Prentice-Hall, 1963.
Taube, M., *Plutonium: A General Survey*, Verlag Chemie, 1974.

4

Some Special Topics

28

complexes of π-acceptor (π-acid) ligands

28-1
Introduction

A characteristic feature of the *d* group transition metal atoms is their ability to form complexes with a variety of neutral molecules such as carbon monoxide, isocyanides, substituted phosphines, arsines and stibines, nitric oxide, and various molecules with delocalized π orbitals, such as pyridine, 2,2′-bipyridine and 1,10-phenanthroline. Very diverse types of complex exist, ranging from binary molecular compounds such as $Cr(CO)_6$ or $Ni(PF_3)_4$ to complex ions such as $[Fe(CN)_5CO]^{3-}$, $[Mo(CO)_5I]^-$, $[Mn(CNR)_6]^+$, and $[Vphen_3]^+$.

In many of these complexes, the metal atoms are in low-positive, zero or even negative *formal* oxidation states. It is a characteristic of the ligands now under discussion that they can stabilize low oxidation states. This property is associated with the fact that these ligands have vacant π orbitals in addition to lone-pairs. These vacant orbitals accept electron density from filled metal orbitals to form a type of π bonding that supplements the σ bonding arising from lone-pair donation. High electron density on the metal atom—of necessity in low oxidation states—can thus be *delocalized onto the ligands*. The ability of ligands to accept electron density into low-lying empty π orbitals is called π acidity. The word acidity is used in the Lewis sense.

The stoichiometries of most complexes of π-acid ligands can be predicted by use of the noble gas formalism. This requires that the number of valence electrons possessed by the metal atom plus the number of pairs of σ electrons contributed by the ligands be equal to the number of electrons in the succeeding noble gas atom. The basis for this rule is the tendency of the metal atom to use its valence orbitals, nd, $(n + 1)s$ and $(n + 1)p$, as fully as possible, in forming bonds to ligands. Although it is of considerable utility in the design of new compounds, particularly of metal carbonyls, nitrosyls and isocyanides, and their substitution products, it is by no means infallible. It fails altogether for the bipyridine and dithiolene type of ligand, and there are significant exceptions even among carbonyls, such as $V(CO)_6$ and $[Mo(CO)_2(diphos)_2]^+$.

Table 28-1 Some Representative Metal Carbonyls and Carbonyl Hydrides

Compound	Color and form	Structure	Comments
A. *Mononuclear Carbonyls*			
$V(CO)_6$	Black crystals; dec. 70°; sublimes in vacuum	Octahedral	Yellow-orange in solution; paramagnetic ($1e^-$)
$Cr(CO)_6$, $Mo(CO)_6$, $W(CO)_6$	Colorless crystals; all sublime in vacuum	Octahedral	Stable to air; dec. 180–200°
$Fe(CO)_5$	Yellow liquid; mp −20° bp 103°	*tbp*	Action of uv gives $Fe_2(CO)_9$
$Ru(CO)_5$	Colorless liquid; mp −22°	*tbp* (by ir)	Very volatile and difficult to prepare
$Ni(CO)_4$	Colorless liquid; mp −25° bp 43°	Tetrahedral	Very toxic; musty smell; flammable; decomposes readily to metal
B. *Polynuclear Carbonyls*			
$Mn_2(CO)_{10}$	Yellow solid mp 151° Subl. 50° (10^{-2} mm)	See *Fig. 28-2*	The Mn—Mn bond is long (2.93 Å) and $Mn_2(CO)_{10}$ is reactive
$Fe_2(CO)_9$	Gold solid mp 100° decomp.	See *Fig. 28-2*	Very insoluble and nonvolatile
$Fe_3(CO)_{12}$	Green-black solid mp 140–150 decomp.	See *Fig. 28-2*	Moderately soluble
$Rh_4(CO)_{12}$	Brick red solid mp 150° decomp. Subl. 65° (10^{-2} mm)	See *Fig. 28-2*	Useful reagent for many carbonyl rhodium compounds
C. *Carbonyl Hydrides*[a]			
$HMn(CO)_5$	Colorless liquid mp −25°	Octahedral	Stable at 25°, weak acid $\tau = 17.5$
$H_2Fe(CO)_4$	Yellow liquid, colorless gas mp −70°	Uncertain	Decomp. −10°. Weak acid $\tau = 21.1$
$H_2Fe_3(CO)_{11}$	Dark red liquid	Uncertain	
$HCo(CO)_4$	Yellow liquid, colorless gas mp −20°	Distorted *tbp*	Decomp. above mp strong acid $\tau = 20$

[a] τ value is position of high-resolution proton magnetic resonance line in parts per million referred to tetramethylsilane reference as 10.00.

CARBON MONOXIDE COMPLEXES

The most important π-acceptor ligand is carbon monoxide. Many carbonyl complexes are of considerable structural interest as well as of importance industrially and in catalytic and other reactions. Carbonyl derivatives of at least one type are known for all of the transition metals. The first metal carbonyls, $Ni(CO)_4$ and $Fe(CO)_5$, were discovered by A. Mond in 1890 and 1891; he developed an industrial process for the isolation of pure nickel based on the formation and subsequent thermal decomposition of the volatile $Ni(CO)_4$.

28-2
Mononuclear Metal Carbonyls

The simplest carbonyls are of the type $M(CO)_x$ (see Table 28-1). The compounds are all hydrophobic, volatile, and soluble to varying degrees in nonpolar liquids.

28-3
Polynuclear Metal Carbonyls

In addition to the ones mentioned above there are numerous polynuclear carbonyls that may be homonuclear, for example, $Fe_3(CO)_{12}$, or heteronuclear, for example, $MnRe(CO)_{10}$. In these compounds there are not only linear M—C—O groups but also either metal—metal bonds alone or M—M bonds plus *bridging carbonyl* groups. The two principal types of bridging group are depicted in *Fig. 28-1*.

Figure 28-1 *The two main types of bridging* CO *groups. (a) Doubly bridging. (b) Triply bridging.*

The doubly bridging types occur fairly frequently and always in conjunction with an M—M bond.

Some important polynuclear carbonyls are listed in Table 28-1 and their structures and those of others are shown in *Fig. 28-2*.

Bridging CO groups very often occur in pairs, as in (28-Ia). Any pair of bridging CO groups can only be regarded as an alternative to a nonbridged arrangement

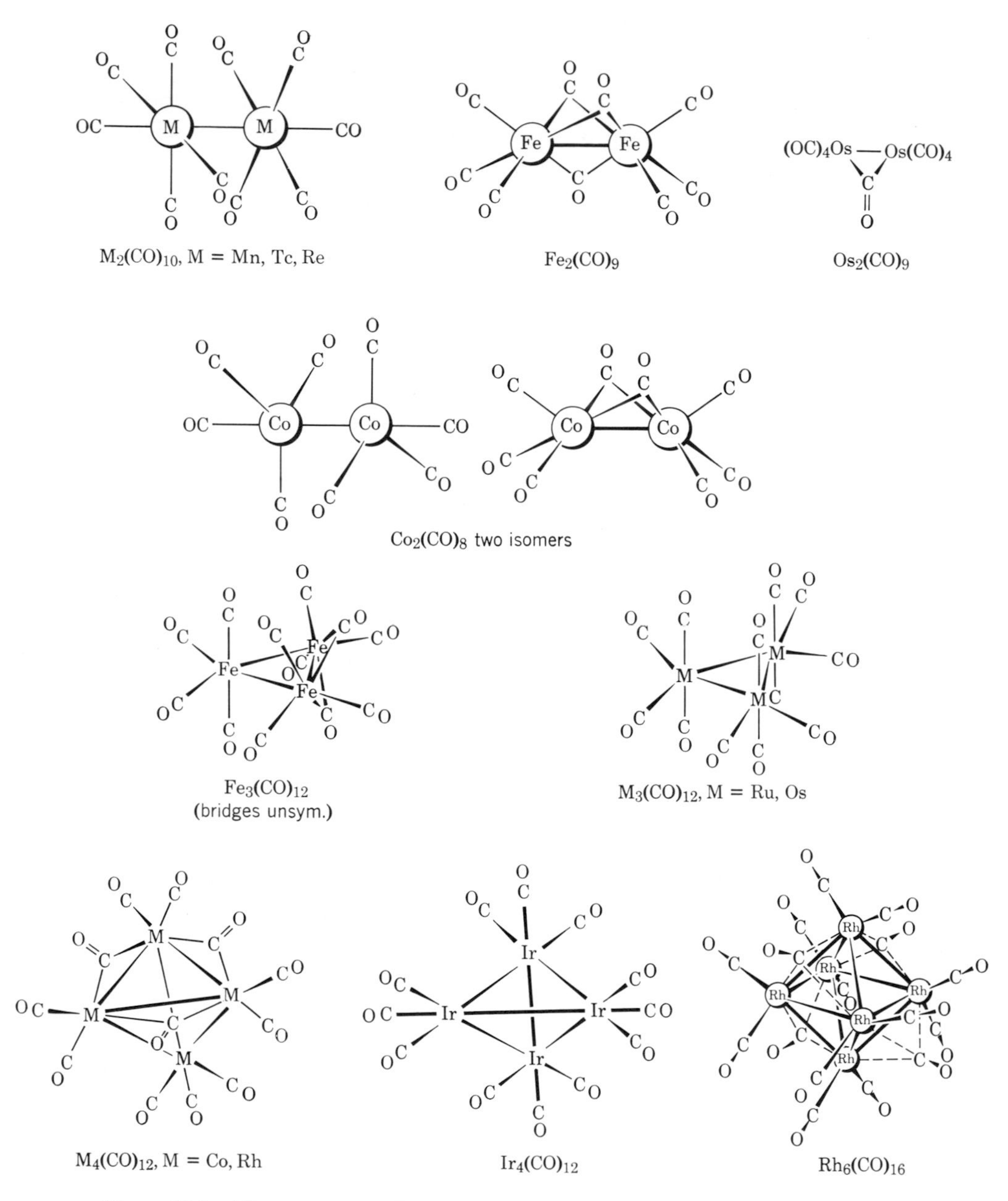

Figure 28-2 *The structures of some polynuclear metal carbonyls. The detailed structure of* $Os_2(CO)_9$ *is not known.*

with two terminal groups, as in (28-Ib):

28-Ia versus 28-Ib

The relative stabilities of the alternatives appear to depend primarily on the size of the metal atoms. The larger the metal atoms the greater is the preference for a nonbridged structure. Thus, in any group the relative stability of nonbridged structures increases as the group is descended. For example, $Fe_3(CO)_{12}$ has two bridging CO's while $Ru_3(CO)_{12}$ and $Os_3(CO)_{12}$ have none. The generalization concerning metal atom size also covers the trend horizontally in the periodic table. Thus, the large Mn atoms form only the nonbridged $(OC)_5Mn—Mn(CO)_5$ molecule whereas the dinuclear cobalt carbonyl, $Co_2(CO)_8$, exists as an equilibrium mixture of the bridged and nonbridged structures.

Carbonyl groups less commonly bridge triangular arrays of three metal atoms (*Fig. 28-1b*) as in $Rh_6(CO)_{16}$ (*Fig. 28-2*).

The presence of bridging CO groups can often be recognized from the infrared spectra of the compounds (see below).

28-4
Stereochemical Nonrigidity in Carbonyls

It is very common for bi- and polynuclear metal carbonyls to undergo rapid intramolecular rearrangements in which CO ligands are scrambled over the two or more metal atoms. These scrambling processes are observed and studied by nuclear magnetic resonance spectroscopy.

In many binuclear compounds the mechanism of scrambling has as its key steps the opening and closing of pairs of bridges, as is illustrated in the following two cases, where Cp represents the C_5H_5 group, which we discuss in detail in Chapter 29.

A more elaborate example is presented by $Rh_4(CO)_{12}$ in which the 12CO ligands move rapidly over the entire tetrahedral skeleton in a series of steps, each involving the concerted opening or closing of a set of three bridges, as shown below:

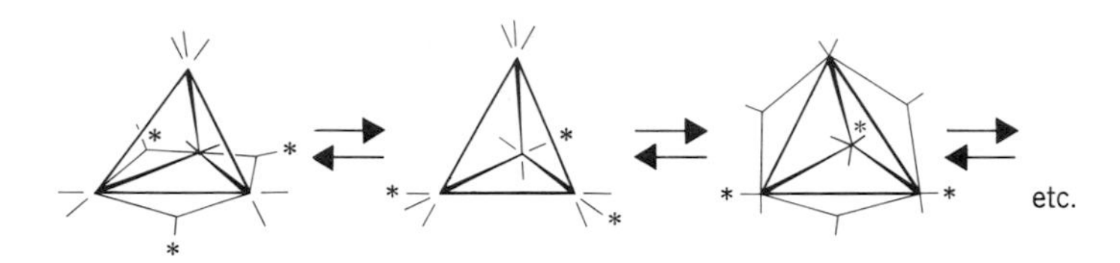

The ease with which these processes proceed in nearly all cases is attributable to the fact that in most polynuclear carbonyls the bridged and nonbridged structures differ very little in energy and, thus, whichever one is the ground state [cf. the $Cp_2Fe_2(CO)_4$ and $Cp_2Mo_2(CO)_6$ cases] the other provides an energetically accessible intermediate for the scrambling. In the examples cited, the rates at which the individual steps occur at room temperature are in the range of 10 to 10^3 per second. Thus, in the course of any ordinary chemical reaction, complete scrambling will occur—many times over.

28-5 Preparation of Metal Carbonyls

Although many metals will react directly with CO when prepared in a highly dispersed form, only $Ni(CO)_4$ and $Fe(CO)_5$ are normally made this way. Finely divided nickel will react at room temperature; an appreciable rate of reaction with iron requires elevated temperatures and pressures.

In general, carbonyls are formed when metal compounds are reduced in the presence of CO. Usually high pressures (200–300 atm) of CO are required. In some cases, CO itself serves as the only necessary reducing agent, for example,

$$Re_2O_7 + 17CO \longrightarrow Re_2(CO)_{10} + 7CO_2$$

but usually an additional reducing agent is needed, typical ones being H_2, metals such as Na, Al, Mg, Cu, or compounds such as a trialkylaluminum or $Ph_2CO^-Na^+$:

$$2CoCO_3 + 2H_2 + 8CO \xrightarrow[120\text{–}150^\circ]{250\text{–}300\text{ atm}} Co_2(CO)_8 + 2CO_2 + 2H_2O$$

$$2Mn(AcAc)_2 + 10CO \xrightarrow{(C_2H_5)_3Al} Mn_2(CO)_{10}$$

$$CrCl_3 + 6CO \xrightarrow{C_6H_5MgBr} Cr(CO)_6$$

The reaction mechanisms are obscure but when Na, Mg, or Al are used reduction to metal probably occurs. When organometallic reducing agents are employed, unstable organo derivatives of the transition metal may be formed as intermediates.

28-6
Bonding in Linear M—C—O Groups

The fact that refractory metals, with high heats of atomization (~400 kJ mol^{-1}), and an inert molecule like CO are capable of uniting to form stable, molecular compounds is quite surprising, especially when the CO molecules retain their individuality. Moreover, the Lewis basicity of CO is *negligible*. However, the explanation lies in the multiple nature of the M—CO bond, for which there is much evidence, some of it semiquantitative.

Although we can formulate the bonding in terms of a resonance hybrid of (28-IIa) and (28-IIb), a molecular-orbital formulation is more detailed and accurate. There is, first, a dative overlap of the filled carbon σ orbital (*Fig. 28-3a*) and,

$$\overset{-}{\text{M}}\text{—}\overset{+}{\text{C}}\equiv\text{O:} \longleftrightarrow \text{M}=\text{C}=\ddot{\text{O}}\text{:}$$

28-IIa 28-IIb

second, a dative overlap of a filled $d\pi$ or hybrid $dp\pi$ metal orbital with an empty antibonding $p\pi$ orbital of the carbon monoxide (*Fig. 28-3b*). This bonding mechanism is synergic, since the drift of metal electrons into CO orbitals will tend to make the CO as a whole negative and, hence, to increase its basicity via the σ orbital of carbon; also the drift of electrons to the metal in the σ bond tends to make the CO positive, thus enhancing the acceptor strength of the π orbitals. Thus, the effects of σ-bond formation strengthen the π bonding and vice versa.

The main lines of physical evidence showing the multiple nature of the M—CO bonds are bond lengths and vibrational spectra. According to the preceding description of the bonding, as the extent of back-donation from M to CO increases, the M—C bond becomes stronger and the C≡O bond becomes weaker. Thus the multiple bonding should be evidenced by shorter M—C and longer C—O bonds as compared with M—C single bonds and C≡O triple bonds, respectively. Although C—O bond lengths are rather insensitive to bond order, for M—C bonds in selected compounds there *is* appreciable shortening consistent with the π-bonding concept.

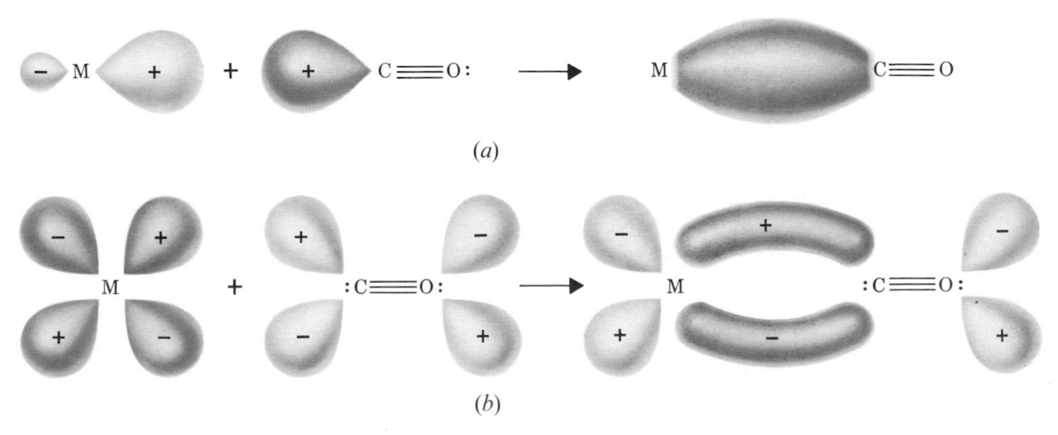

Figure 28-3 *(a) The formation of the metal ← carbon σ bond using an unshared pair on the C atom. (b) The formation of the metal → carbon π bond. The other orbitals on the CO are omitted for clarity.*

28-7
Vibrational Spectra of Metal Carbonyls

Infrared spectra have been widely used in the study of metal carbonyls since the C—O stretching frequencies give very strong sharp bands well separated from other vibrational modes of any other ligands that may also be present.

The CO molecule has a stretching frequency of 2143 cm^{-1}. Terminal CO groups in neutral metal carbonyl molecules are found in the range 2125 to 1850 cm^{-1}, showing the reduction in CO bond orders. Moreover, when changes are made that should increase the extent of M—C back-bonding, the CO frequencies are shifted to even lower values. Thus, if some CO groups are replaced by ligands with low or negligible back-accepting ability, those CO groups that remain must accept more $d\pi$ electrons from the metal to prevent the accumulation of negative charge on the metal atom. Hence, the frequency for $Cr(CO)_6$ is *ca.* 2000 cm^{-1} (exact values vary with phase and solvent) whereas, when three CO's are replaced by amine groups which have essentially no ability to back-accept, as in $Cr(CO)_3(dien)$, dien = $NH(CH_2CH_2NH_2)_2$, there are two CO stretching modes with frequencies of ~1900 and ~1760 cm^{-1}. Similarly, in $V(CO)_6^-$, where more negative charge must be taken from the metal atom, a band is found at ~1860 cm^{-1} corresponding to the one found at ~2000 cm^{-1} in $Cr(CO)_6$. Conversely, a change that would tend to inhibit the shift of electrons from metal to CO π orbitals, such as placing a positive charge on the metal, should cause the CO frequencies to rise, for example,

$Mn(CO)_6^+$, ~ 2090	$Mndien(CO)_3^+$, ~ 2020, ~ 1900
$Cr(CO)_6$, ~ 2000	$Crdien(CO)_3$, ~ 1900, ~ 1760
$V(CO)_6^-$, ~ 1860	

The most important use of infrared spectra of CO compounds is in *structural* diagnosis, whereby bridging and terminal CO groups can be recognized.

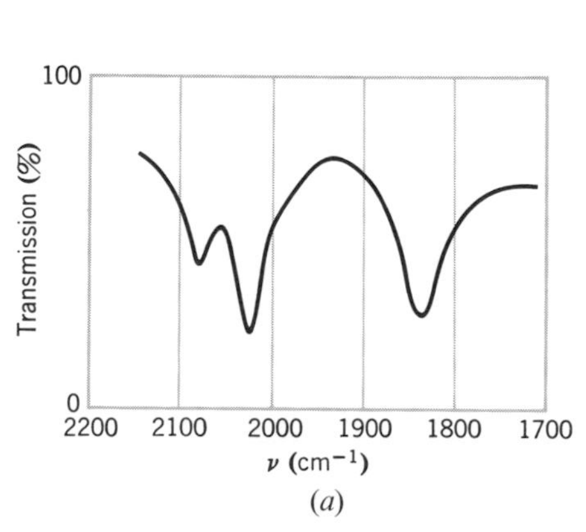

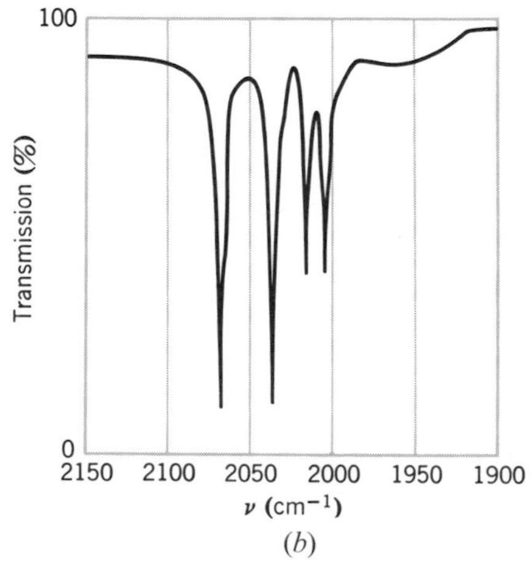

Figure 28-4 *The infrared spectra in the CO stretching region of (a) solid* $Fe_2(CO)_9$, *and (b)* $Os_3(CO)_{12}$ *in solution. Notice the greater sharpness of the solution spectra. The most desirable spectra are those obtained in nonpolar solvents or in the gas phase.*

For terminal M—CO the frequencies of C—O stretches range from 1850 to 2125 cm^{-1} but for bridging CO groups the range is 1750 to 1850 cm^{-1}. *Figure 28-4* shows how these facts may be used to infer structures. Observe that $Fe_2(CO)_9$ has strong bands in both the terminal and the bridging regions. From this alone it could be inferred that the structure must contain both types of CO groups; X-ray study shows that this is true. For $Os_3(CO)_{12}$ several structures consistent with the general rules of valence can be envisioned; some of these would have bridging CO groups, while the actual one (*Fig. 28-2*) does not. The infrared

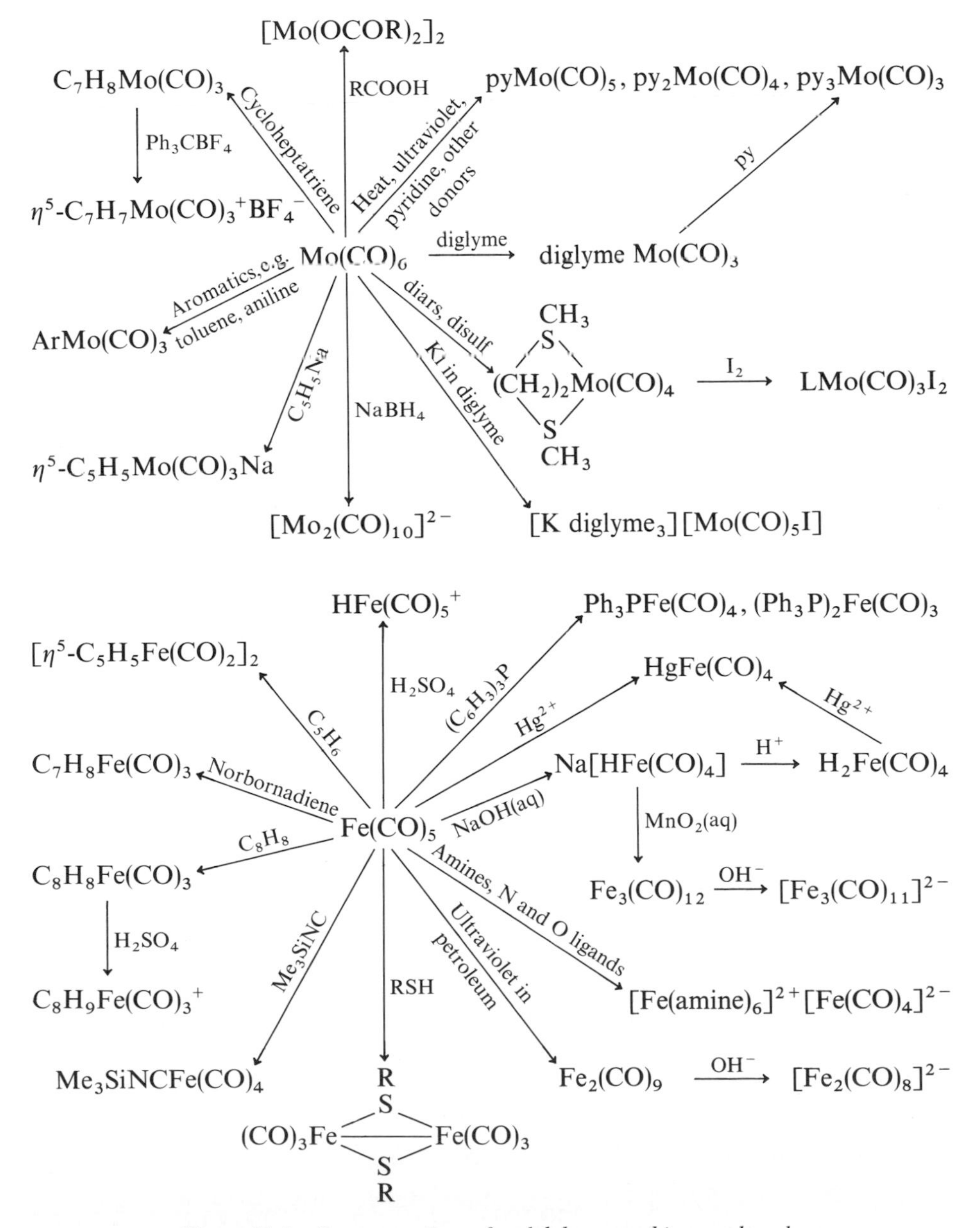

Figure 28-5 *Some reactions of molybdenum and iron carbonyls.*

spectrum alone (*Fig. 28-4b*) shows that no structure with bridging CO groups is acceptable, since there is no absorption band below 2000 cm^{-1}.

In using the *positions* of CO stretching bands to infer the presence of bridging CO groups, certain conditions must be remembered. The frequencies of terminal CO stretches can be quite low if (a) there are ligands present that are good donors but poor π acceptors, or (b) there is a net negative charge on the species. In either case, back-donation to the CO groups becomes very extensive, thus increasing the M—C bond orders, decreasing the C—O bond orders, and driving the CO stretching frequencies down.

28-8 Reactions of Metal Carbonyls

The variety of reactions of the various carbonyls is so large that only a few types can be mentioned. For $Mo(CO)_6$ and $Fe(CO)_5$, *Fig. 28-5* gives an indication of the extensive chemistry that any individual carbonyl typically has.

The most important general reactions of carbonyls are those in which CO groups are displaced by ligands, such as PX_3, PR_3, $P(OR)_3$, SR_2, NR_3, OR_2, RNC, etc., or unsaturated organic molecules, such as C_6H_6 or cycloheptatriene. Derivatives of organic molecules are discussed in Chapter 29.

Another important general reaction is that with bases (OH^-, H^-, NH_2^-) leading to the carbonylate anions that are discussed below.

Substitution reactions may proceed by either thermal or photochemical activation. In some instances, only the photochemical reaction is practical. Generally, the photochemical process involves first expulsion of a CO group after absorption of a photon, followed by entry of the substituent into the coordination system. For example,

$$Cr(CO)_6 \xrightarrow[-CO]{h\nu} Cr(CO)_5 \xrightarrow{+L} Cr(CO)_5L$$

28-9 Carbonylate Anions and Carbonyl Hydrides

Carbonylate anions are formed when carbonyls are treated with aqueous or alcoholic alkali hydroxide or with amines, sulfoxides, or other Lewis bases, when metal—metal bonds are cleaved with sodium, or when certain carbonyls are refluxed with salts in an ether. Illustrative examples are

$$Fe(CO)_5 + 3NaOH(aq) \longrightarrow Na[HFe(CO)_4](aq) + Na_2CO_3(aq) + H_2O$$

$$Co_2(CO)_8 + 2Na/Hg \xrightarrow{THF} 2Na[Co(CO)_4]$$

$$Mn_2(CO)_{10} + 2Li \xrightarrow{THF} 2Li[Mn(CO)_5]$$

$$2Co^{2+}(aq) + 11CO + 12OH^- \xrightarrow{KCN(aq)} 2[Co(CO)_4]^- + 3CO_3^{2-} + 6H_2O$$

The stoichiometries of the simpler carbonylate ions obey the noble-gas formalism. Most of them are readily oxidized by air. The alkali-metal salts are soluble in water, from which they can be precipitated by large cations such as $[Ph_4As]^+$.

There are also many polynuclear species; those of iron have been much studied. They are obtained by reactions such as

$$Fe_2(CO)_9 + 4OH^- \longrightarrow [Fe_2(CO)_8]^{2-} + CO_3^{2-} + 2H_2O$$

$$Fe(CO)_5 + Et_3N \xrightarrow{H_2O,\ 80^\circ} [Et_3NH][HFe_3(CO)_{11}]$$

An important general reaction of carbonylate anions or substituted carbonylate ions is with halogen compounds. By this reaction, metal—carbon or metal—metal bonds can be formed. Typical examples are

$$Mn(CO)_5^- + ClCH_2CH{=}CH_2 = (CO)_5MnCH_2CH{=}CH_2 + Cl^-$$

$$Fe(CO)_4^{2-} + 2Ph_3PAuCl \longrightarrow (Ph_3PAu)_2Fe(CO)_4 + 2Cl^-$$

$$Co(CO)_4^- + Mn(CO)_5Br \longrightarrow (OC)_4CoMn(CO)_5 + Br^-$$

Carbonyl hydrides. In some cases, hydrides corresponding to the carbonylate anions can be isolated. A few of them are listed in Table 28-1 along with their main properties.

Carbonyl hydrides are usually rather unstable. They can be obtained by acidification of the appropriate alkali carbonylates or in other ways. Examples of the preparations are

$$NaCo(CO)_4 + H^+(aq) \longrightarrow HCo(CO)_4 + Na^+(aq)$$

$$Fe(CO)_4I_2 \xrightarrow{NaBH_4 \text{ in THF}} H_2Fe(CO)_4$$

$$Mn_2(CO)_{10} + H_2 \xrightarrow[200^\circ]{200 \text{ atm}} 2HMn(CO)_5$$

$$Co + 4CO + \tfrac{1}{2}H_2 \xrightarrow[150^\circ]{50 \text{ atm}} HCo(CO)_4$$

The hydrides are slightly soluble in water where they behave as acids, ionizing to give the carbonylate ions:

$$HMn(CO)_5 = H^+ + [Mn(CO)_5]^- \qquad pK \sim 7$$

$$H_2Fe(CO)_4 = H^+ + [HFe(CO)_4]^- \qquad pK_1 \sim 4$$

$$[HFe(CO)_4]^- = H^+ + [Fe(CO)_4]^{2-} \qquad pK_2 \sim 13$$

$$HCo(CO)_4 = H^+ + [Co(CO)_4]^- \qquad \text{strong acid}$$

The carbonyl hydrides have sharp M—H stretching bands in the infrared and proton magnetic resonance absorptions at very high τ values, as mentioned in Table 28-1. The hydrogen atom occupies a regular place in the coordination polyhedron and the M—H distances are approximately equal to the values expected from the sum of single-bond covalent radii. A good example is afforded by the structure of $HMn(CO)_5$, shown in *Fig. 28-6*.

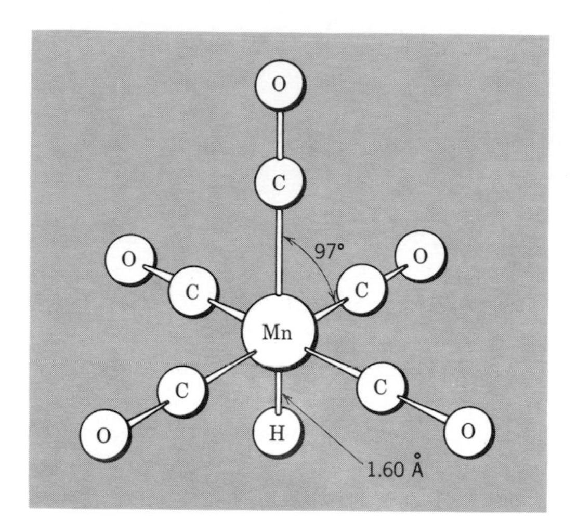

Figure 28-6 *The structure of the* $HMn(CO)_5$ *molecule, showing both the stereochemical activity of the hydrogen atom and the metal-to-hydrogen distance which approximates to the sum of normal covalent radii.*

28-10
Carbonyl Halides and Related Compounds

Carbonyl halides, $M_x(CO)_yX_z$, are known for most of the elements forming binary carbonyls and also for Pd, Pt, Au, Cu^I, and Ag^I. They are obtained either by the direct interaction of metal halides and carbon monoxide, usually at high pressure, or by the cleavage of polynuclear carbonyls by halogens:

$$Mn_2(CO)_{10} + Br_2 \xrightarrow{40^\circ} 2Mn(CO)_5Br \underset{\text{CO, 150 atm}}{\overset{\text{in petrol at } 120^\circ}{\rightleftarrows}} [Mn(CO)_4Br]_2 + 2CO$$

$$RuI_3 + 2CO \xrightarrow{220^\circ} [Ru(CO)_2I_2]_n + \tfrac{1}{2}I_2$$

$$2PtCl_2 + 2CO \longrightarrow [Pt(CO)Cl_2]_2$$

Examples of the halides and some of their properties are listed in Table 28-2. Carbonyl halide anions are also known; they are often derived by reaction of ionic halides with metal carbonyls or substituted carbonyls:

$$M(CO)_6 + R_4N^+X^- \xrightarrow{\text{diglyme}} R_4N^+[M(CO)_5X]^- + CO \qquad M = Cr, Mo, W$$

$$Mn_2(CO)_{10} + 2R_4N^+X^- \longrightarrow (R_4N^+)_2[Mn_2(CO)_8X_2]^{2-} + 2CO$$

Table 28-2 *Some Examples of Carbonyl Halide Complexes*

Compound	Form	Melting point (°C)	Comment
$Mn(CO)_5Cl$	Pale yellow crystals	Sublimes	Loses CO at 120° in organic solvents; can be substituted by pyridine, etc.
$[Re(CO)_4Cl]_2$	White crystals	Dec. >250	Halogen bridges cleavable by donor ligands or by CO (pressure)
$[Ru(CO)_2I_2]_n$	Orange powder	Stable >200	Halide bridges cleavable by ligands
$[Pt(CO)Cl_2]_2$	Yellow crystals	195; sublimes	Hydrolyzed H_2O; PCl_3 replaces CO

Dimeric or polymeric carbonyl halides are invariably bridged through the halogen atoms and not by carbonyl bridges, for example, in (28-III) and (28-IV).

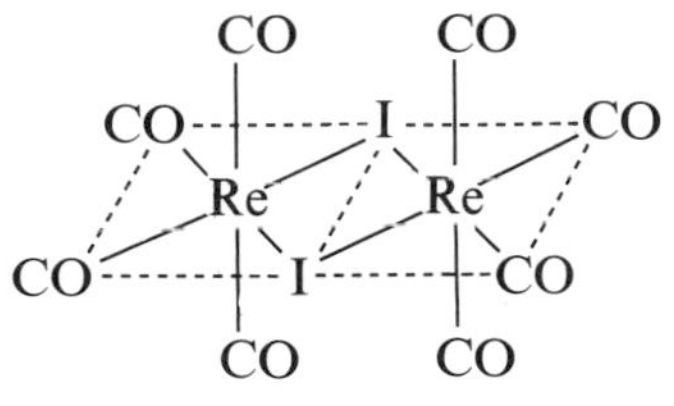

28-III

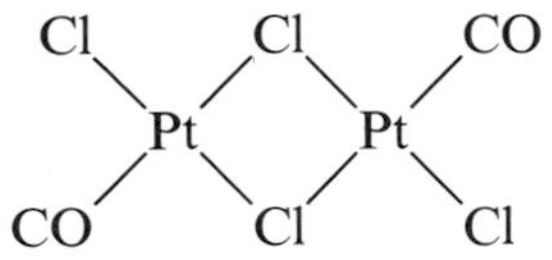

28-IV

The halogen bridges can be broken by numerous donor ligands such as pyridine, substituted phosphines, isocyanides, etc., as in the following reaction:

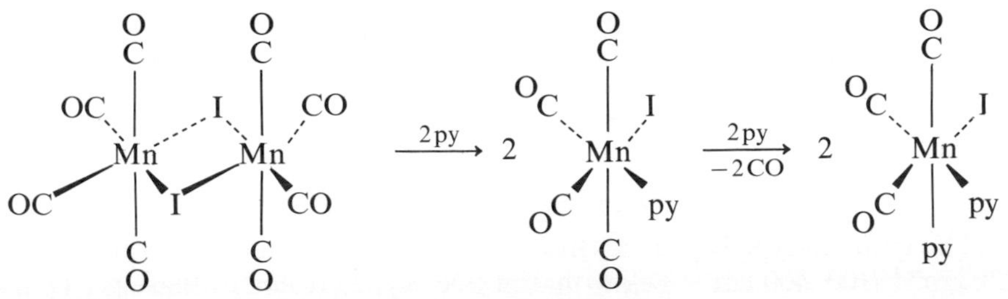

$$\xrightarrow{2py} 2 \qquad \xrightarrow[-2CO]{2py} 2$$

28-V

28-VI

CARBON MONOXIDE ANALOGS

28-11
Isocyanide Complexes

An isocyanide, $R—N\equiv C$:, is very similar electronically to $:O\equiv C:$, and there are many isocyanide complexes stoichiometrically analogous to metal carbonyls. Isocyanides can occupy bridging as well as terminal positions. Examples are such crystalline air-stable compounds as red $Cr(CNPh)_6$, white $[Mn(CNCH_3)_6]I$, and orange $Co(CO)(NO)(CNC_7H_7)_2$, all of which are soluble in benzene.

Isocyanides generally appear to be stronger σ donors than CO, and various complexes such as $[Ag(CNR)_4]^+$, $[Fe(CNR)_6]^{2+}$, and $[Mn(CNR)_6]^{2+}$ are known where π bonding is of relatively little importance; derivatives of this type are not known for CO. However, the isocyanides are capable of extensive back-acceptance of π electrons from metal atoms in low oxidation states. This is indicated qualitatively by their ability to form compounds such as $Cr(CNR)_6$ and $Ni(CNR)_4$, analogous to the carbonyls and more quantitatively by $C\equiv N$ stretching frequencies which, like CO stretching frequencies are markedly lowered when the ligand acts as a π acid.

28-12
Dinitrogen (N_2) Complexes

The fact that CO and N_2 (hereafter called dinitrogen) are isoelectronic has for years led to speculation as to the possible existence of M—NN bonds analogous to M—CO bonds, but it was only in 1965 that the first example, $[Ru(NH)_3)_5N_2]Cl_2$, was reported. Subsequent work has shown that the $[Ru(NH_3)_5N_2]^{2+}$ cation can be obtained in a number of ways, for example,

by reaction of N_2H_4 with aqueous $RuCl_3$;
by reaction of NaN_3 with $[Ru(NH_3)_5(H_2O)]^{3+}$;
by reaction of N_2 with $[Ru(NH_3)_5H_2O]^{3+}$;
by reaction of $RuCl_3(aq)$ with Zn in $NH_3(aq)$.

Of these the direct reaction with N_2 to displace H_2O is perhaps most notable. Despite early reports to the contrary, no effective way of reducing coordinated N_2 into NH_3 has yet been found. However, there are several systems in which reduction of N_2 to NH_3 is catalyzed by low-valent metal atoms, presumably via transient $M—N_2$ complexes.

A *bridging* N_2 ligand is formed in the reaction

$$[Ru(NH_3)_5Cl]^{2+} \xrightarrow[N_2]{Zn/Hg} \{[Ru(NH_3)_5]_2N_2\}^{4+}$$

The terminal-type N_2 ligands have strong infrared bands in the range 1930 to 2230 cm^{-1} (100–400 cm^{-1} below that of free N_2, 2331 cm^{-1}) that may be used diagnostically.

The formation of N_2 complexes by direct uptake of N_2 gas at 1 atm has been observed, especially with tertiary phosphine ligands in reactions such as

$$Co(acac)_3 + 3Ph_3P + N_2 \xrightarrow{Al(iso\text{-}C_4H_9)_3} Co(H)(N_2)(Ph_3P)_3$$

$$FeCl_2 + 3PEtPh_2 + N_2 \xrightarrow{NaBH_4,\ EtOH} FeH_2(N_2)(PEtPh_2)_3$$

$$MoCl_4(PPhMe_2)_2 + N_2 + 2PPhMe_2 \xrightarrow{Na/Hg\ in\ THF} cis\text{-}Mo(N_2)_2(PPhMe_2)_4$$

Several typical compounds containing M—NN groups have been structurally characterized. The three atom chains are essentially linear, the N—N distances are slightly longer than that in the N_2 molecule, and the M—N distances are short enough to indicate some multiple bond character.,

The bonding in M—N_2 groups is similar to that in terminal M—CO groups. The same two basic components, M $\leftarrow$ N_2 σ-donation and M $\rightarrow$ N_2 π-acceptance, are involved. The major quantitative differences, which account for the lower stability of N_2 complexes, arise from small differences in the energies of the MO's of CO and N_2. It appears that N_2 is weaker than CO in both its σ-donor and π-acceptor functions, which accounts for the poor stability of N_2 complexes in general.

28-13 Thiocarbonyl Complexes

The CS molecule, unlike CO, does not exist under ordinary conditions, although it can be made in dilute gas streams by photolysis of CS_2. Nevertheless, CS can be stabilized by complexing and a few compounds are known. Thus $RhCl(PPh_3)_3$ reacts with CS_2 to give $RhCl(\eta^1\text{-}CS_2)(\eta^2\text{-}CS_2)(PPh_3)_2$, which in methanol gives *trans*-$RhCl(CS)(PPh_3)_2$.

Thiocarbonyl complexes have CS stretches in the region 1270 to 1360 cm^{-1}, depending on the oxidation state of the metal, charge on the complex, and the like, whereas the stretch for CS trapped in a matrix at $-190°$ is at 1274 cm^{-1}. The $d\pi$—$p\pi$ bonding is similar to that for the carbonyls.

28-14 Nitric Oxide Complexes

The NO molecule is similar to CO except that it contains one more electron, which occupies a π^* orbital (cf. page 68). Consistent with this similarity, CO and NO form many comparable complexes although, as a result of the presence of the additional electron, NO also forms a class (bent MNO) with no carbonyl analogs.

Linear, Terminal MNO Groups. We have seen that the CO group reacts with a metal atom that presents an empty σ orbital and a pair of filled $d\pi$ orbitals, as is illustrated in *Fig. 28-3*, to give a linear MCO grouping with a C $\rightarrow$ M σ bond and a significant degree of M $\rightarrow$ C π bonding. The NO group engages in an entirely analogous reaction with a metal atom that may be considered, at least formally, to present an empty σ orbital and a pair of $d\pi$ orbitals containing only three electrons. The full set of four electrons for the $Md\pi \rightarrow \pi^*(NO)$ interactions is thus made up of three electrons from M and one from NO. In effect, NO contributes three electrons to the total bonding configuration under circumstances where CO contributes only two. Thus, for purposes of formal electron "bookkeeping," the ligand NO can be regarded as a three-electron donor in the same sense as the ligand CO is considered a two-electron donor. This leads to the following very useful general rules concerning stoichiometry, which may be applied without specifically allocating the difference in the number of electrons to any particular (i.e., σ or π) orbitals:

1. Compounds isoelectronic with one containing an $M(CO)_n$ grouping are those containing $M'(CO)_{n-1}(NO)$, $M''(CO)_{n-2}(NO)_2$, etc., where M′, M″, etc., have atomic numbers that are 1, 2, . . . , etc., less than M. Some examples are: $(\eta^5\text{-}C_5H_5)CuCO$, $(\eta^5\text{-}C_5H_5)NiNO$; $Ni(CO)_4$, $Co(CO)_3NO$, $Fe(CO)_2(NO)_2$, $Mn(CO)(NO)_3$; $Fe(CO)_5$, $Mn(CO)_4NO$.

2. Three CO groups can be replaced by two NO groups. Examples of pairs of compounds so related are:

$Fe(CO)_5$,	$Fe(CO)_2(NO)_2$
$Mn(CO)_4NO$,	$Mn(CO)(NO)_3$
$Co(CO)_3NO$,	$Co(NO)_3$
$[(OC)_3Fe(\mu\text{-}SEt)_2Fe(CO)_3]$	$[(ON)_2Fe(\mu\text{-}SEt)_2Fe(NO)_2]$

Structural data suggest that under comparable circumstances M—CO and M—NO bonds are about equally strong, but in a chemical sense the M—N bonds appear to be stronger, since substitution reactions on mixed carbonyl nitrosyl compounds typically result in displacement of CO in preference to NO. For example, $Co(CO)_3NO$ reacts with R_3P, X_3P, amine and RNC ligands, invariably to yield the $Co(CO)_2(NO)L$ product.

The NO vibration frequencies for linear MNO groups substantiate the idea of extensive M to N π bonding, leading to appreciable population of NO π^* orbitals. Nitric oxide has its unpaired electron in a π^* orbital; the N—O stretching frequency is 1860 cm^{-1}. For typical linear MNO groups in molecules with small or zero charge, the observed frequencies are in the range 1800 to 1900 cm^{-1}. This indicates the presence of approximately one electron pair shared between metal $d\pi$ and NO π^* orbitals.

Bent, Terminal MNO Groups. It has long been known that NO can form single bonds to univalent groups such as halogens and alkyl radicals, affording the bent species

$$X\text{—}\dot{N}\text{=}\ddot{O}: \quad \text{and} \quad R\text{—}\dot{N}\text{=}\ddot{O}:$$

Metal atoms with suitable electron configurations and partial coordination shells may bind NO in a similar way. This type of NO complex is formed when the incompletely coordinated metal ion, L_nM, would have a $t_{2g}{}^6e_g$ configuration, thus being prepared to form one more single σ bond. The M—N—O angles are in, or near, the range 120 to 140°. Typical compounds are $[Co(NH_3)_5NO]Br_2$ and $IrCl_2(PPh_3)_2NO$.

Bridging NO Groups. These are less common than bridging CO groups, but well established cases of both double and triple bridges are known. As in carbonyls, the bridging NO frequencies are at lower frequencies than terminal ones.

Bridging NO groups are also to be regarded as three-electron donors. The doubly bridging ones may be represented as

$$\dot{\underset{\cdot}{N}}::\ddot{\underset{\cdot\cdot}{O}}$$

where the additional electron required to form two metal-to-nitrogen single bonds is supplied by one of the metal atoms. The situation is formally analogous to that for bridging halogen atoms (page 324).

28-15
Donor Complexes of Group V and Group VI Ligands

Trivalent phosphorus, arsenic, antimony, and bismuth compounds, as well as divalent sulfur and selenium compounds, can give complexes with transition metals. These donors are, of course, quite strong Lewis bases and give complexes with Lewis acids such as BR_3 compounds where d orbitals are not involved. However, the donor atoms do have empty $d\pi$ orbitals and back-acceptance into these orbitals is possible, as is shown in *Fig.* 28-7.

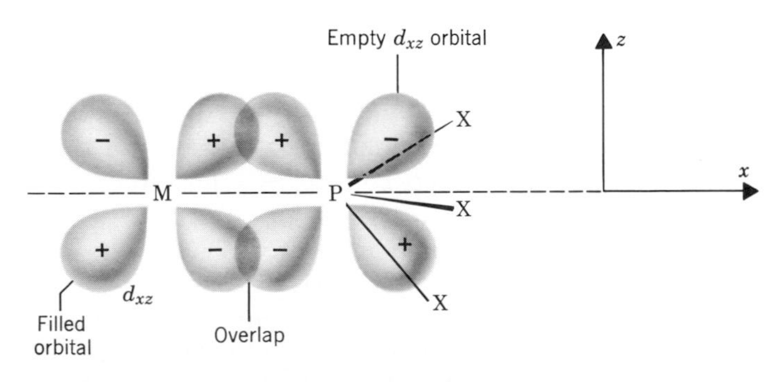

Figure 28-7 *Diagram showing the back-bonding from a filled metal d orbital to an empty phosphorus 3d orbital in the* PX_3 *ligand, taking the internuclear axis as the x axis. An exactly similar overlap occurs in the yz plane using the* d_{yz} *orbitals.*

Based on infrared data an extensive series of ligands involving Group V and Group VI donor atoms can be arranged in the following order of decreasing π acidity:

$$CO \sim PF_3 > PCl_3 \sim AsCl_3 \sim SbCl_3 > PCl_2(OR) > PCl_2R > PCl(OR)_2 > PClR_2 \sim P(OR)_3 > PR_3 \sim AsR_3 \sim SbR_3 \sim SR_2$$

It is noteworthy that ir spectral evidence as well as photoelectron spectroscopy shows that PF_3 is as good or better than CO as a π acid. It is not then surprising that PF_3 forms an extensive group of $M_x(PF_3)_y$ compounds, many of which are analogs of corresponding $M_x(CO)_y$ compounds and some of which, for example, $Pd(PF_3)_4$ and $Pt(PF_3)_4$, actually have no stable carbonyl analogs. Even anions, such as $Co(PF_3)_4^-$, and hydrides, such as $HCo(PF_3)_4$, are known.

The other group V and VI ligands are all capable of replacing some CO groups, to form compounds such as $(R_3P)_3Mo(CO)_3$ or even $(R_3P)_4Mo(CO)_2$, but rarely can they replace all CO groups, because of their inferior π acidity. However, there are some stable molecules such as $Ni(PCl_3)_4$ and $Cr(diphos)_3$.

28-16 Cyanide Complexes

The formation of cyanide complexes is restricted almost entirely to the transition metals of the *d* block and their near neighbors Zn, Cd, and Hg. This suggests that metal—CN π bonding is of importance in the stability of cyanide complexes, and there is evidence of various types to support this. However, the π-accepting tendency of CN^- is much lower than for CO, NO, or RNC. This is, of course, reasonable in view of its negative charge. CN^- is a strong σ donor so that back-bonding does not have to be invoked to explain the stability of its complexes with metals in normal (i.e., II, III) oxidation states. Nonetheless, because of the formal similarity of CN^- to CO, NO, and RNC, it is convenient to discuss its complexes in this chapter.

The majority of cyano complexes have the general formula $[M^{n+}(CN)_x]^{(x-n)-}$ and are anionic, such as $[Fe(CN)_6]^{4-}$, $[Ni(CN)_4]^{2-}$, and $[Mo(CN)_8]^{3-}$. Mixed complexes, particularly of the type $[M(CN)_5X]^{n-}$, where X may be H_2O, NH_3, CO, NO, H, or a halogen, are also well known.

Although bridging cyanide groups might be expected in analogy with those formed by CO, none has been definitely proved. However, linear bridges, M—CN—M, are well known and play an important part in the structures of many crystalline cyanides and cyano complexes. Thus AuCN, $Zn(CN)_2$ and $Cd(CN)_2$ are all polymeric with infinite chains.

The free anhydrous acids corresponding to many cyano anions can be isolated, examples being $H_3[Rh(CN)_6]$ and $H_4[Fe(CN)_6]$. These acids are different from those corresponding to many other complex ions, such as $[PtCl_6]^{2-}$ or $[BF_4]^-$, which cannot be isolated except as hydroxonium (H_3O^+) salts. They are also different from metal carbonyl hydrides in that they contain no metal—hydrogen bonds. Instead, the hydrogen atoms are situated in hydrogen bonds between anions, that is, MCN---H---NCM.

LIGANDS WITH EXTENDED π SYSTEMS

28-17
Bipyridine and Similar Amines

The three ligands bipyridine (bipy) (28-VII), 1,10-phenanthroline (phen) (28-VIII), and terpyridine (terpy) (28-IX) form complexes with a variety of metal atoms in a great range of oxidation states. For metal ions in "normal" oxidation states, the

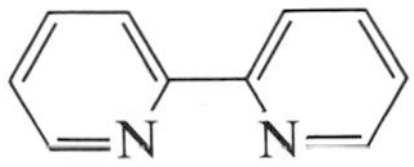

28-VII 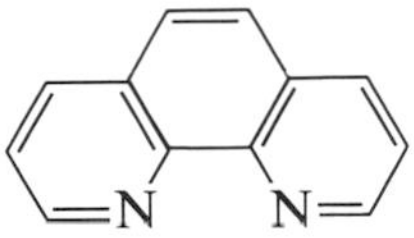 28-VIII 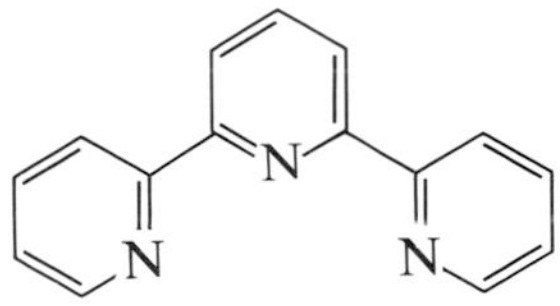 28-IX

interaction of metal $d\pi$ orbitals with the ligand π^* orbitals is significant, but not exceptional. However, these ligands can stabilize metal atoms in low formal oxidation states. In such complexes it is believed that there is extensive occupation of the ligand π^* orbitals, so that the compounds can often be best formulated as having radical anion ligands, L^-.

Complexes of transition-metal ions in "normal" oxidation states can usually be obtained by conventional reactions and then can be reduced with a variety of reagents such as Na/Hg, Mg, or BH_4^-. The most general method employs Li_2bipy:

$$MX_y + yLi_2\text{bipy} + n(\text{bipy}) \xrightarrow{\text{THF}} M(\text{bipy})_n + yLiX + yLi(\text{bipy})$$

The low-valent metal complexes are invariably colored, usually intensely. For those containing transition metals, the bands responsible are believed to be mainly $d \rightarrow \pi^*$ charge-transfer bands.

28-18
1,2-Dithiolene Ligands

The basic types are encompassed by formula (28-X), although several other classes of complexes such as (28-XI), (28-XII) are doubtless related. Moreover, there are many mixed ligand complexes in which 1,2-dithiolene ligands are present along

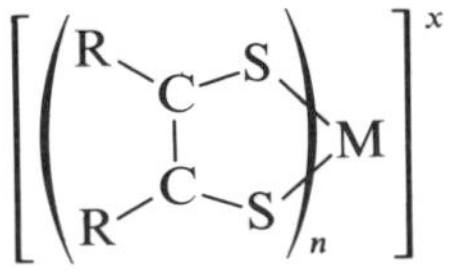

$R = H$, alkyl, C_6H_5, CF_3, CN
$n = 2; x = 0, -1, -2$
$n = 3; x = 0, -1, -2, -3$

28-X

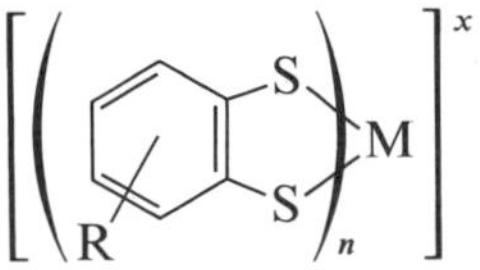

$R =$ alkyl
$n = 2; x = 0, -1, -2$

28-XI

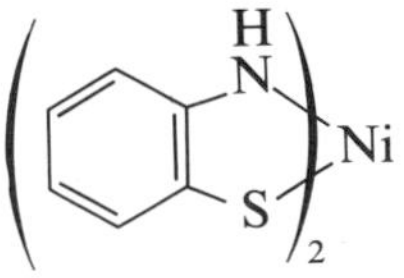

28-XII

with others such as η^5-C_5H_5, NO, olefins, CN^-, O, and the like. In all formulas and structures presented here, lines drawn between atoms are for guidance but do not necessarily indicate valences or bond orders, since the chief property of these systems is the lack of simple descriptions of the bonding.

These compounds, particularly those of types (28-X) and (28-XI), can be prepared by preparative reactions like those shown in the following equations:

$$NiCl_2 + Na_2{}^+[(NC)C(S)C(S)(CN)]^{2-}$$

$$\big\downarrow (C_2H_5)_4N^+$$

$$[(C_2H_5)_4N]^+\left[\left(\begin{matrix} NC-C-S \\ | \\ NC-C-S \end{matrix}\right)_2 Ni\right]^{2-}$$

$$Ni(CO)_4 + 2(C_6H_5)_2C_2 + 4S \longrightarrow \left(\begin{matrix} C_6H_5-C-S \\ | \\ C_6H_5-C-S \end{matrix}\right)_2 Ni$$

$$\swarrow p\text{-}H_2NC_6H_4NH_2 \; / \; (C_2H_5)_4N^+$$

$$[(C_2H_5)_4N]^+\left[\left(\begin{matrix} C_6H_5-C-S \\ | \\ C_6H_5-C-S \end{matrix}\right)_2 Ni\right]^{-}$$

$$Ni(CO)_4 + 2 \begin{matrix} F_3C-C-S \\ \| \quad | \\ F_3C-C-S \end{matrix} \longrightarrow \left(\begin{matrix} F_3C-C-S \\ | \\ F_3C-C-S \end{matrix}\right)_2 Ni$$

$$\big\downarrow \text{spontaneously in acetone} + (C_2H_5)_4N^+$$

$$[(C_2H_5)_4N]^+\left[\left(\begin{matrix} F_3C-C-S \\ | \\ F_3C-C-S \end{matrix}\right)_2 Ni\right]^{-}$$

$$\big\downarrow p\text{-}H_2NC_6H_4NH_2 + (C_2H_5)_4N^+$$

$$[(C_2H_5)_4N]_2\left[\left(\begin{matrix} F_3C-C-S \\ | \\ F_3C-C-S \end{matrix}\right)_2 Ni\right]$$

The most interesting characteristic of the 1,2-dithiolene complexes is their ability, unsurpassed and seldom matched among other complexes, to undergo redox reactions. Examples are:

$$Ni[S_2C_2(CN)_2]_2 \underset{-e}{\overset{+e}{\rightleftharpoons}} Ni[S_2C_2(CN)_2]_2^{-} \underset{-e}{\overset{+e}{\rightleftharpoons}} Ni[S_2C_2CN)_2]_2^{2-}$$

$$[CoL_2]_2^{0} \underset{-e}{\overset{+e}{\rightleftharpoons}} [CoL_2]_2^{1-} \underset{-e}{\overset{+e}{\rightleftharpoons}} [CoL_2]_2^{2-} \underset{-2e}{\overset{+2e}{\rightleftharpoons}} 2[CoL_2]^{2-}$$

$$[L = S_2C_2(CF_3)_2]$$

$$[CrL_3]^{0} \underset{-e}{\overset{+e}{\rightleftharpoons}} [CrL_3]^{1-} \underset{-e}{\overset{+e}{\rightleftharpoons}} [CrL_3]^{2-} \underset{-e}{\overset{+e}{\rightleftharpoons}} [CrL_3]^{3-}$$

$$[L = S_2C_2(CN)_2]$$

The electronic structures of the 1,2-dithiolene complexes have provoked a great deal of controversy. The ring system involved can be written in two extreme forms, (28-XIII) and (28-XIV); the formal oxidation number of the metal differs

28-XIII

28-XIV

by two in these two cases. In MO terms the problem is the extent to which electrons in metal *d* orbitals are delocalized over the ligand. Undoubtedly, considerable delocalization occurs, which accounts for the ability of these complexes to exist with such a range of electron populations.

Study Questions

A

1. Name some π-acid ligands and for each state the nature of its acceptor orbitals.
2. Write the formulas for the mononuclear metal carbonyl molecules formed by V, Cr, Fe, and Ni. Which ones satisfy the noble gas formalism?
3. Why are the simplest carbonyls of the metals Mn, Tc, Re and Co, Rh, Ir groups polynuclear?
4. Explain, with necessary orbital diagrams, how CO, which has negligible donor properties toward simple acceptors such as BF_3 can form strong bonds to transition metal atoms.
5. In what ways can CO be bound to a metal atom?
6. Discuss and explain the trend in CO stretching frequencies in the series $V(CO)_6^{-}$, $Cr(CO)_6$, $Mn(CO)_6^{+}$.
7. Draw the structures of $Fe_2(CO)_9$, $Ru_3(CO)_{12}$, and $Rh_4(CO)_{12}$.
8. Which are the only two metals to react directly with CO under conditions suitable for practical syntheses?
9. What is the general type of reaction used to prepare metal carbonyls? State the main ingredients, the function of each, and some examples.

10. How are the following compounds made? What are their principal physical characteristics?
    (a) $Fe(CO)_5$ from iron powder
    (b) $Co_2(CO)_8$ from hydrated cobalt sulfate
    (c) $Cr(CO)_6$ from hydrated chromic chloride
    (d) $Mn_2(CO)_{10}$ from hydrated manganous chloride
    (e) $Fe_3(CO)_{12}$ from $Fe(CO)_5$.
11. Explain why $Mopy_2(CO)_4$ has two forms, one having a single CO stretching band in the infrared spectrum, the other four.
12. Give the formulas of some simple carbonylate ions and carbonyl hydrides. Do they follow the noble gas rule?
13. In a carbonyl complex with a linear OC—M—CO group, how will the CO stretching frequency change when
    (a) one CO is replaced by triethylamine,
    (b) a positive charge is put on the complex,
    (c) a negative charge is put on the complex?
14. How is N_2 related to CO? Are N_2 complexes more or less stable than CO complexes? What was the first N_2 complex discovered and when?
15. Describe the bonding of NO to a metal in the case where the M—N—O chain is essentially linear; specifically contrast it with the analogous M—C—O bonding in terms of how many electrons are involved.
16. Besides linear M—N—O bonding, what other three kinds are there?
17. Explain why nitric oxide can be regarded as a 3-electron donor ligand. Would you expect the stoichiometries of compounds with the following chelate ligands to be similar to those formed by NO?

O, C—CH_3, ·O· (acetate chelate) CH_2, CH, CH_2 (allyl) CH_2, ←:$N(CH_3)_2$

Would there be any difference in the formal oxidation of the metal for example, in $Mn(CO)_4NO$ and $Mn(CO)_4[CH_2N(CH_3)_2]$?
18. Explain how trialkyl or aryl phosphines can bind to a metal.
19. Which PX_3 ligand is most similar in its bonding ability to CO? Why is it this one rather than another?
20. Discuss the similarities and differences between CN^- and CO as ligands.
21. Why is it that the dithiolene complexes, such as $[Ni\{S_2C_2(CN)_2\}_2]^{0,-1,-2}$ are capable of existing in a whole series of oxidation states while retaining essentially a constant structure?
22. Comment on the differences in the Fe^{II}—Fe^{III} redox potentials

$$[Fe(CN)_6]^{4-} + e = [Fe(CN)_6]^{3-} \qquad E^\circ = 0.36V$$
$$[Fe(H_2O)_6]^{2+} + e = [Fe(H_2O)_6]^{3+} \qquad E^\circ = 0.77V$$
$$[Fe\ phen_3]^{2+} + e = [Fe\ phen_3]^{3+} \qquad E^\circ = 1.12V$$

B

1. In order to have a vanadium carbonyl that satisfies the noble gas formalism, what would be the simplest formula? Why do you think this fails to occur?
2. It is known that in $Mn_2(CO)_{10}$ the carbonyl groups move rapidly from one manganese atom to the other. On the basis of what you find in Sections 28-3 and 28-4 suggest a plausible intermediate for this process.

3. Do you think that carbonyls of the lanthanides are likely to be stable? Whatever your answer, give reasons.
4. Write both bridged and nonbridged structures for $Mn_2(CO)_{10}$ and $Co_2(CO)_8$. The former has CO stretching bands only in the range 2044–1980 cm^{-1} while the latter has bands in the range 2071–2022 cm^{-1} as well as two at 1860 and 1858 cm^{-1}. Which structure is indicated to be correct in each case?
5. What are the formulas of the metal carbonyls which are isoelectronic with $Cr(NO)_4$, $Mn(CO)(NO)_3$, $Mn(CO)_3NO$, $Co(NO)_3$, $Fe(CO)_2(NO)_2$?
6. Write balanced equations for the following reactions
   (a) $Mn_2(CO)_{10}$ is heated with I_2,
   (b) $Mo(CO)_6$ is refluxed with KI in tetrahydrofuran,
   (c) $Fe(CO)_5$ is shaken with aqueous KOH,
   (d) $Ni(CO)_4$ is treated with PCl_3,
   (e) $Co_2(CO)_8$ is treated with NO in petroleum.
7. What is the difference between a π-acid ligand like RNC and a ligand like C_2H_4 that forms π complexes?
8. In a linear group $R_3P—M—CO$, how would the CO frequency change when

$$R = F, CH_3, C_6H_5, -C_6H_4-CH_3, -C_6H_4-F?$$

9. Why is pK_2 for $H_2Fe(CO)_4$ smaller than pK_1 by 9 units? What does this tell us?
10. Put the following ligands in decreasing order of π acidity

$$CH_3CN, (C_2H_5)_2O, PCl_3, As(C_6H_5)_3, CH_3NC, Et_3N.$$

Chapter 28 Study Guide

Supplementary Reading

Abel, E. W. and Stone, F. G. A., "The Chemistry of Transition-metal Carbonyls: Structural Considerations," *Quart. Rev.*, *23*, 325 (1969).

Abel, E. W. and Stone, F. G. A., "The Chemistry of Transition-metal Carbonyls: Synthesis and Reactivity," *Quart. Rev.*, *24*, 498 (1970).

Allen, A. D., "Complexes of Dinitrogen," *Chem. Rev.*, *73*, 11 (1973).

Calderazzo, F., Ercoli, R., and Natta, G., "Metal Carbonyls: Preparation, Structure and Properties" in *Organic Syntheses via Metal Carbonyls*, I. Wender and P. Pino, eds., Interscience-Wiley, 1968.

Eisenberg, R., "Structural Systematics of 1,1- and 1,2-Dithiolato Chelates," *Prog. Inorg. Chem.*, *12*, 295 (1970).

Enemark, J. H. and Feltham, R. D., "Nitric Oxide Complexes," *Coord. Chem. Rev.*, *13*, 339 (1974)

Malatesta, L. and Cenini, S., *Zerovalent Complexes of Metals*, Academic Press, 1974.

McAuliffe, C. A., ed., *Transition Metal Complexes of Phosphorus, Arsenic and Antimony Ligands*, MacMillan, 1973.

29

organometallic compounds

PART I
NONTRANSITION METALS

29-1
General Survey of Types

Organometallic compounds are those in which the *carbon* atoms of organic groups are bound to metal atoms. For example, an alkoxide such as $(C_3H_7O)_4Ti$ is not considered to be an organometallic compound because the organic group is bound to Ti by oxygen, whereas $C_6H_5Ti(OC_3H_7)_3$ is, because a metal to carbon bond is present. The term organometallic is usually rather loosely defined and compounds of elements such as boron, phosphorus, and silicon which are scarcely, if at all metallic, are included in the category. A few general comments on the various types of compound can be made first.

1. *Ionic compounds of electropositive metals.* The organo compounds of highly electropositive metals are usually ionic, insoluble in hydrocarbon solvents, and are very reactive toward air, water, and the like. The stability and reactivity of ionic compounds are determined in part by the stability of the carbanion. Compounds containing unstable anions (e.g., $C_nH_{2n+1}^-$) are generally highly reactive and often unstable and difficult to isolate. Metal salts of carbanions whose stability is enhanced by delocalization of electron density are more stable although still quite reactive; examples are $(C_6H_5)_3C^-Na^+$ and $(C_5H_5^-)_2Ca^{2+}$.

2. *σ-Bonded compounds.* Organo compounds in which the organic residue is bound to a metal atom by a normal 2-electron covalent bond (albeit in some cases with appreciable ionic character) are formed by most metals of lower electropositivity and, of course, by nonmetallic elements. The normal valence rules apply in these cases, and partial substitution of halides, hydroxides, etc., by organic groups occurs as in $(CH_3)_3SnCl$, $(CH_3)SnCl_3$, etc. In most of these compounds, the bonding is predominantly covalent and the chemistry is organic-like, although there are many differences from carbon chemistry due to the following factors. (a) The possibility of using higher *d* orbitals in for example, SiR_4, which is not feasible in CR_4. (b) Donor ability of alkyls or aryls with lone pairs as in PEt_3, SMe_2,

etc. (c) Lewis acidity due to incomplete valence shells as in BR_3 or coordinative unsaturation as in ZnR_2. (d) Effects of electronegativity differences between M—C and C—C bonds.

Transition metals may form simple alkyls or aryls but these are normally less stable than those of main group elements for reasons that we discuss later (Section 29-11). There are numerous compounds in which additional ligands such as CO or PR_3 are present.

3. *Nonclassically bonded compounds.* In many organometallic compounds there is a type of metal to carbon bonding that cannot be explained in terms of ionic or electron-pair σ bonds. One class comprises the alkyls of Li, Be, and Al that have *bridging* alkyl groups. Here, there is electron-deficiency as in boron hydrides, and the bonding is of a similar multicenter type. A second, much larger class comprises compounds of transition metals with alkenes, alkynes, benzene, and other ring systems such as $C_5H_5^-$.

We consider first the organo compounds of the main group elements, including the nonclassically bonded ones, and then turn to the transition metal compounds.

29-2
Synthetic Methods

There are many ways of generating metal to carbon bonds that are useful for both nontransition and transition metals. Some of the more important are as follows.

1. *Direct reactions of metals.* The earliest synthesis, by the English chemist Frankland in 1845, was the interaction of Zn and an alkyl halide. Frankland was, in fact, attempting to synthesize alkyl radicals; his discovery played a decisive part in the development of modern ideas of chemical bonds. Much more useful, however, was the discovery by the French chemist, Grignard, of what are now called Grignard reagents by interaction of magnesium with alkyl or aryl halides in ether:

$$Mg + CH_3I \xrightarrow{\text{ether}} CH_3MgI$$

Direct interactions of alkyl or aryl halides occur also with Li, Na, K, Mg, Ca, Zn, and Cd.

2. *Use of alkylating agents.* The above compounds can be utilized to make other organometallic compounds. The most important and widely used are Grignard and lithium reagents. Aluminum and mercury alkyls and certain sodium derivatives, especially $Na^+C_5H_5^-$, are also useful alkylating agents.

Most nonmetal and metal halides or halide derivatives can be alkylated in ethers, or hydrocarbon solvents, for example,

$$PCl_3 + 3C_6H_5MgCl = P(C_6H_5)_3 + 3MgCl_2$$

$$VOCl_3 + 3(CH_3)_3SiCH_2MgCl = VO(CH_2SiMe_3)_3 + 3MgCl_2$$

$$PtCl_2(PEt_3)_2 + CH_3MgCl = PtCl(CH_3)(PEt_2)_2 + MgCl_2$$

3. *Interaction of metal or nonmetal hydrides with alkenes or alkynes.* One of the best examples for nonmetals, and one that finds wide use in synthesis, is the hydroboration reaction (page 235)

$$\tfrac{1}{2}B_2H_6 + 3\ \gt C{=}C\lt \xrightarrow{\text{ether}} B\left(-\overset{|}{\underset{|}{C}}-\overset{|}{\underset{|}{C}}-\right)_3$$

For transition metals and hydride complexes such reactions are of prime importance in that many catalytic syntheses involving transition metals (page 540) have as an early step the reaction

$$L_nMH + \gt C{=}C\lt = L_nM-\overset{|}{\underset{|}{C}}-\overset{\overset{H}{|}}{\underset{|}{C}}-$$

4. *Oxidative-addition reactions.* The so-called oxad reactions (Section 30-2) where alkyl or aryl halides are added to coordinatively unsaturated transition metal compounds generate metal–carbon bonds; for example:

$$RhCl(PPh_3)_3 + CH_3I = RhClI(CH_3)(PPh_3)_2 + PPh_3$$

5. *Insertion reactions.* Certain "insertion" reactions (Section 30-3) may also allow the generation of bonds to carbon, for example,

$$[(CN)_5Co-Co(CN)_5]^{4-} + HC{\equiv}CH = \left[(CN)_5Co-\overset{H}{C}{=}\underset{H}{C}-Co(CN)_5\right]^{4-}$$

$$SbCl_5 + 2HC{\equiv}CH = Cl_3Sb(CH{=}CHCl)_2$$

The reactions in 3 above can also be regarded as "insertions" into the M—H bond.

29-3
Lithium Alkyls and Aryls

One of the major uses of metallic lithium, industrially and in the laboratory, is for the preparation of organolithium compounds which in their reactions generally resemble Grignard reagents, although they are usually more reactive. They are prepared by direct interaction of the organic halide, usually the chloride (Eq. 29-1), in benzene or petroleum; ethers can be used, but they are attacked slowly by the lithium compounds. Metal–hydrogen exchange (Eq. 29-2), metal–halogen exchange (Eq. 29-3), and metal–metal exchange (Eq. 29-4) may also be used.

$$C_2H_5Cl + 2Li = C_2H_5Li + LiCl \qquad (29\text{-}1)$$

$$n\text{-}C_4H_9Li + \text{Fe} = \text{Fe (Li)} + n\text{-}C_4H_{10} \qquad (29\text{-}2)$$

$$n\text{-}C_4H_9Li + \text{Br, N} = \text{Li, N} + n\text{-}C_4H_9Br \qquad (29\text{-}3)$$

$$2Li + R_2Hg = 2RLi + Hg \qquad (29\text{-}4)$$

n-Butyllithium in hexane, benzene, or ethers is commonly used for such reactions. Methyllithium is also prepared by exchange through the interaction of $n\text{-}C_4H_9Li$ and CH_3I in hexane at low temperatures, whence it precipitates as insoluble white crystals.

Organolithium compounds all react rapidly with oxygen, being usually spontaneously flammable in air, with liquid water and with water vapor. However, lithium bromide and iodide form solid complexes of stoichiometry $RLi(LiX)_{1-6}$ with the alkyls, and these solids are stable in air.

Organolithium compounds are among the very few alkali-metal compounds that have properties—solubility in hydrocarbons or other nonpolar liquids and high volatility—typical of covalent substances. They are generally liquids or low-melting solids. Molecular association is an important feature of the alkyls in both crystals and solutions. Thus in methyllithium (*Fig. 29-1*) the Li atoms are at the corners of a tetrahedron with the alkyl groups centered over the facial planes. The CH_3 group is symmetrically bound to three Li atoms, and this alkyl bridge bonding is of the electron-deficient multicenter type (page 78). Aggregate formation is due principally to the Li—C—Li rather than to Li—Li bonding interactions.

In solutions the nature of the polymerized species depends on the solvent, the steric nature of the organic radical, and temperature. In hydrocarbons MeLi, EtLi, *n*-PrLi, and some others are hexamers, but *tert*-butylithium, which presumably is too bulky, is only tetrameric. In ethers or amines solvated tetramers are formed. There are no aggregates smaller than tetramers.

However, when chelating ditertiary amines, notably tetramethylethylenediamine (TMED), $Me_2NCH_2CH_2NMe_2$, are used, comparatively stable monomeric alkyllithium complexes are obtained. The alkyls and aryls also form complexes with other metal alkyls such as those of Mg, Cd, and Zn. For example,

$$2LiC_6H_5 + Mg(C_6H_5)_2 = Li_2[Mg(C_6H_5)_4]$$

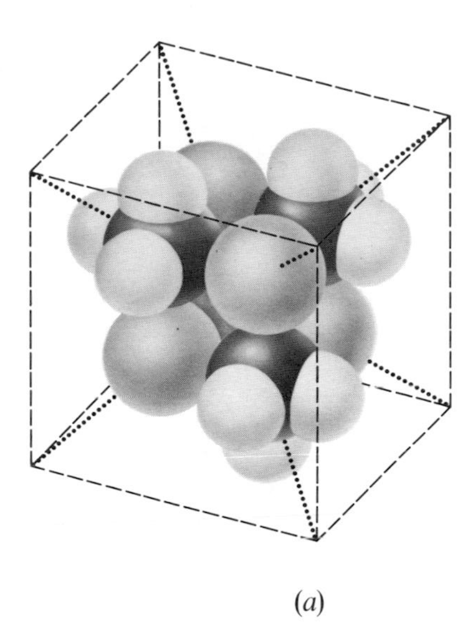

(a)

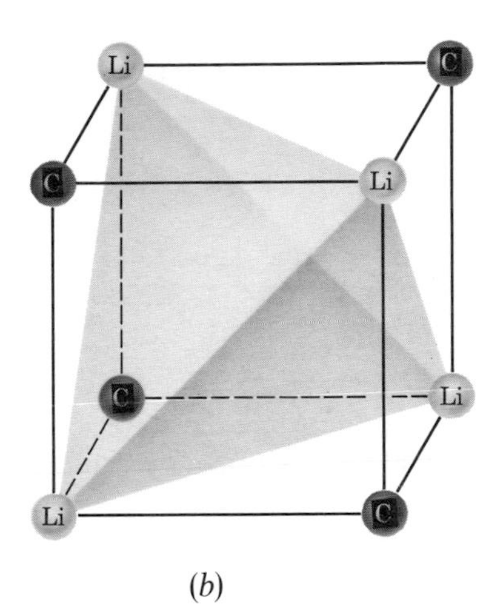

(b)

Figure 29-1 *The structure of* $(CH_3Li)_4$: *(a) showing the tetrahedral* Li_4 *unit with the* CH_3 *groups located symmetrically above each face of the tetrahedron. (Adapted from E. Weiss and E. A. C. Lucken,* J. Organometallic Chem., *1964,* 2, *197.) The structure can also be regarded as derived from a cube (b).*

It is not surprising that there are wide variations in the comparative reactivities of Li alkyls depending on the differences in aggregation and ion-pair interactions. An example is benzyllithium, which is monomeric in tetrahydrofuran and reacts with a given substrate more than 10^4 times as fast as the tetrameric methyllithium. The monomeric TMED complexes mentioned above are also very much more reactive than the corresponding aggregated alkyls. Alkyllithiums can polylithiate acetylenes, acetonitrile, and other compounds; thus, $CH_3C{\equiv}CH$ gives Li_4C_3, which can be regarded as a derivative of C_3^{4-}.

Reactions of lithium alkyls are generally considered to be carbanionic in nature. Lithium alkyls are widely employed as stereospecific catalysts for the polymerization of alkenes, notably isoprene, which gives up to 90% of 1,4-*cis*-polyisoprene; numerous other reactions with alkenes have been studied. The TMED complexes again are especially active: not only will they polymerize ethylene but they will even metallate benzene and aromatic compounds, as well as reacting with hydrogen at 1 atm to give LiH and alkane.

29-4
Organo-sodium and -potassium Compounds

These compounds are all essentially ionic and are not soluble to any appreciable extent in hydrocarbons; they are exceedingly reactive, sensitive to air, and are hydrolyzed vigorously by water.

Most important are the sodium compounds from acidic hydrocarbons such as cyclopentadiene, indene, acetylenes, and the like. These are obtained by inter-reaction with metallic sodium or sodium dispersed in tetrahydrofuran or dimethylformamide.

$$2C_5H_6 + 2Na \longrightarrow 2C_5H_5^-Na^+ + H_2$$
$$RC{\equiv}CH + Na \longrightarrow RC{\equiv}C^-Na^+ + \tfrac{1}{2}H_2$$

29-5
Magnesium

The organic compounds of Ca, Sr, and Ba are highly ionic and reactive and are not useful, but the magnesium compounds are probably the most widely used of all organometallic compounds; they are used very extensively in organic chemistry as well as in synthesis of alkyl and aryl compounds of other elements. They are of the types RMgX (the Grignard reagents) and MgR_2. The former are made by direct interaction of the metal with an organic halide RX in a suitable solvent, usually an ether such as diethyl ether or tetrahydrofuran. The reaction is normally most rapid with iodides, RI, and iodine may be used as an initiator. For most purposes, RMgX reagents are used *in situ*. The species MgR_2 are best made by the dry reaction

$$HgR_2 + Mg(\text{excess}) \longrightarrow Hg + MgR_2$$

The dialkyl or diaryl is then extracted with an organic solvent. Both RMgX, as solvates, and R_2Mg are reactive, being sensitive to oxidation by air and to hydrolysis by water.

The nature of Grignard reagents *in solution* is complex and depends on the nature of the alkyl and halide groups and on the solvent, concentration and temperature. Generally, the equilibria involved are of the type:

$$RMg(\mu\text{-}X)_2MgR \rightleftharpoons 2RMgX \rightleftharpoons R_2Mg + MgX_2$$
$$R_2Mg + MgX_2 \rightleftharpoons R_2Mg(\mu\text{-}X)_2Mg$$

Solvation (not shown) occurs and association is predominantly by halide rather than by carbon bridges, except for methyl compounds where bridging by CH_3 groups may occur.

In dilute solutions and in more strongly donor solvents the monomeric species normally predominate; but in diethyl ether at concentrations greater than 0.1 *M* association gives linear or cyclic polymers. For crystalline Grignard reagents both of the structures $RMgX \cdot nS$ where n the number of solvent molecules,

S, depends on the nature of R, and $R(S)Mg(\mu\text{-}X)_2Mg(S)R$ have been found. The Mg atom is usually tetrahedrally coordinated.

Zinc and *cadmium* compounds are similar to those of magnesium but differ in their reactivities. The lower alkyls of zinc are liquids spontaneously flammable in air. They react vigorously with water.

29-6
Mercury

A vast number of organomercury compounds are known, some of which have useful physiological properties. They are of the types RHgX and R_2Hg. They are commonly made by the interaction of $HgCl_2$ and RMgX, but Hg—C bonds can also be made in other ways discussed below.

The *RHgX compounds* are crystalline solids. When X can form covalent bonds to mercury, for example, Cl, Br, I, CN, SCN, or OH, the compound is a covalent nonpolar substance more soluble in organic liquids than in water. When X is SO_4^{2-} or NO_3^-, the substance is salt-like and presumably quite ionic, for instance, $[RHg]^+NO_3^-$.

The *dialkyls and diaryls* are nonpolar, volatile, toxic, colorless liquids, or low-melting solids. They are unaffected by air or water, presumably because of the low polarity of the Hg—C bond and the low affinity of mercury for oxygen. However, they are photochemically and thermally unstable, as would be expected from the low bond strengths (50–200 kJ mol^{-1}). In the dark, mercury compounds can be kept for months. The decomposition generally proceeds by homolysis of the Hg—C bond and free-radical reactions.

All RHgX and R_2Hg molecules have linear bonds. The principal utility of dialkyl- and diarylmercury compounds, and a very valuable one, is in the preparation of other organo compounds by *interchange* reactions. For example

$$\frac{n}{2}R_2Hg + M \longrightarrow R_nM + \frac{n}{2}Hg$$

This reaction proceeds essentially to completion with the Li and Ca groups, and with Zn, Al, Ga, Sn, Pb, Sb, Bi, Se, and Te, but with In, Tl, and Cd reversible equilibria are established. Partial alkylation of reactive halides can be achieved, for example

$$AsCl_3 + Et_2Hg \longrightarrow EtHgCl + EtAsCl_2$$

Mercury released to the environment, as metal, for example, by losses from electrolytic cells used for NaOH and Cl_2 production, or as compounds such as alkylmercury seed dressings or fungicides, constitutes a serious hazard. This is a result of biological methylation to give highly toxic $(CH_3)_2Hg$ or CH_3Hg^+. Models for vitamin B_{12} such as methylcobaloximes (Chapter 31) which have Co—CH_3 bonds will transfer the CH_3 to Hg^{2+}. There are a number of microorganisms that can perform the same function, possibly by similar routes.

Mercuration and Oxomercuration. An important reaction for the formation of Hg—C bonds, and one that can be adapted to the synthesis of a wide variety of organic compounds, is the addition of mercuric salts, notably the acetate, trifluoroacetate or nitrate to unsaturated compounds.

Mercuration of aromatic compounds occurs as follows:

$$C_6H_6 + Hg(OCOCH_3)_2 \longrightarrow C_6H_5HgOCOCH_3 + CH_3COOH$$

Mercuric salts also react with alkenes in a reversible reaction

$$\gt C{=}C\lt + HgX_2 \rightleftharpoons X{-}\overset{|}{\underset{|}{C}}{-}\overset{|}{\underset{|}{C}}{-}HgX \qquad (29\text{-}5)$$

The reversibility is readily shown by using $Hg(OCOCF_3)_2$, since the latter is soluble in nonpolar solvents; the equilibrium constants for the reaction (29-5) can be measured. In most instances, the reactions must be carried out in an alcohol or other protic medium, where further reaction with the solvent occurs. The reaction is then called *oxomercuration.* For example,

$$\gt C{=}C\lt + Hg(OCOCH_3)_2 + C_2H_5OH$$

$$\downarrow$$

$$C_2H_5O{-}\overset{|}{\underset{|}{C}}{-}\overset{|}{\underset{|}{C}}{-}HgOCOCH_3 + CH_3COOH$$

The evidence that HgX_2 adds across the double bond is usually indirect, often, by observing the products on hydrolysis, for example,

$$CH_2{=}CH_2 + Hg(NO_3)_2 \xrightarrow{OH^-} HOCH_2CH_2Hg^+ + 2NO_3^-$$

In these reactions, *mercurinium ions* of the type (29-I) and (29-II) are believed to be intermediates. In $FSO_3H{-}SbF_5{-}SO_2$ at $-70°$ long-lived mercurinium

$$\underset{\text{29-I}}{\gt C\overset{Hg^{2+}}{\cdots}C\lt} \qquad \underset{\text{29-II}}{\gt C\overset{X-Hg^{+}}{\cdots}C\lt}$$

ions have been obtained by reactions such as

$$CH_3OCH_2CH_2HgCl \xrightarrow{H^+} CH_3OH_2^+ + CH_2CH_2Hg^{2+} + HCl$$

$$C_6H_{10} + Hg(OCOCF_3)_2 \xrightarrow{H^+} C_6H_{10}Hg^{2+}$$

The above type of addition has been used for the synthesis of alcohols, ethers, and amines from alkenes and other unsaturated substances. The additions of HgX_2 are carried out in water, alcohols, or acetonitrile, respectively. The mercury is removed from the intermediate by reduction with sodium borohydride. An example is

$$C_5H_8 \xrightarrow[H_2O,\,THF]{Hg(OCOCH_3)_2} C_5H_8(OH)HgOCOCH_3 \xrightarrow{NaBH_4} C_5H_8(OH)H$$

29-7 Boron

There is a very extensive chemistry of organoboron compounds.

The *trialkyl-* and *aryl-borons* are made from the halides by lithium or Grignard reagents, and by hydroboration. The lower alkyls inflame in air, but the aryls are stable. Like other BX_3 compounds alkyl borons behave as Lewis acids giving adducts, for example, $R_3B \cdot NR_3$. Furthermore, when boron halides are treated with four equivalents of alkylating agent, the trialkyl or triaryl gives an anion BR_4^-. The most important compound is *sodium tetraphenylborate*, $Na[B(C_6H_5)_4]$. This is soluble in water and is stable in weakly acid solution; it gives insoluble precipitates with larger cations such as K^+, Rb^+, or Me_4N^+, that are suitable for gravimetric analysis. There are also di- and monoalkyl compounds such as R_2BX or RBX_2 where X may be halogen, OH, H, etc.

29-8 Aluminum

The alkyls of Al are of great importance because of their industrial use as catalysts for the polymerization of ethylene and propylene (Section 30-9). They are also widely used as reducing and alkylating agents for transition metal complexes.

The alkyls may be prepared by the reactions:

$$2Al + 3R_2Hg \longrightarrow 2R_3Al \text{ (or } [R_3Al]_2) + 3Hg$$

$$RMgCl + AlCl_3 \longrightarrow RAlCl_2,\ R_2AlCl,\ R_3Al$$

More direct methods suitable for large-scale use are

$$AlH_3 + 3C_nH_{2n} \longrightarrow Al(C_nH_{2n+1})_3$$

$$LiAlH_4 + 4C_nH_{2n} \longrightarrow Li[Al(C_nH_{2n+1})_4]$$

Although $(AlH_3)_n$ cannot be made by direct interaction of Al and H_2, nevertheless, in the presence of aluminum alkyl, the following reaction to give the dialkyl hydride can occur:

$$Al + \tfrac{3}{2}H_2 + 2AlR_3 \longrightarrow 3AlR_2H$$

This hydride will then react with olefins:

$$AlR_2H + C_nH_{2n} \longrightarrow AlR_2(C_nH_{2n+1})$$

Thus the direct interaction of Al, H_2, and olefin can be used to give either the dialkyl hydrides or the trialkyls.

Other technically important compounds are the "sesquichlorides" such as $Me_3Al_2Cl_3$ or $Et_3Al_2Cl_3$. These compounds can be made by direct interaction of Al or Mg—Al alloy with the alkyl chloride.

The aluminum lower alkyls are reactive liquids, inflaming in air and exploding with water. All other derivatives are similarly sensitive to air and moisture though not all are spontaneously flammable. Certain aluminum alkyls form reasonably stable dimers. The structure of trimethylaluminum is shown in *Fig. 29-2a*. The alkyl bridge is formed by multicenter bonding, that is, Al—C—Al, 3*c*–2*e* bonds (page 79). Each Al atom supplies an sp^3 hybrid orbital and so does the C atom.

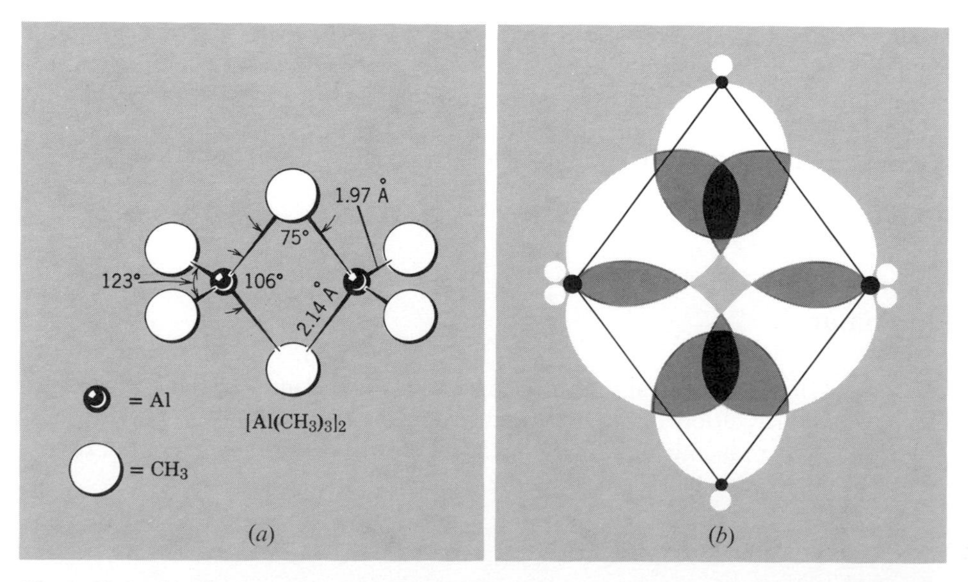

Figure 29-2 *(a) The structure of* $Al_2(CH_3)_6$. *(b) Orbital overlaps in the* Al—C—Al *bonding.*

The bonding situation is shown in *Fig. 29-2b*. A similar description holds true for bridging in $[Be(CH_3)_2]_n$, which is a linear polymer, each Be atom being tetrahedral.

There is no simple explanation why boron trialkyls do not dimerize in a similar way, especially, since hydrogen bridges are very important in the boranes (page 79). The coordinative unsaturation of Al alkyls also means that they behave as Lewis acids, giving adducts such as R_3AlNR_3 or anionic species like $Li[AlEt_4]$. In this respect all of the coordinatively unsaturated alkyls of Group III and Group II elements are similar.

29-9
Silicon, Ge, Sn, and Pb

There is an exceedingly extensive chemistry of the Group IV elements bound to carbon. Some of the compounds, notably silicon—oxygen polymers and alkyl–tin and -lead compounds, are of commercial importance; germanium compounds have no uses.

Essentially all the compounds are of the tetravalent elements. In the divalent state the only well established compounds with σ bonds are the trimethylsilylmethyl derivatives $M[CH(SiMe_3)_2]_2$. Other tin-compounds that might appear to contain Sn^{II} are linear or cyclic polymers of Sn^{IV}.

For all four elements the compounds can generally be designated $R_{4-n}MX_n$, where R is alkyl or aryl and X can be H, Cl, O, COR′, OR′, NR_2', SR′, $Mn(CO)_5$, $W(CO)_3(\eta^5\text{-}C_5H_5)$, etc. The elements can also be incorporated into heterocyclic rings of various types.

For a given class of compound, those with C—Si and C—Ge bonds have higher thermal stability and lower reactivity than those with bonds to Sn and Pb. In catenated compounds, similarly, Si—Si and Ge—Ge bonds are more stable and less reactive than Sn—Sn and Pb—Pb bonds. For example, Si_2Me_6 is very stable, but Pb_2Me_6 blackens in air and decomposes rapidly in CCl_4 although it is fairly stable in benzene.

The bonds to carbon are usually made via interaction of Li, Hg, or Al alkyls or RMgX and the Group IV halide, but there are some special synthetic methods noted below.

Silicon. The organo compounds of Si and Ge are very similar in their properties.

Silicon—carbon bond energies are less than those of C—C bonds but are still quite high, in the region 250 to 335 kJ mol^{-1}. The tetraalkyls and -aryls are hence thermally quite stable; $Si(C_6H_5)_4$, for example, boils unchanged at 530°.

The chemical reactivity of Si—C bonds is generally greater than that of C—C bonds because (a) the greater polarity of the bond, $Si^{\delta+}—C^{\delta-}$, allows easier nucleophilic attack on Si and electrophilic attack on C than for C—C compounds, and (b) displacement reactions at silicon are facilitated by its ability to form 5-coordinate transition states by utilization of *d* orbitals.

Alkyl- and Arylsilicon Halides. These are of special importance because of their hydrolytic reactions. They may be obtained by normal Grignard procedures from $SiCl_4$ or, in the case of the methyl derivatives, by the *Rochow process* in which methyl chloride is passed over heated, copper-activated silicon:

$$CH_3Cl + Si(Cu) \longrightarrow (CH_3)_nSiCl_{4-n}$$

The halides are liquids that are readily hydrolyzed by water, usually in an inert solvent. The silanol intermediates R_3SiOH, $R_2Si(OH)_2$, and $RSi(OH)_3$ can sometimes be isolated, but the diols and triols usually condense under the hydrolysis conditions to *siloxanes* which have Si—O—Si bonds. The exact nature of the products depends on the hydrolysis conditions, and linear, cyclic, and complex cross-linked polymers of varying molecular weights can be obtained. They are often referred to as *silicones*; the commercial polymers usually have $R=CH_3$, but other groups may be incorporated for special purposes.

Controlled hydrolysis of the alkyl halides in suitable ratios can give products of particular physical characteristics. The polymers may be liquids, rubbers, or solids, which have in general high thermal stability, high dielectric strength, and resistance to oxidation and chemical attack.

Examples of simple siloxanes are $Ph_3SiOSiPh_3$ and the cyclic trimer or tetramer $(Et_2SiO)_{3(or\ 4)}$; linear polymers contain $—SiR_2—O—SiR_2—O—$ chains, whereas the cross-linked sheets have the basic unit

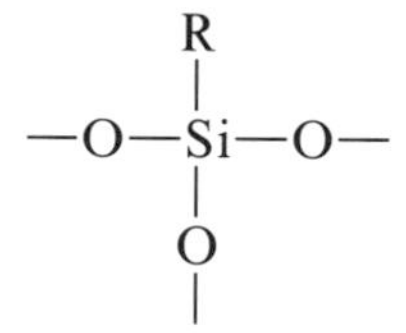

```
        R
        |
  —O—Si—O—
        |
        O
        |
```

Tin. Where the compounds of tin differ from those of Si and Ge, they do so mainly because of a greater tendency of Sn^{IV} to show coordination numbers higher than four and because of ionization to give cationic species.

Trialkyltin compounds, R_3SnX, are always associated in the solid by anion bridging (29-III and 29-IV). The coordination of the tin atom is close to **tbp** with planar $Sn(Me)_3$ groups. In water the perchlorate and some other compounds ionize to give cationic species, for example, $[Me_3Sn(H_2O)_2]^+$.

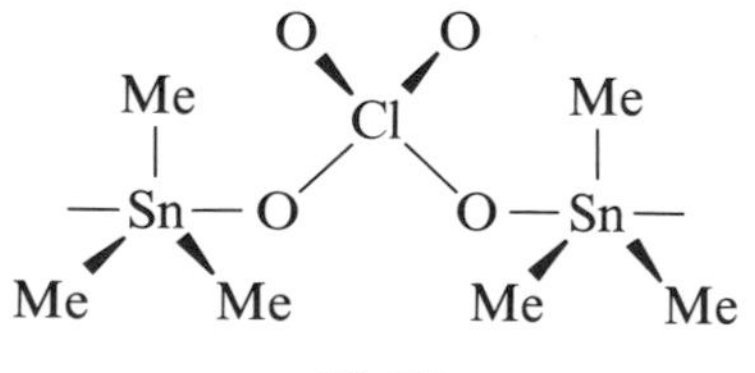

```
            O   O
  Me         \ /         Me              Me  F  Me  F
  |           Cl          |              |  /  ·. | /  ·.
—Sn—O       /   \      O—Sn—          \Sn/      Sn/
 / \                      / \           / \     / \
Me   Me                 Me   Me       Me  Me  Me   Me
```

29-III 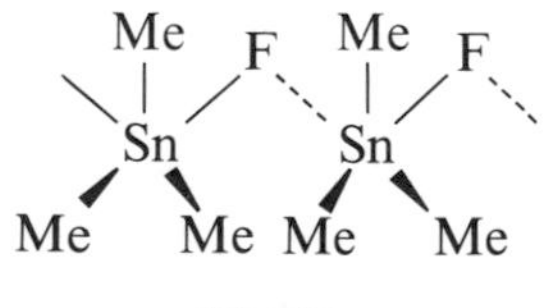 29-IV

Dialkyltin compounds, R_2SnX_2, have behavior similar to that of the trialkyl compounds. Thus the fluoride Me_2SnF_2 is again polymeric, with bridging F atoms, but Sn is octahedral and the Me—Sn—Me group is linear. However, the chloride and bromide have low melting points (90° and 74°) and are essentially molecular compounds. The halides also give conducting solutions in water, and the aqua ion has the linear C—Sn—C group characteristic of the dialkyl species (cf. the linear species Me_2Hg, Me_2Tl^+, Me_2Cd, Me_2Pb^{2+}), probably with four water molecules completing octahedral coordination. The linearity in these species appears to result from maximizing of *s* character in the bonding orbitals of the metal atoms. Organotin *hydrides* are useful reducing agents in organic chemistry and can add to alkenes by free radical reactions to generate other organotin compounds.

Organotin compounds have a number of uses in marine antifouling paints, fungicides, wood preserving, and as catalysts for curing silicone and epoxy resins.

Lead. The most important compounds are $(CH_3)_4Pb$ and $(C_2H_5)_4Pb$, which are made in huge quantities for use as antiknock agents in gasoline. The environmental increase in lead is largely due to the burning of leaded gasolines, and it is quite likely that their use will eventually be phased out.

The commercial synthesis is the interaction of a sodium-lead alloy with CH_3Cl or C_2H_5Cl in an autoclave at 87 to 100°, without solvent for C_2H_5Cl but in toluene at a higher temperature for CH_3Cl. The reaction is complicated and not fully understood, and only a quarter of the lead appears in the desired product:

$$4NaPb + 4RCl \longrightarrow R_4Pb + 3Pb + 4NaCl$$

The required recycling of the lead is disadvantageous and electrolytic procedures have been developed.

The lower alkyls are nonpolar highly toxic liquids. The tetramethyl decomposes around 200° and the tetraethyl around 110° by free radical processes.

29-10
Phosphorus, As, Sb, and Bi

There is an extensive chemistry of organo compounds, especially of phosphorus and arsenic. This was developed largely because of the physiological properties of these compounds. Thus one of the first chemotherapeutic agents, salvarsan, which was discovered by Ehrlich, led to wide study of arylarsenic compounds.

The so-called "organophosphorus" compounds that have anti-cholinesterase activity and are widely used as insecticides do *not* contain P—C bonds, but are P^V derivatives such as phosphates or thionates. For example, parathion is $(EtO)_2P(S)(OC_6H_4NO_2)$.

Most of the genuine organo derivatives are compounds with only three or four bonds to the central atom, although a few R_5M compounds are known. The simplest synthesis is the reaction

$$(O)MX_3 + 3RMgX \longrightarrow (O)MR_3 + 3MgX_2$$

Trimethylphosphine is spontaneously flammable in air, but the higher trialkyls are oxidized more slowly. The R_3MO compounds, which may be obtained from the oxo halides as shown above or by oxidation of the corresponding R_3M compounds, are all very stable. The trialkyl- or alkyl-arylphosphines are usually liquids with an unpleasant odor. The triarylphosphines are white crystalline solids reasonably stable in air. Tertiary phosphines, arsines and stibines are all good π-acid ligands for d-group transition metals (page 489). The oxides, R_3MO, also form many complexes, but they function as simple donors. Trialkyl- and triarylphosphines, -arsines, and -stibines generally react with alkyl and aryl halides to form quaternary salts:

$$R_3M + R'X \longrightarrow [R_3R'M]^+X^-$$

The tetraphenylphosphonium and -arsonium ions are useful for precipitating large anions such as ReO_4^-, ClO_4^-, and complex anions of metals.

An important phosphonium compound is obtained by the reaction

$$PH_3 + 4HCHO + HCl(aq) = [P(CH_2OH)_4]^+Cl^-$$

It is a white crystalline solid, soluble in water, and is used in the flameproofing of fabrics.

Triphenylphosphine, as well as being an important ligand, is utilized in the Wittig reaction for olefin synthesis. This reaction involves the formation of alkylidenetriphenylphosphoranes from the action of butyllithium or other base on the quaternary halide, for example,

$$[(C_6H_5)_3PCH_3]^+Br^- \xrightarrow{n\text{-Butyllithium}} (C_6H_5)_3P{=}CH_2$$

This intermediate reacts very rapidly with aldehydes and ketones to give zwitter-ionic compounds (29-V), which eliminate triphenylphosphine oxides under mild conditions to give olefins (29-VI):

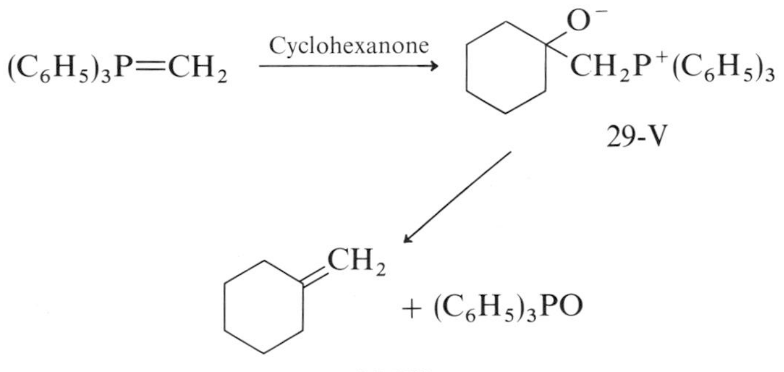

$$(C_6H_5)_3P{=}CH_2 \xrightarrow{\text{Cyclohexanone}} C_6H_{10}(O^-)CH_2P^+(C_6H_5)_3$$

29-V

$$\longrightarrow C_6H_{10}{=}CH_2 + (C_6H_5)_3PO$$

29-VI

Alkylidenephosphoranes such as $Me_3P{=}CH_2$, $Et_3P{=}CH_2$, $Me_2EtP{=}CH_2$ and $Et_3P{=}CHMe$, are all colorless liquids, stable for long periods in an inert atmosphere.

PART II
TRANSITION METALS

For transition metals, σ-bonded alkyls or aryls are stable only under special circumstances. Unstable or labile species with σ bonds to carbon are of great significance particularly in catalytic reactions of alkenes and alkanes induced by transition metals or metal complexes. Transition metal to carbon σ bonds also exist in Nature in Vitamin B_{12} derivatives (Section 31-5).

The unique characteristics of d orbitals allow the binding to metal atoms of unsaturated hydrocarbons and other molecules. The bonding is nonclassical and the metal complexes of alkenes, alkynes, arenes, and the like, have no counterparts elsewhere in chemistry.

29-11
Transition Metal to Carbon σ Bonds

Although the compound $[(CH_3)_3PtI]_4$, which has a structure based on a cube with Pt and I atoms at alternate corners and each Pt bound to three CH_3 groups, was made in 1909 by Pope and Peachy, attempts to prepare compounds such as $(C_2H_5)_3Fe$ by reactions of Grignard reagents with metal halides failed. Although evidence indicated that alkyls were present in solution at low temperatures, complicated decomposition and coupling reactions occurred at ambient temperatures.

About 20 years ago it was found that, provided other ligands such as the η^5-cyclopentadienyl group described later in this Chapter, or those of the π-acid type (Chapter 28) were present, alkyl compounds could be isolated; one example is $CH_3Mn(CO)_5$. It now appears that the principal reason for the stability of these compounds is that the coordination sites required for decomposition reactions to proceed are blocked. The main reason for the instability of most binary alkyls or aryls is that they are coordinatively unsaturated, and there are easy pathways for thermodynamically possible decomposition reactions to occur. Possible decomposition pathways include homolysis of the M—C bond, which generates free radicals, as well as the transfer of a hydrogen atom from carbon to the metal. A particularly common reaction is the transfer from the β-carbon of the alkyl chain (Eq. 29-6)

$$M{-}CH_2{-}CH_2{-}R \rightleftharpoons \left[\begin{array}{l} H \\ | \quad\;\; CHR \\ M \leftarrow \| \\ \quad\;\;\; CH_2 \end{array}\right] \longrightarrow MH + CHR{=}CH_2 \qquad (29\text{-}6)$$

resulting in the elimination of olefin and formation of an M—H bond. The reverse of this reaction, that is, the formation of alkyls by addition of olefins to M—H bonds (cf. page 540) is of very great importance in catalytic reactions discussed in the next Chapter. Once the hydrogen has been transferred to metal, further reaction can occur to give the metal and hydrogen, or the hydrogen can be transferred to the alkene to form alkane. Thus it has been shown that the copper alkyl, $(Bu_3P)CuCH_2C(Me)_2Ph$, decomposes largely by a free radical pathway but that the similar alkyl, $(Bu_3P)CuCH_2CH_2CH_2CH_3$, decomposes by a nonradical pathway involving Cu—H bond formation. The difference is that the latter, but not the former, has a hydrogen atom on the second, β-carbon atom.

There are a number of reasonably thermally stable alkyls that cannot undergo the β-hydride-transfer, alkene-elimination reaction. These have groups such as $-CH_2C_6H_5$, $-CH_2SiMe_3$, $-CH_2CMe_3$, $-CH_2\overset{+}{P}Me_3$ and 1-norbornyl (29-VII).

M

29-VII

Although hydrogen transfer from an α-carbon atom to produce a hydrido carbene intermediate as the first step in decomposition (Eq. 29-7)

$$M{-}CHRR' \longrightarrow \overset{\displaystyle H}{\underset{|}{}}\!\!\!\!\!M{=}CRR' \qquad (29\text{-}7)$$

is possible, this is evidently less favorable than the β-transfer and is rarely observed. Methylmetal compounds, such as $(CH_3)_4Cr$, or the $[(CH_3)_3PtI]_4$ already mentioned, are accordingly much more stable than the homologous ethylmetal compounds. However, even $Ti(CH_3)_4$ decomposes at *ca.* $-50°$, but on addition of ligands such as bipyridine, which leads to coordinative saturation as in $Ti(bipy)(CH_3)_4$, a substantial increase in thermal stability results. This shows again the necessity of having coordination sites on the metal available in order to allow decomposition reactions to proceed. Another striking example of this principle is that substitution–inert complexes (page 143) of Cr^{III}, Co^{III}, and Rh^{III} may have M—C bonds even when H_2O or NH_3 are ligands; one example is $[Rh(NH_3)_5C_2H_5]^{2+}$. Particularly important are the cobalt complexes of the vitamin B_{12} type and their synthetic analogs discussed in Chapter 31. One example is the dimethylglyoxime complex (29-VIII).

Some representative examples of alkyls are given in Table 29-1.

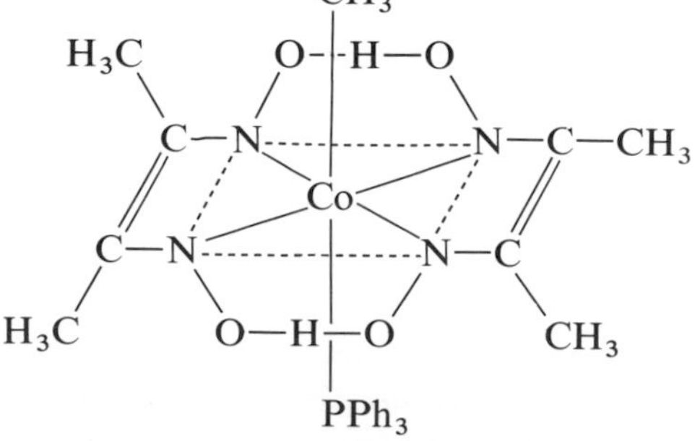

CH_3
H_3C O-·H—O
C—N N—C—CH_3
Co
C—N N—C
H_3C O—H—O CH_3
PPh_3

29-VIII

Table 29-1 Some Binary Transition Metal Alkyls

Compound	Properties
$Ti(CH_2Ph)_4$	Red crystals, mp 70°; tetrahedral
$VO(CH_2SiMe_3)_3$	Yellow needles, mp 75°; has V=O bond
Cr-(1-norbornyl)$_4$	Red brown crystals; tetrahedral; d^2 paramagnetic
$Mo_2(CH_2SiMe_3)_6$	Yellow plates, d. 135°; has Mo—Mo triple bond
$Re(CH_3)_6$	Green crystals; octahedral; d^1 paramagnetic

29-12 Alkene Complexes

About 1830, Zeise, a Danish pharmacist, characterized compounds that had stoichiometries $PtCl_2C_2H_4$ and $K[PtCl_3C_2H_4]$. Although these were the first organo derivatives of transition metals to be prepared, their true nature was fully established only around 1953.

Ethylene and most other alkenes can be bound to transition metals in a wide variety of complexes. The structures of two such are shown in *Fig. 29-3*. The fact that the plane of the olefin, and the C=C axis are perpendicular to one of the expected bond directions from the central metal atom is of key significance. In addition, the expected line of a bond orbital from the metal atom strikes the C=C bond at its midpoint.

The most useful description of the bonding in alkene complexes is illustrated in *Fig. 29-4*. The bonding consists of two interdependent components: (a) overlap of the π-electron density of the alkene with a σ-type acceptor orbital on the metal atom; and (b) a "back-bond" resulting from flow of electron density from filled metal d_{xz} or other $d\pi$—$p\pi$ hybrid orbitals into the π*-*antibonding* orbital on the carbon atoms. It is thus similar to that discussed for the bonding of CO and other

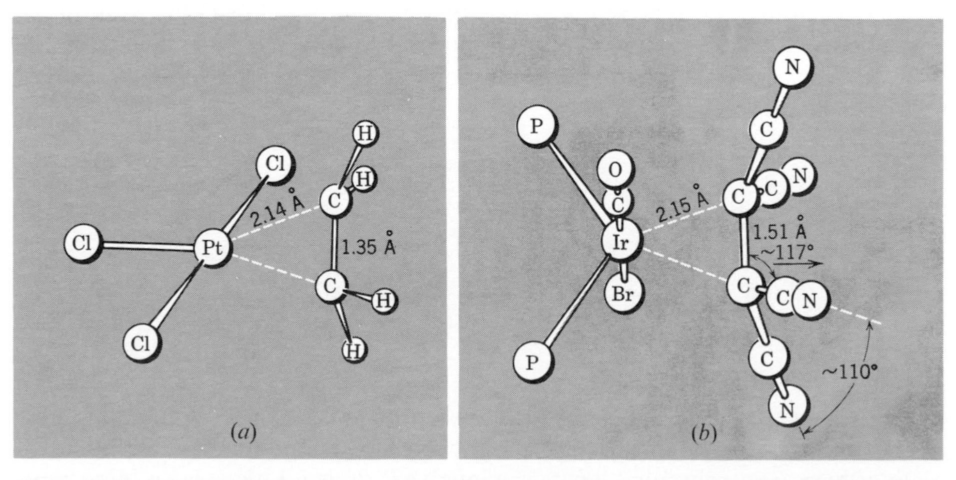

Figure 29-3 *The structures of two monoalkene complexes. (a) The* $[PtCl_3C_2H_4]^-$ *anion of Zeise's salt. (b) The* $(Ph_3P)_2(CO)BrIr[C_2(CN)_4]$ *molecule, with phenyl groups omitted for clarity.*

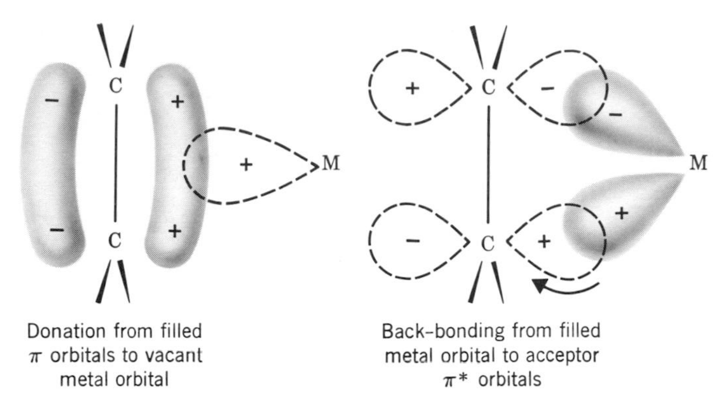

Figure 29-4 *Diagrams showing the molecular orbital view of alkene metal bonding. The donor part of the bond is shown at the left, and the back-bonding part at the right.*

π-acid ligands (Chapter 28) and implies the retention of appreciable "double-bond" character in the olefin. Of course, the donation of π-bonding electrons to the metal σ orbital and the introduction of electrons into the π^*-antibonding orbital both weaken the π bonding in the olefin, and in every case except the anion of Zeise's salt there is significant lengthening of the olefin C—C bond. There appears to be some correlation between lengthening of the bond and the electron-withdrawing power of the substituents of the olefin. This is exemplified by the structures shown in *Fig. 29-3* where the $C_2(CN)_4$ complex has a C—C bond about as long as a normal single bond.

In the extreme of a very long C—C distance the bonding could be formulated as a kind of metallo-cyclopropane ring, involving two $2c$–$2e$ M—C bonds and a C—C single bond. In *Fig. 29-3b* the bond angles at the two olefin carbon atoms are consistent with this view. This representation of the bonding and the MO description are complementary, and there is a smooth gradation of one description into the other.

Alkenes with unconjugated double bonds can form independent linkages to a metal atom. Two representative complexes, of 1,5-cyclooctadiene and norbornadiene, are (29-IX) and (29-X), respectively. Three unconjugated double bonds may be coordinated to one metal atom as in the *trans*, *trans*, *trans*-cyclododecatriene complex (29-XI).

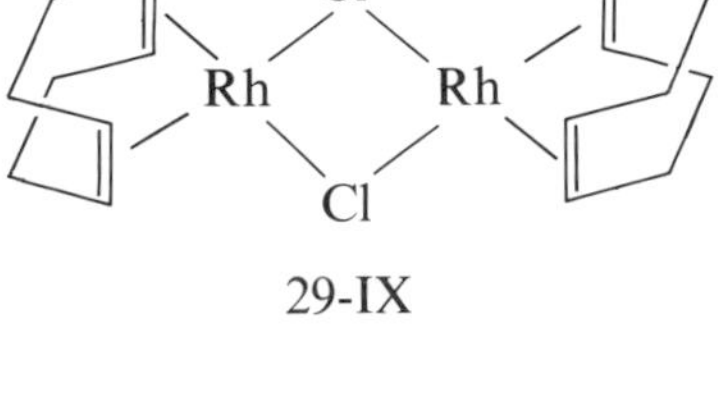

Cl Rh Rh Cl — 29-IX

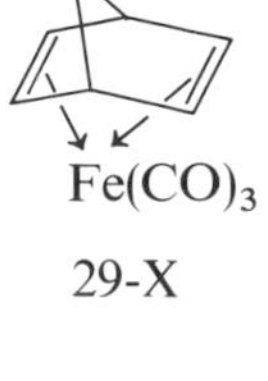

$Fe(CO)_3$ — 29-X

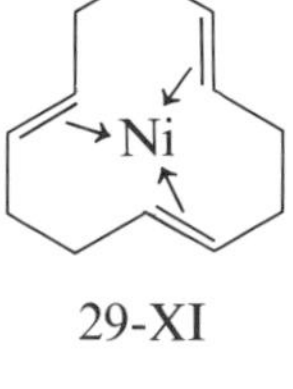

Ni — 29-XI

When two or more *conjugated* double bonds are engaged in bonding to a metal atom the interactions become more complex, though qualitatively the two types of basic, synergic components are involved. Buta-1,3-diene is an important case and shows why it is an oversimplification to treat the bonding as simply collections of separate monoolefin-metal interactions.

Figure 29-5 *Two extreme formal representations of the bonding of a buta-1,3-diene group to a metal atom. Part (a) implies that there are two more or less independent monoolefin-metal interactions, while (b) depicts σ bonds to* C-1 *and* C-4 *coupled with a mono-olefin-metal interaction to* C-2 *and* C-3.

Two extreme formal representations of the bonding of a buta-1,3-diene group to a metal atom are shown in *Fig. 29-5*. The degree to which individual structures approach either of these extremes can be judged by the lengths of the C—C bonds. A short–long–short pattern is indicative of (a) while a long–short–long pattern is indicative of (b). In no case has a pronounced short–long–short pattern been established and the actual variation seems to lie between approximate equality of all three bond lengths and the long–short–long pattern.

From a purely *formal* point of view each double bond in any olefin can be considered as a 2-electron donor. If we have a polyolefin involved, the metal atom usually reacts so as to complete its normal coordination. For example, both $Mo(CO)_6$ and $Fe(CO)_5$ react with cyclohepta-1,3,5-triene to give (29-XII) and (29-XIII) respectively. In (29-XIII) there is one uncoordinated double bond.

$Cr(CO)_3$

29-XII

$Fe(CO)_3$

29-XIII

Cyclooctatetraene with four essentially unconjugated double bonds can bind in several ways depending on the metal system. With $PtCl_2$, it uses its 1- and 5-olefinic linkages, as in (29-XIV), and with $Fe(CO)_3$, which has a predilection for binding 1,3-diolefins, it is bound as in (29-XV).

Cl
Pt
Cl

29-XIV

Fe—CO
OC
CO

29-XV

Synthesis. Olefin complexes are usually synthesized by interaction of metal carbonyls, halides, or occasionally other complexes with the olefin. Some representative examples are

$$Mo(CO)_6 + C_7H_8 \xrightarrow{\text{reflux}} Mo(CO)_3C_7H_8$$

$$RhCl_3(aq) + C_2H_4 \xrightarrow{25^\circ \text{ in ethanol}} [(C_2H_4)_2RhCl]_2$$

Some of the earliest studies were made with the Ag^+ ion and in solutions we have equilibria of the type

$$Ag^+(aq) + \text{olefin} = [\text{Ag olefin}]^+$$

The interaction of hydrocarbons with Ag^+ ions sometimes gives crystalline precipitates that are useful for purification of the olefin. Thus cyclooctatetraene or bicyclo-2,5-heptadiene, when shaken with aqueous silver perchlorate (or nitrate), give white crystals of stoichiometry olefin·$AgClO_4$ or 2 olefin·$AgClO_4$, depending on the conditions. Benzene also gives crystalline complexes with $AgNO_3$, $AgClO_4$, or $AgBF_4$. In $[C_6H_6 \cdot Ag]^+ClO_4$ the metal ion is asymmetrically located with respect to the ring.

29-13
Notation in Alkene and Other Related Complexes

In addition to alkene complexes there are more complex systems in which allylic and other delocalized π systems are bound to metals. Some systematic notation is required to designate the number of carbon atoms that are bound to the metal. This is done by use of the term *hapto* (from the Greek, to fasten), for example, *trihapto-*, *tetrahapto-*, *pentahapto-*, etc., designated as η^3-, η^4-, η^5-, etc. If necessary, the bound carbon atoms may be specified by numbers, using a conventional naming and numbering scheme for the organic group. The formulas (29-XVI), (29-XVII), and (29-XVIII) should make clear the use of the notation.

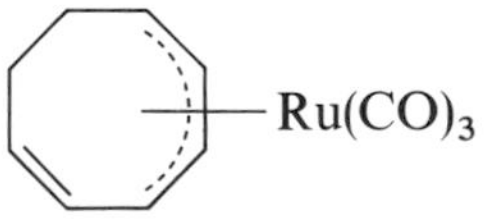

Ru(CO)$_3$

29-XVI

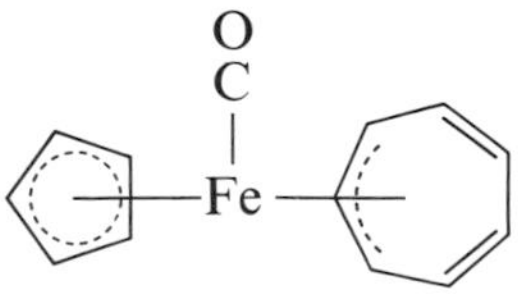

O C Fe

29-XVII

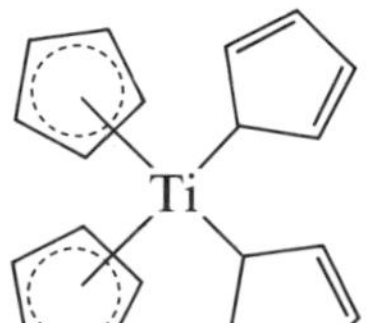

Ti

29-XVIII

(XVI) 1-4-*tetrahapto*,1,3,5-cyclooctatrienetricarbonylruthenium
(XVII) (*pentahapto*cyclopentadienyl)(1-3-*trihapto*cycloheptatrienyl)carbonyliron
(XVIII) $(\eta^5\text{-}C_5H_5)_2(\eta^1\text{-}C_5H_5)_2Ti$

29-14
Compounds of Delocalized Carbocyclic Groups

In 1951 a compound of formula $(C_5H_5)_2Fe$ was reported and was subsequently shown to have a unique "sandwich" structure (29-XIX) in which the metal lies between two planar cyclopentadienyl rings. Many η^5-C_5H_5 compounds are now known. Some have only one η^5-C_5H_5 ring as in (29-XX), others have two rings but with these at an angle as in (29-XXI).

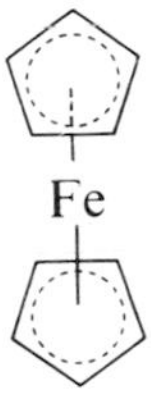

Fe
29-XIX

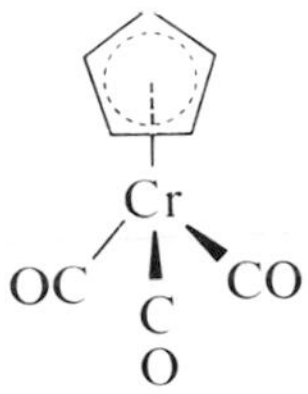

Cr OC CO C O
29-XX

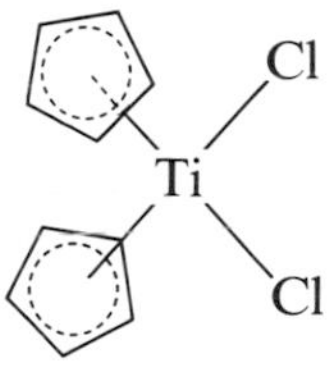

Ti Cl Cl
29-XXI

Other symmetric ring systems now known to form complexes are C_3Ph_3, C_4H_4, C_6H_6, C_7H_7, and C_8H_8. There is a formalism of describing these ring systems as if they assume the charge required to achieve an aromatic electron configuration. The "magic numbers" for aromaticity are 2, 6, and 10 so that these carbocycles can be written as:

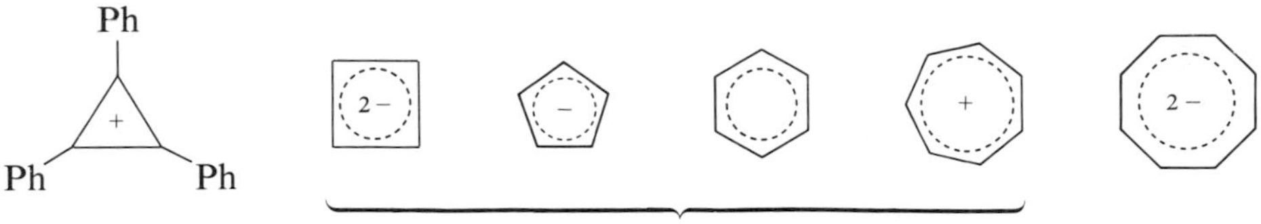

Ph Ph Ph + 2e

2− − + 6e

2− 10e

The charges may be used in assigning formal oxidation numbers to the metal atoms in the complexes. Thus $(\eta^5$-$C_5H_5)_2Fe$ can be regarded as formed from the cyclopentadienide ion $C_5H_5^-$ and Fe^{2+} so that the compound contains Fe^{II}. In the benzene compound $C_6H_6Cr(CO)_3$ chromium has the formal oxidation state 0 as in $Cr(CO)_6$.

Examples of carbocyclic complexes are (29-XXII to 29-XXV).

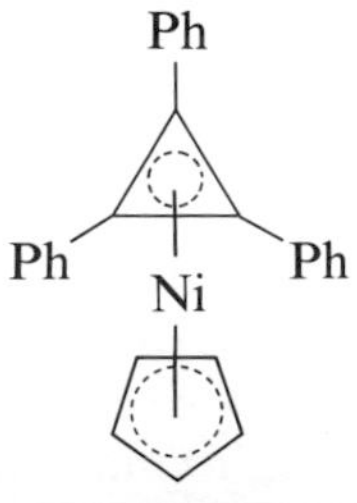

Ph Ph Ph Ni
29-XXII

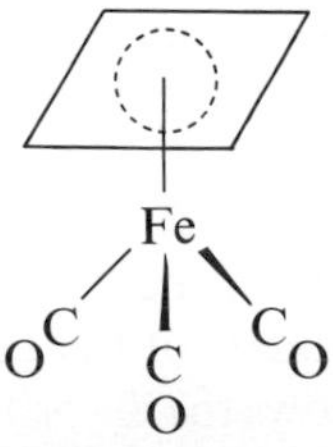

Fe OC CO C O
29-XXIII

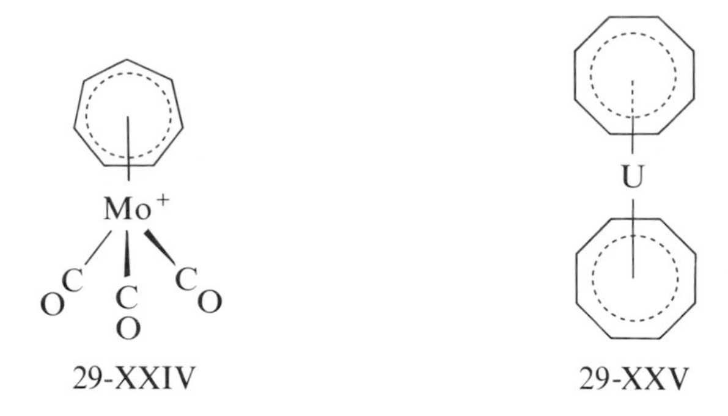

29-XXIV 29-XXV

Cyclopentadienyls. Cyclopentadiene is a weak acid (pK_a ~ 20) and with strong bases forms the cyclopentadienide ion $C_5H_5^-$. The general method for synthesizing metal complexes is the reaction of this ion with a metal halide or other complex, for example,

$$C_5H_6 + Na \xrightarrow{\text{THF}} C_5H_5^- + Na^+ + \tfrac{1}{2}H_2 \text{ (main reaction)}$$

$$2C_5H_5^- + NiCl_2 \xrightarrow{\text{THF}} (\eta^5\text{-}C_5H_5)_2Ni + 2Cl^-$$

Two other methods are (a) the use of C_5H_5Tl, which is insoluble in water, stable, and easily stored:

$$C_5H_6 + TlOH \xrightarrow{H_2O} C_5H_5Tl\downarrow + H_2O$$

$$FeCl_2 + 2TlC_5H_5 \xrightarrow{\text{THF}} 2TlCl + (\eta^5\text{-}C_5H_5)_2Fe$$

and (b), the use of a strong organic base as proton acceptor,

$$2C_5H_6 + CoCl_2 + 2Et_2NH \xrightarrow[\text{amine}]{\text{in excess}} (\eta^5\text{-}C_5H_2)_2Co + 2Et_2NH_2Cl$$

Since the $C_5H_5^-$ anion acts as a uninegative ligand, the dicyclopentadienyl compounds are of the type $(\eta^5\text{-}C_5H_5)_2MX_{n-2}$ where the oxidation state of the metal M is n and X is a uninegative ion. When $n = 2$ we obtain neutral molecules like $(\eta^5\text{-}C_5H_5)_2Fe^{II}$. When $n = 3$, we may obtain a cation like $[(\eta^5\text{-}C_5H_5)_2Co^{III}]^+$ or, when $n = 4$, a halide like $(\eta^5\text{-}C_5H_5)_2Ti^{IV}Cl_2$. Some typical η^5-cyclopentadienyl compounds are given in Table 29-2. The C—C distance and bond order in $\eta^5\text{-}C_5H_5$ rings are similar to C—C distances in benzene. For two compounds, $(\eta^5\text{-}C_5H_5)_2Fe$, which has been given the trivial name *ferrocene*, and $(\eta^5\text{-}C_5H_5)Mn(CO)_3$ or *cymantrene*, aromatic-like reactions can be carried out. These compounds will survive the reaction conditions, but other $\eta^5\text{-}C_5H_5$ compounds are decomposed. Typical reactions are Friedel–Crafts acylation, metallation by butyllithium, sulfonation, etc. Indeed, there is an extensive "organic chemistry" of these molecules.

Bonding in η^5-C_5H_5 Metal Compounds. An MO treatment gives a good description of the bonding. The main source of bonding is the overlap between the d_{xz} and d_{yz} orbitals of the metal atom with the $p\pi$ orbitals of the ring, or rings. *Figure 29-6* shows the d_{xz} orbital overlaps; that of the d_{yz} orbital is precisely equivalent but occurs in the plane perpendicular to that shown. Because these two

Table 29-2 Some Di-η^5-cyclopentadienylmetal Compounds

Compound	Appearance; mp (°C)	Unpaired electrons	Other properties[a]
$(\eta^5\text{-}C_5H_5)_2Fe$	Orange crystals; 174	0	Oxidized by Ag^+(aq) or dil. HNO_3 to blue cation $\eta^5\text{-}Cp_2Fe^+$. Stable thermally to >500°
$(\eta^5\text{-}C_5H_5)_2Cr$	Scarlet crystals; 173	2	Very air-sensitive
$(\eta^5\text{-}C_5H_5)_2Co^+$	Yellow ion and salts	0	Forms numerous salts and a stable strong base (absorbs CO_2 from air); thermally stable to ~400°
$(\eta^5\text{-}C_5H_5)_2TiCl_2$	Bright red crystals; 230	0	C_6H_5Li gives $\eta^5\text{-}Cp_2Ti(C_6H_5)_2$; reducible to $\eta^5\text{-}Cp_2TiCl$
$(\eta^5\text{-}C_5H_5)_2WH_2$	Yellow crystals; 163	0	Moderately stable in air, soluble benzene, etc.; soluble in acids giving $\eta^5\text{-}Cp_2WH_3^+$ ion

[a] Cp = C_5H_5.

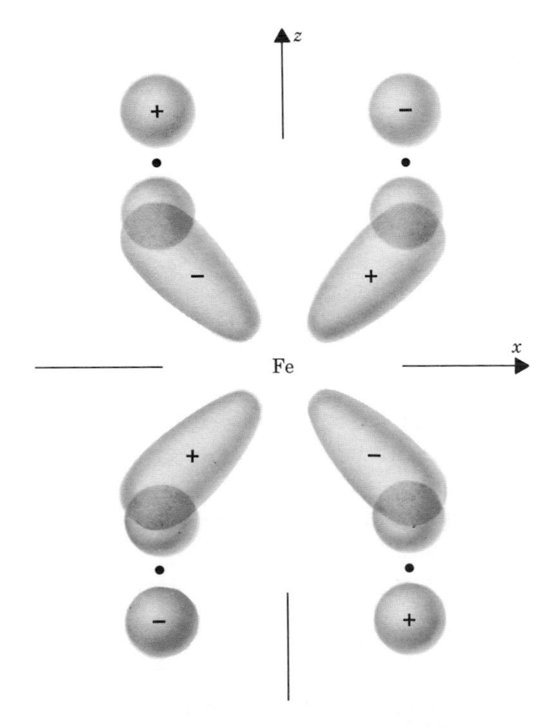

Figure 29-6 *A sketch showing how the d_{xz} orbital overlaps with a ring π orbital to give a delocalized metal-ring bond. The view is a cross-section taken in the xz plane.*

mutually perpendicular bonds are equivalent, the rings are able to rotate freely about the axis from the metal to the ring center.

In the compounds with only one η^5-C_5H_5 ring the lobes of the *d* orbitals not involved in bonding to the ring can overlap with suitable orbitals in other ligands such as CO, NO, R_3P, etc. Observe that only in the neutral compounds and $(\eta^5\text{-}C_5H_5)_2M^+$ are the rings parallel; in other compounds such as (29-XXI) the rings are at an angle.

Benzenoid—Metal Complexes. Of other carbocycles, those containing benzene and substituted benzenes are the most important. Curiously, the first $(\eta^6\text{-}C_6H_6)M$ compounds were prepared as long ago as 1919, but their true identities were recognized only in 1954. A series of chromium compounds was obtained by Hein from the reaction of $CrCl_3$ with C_6H_5MgBr; they were formulated as "polyphenylchromium" compounds, namely, $(C_6H_5)_nCr^{0,1+}$ where $n = 2$, 3, or 4. They actually contain "sandwich"-bonded C_6H_6 and C_6H_5—C_6H_5 groups as, for example, in 29-XXVI.

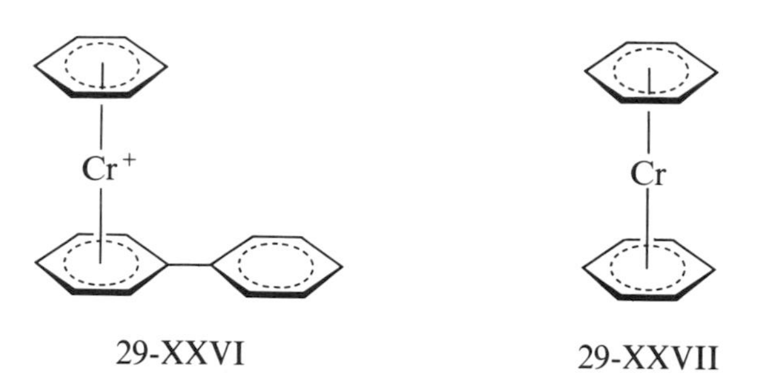

29-XXVI 29-XXVII

The prototype neutral compound, *dibenzenechromium*, $(C_6H_6)_2Cr$ (29-XXVII), has also been obtained from the Grignard reaction of $CrCl_3$. A more effective method applicable to other metals is the direct interaction of an aromatic hydrocarbon and a metal halide in presence of Al powder as a reducing agent and halogen acceptor plus $AlCl_3$ as a Friedel–Crafts-type activator. Although the neutral species are formed directly in the case of chromium, the usual procedure is to hydrolyze the reaction mixture with dilute acid which gives the cations $(C_6H_6)_2Cr^+$, $(\text{mesitylene})_2Ru^{2+}$, etc. These cations may then be reduced to the neutral molecules.

Dibenzenechromium, which forms dark brown crystals, is much more sensitive to air than is ferrocene, with which it is isoelectronic; it does not survive the reaction conditions of aromatic substitution. As with the η^5-C_5H_5 compounds complexes with only one arene ring can be prepared:

$$C_6H_5CH_3 + Mo(CO)_6 \xrightarrow{\text{reflux}} C_6H_5CH_3Mo(CO)_3 + 3CO$$

$$C_6H_6 + Mn(CO)_5Cl + AlCl_3 \longrightarrow C_6H_6Mn(CO)_3{}^+AlCl_4{}^-$$

The cyclooctatetraenyl ion, $C_8H_8{}^{2-}$ forms similar sandwich compounds with actinides, for example, $(\eta^8\text{-}C_8H_8)_2U^{IV}$ (29-XXV). It appears that *f* orbitals are involved in the bonding here.

29-15
Alkyne Complexes

In acetylenes there are *two* π-bonds at 90° to each other and both can be bound to a metal as in (29-XXVIII). The Co atoms and the acetylene carbon atoms form a distorted tetrahedron, and the C_6H_5 (or other groups) on the acetylene are bent away as shown.

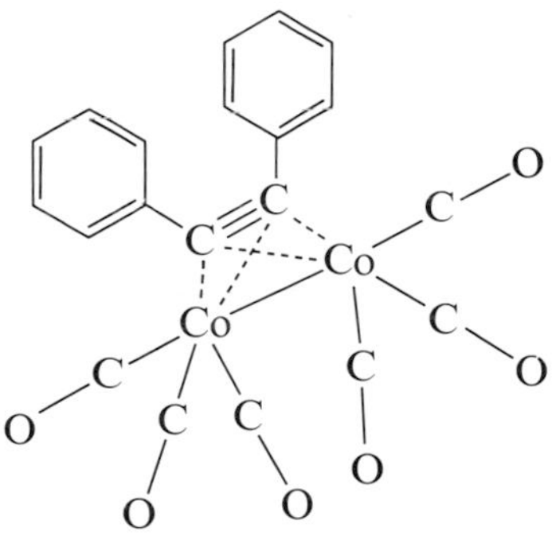
C
C
Co
Co
C
O
C
O
C
O
C
O
C
O
C
O

29-XXVIII

There are also complexes where the alkyne is coordinated to only one metal atom and serves simply as the equivalent of an olefin or carbon monoxide ligand. Thus we have the reactions:

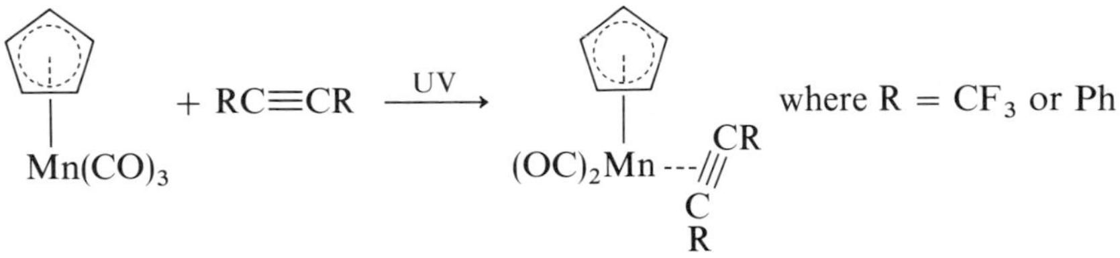
Mn(CO)3
+ RC≡CR
UV
(OC)2Mn
CR
C
R
where R = CF3 or Ph

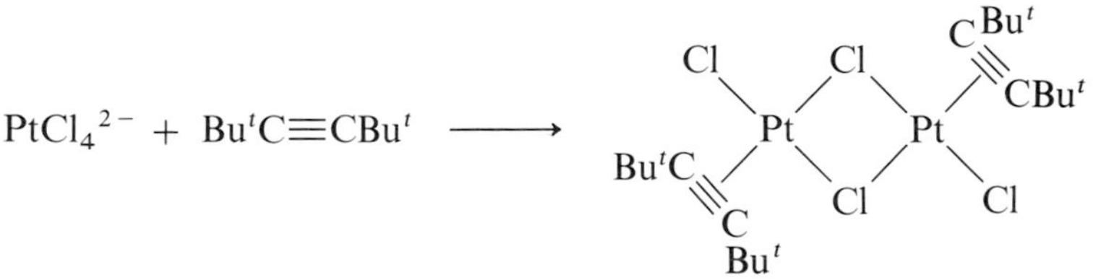
PtCl4 2−
+ Bu'C≡CBu'
Cl
Cl
CBu'
CBu'
Pt
Pt
Bu'C
Cl
Cl
C
Bu'

A third way of bonding, notably in Pt, Pd, and Ir complexes is that shown in *Fig. 29-7*. In these the C—C stretching frequency is lowered considerably, to the

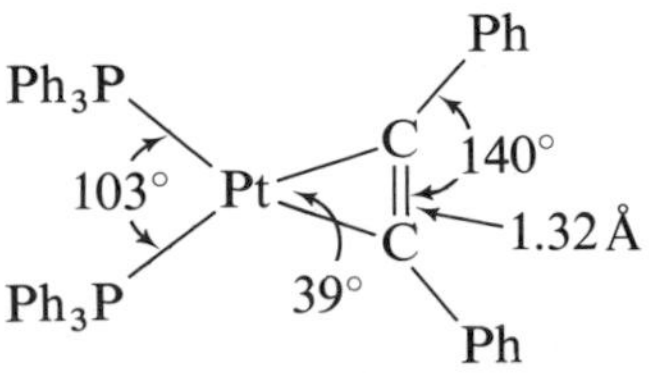
Ph
Ph3P
C
140°
103°
Pt
C
1.32 Å
Ph3P
39°
Ph

Figure 29-7 *The structure of* $(Ph_3P)_2Pt(PhC_2Ph)$, *in which diphenylacetylene is most simply formulated as a divalent, bidentate ligand.*

range 1750 to 1770 cm^{-1}, indicative of a C—C double bond. The C—C bond length of 1.32 Å is consistent with this view, as is the large distortion from linearity.

Finally, many important reactions of acetylenes, especially with metal carbonyls, involve incorporation of the acetylenes into rings, thus producing species with new organic ligands bound to the metals. Some examples are the following:

$$Fe(CO)_5 + 2C_2H_2 \longrightarrow (C_4H_4CO)Fe(CO)_3 + [(C_5H_4O)Fe(CO)_2]_2$$

$$(\eta^5\text{-}C_5H_5)Co(CO)_2 + 2R_2C_2(R = CH_3 \text{ or } CF_3) \longrightarrow (\eta^5\text{-}C_5H_5)Co(C_4R_4CO)$$

$$Fe(CO)_5 + 2C_2(CH_3)_2 \xrightarrow{h\nu} (C_6(CH_3)_4O_2)Fe(CO)_3$$

29-16 Allyl Complexes

The allyl group $CH_2{\cdots}CH{\cdots}CH_2$ can be bound to metals in a delocalized (trihapto) fashion and can be considered to behave as a 3-electron donor ligand. A representative structure is shown in *Fig. 29-8*. The allyl group can also be bound by a σ bond and transformation between σ or η^1 allyls and π or η^3 allyls may be quite facile:

$$M{-}CH_2{-}CH{=}CH_2 \rightleftharpoons M{-}(\eta^3\text{-}CH_2CHCH_2)$$

σ-allyl π or η^3-allyl

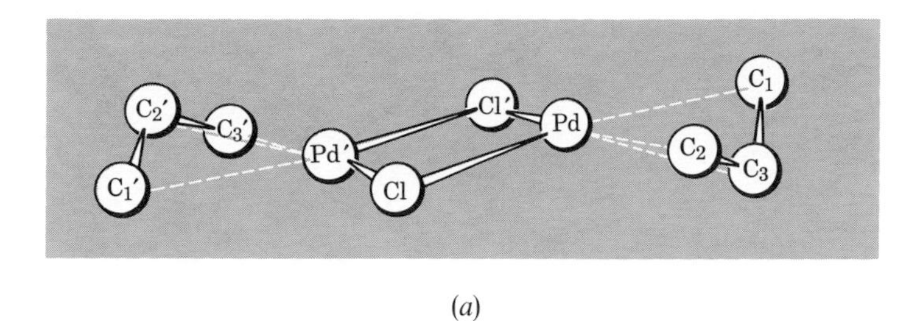

(a)

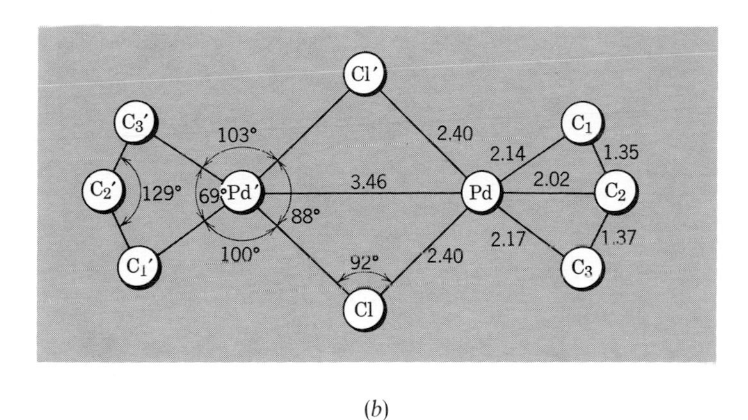

(b)

Figure 29-8 *Structure of allylpalladium(II) chloride dimer. (a) side view, and (b) top view. (Reprinted by permission from* J. Organometallic Chem., *1965*, 3, *43*.)

Allyl complexes may be obtained from allyl Grignard reagents or from allyl chloride as in the reactions:

$$NiCl_2 + 2C_3H_5MgBr \xrightarrow[-10^\circ]{\text{Ether}} Ni(\eta^3\text{-}C_3H_5)_2 + 2MgX_2$$

$$Na^+\left[(C_5H_5)Mo(CO)_3\right]^- + C_3H_5Cl \longrightarrow (C_5H_5)Mo(CO)_3(H_2C{-}CH{=}CH_2) \xrightarrow[\text{uv}]{\Delta\text{ or}} (C_5H_5)Mo(CO)_2(\eta^3\text{-}C_3H_5)$$

They can also be obtained by protonation of butadiene complexes, for example,

$$(C_4H_6)Fe(CO)_3 + HCl \longrightarrow (\eta^3\text{-}CH_2CHCHCH_3)Fe(CO)_3Cl$$

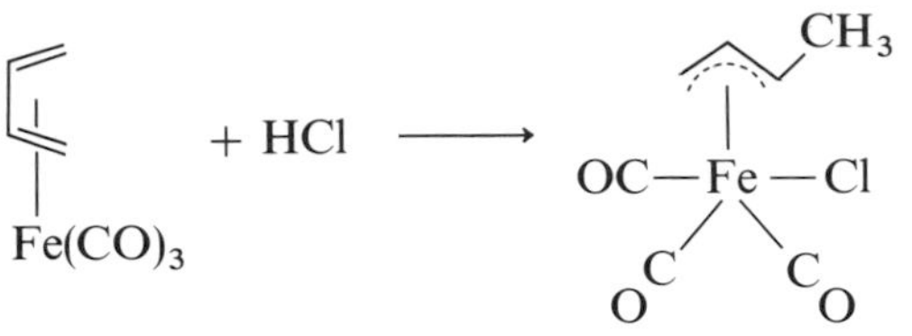

Allyl complexes play an important role in many catalytic reactions, particularly those involving conjugated alkenes.

29-17
Carbene Complexes

Although carbenes, $:CR_2$, are short-lived in the free state, many organic reactions proceed by way of carbene intermediates. An increasing number of compounds are now known in which carbenes are "stabilized" by binding to a transition metal. A carbene could be regarded as a 2-electron donor comparable to CO, since there is a lone pair present and the carbon atom is formally divalent. General methods of synthesis involve attacks of nucleophilic reagents on coordinated CO or RNC, for example,

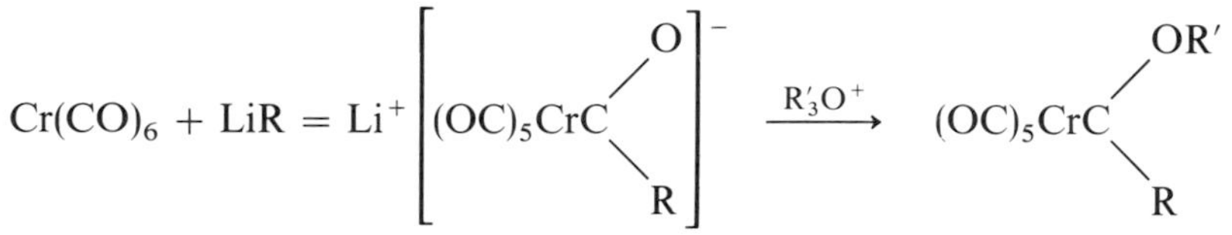

$$Cr(CO)_6 + LiR = Li^+\left[(OC)_5CrC{\textstyle{O \atop R}}\right]^- \xrightarrow{R'_3O^+} (OC)_5CrC{\textstyle{OR' \atop R}}$$

Reactions of certain electron-rich olefins can also lead to cleavage of the C=C bond and formation of carbene complexes, for example,

$$(PhNCH_2CH_2NPh)C{=}C(PhNCH_2CH_2NPh) + [PtCl_2(PEt_3)]_2 \xrightarrow[\text{xylene}]{\text{reflux}} Et_3P\text{—}PtCl_2\text{—}C(NPhCH_2CH_2NPh)$$

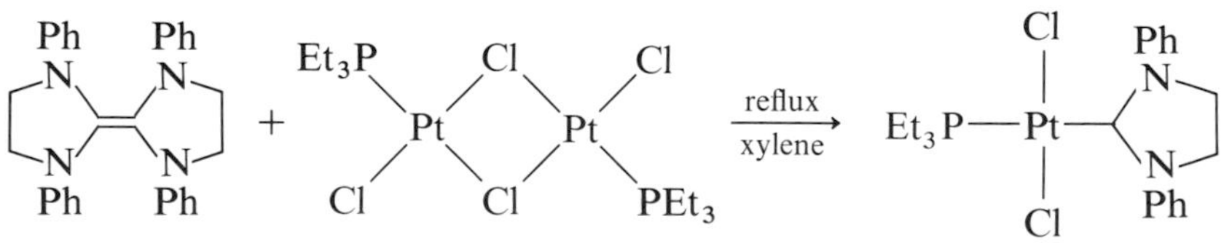

Structural studies on carbenes, for example, (29-XXIX) and (29-XXX) show that the M—CXX′ skeleton is always *planar* while the M—C distances indicate multiple bonding as in the bonding of π-acid ligands.

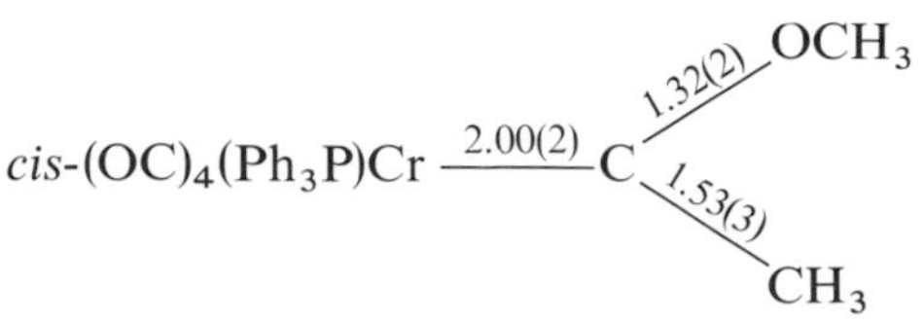

cis-$(OC)_4(Ph_3P)Cr\overset{2.00(2)}{\text{—}}C(\overset{1.32(2)}{\text{—}}OCH_3)(\overset{1.53(3)}{\text{—}}CH_3)$

29-XXIX

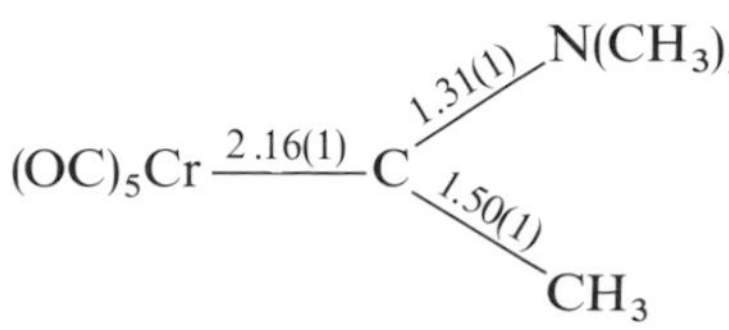

$(OC)_5Cr\overset{2.16(1)}{\text{—}}C(\overset{1.31(1)}{\text{—}}N(CH_3)_2)(\overset{1.50(1)}{\text{—}}CH_3)$

29-XXX

Study Questions

A

1. Give a definition of an organometallic compound.
2. What are the three broad classes of organometallic compounds? Cite an example of each.
3. Describe at least three important general methods for preparing organometallic compounds.
4. What is the most characteristic structural feature of lithium alkyls?
5. Why are the tetramethylethylenediamine complexes of the lithium alkyls more reactive than the alkyls themselves?
6. What is a formula for a Grignard reagent and how are Grignard reagents prepared? What is the other general type of organomagnesium compound?
7. What sorts of species are believed to actually exist, in equilibrium, in a Grignard reagent in diethyl ether solution?
8. Give an example of a metal interchange reaction involving an organomercury compound.
9. Indicate with sketches the structures of the following:

$$LiCH_3,\ MgCH_3Br[O(C_2H_5)_2]_2,\ Hg(CH_3)_2,\ Al(CH_3)_3,\ (CH_3)_3SnF.$$

10. Write an equation illustrating each of the terms: hydroboration reaction, mercuration, oxymercuration.
11. How would you prepare each of the following: $NaBPh_4$ (from BCl_3); cyclopropylmercury bromide (from mercury); diethylzinc (from zinc); trimethylaluminum (from aluminum)?
12. What are siloxanes? Silicones? How are they made?
13. What is an alkylidenetriphenylphosphorane (Wittig reagent)? How are they made and what are they used for?
14. Why is $Ti(bipy)(CH_3)_4$ much more thermally stable than $Ti(CH_3)_4$?
15. Explain mechanistically why transition metal alkyls which have a β hydrogen atom are usually unstable, whereas analogous compounds in which the alkyls do not have β-hydrogen atoms generally are stable.
16. Besides β-hydrogen transfer, which is another important mode of decomposition of some metal alkyls?
17. Describe the structure of the anion, $[PtCl_3C_2H_4]^-$, in Zeise's salt, emphasizing the significant features on which an understanding of the metal-olefin bonding must be based.
18. Show with drawings the two important types of orbital overlap that explain the metal-olefin bonding in $[PtCl_3C_2H_4]^-$.
19. Show with drawings the expected structures of the following cyclooctatetraene (COT) complexes: $(COT)Cr(CO)_3$, $(COT)Fe(CO)_3$, $(COT)PtCl_2$.
20. Give the formal names for the following:

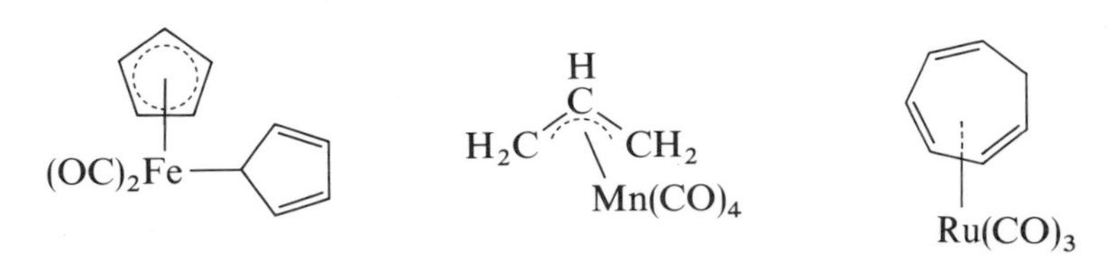

21. Write equations for a two-step preparation of $(\eta^5\text{-}C_5H_5)_2Ni$ from C_5H_6, Na, and $NiCl_2$.

22. List the five symmetric ring systems that are known to form carbocyclic complexes of the type $[(RC)_n]ML_x$.
23. How may alkynes be bound to transition metals?
24. State two ways of obtaining η^3-allyl complexes.
25. Show how carbene complexes can be obtained from metal carbonyls.

B

1. What are mercurinium ions and what part do they play in oxomercuration reactions?
2. Describe the bonding in (a) dimethylberyllium, and (b) trimethylaluminum in terms of multicenter bonds.
3. Discuss the mechanism of the synthesis of alkenes from aldehydes or ketones by use of the Wittig reaction.
4. Explain the following observations
   (a) Although the M—C force constants and presumably bond strengths are comparable, $PbMe_4$ begins to decompose by radical formation at *ca.* 200°, while $TiMe_4$ is unstable above −80°.
   (b) Alkyl halides R′X react with phosphate esters $P(OR)_3$ to give dialkylphosphonates $O{=}P(OR)_2R'$ and RX.
   (c) The compound $(CH_3)_2PBH_2$ is trimeric and is extraordinarily stable and inert.
   (d) At −75° the proton resonance spectrum of trimethylaluminum shows two resonances in the ratio 2:1 but at 25° only one peak at an average position is found.
5. Consider the interaction of a hydrido species L_nMH with hex-1-ene in benzene solution at 25° C.
   (a) Why and how are *cis*- and *trans*-hex-2-enes formed?
   (b) If L_nMD was used where would the deuterium finish up?
6. Assuming H transfer from CH_3 to metal is a plausible first step in decomposition of a methyl compound, write a mechanism for decomposition of $Ti(CH_3)_4$ in petroleum solution.
7. The interaction of $Na_2Fe(CO)_4$ with $(CH_3)_2NCH_2I$ gives the carbene complex $(CO)_4FeCHN(CH_3)_2$. Write a plausible reaction sequence.
8. The interaction of $(\eta^5\text{-}C_5H_5)_2Co^+$ salts with sodium borohydride as H^- source gives a red diamagnetic hydrocarbon-soluble product, $C_{10}H_{11}Co$. This has no band in the infrared around 2000 cm^{-1} and the nmr spectra is quite complex. The interaction of $(\eta^5\text{-}C_5H_5)_2Co$ with CH_3I gives $(\eta^5\text{-}C_5H_5)_2CoI$ and $C_{11}H_{13}Co$. Explain these reactions.
9. Compare the bonding of C_2H_4 and O_2 in the compounds $(Ph_3P)_2Pt(O_2)$ and $(Ph_3P)_2Pt(C_2H_4)$.

Chapter 29
Study Guide

Purpose and Scope. Although organometallic chemistry is a subarea of both inorganic and organic chemistry, it is such a large field that it could well be considered a full-fledged branch of chemistry. It draws on both inorganic and organic chemistry; yet, the whole is greater than the sum of its parts. In this one Chapter only a few of the most salient points can be covered. There are many new concepts of structure and bonding here that should be studied very carefully. Also, this Chapter is an indispensable prerequisite to the study of Chapter 30.

Supplementary Reading

Baker, R., "Pi-Allyl Complexes in Organic Synthesis," *Chem. Rev.*, *73*, 487 (1973).
Becker, E. I. and Tsutsui, M., *Organometallic Reactions*, Vols. 1–4, Wiley, 1970–1972.
Heck, R. F., *Organotransition Metal Chemistry; A Mechanistic Approach*, Academic Press, 1974.
Herberhold, M., *Metal Pi Complexes*, Elsevier, 1972.
Jolly, P. W. and Wilke, G., *The Organic Chemistry of Nickel*, Vol. 1, Academic Press, 1974.
King, R. B., *Transition Metal Organometallic Chemistry: An Introduction*, Academic Press, 1969.
Kosolapoff, G. M. and Maier, L., *Organophosphorus Compounds*, Wiley, 1972.
Maitlis, P. M., *The Organic Chemistry of Palladium*, Academic Press, 1971.
Matteson, D. E., *Organometallic Reaction Mechanisms of Nontransition Elements*, Academic Press, 1974.
Mole, T., and Jeffery, E. A., *Organoaluminum Compounds*, Elsevier, 1972.
Sawyer, A. K., *Organotin Compounds*, Vols. 1–3, Dekker, 1974.
Stone, F. G. A. and West, R., eds., *Advances in Organometallic Chemistry*, Vols. 1–12, Academic Press, 1963–1974.
Wakefield, B. J., *Organolithium Compounds*, Pergamon, 1974.

30

organometallic compounds in homogeneous catalytic reactions

The use of transition-metal complexes for the conversion of unsaturated hydrocarbons into polymers, alcohols, ketones, carboxylic acids, and the like, has generated an extensive patent as well as scientific-journal literature. The discovery of the low-pressure polymerization of ethylene and propene by Ziegler and Natta led to the wide use of aluminum alkyls as alkylating agents and reductants for metal complexes. Also the discovery by Smidt of palladium-catalyzed oxidation of alkenes stimulated an enormous growth in the use of palladium complexes for a variety of catalytic and stoichiometric reactions of organic compounds.

Before we discuss specific catalytic reactions, we must consider a number of stoichiometric ones that are important in themselves as well as for their relevance, actual or potential, to catalyses. Although the principles discussed below have some applicability to heterogeneous catalysis, we shall not discuss those processes. For separation of the products from reactants and catalyst, heterogeneous systems have great practical advantage over homogeneous ones, but it is often difficult to obtain from studies of heterogeneous systems the same insight into reaction mechanisms as can sometimes be obtained from homogeneous systems.

STOICHIOMETRIC REACTIONS

The transfer of atoms or groups of atoms from a metal atom to a ligand and vice versa is one of the fundamental processes involved in catalysis by metals. The metal atom or its ligand may be attacked by electrophilic or nucleophilic reagents, but it may be difficult to prove that an attacking reagent is bound to a metal before transfer to a ligand occurs. In certain cases, however, the distinction between direct attack on a ligand and coordination prior to transfer can be made.

30-1 Coordinative Unsaturation

If two substances *A* and *B* are to react at a metal atom contained in a complex in solution, then clearly there must be vacant sites for their coordination. In heterogeneous reactions, the surface atoms of metals, metal oxides, halides, and the like, are necessarily coordinatively unsaturated; but, when even intrinsically coordinatively unsaturated complexes such as square d^8 species are in solution, solvent molecules will occupy the remaining sites, and these will have to be displaced by reacting molecules. Thus:

$$\text{M}(\text{L}_1)(\text{L}_2)(\text{L}_3)(\text{L}_4)(\text{Solv.})_2 + \text{A} + \text{B} \longrightarrow \text{M}(\text{L}_1)(\text{L}_2)(\text{L}_3)(\text{L}_4)(\text{A})(\text{B})$$

In 5- or 6-coordinated metal complexes, coordination sites may be made available by dissociation of one or more ligands either thermally or photochemically. Examples of thermal dissociation are

$$RhH(CO)(PPh_3)_3 \xrightleftharpoons{-PPh_3} RhH(CO)(PPh_3)_2 \xrightleftharpoons{-PPh_3} RhH(CO)(PPh_3)$$

$$Ni[P(o\text{-tolyl})_3]_4 \xrightleftharpoons{-PR_3} Ni[P(o\text{-tolyl})_3]_3$$

The iridium analog of the first complex, namely, $IrH(CO)(PPh_3)_3$, does not catalyze the reactions that the Rh species does at 25°, but will do so when dissociation is induced either by heat or by ultraviolet irradiation.

Dissociation may also be promoted by a change in oxidation state as in oxidative addition reactions discussed below.

30-2
The Acid-Base Behavior of Metal Atoms in Complexes

Protonation and Lewis Base Behavior. In electron-rich complexes the metal atom may have substantial nonbonding electron density located on it and, consequently, may be attacked by the proton or by other electrophilic reagents. An example is $(\eta^5\text{-}C_5H_5)_2ReH$ which is a base comparable in strength to ammonia:

$$H_3N + H^+ \rightleftharpoons H_4N^+$$

$$(\eta^5\text{-}C_5H_5)_2HRe + H^+ \rightleftharpoons (\eta^5\text{-}C_5H_5)_2H_2Re^+$$

Many metal carbonyls and phosphine or phosphite complexes can be protonated, and in some cases the salts can be isolated:

$$Fe(CO)_5 + H^+ \rightleftharpoons FeH(CO)_5^+$$

$$Ni[P(OEt)_3]_4 + H^+ \rightleftharpoons NiH[P(OEt)_3]_4^+$$

$$Ru(CO)_3(PPh_3)_2 + H^+ \rightleftharpoons [RuH(CO)_3(PPh_3)_2]^+$$

$$Os_3(CO)_{12} + H^+ \rightleftharpoons [HOs_3(CO)_{12}]^+$$

The acidification of carbonylate anions (page 483) can be regarded similarly, for example,

$$Mn(CO)_5^- + H^+ \rightleftharpoons HMn(CO)_5$$

The Oxidative Addition or Oxad Reaction. Coordinatively unsaturated compounds, whether transition metal or not, can generally add neutral or anionic nucleophiles, for example,

$$PF_5 + F^- \rightleftharpoons PF_6^-$$

$$TiCl_4 + 2OPCl_3 \longrightarrow TiCl_4(OPCl_3)_2$$

but it is important to note that, even when they are electron-rich, coordinatively unsaturated species may show Lewis acid as well as Lewis base behavior, for example,

$$trans\text{-}IrCl(CO)(PPh_3)_2 + CO \rightleftharpoons IrCl(CO)_2(PPh_3)_2$$

$$PdCl_4^{2-} + Cl^- \rightleftharpoons PdCl_5^{3-}$$

When a complex behaves *simultaneously* as acid and base we have the so-called oxidative addition reaction, which can be written generally as

$$L_nM + XY \longrightarrow L_n(X)(Y)M$$

The formal oxidation number of the metal atom increases by two units. The reverse reaction can be termed reductive elimination.

For oxad reactions to proceed, we must have (a) nonbonding electron density on the metal M, (b) two vacant coordination sites on the complex L_nM to allow formation of two new bonds to X and Y, and (c) a metal M with its stable oxidation states separated by two units.

Many reactions of compounds even of nonmetals, not usually thought of as additions, may be so designated, for example,

$$(CH_3)_2S + I_2 \rightleftharpoons (CH_3)_2SI_2$$

$$PF_3 + F_2 \rightleftharpoons PF_5$$

$$SnCl_2 + Cl_2 \rightleftharpoons SnCl_4$$

For transition metals, the most common reactions are those involving complexes of metals with the d^8 and d^{10} electron configurations, notably, Fe^O, Ru^O, Os^O; Rh^I, Ir^I; Ni^O, Pd^O, Pt^O and Pd^{II} and Pt^{II}. An especially well-studied complex is the square *trans*-$IrCl(CO)(PPh_3)_2$ (page 436) which undergoes reactions such as

$$trans\text{-}Ir^ICl(CO)(PPh_3)_2 + HCl \rightleftharpoons Ir^{III}HCl_2(CO)(PPh_3)_2$$

It will be noted that in additions of molecules such as H_2, HCl, or Cl_2, *two* new bonds to the metal are made and the H—H, H—Cl or Cl—Cl bond is broken. However, molecules that contain multiple bonds may be added oxidatively without cleavage to form new complexes that have three-membered rings (cf. pages 307 and 521), for example,

$$trans\text{-}IrCl(CO)(PPh_3)_2 + O_2 \rightleftharpoons IrCl(CO)(PPh_3)_2(O_2)$$

$$Pt(PPh_3)_4 + (CF_3)_2CO \longrightarrow (Ph_3P)_2Pt[OC(CF_3)_2] + 2PPh_3$$

The latter reaction also provides an example of the situation where the most stable coordination number in the oxidized state would be exceeded, so that expulsion of one or more ligands may occur.

In Table 30-1 we list types of molecules that add to at least one complex.

Table 30-1 Some Substances that Can Be Added to Complexes in Oxidative Addition Reactions

Atoms separate	Atoms remain connected
H_2	O_2
HX (X = Cl, Br, I, CN, RCOO, ClO_4)	SO_2
H_2S, C_6H_5SH	$CF_2{=}CF_2$, $(CN)_2C{=}C(CN)_2$
RX	$RC{\equiv}CR'$
RCOX R = Me, Ph, CF_3, etc.,	RNCS
RSO_2X	RNCO
R_3SnX X = Cl, Br, I	$RN{=}C{=}NR'$
R_3SiX	$RCON_3$
Cl_3SiH	$R_2C{=}C{=}O$
Ph_3PAuCl	CS_2
HgX_2, CH_3HgX (X = Cl, Br, I)	$(CF_3)_2CO$, $(CF_3)_2CS$, CF_3CN
C_6H_6	

Oxad reactions can be considered as equilibria:

$$L_nM^n + XY \rightleftharpoons L_nM^{n+2}XY$$

Whether the equilibrium lies on the reduced or the oxidized side depends very critically on (a) the nature of the metal and its ligands; (b) the nature of the added molecule XY and of the M—X and M—Y bonds so formed; and (c) on the medium in which the reaction is conducted. When the XY molecule adds without severance of X from Y the two new bonds to the metal are *necessarily* in *cis*-positions, but when X and Y are separated the product may be one or more of several isomers with either *cis* or *trans* MX and MY groups, for example,

$$ML_1L_2L_3L_4 + XY$$

$$\downarrow$$

$$MXY(L_1L_2L_3L_4)\ (X \text{ and } Y \text{ trans}),\quad MXY(L_1L_2L_3L_4)\ (X \text{ trans to } L_4),\quad MXY(L_1L_2L_3L_4)\ (X \text{ trans to } L_3),\quad \text{etc.}$$

The final product will be the isomer or isomer mixture that is the most stable thermodynamically under the reaction condition. The ligands, solvent, temperature, pressure, and the like, will have a decisive influence on this. The nature of the *final* product does not necessarily give a guide to the *initial* product of the reaction, since isomerization of the initial product may also occur. The following observations have been made:

1. When solid *trans*-$IrCl(CO)(PPh_3)_2$ reacts with HCl gas the product has H and Cl in *cis*-positions.

2. The addition of HCl or HBr to *trans*-$IrCl(CO)(PPh_3)_2$ in nonpolar solvents such as benzene also gives *cis*-addition. If wet solvents or polar solvents, such as dimethylformamide, are used *cis*-*trans*-mixtures are obtained.

In polar media HCl or HBr will be dissociated and protonation of a square complex will produce first a cationic 5-coordinate complex which may then isomerize by an intramolecular mechanism (page 157). Coordination of halide ion would finally give the oxad product:

$$MXL_3 + H^+(\text{solv}) \longrightarrow MHXL_3^+$$

$$MHXL_3^+ + Cl^-(\text{solv}) \longrightarrow MHClXL_3$$

Study of the mechanisms of the oxad reactions, although still incomplete, suggest that there are the following pathways.

1. A purely ionic mechanism, as just noted, especially in polar solvents.

2. An S_N2 attack of the type common in organic chemistry. A transition metal nucleophile attacks an alkyl halide, namely,

$$L_nM\!: \curvearrowright CR^1R^2R^3X \longrightarrow L_nM^{\delta+}\cdots C(R^1)(R^2)(R^3)\cdots X^{\delta-}$$

$$\downarrow$$

$$L_nMX(CR^1R^2R^3) \longleftarrow [L_nM{-}CR^1R^2R^3]^+ + X^-$$

3. Several oxad reactions are free radical in nature and can be initiated by free radical sources such as peroxides, azoisobutyronitrile, and the like.

4. Under nonpolar conditions, particularly with molecules that have little or no polarity—such as hydrogen—one-step *concerted processes* give products with the new bonds formed in *cis*-positions, namely,

$$\oplus X{-}Y \ominus \quad L{-}Ir{-}L \longrightarrow (X)(Y)Ir(L)(L)$$

The fact that many of the d^8 complexes react with molecular hydrogen might seem surprising in view of the high energy (*ca.* 450 kJ mol^{-1}) of the H—H bond. The attack on the H_2 molecule probably results from electron density of the metal entering the hydrogen σ-antibonding orbital, thus leading to bond weakening. Where two coordination sites are available on the metal atom two *cis*-M—H bonds result (cf. page 542). An alternative reaction is heterolysis of H_2 by the removal of H^+ in the presence of strong bases, for example,

$$RuCl_2(PPh_3)_3 + H_2 + Et_3N \longrightarrow RuHCl(PPh_3)_3 + Et_3NH^+Cl^-$$

30-3
Migration of Atoms or Groups from Metal to Ligand; the "Insertion Reaction"

The concept of "insertion" is of wide applicability in chemistry when defined as a reaction whereby any atom or group of atoms is inserted between two atoms initially bound together:

$$L_nM{-}X + YZ \longrightarrow L_nM{-}(YZ){-}X$$

Some representative examples are

$$R_3SnNR_2 + CO_2 \longrightarrow R_3SnOC(O)NR_2$$

$$Ti(NR_2)_4 + 4CS_2 \longrightarrow Ti(S_2CNR_2)_4$$

$$R_3PbR' + SO_2 \longrightarrow R_3PbOS(O)R'$$

$$[(NH_3)_5RhH]^{2+} + O_2 \longrightarrow [(NH_3)_5RhOOH]^{2+}$$

$$(CO)_5MnCH_3 + CO \longrightarrow (CO)_5MnCOCH_3$$

For transition metals, detailed studies have been made on the "insertion" of CO into metal-to-carbon bonds, but insertions into M—H, M—N, and M—O bonds are also known.

Mechanistic studies using $CH_3Mn(CO)_5$ with ^{14}CO as tracer, have shown that (a) the CO molecule that becomes the acyl-carbonyl is *not* derived from external CO but is one already coordinated to the metal atom, and (b) the incoming CO is added cis to the acyl group, that is,

$$CH_3Mn(CO)_5 + {}^{14}CO \longrightarrow (^{14}CO)(CO)_4Mn{-}C(O){-}CH_3$$

(c) the conversion of alkyl into acyl can be effected by addition of ligands other than CO, for example,

$$CH_3Mn(CO)_5 + PPh_3 \longrightarrow (Ph_3P)(CO)_4Mn{-}C(O){-}CH_3$$

Kinetic studies of the reactions show that the first step involves an equilibrium between the octahedral alkyl and a 5-coordinate acyl species:

$$CH_3Mn(CO)_5 \rightleftharpoons CH_3COMn(CO)_4$$

The incoming ligand ($L = CO, Ph_3P$, etc.) then adds to the 5-coordinate species:

$$CH_3COMn(CO)_4 + L \longrightarrow CH_3COMn(CO)_4L$$

Since 5-coordinate species can undergo intramolecular rearrangements (page 157), more than one isomer of the final product may be formed.

The insertion reaction is thus best considered as an *alkyl migration* to a coordinated carbon monoxide ligand in a *cis*-position, and the migration probably proceeds through a three-center transition state:

$$\mathrm{M{-}CO\;(M{-}CR_3)} \longrightarrow \mathrm{M\cdots CO\;(bridged\;by\;CR_3)} \longrightarrow \mathrm{M{-}C({=}O)CR_3}$$

30-4 Reactions of Coordinated Ligands

Hydride ion addition to certain η^5-C_5H_5 compounds produces cyclopentadiene complexes while addition to arene complexes gives η^5-cyclohexadienyls.

(a) $(C_5H_5\text{-}H_\alpha H_\beta)Fe(CO)_2PPh_3 \underset{+H^-}{\overset{-H^-}{\rightleftharpoons}} [\eta^5\text{-}C_5H_5Fe(CO)_2PPh_3]^+$

(b) $[C_6H_6Mn(CO)_3]^+ + H^- \longrightarrow (\eta^5\text{-}C_6H_5H_2)Mn(CO)_3$

Similar hydride transfers occur with certain complex alkyls, where conversion into the olefin complex can be achieved by abstraction of H^- by triphenylmethyl tetrafluoroborate:

$$\eta^5\text{-}C_5H_5(CO)_2Fe{-}CHRCH_2R'$$

$$\underset{BH_4^-}{\uparrow}\;\;\underset{Ph_3C^+BF_4^-}{\downarrow}$$

$$\left[\eta^5\text{-}C_5H_5(CO)_2Fe\cdots\begin{matrix}HRC\\ \| \\ HR'C\end{matrix}\right]^+ BF_4^- + CHPh_3$$

A good example of a protonation reaction is

$$\eta^5\text{-}C_5H_5(CO)_3Mo{-}CH_2{-}CH{=}CH_2 \xrightarrow{H^+} \left[\eta^5\text{-}C_5H_5(CO)_3Mo \cdots \begin{matrix} H \quad CH_3 \\ \diagdown \; \diagup \\ C \\ \| \\ C \\ \diagup \; \diagdown \\ H \quad H \end{matrix}\right]^+$$

Nucleophilic Attack. There are numerous reactions involving anions or bases such as OH^-, OR^-, $OCOR^-$, N_3^-, R^-, NR_3, N_2H_4, etc., and the ligands attacked may be CO, NO, RCN, RNC, alkenes, etc. It is not always certain that attack is direct, and prior coordination may well occur so that then the reactions could be considered as intramolecular transfers.

Examples long known are attacks on coordinated NO and CO by OH^- ion:

$$[Fe(CN)_5NO]^{2-} + HO^- \xrightarrow{\text{Slow}} \left[Fe(CN)_5N\begin{matrix} \diagup O \\ \diagdown OH \end{matrix}\right]^{3-}$$

$$\Big\downarrow \text{Fast} \mid HO^-$$

$$[Fe(CN)_5NO_2]^{4-} + H_2O$$

$$Fe(CO)_5 + HO^- \longrightarrow \left[(CO)_4Fe{-}C\begin{matrix} /\!\!/ O \\ \diagdown OH \end{matrix}\right]^-$$

$$\Big\downarrow HO^-$$

$$(CO)_4FeH^- + HCO_3^-$$

The attack of alkoxide ions on CO gives M—COOR groups and has been observed for complexes of Mn, Re, Fe, Ru, Os, Co, Rh, Ir, Pd, Pt, and Hg, for example,

$$[Ir(CO)_3(PPh_3)_2]^+ \underset{H^+}{\overset{CH_3O^-}{\rightleftarrows}} Ir(CO)_2(COOCH_3)(PPh_3)_2$$

The reaction is important in the synthesis of carboxylic acids and esters from alkenes, CO, and water or alcohols. Similar attacks on CO by OH^- or H_2O are involved in reductions of Co^{2+} or Rh^{3+} by CO to form CO_2.

Coordinated carbon monoxide can also be attacked by lithium alkyls or dialkylamides, for example,

$$LiCH_3 + W(CO)_6 \longrightarrow Li^+[(CO)_5W{-}C(O)CH_3]^-$$

These anions may in turn be converted into coordinated carbenes (see page 524).

Alkene and dienyl complexes are also attacked by nucleophiles, for example,

$$(\text{diene})PtCl_2 + 2ROH \longrightarrow [(\text{RO-alkenyl})Pt(\mu\text{-}Cl)_2Pt(\text{alkenyl-OR})] + 2HCl$$

Nitrile complexes with aromatic amines and alcohols give complexes of amidines and imidate esters, respectively, for example,

$$(CH_3CN)_2ReCl_4 + 2PhNH_2 \longrightarrow [PhHN(H_3C)C{=}N(H)]ReCl_4[N(H){=}C(NHPh)CH_3]$$

Isocyanide complexes on the other hand are attacked to form carbene complexes (page 524), for example,

$$(Et_3P)Cl_2PtCNPh + EtOH \longrightarrow (Et_3P)Cl_2Pt{=}C(OEt)(NHPh)$$

Intramolecular Hydrogen Transfer. Groups may be transferred from the metal to ligand by insertion reactions, but a special case of transfer reactions is one between certain ligands and the metal in which a hydrogen atom is initially transferred and is then subsequently lost. Such reactions are especially important for triarylphosphines and triarylphosphites. An example is the following:

$$(Ph_3P)_3IrCl \longrightarrow (Ph_3P)_2IrHCl(C_6H_4PPh_2) \xrightarrow{-HCl} (Ph_3P)_2Ir(C_6H_4PPh_2)$$

The formation of M—C bonds in this way also occurs in reactions with azobenzene, and *N,N*-dimethylbenzylamine, for example,

$$2\ PhN{=}NPh + 2PdCl_4^{2-} \xrightarrow{-2HCl} [(C_6H_4N{=}NPh)Pd(\mu\text{-}Cl)]_2 + 4Cl^-$$

30-5 Reactions of Coordinated Molecular Oxygen

We observed above that molecular oxygen can add to certain complexes without breaking the O—O bond. Oxygen also gives reversible reactions with Schiff base complexes of Co (page 406) and hemoglobin (page 553).

Coordinated oxygen may be kinetically more reactive than the free molecule. The mechanisms of attacks on coordinated oxygen are not too well understood. In many cases the reactions involve free radicals. However, for some tertiary phosphine complexes the reaction proceeds through peroxo intermediates, which may be isolated, for example, the platinum peroxocarbonate:

$$(Ph_3P)_2Pt(O_2) + CO_2 \longrightarrow (Ph_3P)_2Pt(O{-}O{-}C({=}O){-}O)$$

The mechanism of oxidation of SO_2 by $IrCl(CO)O_2(PPh_3)_2$ to give a sulfato complex has been shown by use of ^{18}O tracer to be similar, namely,

$$Ir(O^*_2) \xrightarrow{SO_2} Ir(O^*{-}O^*{-}S({=}O){-}O) \longrightarrow Ir(O^*{-}S({=}O^*)({=}O){-}O)$$

CATALYTIC REACTIONS OF ALKENES

The term catalyst is ambiguous and requires careful use. In heterogeneous reactions, where for example a gas mixture is passed over a solid that evidently undergoes no change, the term, meaning a substance added to accelerate a reaction, may have some point. However, homogeneous catalytic reactions in solution commonly proceed by way of linked chemical reactions involving different metal species. The concept of one particular species being "the catalyst," even if it is the one added to initiate or accelerate the reaction, has no validity. It is necessary to think in terms of intermediates involved in the various chemical reactions of a *catalytic cycle.*

Observe that the cycles involve fundamental changes in both the oxidation state and coordination number of the metal atom in complexes consequent on coordination, loss of reactants or products as well as oxad or transfer reactions.

30-6 Isomerization

Many transition-metal ions and complexes, especially those of Group VIII metals, promote double-bond migration—that is, isomerization—in alkenes, to give the thermodynamically most stable isomeric mixture. Thus 1-alkenes give (*cis* + *trans*)-2-alkenes. The isomerization involves the transfer of a hydrogen atom from the metal to the coordinated alkene, thus forming an alkyl. This reaction is characteristic for many transition-metal hydrido species; and, in addition, many complexes that do not have M—H bonds, for example, $(Et_3P)_2NiCl_2$, will isomerize alkenes provided that a source of hydride ion such as molecular hydrogen is present.

The first step in the reaction must be the coordination of the alkene:

$$L_nMH + RCH{=}CH_2 \rightleftharpoons L_nMH(RCH{=}CH_2)$$

which is followed by hydride transfer to form an alkyl group:

H H H \ / | C M---|| C / \ H R ⇌ H H \ / H---C ⁞ || M---C / \ H R ⇌ H H H \ | / C / M—C / \ H R

This reaction, in the reverse direction, has been described previously as one of the major routes for the decomposition of metal alkyls (page 511).

Although it has not been proved that the hydrogen atom on the metal transfers to the second carbon of the alkyl chain, mainly because such reactions are very readily reversible, thus leading to a scrambling of all hydrogen atoms, when fluoroolefins are used this can be proved because stable products are obtained:

$$RhH(CO)(PPh_3)_3 + C_2F_4 \longrightarrow Rh(CF_2CF_2H)(CO)(PPh_3)_2 + PPh_3$$

Notice that in the hydrido alkene intermediate two coordination sites are involved on the metal, but in the alkyl only one is involved.

With alkenes other than ethylene, there is the possibility of addition of M—H to the double bond in either the Markownikoff or the anti-Markownikoff sense just as with the addition of any other X—H reagent. Thus we may have reactions that give either product (A) or product (B). Since in the anti-Markownikoff addition the H atom is transferred from the metal to the β-carbon of the chain,

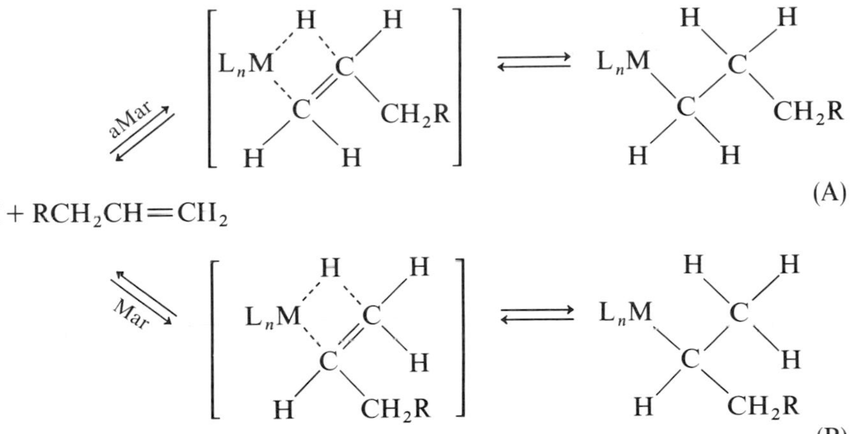

$$L_nMH + RCH_2CH{=}CH_2 \underset{\text{Mar}}{\overset{\text{aMar}}{\rightleftharpoons}} \left[L_nM\cdots H \cdots C(H)(H)\!=\!C(H)(CH_2R) \right] \rightleftharpoons L_nM{-}CH_2{-}CH_2CH_2R \quad \text{(A)}$$

$$\left[L_nM\cdots H\cdots C(H)(CH_2R)\!=\!C(H)(H) \right] \rightleftharpoons L_nM{-}CH(CH_2R){-}CH_3 \quad \text{(B)}$$

aMar = anti-Markownikoff
Mar = Markownikoff

to give the primary alkyl derivative (A), the reverse reaction must re-form the original alkene. There is thus no isomerization in this case; observe, however, that because of rotation about the C—C bond the same H atom need not necessarily be removed, so that hydrogen atom exchange can occur. On the other hand, there are two possibilities for the secondary alkyl derivative (B); if the H atom is transferred from the CH_3 group, again the original l-alkene is formed, but if it is transferred from the methylene of the CH_2R group, then a 2-alkene is formed. Thus isomerizations can occur only following initial Markownikoff addition, and it should be noted that either *cis*- or *trans* 2-alkenes, or both may be formed.

30-7
Hydrogenation of Alkenes

Molecular hydrogen reacts with many substances at room temperature and atmospheric pressure—aqueous ions such as Ag^+ or MnO_4^- and complexes such as Cu^{2+} in quinoline or Co^{2+} in aqueous NaCN solution. Recently, useful catalysts for the reduction of unsaturated compounds such as alkenes or alkynes have been developed from these systems. The most successful of them uses the complex $RhCl(PPh_3)_3$ in benzene or ethanol-benzene solutions. Because of differences in rates of hydrogenation depending on the nature of groups at the double bond, selective reductions are possible, for example,

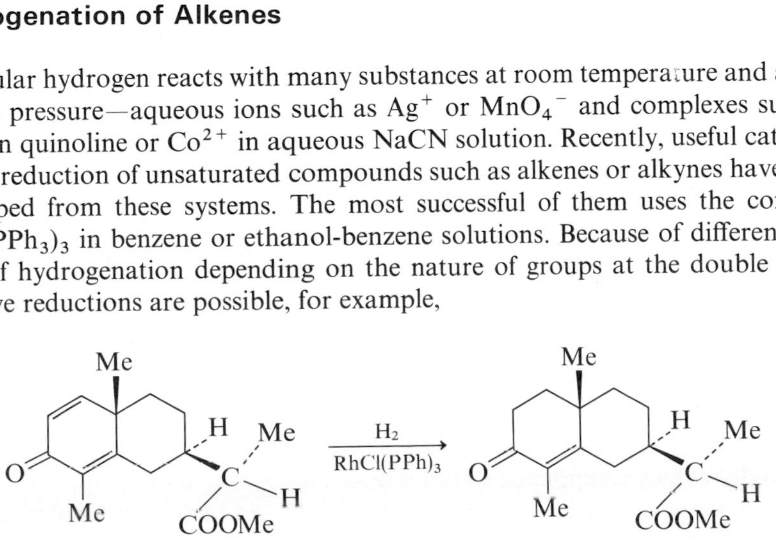

$$\xrightarrow[\text{RhCl(PPh)}_3]{H_2}$$

(Me, H, Me, C, H, COOMe, O, Me — reactant and product structures)

Furthermore, in contrast to heterogeneous catalysis, where scattering of deuterium throughout the molecule usually results, selective addition of D_2 to a double bond occurs. Finally, asymmetric hydrogenation has been achieved by use of complexes with phosphines that are optically active either at the phosphorus atom or at a carbon atom on the group attached to P. This technique is used to synthesize from styrenes D or L-α-amino acids some of which have important physiological properties, for example, L-Dopa which is used in treatment of Parkinson's disease.

The mechanism of hydrogenations using $RhCl(PPh_3)_3$ appears to involve the following cycle (where $P = PPh_3$):

$$RhCl(P)_3 + H_2 \rightleftharpoons RhClH_2(P)_3 \underset{+P}{\overset{-P}{\rightleftharpoons}} RhClH_2(P)_2$$

$$RhClH_2(P)_2 \underset{}{\overset{>C=C<}{\rightleftharpoons}} RhClH_2(P)_2(>C=C<) \rightleftharpoons RhClH(P)_2(-C-C<H) \xrightarrow{-\text{Alkane}} RhCl(P)_2$$

$$RhCl(P)_2 \underset{-P}{\overset{+P}{\rightleftharpoons}} RhCl(P)_3$$

There are a number of other similar systems that can reduce not only C=C and C≡C but also >C=O, —N=N—, and —CH=N—. One employs $[RhH_2(PR_3)_2S_2]^+$ where *S* is a solvent molecule or unsaturated substrate, and another $RhCl_3py_3$ in dimethylformamide plus $NaBH_4$.

30-8
Carbonylation Reactions

The hydroformylation reaction is the addition of H_2 and CO (or formally of H and the formyl group, HCO) to an alkene, usually a 1-alkene, to form an aldehyde, which may be further reduced to the alcohol:

$$RCH{=}CH_2 + H_2 + CO \longrightarrow RCH_2CH_2CHO \xrightarrow{H_2} RCH_2CH_2CH_2OH$$

Originally cobalt compounds were used as catalysts at temperatures of *ca.* 150° and >200 atm pressure, and some three million tons a year of alcohols, usually

C_7—C_9, are produced in this way. The process ordinarily gives both straight- and branched-chain products in the ratio *ca.* 3:1, but considerable efforts have been made to improve the yield of the linear product.

The cobalt system is difficult to study and further information on the catalyst cycle is provided by use of $RhH(CO)(PPh_3)_3$. This complex is catalytically active even at 25° and 1 atm pressure and, in contrast to the cobalt system, produces only aldehyde. On use of high concentrations of PPh_3, high yields of linear aldehyde can be obtained with little or no loss of alkene as alkane (which is a disadvantage of the cobalt systems). The reaction cycle is shown in *Fig. 30-1*. The initial step is associative attack of the alkene on the species $RhH(CO)_2(PPh_3)_2$ [(A) in *Fig. 30-1*,] which leads to the alkyl complex, (B). The latter then undergoes CO insertion to form the acyl derivative (C), which subsequently adds hydrogen oxidatively to give the dihydridoacyl complex (D).

The last of these steps, which is the only one in the cycle that involves a change in oxidation state of the metal, is probably rate-determining. The final steps are another H transfer to the carbon atom of the acyl group in (D), followed by loss of aldehyde and regeneration of the 4-coordinate species (E).

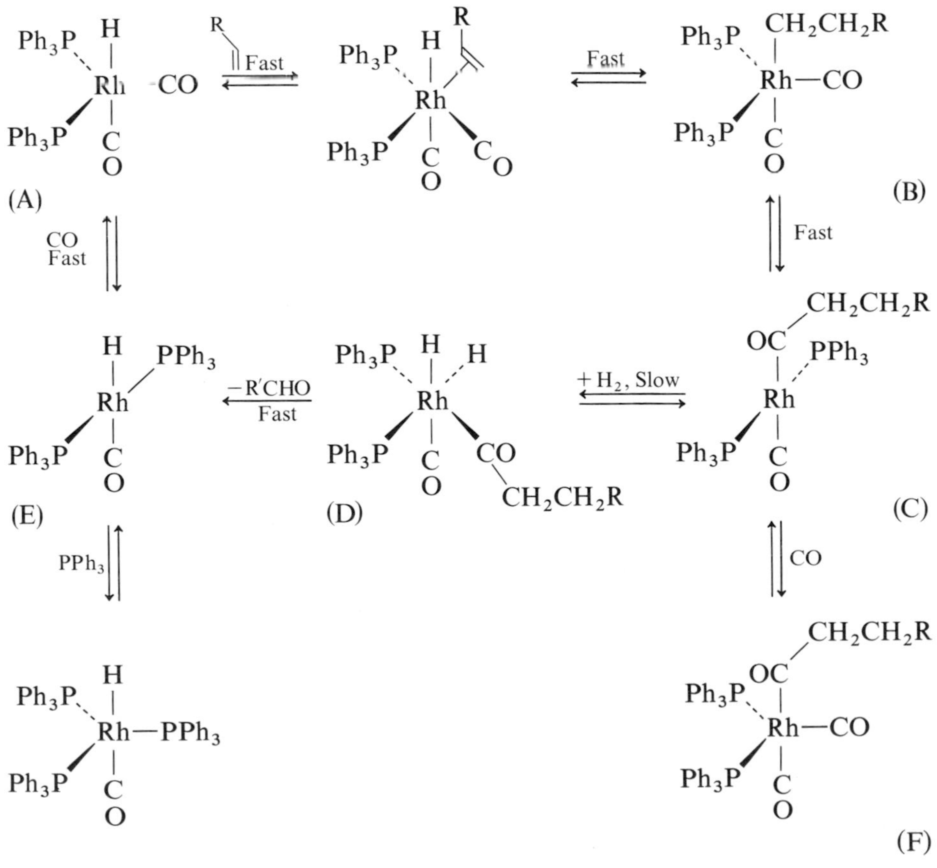

(A) $RhH(CO)_2(PPh_3)_2$ ⇌ (R–CH=CH₂, Fast) ⇌ (Fast) (B) $Rh(CH_2CH_2R)(CO)_2(PPh_3)_2$

(A) ⇌ (CO, Fast) (E) $RhH(CO)(PPh_3)_2$ ← (−R′CHO, Fast) (D) $RhH_2(COCH_2CH_2R)(CO)(PPh_3)_2$ ⇌ (+H_2, Slow) (C) $Rh(COCH_2CH_2R)(CO)(PPh_3)_2$

(B) ⇌ (Fast) (C); (E) ⇌ (PPh_3) $RhH(CO)(PPh_3)_3$; (C) ⇌ (CO) (F) $Rh(COCH_2CH_2R)(CO)_2(PPh_3)_2$

Figure 30-1 *Catalytic cycle for the hydroformylation of alkenes involving triphenylphosphine rhodium complex species. The configurations of the complexes are not known with certainty.*

The high PPh_3 concentrations that are essential to provide high yields (>95%) of linear aldehyde are probably required to suppress dissociation and the formation of monophosphine species, and thus to force the associative attack of olefin on the bisphosphine species (A), for which the specificity of anti-Markownikoff addition is high.

Another important commercial process is the carbonylation of methanol to give acetic acid. Cobalt was used originally but high temperatures and pressures were required; with rhodium much milder conditions can be used. They key to the reaction is the presence of iodide, which reacts to give methyl iodide

$$CH_3OH + HI \rightleftharpoons CH_3I + H_2O$$

The CH_3I then oxidatively adds:

$$L_nRh^I + CH_3I \rightleftharpoons L_nRh(CH_3)\text{—}I$$

which is followed by CO insertion and hydrolysis:

$$L_nRh(CH_3)\text{—}I \xrightarrow{CO} L_nRh(C(=O)CH_3)\text{—}I \xrightarrow{H_2O} L_nRh^I + HI + CH_3COOH$$

Hydrosilylation of Alkenes. The hydrosilylation reaction (Speier reaction) of alkenes is

$$RCH{=}CH_2 + HSiR_3 \longrightarrow RCH_2CH_2SiR_3$$

The commercial reaction uses hexachloroplatinic acid as the catalyst, but phosphine complexes of cobalt, rhodium, palladium, or nickel can also be used. The addition of silanes to *trans*-$IrCl(CO)(PPh_3)_2$ provides a model for the first step, namely, the oxidative addition

$$IrCl(CO)(PPh_3)_2 + R_3SiH \longrightarrow IrHCl(SiR_3)(CO)(PPh_3)_2$$

In this case there is no vacant site on the metal, and the process ends here. In actual catalysts there must be an open coordination site to which the alkene is added. This will be followed by M—H addition across the double bond to form an alkyl group, which can then combine with SiR_3 in a reductive elimination to liberate the product.

30-9
Alkene Polymerization and Oligomerization

Ziegler–Natta polymerization. Hydrocarbon solutions of $TiCl_4$ in presence of triethylaluminum polymerize ethylene at 1 atm pressure.

The Ziegler–Natta system is heterogeneous, and the active metal species is a fibrous form of $TiCl_3$ formed *in situ* from $TiCl_4$ and $AlEt_3$, but preformed $TiCl_3$ can be used. The second function of the aluminum alkyl appears to be replacement of one of the chloride ions at the $TiCl_3$ surface by an alkyl radical derived from it; the surface Ti atom has one of its 6-coordination sites vacant. An ethylene molecule then becomes bound at the vacant site. The alkyl group is then transferred to the coordinated ethylene. A further molecule of ethylene is then bound to the vacant site and the process is repeated. The mechanism is then as follows:

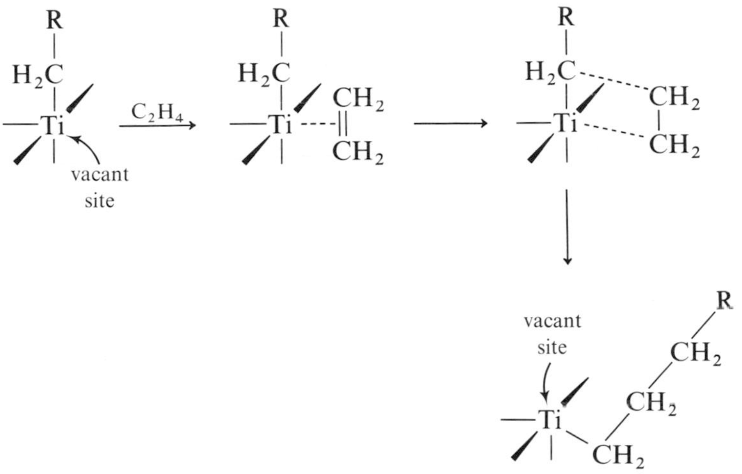

R
H_2C
Ti
vacant site
$\xrightarrow{C_2H_4}$
R
H_2C
Ti ---- CH_2=CH_2
⟶
R
H_2C-----CH_2
Ti-------CH_2
↓
vacant site
Ti—CH_2—CH_2—CH_2—R

The stereoregular polymerization of propene may arise because of the nature of the sterically hindered surface sites on the $TiCl_3$ lattice.

An important extension of Ziegler–Natta polymerization is the copolymerization of styrene, butadiene, and a third component such as dicyclopentadiene or 1,4-hexadiene to give synthetic rubbers. Vanadyl halides instead of titanium halides are then the preferred catalysts.

30-10 Palladium-Catalyzed Reactions

It was long known that ethylene compounds of palladium, for example, $[C_2H_4PdCl_2]_2$, are rapidly decomposed in aqueous solution to form acetaldehyde and Pd metal. The conversion of this stoichiometric reaction into a cyclic one (Wacker process) required the linking together of the known individual reactions:

$$C_2H_4 + PdCl_2 + H_2O \longrightarrow CH_3CHO + Pd + 2HCl$$

$$Pd + 2CuCl_2 \longrightarrow PdCl_2 + 2CuCl$$

$$2CuCl + 2HCl + \tfrac{1}{2}O_2 \longrightarrow 2CuCl_2 + H_2O$$

$$C_2H_4 + \tfrac{1}{2}O_2 \longrightarrow CH_3CHO$$

The oxidation of ethylene by palladium(II)–copper(II) chloride solution is essentially quantitative and only low Pd concentrations are required; the process can proceed either in one stage or in two stages; in the latter the reoxidation by O_2 is done separately.

Since the reaction proceeds in Pd^{II} solutions with a chloride ion concentration $>0.2M$, the metal is most likely present as $[PdCl_4]^{2-}$. The following reactions then occur:

$$[PdCl_4]^{2-} + C_2H_4 \rightleftharpoons [PdCl_3(C_2H_4)]^- + Cl^- \quad \text{(fast)}$$

$$[PdCl_3(C_2H_4)]^- + H_2O \rightleftharpoons [PdCl_2(H_2O)(C_2H_4)] + Cl^-$$

$$[PdCl_2(H_2O)C_2H_4] + H_2O \rightleftharpoons [PdCl_2(OH)(C_2H_4)]^- + H_3O^+$$

The *trans*-isomer of this hydroxo species is doubtless more stable than the *cis*-isomer (see trans-effect, page 150), but kinetically significant amounts of the latter will be present so that further reaction occurs by *cis*-transfer. This transfer may well be Cl^- ion- or solvent-assisted in a 5-coordinate solvated species, for example,

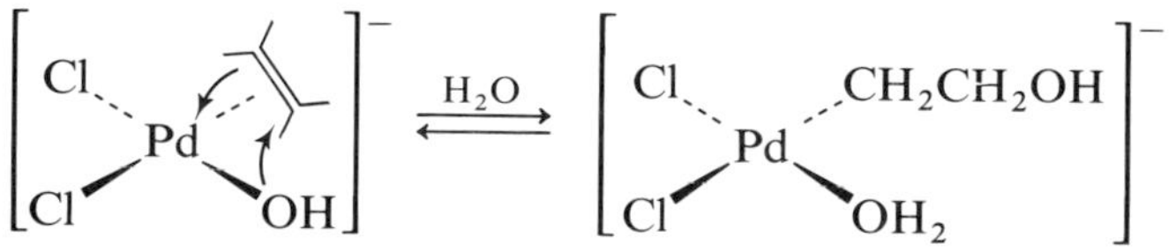

$$[Cl_2Pd(C_2H_4)(OH)]^- \underset{}{\overset{H_2O}{\rightleftharpoons}} [Cl_2Pd(CH_2CH_2OH)(OH_2)]^-$$

This reaction is followed by three further steps, namely (a) a fast hydrogen transfer from the β-carbon of the chain to the metal, (b) hydride transfer from metal to the α-CH_2 group (as in hydrogenation), and finally (c) reductive elimination of Pd metal:

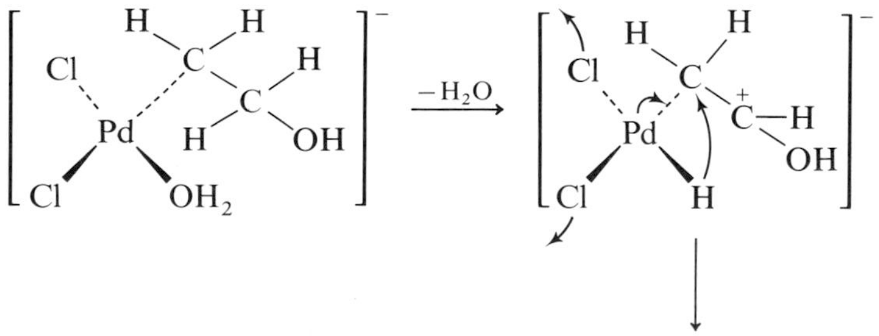

$$[Cl_2Pd(OH_2)(CH_2CH(H)OH)]^- \xrightarrow{-H_2O} [Cl_2Pd(H)(CH_2\overset{+}{C}HOH)]^-$$

$$\downarrow$$

$$CH_3CHO + H^+ \longleftarrow CH_3CHOH^+ + Pd^0 + 2Cl^-$$

The sequence accounts not only for the rate laws and dependence on (inhibition by) Cl^- and H^+ but also for results of deuteration studies which show that in D_2O no deuterium is incorporated into the acetaldehyde.

The mechanism for the oxidation of Pd metal by Cu^{II} chloro complexes is not well understood, but electron transfer via halide bridges (cf. page 156) is probably involved. The extremely rapid air-oxidation of Cu^I chloro complexes is better known and probably proceeds through an initial oxygen complex:

$$CuCl_2^- + O_2 \rightleftharpoons ClCuO_2 + Cl^-$$

followed by formation of radicals such as O_2^-, OH, or HO_2:

$$ClCuO_2 + H_3O^+ \longrightarrow CuCl^+ + HO_2 + H_2O$$

The reactivity of palladium complexes in other systems has been extensively studied, and there are now many catalytic reactions involving alkenes, arenes, CO, acetylenes, and the like. Extensions of the Wacker process using media other than water are known; thus in acetic acid, vinyl acetate is obtained while in alcohols, vinyl ethers are obtained. Also with alkenes other than ethylene, ketones may be obtained. For example, propene gives acetone.

Study Questions

A

1. What is meant by a coordinatively unsaturated species? Give two examples, and explain how these species may arise in solutions beginning with coordinatively saturated ones.
2. Define the term oxidative addition (oxad) reaction. What conditions must be met for such a reaction to occur? What is the reverse of such a reaction called?
3. Draw plausible structures for the reaction products of $IrCl(CO)(PR_3)_2$ with H_2, CH_3I, PhNCS, CF_3CN, $(CF_3)_2CO$, O_2.
4. How can one account for the low activation energy for oxidative addition of H_2, with its very strong H—H bond?
5. What is an insertion reaction? Give two real examples.
6. Describe the actual pathway for the reaction of PEt_3 with $CH_3Mn(CO)_5$ to give $CH_3COMn(CO)_4PEt_3$.
7. Complete the following equations and show with diagrams the structures of the principal products:
   (a) $Ru(CO)_3(PPh_3)_2 + HBF_4 \longrightarrow$
   (b) $Ir(CO)_3(PPh_3)_2^+ + CH_3O^- \longrightarrow$
   (c) $W(CO)_6 \xrightarrow{LiCH_3} [A] \xrightarrow{Me_3O^+} [B]$
   (d) $[Fe(CN)_5NO]^{2-} + 2OH^- \longrightarrow$
   (e) $ReCl_4(NCCH_3) + 2PhNH_2 \longrightarrow$
8. Show the steps by which a hydrido complex can cause isomerization of 1-alkenes to 2-alkenes? Is this generally stereospecific?
9. Write a balanced equation showing the overall (net) reaction in each of the following processes: hydroformylation; hydrosilylation; the Ziegler–Natta process; the Wacker process for synthesis of acetaldehyde.
10. Outline the main steps by which Ziegler–Natta polymerization proceeds.
11. Outline the mechanism of the Wacker process.

B

1. Write a plausible mechanism for the reaction of $Ti(NEt_2)_4$ with CS_2 to give $Ti(S_2CNEt_2)_4$.
2. Give a plausible catalytic cycle to account for the conversion of ethylene to propionaldehyde employing $RhH(CO)(PPh_3)_3$ as the catalyst.

3. Suggest a catalytic cycle to account for the action of

$$[Rh(PEtPh_2)_2(CH_3OH)_2]^+PF_6^-$$

in methanol as a catalyst for hydrogenation of but-1-ene by H_2 at 25°C and 1 atm pressure.
4. The complex $Ni[P(OEt)_3]_4$ in acidic solution is used in the synthesis of hexa-1,4-diene from ethylene and butadiene. Suggest a plausible catalytic cycle.
5. $Ni[P(OEt)_3]_4$ is also used to catalyze the process

$$CH_2{=}CH{-}CH{=}CH_2 + HCN \longrightarrow NC(CH_2)_4CN$$

Again, suggest a sensible sequence of steps.
6. Suggest a mechanism for the following, socalled 1,3-insertion, reaction.

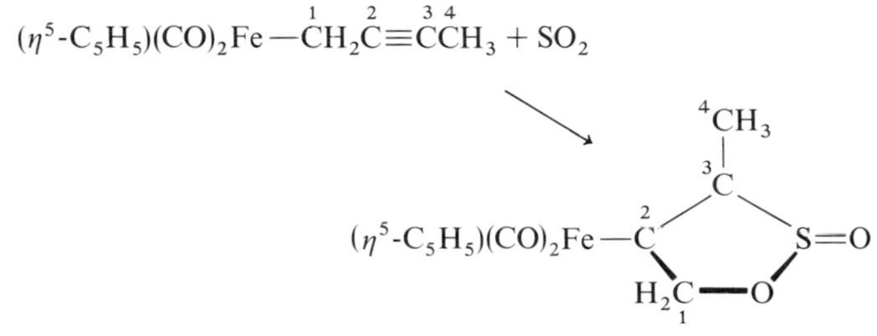

$$(\eta^5\text{-}C_5H_5)(CO)_2Fe{-}\overset{1}{C}H_2\overset{2}{C}{\equiv}\overset{3}{C}\overset{4}{C}H_3 + SO_2 \longrightarrow$$

$$(\eta^5\text{-}C_5H_5)(CO)_2Fe{-}\overset{2}{C}\ \text{(ring: } \overset{2}{C}{=}\overset{3}{C}(\overset{4}{C}H_3){-}S({=}O){-}O{-}\overset{1}{C}H_2{-}\overset{2}{C}\text{)}$$

7. It has been proved that the alkyl group retains its configuration when CO insertion to produce the acyl occurs for $(\eta^5\text{-}C_5H_5)(CO)_2Fe{-}CHD{-}CHD{-}C(CH_3)_3$. Propose a mechanism that accounts for this.
8. Write a mechanism for the conversion of butadiene to *trans-trans-trans*-cyclododecatriene, using a Ni(O) species.
9. $RhH(CO)(PR_3)_3$ in benzene under pressure of ethylene reacts with benzoyl chloride to give propiophenone, $PhC(O)C_2H_5$. Suggest a mechanism.

Chapter 30
Study Guide

Scope and Purpose. The purpose of this chapter is to present a few of the basic principles and processes involved in the practical use of organometallic compounds as homogeneous catalysts. The large-scale synthesis of useful organic compounds, including polymers, from simpler and cheaper starting materials is one of the major activities of the worldwide chemical industry. A great deal of industrial chemical research is concerned with the discovery and improvement of these processes. In addition to the practical importance of the field, it offers fascinating problems from the purely scientific point of view.

Supplementary Reading

Advances in Chemistry Series, No. 70, Homogeneous Catalysis: Industrial Applications and Implications, American Chemical Society, 1968.
Bird, C. W., *Transition Metal Intermediates in Organic Synthesis*, Academic Press, 1967.

Candlin, J. P., Taylor, A. K., and Thompson, D. T., *Reactions of Transition Metal Complexes*, Elsevier, 1968.

Collman, J. P., "Patterns of Organometallic Reactions Related to Homogeneous Catalysis," *Accts. Chem. Research*, *1*, 136 (1968).

James, B. R., *Homogeneous Hydrogenation*, Wiley, 1973.

Martell, A. E. and Taqui Khan, M. M., *Homogeneous Catalysis by Metal Complexes*, Vols. 1 and 2, Academic Press, 1974.

Rylander, P. N., *Organic Synthesis with Noble Metal Catalysts*, Academic Press, 1973.

Schrauzer, G. N., ed., *Transition Metals in Homogeneous Catalysis*, Dekker, 1971.

31

metals in biological systems

Biochemistry is not merely an elaboration of organic chemistry. The chemistry of life involves, in an essential and indispensible way, many of the chemical elements, including metals. The importance of sodium, calcium, and iron has long been recognized but many others, especially Cu, Zn, Mn, Mo, and Co are necessary to life. In this chapter we survey the major aspects of metals in biological systems or what is sometimes called, bioinorganic chemistry.

METALLOPORPHYRINS

One of the most important ways in which metal ions are involved is in complexes with a type of macrocylic ligand called a *porphyrin*. Porphyrins are derivatives of *porphine*: they differ in the arrangement of substituents around the periphery. The porphin molecule is shown in *Fig. 31-1a*, and the two most important metal complexes of porphyrins, *chlorophyll* and *heme*, are shown in *Figs.* 31-1*b* and *c*. In the complexes the inner hydrogen atoms have been displaced by the metal ions.

31-1
Chlorophyll

There are several very similar but not identical chlorophyll molecules. Green plants contain two and various algae contain others. Notice that in *Fig. 31-1b* the basic porphin system has been modified in two ways. In pyrrole ring IV, one of the double bonds has been *trans*-hydrogenated, and a cyclopentanone ring has been fused to the side of pyrrole ring III. Nevertheless, the fundamental properties of the porphin system are retained.

Photosynthesis is a complex sequence of processes in which solar energy is first absorbed and ultimately—in a series of redox reactions, some of which proceed in the dark—used to drive the overall endothermic process of combining water and carbon dioxide to give glucose; molecular oxygen is released simultaneously:

$$6CO_2 + 6H_2O = C_6H_{12}O_6 + 6O_2$$

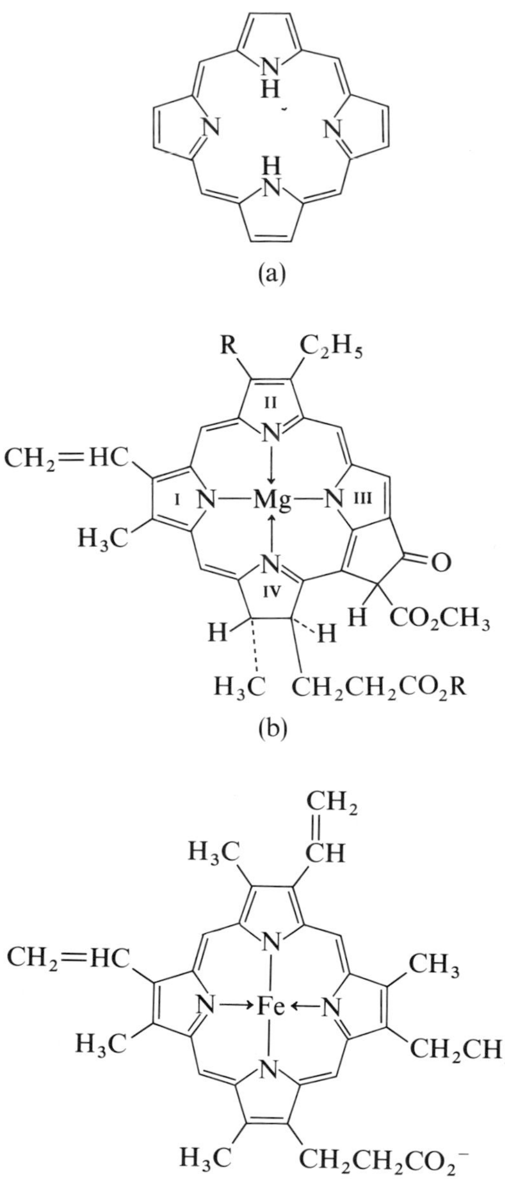

(a)

(b)

(c)

Figure 31-1 (*a*) *The prototype porphine molecule.* (*b*) *One of the chlorophyll molecules.* (*c*) *The heme group.*

The function of the chlorophyll molecules in the chloroplast is to absorb photons in the red part of the visible spectrum (near 700 nm) and pass this energy of excitation on to other species in the reaction chain. The ability to absorb the light is due basically to the conjugated polyene structure of the porphine ring system. The role of the magnesium ion is, at least, twofold. (1) It helps to make the entire molecule rigid so that energy is not too easily lost thermally, that is, degraded to molecular vibrations. (2) It enhances the rate at which the short-lived singlet excited state initially formed by photon absorption is transformed into the corresponding triplet state, which has a longer lifetime and thus can transfer its excitation energy into the redox chain.

At an early stage of the electron-transfer sequence which leads ultimately to the release of molecular oxygen, a manganese complex, of unknown composition, undergoes reversible redox reactions. At still other stages, iron-containing substances, called cytochromes and ferredoxins, and a copper-containing substance, called plastocyanin, also participate. Thus, photosynthesis requires the participation of complexes of no less than four metallic elements.

THE BIOINORGANIC CHEMISTRY OF IRON

Iron is certainly the most widespread of the heavy or transition metals in living systems. Its compounds participate in a variety of activities. The two main functions of iron-containing materials are: (1) transport of oxygen, and (2) mediation in electron-transfer chains. So much iron is required for these purposes that there is also a chemical system to store and transport iron. We turn first to compounds in which the iron is present as heme, the porphyrin complex depicted in *Fig. 31-1c*.

31-2 Heme Proteins

The heme group functions in all cases in intimate association with a protein molecule. The chief heme proteins are

1. Hemoglobins.
2. Myoglobins.
3. Cytochromes.
4. Enzymes such as catalase and peroxidase.

Hemoglobin and Myoglobin. These are closely related. Hemoglobin has a molecular weight of 64,500 and consists of four subunits, each containing one heme group. Myoglobin is very similar to one of the subunits of hemoglobin, one of which is shown in *Fig. 31-2*. Hemoglobin has two functions. (1) It binds oxygen molecules to its iron atoms and transports them from the lungs to muscles where they are delivered to myoglobin molecules. These store the oxygen until it is required for metabolic action. (2) The hemoglobin then uses certain amino groups to bind carbon dioxide and carry it back to the lungs.

The heme group is attached to the protein in both hemoglobin and myoglobin through a coordinated histidine-nitrogen atom, F8, shown in *Fig. 31-2*. The position *trans* to the histidine-nitrogen atom is occupied by a water molecule in the deoxy species or O_2 in the oxygenated species. The structure of the Fe—O_2 grouping is still unknown, but changes in the oxidation state of iron and the introduction of O_2 (and other ligands) cause important changes in the structure of heme, as we describe below.

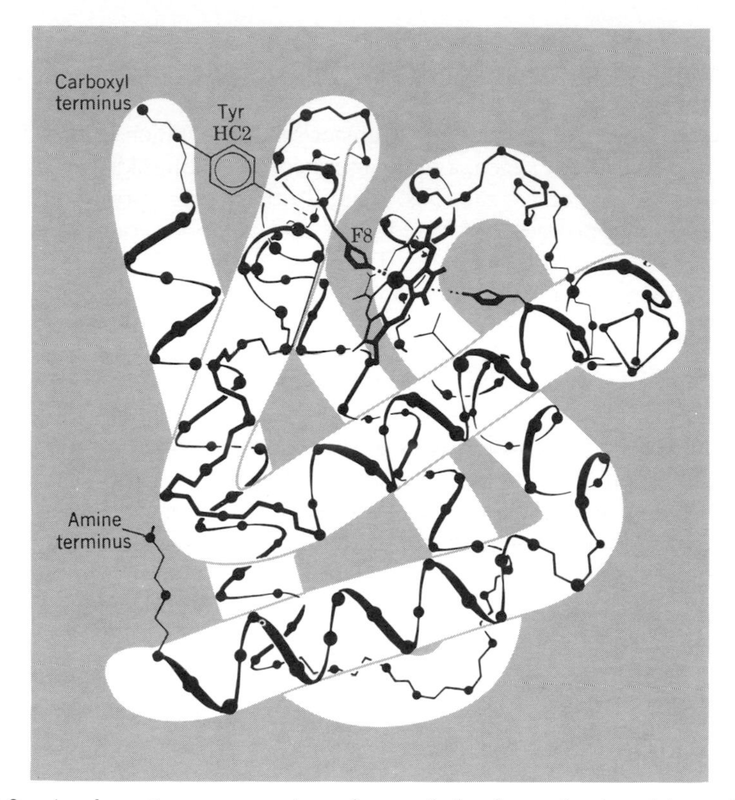

Figure 31-2 *A schematic representation of one of the four subunits of hemoglobin. The continuous black band represents the peptide chain and the various sections of helix are evident. Dots on the chain represent the α-carbon atoms. The heme group can be seen at upper right center with the iron as a large dot. The coordinated histidine side chain is labelled* F8 *(meaning the 8th residue of the* F *helix). (This figure was adapted from one kindly supplied by M. Perutz.)*

Hemoglobin is not simply a passive container for oxygen but an intricate molecular machine. This may be appreciated by comparing its affinity for O_2 to that of myoglobin. For myoglobin (Mb) we have the following simple equilibrium:

$$Mb + O_2 = MbO_2 \qquad K = \frac{[MbO_2]}{[Mb][O_2]}$$

If f represents the fraction of myoglobin molecules bearing oxygen and P represents the equilibrium partial pressure of oxygen, then

$$K = \frac{f}{(1 - f)P} \quad \text{and} \quad f = \frac{KP}{1 + KP}$$

This is the equation for the hyperbolic curve labeled Mb in *Fig. 31-3.* Hemoglobin with its four subunits has more complex behavior; it approximately follows the equation

$$f = \frac{KP^n}{1 + KP^n}; \qquad n \approx 2.8$$

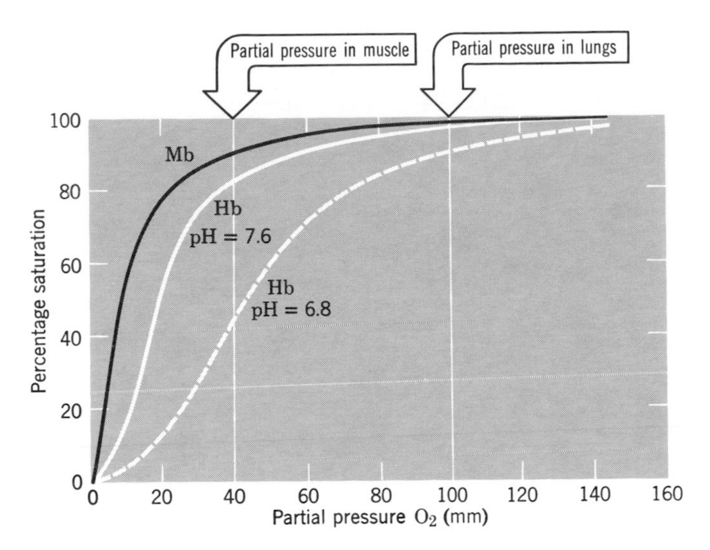

Figure 31-3 *The oxygen-binding curves for myoglobin* (Mb) *and hemoglobin* (Hb), *showing also the* pH-*dependence* (*Bohr effect*) *for the latter.*

where the exact value of n depends on pH. Thus, for hemoglobin (Hb) the oxygen-binding curves are sigmoidal as is shown in *Fig. 31-3*. The fact that n exceeds unity can be ascribed physically to the fact that attachment of O_2 to one heme group increases the binding constant for the next O_2, which in turn increases the constant for the next one, and so on.

Although Hb is about as good an O_2 binder as Mb at high O_2 pressure, it is much poorer at the lower pressures prevailing in muscle and, hence, passes on its oxygen to the Mb as required. Moreover, the need for O_2 will be greatest in tissues that have already consumed oxygen and simultaneously have produced CO_2. The CO_2 lowers the pH, thus causing the Hb to release even more oxygen to the Mb. The pH-sensitivity (called the Bohr effect), as well as the progressive increase of the O_2 binding constants in Hb, is due to interactions between the subunits; Mb behaves more simply because it consists of only one unit. It is clear that each of the two is essential in the complete oxygen-transport process. Carbon monoxide, PF_3, and a few other substances are toxic because they become bound to the iron atoms of Hb more strongly than O_2; their effect is one of competitive inhibition.

The way in which interactions between the four subunits in Hb give rise to both the cooperativity in oxygen binding and to the Bohr effect (pH dependence), both of which are so essential to the role played by Hb is now partly understood. The mechanism is very intricate, but one essential feature depends directly on the coordination chemistry involved. Deoxyhemoglobin has a high-spin distribution of electrons, with one electron occupying the $d_{x^2-y^2}$ orbital which points directly toward the four porphyrin nitrogen atoms. The presence of this electron in effect increases the radius of the iron atom in these directions by repelling the lone pair electrons of the nitrogen atoms. The result is that the iron atom actually lies about 0.7 to 0.8 Å out of the plane of these nitrogen atoms, in order that it not be in too close contact with them. The iron atom is also coordinated by a nitrogen atom on the imidazole ring of the amino acid histidine, labeled F-8 in *Fig. 31-2*.

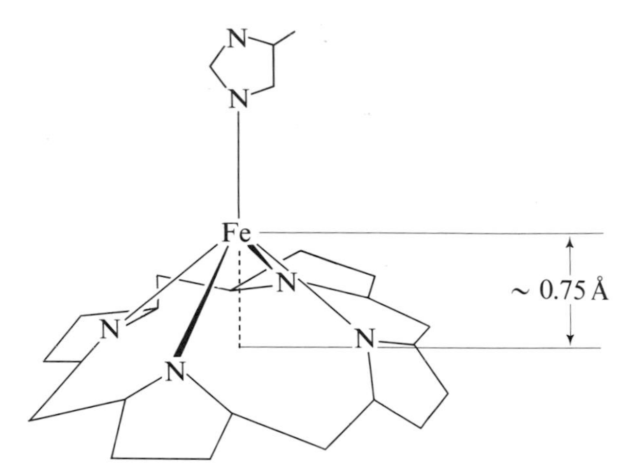

Figure 31-4a *The 5-coordinate, high-spin* Fe(II) *atom in deoxyhemoglobin.*

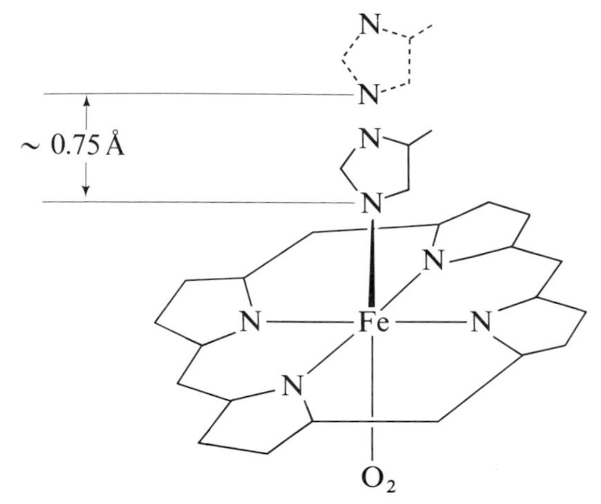

Figure 31-4b *The 6-coordinate, low-spin iron in oxyhemoglobin. The distance which the side-chain of histidine* F-8 *has moved is indicated.*

Thus the iron atom in deoxyhemoglobin has square pyramidal coordination, as is shown in *Fig. 31-4a*.

When an oxygen molecule becomes bound to the iron atom, it occupies a position opposite to the imidazole-nitrogen atom. The presence of this sixth ligand alters the strength of the ligand field, and the iron atom goes into a low-spin state, in which the six d electrons occupy the d_{xy}, d_{yz} and d_{zx} orbitals. The $d_{x^2-y^2}$ orbital is then empty and the previous effect of an electron occupying this orbital in repelling the porphyrin nitrogen atoms vanishes. The iron atom is thus able to slip into the center of an approximately planar porphyrin ring and an essentially octahedral complex is formed, as shown in *Fig. 31-4b*.

As the iron atom moves, it pulls the imidazole side chain of histidine F8 with it, thus moving that ring about 0.75 Å. This shift is then transmitted to other parts of the protein chain to which F8 belongs and, in particular, a large movement of the phenolic side chain of tyrosine HC2 is produced. From here various shifts of atoms in the neighboring subunit are caused, and these shifts influence the oxygen-binding capability of the heme group in that subunit. Thus the movement of the iron atom of the heme group in one subunit of hemoglobin acts as a kind of "trigger," which sets into motion extensive structural changes in other subunits.

One of the still unsolved problems about oxygen binding by hemoglobin concerns the structure of the Fe—O_2 grouping. Three possibilities are shown in *Fig. 31-5*. The linear one has no precedent and is least probable. The side-on arrangement is similar to the structure found in some simple O_2 complexes involving other metals, such as $(PPh_3)_2(CO)ClIrO_2$ (see page 307). However, the bent chain appears most probable, since O_2 is isoelectronic with NO^-, which forms complexes with bent Co^{III}—N—O chains (page 488), and there is one fairly good model compound, an iron(II) porphyrin complex of O_2, in which this arrangement has been found.

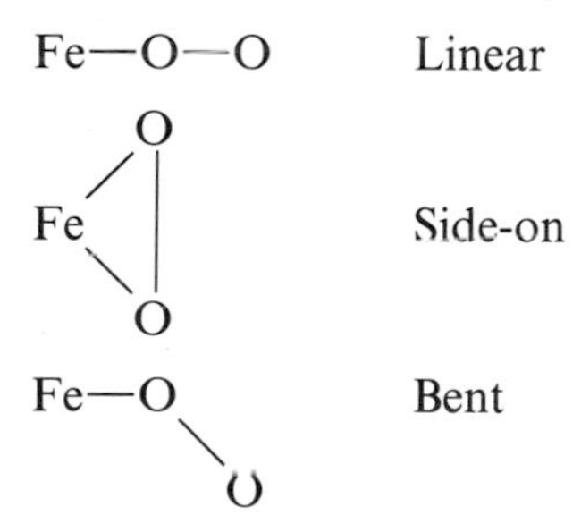

Figure 31-5 *Three conceivable O_2-to-iron binding geometries for hemoglobin or myoglobin.*

Cytochromes. These are heme proteins, found in both plants and animals, which serves as electron carriers. They accept an electron from a slightly better reducing agent and pass it on to a slightly better oxidizing agent. In the cytochromes the heme iron is also coordinated by a nitrogen atom of an imidazole ring on one side of the porphyrin-ring plane but, in addition, it is coordinated on the other side by a sulfur atom of a methionine residue in a different part of the protein chain. Thus its potential oxygen-binding capacity is shut off.

Heme-Containing Enzymes. Both catalase and peroxidase catalyze the decomposition of hydrogen peroxide:

$$2H_2O_2 \longrightarrow 2H_2O + O_2 \quad \text{(catalase)}$$

$$H_2O_2 + AH_2 \longrightarrow 2H_2O + A \quad \text{(peroxidase; } AH_2 = \text{a co-enzyme)}$$

31-3
Nonheme Iron Proteins

These are proteins that contain strongly bound, functional iron atoms but no porphyrins. The iron atoms are bound by sulfur atoms. These proteins all participate in electron-transfer sequences.

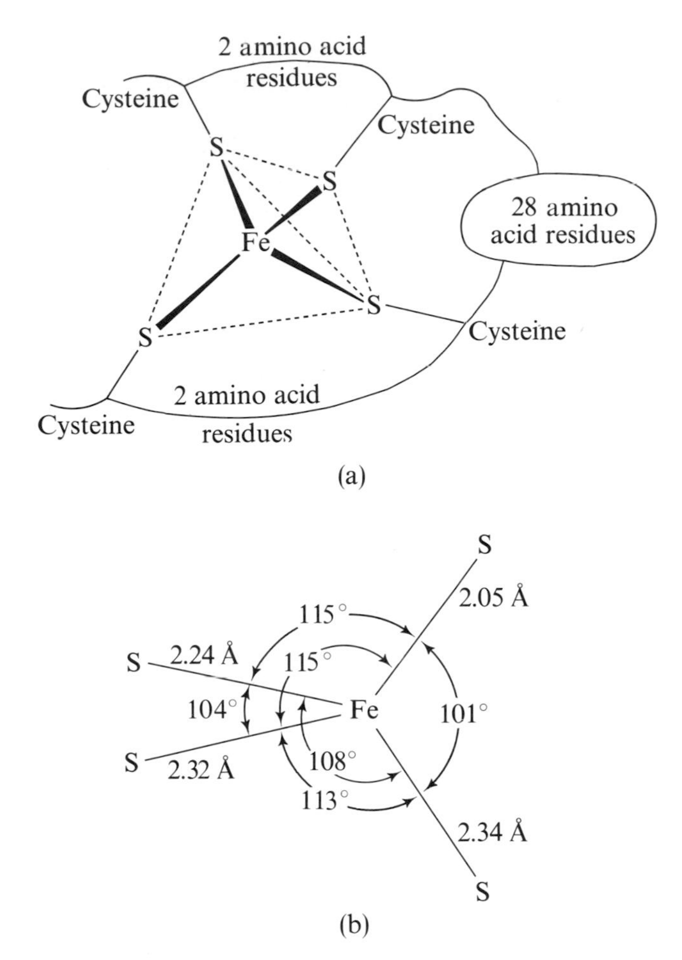

Figure 31-6 *Rubredoxin. (a) The location of the iron atom in the molecule. (b) The bond lengths and angles about the* FeS_4 *"tetrahedron." (According to published data in 1973.)*

Rubredoxins. These substances participate in a number of biological redox reactions, especially in anaerobic bacteria. A presumably typical one is that isolated from the bacterium *C. pasturianum.* It has a molecular weight of about 6000 and consists of a single peptide chain of 53 amino acid residues plus one iron atom. An X-ray crystallographic study has shown that the iron atom (in the ferric state) is coordinated by four sulfur atoms of cysteine residues, spaced as shown in *Fig. 31-6a.* The coordination is roughly tetrahedral, but severe distortions exist, as can be seen in *Fig. 31-6b.*

Ferredoxins. These are relatively small proteins (molecular weights of 6,000–12,000) which contain sulfur-bound iron atoms and, like rubredoxin, participate in electron-transfer chains. Indeed, rubredoxin might logically be considered as one type of ferredoxin. The other types, to which the name is conventionally applied, contain two, four, or eight atoms of iron per molecule. Those with two iron atoms are not so well understood as the others; they probably contain the structure unit

```
cys-S      S      S-cys
     \   /   \   /
      Fe       Fe
     /   \   /   \
cys-S      S      S-cys
```

where cys-S or S-cys represents the —CH_2S^- side chain of a cysteine residue of the protein.

The structure of one of the 8-Fe ferredoxins (from *P. aerogenes*) is known. It has two widely separated (12 Å) Fe_4S_4 groups. Each of these is held in its place in the molecule by bonds from cys-S groups to each of the iron atoms, as is shown in *Fig. 31-7.* It seems probable that the 4-Fe ferredoxins contain one such unit, although this has not been proved. The function of this Fe_4S_4 group is to serve as a source or sink for electrons.

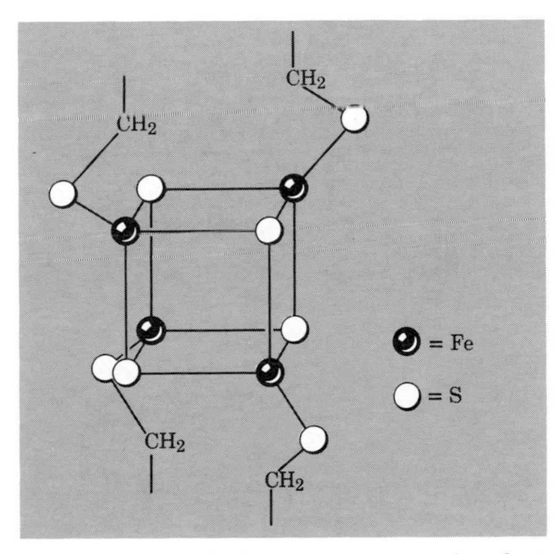

Figure 31-7 *The* Fe_4S_4(S-cys)$_4$ *unit which occurs in certain ferredoxins and in* HiPIP. *The central* Fe_4S_4 *group consists of two concentric tetrahedra which, contrary to this idealized drawing are of different sizes, the* Fe_4 *one being smaller.*

High-Potential Iron Proteins. These proteins, called HiPIP's, have redox potentials about 0.75 V more positive than the ferredoxins. However, in at least one case, and possibly in all, they contain the same Fe_4S_4(S-cys)$_4$ unit (see *Fig. 31-7*). How, then, does one account for the great difference in redox potentials? The basic reason is probably as follows. The Fe_4S_4 cage, C, can exist in three oxidation states, C, C^+, and C^-. In reduced HiPIP and oxidized ferredoxin, we have C. The two redox reactions, then, are

$$\text{HiPIP:} \quad C^+ + e = C \quad E_0 \approx 0.35\text{ V}$$

$$\text{Ferredoxin:} \quad C + e = C^- \quad E_0 \approx -0.40\text{ V}$$

There is support for this hypothesis from detailed structural comparisons of the Fe_4S_4 cages in the oxidized ferredoxin and in oxidized and reduced HiPIP, and from studies of an excellent model system in which the Fe_4S_4 cage is attached to four $C_6H_5CH_2S^-$ groups instead of to the four cys-S groups (see page 568).

31-4 Iron Supply and Transport

Iron metabolism requires provision for storing and transporting iron. In man and in many other higher animals the storage materials are *ferritin* and *hemosiderin.* These are present in liver, spleen, and bone marrow. Ferritin is a water-soluble, crystalline substance consisting of a roughly spherical protein sheath, of ~75 Å inside diameter and ~120 Å outside diameter, which in turn is built up of ~20 subunits. Within this sheath is a micelle of colloidal Fe_2O_3—H_2O—phosphate. Up to 23% of the dry weight may be iron; the protein portion alone, called apoferritin, is stable, crystallizes, and has a molecular weight of about 450,000. Hemosiderin contains even larger proportions of "iron hydroxide," but its constitution is variable and ill-defined in comparison with that of ferritin.

Transferrin is a protein that binds ferric iron very strongly and transports it from ferritin to red cells and *vice versa.* Iron passes between ferritin and transferrin as Fe^{2+}, but the details of the redox process are obscure.

In microorganisms, iron is transported by substances called *ferrichromes* and *ferrioxamines.* The former are trihydroxamic acids in which the three hydroxamate groups are on three side chains of a cyclic hexapeptide. The latter have the three hydroxamate groups as part of the peptide chain, which may be cyclic or acyclic. Typical structures are shown in *Fig. 31-8.*

The importance of these compounds derives from their exceptional ability to chelate Fe(III) and then pass through cell membranes, thus carrying iron from inorganic sources, such as $Fe_2O_3 \cdot xH_2O$, to points of need in the cells.

THE BIOINORGANIC CHEMISTRY OF COBALT

31-5 Vitamin B_{12}

The best-known biological function of cobalt is its intimate involvement in the coenzymes related to vitamin B_{12}, the basic structure of which is shown in *Fig. 31-9.* This structure is not as overwhelming as it might seem at first glance. It consists of four principal components.

1. A cobalt atom.
2. A macrocyclic ligand called the *corrin* ring, which bears various substituents. The essential corrin-ring system is shown in bold lines. It resembles the porphine ring, but differs in various ways, notably in the absence of one methine, =CH—, bridge between a pair of pyrrole rings.
3. A complex organic portion consisting of a phosphate group, a sugar, and an organic base, the latter being coordinated to the cobalt atom.

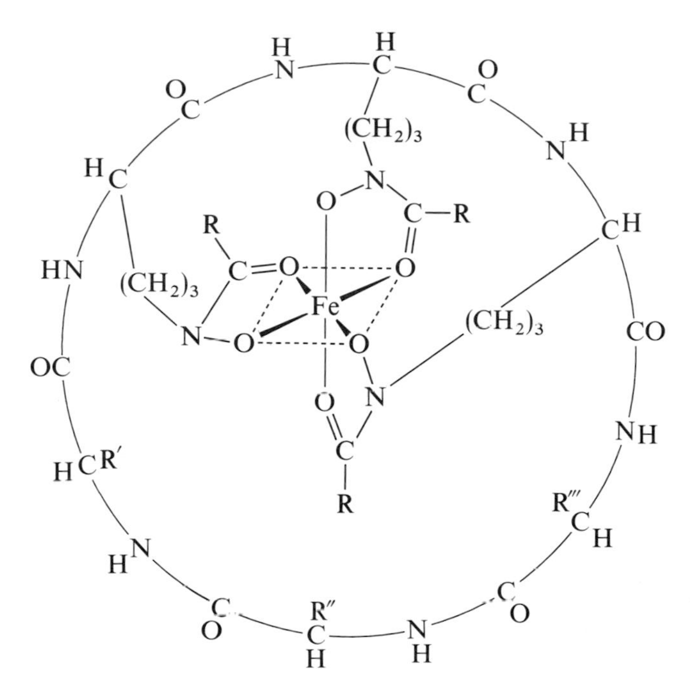

Figure 31-8a *Diagrammatic structure of a typical ferrichrome.*

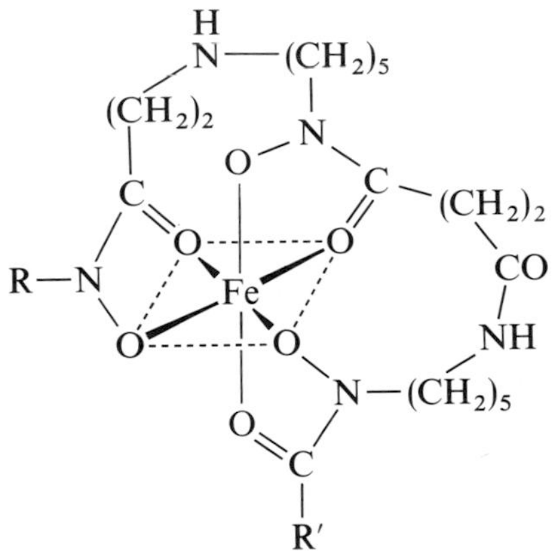

Figure 31-8b *Typical structure of an acyclic ferrioxamine.*

4. A sixth ligand, X, may be coordinated to the cobalt atom. The nature of this ligand can be varied and when the cobalt atom is reduced to the oxidation state +1, it is evidently absent.

The entire entity shown in *Fig. 31-9*, but neglecting the ligand X, is called *cobalamin.*

The term, vitamin B_{12}, refers to cyanocobalamin, which has cobalt in the +3 oxidation state and CN^- as the ligand X. The CN^- ligand is introduced during the isolation procedure and is not present in any active form of the vitamin. In the biological system, the ligand X is likely to be H_2O much of the time, but another

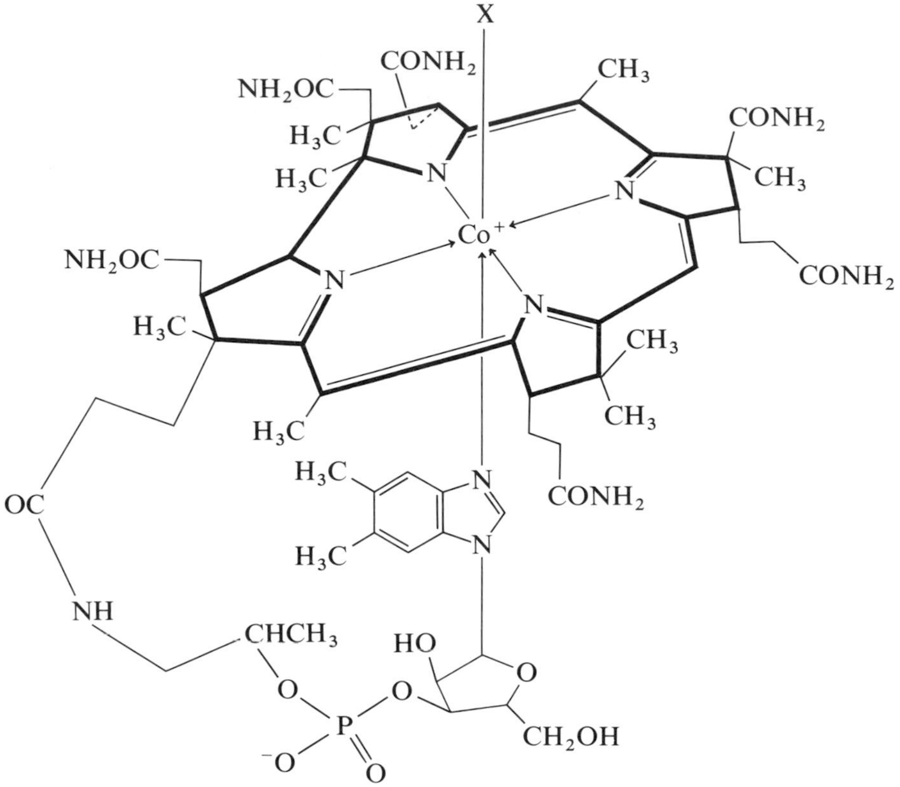

Figure 31-9 The structure of cobalamin. The corrin ring is shown in heavy lines.

possibility, which has been identified by actual isolation of the complex, the 5′-deoxyadenosyl radical, is shown in *Fig. 31-10.* The particular coenzyme in which this is found was the first organometallic compound to be observed in a living system.

The B_{12} coenzymes act in concert with a number of enzymes, but much remains to be done to elucidate their precise role. The best-studied systems involve the dioldehydrases, where the following reactions are catalyzed:

$$RCHOHCH_2OH \longrightarrow RCH_2CHO + H_2O \; (R = CH_3 \text{ or } H)$$

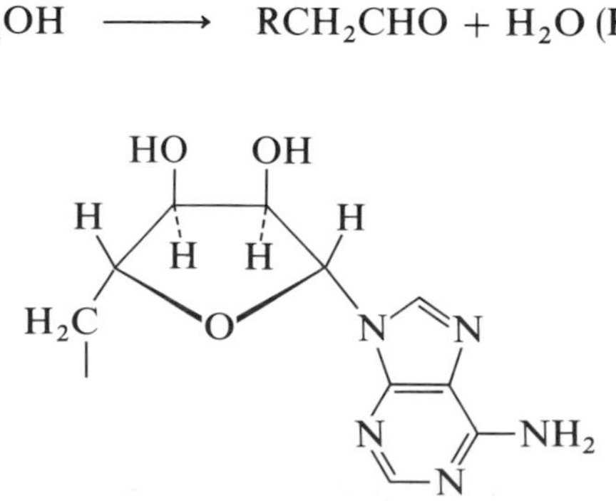

Figure 31-10 The 5′-deoxyadenosyl group which may constitute the ligand denoted X in Fig. 31-9.

From studies of the nonenzymic chemistry of B_{12} coenzymes and of model systems noted below, a body of knowledge about basic B_{12} chemistry has been built up. Some of this chemistry undoubtedly plays a role in its activities as a coenzyme. The cobalamins can be reduced in neutral or alkaline solution to give Co(II) and Co(I) species, often called B_{12r} and B_{12s}, respectively. The latter is a powerful reducing agent, decomposing water to give hydrogen and B_{12r}. These reductions can apparently be carried out in vivo by reduced ferredoxin. When cyano or hydroxo cobalamin is reduced, the ligand, CN^- or OH^-, is lost, and the Co(I) complex is 5-coordinate. There is considerable evidence that these 5-coordinate Co(I) species react with adenosine triphosphate in presence of a suitable enzyme to generate the B_{12} coenzyme.

In nonenzymic systems, rapid reaction of B_{12s} occurs with alkyl halides, acetylenes, and the like, as shown below, where [Cb] represents the cobalamine group.

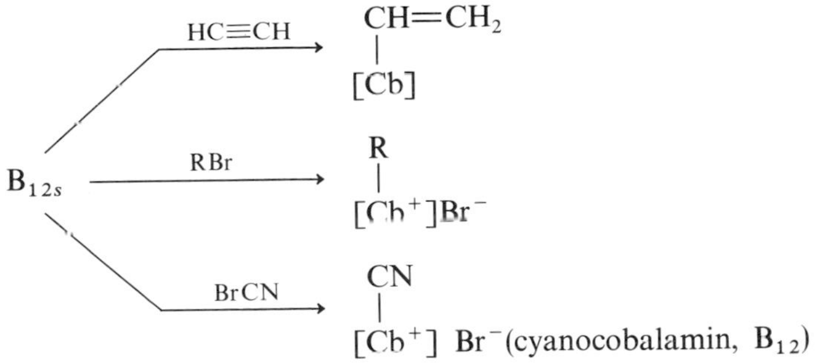

$$B_{12s} \xrightarrow{HC\equiv CH} \underset{[Cb]}{\overset{CH=CH_2}{|}}$$

$$B_{12s} \xrightarrow{RBr} \underset{[Cb^+]Br^-}{\overset{R}{|}}$$

$$B_{12s} \xrightarrow{BrCN} \underset{[Cb^+]\ Br^-(\text{cyanocobalamin, } B_{12})}{\overset{CN}{|}}$$

Methylcobalamin has an extensive chemistry, some of which is involved in the metabolism of methane-producing bacteria. It transfers CH_3 groups to Hg^{II}, Tl^{III}, Pt^{II}, and Au^{I}. It is, evidently, in this way that certain bacteria accomplish their unfortunate feat of converting relatively harmless elemental mercury, which collects in sea or lake bottoms, into the exceeding toxic methylmercury ion, CH_3Hg^+.

METALLOENZYMES

31-6 Survey and Definitions

Enzymes are large protein molecules so built that they can bind at least one reactant, (called the substrate) and catalyze a biochemically important reaction. They are extremely efficient as catalysts, typically causing rates to increase 10^6 times or more compared to the uncatalyzed rate, or even to the rate attained with conventional, nonenzymic catalysts.

Some enzymes incorporate one or more metal atoms in their normal structure. The metal ion does not merely participate during the time that the enzyme-substrate complex exists, but is a permanent part of the enzyme. The metal atom, or at least one of the metal atoms when two or more are present, occurs at or very near to the active site (the locus of the bound, reacting substrate) and plays a role in the activity of the enzyme. Such enzymes are called *metalloenzymes*, and at least 50 have been identified.

The following metals have been found in metalloenzymes: Ca, Mn, Fe, Cu, Zn, and Mo. Although Co^{2+} can often be made to replace Zn^{2+} in zinc metalloenzymes, with retention or even enhancement of activity, the actual presence of Co^{2+} in the native enzymes is at best rare. The most commonly used metals are Zn, Fe, and Cu.

31-7 Zinc Metalloenzymes

No less than 20 of these are known. Two of the most important or, at least, best studied, are the following.

Carbonic Anhydrase (MW = 30,000; 1 Zn). This enzyme occurs in red blood cells and catalyzes the dehydration of the bicarbonate ion and the hydration of CO_2:

$$OH^- + CO_2 = HCO_3^-$$

These reactions would otherwise proceed too slowly (see page 260) to be compatible with physiological requirements.

Carboxypeptidase (MW = 34,300; 1 Zn). This enzyme in the pancreas of mammals catalyzes the hydrolysis of the peptide bound at the carboxyl end of a peptide chain,

$$\text{—R''CH—CONH—CHR'—CONH—CHRCOO}^- + H_2O \longrightarrow$$

$$\text{—R''CH—CONH—CHR'—CO}_2^- + H_3N^+\text{CHRCOO}^-$$

The enzyme has a particular preference for substrates in which the side chain R is aromatic, that is, $—CH_2C_6H_5$ or $—CH_2C_6H_4OH$.

Its structure and mechanism of action have been partly elucidated. The zinc ion is bound in a distorted tetrahedral environment, with two histidine-nitrogen atoms, one glutamate carboxyl oxygen atom and a water molecule as ligands. The binding of the substrate probably occurs as is shown in *Fig. 31-11a.* Notice that the carbonyl oxygen atom of the peptide linkage which is to be broken has replaced the water molecule in the coordination sphere of the zinc ion.

The key step in a possible, but speculative, mechanism is shown in *Fig. 31-11b.* Once the peptide bond has been broken with formation of the acid anhydride, rapid hydrolysis of the anhydride would occur, as in *Fig. 31-11c.* The products would then vacate the active site, leaving it ready to bind another molecule of substrate and repeat the cycle.

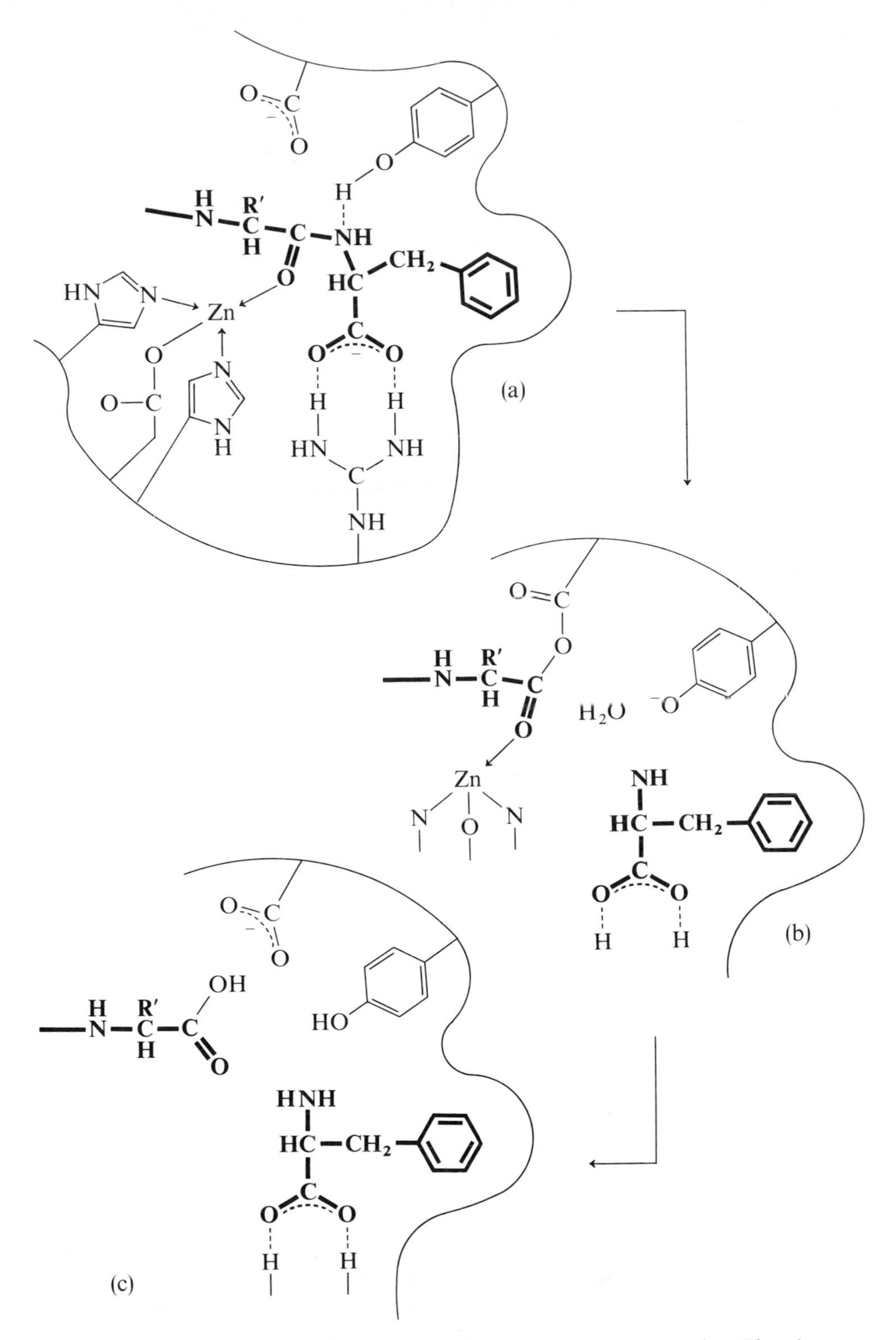

Figure 31-11 *(a) A proposed mode of binding of the substrate in carboxypeptidase. The substrate is shown in heavy type and lines. The curved line schematically defines the "surface" of the enzyme molecule. (b) A possible first step in the mechanism, wherein a carboxyl side chain attacks the carbonyl carbon atom, forming an anhydride. (c) The next steps would be hydrolysis of the intermediate anhydride and dissociation of the products from the active site.*

31-8
Copper Metalloenzymes and Other Metalloproteins

More than 15 of these have been isolated, but in no case is structure or function well understood in a chemical sense. The copper enzymes are mostly oxidases, that is enzymes which catalyze oxidations. Examples are: (1) *Ascorbic acid oxidase* (MW $\approx$ 140,000; 8 Cu) is widely distributed in plants and microorganisms. It catalyzes oxidation of ascorbic acid (vitamin C) to dehydroascorbic acid. (2) *Cytochrome oxidase*, the terminal electron acceptor in the oxidative pathway of cell mitochondria. This enzyme also contains heme. (3) Various *tyrosinases*, which catalyze the formation of pigments (melanins) in a host of plants and animals.

In many lower animals, such as crabs and snails, the oxygen-carrying molecule is a copper-containing protein *hemocyanin* (although it contains no heme groups), which has a very high molecular weight, and appears to bind one molecule of O_2 per two Cu atoms.

NITROGEN FIXATION

Elemental nitrogen, N_2, is relatively unreactive. In order to "fix" nitrogen, that is, make N_2 react with other substances to produce nitrogen compounds, it is generally necessary to use energy-rich conditions. High temperatures or electrical discharges can supply the necessary activation energy. It has long been a problem that primitive bacteria and some blue-green algae can fix nitrogen under mild conditions, that is, at ambient pressure and temperature. Metalloenzymes play a key role in this process.

31-9
Bacterial Nitrogenase Systems

Our more detailed chemical information comes mainly from studies of free-living soil bacteria. These can be cultured in the laboratory and essential components isolated and purified. Biological nitrogen fixation is reductive. An important fact, established using $^{15}N_2$, is that the first recognizable product is always NH_3. Apparently all intermediates remain bound to the enzyme system.

It has been known since 1930 that molybdenum is essential for bacterial nitrogen fixation, since this function can be turned off and on by removing and then restoring molybdenum to the environment. Magnesium and iron are also essential.

In 1960 the first active cell-free extracts were prepared and since then, *nitrogenases*, as the enzymes are called, have been obtained in fairly pure condition from several bacteria. In each case, the nitrogenase can be separated into two proteins, one with a molecular weight of about 250,000, and the other around 70,000. Neither of these separately is active, but on mixing them activity is observed immediately. The first one contains 1 (or possibly 2) atoms of molybdenum and about 15 atoms of iron, and has a higher than average sulfur content, suggesting that the metals may be coordinated by sulfur. The second, smaller component of nitrogenase contains two atoms of iron, two "labile" or "inorganic" sulfur atoms (i.e., not part of the amino acids cysteine or methionine) but no molybdenum.

There has been much speculation about the roles played by these components of nitrogenase. The general idea is that the molybdenum atom or atoms bind the dinitrogen and that the iron atoms participate in one or more redox chains that supply the electrons needed to reduce it. When the system operates in vitro it requires considerable amounts of adenosine triphosphate (ATP) and something like 440 kJ (105 kcal) of free energy are dissipated per mole of N_2 fixed. This is puzzling, since the actual overall reaction is slightly exothermic.

$$N_2(g) + 3H_2(g) = 2NH_3(g) \qquad \Delta H^\circ_{298} = -46\ \text{kJ mol}^{-1}$$

The nitrogenase system even more efficiently reduces acetylene, exclusively to ethylene, and this constitutes a convenient means of monitoring nitrogenase activity.

31-10
Synthetic Nitrogen Fixation Systems

What little is known about the operation of the natural nitrogenase systems has stimulated efforts to concoct synthetic ones. An obvious development based on the biochemical clues is an aqueous system containing $MoO_4{}^{2-}$, Fe^{2+}, and SH^- ions, organic thiols, and a reducing agent, such as $BH_4{}^-$. Such mixtures reduce acetylene quite effectively and show feeble activity toward N_2. Other systems containing titanium or vanadium and reducing agents such as aluminum alkyls may react with nitrogen and on hydrolysis give ammonia, but they bear no noticeable relationship to the natural systems.

MODEL SYSTEMS

Living systems are invariably complex and difficult to study in the same way that chemists can study and interpret simpler chemical systems. Thus there is always a desire to find a simple system that "models" the essential features of the complex natural one, but lends itself to more convenient study. The construction of model systems, usually poor ones, has been particularly popular in bioinorganic chemistry.

The problem with models is to judge correctly how far they can be trusted to replicate the true behavior of the real system. No model can give more than a partial picture of how the real one behaves. Some properties of the model may be entirely unrelated to the behavior of the natural one, and thus potentially misleading. Two model systems will be discussed in illustration.

31-11
Cobalamin Models

In 1964 it was reported that many of the chemical reactions of the cobalt atom in cobalamin are simulated by bis(dimethylglyoximato) cobalt complexes, an example being that in *Fig. 31-12*. A broad similarity to cyanocobalamin is obvious. These

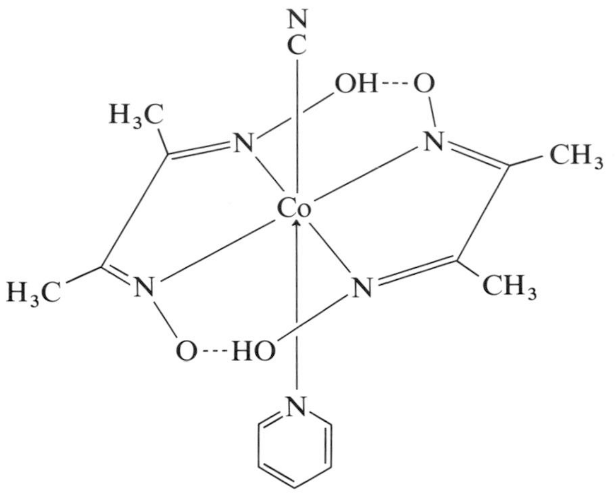

Figure 31-12 *A bis(dimethylglyoximato)cobalt complex, a "cobaloxime," which is a model for cyanocobalamine, vitamin* B_{12}.

dimethylglyoximato complexes have been given the short name *cobaloximes*, by analogy to the cobalamins. Actually, many compounds in which a strong, planar ligand system encircles the cobalt can also, in varying degrees, act similarly, but the cobaloximes have proved to be the most useful models. They are capable of undergoing reduction to the Co(II) and Co(I) state in aqueous media. The Co(I) complex behaves in many respects like vitamin B_{12s}.

Interestingly, cobalt porphyrins are less acceptable models, since they cannot be reduced to the Co(I) state under the conditions in which vitamin B_{12s} is obtained. This inability of the porphyrin ligand to stabilize the Co(I) species may be the reason why the corrin system developed during the course of evolution.

The cobaloxime with Co(I) undergoes essentially all of the same reactions as vitamin B_{12s} (see, for example, those mentioned on page 563), and study of the cobaloxime chemistry has contributed to our understanding of the vitamin B_{12} coenzymes. Of course, study of the coenzymes themselves is essential, and the data on the models can be considered only as auxiliary to that main task.

31-12
Ferredoxin and HiPIP Models

The complex Fe_4S_4(S-cys)$_4$ group which functions as the electron donor-acceptor system in HiPIP and the 4-Fe and 8-Fe ferredoxins needs detailed study if it is to be fully understood. However, accurate structural and physical data are difficult to obtain for the natural systems, in which the Fe_4S_4 cage is embedded in the large protein molecule. R. H. Holm and his coworkers have shown that it is possible to synthesize models that contain the Fe_4S_4 cage surrounded by mercaptide ions, such as CH_3S^- or $C_5H_5CH_2S^-$, in place of the cysteinyl groups. A typical compound is the crystalline $[(C_2H_5)_4N]_2[Fe_4S_4(SCH_2C_6H_5)_4]$.

This model has already provided information on structural and magnetic properties that is virtually inaccessible on the natural systems. For example, the magnetic properties of the model could readily be studied from 4 to 300 K, whereas the natural systems had been studied with much lower accuracy only from 5 to 50° C.

Study Questions

1. Name four transition metals and two nontransition metals that play important functional roles in biological processes.
2. Draw the structure of porphin and explain how the structures of heme and chlorophyll are related to it.
3. What role does the magnesium ion play in the functioning of chlorophyll?
4. What is a heme protein? Name at least three.
5. What are the functions of hemoglobin and myoglobin? What are the similarities and differences in their structures?
6. What changes occur in the heme groups of hemoglobin in going from deoxy- to oxyhemoglobin?
7. What prosthetic group (i.e., nonprotein group) is found in bacterial rubredoxin?
8. What is the structure of the redox center of HiPIP and of the 4-Fe and 8-Fe ferredoxins?
9. What functions do ferrichromes and ferrioxamines have? What are their chief chemical features?
10. State the main components of cobalamin. How do B_{12}, B_{12r} and B_{12s} differ?
11. What does an enzyme do? Specify the two characteristics of a metalloenzyme.
12. What role does the zinc ion appear to play in the action of carboxypeptidase?
13. What are the known facts about the chemical nature of nitrogenase?
14. What is a cobaloxime, and of what interest are cobaloximes?

Chapter 31
Study Guide

Scope and Purpose. This chapter introduces one of the newest and most rapidly expanding areas of research in all of chemistry. Space permits only brief discussions of a few aspects in which activity is already intense and productive. This is a field that both instructor and student will have to follow in the literature if they are to keep up to date.

The major message, quite aside from any specific facts, is that biochemistry involves some 20 or more other elements besides those traditionally treated by organic chemists (C, H, N, O, P, halogens), including many transition metals. Though these other elements tend to have more limited roles, life processes require them just as surely as they require proteins, carbohydrates and lipids. An understanding of their roles is therefore essential to a complete understanding of life processes.

Supplement Reading. The only comprehensive survey of the field is *Inorganic Biochemistry*, G. L. Eichhorn, ed., Elsevier Scientific Publishing Company, New York, 1973. This two-volume work consists of 34 chapters, each on a specialized topic. A large number of references to the original literature are given.

Other books of value are: *Metalloproteins*, by B. L. Vallee and W. E. C. Wacker, Volume V of *The Proteins*, H. Neurath, ed., 2nd ed., Academic Press, 1970. *The Inorganic Chemistry of Biological Processes*, by M. N. Hughes, Wiley, 1972. *The Biochemistry of Copper*, J. Peisach,

P. Aisen, and W. E. Blumberg, eds., Academic Press, 1966. *The Chemistry and Biochemistry of Nitrogen Fixation*, J. R. Postgate, ed., Plenum Press, 1971. S. J. Lippard, ed., *Current Research Topics in Bioinorganic Chemistry*, Vol. 18 of *Progress in Inorganic Chemistry*, Wiley-Interscience, 1973.

Among the many articles in the recent review literature, the following are recommended.

Crichton, R. R., "Structure and Function of Ferritin," *Angewandte Chem. Internat. Ed.* (English), *12*, 57 (1973).

Doig, M. T., Heyl, M. G., and Martin, D. F., "Lithium and Mental Health," *J. Chem. Educ.*, *50*, 343 (1973).

Fleischer, E. B., "The Structure of Porphyrins and Metallophyrins," *Accts. Chem. Research*, *3*, 105 (1970).

Holm, R. H., Iron-sulphur Clusters in Natural and Synthetic Systems, *Endeavour*, *34*, 38 (1975).

Mason, R. and Zubieta, J. A., "Iron-Sulfur Proteins: Structural Chemistry of Their Chromophores and Related Systems," *Angewandte Chem. Internat. Ed.* (English), *12*, 390 (1973).

Schrauzer, G. N., "Organocobalt Chemistry of Vitamin B_{12} Model Compounds (Cobaloximes)," *Accts. Chem. Research*, *1*, 97 (1968).

Siegel, H. and McCormick, D. B., "On the Discriminating Behavior of Metal Ions and Ligands with regard to their Biological Significance," *Accts. Chem. Research*, *3*, 201 (1970).

Williams, R. J. P., "The Biochemistry of Sodium, Potassium, Magnesium and Calcium," *Quart. Rev.*, *24*, 331 (1970).

index